Advanced Lectures in Mathematics (ALM)

ALM 1: Superstring Theory

ALM 2: Asymptotic Theory in Probability and Statistics with Applications

ALM 3: Computational Conformal Geometry

ALM 4: Variational Principles for Discrete Surfaces

ALM 6: Geometry, Analysis and Topology of Discrete Groups

ALM 7: Handbook of Geometric Analysis, No. 1

ALM 8: Recent Developments in Algebra and Related Areas

ALM 9: Automorphic Forms and the Langlands Program

ALM 10: Trends in Partial Differential Equations

ALM 11: Recent Advances in Geometric Analysis

ALM 12: Cohomology of Groups and Algebraic K-theory

ALM 13: Handbook of Geometric Analysis, No. 2

ALM 14: Handbook of Geometric Analysis, No. 3

ALM 15: An Introduction to Groups and Lattices: Finite Groups and Positive Definite Rational Lattices

ALM 16: Transformation Groups and Moduli Spaces of Curves

ALM 17: Geometry and Analysis, No. 1

ALM 18: Geometry and Analysis, No. 2

ALM 19: Arithmetic Geometry and Automorphic Forms

ALM 20: Surveys in Geometric Analysis and Relativity

ALM 21: Advances in Geometric Analysis

ALM 22: Differential Geometry: Under the Influence of S.-S. Chern

ALM 23: Recent Developments in Geometry and Analysis

ALM 24: Handbook of Moduli, Volume I

ALM 25: Handbook of Moduli, Volume II

ALM 26: Handbook of Moduli, Volume III

Advanced Lectures in Mathematics
Volume 24

Handbook of Moduli

Volume I

edited by

Gavril Farkas · Ian Morrison

International Press
www.intlpress.com

高等教育出版社
HIGHER EDUCATION PRESS

Advanced Lectures in Mathematics, Volume 24
Handbook of Moduli, Volume I

Volume Editors:
Gavril Farkas (Humboldt-Universität, Berlin)
Ian Morrison (Fordham University, New York)

2010 Mathematics Subject Classification. Primary: 14D20. Secondary: 14H10, 14K10, 11G15, 14N35, 14E30, 14D15.

ISBN: 978-1-57146-257-2

Printed in the United States of America.

17 16 15 14 1 2 3 4 5 6 7 8 9

The Handbook of Moduli *is dedicated to the memory of Eckart Viehweg, whose untimely death precluded a planned contribution, and to David Mumford, who first proposed the project, for all that they both did to nurture its subject; and to Angela Ortega and Jane Reynolds for everything that they do to sustain its editors.*

Contents

Volume I

Volume II

Volume III

Preface

Gavril Farkas and Ian Morrison

The title of these volumes might lead unwary readers to expect an encyclopedic survey for experts in the study of moduli problems in algebraic geometry. What they will discover is rather different. Our aims here are, first, to clarify the audience that we hope the Handbook *will* serve and the approach it *does* takes to its subject and, second, to thank all those who have assisted us in helping it realize these aims.

To begin with, a bit of history. The idea for a Handbook of Moduli originated in a discussion between David Mumford and Lizhen Ji at Michigan in 2006. Lizhen and David produced a draft table of contents that was circulated at the Symposium marking David's retirement from Brown in 2007. The Handbook was originally to have been edited by Ching-Li Chai and Amnon Neeman, but the demands of their work with Takahiro Shiota as editors of the second volume of Mumford's Collected Papers took priority and, at their urging, we agreed to take over editorship in the spring of 2009.

We quickly reached the conclusion that what was needed for many topics was not a discussion of the latest results aimed at specialists, but a survey aimed at a broad community of producers (and even some consumers from cognate areas) of algebraic geometry, most of whom had little prior familiarity with the area. Our goal became a Handbook that would introduce the techniques, examples and results essential to each topic, and say enough about recent developments to prepare the reader to tackle the primary literature in the area. We particularly sought to elicit contributions that illustrated "secret handshakes", yogas and heuristics that experts use privately to guide intuition or simplify calculation but that are replaced by more formal arguments, or simply do not appear, in articles aimed at other specialists.

For many topics, the Handbook succeeds much better than we dared to hope. The credit is due entirely to the hard work of the Handbook's many authors in producing articles that conformed to the goals we had set. Again and again, we were delighted to find that authors, instead of taking the easy course of cutting and pasting from earlier surveys and primary references, had made the substantially greater effort to write the original treatments needed to bridge gaps in the literature and make important problems accessible to a wide audience for the first time.

We expect that they will reap a just reward and that their articles will be widely read and referenced. Here we want to offer them not only our sincerest thanks, but also those of the Handbook's readers, for their exceptional generosity. Many

Handbook articles were also improved by extensive and thoughtful referees' reports. We are grateful for all work that the referees did to improve the Handbook and take this opportunity to thank them collectively on behalf of the contributors.

We must, however, disclaim that the Handbook's coverage is often incomplete, in extreme cases, non-existent. The blame for these gaps is mostly ours. When we solicited contributions to the Handbook, each invitation was accompanied by a suggested topic, and we selected contributors who we thought would be able to cover their topics in the spirit discussed above. The results reflect both our knowledge and taste—of topics and of experts in them—and also, in some cases, our ignorance.

In some areas, we found it easy to produce candidate contributor–topic pairs, and to recruit the contributors we had identified. The Handbook's discussion of, for example, moduli spaces of curves is, therefore, particularly complete—some will say, not without a certain justice, excessive.

In other areas, we had more difficulty both in identifying and in enlisting candidates. A few of the more obvious gaps arose when authors who had accepted our invitation backed out after it was too late to find replacements. A more deeply felt loss—one that impacts the whole subject of moduli—was the untimely death of Eckart Viehweg, who had been one of the first to agree to contribute.

We also omitted a few topics as a courtesy to the authors of monographs devoted to them that we knew to be in preparation, others because papers treating them in the spirit we were seeking had recently appeared, and yet others because we felt that they were developing so rapidly that any contribution dealing with them would have a limited shelf-life. In hindsight, not all of these decisions were well taken.

As a result, the Handbook's treatment of moduli has some major lacunae (mirror symmetry, wall crossing formulae) and there are other topics (moduli of sheaves and bundles) which are discussed but not in the depth that their importance merits. We apologize to readers who may have hoped to find more about these subjects in the Handbook, and (with Lizhen's encouragement) we challenge experts who feel that their areas deserve a fuller exposition to offer him proposals for additional Handbook volumes devoted to them.

The Handbook also benefitted from the efforts of many other colleagues. Amnon Neeman showed considerable doggedness in recruiting us to succeed him and Ching-Li as editors. Scott Wolpert provided valuable advice on the cat-herding elements of the editor's job. Dave Bayer helped enormously in setting up the final production process both to automate complex and error prone operations and to prevent inconsistencies between the LaTeX installations on our home systems and those at Higher Education Press.

Brian Bianchini, International Press' book production manager, ensured that we had the resources we needed throughout the Handbook's growth from the single volume originally projected to the present three. The Advanced Mathematics series editor, Lizhen Ji, was always ready to answer our questions, help with practical difficulties, and adjust his schedule for the series to adapt to changes in ours. Liping Wang and her production staff at the Higher Education Press were unfailingly accommodating and helpful to us in resolving LaTeX issues—even reTeXing several submissions to bring them into conformity with the Handbook style—and made every effort to ensure that the appearance of the Handbook volumes was up to the standard of their contents.

To all of them, and to many others who provided more informal help, we here offer our sincerest thanks.

Humboldt Universität, Institut für Mathematik, Unter den Linden 6, 10099 Berlin
E-mail address: farkas@mathematik.hu-berlin.de

Department of Mathematics, Fordham University, Bronx, NY 10458
E-mail address: morrison@fordham.edu

Logarithmic geometry and moduli

Dan Abramovich, Qile Chen, Danny Gillam, Yuhao Huang, Martin Olsson, Matthew Satriano, and Shenghao Sun

Abstract. We discuss the role played by logarithmic structures in the theory of moduli.

Contents

1. Introduction

Logarithmic structures in algebraic geometry

It can be said that Logarithmic Geometry is concerned with a method of finding and using "hidden smoothness" in singular varieties. The original insight comes from consideration of de Rham cohomology, where logarithmic differentials can reveal such hidden smoothness. Since singular varieties naturally occur "at the boundary" of many moduli problems, logarithmic geometry was soon applied in the theory of moduli.

Foundations for this theory were first given by Kazuya Kato in [27], following ideas of Fontaine and Illusie. The main body of work on logarithmic geometry

2000 *Mathematics Subject Classification.* Primary 14A20; Secondary 14Dxx.
Key words and phrases. moduli, logarithmic structures.

has been concerned with deep applications in the cohomological study of p-adic and arithmetic schemes. This gave the theory an aura of "yet another extremely complicated theory". The treatments of the theory are however quite accessible. We hope to convince the reader here that the theory is simple enough and useful enough to be considered by anybody interested in moduli of singular varieties, indeed enough to be included in a Handbook of Moduli.

Normal crossings and logarithmic smoothness

So what is the original insight? Let X be a nonsingular irreducible complex variety, S a smooth curve with a point s and $f : X \to S$ a dominant morphism smooth away from s, in such a way that the fiber $f^{-1}s = X_s = Y_1 \cup \ldots \cup Y_m$ is a reduced simple normal crossings divisor. Then of course $\Omega_{X/S} = \Omega_X / f^*\Omega_S$ fails to be locally free at the singular points of f. But consider instead the sheaves $\Omega_X(\log(X_s))$ of differential forms with at most logarithmic poles along the Y_i, and similarly $\Omega_S(\log(s))$. Then there is an injective sheaf homomorphism $f^*\Omega_S(\log(s)) \to \Omega_X(\log(X_s))$, and *the quotient sheaf* $\Omega_X(\log(X_s))/\Omega_S(\log(s))$ *is locally free.*

So in terms of logarithmic forms, *the morphism* f *is as good as a smooth morphism.*

There is much more to be said: first, this $\Omega_X(\log(X_s))/\Omega_S(\log(s))$ can be extended to a logarithmic de Rham complex, and its hypercohomology, while not recovering the cohomology of the singular fibers, does give rise to the limiting Hodge structure. So it is evidently worth considering.

Second, the picture is quite a bit more general, and can be applied to all toric and toroidal maps between toric varieties or toroidal embeddings (with a little caveat about the characteristic of the residue fields). So there is some flexibility in choosing $X \to S$.

The search for a structure

Since we are considering moduli, then as soon as we consider $X \to S$ as above we must also consider the normal crossings fiber $X_s \to \{s\}$. But what structure should we put on this variety? The notion of differentials with logarithmic poles along X_s is not in itself intrinsic to X_s. Also the normal crossings variety X_s is not in itself toric or toroidal, so a new structure is needed to incorporate it into the picture.

One is tempted to consider varieties which are assembled from nice varieties by some sort of gluing, as normal crossings varieties are. But already normal crossings varieties do not give a satisfactory answer in general, because their deformation spaces have "bad" components. Here is a classical example: consider a smooth projective variety Z such that $\mathrm{Pic}^0(Z)$ is nontrivial. Let L be a line bundle on Z and set $Y = \mathbb{P}(\mathcal{O} \oplus L)$, with zero section $Z \subset Y$. Let X be the blowing up of $Z \times 0 \subset Y \times \mathbb{A}^1$. We have a flat morphism $f : X \to \mathbb{A}^1$ with fiber $X_0 = f^{-1}(0) \simeq Y \cup Y$, where the two copies of Y are glued with the zero section of one attached to the ∞ section of the other.

So clearly X_0 is a normal crossings variety with a nice smoothing to a copy of Y. But there are other deformations: the variety $Y \cup Y$ also deforms to $Y \cup Y'$ where $Y' = \mathbb{P}(\mathcal{O} \oplus L')$ and L' a deformation of the line bundle L. And it is not hard to see that $Y \cup Y'$ does not have a smoothing. Ideally one really does not want to see this deformation $Y \cup Y'$ in the picture - and ideally X_0 should have a natural structure whose deformation space excludes $Y \cup Y'$ automatically.

Such a structure was proposed by Friedman in [10], where the notion of *d-semistable varieties* was introduced. This structure is somewhat subtle, and while it solves the issue in this case, it is not quite as flexible as one could wish. As we will see in Section 5, logarithmic structures subsume d-semistability and do provide an appropriate flexibility.

Organization of this chapter

In this chapter we briefly describe logarithmic structures and indicate where they can be useful in the study of moduli spaces. Section 2 gives the basic definitions of logarithmic structures, and Section 3 discusses logarithmic differentials and log smooth deformations, which are important in considering moduli spaces.

Section 4 gives the first example where logarithmic geometry fits well with moduli spaces: the moduli space of stable curves is the moduli space of log smooth curves. The issue of d-semistability does not arise since a nodal curve is automatically d-semistable. So the theory for curves is simple. Turning to higher dimensions, Section 5 shows how d-semistability can be described using logarithmic structures.

If one is to enlarge algebraic geometry to include logarithmic structures, the task of generalizing the techniques of algebraic geometry to logarithmic structure can certainly seem daunting. In Section 6 we show how to encode logarithmic structure in terms of certain algebraic stacks. This allows us to reduce various constructions to the case of algebraic stacks. (One can argue that the theory of stacks is not simple either, but at least in the theory of moduli they have come to be accepted, with some exceptions [34].)

In Section 7 we make use of logarithmic stacks to describe the complexes which govern deformations and obstructions for logarithmic structures even in the non-smooth case. This comes in handy later. For instance, even when studying moduli of log-smooth schemes, the moduli spaces tend to be singular, and their cotangent complexes are necessary ingredients in constructing virtual fundamental classes.

Section 8 describes a beautiful construction, similar to polar coordinates, in which families of complex log smooth varieties give rise canonically to families of topological manifolds. Differential geometers have used polar coordinates on nodal curves to "make space" for monodromy to act by Dehn twists. Rounding (using Ogus's terminology) is a magnificent way to generalize this.

The immediate implications of logarithmic structures for De Rham cohomology and Hodge structures are described in Section 9.

We conclude by describing three applications, where logarithmic structures serve as the proverbial "magic powder" (term suggested by Kato and Ogus) to clarify or remove unwanted behavior from moduli spaces.

Section 10 describes a number of cases where the main irreducible component of a moduli space can be separated from other "unwanted" components by sprinkling the objects with a bit of logarithmic structure.

In Section 11 we introduce twisted curves, a central object of orbifold stable maps, and show how logarithmic structures give a palatable way to construct the moduli stack of twisted curves.

Section 12 gives background for the work of B. Kim, in which Jun Li's moduli space of relative stable maps, with its obstruction theory and virtual fundamental class, is beautifully simplified using logarithmic structures.

Notation

Following the lead of Ogus [45], we try whenever possible to denote a logarithmic scheme by a regular letter (such as X) and the underlying scheme by $\underline{X}$. When this is impossible we write X for the underlying scheme and $(X, \mathcal{M}_X)$ for a logarithmic scheme over it.

Acknowledgements

This chapter originated from lectures given by Olsson at the School and Workshop on Aspects of Moduli, June 15-28, 2008 at the De Giorgi Center at the Scuola Normale Superiore in Pisa, Italy. The material was revisited and expanded in our seminar during the Algebraic Geometry program at MSRI, 2009. We thank the De Giorgi Center, MSRI, their staff and program organizers for providing these opportunities. Thanks are due to Arthur Ogus and Phillip Griffiths, who lectured on two topics at the MSRI seminar. While no new material is intended here, we acknowledge that research by Abramovich, Gillam and Olsson is supported by the NSF.

2. Definitions and basic properties

In this section we introduce the basic definitions of logarithmic geometry in the sense of [27]. Good introductions are given in [27] and [45]. Further technique is developed in [12].

Logarithmic structures

The basic definitions are as follows:

Definition 2.1. A *monoid* is a commutative semi-group with a unit. A morphism of monoids is required to preserve the unit element. We use Mon to denote the category of Monoids.

Definition 2.2. Let $\underline{X}$ be a scheme. A *pre-logarithmic structure* on $\underline{X}$ is a sheaf of monoids $\mathcal{M}_X$ on the étale site $\underline{X}_{ét}$ combined with a morphism of sheaves of monoids: $\alpha : \mathcal{M}_X \longrightarrow \mathcal{O}_{\underline{X}}$, called the structure morphism, where we view $\mathcal{O}_{\underline{X}}$ as a monoid under multiplication. A pre-log structure is called a *log structure* if $\alpha^{-1}(\mathcal{O}^*_{\underline{X}}) \cong \mathcal{O}^*_{\underline{X}}$ via α. The pair $(\underline{X}, \mathcal{M}_X)$ is called a *log scheme*, and will be denoted by X.

Note that, given a log structure $\mathcal{M}_X$ on $\underline{X}$, we can view $\mathcal{O}^*_{\underline{X}}$ as a subsheaf $\mathcal{M}_X$.

Definition 2.3. Given a log scheme X, the quotient sheaf $\overline{\mathcal{M}}_X = \mathcal{M}_X/\mathcal{O}^*_{\underline{X}}$ is called the *characteristic of the log structure* $\mathcal{M}_X$.

Definition 2.4. Let $\mathcal{M}$ and $\mathcal{N}$ be pre-log structures on $\underline{X}$. A *morphism* between them is a morphism $\mathcal{M} \to \mathcal{N}$ of sheaves of monoids which is compatible with the structure morphisms.

How should one think of such a beast? There are two extreme cases:

(1) If an element $m \in \mathcal{M}$ has $\alpha(m) = x \neq 0$, one often thinks of m as some sort of partial data of a "branch of the logarithm of x". Evidently no data is added if x is invertible, but some is added otherwise. In particular, we will see later that m permits us to take the logarithmic differential dx/x of x.
(2) If $\alpha(m) = 0$ it is often the case that it m comes by restricting the log structure of an ambient space, and serves as the "ghost" of a logarithmic cotangent vector coming from that space. So the log structure "remembers" deformations that are lost when looking at the underlying scheme.

The log structure associated to a pre-log structure

We have a natural inclusion

$$i : (\text{log structures on } \underline{X}) \hookrightarrow (\text{pre-log structures on } \underline{X})$$

by viewing a log structure as a pre-log structure. We now construct a left adjoint.

Let $\alpha : \mathcal{M} \to \mathcal{O}_{\underline{X}}$ be a pre-log structure on $\underline{X}$. We define the *associated log structure* $\mathcal{M}^a$ to be the push-out of

$$\begin{array}{ccc} \alpha^{-1}(\mathcal{O}^*_{\underline{X}}) & \longrightarrow & \mathcal{M} \\ \downarrow & & \\ \mathcal{O}^*_{\underline{X}} & & \end{array}$$

in the category of sheaves of monoids on $\underline{X}_{ét}$, endowed with

$$\mathcal{M}^a \to \mathcal{O}_{\underline{X}} \quad (a, b) \mapsto \alpha(a)b \ (a \in \mathcal{M}, b \in \mathcal{O}^*_{\underline{X}}).$$

In this way, we obtain a functor a : (pre-log structures on $\underline{X}$) $\to$ (log structures on $\underline{X}$). From the universal property of push-out, any morphism of pre-log structure from a pre-log structure $\mathcal{M}$ to a log structure on $\underline{X}$ factor through $\mathcal{M}^a$ uniquely.

Lemma 2.5. [45, 1.1.5] *The functor a is left adjoint to i.*

Example 2.6. The category of log structures on $\underline{X}$ has an initial object, called the trivial log structure, given by the inclusion $\mathcal{O}_{\underline{X}}^* \hookrightarrow \mathcal{O}_{\underline{X}}$. It also has a final object, given by the identity map $\mathcal{O}_{\underline{X}} \to \mathcal{O}_{\underline{X}}$. Trivial log structures are quite useful as they make the category of schemes into a full subcategory of the category of log schemes (see Definition 2.9). The final object is rarely used since it is not fine, see Definition 2.16.

Example 2.7. Let $\underline{X}$ be a regular scheme, and $D \subset \underline{X}$ a divisor. We can define a log structure $\mathcal{M}$ on $\underline{X}$ associated to the divisor D as

$$\mathcal{M}(U) = \{ g \in \mathcal{O}_{\underline{X}}(U) : g|_{U \setminus D} \in \mathcal{O}_{\underline{X}}^*(U \setminus D) \} \subset \mathcal{O}_{\underline{X}}(U).$$

The case where D is a normal crossings divisor is special - we will see later that it is *log smooth*.

Note that the concept of normal crossing is local in the étale topology. This is one reason we use the étale topology instead of the Zariski topology.

Example 2.8. Let P be a monoid, R a ring, and denote by R[P] the monoid algebra. Let $\underline{X} = \operatorname{Spec} R[P]$. Then $\underline{X}$ has a canonical log structure associated to the canonical map $P \to R[P]$. We denote by Spec $(P \to R[P])$ the log scheme with underlying $\underline{X}$, and the canonical log structure.

The inverse image and the category of log schemes

Let $f : \underline{X} \to \underline{Y}$ be a morphism of schemes. Given a log structure $\mathcal{M}_Y$ on $\underline{Y}$, we can define a log structure on $\underline{X}$, called the inverse image of $\mathcal{M}_Y$, to be the log structure associated to the pre-log structure $f^{-1}(\mathcal{M}_Y) \to f^{-1}(\mathcal{O}_{\underline{Y}}) \to \mathcal{O}_{\underline{X}}$. This is usually denoted by $f^*(\mathcal{M}_Y)$. Using the inverse image of log structures, we can give the following definition.

Definition 2.9. A *morphism of log schemes* $X \to Y$ consists of a morphism of underlying schemes $f : \underline{X} \to \underline{Y}$, and a morphism $f^\flat : f^*\mathcal{M}_Y \to \mathcal{M}_X$ of log structures on $\underline{X}$.

We denote by LSch the category of log schemes.

Example 2.10. In Example 2.8, the log structure on Spec $(P \to R[P])$ can be viewed as the inverse image of the log structure on Spec $(P \to \mathbb{Z}[P])$ via the canonical map Spec $(R[P]) \to$ Spec $(\mathbb{Z}[P])$.

Example 2.11. Let k be a field, $\underline{Y}$=Spec $k[x_1, \cdots, x_n]$, $D = V(x_1 \cdots x_r)$. Note that D is a normal crossing divisor in $\underline{Y}$. By Example 2.7, we have a log structure $\mathcal{M}_Y$ on $\underline{Y}$ associated to the divisor D. In fact, $\mathcal{M}_Y$ can be viewed as a submonoid of $\mathcal{O}_{\underline{Y}}$ generated by $\mathcal{O}_{\underline{Y}}^*$ and $\{x_1, \cdots, x_r\}$.

Consider the inclusion $j : p = \operatorname{Spec} k \hookrightarrow \underline{Y}$ sending the point to the origin of $\underline{Y}$. Then $j^*\mathcal{M}_Y = k^* \oplus \mathbb{N}^r$, and the structure map $j^*\mathcal{M} \longrightarrow \mathcal{O}_{\underline{X}}$ is given by $(a, n_1, \cdots, n_r) \mapsto a \cdot 0^{n_1 + \cdots + n_r}$, where we define $0^0 = 1$ and $0^n = 0$ if $n \neq 0$. Such point with the log structure above is call a *logarithmic point*; when $r = 1$ we call it the *standard logarithmic point*.

Charts of log structures

Definition 2.12. Let X be a log scheme, P a monoid, and consider the constant sheaf P_X. A *chart* for $\mathcal{M}_X$ is a morphism $P_X \to \mathcal{M}_X$, such that the induced map of log structures $P^a \to \mathcal{M}_X$ is an isomorphism, where P^a is the log structure associated to the pre-log structure given by $P_X \to \mathcal{M}_X \to \mathcal{O}_{\underline{X}}$.

In fact, a chart of $\mathcal{M}_X$ is equivalent to a morphism

$$f : X \to \mathrm{Spec}\,(P \to \mathbb{Z}[P]),$$

such that $f^\flat$ is an isomorphism. In general, we have the following:

Lemma 2.13. [45, 1.1.9] *The morphism*

$$\mathrm{Hom}_{\mathrm{LSch}}(X, \mathrm{Spec}\,(P \to \mathbb{Z}[P])) \;\to\; \mathrm{Hom}_{\mathrm{Mon}}(P, \Gamma(X, \mathcal{M}_X))$$

associating to f *the composition*

$$P \longrightarrow \Gamma(\underline{X}, P_X) \xrightarrow{\Gamma(f^\flat)} \Gamma(\underline{X}, \mathcal{M}_X)$$

is an isomorphism.

We can also consider charts for log morphisms.

Definition 2.14. Let $f : X \to Y$ be a morphism of log schemes. A *chart* for f is a triple $(P_X \to \mathcal{M}_X, Q_Y \to \mathcal{M}_Y, Q \to P)$ where P_X and Q_Y are the constant sheaves associated to the monoids P and Q, which satisfy the following conditions:

(1) $P_X \to \mathcal{M}_X$ and $Q_Y \to \mathcal{M}_Y$ are charts of $\mathcal{M}_X$ and $\mathcal{M}_Y$;
(2) the morphism of monoids $Q \to P$ makes the following diagram commutative:

$$\begin{array}{ccc} Q_X & \longrightarrow & P_X \\ \downarrow & & \downarrow \\ f^*\mathcal{M}_Y & \longrightarrow & \mathcal{M}_X. \end{array}$$

Fine log structures

Arbitrary log structures are too wild to manipulate; they are roughly analogous to arbitrary ringed spaces: both notions are useful for general constructions, but a narrower, more geometric category is desirable. In Definition 2.16 below we introduce the notion of fine log structures. Continuing the analogy above, these are well-behaved log structures analogous to noetherian schemes, in the sense that you can do geometry on them.

Given a monoid P, we can associate a group

$$P^{gp} := \{(a, b) | (a, b) \sim (c, d) \text{ if } \exists s \in P \text{ such that } s + a + d = s + b + c\}.$$

Note that any morphism from P to an abelian group factors through P^{gp} uniquely.

Definition 2.15. P is called *integral* if $P \to P^{gp}$ is injective. It is called *saturated* if it is integral and for any $p \in P^{gp}$, if $n \cdot p \in P$ for some positive integer n then $p \in P$.

Definition 2.16. A log scheme X is said to be *fine,* if étale locally there is a chart $P \to \mathcal{M}_X$ with P a finitely generated integral monoid. If moreover P can be chosen to be saturated, then X is called a *fine and saturated* (or *fs*) log structure. This is equivalent to saying that for every geometric point $\bar{x} \to \underline{X}$ the monoid $\overline{\mathcal{M}}_{\bar{x},X}$ is finitely generated and saturated as in Definition 2.15. Finally if $P \simeq \mathbb{N}^k$ we say that the log structure is *locally free.*

In the following, we will focus on fine log schemes.

3. Differentials, smoothness, and log smooth deformations

The main reference in this section is [27].

Logarithmic differentials

In [14] Grothendieck defines a derivation as the difference of infinitesimal liftings of a section. We can do the same thing with logarithmic schemes. First, we need a concept of infinitesimal extension, which requires the following definition.

Definition 3.1. A morphism $f : X \to Y$ of log schemes is called *strict* if $f^\flat : f^*\mathcal{M}_Y \to \mathcal{M}_X$ is an isomorphism. It is called a *strict closed immersion* [1] if it is strict and the underlying map $\underline{X} \to \underline{Y}$ is a closed immersion in the usual sense.

Let us consider a commutative diagram of solid arrows of log schemes:

$$\begin{array}{ccc} T_0 & \xrightarrow{\ \phi\ } & X \\ {\scriptstyle j}\big\downarrow & \overset{g_1}{\nearrow}\ \underset{g_2}{\nearrow} & \big\downarrow{\scriptstyle f} \\ T_1 & \xrightarrow[\ \psi\ ]{} & Y \end{array}$$

where j is a strict closed immersion defined by an ideal J with $J^2 = 0$. Note that T_0 and T_1 have the same underlying topological space, and isomorphic étale sites. Then we have the following commutative diagram of sheaves of algebras:

$$\begin{array}{ccc} \mathcal{O}_{\underline{T}_0} & \longleftarrow & \phi^{-1}\mathcal{O}_{\underline{X}} \\ \big\uparrow & \overset{g_1^\#}{\swarrow}\ \underset{g_2^\#}{\swarrow} & \big\uparrow \\ \mathcal{O}_{\underline{T}_1} & \longleftarrow & \psi^{-1}\mathcal{O}_{\underline{Y}} \end{array}$$

Then $g_1^\# - g_2^\#$ is a derivation $\partial_{g_1 - g_2} : \phi^{-1}\mathcal{O}_{\underline{X}} \to J$ in the usual sense. We also have a commutative diagram given by the log structures:

[1] The term used in [27] is *an exact closed immersion.*

$$\begin{array}{ccc} \mathcal{M}_{T_0} & \xleftarrow{\quad} & \phi^{-1}\mathcal{M}_X \\ \uparrow & \swarrow^{g_1^\flat} \quad \swarrow^{g_2^\flat} & \uparrow \\ \mathcal{M}_{T_1} & \xleftarrow{\quad} & \psi^{-1}\mathcal{M}_Y \end{array}$$

Note that we have an "exact sequence" of multiplicative monoids

$$\mathbf{1} \to (1+J) \to \mathcal{M}_{T_1} \to \mathcal{M}_{T_0} \to \mathbf{1},$$

by which we mean that the group $1+J$ acts freely on $\mathcal{M}_{T_1}$ with quotient $\mathcal{M}_{T_0}$. Hence we obtain a morphism $D_{g_1-g_2} : \phi^{-1}\mathcal{M}_X \to J$ such that for every $m \in \phi^{-1}\mathcal{M}_X$ we have $(g_1^\flat - g_2^\flat)(m) = 1 + D_{g_1-g_2}(m)$. It is not hard to check that it is a monoid homomorphism: $D_{g_1-g_2}(m \cdot n) = D_{g_1-g_2}(m) + D_{g_1-g_2}(n)$ for any $m, n \in \phi^{-1}(\mathcal{M}_X)$. By the definition of log structures, we also have

(1) $\alpha(m)D_{g_1-g_2}m = \partial_{g_1-g_2}(\alpha(m)), \forall m \in \phi^{-1}\mathcal{M}_X$;
(2) $D_{g_1-g_2}|_{\psi^{-1}\mathcal{M}_Y} = 0$.

Remark 3.2. (1) Since the log structure contains all the invertible elements in the structure sheaf, the map $D_{g_1-g_2}$ determines $\partial_{g_1-g_2}$.
(2) The above properties show that $D_{g_1-g_2}$ behaves like "d log". This is one of the reasons for the name "logarithmic structure".

Summarizing the above discussion gives the following definitions:

Definition 3.3. [44],[45, Definition 1.1.1] Consider a morphism $f : X \to Y$ of fine log schemes. Let I be an $\mathcal{O}_{\underline{X}}$-module. A *log derivation of* X *over* Y *to* I is a pair (∂, D) where $\partial \in \mathcal{D}er_{\underline{Y}}(\underline{X}, I)$ and $D : \mathcal{M}_X \to I$ is an additive map such that the following conditions hold:

(1) $D(ab) = D(a) + D(b)$ for $a, b \in \mathcal{M}_X$;
(2) $\alpha(a)D(a) = \partial(\alpha(a))$, for $a \in \mathcal{M}_X$.
(3) $D(a) = 0$, for $a \in f^{-1}\mathcal{M}_Y$.

The sheaf $\mathcal{D}er_Y(X, I)$ of log derivations of X over Y to I is the sheaf of germs of pairs (∂, D). The sheaf $\mathcal{D}er_Y(X, \mathcal{O}_{\underline{X}})$ is usually denoted by $T_{X/Y}$, and is called *the logarithmic tangent sheaf of* X *over* Y.

As an analogue of differentials of usual schemes, we have the following result:

Proposition 3.4. [45, IV.1.1.6] *There exists an* $\mathcal{O}_{\underline{X}}$*-module* $\Omega^1_{X/Y}$ *with a universal derivations* $(\partial, D) \in \mathcal{D}er_Y(X, \Omega^1_{X/Y})$*, such that for any* $\mathcal{O}_{\underline{X}}$*-module* I*, the canonical map*

$$\mathcal{H}om_{\mathcal{O}_{\underline{X}}}(\Omega^1_{X/Y}, I) \to \mathcal{D}er_Y(X, I), \quad u \mapsto (u \circ \partial, u \circ D)$$

is an isomorphism of $\mathcal{O}_{\underline{X}}$*-modules. In fact, we have the following construction:*

$$\Omega^1_{X/Y} = \Omega_{\underline{X}/\underline{Y}} \oplus (\mathcal{O}_{\underline{X}} \otimes_{\mathbb{Z}} \mathcal{M}_X^{gp})/\mathcal{K}$$

where $\mathcal{K}$ *is the* $\mathcal{O}_{\underline{X}}$*-module generated by local sections of the following forms:*

(1) $(d\alpha(a),0) - (0, \alpha(a) \otimes a)$ *with* $a \in \mathcal{M}_X$;
(2) $(0,1 \otimes a)$ *with* $a \in \mathrm{Im}(f^{-1}(\mathcal{M}_Y) \to \mathcal{M}_X)$.

The universal derivation (∂, D) *is given by* $\partial : \mathcal{O}_{\underline{X}} \xrightarrow{d} \Omega_{\underline{X}/\underline{Y}} \to \Omega^1_{X/Y}$ *and* $D : \mathcal{M}_X \to \mathcal{O}_{\underline{X}} \otimes_{\mathbb{Z}} \mathcal{M}_X^{gp} \to \Omega^1_{X/Y}$.

Definition 3.5. Given a morphism $f : X \to Y$ of log schemes, the $\mathcal{O}_{\underline{X}}$-module $\Omega^1_{X/Y}$ is called *the sheaf of logarithmic differentials.* Sometimes we use the short notation Ω^1_f for $\Omega^1_{X/Y}$.

Note that $\mathcal{H}om(\Omega^1_{X/Y}, \mathcal{O}_{\underline{X}}) \cong T_{X/Y}$.

Remark 3.6. If we consider only fine log structures, and assume that $\underline{Y}$ is locally noetherian and $\underline{X}$ locally of finite type over $\underline{Y}$, then both $\mathcal{D}er_Y(X, I)$ and $\Omega^1_{X/Y}$ in the definitions above are coherent sheaves. The proof of this can be found in [45, IV.1.1]

Example 3.7. Consider $R = k[x_1, \cdots, x_n]/(x_1 \cdots x_r)$, where k is a field. Denote $\underline{X} = \mathrm{Spec}\, R$. Let $\mathcal{M}_X$ be the log structure on $\underline{X}$ given by $\mathbb{N}^r \to R$, $e_i \mapsto x_i$, where e_i is the standard generator of the monoid $\mathbb{N}^r$. Let $Y = \mathrm{Spec}\ (\mathbb{N} \to k)$ be the logarithmic point described in 2.11. Now we can define a morphism $f : X \to Y$ by the following diagram:

$$\begin{array}{ccc} \mathbb{N}^r & \longrightarrow & R \\ {\scriptstyle\Delta}\uparrow & & \uparrow \\ \mathbb{N} & \longrightarrow & k \end{array}$$

where $\Delta : e \mapsto e_1 + \cdots + e_r$, and e is the standard generator of $\mathbb{N}$. Then it is easy to see that $\mathcal{D}er_Y(X, \mathcal{O}_X)$ is a free $\mathcal{O}_{\underline{X}}$-module generated by

$$x_1 \frac{\partial}{\partial x_1}, \cdots, x_r \frac{\partial}{\partial x_r}, \frac{\partial}{\partial x_{r+1}}, \cdots, \frac{\partial}{\partial x_n},$$

with a relation $x_1 \frac{\partial}{\partial x_1} + \cdots + x_r \frac{\partial}{\partial x_r} = 0$. The sheaf Ω^1_f is a free $\mathcal{O}_X$-module generated by the logarithmic differentials:

$$\frac{dx_1}{x_1}, \cdots, \frac{dx_r}{x_r}, dx_{r+1}, \cdots, dx_n$$

with a relation $\frac{dx_1}{x_1} + \cdots + \frac{dx_r}{x_r} = 0$.

Example 3.8. Let $h : Q \to P$ be a morphism of fine monoids. Denote $X = \mathrm{Spec}\ (P \to \mathbb{Z}[P])$ and $Y = \mathrm{Spec}\ (Q \to \mathbb{Z}[Q])$. Then we have a morphism $f : X \to Y$ induced by h. A direct calculation shows that $\Omega^1_f = \mathcal{O}_{\underline{X}} \otimes \mathbf{Cok}(h^{gp})$. This can also be seen from the universal property of the sheaf of logarithmic differentials.

Logarithmic smoothness

Let us go back to the following cartesian diagram of log schemes:

(3.9)
$$\begin{array}{ccc} T_0 & \xrightarrow{\phi} & X \\ {\scriptstyle J}\downarrow{\scriptstyle j} & & \downarrow{\scriptstyle f} \\ T_1 & \xrightarrow[\psi]{} & Y \end{array}$$

where j is a strict closed immersion defined by J with $J^2 = 0$. As in the usual case, we can define log smoothness by the infinitesimal lifting property.

Definition 3.10. A morphism $f : X \to Y$ of fine log schemes is called *log smooth* (resp. *étale*) if the underlying morphism $\underline{X} \to \underline{Y}$ is locally of finite presentation and for any commutative diagram (3.9), étale locally on T_1 there exists a (resp. there exists a unique) morphism $g : T_1 \to X$ such that $\phi = g \circ j$ and $\psi = f \circ g$.

We have the following useful criterion for smoothness from [27, Theorem 3.5].

Theorem 3.11. *(K.Kato) Let $f : X \to Y$ be a morphism of fine log schemes. Assume we have a chart $Q \to \mathcal{M}_Y$, where Q is a finitely generated integral monoid. Then the following are equivalent:*

(1) *f is log smooth (resp. log étale);*
(2) *étale locally on X, there exists a chart $(P_X \to \mathcal{M}_X, Q_Y \to \mathcal{M}_Y, Q \to P)$ extending the chart $Q_Y \to \mathcal{M}_Y$, satisfying the following properties.*
  (a) *The kernel and the torsion part of the cokernel (resp. the kernel and the cokernel) of $Q^{gp} \to P^{gp}$ are finite groups of order invertible on X.*
  (b) *The induced morphism from $\underline{X} \to \underline{Y} \times_{\mathrm{Spec}\ \mathbb{Z}[Q]} \mathrm{Spec}\ \mathbb{Z}[P]$ is étale in the classical sense.*

Remark 3.12. (1) We can require $Q^{gp} \to P^{gp}$ in (a) to be injective, and replace the requirement of $\underline{X} \to \underline{Y} \times_{\mathrm{Spec}\ \mathbb{Z}[Q]} \mathrm{Spec}\ \mathbb{Z}[P]$ be étale in (b) by requiring it to be smooth without changing the conclusion of the Theorem 3.11.

(2) The arrow in (b) shows that a log smooth morphism is "locally toric" relative to the base. Consider the case Y is a log scheme with underlying space given by Spec $\mathbb{C}$ with the trivial log structure, and $X = \mathrm{Spec}(P \to \mathbb{C}[P])$ where P is a fine, saturated and torsion free monoid. Then $\underline{X}$ is a toric variety with the action of Spec $\mathbb{C}[P^{gp}]$. According to the theorem, X is log smooth relative to Y, though the underlying space might be singular. These singularities are called toric singularities in [28]. This is closely related to the classical notion of toroidal embeddings, see [32].

Example 3.13. Using the theorem, we can check directly that the morphism f in example 3.7 is log smooth, but the underlying map has normal crossing singularities. We will see later that one of the major advantages of log structures is in dealing with the normal crossing singularities.

Let $X \xrightarrow{f} Y \xrightarrow{g} Z$ be morphisms of fine log schemes. Consider the sheaves of log differentials Ω^1_g and $\Omega^1_{g\circ f}$, with their universal derivations (∂_g, D_g) and $(\partial_{g\circ f}, D_{g\circ f})$ respectively. We have a canonical map $f^*\Omega^1_g \to \Omega^1_{g\circ f}$ induced by

$$f^*(\partial_g u) \mapsto \partial_{g\circ f} f^*(u) \quad \text{and} \quad f^*(D_g v) \mapsto D_{g\circ f} f^\flat(v),$$

where $u \in \mathcal{O}_Y$ and $v \in \mathcal{M}_Y$. Denote by (∂_f, D_f) the universal derivation associated to Ω^1_f. Similarly, we have a canonical map $\Omega^1_{g\circ f} \to \Omega^1_f$ induced by

$$\partial_{g\circ f} u' \mapsto \partial_f u' \quad \text{and} \quad D_{g\circ f} v' \mapsto D_f v',$$

where $u' \in \mathcal{O}_X$ and $v' \in \mathcal{M}_X$. The following proposition shows that log differentials behave like usual differentials, especially for log smooth morphisms.

Proposition 3.14. (1) *The sequence* $f^*\Omega^1_g \to \Omega^1_{g\circ f} \to \Omega^1_f \to 0$ *is exact.*

(2) *If* f *is log smooth, then* Ω^1_f *is a locally free* $\mathcal{O}_X$*-module, and we have the following exact sequence:* $0 \to f^*\Omega^1_g \to \Omega^1_{g\circ f} \to \Omega^1_f \to 0$.

(3) *If* $g \circ f$ *is log smooth and the sequence in (2) is exact and splits locally, then* f *is log smooth.*

A proof can be found in [45, Chapter IV].

Logarithmic smooth deformation

Having discussed log smoothness, a natural thing to do is to develop log smooth deformations. In many cases, we would require this to be a flat deformation for the underlying space. Unfortunately log smoothness does not imply flatness, so we need the following definition.

Definition 3.15. A map of fine monoids $h : Q \to P$ is called *integral* if the induced map on monoid algebra $\mathbb{Z}[Q] \to \mathbb{Z}[P]$ is flat.

Definition 3.16. A morphism $f : X \to Y$ of fine log schemes is called *integral* if for every geometric point $\bar{x} \in X$, the map of characteristic monoids $h : f^{-1}(\overline{\mathcal{M}}_Y)_{\bar{x}} \to \overline{\mathcal{M}}_{X,\bar{x}}$ is integral.

Remark 3.17. (1) If f is integral, then étale locally we have a chart $(P_X \to \mathcal{M}_X, Q_Y \to \mathcal{M}_Y, Q \xrightarrow{h} P)$ such that h is integral.

(2) If $h : Q \to P$ is integral of integral monoids, then for any integral monoid Q', the push-out of $P \leftarrow Q \to Q'$ in the category of monoids is integral. Thus integral morphisms are stable under base change by integral log schemes.

(3) Given a morphism $h : Q \to P$ of integral monoids, there is an explicit criterion, which looks complicated, but sometimes is useful for checking integrality of h directly: if $a_1, a_2 \in Q$, $b_1, b_2 \in P$ and $h(a_1)b_1 = h(a_2)b_2$, then there exist $a_3, a_4 \in Q$ and $b \in P$ such that $b_1 = h(a_3)b$, $b_2 = h(a_4)b$ and $a_1a_3 = a_2a_4$. This comes essentially from the equational criterion for flatness.

Now we have the following fact from [27, 4.5].

Proposition 3.18. *If* f *is a log smooth and integral morphism of fine log schemes, then* $\underline{f}$ *the underlying map is flat in the usual sense.*

Now let us consider the following deformation problem. We are given a log smooth integral morphism $f_0 : X_0 \to B_0$ of fine log schemes, and a strict closed immersion $j : B_0 \to B$ defined by an ideal J with $J^2 = 0$. We want to find a log smooth lifting $f : X \to B$ fitting in the following cartesian diagram:

$$\begin{array}{ccc} X_0 & \hookrightarrow & X \\ \downarrow & & \downarrow \\ B_0 & \hookrightarrow & B. \end{array}$$

Remark 3.19. Since f_0 is integral, and $\mathcal{M}_X/(1+J) \cong \mathcal{M}_{X_0}$, it is not hard to show that the lifting f is automatically integral and hence flat. Without the integrality assumption a lifting of a log smooth morphism might fail to be flat or log smooth - see [24, Remark 8.2].

We have the following theorem for log smooth deformations.

Theorem 3.20. [27, 3.14]*With the notation as above, we have:*

(1) *There is a canonical obstruction* $\eta \in H^2(\underline{X_0}, T_{X_0/B_0} \otimes J)$ *such that* $\eta = 0$ *if and only if there exists a log smooth lifting.*
(2) *If* $\eta = 0$*, the set of log smooth deformations form a torsor under* $H^1(\underline{X_0}, T_{X_0/B_0} \otimes J)$.
(3) *The automorphism group of any deformation is given by* $H^0(\underline{X_0}, T_{X_0/B_0} \otimes J)$.

The theorem can be proved in a manner similar to the case of usual deformation theory as in [15, Exposé 3]. Another proof using the logarithmic cotangent complex can be found in [49, Thm 5.6], which we will discuss later.

4. Log smooth curves and their moduli

Now that we have reviewed some of the foundations, we can discuss the first application to the study of moduli spaces: F. Kato's interpretation of $\overline{\mathcal{M}}_{g,n}$ as a moduli space for log curves. The general philosophy is expressed by F. Kato in the introduction to [26]:

Philosophy. Since log smoothness includes some degenerating objects like semistable reductions, etc., the moduli space of log smooth objects should be already compact, once its existence has been established.

Along the lines of this philosophy, to compactify $\mathcal{M}_{g,n}$, we want to introduce a notion of *log curve* which extends the notion of smooth curve. Following F. Kato, we do so after some preliminaries.

Relative characteristic sheaves

Recall from Definition 2.3 that the characteristic $\overline{\mathcal{M}}_X$ of a log scheme X is defined as $\mathcal{M}_X/\mathcal{O}_{\underline{X}}^*$. In the study of log curves, the following relative notion of characteristic plays an important role.

Definition 4.1. Given a morphism $f : X \to Y$ of log schemes, the *relative characteristic* $\overline{\mathcal{M}}_{X/Y}$ is defined as the quotient $\mathcal{M}_X/\operatorname{im}(f^*\mathcal{M}_Y \to \mathcal{M}_X)$.

Example 4.2. Let $f : X \to Y$ be the morphism from Example 3.7. Then the relative characteristic $\overline{\mathcal{M}}_{X/Y}$ is the cokernel in the category of integral monoids of the diagonal map $\Delta : \mathbb{N} \to \mathbb{N}^2$, which is $\mathbb{Z}$.

Lemma 4.3 ([26, Lemma 1.6]). *If $f : X \to Y$ is an integral morphism of fine log schemes, then $\overline{\mathcal{M}}_{X/Y,\bar{x}} = 0$ if and only if f is strict in an étale neighborhood of x.*

As the following example illustrates, the integrality assumption on f is necessary.

Example 4.4. Let P be the monoid on three generators x, y, and z subject to the relation $x + z = 2y$. We have an injection

$$i : P \longrightarrow \mathbb{N}^2$$

sending x to (2,0), y to (1,1), and z to (0,2). Let $X = \operatorname{Spec} k[\mathbb{N}^2]$ and $Y = \operatorname{Spec} k[P]$ with their canonical log structures. Then it follows from [27, Prop 3.4] that the morphism of log schemes $f : X \to Y$ induced by i is log étale, but $\underline{f}$ is not flat, and hence, by [27, Cor 4.5], f is not integral. It is easy to check that the *non-integral* quotient

$$\mathcal{M}_{X,0}/\operatorname{im}(f^*\mathcal{M}_{Y,0}) = \mathbb{N}^2/P \simeq \mathbb{Z}/2,$$

so $\overline{\mathcal{M}}_{X/Y,0} = 0$, but f is not strict.

Log curves

Definition 4.5. A *log curve* is a log smooth integral morphism $f : X \to S$ of fs log schemes such that the geometric fibers of $\underline{f}$ are reduced connected 1-dimensional schemes.

We require f to be integral so that, by [27, Cor 4.5], $\underline{f}$ is flat. The reason for the fs assumption is to avoid cusps or worse singularities, as the following example shows.

Example 4.6. If $X = \operatorname{Spec} k[\mathbb{N} - \{1\}]$ is given its canonical log structure and $S = \operatorname{Spec} k$ is given the trivial log structure, then $X \to S$ is log smooth and integral; however, $\underline{X} = \operatorname{Spec} k[x, y]/(y^2 - x^3)$ has a cusp.

It is a remarkable fact that by endowing our curves with log structures as in Definition 4.5, this is enough to control the singularities of the curve.

Theorem 4.7 ([26, Thm 1.3]). *If* k *is a separably closed field and* $f : X \to S$ *is a log curve with* $\underline{S} = \operatorname{Spec} k$, *then* $\underline{X}$ *has at worst nodal singularities. Moreover, if* $r_1, \ldots, r_\ell$ *are the nodes of* $\underline{X}$, *then there exist smooth points* $s_1, \ldots, s_n$ *of* $\underline{X}$ *such that*

$$\overline{\mathcal{M}}_{X/S} = \mathbb{Z}_{r_1} \oplus \cdots \oplus \mathbb{Z}_{r_\ell} \oplus \mathbb{N}_{s_1} \oplus \ldots \mathbb{N}_{s_n};$$

here M_x *denotes the skyscraper sheaf for a monoid* M *supported at a point* $x \in \underline{X}$.

The reader should think of the s_i in the above theorem as marked points. So we can already see how n-pointed curves emerge naturally from the log geometry perspective. The stalk $\mathbb{Z}$ of the relative characteristic at a node can be seen in Example 3.7, taking $n = r = 2$.

Example 4.8. Consider the closed subscheme $\underline{X}$ of $\mathbb{P}^2_k \times_k \mathbb{A}^1_k$ defined by $xz = ty$, where t is the coordinate of $\mathbb{A}^1_k$ and x, y, and z are the coordinates of $\mathbb{P}^2_k$. Then $\underline{X}$ has a natural log structure $\mathcal{M}_X$. For example, on the locus where z is invertible, $\underline{X}$ is given by $\operatorname{Spec} k[P_z]$ with P_z a monoid on five generators a, b, c, c', u subject to the relations $c + c' = 0$ and $a + c = b + u$; here $\mathcal{M}_X$ is given by the canonical log structure associated to P_z. Then the projection

$$X \longrightarrow \mathbb{A}^1_k$$

is a log curve, where X is given the log structure above and $\mathbb{A}^1_k$ is given the log structure defined by the divisor $t = 0$. We see that every fiber above $t \neq 0$ is isomorphic to $\mathbb{P}^1_k$ with log structure given by the divisor at 0 and ∞; the fiber above $t = 0$ is nodal. The n in Theorem 4.7 is equal to 2 for all geometric fibers.

Since our goal is to give a log geometric description of $\overline{\mathcal{M}}_{g,n}$, we would like to express the stability condition purely in terms of log geometry. The following proposition provides the key.

Proposition 4.9 ([26, Prop 1.13]). *With notation as in Theorem 4.7, there is a natural isomorphism*

$$\Omega^1_{X/S} \longrightarrow \omega_{\underline{X}}(s_1 + \cdots + s_n),$$

where $\omega_{\underline{X}}$ *is the dualizing sheaf of* $\underline{X}$.

We therefore make the following definition.

Definition 4.10. Let $f : X \to S$ be a log curve and for all geometric points $\bar{t}$ of $\underline{S}$, let $\ell(\bar{t})$ and $n(\bar{t})$ be such that

$$\overline{\mathcal{M}}_{X_{\bar{t}}/\bar{t}} = \mathbb{Z}_{r_1} \oplus \cdots \oplus \mathbb{Z}_{r_{\ell(\bar{t})}} \oplus \mathbb{N}_{s_1} \oplus \ldots \mathbb{N}_{s_{n(\bar{t})}}.$$

We say f is *of type* (g, n) if $\underline{f}$ is proper, $\underline{X}$ has genus g, and $n(\bar{t}) = n$ for all $\bar{t}$. We say f is *stable of type of* (g, n) if it of type (g, n) and

$$H^0(\underline{X}_{\bar{t}}, T_{X_{\bar{t}}/\bar{t}}) = 0$$

for all geometric points $\bar{t}$ of $\underline{S}$.

It is, in fact, true ([26, Prop 1.7]) that if $f : X \to S$ is a log curve of type (g, n), then the s_i in each geometric fiber fit together to yield n sections σ_i of $\underline{f}$. It follows then that every stable log curve of type (g, n) is an n-pointed stable curve of genus g in the classical sense.

Log structures on stable curves

Having now shown that every log curve is naturally a pointed nodal curve, we shift gears and ask the following question: given a stable genus g curve $\underline{f} : \underline{X} \to \underline{S}$ with n marked points, how many log structures can we put on $\underline{X}$ and $\underline{S}$ so that the associated morphism of log schemes is a log curve with relative characteristic supported on our given n marked points? We begin by sketching the construction of a canonical such log structure.

Lemmas 2.1 and 2.2 of [26] show that to endow $\underline{X}$ and $\underline{S}$ with log structures as desired, it is enough to consider the case when $\underline{S} = \operatorname{Spec} A$ and A is strict Henselian. For every node r_i of the closed fiber of $\underline{f}$, we can find an étale neighborhood U_i of the points specializing to r_i and a diagram

$$\begin{array}{ccc} U_i & \xrightarrow{\psi_i} & \operatorname{Spec} A[x, y, t]/(xy - t) \\ \downarrow & & \downarrow \pi \\ S & \xrightarrow{\varphi_i} & \operatorname{Spec} A[t] \end{array}$$

which is cartesian. Let $t_i \in A$ be the image of t under the morphism induced by φ_i. Endowing $\operatorname{Spec} A[t]$ with the log structure associated to the morphism $\mathbb{N} \to A[t]$ sending 1 to t, and $\operatorname{Spec} A[x, y, t]/(xy - t)$ with the log structure associated to $\mathbb{N}^2 \to A[x, y, t]/(xy - t)$ sending e_1 (resp. e_2) to x (resp. y), we see that (π, Δ) is a morphism of log schemes, where $\Delta : \mathbb{N} \to \mathbb{N}^2$ is the diagonal map. Pulling back these log structures under φ_i (resp. ψ_i), we obtain log structures $\mathcal{L}_i$ (resp. $\mathcal{M}_i'$) on S (resp. U_i). Away from the points specializing to r_i, we define a log structure $\mathcal{M}_i''$ as the pullback of $\mathcal{L}_i$. The log structures $\mathcal{M}_i'$ and $\mathcal{M}_i''$ glue to yield a log structure $\mathcal{M}_i$ on X. Let $\mathcal{N}$ be the log structure on X associated to the divisor defined by the marked points. Writing ℓ for the number of nodes, we let

$$\mathcal{M}_X = \mathcal{M}_1 \oplus_{\mathcal{O}_X^*} \cdots \oplus_{\mathcal{O}_X^*} \mathcal{M}_\ell \oplus_{\mathcal{O}_X^*} \mathcal{N}$$

and

$$\mathcal{M}_S = \mathcal{L}_1 \oplus_{\mathcal{O}_S^*} \cdots \oplus_{\mathcal{O}_S^*} \mathcal{L}_\ell .$$

It is not difficult to see that with these definitions, we have endowed f with the structure of a log curve.

Moreover, a detailed analysis of the proof of Theorem 4.7 shows that this log structure we have just constructed is "minimal" among all possible log structures giving $\underline{X}/\underline{S}$ the structure of a log curve (see 1.8 and Thm 2.3 of [26]):

Theorem 4.11. *Let $\underline{X}/\underline{S}$ be a stable genus g curve with n marked points and let X/S be the log curve obtained by endowing $\underline{X}/\underline{S}$ with the canonical log structure above. If X'/S' is a log curve and $\underline{a} : \underline{S}' \to \underline{S}$ and $\underline{b} : \underline{X}' \to \underline{X}$ are morphisms such that $\underline{X}' \simeq \underline{X} \times_{\underline{S}} \underline{S}'$ and such that the divisors of marked points in $\underline{X}'$ are sent scheme-theoretically to the divisors of marked points in $\underline{X}$, there are unique morphisms a and b of log schemes extending the morphisms $\underline{a}$ and $\underline{b}$ above such that*

$$\begin{array}{ccc} X' & \xrightarrow{b} & X \\ \downarrow & & \downarrow \\ S' & \xrightarrow{a} & S \end{array}$$

is cartesian in the category of fs log schemes.

Definition 4.12. A log curve X/S is called *basic* if it satisfies the universal property in Theorem 4.11.

We note that one often meets non-basic log curves. For instance, consider a one-parameter family of curves $X \to S$ over a smooth curve S with smooth total space X. Assume that the generic fiber is a smooth curve of genus 2, but the special fiber over $s \in S$ is the union of two rational curves glued at three nodes. We can give S the log structure associated to the divisor s, and give X the log structure associated to the normal crossings divisor X_s. Then $X \to S$ is log smooth but not basic, since $\overline{\mathcal{M}}_{S,s} = \mathbb{N}$ but the fiber has three nodes. Intuitively, the log structure $\overline{\mathcal{M}}_{S,s}$ remembers the particular direction of deformation of the nodes, whereas the basic log structure does not give preference to any particular direction of deformation.

Moduli

Before discussing moduli of log smooth curves, we begin with some generalities about log structures and stacks. Note that the definition of a log structure is not particular to schemes; indeed, Definition 2.2 makes sense for any ringed topos. We can therefore define log structures on the étale site of a Deligne-Mumford stack or the lisse-étale site of an Artin stack. The notions of fine and fs log structures carry over to this setting as well, so one can speak of fine (or fs) log algebraic stacks.

There is another equivalent way to talk about log structures on stacks; namely, if $\mathfrak{X}$ is a stack over the category of schemes, and $\mathcal{M}_{\mathfrak{X}}$ is a log structure on $\mathfrak{X}$, then $\mathfrak{X}$ can naturally be viewed as a stack over the category of log schemes. For concreteness, say $\mathfrak{X}$ is a stack over the category of schemes with the étale topology. Then we obtain a category $\tilde{\mathfrak{X}}$ fibered over the category of log schemes by defining $\tilde{\mathfrak{X}}(S, \mathcal{M}_S)$ to be the category whose objects are morphisms

$$f : (S, \mathcal{M}_S) \to (\mathfrak{X}, \mathcal{M}_{\mathfrak{X}})$$

of log stacks and whose morphisms g from $f \in \tilde{\mathfrak{X}}(S, \mathcal{M}_S)$ to $f' \in \tilde{\mathfrak{X}}(S', \mathcal{M}_{S'})$ are given by diagrams

$$\begin{array}{ccc} (S, \mathcal{M}_S) & \xrightarrow{\quad g \quad} & (S', \mathcal{M}_{S'}) \\ & {\scriptstyle f}\searrow \quad \swarrow{\scriptstyle f'} & \\ & (\mathfrak{X}, \mathcal{M}_{\mathfrak{X}}). & \end{array}$$

One checks that $\tilde{\mathfrak{X}}$ is a stack over the category of log schemes where coverings are given by surjective strict étale morphisms.

Conversely, given any stack $\mathcal{Y}$ over the category of log schemes with the strict étale topology, we obtain a log stack $(\mathcal{Y}', \mathcal{M}_{\mathcal{Y}'})$ over the category of schemes with the étale topology by letting $\mathcal{Y}'(S)$ be the category of pairs $(\mathcal{M}_S, \xi)$ where ξ is an object of $\mathcal{Y}(S, \mathcal{M}_S)$. The log structure $\mathcal{M}_{\mathcal{Y}'}$ is then defined by the following property: if $f : S \to \mathcal{Y}$ is a morphism which corresponds to the pair $(\mathcal{M}_S, \xi)$, then $f^*\mathcal{M}_{\mathcal{Y}'} = \mathcal{M}_S$.

It is however important to note that these *two procedures are not inverse to each other*. If we start with a logarithmic stack $(\mathfrak{X}, \mathcal{M}_{\mathfrak{X}})$, and take the result $(\tilde{\mathfrak{X}})'$ of the composite operation, we do not get $\mathfrak{X}$ but rather the stack $LOG_{\mathfrak{X}}$ described in section 6. In order to recover the stack $\mathfrak{X}$ over schemes from a stack $\tilde{\mathfrak{X}}$ over log schemes, it is necessary to distinguish objects similar to the basic log curves of Definition 4.12. Rather than launch into a premature categorical discussion, let us see how this works for log curves. The issue is revisited in sections 11 (with Olsson's terminology of *special log structures*) and 12 (with Kim's terminology of *minimal log structures*).

Let $\overline{\mathcal{M}}_{g,n}^{log}$ be the stack over the category of fs log schemes with the strict étale topology where $\overline{\mathcal{M}}_{g,n}^{log}(S, \mathcal{M}_S)$ is the category of stable log curves of type (g, n) over $(S, \mathcal{M}_S)$. Let $\mathcal{M}_{g,n}^{bas}$ be the sub-stack of basic stable log curves of type (g, n). By the above discussion, we obtain a log stack over the category of schemes with the étale topology, which we again denote by $(\mathcal{M}_{g,n}^{bas}, \mathcal{M}_{\mathcal{M}_{g,n}^{bas}})$.

Note that the Deligne-Mumford compactification $\overline{\mathcal{M}}_{g,n}$ carries a natural log structure $\mathcal{M}_{\overline{\mathcal{M}}_{g,n}}$ coming from the simple normal crossing divisor at the boundary. It follows from [9, §2] that $\mathcal{M}_{\overline{\mathcal{M}}_{g,n}}$ can also be described as the log structure which assigns to each stable curve $\underline{X}/\underline{S}$ the basic log structure obtained on $\underline{S}$. The discussion following Definition 4.5 above shows that we have a natural morphism

$$F : (\mathcal{M}_{g,n}^{bas}, \mathcal{M}_{\mathcal{M}_{g,n}^{bas}}) \longrightarrow (\overline{\mathcal{M}}_{g,n}, \mathcal{M}_{\overline{\mathcal{M}}_{g,n}})$$

and from Theorem 4.11 we see

Theorem 4.13 ([26, §4]). *The morphism* F *is an equivalence of log stacks.*

4.14. Back to the big picture

We end by mentioning a type of converse to the philosophic principal mentioned at the beginning of this section. We have seen that since log smoothness includes degenerate objects, log geometry can naturally lead to compactifications;

however, it is also generally true that we do not end up with "too many" degenerate objects.

Philosophy. Log geometry controls degenerations.

In higher dimensions, compactifications tend to have unwanted extra components. Log geometry helps to cut down on these components. Let us give some inkling of an idea as to why this should be true. Suppose $\mathfrak{X}$ is an algebraic stack which is irreducible. Suppose we can find a proper algebraic stack $\bar{\mathfrak{X}}$ with a fine log structure and an open immersion $i : \mathfrak{X} \to \bar{\mathfrak{X}}$ such that $\mathfrak{X}$ is the trivial locus of $\bar{\mathfrak{X}}$.

As we now explain, log geometry provides us with a good method of trying to show that $\bar{\mathfrak{X}}$ is irreducible as well. If k is separably closed and $x : \operatorname{Spec} k \to \bar{\mathfrak{X}}$ is a morphism, then pulling back the log structure on $\bar{\mathfrak{X}}$ endows $\operatorname{Spec} k$ with a fine log structure. By [27, Lemma 2.10], it follows that this is the log structure associated to a morphism of monoids $P \to k$, where P is fine. Hence, we have a strict closed immersion of log schemes $j : \operatorname{Spec} k \to \operatorname{Spec} k[[P]]$, where $\operatorname{Spec} k[[P]]$ is given its canonical log structure. Note that the generic point $\operatorname{Spec} K$ of $\operatorname{Spec} k[[P]]$ carries the trivial log structure. Therefore, if x factors as a morphism of log stacks through j, then we automatically obtain a commutative diagram

$$\begin{array}{ccccc} \mathfrak{X} & & \xrightarrow{\quad i \quad} & & \bar{\mathfrak{X}} \\ \uparrow{\scriptstyle y} & & & \nearrow & \uparrow{\scriptstyle x} \\ \operatorname{Spec} K & \longrightarrow & \operatorname{Spec} k[[P]] & \xleftarrow[j]{} & \operatorname{Spec} k \end{array}$$

and hence x is the specialization of the point y of $\mathfrak{X}$. We see then that the log structure on $\operatorname{Spec} k$ obtained from $\bar{\mathfrak{X}}$ somehow serves as a compass telling us which way to look in order to find a family degenerating to our given point x of $\bar{\mathfrak{X}}$.

5. Friedman's concept of d-semistability and log structures

The main references here are [10, 24, 47].

Convention. Throughout this section, every scheme is over an algebraically closed field. The notation X will be reserved for a *normal crossing variety*. By this we mean a variety for which every closed point $x \in X$ has an étale neighborhood $x \to U$ which admits an étale map $U \to \operatorname{Spec} \dfrac{k[x_1, \cdots, x_n]}{(x_1 \cdots x_r)}$ such that x maps to the point with coordinates $(0, \cdots, 0)$. By a standard neighborhood of $x \in X$, we mean an étale neighborhood of x as above. The notation $\mathcal{M}_k$ denotes the log structure of the standard log point $\mathbb{N} \to k, n \mapsto 0^n$, see Example 2.11.

Introduction

To study the geometry of a normal crossing variety X, e.g. to study the deformation theory of X, one would like to ask the following questions:

Question 5.1. Can we embed X into another smooth variety $i : X \to \mathscr{X}$ as a normal crossing divisor?

Question 5.2. Can we find a semi-stable smoothing of X, i.e. embed $X \to \operatorname{Spec} k$ in a flat family over a curve $\mathscr{X} \to C$, in such a way that there exists a diagram

$$\begin{array}{ccccc} X & \longrightarrow & \mathscr{X} & \longleftarrow & \mathscr{X}^* = \mathscr{X} \backslash X \\ \downarrow & & \downarrow & & \downarrow {\scriptstyle f^*} \\ \operatorname{Spec} k & \xrightarrow{0} & C & \longleftarrow & C^* = C \backslash 0 \end{array}$$

where $\mathscr{X}$ is smooth, the squares are cartesian and f^* is smooth?

The answer to these questions are not always yes, because the existence of such maps would imply the existence of certain log structures on X, which in turn would imply intrinsic condition on X. so their existence is not guaranteed.

Example 5.3 (log structure of embedding type). If we can find an embedding as in Question 5.1 above, then $i^*(j_* \mathcal{O}^{\times}_{\mathscr{X}^*}) \to \mathcal{O}_X$ defines a log structure $\mathcal{M}_X$ on X, which étale locally has a chart $\mathbb{N}^r \to \operatorname{Spec} \dfrac{k[x_1, \cdots, x_n]}{(x_1 \cdots x_r)}$ sending the element e_i in the standard basis of $\mathbb{N}^r$ to x_i. This is called a *log structure of embedding type.*

Example 5.4 (log structure of semi-stable type). If we can find a semi-stable smoothing as in Question 5.2 above, then what we have in this case is not only a log structure of embedding type on X, but also a morphism (of sheaf of monoids) $f^\flat : f^*\mathcal{M}_k \to \mathcal{M}_X$, which makes X a log smooth variety over the standard log point $(\operatorname{Spec} k, \mathcal{M}_k)$. Étale locally a chart for the log structure on X can be put in the form

$$\Big(\operatorname{Spec} \frac{k[x_1, \cdots, x_n]}{(x_1 \cdots x_r)}, \mathbb{N}^r, e_i \mapsto x_i, i = 1, \cdots, r\Big).$$

Modulo the units in the monoid, the morphism of quotient monoids induced by $f^\flat : f^*\mathcal{M}_k \to \mathcal{M}_X$ is just the diagonal $\Delta : \mathbb{N} \to \mathbb{N}^r$. Such a pair $(\mathcal{M}_X, f^\flat : f^*\mathcal{M}_k \to \mathcal{M}_X)$ is called a *log structure of semi-stable type on* X.

Remark 5.5. The log structure $\mathcal{M}_k$ on $\operatorname{Spec} k$ can be defined by the pullback of the log structure $(\mathcal{M}_0 = j_* \mathcal{O}^{\times}_{C^*} \to \mathcal{O}_C)$ on C, i.e. the log structure defined by the divisor 0 of C. We have an isomorphism $\overline{\mathcal{M}_k} := \mathcal{M}_k / k^\times \cong (\mathcal{M}_0 / \mathcal{O}^{\times}_C)_0 \cong \mathbb{N}$, where the second isomorphism assigns each function to its vanishing order at the point 0 in $\mathbb{N}$. This gives a geometric interpretation of the standard log point.

Concerning the existence of such a log structure on the normal crossing variety X, we have the following theorems ([24], Sec.11):

Theorem 5.6. *Let X be a normal crossing variety over the spectrum of an algebraically closed field, then X can be equipped with a log structure of embedding type iff there exists a line bundle $\mathcal{L}$ on X such that*

$$\mathcal{E}xt^1_X(\Omega^1_X, \mathcal{O}_X)|_D \cong \mathcal{L}|_D$$

where D is the non-smooth locus of X.

Remark 5.7. It is not hard to see $\mathcal{E}xt^1_X(\Omega^1_X, \mathcal{O}_X)|_D$ is a line bundle on D.

Definition 5.8. (see [10, Def. 1.13 and Prop. 2.3]) Let X and D as before. If $\mathcal{E}xt^1_X(\Omega^1_X, \mathcal{O}_X)|_D$ is a trivial on D, then we say that X is *d-semistable.*

Theorem 5.9 (d-semistability). *Let X be a normal crossing variety over the spectrum of an algebraically closed field, then X can be equipped with a log structure over the standard log point, such that the structure morphism is log smooth if and only if X is d-semistable.*

Generalization of these theorems can be found in [47], Section 3.

Corollary 5.10. *If X has a semistable smoothing as in Question 5.2, then X is d-semistable.*

In fact, we can put a log structure on X such that it is log smooth over the standard log point by Example 5.4, so we can apply Theorem 5.9

Remark 5.11. Being d-semistable is not equivalent to having a semi-stably smoothing. See [54, Section 3] for counterexamples.

Example 5.12 (a normal crossing variety that is not d-semistable [10]). Let X be the subvariety of $\mathbf{P}^3$ defined by the product of 4 linear equations $f = L_1L_2L_3L_4 = 0$. It is a normal crossing variety provided the four planes has no points in common. Then D is defined by the homogeneous ideal $(L_iL_j | 1 \leqslant i < j \leqslant 4)$, and it is not hard to calculate that $\mathcal{E}xt^1_X(\Omega^1_X, \mathcal{O}_X)|_D \cong \mathcal{O}_D(4)$, which is not trivial. So X is not d-semistable.

Corollary 5.13. *The X in Example 5.12 does not admit a semi-stably smoothing.*

Remark 5.14. If we put X in the 1−dimensional family $\mathscr{X}$ defined by $f + tg = 0$ with parameter t, where g is a smooth quartic, then X is the fiber over $t = 0$ and for generic g, $\mathscr{X}$ over $(t \neq 0)$ is smooth. But the whole space of this family is not smooth: in fact, for a generic g, this family is singular at the 24 points of $D \cap \{g = 0\}$. However, a single blowing up at such a point gives a $\mathbf{P}^1 \times \mathbf{P}^1$. Contract along either ruling gives back a family with parameter t which has one less singularity. If we do this process to all 24 points, we will get a family which is a semi-stable smoothing of $\tilde{X}$, the blowing-up of X at those 24 points. $\tilde{X}$ is d-semistable. For details, see [10, Rem. 1.14].

Refined analysis of the existence of log structures

Continuing with the notation X, D above, let us analyze the situation. We want to break the job of finding a suitable log structure over the standard log point into 2 steps:

(1) Put a log structure of embedding type on X.
(2) See if it is possible to make that log structure semi-stable.

We will see that two related obstructions arise naturally, where the vanishing of the first corresponds to the first step, and the vanishing of the second, which means precisely being d-semistable, allows us to do the second step.

Since étale locally a log structure of embedding type always exists, let us consider the stack $\mathcal{G}$, which to each $U \in X_{\acute{e}t}$ associates the groupoid of log structure of embedding type on U. Using Artin's approximation theorem one can show that any two elements of $\mathcal{G}(U)$ are locally isomorphic, which means $\mathcal{G}$ is a gerbe. Since $\mathrm{Aut}\,\mathcal{G} \cong \mathcal{K}$, where $\mathcal{K}$ is the kernel of the restriction map $\mathcal{O}_X^\times \to \mathcal{O}_D^\times$, we have:

Proposition 5.15. *There is an obstruction* η *in* $H^2(X_{\acute{e}t}, \mathcal{K})$ *whose vanishing is equivalent to the existence of a log structure of embedding type on* X.

For the calculation of this obstruction, we state the following result (See [24], Sec.11 and [47], Sec.3):

Proposition 5.16. *In the long exact sequence of cohomology associated to the short exact sequence* $1 \to \mathcal{K} \to \mathcal{O}_X^\times \to \mathcal{O}_D^\times \to 1$, *the line bundle* $\mathcal{E}xt_X^1(\Omega_X^1, \mathcal{O}_X)|_D$ *maps to* $-\eta \in H^2(X_{\acute{e}t}, \mathcal{K})$.

Combining these results with the exactness of the long exact sequence , we get Theorem 5.6.

Remark 5.17. General theory tells us if $\eta = 0$, then the set of all log structure of embedding type on X is naturally a torsor under $H^1(X_{\acute{e}t}, \mathcal{K})$. In general the set of all log structure of embedding type is only a pseudo torsor under this group.

Suppose $\eta = 0$, then we can put a log structure of embedding type on X, which maps to $(\mathrm{Spec}\,k, k^\times)$. To go one step further, i.e., to make it a log structure of semistable type, we need a morphism of monoids $f^*\mathcal{M}_k \to \mathcal{M}_X$, such that in an standard neighborhood of $x \in X$, a chart of the morphism is given by the diagonal $\Delta : \mathbb{N} \to \mathbb{N}^r$, where r is the number of irreducible components passing through x.

Since $\mathcal{M}_k \cong \mathbb{N} \oplus k^\times$ (non-canonically), and the image of $k^\times \subset \mathcal{M}_k$ is determined by the underlying morphism of schemes. To give the morphism wanted from $\mathcal{M}_k$ to $\mathcal{M}_X$, we only have to specify the image of an element in $\mathcal{M}_k$ having vanishing order 1.

Now the question becomes a lifting problem for morphism of sheaves in monoids:

$$\begin{array}{ccc} & & \mathcal{M}_X \\ & \nearrow & \big\downarrow \beta \\ \mathbb{N} \cong f^{-1}\overline{\mathcal{M}_k} & \xrightarrow{\overline{f^\flat}} & \overline{\mathcal{M}_X} \end{array}$$

where étale locally $\overline{f^\flat}$ is the diagonal Δ. To lift $\overline{f^\flat}$ it is equivalent to lift the element $\overline{f^\flat}(1)$.

Consider the sheaf of all the possible local liftings of $\overline{f^\flat}(1)$, $T = \beta^{-1}(\overline{f^\flat}(1))$, then T is a torsor under $\mathcal{O}_X^\times$. To find a lifting of $\overline{f^\flat}(1)$, it is equivalent to find a global section of T, i.e. a trivialization of T.

It seems like we have got an obstruction of finding a log structure on X which is semi-stable in $H^1(X_{\acute{e}t}, \mathcal{O}_X) = \mathrm{Pic}(X)$. This is, however, not quite true. In fact, what we got is for each log scheme X with a log structure of embedding type, the obstruction of making it semi-stable. And our original question (on the existence of log structure of semi-stable type) allows some ambiguity of choosing the log structure of embedding type $\mathcal{M}_X$ on X. As we said in remark 5.17, in this case the set of all log structure of embedding type on X is an $H^1(X_{\acute{e}t}, \mathcal{K})$-torsor, which implies:

Proposition 5.18. *If $\eta = 0$, i.e. there exists a log structure of embedding type on X. Then there is an obstruction for finding a log structure of semi-stable type on X, $\eta' \in H^1(X_{\acute{e}t}, \mathcal{O}_X^\times)/H^1(X_{\acute{e}t}, \mathcal{K})$, whose vanishing is equivalent to the existence of such a log structure.*

By the long exact sequence of cohomology, $H^1(X_{\acute{e}t}, \mathcal{O}_X^\times)/H^1(X_{\acute{e}t}, \mathcal{K})$ embeds into $H^1(D_{\acute{e}t}, \mathcal{O}_D^\times) \cong \mathrm{Pic}(D)$. For the calculation of η' as an element of $\mathrm{Pic}(D)$, we state the following proposition (See [47]).

Proposition 5.19. *We have* $-\eta' = [\mathcal{E}xt^1_X(\Omega^1_X, \mathcal{O}_X)|_D] \in \mathrm{Pic}(D)$.

Combining these two proposition, we get Theorem 5.9.

6. Stacks of logarithmic structures

The main reference of this section is [46].

A motivating example

Before introducing the stack LOG_S classifying fine log structures on schemes over a fine log scheme S, constructed by Olsson in [46], let us look at an example, which will give the local covers of LOG_S.

Definition 6.1. Let $\underline{X}$ be a scheme and $r \geqslant 1$ an integer. A *Deligne-Faltings log structure of rank* r *on* $\underline{X}$ (abbreviated as a *DF log structure of rank* r) is the following data:

- a sequence $L_1, \cdots, L_r$ of line bundles on $\underline{X}$, and
- a morphism $s_i : L_i \to \mathcal{O}_X$ of line bundles, for each i.

Consider the following three categories fibered in groupoids over the category of schemes:

(1) the category of triples $(\underline{X}, L, s : L \to \mathcal{O}_{\underline{X}})$ consisting of a scheme $\underline{X}$ and a DF log structure of rank 1 on $\underline{X}$;
(2) the category of pairs $(X, \beta : \mathbb{N} \to \overline{\mathcal{M}}_X)$ consisting of a fine log scheme X and a morphism of sheaves of monoids β that étale locally lifts to a chart: $\widetilde{\beta} : \mathbb{N} \to \mathcal{M}_X$;
(3) the quotient stack $[\mathbb{A}^1/\mathbb{G}_m]$, where the quotient is formed with respect to the multiplication action of $\mathbb{G}_m$ on $\mathbb{A}^1$.

Lemma 6.2. *([27], complement 1). These three categories fibered in groupoids are equivalent.*

Let us sketch the proof. Given a DF log structure $(L, s : L \to \mathcal{O}_{\underline{X}})$ of rank 1 on $\underline{X}$, define a sheaf of monoids $\mathcal{M}'$ on $\underline{X}$ to be

$$\coprod_{n \geqslant 0} \underline{\mathrm{Isom}}(\mathcal{O}_{\underline{X}}, L^{\otimes n}),$$

the sheafification of the presheaf that takes U to $\coprod_{n \geqslant 0} \mathrm{Isom}(\mathcal{O}_U, (L|_U)^{\otimes n})$. It comes with a natural morphism of sheaves of monoids $\mathcal{M}' \to \mathbb{N}$, where the monoid structure on $\mathcal{M}'$ is induced by

$$(n, a : \mathcal{O} \to L^{\otimes n}) \cdot (m, b : \mathcal{O} \to L^{\otimes m}) = (n + m, a \otimes b).$$

The map $s : L \to \mathcal{O}_{\underline{X}}$ induces a morphism

$$\underline{\mathrm{Isom}}(\mathcal{O}_{\underline{X}}, L^{\otimes n}) \xrightarrow{\otimes s} \underline{\mathrm{Hom}}(\mathcal{O}_{\underline{X}}, \mathcal{O}_{\underline{X}}) = \mathcal{O}_{\underline{X}}$$

of sheaves, hence giving a pre-log structure on $\mathcal{M}'$:

$$\mathcal{M}' \to \mathcal{O}_{\underline{X}}.$$

We take $\mathcal{M}_X$ to be the log structure associated to this pre-log structure $\mathcal{M}'$. Note that $\mathcal{M}'/\mathcal{O}^*_{\underline{X}} \cong \mathbb{N}$, and we define $\beta : \mathbb{N} \to \overline{\mathcal{M}}_X$ to be the composite

$$\beta : \mathbb{N} \cong \mathcal{M}'/\mathcal{O}^*_{\underline{X}} \to \mathcal{M}_X/\mathcal{O}^*_{\underline{X}}.$$

Locally the line bundle L is trivial, and one can choose a trivialization of L, which gives trivialization of all $L^{\otimes n}$. Sending $n \in \mathbb{N}$ to this trivialization defines a section $\mathbb{N} \to \mathcal{M}'$, and hence a section $\widetilde{\beta} : \mathbb{N} \to \mathcal{M}' \to \mathcal{M}_X$. One can check that this is a chart.

Conversely, given a fine log structure $(\mathcal{M}_X, \mathcal{M}_X \xrightarrow{\alpha} \mathcal{O}_X)$ on $\underline{X}$ with a morphism $\beta : \mathbb{N} \to \overline{\mathcal{M}}_X$ that étale locally lifts to a chart $\widetilde{\beta} : \mathbb{N} \to \mathcal{M}_X$, we have a section $\beta(1)$ of $\overline{\mathcal{M}}_X$, and its inverse image under $\pi : \mathcal{M}_X \to \overline{\mathcal{M}}_X$ is an $\mathcal{O}^*_X$-torsor, which corresponds to a line bundle L. The composition

$$\pi^{-1}(\beta(1)) \subset \mathcal{M}_X \xrightarrow{\alpha} \mathcal{O}_X$$

gives a morphism of line bundles $s : L \to \mathcal{O}_X$.

Giving a morphism $\underline{X} \to [\mathbb{A}^1/\mathbb{G}_m]$ is equivalent to giving a $\mathbb{G}_m$-torsor (namely a line bundle L) with a $\mathbb{G}_m$-equivariant morphism to $\mathbb{A}^1$:

$$\begin{array}{ccc} Y & \longrightarrow & \mathbb{A}^1 \\ \downarrow & & \downarrow \\ X & \longrightarrow & [\mathbb{A}^1/\mathbb{G}_m]. \end{array}$$

This diagram is equivalent to the following one

$$\begin{array}{ccc} Y & \xrightarrow{s} & \mathbb{A}^1_X \\ \downarrow & & \downarrow \\ X & \longrightarrow & [\mathbb{A}^1/\mathbb{G}_m]_X, \end{array}$$

and the top arrow is $\mathbb{G}_m$-equivariant, namely a morphism of line bundles $s : L \to \mathcal{O}_X$. This finishes the proof.

In fact, in the three fibered categories, one can replace $\mathbb{N}$ by $\mathbb{N}^r$, and rank 1 DF-log structure by rank r DF-log structure, and replace $[\mathbb{A}^1/\mathbb{G}_m]$ by $[\mathbb{A}^r/\mathbb{G}_m^r]$ (which is equivalent to $[\mathbb{A}^1/\mathbb{G}_m]^r$), and they are still equivalent.

More generally, let P be a fine monoid and S a scheme, and let S[P] be the product $S \times_{\mathrm{Spec}\,\mathbb{Z}} \mathrm{Spec}\,\mathbb{Z}[P]$, which has a fine log structure coming from the chart $P \to \mathbb{Z}[P]$ (2.8). For an affine S-scheme Spec R, the set of R-points S[P](R) is the set of monoid homomorphisms $\mathrm{Hom}_{\mathrm{mon}}(P, R)$, where R is regarded as a multiplicative monoid. Let P^{gp} be the group associated to P. For any affine S-scheme Spec R, the group $\mathrm{Hom}_{\mathrm{mon}}(P^{gp}, R) = \mathrm{Hom}_{\mathrm{gp}}(P^{gp}, R^*)$ acts on the set $\mathrm{Hom}_{\mathrm{mon}}(P, R)$ by pointwise multiplication. This induces an action of the S-group scheme $S[P^{gp}]$ on S[P]. When $S = \mathrm{Spec}\, k$ for a field k and P is saturated and torsion-free, the k-group variety $S[P^{gp}]$ is a torus, and S[P] is a toric variety with respect to this torus action.

We have the following.

Lemma 6.3. *([46], 5.14, 5.15) The following two categories fibered in groupoids over the category of S-schemes are equivalent:*

(1) *the category of pairs $(X, \beta : P \to \overline{\mathcal{M}}_X)$ consisting of a fine log scheme X with a morphism $\underline{X} \to S$ and a morphism of sheaves of monoids β that fppf locally lifts to a chart: $\widetilde{\beta} : P \to \mathcal{M}_X$;*

(2) *the quotient stack $\mathcal{S}_P := [S[P]/S[P^{gp}]]$.*

If in addition P is fs, then one can replace "fppf" by "étale".

6.4. In fact, the action of $S[P^{gp}]$ on S[P] extends to an action on the log structure on S[P], and so this log structure descends to a log structure $\mathcal{M}_{\mathcal{S}_P}$ on the stack $\mathcal{S}_P$ (cf. Section 4 for the definition of log structures on stacks), and there is a natural morphism $\pi_P : P \to \overline{\mathcal{M}}_{\mathcal{S}_P}$ of sheaves of monoids that fppf locally lifts to a chart. This is the universal pair $(\mathcal{M}_{\mathcal{S}_P}, \pi_P)$ on $\mathcal{S}_P$ that induces the equivalence in (6.3) above.

Moreover, for a morphism $h : Q \to P$ of fine monoids, the induced morphism

$$S[h] : S[P] \to S[Q]$$

is compatible with the actions of $S[P^{gp}]$, $S[Q^{gp}]$ and the homomorphism $S[h^{gp}] : S[P^{gp}] \to S[Q^{gp}]$, hence it descends to a morphism

$$S(h) : \mathcal{S}_P \to \mathcal{S}_Q$$

of S-stacks. The map $h : Q \to P$, regarded as a morphism of constant sheaves, induces a morphism $S(h)^*\mathcal{M}_{\mathcal{S}_Q} \to \mathcal{M}_{\mathcal{S}_P}$ of log structures, making $S(h)$ into a morphism of S-log stacks.

The stack of log structures

Now we can discuss the stack LOG_S parameterizing fine log structures.

Let S be a fine log scheme. Define LOG_S to be the category with

- objects: morphisms $X \to S$ of fine log schemes, and
- morphisms: strict morphisms $X \to Y$ over S.

With the functor $(X \to S) \mapsto (\underline{X} \to \underline{S})$ from $LOG_S \to Sch_{\underline{S}}$, this defines a fibered category over $\underline{S}$. One of the main results in [46] is the following.

Theorem 6.5. *([46], 1.1) LOG_S is an algebraic stack locally of finite presentation over $\underline{S}$.*

Here for an algebraic stack we use a slightly different definition from [36, 4.1]. Namely the first axiom there that the diagonal is representable, separated and quasi-compact, is replaced by that the diagonal is representable and of finite presentation. In fact the stack LOG_S is not quasi-separated [46, 3.17].

Here are two basic properties of LOG_S.

Proposition 6.6. *([46], 3.19, 3.20) (1). The natural map $i_S : \underline{S} \to LOG_S$ corresponding to the identity morphism $S \to S$ is an open immersion;*

(2). The 2-functor

$$S \mapsto LOG_S : \{\text{fine log schemes}\} \to \{\text{algebraic stacks}\}$$

preserves fiber product. More precisely, if

$$\begin{array}{ccc} X' & \longrightarrow & X \\ \downarrow & & \downarrow \\ S' & \longrightarrow & S \end{array}$$

is a Cartesian square of fine log schemes, then the induced diagram

$$\begin{array}{ccc} LOG_{X'} & \longrightarrow & LOG_X \\ \downarrow & & \downarrow \\ LOG_{S'} & \longrightarrow & LOG_S \end{array}$$

is a 2-Cartesian square of algebraic stacks.

What is LOG_S good for?

One can use this stack LOG_S to reinterpret many concepts in log geometry. Note that for a morphism $f : X \to S$ of fine log schemes, the induced morphism $\mathrm{LOG}(f) : \mathrm{LOG}_X \to \mathrm{LOG}_S$ is faithful, hence representable.

Definition 6.7. Let P be a property of representable morphisms of algebraic stacks. Then we say that $f : X \to S$ *has property* LOG(P) if $\mathrm{LOG}(f) : \mathrm{LOG}_X \to \mathrm{LOG}_S$ has property P. We say that f *has property weak* LOG(P) if the map $\underline{X} \to \mathrm{LOG}_S$ corresponding to the given morphism $f : X \to S$ has property P.

6.8. Caution: The diagram

$$\begin{array}{ccc} \underline{X} & \xrightarrow{i_X} & \mathrm{LOG}_X \\ {\scriptstyle \underline{f}}\downarrow & & \downarrow{\scriptstyle \mathrm{LOG}(f)} \\ \underline{S} & \xrightarrow[i_S]{} & \mathrm{LOG}_S \end{array}$$

does not necessarily commute. It commutes if and only if f is strict. In [49, Section 2] a device is introduced in order to fix this issue, using stacks of diagrams of logarithmic structures.

Recall from (3.10) the notion of log smoothness and log étaleness.

Theorem 6.9. *For a morphism $f : X \to S$ of fine log schemes, f is LOG smooth (resp. LOG étale) if and only if f is log smooth (resp. log étale), if and only if f is weakly LOG smooth (resp. weakly LOG étale).*

This is part of [46, 4.6].

6.10. Another application is the following. Consider the deformation problem for a log smooth integral morphism $f_0 : X_0 \to B_0$ of fine log schemes and a strict square-zero thickening $B_0 \to B$ defined by an ideal $J \subset \mathcal{O}_{\underline{B}}$. In (3.20) we gave the relation between this deformation problem and the cohomology groups of the log tangent bundle T_{X_0/B_0}. The stack LOG_S provides another way of thinking of this problem.

By (6.9), the log smooth morphism $f_0 : X_0 \to B_0$ induces a representable smooth morphism $\underline{X_0} \to \mathrm{LOG}_{B_0}$, denoted $\mathcal{L}_{f_0}$, and the deformation problem

$$\begin{array}{ccc} X_0 & \hookrightarrow\cdots & X \\ {\scriptstyle f_0}\downarrow & & \vdots\,{\scriptstyle f} \\ B_0 & \hookrightarrow & B \end{array}$$

is equivalent to the following

$$\begin{array}{ccc} \underline{X_0} & \dashrightarrow & \underline{X} \\ {\scriptstyle \mathcal{L}_{f_0}}\downarrow & & \vdots\, {\scriptstyle \mathcal{L}_f} \\ \mathrm{LOG}_{B_0} & \longrightarrow & \mathrm{LOG}_B. \end{array}$$

The solution to this deformation problem is the cohomology groups of the *ordinary tangent bundle* $T_{\underline{X_0}/\mathrm{LOG}_{B_0}}$, therefore, theorem (3.20) holds with T_{X_0/B_0} replaced by $T_{\underline{X_0}/\mathrm{LOG}_{B_0}}$. In fact, we have $\Omega^1_{X_0/B_0} \cong \Omega_{\underline{X_0}/\mathrm{LOG}_{B_0}}$ (cf. ([49], 3.8)).

See Section 7 for the general deformation theory of log schemes.

Local structure of LOG_S

For a fine log scheme S, the relation between the quotient stacks $\mathcal{S}_P$ and LOG_S is that, the stack LOG_S can be covered by the relative versions of the $\mathcal{S}_P$'s.

Let $u : U \to S$ be a strict morphism of fine log schemes, such that the underlying morphism $\underline{u}$ is étale. We will just say that u is an étale strict morphism, if there is no confusion. Let $\beta : Q \to \mathcal{M}_U$ be a chart, and let $h : Q \to P$ be a morphism of fine monoids. The chart β induces a strict morphism $U \to \underline{S}[Q]$, which we also denote by β.

Let $\mathcal{S}_P$ be the quotient stack $[\underline{S}[P]/\underline{S}[P^{gp}]]$ with the natural fine log structure $\mathcal{M}_{\mathcal{S}_P}$ in (6.4), and let $\underline{\mathcal{S}_P}$ be the underlying stack. Consider the 2-commutative diagram

$$\begin{array}{ccccc} \underline{\mathcal{S}_P} \times_{\underline{\mathcal{S}_Q}} \underline{U} & \xrightarrow{pr_2} & \underline{U} & \xrightarrow{\beta} & \underline{S[Q]} \\ {\scriptstyle pr_1}\downarrow & & & & \downarrow{\scriptstyle \pi} \\ \underline{\mathcal{S}_P} & & \xrightarrow[\underline{S}(h)]{} & & \underline{\mathcal{S}_Q}. \end{array}$$

Let Z be the $\underline{U}$-log stack with underlying stack $\underline{Z} = \underline{\mathcal{S}_P} \times_{\underline{\mathcal{S}_Q}} \underline{U}$ and the inverse image log structure $\mathcal{M}_Z = pr_1^*\mathcal{M}_{\mathcal{S}_P}$. Applying pr_1^* to the morphism of log structures $\underline{S}(h)^*\mathcal{M}_{\mathcal{S}_Q} \to \mathcal{M}_{\mathcal{S}_P}$ (cf. (6.4)), noting that $\pi \circ \beta : U \to \mathcal{S}_Q$ is strict, we obtain a morphism $pr_2^*\mathcal{M}_U \to \mathcal{M}_Z$, making pr_2 into a morphism of log stacks $pr_2 : Z \to U$. This gives a morphism

$$\underline{Z} \to \mathrm{LOG}_U \to \mathrm{LOG}_S.$$

Proposition 6.11. *([46], 5.25) For any fine log scheme* S*, the natural morphism*

$$\coprod_{(U,\beta,h)} \underline{\mathcal{S}_P} \times_{\underline{\mathcal{S}_Q}} \underline{U} \to \mathrm{LOG}_S$$

is a representable étale surjection, where the disjoint union is taken over the isomorphism classes of all triples (U, β, h) *consisting of an étale strict morphism* $U \to S$*, a chart* $Q \xrightarrow{\beta} \mathcal{M}_U$*, and a morphism* $h : Q \to P$ *of fine monoids, for some fine monoids* P *and* Q*.*

7. Log deformation theory in general

The main reference here is [49].

As is well known, the general deformation theory of schemes and morphisms of schemes is not as easy as in the smooth case. To understand deformation theory of general morphisms, one has to use the full power of the cotangent complex, see [19, 20]. In log geometry, one can generalize it to get a reasonable theory of logarithmic cotangent complex.

This log cotangent complex will be compatible with the usual cotangent complex when the morphism in question is strict and is also compatible with log smooth deformation theory for log smooth morphisms. Basically, this is an application of the deformation theory of representable morphisms to algebraic stacks ([50]) to the classifying morphisms from the underlying scheme of X to the stack LOG_Y ([46], see also Section 6 of this chapter.)

We will focus on the category of fine log schemes.

Remark 7.1. We will work with the category $D'(\underline{X}_{\acute{e}t})$ and similar categories, and one can talk about distinguished triangles and Ext's in these categories. For relevant definitions, see [49] and [50].

Our presentation here follows [49]. An alternative approach to the cotangent complex due to Gabber is also explained in [49, Section 8].

The Log cotangent complex

In Section 6, an Artin stack LOG_Y is defined for a log scheme Y. It has the property that morphisms of log schemes $X \to Y$ are equivalent to morphisms $\underline{X} \to LOG_Y$. Thus one may interpret deformations of a morphism of log schemes $X \to Y$ as deformations of the associated representable morphism $\underline{X} \to LOG_Y$. In [50], the deformation theory of representable morphisms of stacks was studied in detail. As an application of this theory, one makes the following definition:

Definition 7.2. For a morphism of log schemes $f : X \to Y$, the *logarithmic cotangent complex of* f is the complex $L_f = L_{\underline{X}/LOG_Y}$, where the right hand side is the cotangent complex of the morphism $\underline{X} \to LOG_Y$ defined in Section 6.

Remark 7.3. One should think about L_f as an object of the category $D'_{qcoh}(\underline{X}_{\acute{e}t})$. In the above definition, the right hand side is an object of the category $D'_{qcoh}(\underline{X}_{lis-\acute{e}t})$. As the restriction functor $D'_{qcoh}(\underline{X}_{lis-\acute{e}t}) \to D'_{qcoh}(\underline{X}_{\acute{e}t})$ is an equivalence of categories, no information of the cotangent complex would lost.

Basic properties

For every morphism of fine log schemes $f : X \to Y$ the log cotangent complex is a projective system

$$L_f = (\cdots \to L_f^{\geqslant -n-1} \to L_f^{\geqslant -n} \to \cdots \to L_f^{\geqslant 0})$$

where each $L_f^{\geqslant -n}$ is an essentially constant ind-object in $D^{[-n,0]}(\mathcal{O}_X)$ (The derived category of $\mathcal{O}_X$-modules supported in $[-n,0]$).

The log cotangent complex L_f has the following properties:

(1) For any $n \geqslant 0$, the natural map $\tau_{\geqslant -n} L_f^{\geqslant -n-1} \to L_f^{\geqslant -n}$ is an isomorphism.
(2) If f is strict, then the system $(\tau_{\geqslant -n} L_{f'})$ represents L_f, where $L_{f'}$ is the usual cotangent complex of the underlying morphism of schemes f'.
(3) If $f : X \to Y$ is log smooth, then the sheaf of log differentials $\Omega^1_{X/Y}$ represents L_f.
(4) If

$$\begin{array}{ccc} X' & \xrightarrow{a} & X \\ {\scriptstyle g}\downarrow & & \downarrow{\scriptstyle f} \\ Y' & \xrightarrow{b} & Y \end{array}$$

is a commutative diagram of fine log schemes, then there is a natural map

$$a^* L_f \to L_g$$

which is an isomorphism if the square above is cartesian and f is log flat. Furthermore, if the composite $X' \to Y' \to Y$ satisfies the condition (T) below, then the map

$$g^* L_b \oplus a^* L_f \to L_{bg}$$

is also an isomorphism.
(5) Given a composite

$$X \xrightarrow{f} Y \xrightarrow{g} Z$$

satisfying condition (T) below, there is a natural map

$$L_f \to f^* L_g[1]$$

making the resulting triangle

(7.4) $$f^* L_g \to L_{gf} \to L_f \to f^* L_g[1]$$

distinguished.

Remark 7.5. In (4) and (5) above, f^*, a^*, g^* should be understood in the derived sense.

Remark 7.6. One might hope for a theory of log cotangent complex in which every triangle (7.4) is distinguished. This is unfortunately not the possible — an example due to W. Bauer is given in [49, Section 7].

On the other hand, Gabber has shown (see [49, Section 8]) that if one loosens the requirement 3, then one can obtain a theory of log cotangent complexes for which one has a distinguished triangle (7.4) for all composites $X \xrightarrow{f} Y \xrightarrow{g} Z$.

The Condition (T) mentioned above is the following:
There exists a family of commutative diagrams

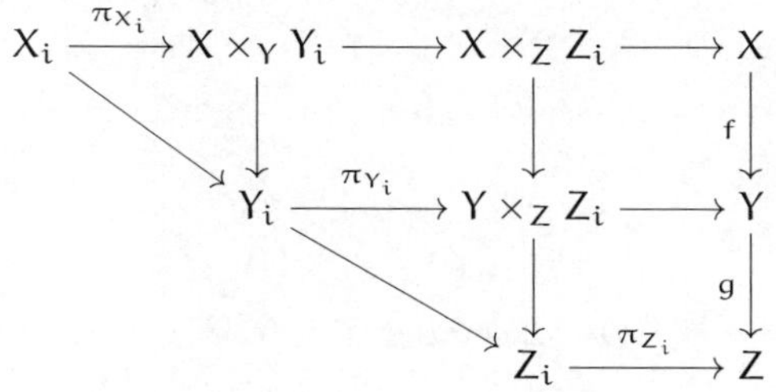

such that

(1) The underlying schemes of X_i, Y_i, Z_i are all affine.
(2) The π's are all strict, and their underlying morphisms are flat and locally of finite presentation.
(3) The underlying family of morphisms of schemes of $\{X_i \to X\}$ is jointly surjective.
(4) There exists charts

$$\beta_{X_i} : Q_{X_i} \to \mathcal{M}_{X_i}, \beta_{Y_i} : Q_{Y_i} \to \mathcal{M}_{Y_i}, \beta_{Z_i} : Q_{Z_i} \to \mathcal{M}_{Z_i}$$

and injective maps

$$Q_{Z_i} \to Q_{Y_i} \to Q_{X_i}$$

compatible with the morphisms f_i, g_i and

$\mathrm{Tor}^j_{\mathcal{O}_{Z_i} \otimes_{\mathbb{Z}[Q_{Z_i}]} \mathbb{Z}[Q_{Y_i}]}(\mathcal{O}_{Z_i} \otimes_{\mathbb{Z}[Q_{Z_i}]} \mathbb{Z}[Q_{X_i}], \mathcal{O}_{Y_i}[G]) = 0$ for all $j > 0$.

Here $G := \mathrm{Coker}(Q^{gp}_{Z_i} \to Q^{gp}_{Y_i})$ and $\mathcal{O}_{Y_i}[G]$ is viewed as an $\mathcal{O}_{Z_i} \otimes_{\mathbb{Z}[Q_{Z_i}]} \mathbb{Z}[Q_{Y_i}]$-algebra via the map

$$\mathcal{O}_{Z_i} \otimes_{\mathbb{Z}[Q_{Z_i}]} \mathbb{Z}[Q_{Y_i}] \to \mathcal{O}_{Y_i}[G], t \otimes e_q \mapsto g_i^*(t)\beta_{Q_{Y_i}}(q) \cdot \bar{q}$$

where $\bar{q}$ denotes the image of q in G.

Deformation theory of Log schemes in general

In this section, we explain the relation between the log cotangent complex and deformation theory of log schemes. Let $f : X \to Y$ be a morphism of fine log schemes and let I be a quasi-coherent sheaf on $\underline{X}$. Define a Y-extension of X by I to be a commutative diagram of log schemes

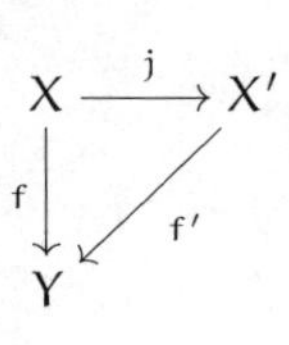

where j is an strict closed immersion defined by a square-zero ideal, together with an isomorphism $\epsilon_j : I \cong \mathrm{Ker}(\mathcal{O}_{X'} \to \mathcal{O}_X)$. The set of Y-extensions of X by I forms, in a natural way, a category $\underline{\mathrm{Exal}}_Y(X, I)$. Let $\mathrm{Exal}_Y(X, I)$ be the set of isomorphism classes of this category.

There is a tautological equivalence of categories (see [50, Problem 1] for the meaning of the right hand side):

$$\underline{\mathrm{Exal}}_Y(X, I) \cong \underline{\mathrm{Exal}}_{\mathrm{LOG}_Y}(\underline{X}, I).$$

Hence, by [50, Theorem 1.1] and our definition of L_f, we obtain the following result:

Theorem 7.7. ([49, Theorem 5.2]) *There is a natural bijection*

$$\mathrm{Exal}_Y(X, I) \cong \mathrm{Ext}^1(L_f, I).$$

It is precisely the theorem above that guarantees that general deformation theory is controlled by our logarithmic cotangent complex.

Definition 7.8. Let $j_0 : Y_0 \hookrightarrow Y$ be an strict closed immersion of fine log schemes defined by a square-zero ideal $I \subset \mathcal{O}_Y$, and let $f_0 : X_0 \to Y_0$ be a LOG flat morphism (Definition 6.7). A *log flat deformation of* X_0 *to* Y is a cartesian square

$$\begin{array}{ccc} X_0 & \xrightarrow{j} & X \\ {\scriptstyle f_0}\downarrow & & \downarrow{\scriptstyle f} \\ Y_0 & \xrightarrow{j_0} & Y \end{array}$$

with f LOG flat.

To give a log flat deformation as above is equivalent to give a 2-commutative diagram

$$\begin{array}{ccc} \underline{X_0} & \xrightarrow{j} & \underline{X} \\ {\scriptstyle \mathcal{L}_{f_0}}\downarrow & & \downarrow{\scriptstyle \mathcal{L}_f} \\ \mathrm{LOG}_{Y_0} & \xrightarrow{j_0} & \mathrm{LOG}_Y \end{array}$$

with $\mathcal{L}_f$ flat. Thus from ([50], 1.4) we obtain the following:

Theorem 7.9. *Let* J *denote the ideal of* LOG_{Y_0} *in* LOG_Y. *Then*

(1) *There exists a canonical class* $o \in \mathrm{Ext}^2(L_{f_0}, \mathcal{L}^*_{f_0}J)$ *whose vanishing is equivalent to the existence of a log flat deformation of* X_0 *to* Y.
(2) *If* $o = 0$, *then the set of isomorphism classes of log flat deformations of* X_0 *to* Y *is naturally a torsor under* $\mathrm{Ext}^1(L_{f_0}, \mathcal{L}^*_{f_0}J)$.
(3) *The automorphism group of any log flat deformation of* X_0 *to* Y *is canonically isomorphic to* $\mathrm{Ext}^0(L_{f_0}, \mathcal{L}^*_{f_0}J))$.

This theorem gives an answer to the question of general deformation theory of log schemes. In case f_0 and f are log smooth and integral, we have that $L_{f_0} = \Omega^1_{X_0/Y_0}$, and this theorem reduces to Theorem 3.20.

Deformations of morphisms

As in the case of usual schemes, once one understands deformations of log schemes, one obtains a solution to the related problem of deformations of a log morphisms.

We are given a commutative diagram of solid arrows

(7.10)

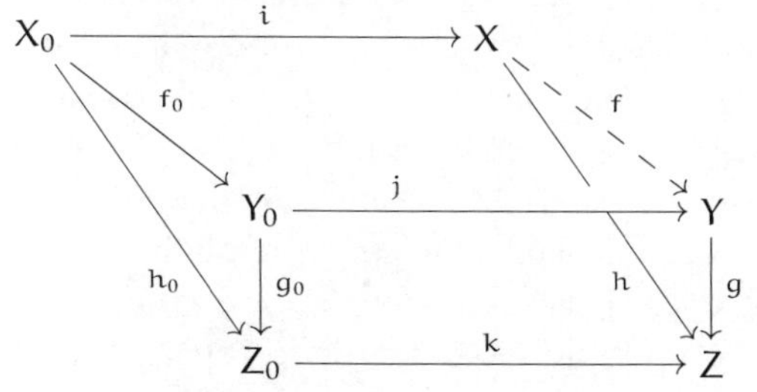

where i, j, k are strict closed immersion defined by square-zero ideal sheaves I, J, K living on X, Y, Z respectively. The question is to find a dotted arrow f fitting in the diagram. To nail down f, we need some more data.

The morphisms h and g induce morphisms $w : h_0^*K \to I$ and $v : g_0^*K \to J$. Assume given a morphism $u : f_0^*J \to I$ such that the composite

$$h_0^*K = f_0^*g_0^*K \to f_0^*J \to I$$

is equal to w.

What we want to find is an f fitting in the diagram (7.10) such that the morphism $f_0^*J \to I$ induced by f is equal to u. This can also be solved by the logarithmic cotangent complex.

Theorem 7.11. *In the situation above, assume in addition that u induces a map $\mathcal{L}^*_{f_0} J' \to I$, where J' is the ideal of LOG_{Y_0} in LOG_Y. Then there is a canonically defined class $o \in \mathrm{Ext}^1(f_0^*L_{Y_0/Z_0}, I)$ that vanishes iff there exists an f fitting in the diagram (7.10) such that the morphism $f_0^*J \to I$ induced by f is equal to u. If $o = 0$, then the set of such maps f is a torsor under the group $\mathrm{Ext}^0(f_0^*L_{Y_0/Z_0}, I)$.*

For a proof of this theorem, see [49], Theorem 5.9.

8. Rounding

The main reference for this section is [30]. We have benefitted from a lecture by A. Ogus, who reported on results in [43].

What is rounding?

The process of "rounding", in its most basic form, produces a manifold *with corners* from a *smooth* analytic space with a normal crossings divisor. So the corners are not rounded but rather the opposite: they are created. On the other hand, these corners are rather round and shapely. Also, anybody who has seen the construction under any name and hears the name "rounding" immediately knows what this is about. Evidently then, even though "rounding" might be something of a misnomer, it is a very good name. The origin of the name seems to be in work of Kajiwara, Nakayama and Ogus [23, 43].

In various moduli problems, the rounding of the moduli space often has a more natural topological interpretation than the moduli space itself. A good example is the Deligne-Mumford moduli space $\overline{\mathcal{M}}_{g,n}$ of marked nodal curves, whose boundary is a normal crossings divisor. The "interior" $\mathcal{M}_{g,n}$ can be described topologically as a quotient of Teichmuller space by the appropriate mapping class group. There is a natural generalization of Teichmuller space involving 2-manifolds decorated with circles, due to Harvey [18]; the analogous quotient yields a topological description of the *rounding* of $\overline{\mathcal{M}}_{g,n}$, rather than the moduli space $\overline{\mathcal{M}}_{g,n}$ itself.

Another example is that of twisted curves, discussed in Section 11. A twisted curve is an algebraic stack with a log structure, so it is a bit exotic. But its rounding is a good old topological space. A similar example occurs in current work of one of us (Gillam): the relative Hilbert stack of a marked Riemann surface can be defined algebraically using Jun Li style expansions, but it is not representable (except in some trivial cases). However its rounding is a topological space (even a manifold with corners) which is relatively easy to describe. A similar phenomenon occurs in many moduli problems involving expansions, discussed briefly below.

The topological preeminence of the rounding of moduli spaces might ultimately be traced back to the preference in topology for operations involving real codimension one subspaces (e.g. connected sum of manifolds) as opposed to the algebra-geometric preference for complex codimension one operations (e.g. pushout of two smooth varieties along a common divisor). From this point of view, one might think of log geometry as an attempt to speak algebraically about various "real codimension one" phenomena.

The oriented real blowup

The most general rounding operation is the *Kato–Nakayama logarithmic space* associated to a log analytic space [30]. In the basic example of a smooth analytic space with log structure from a normal crossings divisor, the Kato–Nakayama space can be described in terms of *oriented real blowup*, which is a relatively simple rounding operation that can be described as follows.

Suppose X is a topological space, $\pi : L \to X$ is a complex line bundle, and $s : X \to L$ is a section of π. Locally on X we can choose a trivialization $(\pi, \phi) : L \to$

$X \times \mathbb{C}$ and consider the subspace

$$B_{L,s,\phi} X := \left\{ l \in L \ : \ |\phi(l)| \cdot (\phi s \pi)(l) \ = \ \phi(l) \cdot |(\phi s \pi)(l)| \right\}$$

of L. A continuous function $u : X \to \mathbb{C}^*$ yields a new trivialization $(\pi, u \cdot \phi)$, where $(u \cdot \phi)(l) := (u\pi)(l)\phi(l)$. The key observation is that $B_{L,s,\phi} X = B_{L,s,u\cdot\phi} X$, so the subspace $B_{L,s,\phi} X$ is independent of the choice of ϕ, hence one can define a subspace $B_{L,s} X \subseteq L$ by defining it locally on X using a trivialization, then gluing the locally defined subspaces. From the local picture using a trivialization, it is clear that the subspace $B_{L,s} X$ contains the zero section and $L|_{Z(s)}$ (where $Z(s) \subseteq X$ is the zero locus of s) and is invariant under the $\mathbb{R}_{>0}$ action on L inherited from the full $\mathbb{C}^*$ scaling action. We let $B^*_{L,s} X$ be the complement of the zero section in $B_{L,s} X$ and we call

$$\mathrm{Blo}_{L,s} X := (B^*_{L,s} X)/\mathbb{R}_{>0}$$

the *oriented real blowup* of X along (L, s).

The space $\mathrm{Blo}_{L,s} X$ is a closed subspace of the oriented circle bundle $S^1 L := L^*/\mathbb{R}_{>0}$ associated to L and is, in particular, proper over X. The projection $\tau : \mathrm{Blo}_{L,s} X \to X$ is an isomorphism away from $Z(s)$ and $\tau^{-1}(Z(s))$ is oriented circle bundle $S^1 L|_{Z(s)}$. The spaces $B_{L,s} X$ and $\mathrm{Blo}_{L,s} X$ are natural under pulling back line bundles and sections.

If X is an analytic space and $D \subseteq X$ is a Cartier divisor, then D determines a line bundle $\mathcal{O}_X(D)$ together with a section s whose zero locus is D. In this situation, we will write $B_D X$, $\mathrm{Blo}_D X$, etc. and speak of the *oriented real blowup of* X *along* D. The space $\mathrm{Blo}_D X$ inherits a differentiable structure from its inclusion in $S^1\mathcal{O}_X(D)$.

The basic example to keep in mind is the oriented real blowup $\mathrm{Blo}_0 \mathbb{C}$ of the complex plane $\mathbb{C}$ at the origin. The origin is the zero locus of the identity map $\mathrm{Id} : \mathbb{C} \to \mathbb{C}$, hence

$$\begin{aligned} \mathrm{Blo}_0 \mathbb{C} &= \left\{(z, Z) \in \mathbb{C} \times \mathbb{C}^* \ : \ |z|Z = z|Z|\right\} / \mathbb{R}_{>0} \\ &= \left\{(z, Z) \in \mathbb{C} \times S^1 \ : \ |z|Z = z\right\} \\ &\cong \mathbb{R}_{\geqslant 0} \times S^1, \end{aligned}$$

where the last isomorphism from $\mathbb{R}_{\geqslant 0} \times S^1$ is given by $(\lambda, Z) \mapsto (\lambda Z, Z)$. Evidently $\mathrm{Blo}_0 \mathbb{C}$ is a half-infinite annulus whose boundary S^1 is the exceptional locus of $\tau : \mathrm{Blo}_0 \mathbb{C} \to \mathbb{C}$ (the fiber over the origin).

The Kato–Nakayama space

Let $(X, \mathcal{M}_X)$ be a fine and saturated logarithmic analytic space.

Definition 8.1. [[30, 1.2]] We define its *canonical rounding*, or *Kato-Nakayama space*, denoted $X^{\log}$, as the space whose points are pairs (x, F) where $x \in X$ and $F : \mathcal{M}_{X,x} \to S^1$ is a monoid homomorphism satisfying $F(u) = u(x)/|u(x)|$ for every $u \in \mathcal{O}^*_{X,x} \subseteq \mathcal{M}_{X,x}$.

This space has a natural topology. Let us describe the topology in the special case where $\mathcal{M}_X$ is the canonical log structure associated to a Cartier divisor $D \subseteq X$, see Example 2.7. Locally on X we can find $f_1, \ldots, f_n \in \mathcal{M}_X(X)$ which, together with the units, generate $\mathcal{M}_X$. The map

$$X^{\log} \to X \times (S^1)^n$$
$$(x, F) \mapsto (x, F(f_{1,x}), \ldots, F(f_{n,x}))$$

is then easily seen to be a monomorphism onto a closed subset of $X \times (S^1)^n$, so we give $X^{\log}$ the subspace topology so that this is a closed embedding. Since one can check easily that this topology does not depend on the choice of generators $f_1, \ldots, f_n$, the locally defined topologies glue to a topology on $X^{\log}$ making the projection $\tau : X^{\log} \to X$ given by $\tau(x, F) := x$ a proper map.

8.2. Relating Kato–Nakayama spaces to oriented blowups

There is a morphism

$$\phi : X^{\log} \to \mathrm{Blo}_D X$$
$$(x, F) \mapsto (x, f \mapsto F(\bar{f}))$$

of topological spaces over X which requires a little explanation. Here $f \in S^1\mathcal{O}_X(-D)|_x$ is in the circle bundle associated to the fiber $\mathcal{O}_X(-D)|_x \cong \mathbb{C}$ and $\bar{f} \in \mathcal{O}_{X,x}(-D)$ is a lifting of f to the stalk (one shows that any such $\bar{f}$ is actually in $\mathcal{M}_{X,x}$ and that $F(\bar{f})$ does not depend on this choice of lifting $\bar{f}$). If we use the identification

$$S^1\mathcal{O}_X(D) = \mathrm{Hom}_{S^1}(S^1\mathcal{O}_X(-D), S^1),$$

then we can think of ϕ as a map from $X^{\log}$ to $S^1\mathcal{O}_X(D)$; one then shows that this ϕ is continuous and that ϕ factors through $\mathrm{Blo}_D X \subseteq S^1\mathcal{O}_X(D)$.

When X is a smooth analytic space and D is a smooth divisor, the map ϕ is easily seen to be an isomorphism since one can reduce to the case $(X, D) = (\mathbb{C}, 0)$ on formal grounds. Slightly more generally, if X is smooth, but D is only a normal crossings divisor, then locally we can write D as a union of smooth divisors $D_1, \ldots, D_i$ which look like the first i coordinate hyperplanes in $\mathbb{C}^n$ ($n = \dim X$), and we can define a variant of the oriented real blowup

$$\mathrm{Blo}'_D X := (\mathrm{Blo}_{D_1} X) \times_X \cdots \times_X (\mathrm{Blo}_{D_i} X)$$

and a map $\phi : X^{\log} \to \mathrm{Blo}'_D X$. In this local picture, the log structure $\mathcal{M}_X$ is the direct sum (in the category of log structures) of the log structures $\mathcal{M}^j_X$ from $D_1, \ldots, D_i$, the associated Kato–Nakayama space $X^{\log}$ is the fibered product over X of the $(X, \mathcal{M}^j_X)^{\log}$, and ϕ is just the fibered product over X of the previously constructed isomorphisms $\phi_j : (X, \mathcal{M}^j_X)^{\log} \to \mathrm{Blo}_{D_j} X$. The locally defined variants can be glued to define a global variant $\mathrm{Blo}'_D X$ of the oriented real blowup and an isomorphism $\phi : X^{\log} \cong \mathrm{Blo}'_D X$ of topological spaces over X.

8.3. Topology, cohomology, and the Kato–Nakayama space

Locally, if $X = \mathbb{C}^n$ and D is the union of the first i coordinate hyperplanes, then D is the zero locus of $(z_1, \dots, z_n) \mapsto z_1 \cdots z_i \in \mathbb{C}$ and we have

$$\begin{aligned} \mathrm{Blo}_D X &= \{(z_1, \dots, z_n, Z) \in \mathbb{C}^n \times S^1 : |z_1 \cdots z_i| Z = z_1 \cdots z_i\} \\ \mathrm{Blo}'_D X &= \{(\overline{z}, \overline{Z}) \in \mathbb{C}^n \times (S^1)^i : |z_j| Z_j = z_j \text{ for } j = 1, \dots, i\}. \end{aligned}$$

In the general normal crossings divisor situation, the fiber of $\tau : X^{\log} \to X$ over a point $x \in X$ is naturally identified with

$$S^1 N_{D_1/X}|_x \times \cdots \times S^1 N_{D_i/X}|_x,$$

where $D_1, \dots, D_i$ are the branches of D containing x. When $\dim X = n$, a point $y \in \tau^{-1}(x)$ has a neighborhood diffeomorphic to a neighborhood of the origin in $\mathbb{R}^i_{\geqslant 0} \times \mathbb{R}^{2n-i}$. (Note $i \leqslant n$, so the depth of the corners in a Kato–Nakayama space is somewhat constrained.) Recall that the *topology* near the origin only depends on whether $i > 0$, but the differentiable structure depends on the actual value of i. In particular, the topological boundary of the manifold $X^{\log}$ is given by $\tau^{-1}(D)$, and this manifold with boundary is homotopy equivalent to its interior, so $H^*(X^{\log}) = H^*(X \setminus D)$.

The topology of morphisms of Kato–Nakayama spaces is also very nice, as shown by the following beautiful general result, see [43, Theorem 0.3]:

Theorem 8.4. *Let $f : X \to Y$ be a log smooth and integral morphism of fine log analytic spaces. Then the associated map $X^{\log} \to Y^{\log}$ is a topological submersion.*

In fact, Nakayama and Ogus prove a more general result, replacing integrality by K. Kato's notion of *exact morphisms*, see [27, Definition 4.6].

The fact that the topology is nice suggests that one expects nice cohomological implications. This is indeed the original motivation leading Kato and Nakayama to define $X^{\log}$, see [30, Theorem 0.2 (1)]:

Theorem 8.5. *Let (X, M) be a fine and saturated log scheme with X of finite type over $\mathbb{C}$. Let F be a constructible sheaf on the log-étale site $X_{\text{log-ét}}$, and let $F^{\log}$ be its pullback to the topological space $X^{\log}$. Then for all $q \in \mathbb{Z}$ we have*

$$H^q(X_{\textit{log-ét}}, F) = H^q(X^{\log}, F^{\log}).$$

Very strong results hold true for de Rham cohomology. In fact the Kato–Nakayama space is a model for the log de Rham cohomology of X in the sense that

$$H^*(X^{\log}, \mathbb{C}) = \mathbb{H}^*(X, \wedge^\bullet \Omega_{(X,M)})$$

under mild assumptions on X. We discuss this in Section 9 below.

Kato–Nakayama spaces of expanded pairs

Given a pair (X, D) consisting of a smooth variety X over $\mathbb{C}$ with a smooth divisor $D \subseteq X$, the notion of an *expanded pair* $t : \mathcal{X} \to B$ over a base B arises in various relative curve counting theories. The fiber of t over a point $b \in B$ always looks like

$$X[n]_0 = X \coprod_D \Delta_1 \coprod_D \cdots \coprod_D \Delta_n$$

(for an appropriate n), where $\Delta_i = \mathbb{P}(N_{D/X} \oplus \mathcal{O}_D)$ is a $\mathbb{P}^1$ bundle over X. Both $\mathcal{X}$ and B have natural log structures making t a log smooth map of log schemes. The fiber of $t^{\log} : \mathcal{X}^{\log} \to B^{\log}$ over a point $c \in \tau_B^{-1}(b)$ looks like

$$X^{\log} \coprod_{c_1 : S^1 N_{D/X} \cong S^1 N_{D/\Delta_1}} \Delta_1^{\log} \cdots \coprod_{c_n : S^1 N_{D/\Delta_{n-1}} \cong S^1 N_{D/\Delta_n}} \Delta_n^{\log},$$

where the choice of $c \in \tau^{-1}(b) \cong (S^1)^n$ determines the choice of orientation reversing S^1 bundle isomorphisms $c_1, \ldots, c_n$. Here each Δ_j has the log structure from the two copies of D, and $\Delta_j^{\log}$ is a cylinder bundle over D (better: an I-bundle over $D^{\log}$).

The action of $(\mathbb{C}^*)^n$ on $X[n]_0$ given by scaling the fibers of the $\mathbb{P}^1$ bundles Δ_i is an action by isomorphisms of log schemes, so it lifts to an action on Kato–Nakayama spaces. This lifted action is nontrivial on $B^{\log}$ as the $(S^1)^n$ factor of $(\mathbb{C}^*)^n$ acts simply transitively on $\tau_B^{-1}(b) \cong (S^1)^n$. In the usual moduli problems involving expansions, the isotropy group of a point b involves elements of $(\mathbb{C}^*)^n$ such that the induced action on $X[n]_0$ respects a map from a curve to $X[n]_0$, a subscheme of $X[n]_0$, etc., and this isotropy is usually required to be finite to have a good moduli problem. Since $G \cap \mathbb{R}_{>0} = \{\mathrm{Id}\}$ for any finite subgroup G of $\mathbb{C}^*$, the Kato–Nakayama space of the moduli problem is often representable even if the moduli problem itself is not. This is always the case for moduli problems involving, say, quotients of sheaves on $X[n]_0$ pulled back from X, since these quotients themselves have no automorphisms and the only isotropy comes from the subgroup of $(\mathbb{C}^*)^n$ preserving the quotient.

9. Log de Rham and Hodge structures

The main references of this section are [31, 30, 29]. This section owes much to a lecture by Phillip Griffiths [13].

Moduli spaces of polarized Hodge structures

We assume the reader to be familiar with some basic concepts of Hodge theory. First of all, we briefly summarize the classical theory of the moduli spaces of polarized Hodge structures.

9.1. The moduli space $M_h = \Gamma \backslash D_h$. Let n be an integer, and let h be a sequence of non-negative integers $(h^{n,0}, h^{n-1,0}, \cdots, h^{0,n})$ satisfying $h^{p,q} = h^{q,p}$, called the *Hodge numbers*. Let $H_{\mathbb{Z}}$ be a free abelian group of rank $\sum h^{p,n-p}$, with a non-degenerate bilinear form $Q : H_{\mathbb{Z}} \otimes H_{\mathbb{Z}} \to \mathbb{Z}$, which is symmetric (resp. anti-symmetric) if n is even (resp. odd). Let $G_{\mathbb{Z}}$ be the group functor $\mathrm{Aut}(H_{\mathbb{Z}}, Q)$ on commutative rings, sending a ring R to the group of automorphisms on the free R-module $H_R := H_{\mathbb{Z}} \otimes R$ preserving the bilinear form Q. It is an affine group scheme of finite type over $\mathbb{Z}$ (which is clear if we write down the matrix representing the bilinear form Q with respect to some basis of $H_{\mathbb{Z}}$). Let Γ be an arithmetic subgroup of $G_{\mathbb{Z}}(\mathbb{Z})$.

The set of Hodge structures of weight n on $H_{\mathbb{R}}$ with prescribed Hodge numbers h, such that Q induces a *polarization* on $H_{\mathbb{R}}$ (i.e. it induces a morphism $H_{\mathbb{R}} \otimes H_{\mathbb{R}} \to \mathbb{R}(-n)$ of Hodge structures, and the bilinear form $Q_C(u, v) := Q(u, Cv)$, where C is the Weil operator, is symmetric and positive definite), is parameterized by the homogeneous space $D_h = G_{\mathbb{R}}/K$, where K is the stabilizer group of a fixed polarized Hodge structure F_0 on $H_{\mathbb{R}}$. See for instance ([8]) for these concepts.

This homogeneous space $D = D_h = G_{\mathbb{R}}/K$ has a complex structure defined as follows. It is clear that $Q : H_{\mathbb{R}} \otimes H_{\mathbb{R}} \to \mathbb{R}(-n)$ is a morphism of Hodge structures if and only if $Q(F^p, F^{n-p+1}) = 0$ for all p. Let $f^p = \sum_{r \geqslant p} h^{r,n-r}$, and let $D^\vee$, the *compact dual of* D, be the subspace of the product of the Grassmannians $\prod_p \mathrm{Gr}(f^p, H_{\mathbb{R}})$ consisting of all flags $F^\bullet$:

$$\cdots \subset F^{p+1} \subset F^p \subset \cdots$$

such that $Q(F^p, F^{n-p+1}) = 0$. Then $D^\vee \simeq G_{\mathbb{C}}/P$, where P is a parabolic subgroup preserving a fixed flag. This gives $D^\vee$ a complex structure. We see that $D \subset D^\vee$ is the locus of flags satisfying

(i) $F^p \cap \overline{F}^{n-p+1} = 0$ (so that $F^p \oplus \overline{F}^{n-p+1} \cong H_{\mathbb{C}}$) for all p, and

(ii) $Q(\overline{u}, Cu) > 0$ for all $u \neq 0$ in $H_{\mathbb{C}}$.

They are both open conditions, so $D \subset D^\vee$ is an open complex submanifold. The group Γ acts on D_h properly discontinuously, and the quotient $M_h = \Gamma \backslash D_h$ is the moduli space of *Γ-equivalence classes of Q-polarized Hodge structures on $H_{\mathbb{C}}$ with Hodge type* h. See ([31], 0.3.6, 0.3.7).

Variations of Hodge structures.

Definition 9.2. ([8], 3.11, 3.12) Let S be a complex manifold. A *variation of Hodge structures $\mathscr{H}$ of weight n on* S is given by

- a local system $\mathscr{H}_{\mathbb{Z}}$ of free abelian groups of finite rank on S;
- a finite decreasing filtration $F^\bullet \mathscr{H}_{\mathscr{O}}$ of the vector bundle $\mathscr{H}_{\mathscr{O}} := \mathscr{H}_{\mathbb{Z}} \otimes_{\mathbb{Z}} \mathscr{O}_S$ by holomorphic sub-bundles,

such that the following conditions are satisfied:

1) (Griffiths transversality) the natural flat connection $\nabla = d \otimes \mathrm{id}_{\mathscr{H}_{\mathscr{O}}} : \mathscr{H}_{\mathscr{O}} \to \Omega^1_S \otimes \mathscr{H}_{\mathscr{O}}$ takes $F^p \mathscr{H}_{\mathscr{O}}$ into $\Omega^1_S \otimes F^{p-1} \mathscr{H}_{\mathscr{O}}$, for every p;

2) for each point $s \in S$, the fiber $F^\bullet(s)$ over s is a Hodge structure of weight n.

A *polarization* of the variation of Hodge structures $\mathscr{H}$ is a locally constant bilinear form

$$\mathscr{Q} : \mathscr{H}_{\mathbb{Z}} \otimes \mathscr{H}_{\mathbb{Z}} \to \mathbb{Z}$$

such that on each fiber over $s \in S$, it induces a polarization of the fiber Hodge structure.

Suppose we have a polarized family of Hodge structures $(\mathscr{H}, \mathscr{Q} : \mathscr{H}_{\mathbb{Z}} \otimes \mathscr{H}_{\mathbb{Z}} \to \mathbb{Z})$ of weight n on S (not necessarily a variation of Hodge structures), and a global section of the sheaf

$$\Gamma\backslash\underline{\mathrm{Isom}}((\mathscr{H}_{\mathbb{Z}}, \mathscr{Q}), (H_{\mathbb{Z}}, Q)),$$

where $H_{\mathbb{Z}}$ is regarded as a constant sheaf on S. If the monodromy group of this family of Hodge structures on S is contained in Γ, i.e. the image of the homomorphism

$$\pi_1(S) \to G_{\mathbb{Z}}(\mathbb{Z})$$

is contained in Γ, then there is a well-defined function

$$\varphi : S \to M_h$$

inducing this family of Hodge structures. This map is locally liftable to D_h.

If $f : X \to S$ is a projective smooth morphism between quasi-projective complex algebraic manifolds, with a relative hyperplane section $\eta \in H^0(S, R^2 f_* \mathbb{Z})$, then the family of the primitive parts $P^n(X_s, \mathbb{Z})$ of the cohomology groups $H^n(X_s, \mathbb{Z})$ modulo torsion form a polarized variation of Hodge structures of weight n on S, and it induces a map

$$\varphi : S \to M_h$$

if Γ contains the monodromy group. The map φ is called the *period map*. To be precise, the family of $H^n(X_s, \mathbb{C})$'s are the stalks of $R^n f_*(f^{-1}\mathscr{O}_S)$, and the Hodge filtration on $R^n f_*(f^{-1}\mathscr{O}_S)$ is given by the degenerate spectral sequence

$$E_1^{pq} = R^q f_* \Omega^p_{X/S} \Longrightarrow R^{p+q} f_*(f^{-1}\mathscr{O}_S),$$

which is induced from the natural quasi-isomorphism (the relative holomorphic Poincaré lemma)

$$f^{-1}\mathscr{O}_S \longrightarrow \Omega^\bullet_{X/S}.$$

Since η is a global section, the primitive part forms a family of sub-Hodge structures on S. Griffiths proved that the period map is holomorphic, and the family of Hodge structures is a variation of Hodge structures. See ([8], 3) for more detail.

For instance, one can take S to be the moduli space A_g (resp. M_g) of principally polarized abelian varieties of dimension g (resp. projective smooth curves of genus g), and take X to be the universal family of such objects over S, and $n = 1$. In this case, the local system $\mathscr{H}_{\mathbb{Z}} = R^1 f_* \mathbb{Z}$ is torsion-free and is equal to its primitive part, and the filtration $F^\bullet \mathscr{H}_{\mathscr{O}}$ is given by $F^0 \mathscr{H}_{\mathscr{O}} = \mathscr{H}_{\mathscr{O}}$, $F^1 \mathscr{H}_{\mathscr{O}} = R^0 f_* \Omega_{X/S}$ and $F^2 \mathscr{H}_{\mathscr{O}} = 0$. This variation of Hodge structures (and its tensor powers) are of great arithmetic interest.

Logarithmic Hodge structures

Consider the following situation. Let $f : X \to S$ be a family of projective manifolds, where S is the complement of a normal crossing divisor D in a compact manifold $\overline{S}$, and suppose one can extend the family f to a family $\overline{f} : \overline{X} \to \overline{S}$ which is log smooth (here $\overline{S}$ has the log structure induced by the divisor D (2.7)). Then one can ask if it is possible to enlarge the moduli space M_h to some $\overline{M_h}$ so that the period map extends to $\overline{\varphi} : \overline{S} \to \overline{M_h}$.

To study the degenerations of Hodge structures, Kato and Usui introduced the notion of logarithmic Hodge structures [31].

Log de Rham complex. Let $(X, \alpha : M_X \to \mathcal{O}_X)$ be an fs log analytic space over $\mathbb{C}$, and let $X^{\log}$ be the Kato-Nakayama space (Definition 8.1), with $\tau = \tau_X : X^{\log} \to X$ the natural proper map. Over the open set $X^* \subset X$ where the log structure is trivial, the map τ is a homeomorphism, and the section $j^{\log} : X^* \hookrightarrow X^{\log}$ is a homotopy equivalence. For $x \in X$, the fiber $\tau^{-1}(x)$ is a compact torus $(S^1)^m$, where m is the rank of $\overline{M}^{gp}_{X,x}$. For instance, let Δ be the open unit disk $\{|z| < 1\}$ equipped with the log structure induced by the center $\{z = 0\}$ (cf. 2.7). Then $\Delta^{\log}$ is homeomorphic to $[0,1) \times S^1$. See Sections 8.2 and 8.3 for more examples.

One can define a sheaf of rings $\mathcal{O}_{X^{\log}}$ on $X^{\log}$. Roughly speaking, this is the subsheaf of rings of $j^{\log}_* \mathcal{O}_{X^*}$ on $X^{\log}$ generated over $\tau^{-1}\mathcal{O}_X$ by "$\log(q)$", for all $q \in M^{gp}_X$. See ([30], Section 3) for the precise definition.

For example, if $x \in X$ and $y \in \tau^{-1}(x)$, and the free abelian group $\overline{M}^{gp}_{X,x}$ has rank m and is generated by $f_1, \cdots, f_m \in M^{gp}_{X,x}$, then the stalk $\mathcal{O}_{X^{\log},y}$ is isomorphic to the polynomial ring $\mathcal{O}_{X,x}[\log(f_1), \cdots, \log(f_m)]$. This shows that in general, $(X^{\log}, \mathcal{O}_{X^{\log}})$ is not a locally ringed space.

For a morphism $f : X \to Y$ of fs log analytic spaces, one can define the *sheaf of relative log differentials* $\Omega^1_{X/Y}$ in the same way as Definition 3.5, namely it is the sheaf representing the functor of Y-log derivations (Definition 3.3), where we use the sheaf $\mathcal{O}_X$ of holomorphic functions as the structure sheaf. The explicit description in Proposition 3.4 still applies. If (∂, D) is the universal Y-log derivation of X to $\Omega^1_{X/Y}$, the morphism $D : M_X \to \Omega^1_{X/Y}$ is also written as $d\log$, and it can be extended by linearity to M^{gp}_X. In the explicit description

$$\Omega^1_{X/Y} = (\Omega_{\underline{X}/\underline{Y}} \oplus \mathcal{O}_{\underline{X}} \otimes_{\mathbb{Z}} M^{gp}_X)/\mathcal{K},$$

$d\log(a)$ is the image of $0 \oplus (1 \otimes a)$, for a local section a of M^{gp}_X.

The sheaf $\Omega^1_{X/Y}$ is an analytic coherent $\mathcal{O}_X$-module. For an integer $r \geqslant 1$, let $\Omega^r_{X/Y}$ be the r-th exterior power of $\Omega^1_{X/Y}$. The derivation $\partial : \mathcal{O}_{\underline{X}} \to \Omega^1_{X/Y}$ can be prolonged to a complex $\Omega^\bullet_{X/Y}$:

$$\mathcal{O}_{\underline{X}} \xrightarrow{\partial} \Omega^1_{X/Y} \xrightarrow{d} \Omega^2_{X/Y} \xrightarrow{d} \cdots \longrightarrow \Omega^r_{X/Y} \longrightarrow \cdots$$

by imposing that $d(d\log(a)) = 0$ for $a \in M_X^{gp}$. This is a complex of $\underline{f}^{-1}\mathscr{O}_{\underline{Y}}$-modules, called the *relative log de Rham complex on* X *with respect to* f.

For any sheaf F of $\mathscr{O}_X$-modules, define

$$\tau^*F := \tau^{-1}F \otimes_{\tau^{-1}\mathscr{O}_X} \mathscr{O}_{X^{\log}}$$

as a sheaf on $X^{\log}$. For an integer $r \geqslant 1$, define

$$\Omega^r_{X^{\log}/Y^{\log}} := \tau^*\Omega^r_{X/Y}.$$

The structure sheaf $\mathscr{O}_{X^{\log}}$ comes with a natural derivation $d : \mathscr{O}_{X^{\log}} \to \Omega^1_{X^{\log}/Y^{\log}}$ (see ([30], 3.5)), which can be prolonged to a complex $\Omega^\bullet_{X^{\log}/Y^{\log}}$ of $(f^{\log})^{-1}\mathscr{O}_{Y^{\log}}$-modules

$$\mathscr{O}_{X^{\log}} \xrightarrow{d} \Omega^1_{X^{\log}/Y^{\log}} \xrightarrow{d} \Omega^2_{X^{\log}/Y^{\log}} \xrightarrow{d} \cdots \longrightarrow \Omega^r_{X^{\log}/Y^{\log}} \longrightarrow \cdots,$$

which can be called the *relative log de Rham complex on* $X^{\log}$ *with respect to* f.

When Y is a point with trivial log structure, we denote $\Omega^\bullet_{X/Y}$ (resp. $\Omega^\bullet_{X^{\log}/Y^{\log}}$) by $\Omega^\bullet_X$ (resp. $\Omega^\bullet_{X^{\log}}$), and call it the *absolute log de Rham complex on* X (resp. $X^{\log}$).

Under mild conditions, F. Kato proved the relative log Poincaré lemma and the logarithmic analogue of the de Rham theorem. We state in the following a weaker version. See [25] for the more general version. We say an fs log analytic space is *log smooth* if it is so over a point with trivial log structure $(\mathrm{pt}, \mathbb{C}^*)$.

Theorem 9.3. *([25], 3.4.2, 3.2.5) Let* $f : X \to Y$ *be a log smooth morphism of fs log analytic spaces, with* Y *log smooth. Assume that the induced morphism* $f^{-1}\overline{M}_Y \to \overline{M}_X$ *is injective, that the stalk* $\overline{M}_{X/Y,x}$ *of the relative characteristic (4.1) is torsion-free for every* $x \in X$*, and that* f *is exact ([27], 4.6). Then there is a natural quasi-isomorphism*

$$(f^{\log})^{-1}\mathscr{O}_{Y^{\log}} \longrightarrow \Omega^\bullet_{X^{\log}/Y^{\log}}.$$

Corollary 9.4. *([25], 4.1.5) Let* $f : X \to Y$ *be as in Theorem 9.3 above, and assume in addition that* $\underline{f}$ *is proper. Then there is a natural quasi-isomorphism*

$$\tau_Y^* Rf_*\Omega^\bullet_{X/Y} \longrightarrow Rf_*^{\log}\underline{\mathbb{Z}} \otimes_{\mathbb{Z}} \mathscr{O}_{Y^{\log}}.$$

These results apply in particular to semi-stable degenerations. Also, the absolute log Poincaré lemma was proved earlier by Kato and Nakayama ([30], Theorem 3.8).

Log variations of polarized Hodge structures. For $y \in X^{\log}$ and $x = \tau(y) \in X$, let $\mathrm{sp}(y)$ be the set of all ring homomorphisms $s : \mathscr{O}_{X^{\log},y} \to \mathbb{C}$ that extend the evaluation map $\mathrm{ev}_x : \mathscr{O}_{X,x} \to \mathbb{C}$. Since $\mathscr{O}_{X^{\log},y}$ is isomorphic to the polynomial ring over $\mathscr{O}_{X,x}$ generated by log of a basis for $\overline{M}_{X,x}$, if we fix an $s_0 \in \mathrm{sp}(y)$, then we have a bijection:

$$s \mapsto (f \mapsto s(\log(f)) - s_0(\log(f))) : \mathrm{sp}(y) \xrightarrow{\sim} \mathrm{Hom}_{\mathrm{group}}(\overline{M}^{gp}_{X,x}, \mathbb{C}),$$

where $\mathbb{C}$ is viewed as an additive group.

Definition 9.5. Let X be an fs log analytic space. A *log variation of polarized Hodge structures of weight* n (abbreviated as *LVPHS*) on X is given by

- a local system of free abelian groups of finite rank $\mathscr{H}_{\mathbb{Z}}$ on $X^{\log}$,
- a bilinear form $\mathscr{Q} : \mathscr{H}_{\mathbb{Z}} \otimes \mathscr{H}_{\mathbb{Z}} \to \mathbb{Z}$,
- a finite decreasing filtration $F^\bullet \mathscr{H}_{\mathscr{O}}$ of $\mathscr{H}_{\mathscr{O}} := \mathscr{H}_{\mathbb{Z}} \otimes_{\mathbb{Z}} \mathscr{O}_{X^{\log}}$ by $\mathscr{O}_{X^{\log}}$-submodules,

such that the following conditions are satisfied:

1) there exist a locally free $\mathscr{O}_X$-module $\mathscr{E}$ and a finite decreasing filtration $F^\bullet\mathscr{E}$ by $\mathscr{O}_X$-submodules, such that $\mathrm{Gr}_p(\mathscr{E})$ is locally free and

$$F^p \mathscr{H}_{\mathscr{O}} \cong \tau^* F^p \mathscr{E}$$

for each p;

2) for $y \in X^{\log}$ and $x = \tau(y) \in X$, let $s \in \mathrm{sp}(y)$ and let $f_1, \cdots, f_r \in M_{X,x} - \mathscr{O}^*_{X,x}$ generate the monoid $\overline{M}_{X,x}$. If the real numbers $|\exp(s(\log(f_i)))|$ are sufficient small for all i, then $(\mathscr{H}_{\mathbb{Z},y}, \mathscr{Q}, F^\bullet(s))$ is a polarized Hodge structure of weight n;

3) (Griffiths transversality) the connection $\nabla = d \otimes \mathrm{id}_{\mathscr{H}_{\mathbb{Z}}} : \mathscr{H}_{\mathscr{O}} \to \Omega^1_{X^{\log}} \otimes_{\mathscr{O}_{X^{\log}}} \mathscr{H}_{\mathscr{O}}$ takes $F^p \mathscr{H}_{\mathscr{O}}$ into $\Omega^1_{X^{\log}} \otimes F^{p-1} \mathscr{H}_{\mathscr{O}}$.

Here $F^\bullet(s)$, the *specialization of* $F^\bullet$ *at* s, is the decreasing filtration of $\mathscr{H}_{\mathbb{C},y} := \mathbb{C} \otimes_{\mathbb{Z}} \mathscr{H}_{\mathbb{Z},y}$ defined by $F^p(s) = \mathbb{C} \otimes_{s,\mathscr{O}_{X^{\log},y}} (F^p \mathscr{H}_{\mathscr{O}})_y$. For a fixed point $y \in X^{\log}$, the family $(\mathscr{H}_{\mathbb{Z},y}, \mathscr{Q}, F^\bullet(s))_{s \in \mathrm{sp}(y)}$ is called a *polarized log Hodge structure of weight* n on the log point $(x, M_{X,x})$; this is the same as a log variation of polarized Hodge structures of weight n on the log point $(x, M_{X,x})$.

9.6. Variant. The definition we gave here follows ([31], 0.2.19), except that our polarization $\mathscr{Q}$ is integral. This definition differs slightly from the one as in ([41], 5.3) and ([29], 2.3), and is weaker. The main difference is that, in *loc. cit.* the locally free $\mathscr{O}_X$-module $\mathscr{E}$ with its filtration $F^\bullet\mathscr{E}$ is part of the data of the definition, and the flat connection is for $\mathscr{E}$ on X (namely $\nabla : \mathscr{E} \to \Omega^1_X \otimes \mathscr{E}$) and is required to satisfy the Griffiths transversality.

9.7. LVPHS from geometry. Log variations of polarized Hodge structures arise from geometry in the following way. Let $f : X \to Y$ be a projective vertical log smooth morphism between log smooth fs log analytic spaces, and we fix a line bundle on $\underline{X}$ which is relatively very ample over $\underline{Y}$. Here "vertical" means $f^{-1}(Y^*) = X^*$. By a theorem of Kajiwara and Nakayama ([23], 0.3), for every integer n, the sheaf $R^n f^{\log}_* \underline{\mathbb{Z}}$ is a local system on $Y^{\log}$. We take $\mathscr{H}_{\mathbb{Z}}$ to be $R^n f^{\log}_* \underline{\mathbb{Z}}$ modulo torsion, take $\mathscr{Q}$ to be the pairing induced by the fixed line bundle (which is obtained in the same way as ([29], 8.2) where we replace all local systems of $\mathbb{Q}$-vector spaces by the integral lattices in them), take $\mathscr{E}$ to be $R^n f_*(\Omega^\bullet_{X/Y})$, with filtration $F^p\mathscr{E} = R^n f_*(\Omega^{\geqslant p}_{X/Y}) \subset \mathscr{E}$, and take $F^p \mathscr{H}_{\mathscr{O}}$ to be $\tau^* F^p \mathscr{E}$. Then by a theorem of Kato, Matsubara and Nakayama ([29], Theorem 8.1), this is a log variation of polarized Hodge structures on Y, even in the stronger sense ([29], 2.3). See ([29], Theorem 8.1) for a more general version with coefficients. The special case when Y is the unit disk $\{|z| < 1\}$ in the complex plane with log structure induced by the divisor $\{z = 0\}$ and f is family of projective manifolds with semi-stable reduction over $\{z = 0\}$ was proved earlier by Matsubara ([41], Theorem C), except for the polarization part.

9.8. Relation to Deligne's canonical extensions. Suppose that $\underline{X}$ and $\underline{Y}$ are smooth, that the log structure on Y is induced by a normal crossing divisor $D \subset Y$, and that $f : X \to Y$ is a morphism of semi-stable reduction. Let $f' : X^* \to Y^*$ be the restriction of f. Then the flat connection $(R^n f'_* \Omega^\bullet_{X^*/Y^*}, \nabla')$ on Y^* has unipotent local monodromy around each component of D, and the flat connection $(R^n f_* \Omega^\bullet_{X/Y}, \nabla)$ on Y is its canonical extension in the sense of Deligne [7].

Kato-Usui spaces

We fix $n, h, H_{\mathbb{Z}}, Q, G_{\mathbb{Z}}, D$ and $D^\vee$ as in (9.1). For a ring R, let $\mathfrak{g}_R = \mathrm{Lie}(G_R)$. A subset $\sigma \subset \mathfrak{g}_{\mathbb{R}}$ is called a *nilpotent cone* if it is a cone

$$\sigma = \sum_{i=1}^{n} \mathbb{R}_{\geqslant 0} N_i$$

generated by mutually commutative nilpotent operators $N_i \in \mathfrak{g}_{\mathbb{R}} \subset \mathrm{End}(H_{\mathbb{R}})$. It is called a *rational nilpotent cone* if it can be generated by nilpotent operators in $\mathfrak{g}_{\mathbb{Q}}$. Let Γ be a *neat subgroup* of $G_{\mathbb{Z}}(\mathbb{Z})$, i.e. for every element $\gamma \in \Gamma$, its eigenvalues on $H_{\mathbb{C}}$ generate a torsion-free subgroup of $\mathbb{C}^*$.

Nilpotent orbits

Definition 9.9. Let $\sigma = \sum_i \mathbb{R}_{\geqslant 0} N_i$ be a nilpotent cone. A subset $Z \subset D^\vee$ is called a *σ-nilpotent orbit*, if there exists an $F_0 \in D^\vee$ such that

- $Z = \exp(\sum_i \mathbb{C} N_i) F_0$,
- $N F_0^p \subset F_0^{p-1}$ for all $p \in \mathbb{Z}$ and $N \in \sigma$,
- $\exp(\sum_i z_i N_i) F_0 \in D$ if $\mathrm{Im}(z_i) \gg 0$ for all i.

We also call the pair (σ, Z) a *nilpotent orbit.*

Let Σ be a *fan* in $\mathfrak{g}_{\mathbb{Q}}$, i.e. Σ is a non-empty set of rational nilpotent cones in $\mathfrak{g}_{\mathbb{R}}$ such that

- if $\sigma \in \Sigma$, then all faces of σ are in Σ,
- for $\sigma, \sigma' \in \Sigma$, the intersection $\sigma \cap \sigma'$ is a face of both σ and σ',
- for every $\sigma \in \Sigma$, we have $\sigma \cap (-\sigma) = 0$.

One can then define the set $D_{h,\Sigma}$ (or just D_Σ, if there is no confusion) of *nilpotent orbits in the directions in* Σ to be the set of nilpotent orbits (σ, Z) where $\sigma \in \Sigma$. There is a natural injection

$$F \mapsto (0, \{F\}) : D \hookrightarrow D_\Sigma.$$

The moduli space M_Σ Let Σ be a fan in $\mathfrak{g}_\mathbb{Q}$ and let $\Gamma \subset G_\mathbb{Z}(\mathbb{Z})$ be a subgroup. Then we say that Γ *is compatible with* Σ if for every $\gamma \in \Gamma$ and $\sigma \in \Sigma$, we have $\mathrm{Ad}(\gamma)(\sigma) \in \Sigma$. In this case, there is an action of Γ on D_Σ given by

$$(\sigma, Z) \overset{\gamma}{\longmapsto} (\mathrm{Ad}(\gamma)(\sigma), \gamma Z).$$

We say that Γ *is strongly compatible with* Σ if every cone $\sigma \in \Sigma$ is generated by elements in $\log \Gamma$. Kato and Usui showed that when Γ is strongly compatible with Σ and the arithmetic subgroup Γ is neat, the quotient set $M_\Sigma := \Gamma\backslash D_\Sigma$ can be given the structure of a log locally ringed space over $\mathbb{C}$, in fact a *log manifold* (see ([31], 3.5.7)). Roughly speaking, a log manifold is a log locally ringed space over $\mathbb{C}$, which is locally isomorphic to the "zero locus" of some log differential forms on a log smooth analytic space.

Informally speaking, Kato and Usui proved the following. First, there is a one-to-one correspondence between D_Σ and the set of polarized log Hodge structures of the given type. Second, if $\overline{X} \to \overline{S}$ is a log smooth family extending the projective smooth family $X \to S$, where $S \subset \overline{S}$ is the complement of a normal crossing divisor, then the period map extends to $\overline{S} \to M_\Sigma$. We briefly explain the first part in the following.

We shall show how to get a nilpotent orbit from a polarized log Hodge structure on a log point ([31], 0.4.24). Let x be an fs log point with log structure M_x. Then $\overline{M}_x$ is a sharp fs monoid and $\overline{M}_x^{gp}$ if a free abelian group of finite rank, say r. Fix $y \in x^{\log}$. We have $x^{\log} = \mathrm{Hom}(\overline{M}_x^{gp}, S^1) \simeq (S^1)^r$ and hence $\pi_1(x^{\log}) = \mathrm{Hom}(\overline{M}_x^{gp}, \mathbb{Z}) \simeq \mathbb{Z}^r$. Let $\pi_1^+(x^{\log}) \subset \pi_1(x^{\log})$ be the subset consisting of those homomorphisms $a : \overline{M}_x^{gp} \to \mathbb{Z}$ that take $\overline{M}_x$ into $\mathbb{N}$; this subset is an fs monoid.

Let $(H_\mathbb{Z}, Q, F^\bullet H_\mathcal{O})$ be a polarized log Hodge structure on x. Let $(h_i)_{i=1}^n$ be a family of generators for $\pi_1^+(x^{\log})$ and fix an $s_0 \in sp(y)$. Let $z_1, \cdots, z_r$ be complex numbers, and let $s \in sp(y)$ be such that

$$s\Big(\frac{\log(f)}{2\pi i}\Big) - s_0\Big(\frac{\log(f)}{2\pi i}\Big) = \sum_{i=1}^{r} z_i h_i(f), \quad \text{for } f \in \overline{M}_x^{gp}.$$

Let $N_i : H_{\mathbb{Q},y} \to H_{\mathbb{Q},y}$ be the logarithm of h_i. Then we have

$$F(s) = \exp\Big(\sum_{i=1}^{n} z_i N_i\Big) F(s_0),$$

which shows that $(F(s))_{s\in sp(y)}$ is an orbit of filtrations under $\exp(\sigma \otimes \mathbb{C})$ for $\sigma = \sum_i \mathbb{R}_{\geqslant 0} N_i$. Moreover, the condition 2) in Definition 9.5 implies that $F(s) \in D$ if $\mathrm{Im}(z_i) \gg 0$ for all i, and the condition 3) in Definition 9.5 implies that $NF(s_0)^p \subset F(s_0)^{p-1}$ for all $p \in \mathbb{Z}$ and $N \in \sigma$. In other words, the family $(F(s))_s$ is a σ-nilpotent orbit.

10. The main component of moduli spaces

Moduli: compactness and main components

In Section 4, we gave an overview of F. Kato's work [26] in which he uses log geometry to compactify the moduli space $\mathcal{M}_{g,n}$ of curves. Specifically, he shows that the moduli space of log smooth curves agrees with the Deligne-Mumford compactification $\overline{\mathcal{M}}_{g,n}$. The key philosophic idea in that section was that since moduli spaces of log smooth objects already includes degenerate objects, it is reasonable to expect that such a moduli space is a compactification of the moduli of objects with trivial log structure.

While the Deligne–Mumford space of stable curves $\overline{\mathcal{M}}_{g,n}$ turns out to be irreducible, it is an unfortunate fact of life that if $\mathfrak{X}$ is a moduli space of higher dimensional objects, then moduli-theoretic "compactifications" $\bar{\mathfrak{X}}$ of $\mathfrak{X}$ tend to have many irreducible components. If $\mathfrak{X}$ is irreducible, then it sits entirely within one of these many components of $\bar{\mathfrak{X}}$ and so it is natural then to ask if this "main component" can itself be given a moduli interpretation.

In Section 4.14 we stated a second philosophic principle: log geometry controls degenerations; that is, moduli of log smooth objects does not incorporate "too many" degenerate objects. This provides a type of converse to the aforementioned philosophy that moduli of log smooth objects should be compact. As explained in Section 4.14, the log structure gives us a fighting chance to show that our moduli space is irreducible (although it is of course too naïve to expect that moduli of log smooth objects is always irreducible). Combining the two principles, one may hope that if $\mathfrak{X}$ is an irreducible moduli space and $\bar{\mathfrak{X}}$ a moduli-theoretic compactification, then by appropriately incorporating log structures into the objects parameterized by $\bar{\mathfrak{X}}$, one will isolate the main component.

This technique of using log geometry to isolate the main component of a moduli space has been carried out by M. Olsson in several different settings. In [53], Olsson gives a moduli interpretation to the normalization of the main component of the toric Hilbert scheme. In [52], he isolates the normalization of the main component of V. Alexeev's compactification of the moduli space of principally polarized abelian varieties given in [3]; he further constructs a moduli-theoretic irreducible compactification of the moduli space of abelian varieties with higher degree polarization. In [48], he gives an irreducible modular compactification of the moduli space of polarized K3 surfaces.

Example: the toric Hilbert scheme

Our goal in this section is to explain the technique of isolating the main component of a moduli space by following Olsson's work [53]. We begin with the definition of the toric Hilbert scheme. Let k be a field and let P and Q be finitely-generated integral monoids with Q sharp and P^{gp} and Q^{gp} torsion-free.

Fix a surjective morphism $\pi : P \to Q$. This yields a closed immersion from $A_Q := \operatorname{Spec} k[Q]$ to $A_P := \operatorname{Spec} k[P]$, which is T_Q-equivariant, where T_Q (resp. T_P) denotes the torus associated to Q^{gp} (resp. P^{gp}). Consider the functor $\mathcal{H}$ whose S-valued points are diagrams

$$\begin{array}{ccc} Z & \xrightarrow{i} & A_{P,S} \\ & \searrow^{g} & \downarrow \\ & & S \end{array}$$

where i is a T_Q-invariant closed immersion and for every $q \in Q^{gp}$, the q-eigenspace of $g_*\mathcal{O}_Z$ is a finitely-presented projective $\mathcal{O}_S$-module of rank 1 if $q \in Q$ and rank 0 otherwise. By [16, Thm 1.1], this functor is representable by a quasi-projective scheme, which we call the *toric Hilbert scheme*.

Given a closed subscheme Z of $A_{P,S}$ as above, we can move Z by the action of T_P on $A_{P,S}$. This yields an action of T_P on $\mathcal{H}$. Since Z is T_Q-invariant, this action factors through $T_K = T_P/T_Q$, where K denotes the kernel of π^{gp}. We therefore obtain a map

$$T_K \longrightarrow \mathcal{H}$$

by letting $u \in T_K$ act on the distinguished point of $\mathcal{H}(k)$ given by the closed immersion $A_Q \to A_P$. By [6, 3.6(2)], this map is an open immersion. Therefore the normalization $\mathcal{S}$ of the scheme-theoretic closure of its image is a normal toric variety, and hence carries a natural fs log structure $\mathcal{M}_\mathcal{S}$ which makes it log smooth over $\operatorname{Spec} k$ (endowed with the trivial log structure). The goal of [53] is to give a moduli-theoretic interpretation of $(\mathcal{S}, \mathcal{M}_\mathcal{S})$.

Consider the functor $\mathcal{H}^{log}$ on the category of fs log schemes over k whose $(S, \mathcal{M}_S)$-valued points are given by diagrams

$$\begin{array}{ccc} (Z, \mathcal{M}_Z) & \xrightarrow{i} & (A_P, \mathcal{M}_{A_P}) \times (S, \mathcal{M}_S) \\ & \searrow^{g} & \downarrow \\ & & (S, \mathcal{M}_S) \end{array}$$

where the underlying maps on schemes defines a point of $\mathcal{H}(S)$, where g is log smooth and integral, and where the map

$$P \longrightarrow \mathcal{M}_{(Z,\mathcal{M}_Z)/(S,\mathcal{M}_S)} := \operatorname{coker}(g^*\mathcal{M}_S \to \mathcal{M}_Z)$$

induced by i factors through Q. The main theorem of [53] is then

Theorem 10.1 ([53, Thm 1.6]). *The functor $\mathcal{H}^{log}$ is representable by $(\mathcal{S}, \mathcal{M}_\mathcal{S})$.*

We explain briefly how Olsson obtains a natural morphism

$$F : \mathcal{H}_\mathcal{S} \longrightarrow \mathcal{H}^{log}$$

which he then shows is an equivalence; here $\mathcal{H}_{\mathcal{S}}$ denotes the functor of points of $(\mathcal{S}, \mathcal{M}_{\mathcal{S}})$. Olsson obtains F by showing that the pullback to $\mathcal{S}$ of the universal family over $\mathcal{H}$ yields a point of $\mathcal{H}^{\log}(\mathcal{S}, \mathcal{M}_{\mathcal{S}})$. Explicitly, if $i : \mathcal{Z} \to A_P \times \mathcal{S}$ is the pullback of the universal family, he constructs a log structure on $\mathcal{Z}$ as follows. Since $\mathcal{S}$ is a toric variety with torus T_K, it can be covered by open affines of the form Spec k[L] with L a submonoid of K whose associated group is K. Olsson proves ([53, 2.4]) that over such an open affine, the closed immersion i is given by

$$\mathcal{Z} \times_{\mathcal{S}} \operatorname{Spec} k[L] = \operatorname{Spec} k[E_L] \longrightarrow \operatorname{Spec} k[P \oplus L] = A_P \times \operatorname{Spec} k[L],$$

where E_L is the image of $P \oplus L$ in P^{gp} under the map $(p, \ell) \mapsto p + \ell$.

Example 10.2. *Let* $\pi : \mathbb{N}^2 \to \mathbb{N}$ *send* e_1 *to* 2 *and* e_2 *to* 1. *Then the kernel* K *of* π^{gp} *is all integer multiples of* $2e_2 - e_1$. *Let* L *be the submonoid of* K *consisting of the non-negative multiples of* $e_1 - 2e_2$ *and let* M *be the submonoid consisting of the non-positive multiples. Then* E_L *is generated by* e_1, e_2, *and* $2e_2 - e_1$; *hence,*

$$\operatorname{Spec} k[E_L] \simeq k[x, y, z]/(xy - z^2).$$

We see that E_M *is freely generated by* e_2 *and* $e_1 - 2e_2$, *so* $\operatorname{Spec} k[E_M] \simeq \mathbb{A}^2_k$.

We see then that $\mathcal{Z} \times_{\mathcal{S}} \operatorname{Spec} k[L]$ carries a natural log structure. These log structures glue to give a log structure on $\mathcal{Z}$ (by [53, Lemma 2.8]) which makes i a closed immersion of log schemes. This therefore yields F above.

Example: moduli of K3 surfaces

In this section we discuss Olsson's work [48] on the moduli of K3 surfaces. Recall that a surface $\underline{X}/\underline{S}$ is called *K3* if for all geometric points s of $\underline{S}$, the canonical divisor $K_{\underline{X}_s}$ is trivial and $H^1(\underline{X}_s, \mathcal{O}_{\underline{X}_s}) = 0$. In his thesis, R. Friedman constructs partial compactifications of the coarse space of the moduli of polarized K3 surfaces over $\mathbb{C}$. The key notion in his construction is that of a combinatorial K3 surface, which we now recall (see [11, p.2]).

Definition 10.3. If k is an algebraically closed field, then a *combinatorial K3 surface* over $\underline{S} = \operatorname{Spec} k$ is a k-scheme $\underline{X}$ with normal crossings (see Section 5) which is d-semistable, has trivial dualizing sheaf, and satisfies one of the following: $\underline{X}$ is a smooth K3 surface, $\underline{X}$ is a chain of elliptic ruled surfaces with rational surfaces on either end, or $\underline{X}$ is a union of rational surfaces where the double curves on each component form a cycle of rational curves with the dual graph of $\underline{X}$ a triangulation of S^2.

In light of the d-semistability condition and our discussion in Section 5, we should expect to be able to "explain" Friedman's moduli from the point of view of log geometry. With this in mind, we give the following definition of a log K3 surface (see [48, Def 5.1]).

Definition 10.4. A morphism $f : X \to S$ of log algebraic spaces is a *log K3 surface* if f is log smooth and integral, $\underline{f}$ is proper, the cokernel of $f^*\mathcal{M}_S \to \mathcal{M}_X$ in the category of integral sheaves of monoids is a sheaf of groups, and for every geometric point s of $\underline{S}$, we have $\Omega^2_{X_s/s} = \mathcal{O}_{\underline{X}_s}$, $H^1(\mathcal{O}_{\underline{X}_s}) = 0$, and $\underline{X}_s$ is a normal crossing variety. The log K3 surface is called *stable* if it has no infinitesimal automorphisms.

As explained in [48, Rmk 5.3], if X/S is a log K3 surface, then for every geometric point s of $\underline{S}$, the fiber $\underline{X}_s$ is a combinatorial K3 surface.

Now that we have a definition of log K3 surfaces, we discuss the logarithmic counterpart to the polarization. In order to ease the exposition, in what follows we will always assume that our log K3 surfaces satisfy the technical *special* condition defined in [48, Def 2.7]. Olsson introduces a notion of logarithmic Picard functor which generalizes the usual Picard functor in the case of trivial log structure (see Definition 4.5 and Corollary 5.6 of [48]):

Definition 10.5. If X/S is a log K3 surface, then the *log Picard functor* $\underline{\mathrm{Pic}}(X/S)$ is the sheafification of the presheaf associating to any $\underline{S}$-scheme $\underline{T}$ the isomorphism classes of $\mathcal{M}^{gp}_{X_T}$-torsors on $\underline{X}_{\underline{T},\text{ét}}$.

A polarization on a log K3 surface X/S is then defined to be ([48, Def 5.7]) a morphism $\lambda : \underline{S} \to \underline{\mathrm{Pic}}(X/S)$ such that on each geometric fiber $\underline{X}_s$, there is a line bundle $\mathcal{L}$ which lifts λ_s to $H^1(\underline{X}_s, \mathcal{O}^*_{\underline{X}_s})$ and satisfies the following. There is some $N > 0$ such that $\mathcal{L}^N$ is generated by global sections and the map defined by $\mathcal{L}^N$ only contracts finitely many curves to points.

With these definitions in place, Olsson fixes a positive integer k and considers the stack $\mathbb{M}_{2k}/\mathbb{Q}$ whose fiber over $\underline{T}$ is the groupoid of triples $(\mathcal{M}_T, X/(\underline{T}, \mathcal{M}_T), \lambda)$ where $\mathcal{M}_T$ is a log structure on $\underline{T}$ and $(X/(\underline{T}, \mathcal{M}_T), \lambda)$ is a stable polarized log K3 surface such that $\lambda_t^2 = 2k$ for every geometric point t of $\underline{T}$. Note that $\mathbb{M}_{2k}$ carries a natural log structure given by base log structure $\mathcal{M}_T$ in each fiber. One of the main results of [48] is then:

Theorem 10.6 ([48, Thm 6.2]). *The stack $\mathbb{M}_{2k}$ is smooth, log smooth, Deligne-Mumford, and contains an open sub-stack $\mathbb{M}^{sm}_{2k}$ parameterizing polarized* K3 *surfaces in the classical sense. The compliment of $\mathbb{M}^{sm}_{2k}$ in $\mathbb{M}_{2k}$ is a smooth divisor and the induced log structure agrees with the natural one on $\mathbb{M}_{2k}$.*

11. Twisted curves and log twisted curves

Twisted curves are a central object in the theory of *twisted stable maps* [2, 5, 1]: in order to have a complete moduli space of stable maps $C \to X$ of type Γ, where X is a proper tame stack with projective coarse moduli space and $\Gamma = (g, n, \beta)$ are the relevant discrete data, one must allow the curve C itself to be a certain type of stack, called *twisted curve*.

The original treatments of twisted curves relied on ad-hoc methods. The more recent approach of [1] relies on a method introduced in [51], which uses a construction with logarithmic structures.

Twisted curves

For simplicity we will stick with the case of Deligne–Mumford stacks.

First consider the geometric objects: fix an algebraically closed field k.

Definition 11.1. A *twisted curve* over k is a tame, purely 1-dimensional Deligne–Mumford stack $\mathcal{C}/k$, with at most nodes as singularities, satisfying the following conditions:

(1) Let $\pi : \mathcal{C} \to C$ be the morphism to the coarse moduli space. Then $\mathcal{C}^{sm} = \pi^{-1}C^{sm}$, and $\pi : \mathcal{C} \to C$ is an isomorphism over a dense open subset of C.
(2) Consider a node $\bar{x} \to C$, where the strictly henselian local ring $\mathcal{O}_{C,\bar{x}}$ is the strict henselization of $k[x, y]/(xy)$. Then

$$\mathcal{C} \times_C \operatorname{Spec} \mathcal{O}_{C,\bar{x}} \simeq \left[\operatorname{Spec} \mathcal{O}_{C,\bar{x}}[z, w]/(zw, z^m - x, w^m - y) \Big/ \mu_m \right],$$

where $\zeta \in \mu_m$ acts by $(z, w) \mapsto (\zeta z, \zeta^{-1} w)$.

An action such as (2) above is called *balanced* - it is crucial to our discussion of log structures below. The Deligne–Mumford assumption implies that the integer m is prime to the residue characteristic of $\bar{x}$. Note that $\mathcal{C}$ may have a stack structure at isolated smooth points as well - such points will behave like $[\mathbb{A}^1/\mu_a]$, where μ_a acts by multiplication, and again a is prime to the characteristic.

Over a general base S twisted curves are detected by their geometric fibers: a *twisted curve* $\mathcal{C} \to S$ is a flat, tame Deligne–Mumford stack locally of finite presentation, all of whose geometric fibers are twisted curves as in the definition above.

The *genus* of $\mathcal{C}$ is simply the genus of C. One typically needs to consider n-pointed twisted curves, where the markings are described in families as follows:

Definition 11.2. An *n-pointed twisted curve* $\mathcal{C}/S$ marked by disjoint closed substacks $\{\Sigma_i\}_{i=1}^n$ in $\mathcal{C}$ is assumed to satisfy the following:

(1) the Σ_i are contained in the smooth locus $\mathcal{C}^{sm}$,
(2) each Σ_i is a tame étale gerbe over S, and
(3) $\mathcal{C}_{gen} := \mathcal{C}^{sm} \smallsetminus \cup_i \Sigma_i \longrightarrow C$ is an open embedding.

Remark 11.3. When $S = \operatorname{Spec} k$ where $k = \bar{k}$, then $\Sigma_i = B\mu_{a_i}$, and moreover a_i is locally constant in families.

Remark 11.4. When $(\mathcal{C}/S, \{\Sigma_i\})$ is an n-pointed twisted curve, then the coarse moduli space of Σ_i is isomorphic to S. This means that the composite morphism $\Sigma_i \to \mathcal{C} \to C$ factors through a section $p_i : S \to C$. It follows that $(C, \{p_i\})$ is an n-pointed curve in the usual sense. This gives a functor

$$(\mathcal{C}/S, \{\Sigma_i\}) \mapsto (C, \{p_i\})$$

One can ask oneself: what does one need in order to recover a twisted n-pointed curve $(\mathcal{C}/S, \{\Sigma_i\})$ from a usual n-pointed curve $(C, \{p_i\})$? In other words, can we enrich the functor above to something like

$$(\mathcal{C}/S, \{\Sigma_i\}) \mapsto (C, \{p_i\}) + ?$$

which is nice and explicit and actually an equivalence of categories?

The stack structure at the marking definitely needs the data of the integers a_i, but in fact this is all that is necessary for the markings: near p_i, the curve $\mathcal{C}$ is canonically isomorphic to the root stack $C(\sqrt[a_i]{p_i})$. If x is a local generator of the ideal of p_i, then Zariski locally we have

$$\mathcal{C} \simeq \left[\operatorname{Spec} \mathcal{O}_C[z]/(z^{a_i} - x) \Big/ \mu_{a_i}\right].$$

The story is a bit more interesting at a node. It has to be - a twisted curve $\mathcal{C}$ with a node of index $m > 1$ has "ghost" automorphisms in μ_m which are not detectible on the coarse curve C: using the local coordinates given in Definition 11.1 (2), the μ_m action

$$(z, w) \mapsto (\zeta z, w)$$

on $\operatorname{Spec} \mathcal{O}_{C,\bar{x}}[z, w]/(zw, z^m - x, w^m - y)$ commutes with the action defining the quotient stack. It therefore descends to a nontrivial action on $\mathcal{C}$ which becomes trivial on the coarse moduli space C.

Log twisted curves

Let X be a Deligne–Mumford stack. Recall from Definition 2.16 that a fine log structure M on X is said to be *locally free* if for every geometric point $\bar{x} \to X$ we have that the characteristic sheaf $\overline{M}_{\bar{x}}$ is isomorphic to $\mathbb{N}^r$ for some r.

In this situation we say that a morphism of sheaves of monoids $M \to M'$ is *simple* if for every geometric point $\bar{x} \to X$ we can identify the map as the diagonal map

$$\begin{array}{ccc} \overline{M}_{\bar{x}} & \longrightarrow & \overline{M}'_{\bar{x}} \\ \downarrow \simeq & & \downarrow \simeq \\ \mathbb{N}^r & \xrightarrow{(m_1, \dots, m_r)} & \mathbb{N}^r \end{array}$$

where all m_i are prime to the characteristic of the field.

Definition 11.5. An *n-pointed log twisted curve over* S is the data

$$(C/S, \{\sigma_i, a_i\}, \ell : M_S \to M'_S)$$

where

- $(C, \{\sigma_i\})/S$ is an n-pointed nodal curve.
- M_S is the canonical log structure coming from the family $(C, M_C) \to (S, M_S)$ (see Section 4).

- $a_i : S \to \mathbb{Z}_{>0}$ are locally constant, with $a_i(s)$ invertible in the residue field $k(s)$.
- $\ell : M_S \to M'_S$ is a simple morphism.

And we have the following:

Theorem 11.6 ([51, Theorem 1.8]). *The fibered category of* n*-pointed twisted curves is naturally equivalent to the stack of* n*-pointed log twisted curves.*

The picture is as follows: we have already noted that we can replace a marking p_i by a stacky marking just using the data a_i. Now the j-th node which looks like $\operatorname{Spec} \mathcal{O}_S[x,y]/(xy-t)$ needs to be replaced by $\left[\operatorname{Spec} \mathcal{O}_S[z,w]/(zw - t^{1/m_j}) \;/\; \mu_{m_j}\right]$, and the data is encoding by dividing the j-th generator of $\mathbb{N}^r$ by m_j.

Remark 11.7. We can decompose the stack according to a_i:

$$\begin{array}{ccccc} \coprod_{\underline{a}} \mathcal{M}^{tw}_{g,n,\underline{a}} & = & \mathcal{M}^{tw}_{g,n} & = & \text{stack of } n\text{-pointed twisted curves} \\ & & \downarrow & & \\ & & \mathcal{S}_{g,n} & = & \text{stack of } n\text{-pointed nodal curves} \end{array}$$

and it can be deduced form the theorem that $\mathcal{M}^{tw}_{g,n,\underline{a}}$ is obtained from $\mathcal{S}_{g,n}$ using a root construction applied to the boundary divisor of $\mathcal{S}_{g,n}$, see [51, Remark 1.10].

In fact we have to apply all possible roots, accounting for all possible twisting of nodes, and glue together, so $\mathcal{M}^{tw}_{g,n,\underline{a}}$ is highly non-separated.

Below we sketch the main ideas in proving this. We stress that the assumption that our twisted curves are *balanced* is crucial - the case of unbalanced curves has not been treated.

From twisted curves to log twisted curves

Fix a twisted curve $f : \mathcal{C} \to S$.

We can follow F. Kato [26], giving log structures on nodal curves: consider all possible triples

$$(M_S, M_{\mathcal{C}}, f^\flat : f^{-1}M_S \to M_{\mathcal{C}})$$

such that

1. $(\mathcal{C}, M_{\mathcal{C}}) \to (S, M_S)$ is log smooth;
2. $M_{\mathcal{C}}, M_S$ are locally free; and
3. for all geometric points $\bar{x} \to \mathcal{C}$ mapping to nodes we have

$$\begin{array}{ccc} \overline{M}_{\mathcal{C},\bar{x}} & \xrightarrow{\sim} & \mathbb{N}^{r-1} \oplus \mathbb{N}^2 \\ \uparrow & & \uparrow{\scriptstyle \mathrm{id}\oplus\Delta} \\ \overline{M}_{S,f(\bar{x})} & \xrightarrow{\sim} & \mathbb{N}^{r-1} \oplus \mathbb{N} \end{array}$$

If $S = \operatorname{Spec} k$ we get a natural map

$$\mathbb{N}^{\text{number of nodes}} \to \overline{M}_S.$$

We say that $(M_{\mathcal{C}}, M_S, f^\flat)$ is *special* if for every geometric point $\bar{s} \to S$ this map is an isomorphism.

A result [51, Theorem 3.6] analogous to F. Kato's [26, Theorem 2.1] says that there is a unique special triple $(M_{\mathcal{C}}, M'_S, f^\flat)$ associated to $f : \mathcal{C} \to S$. Analyzing the coarse moduli space we obtain a unique diagram

$$\begin{array}{ccc} (\mathcal{C}, M_{\mathcal{C}}) & \longrightarrow & (C, M_C) \\ \downarrow & & \downarrow \\ (S, M'_S) & \xrightarrow{(\mathrm{id},\ell)} & (S, M_S) \end{array}$$

where $\ell : M_S \to M'_S$ is simple. In particular we obtain a log twisted curve

$$(C/S, \{\sigma_i, a_i\}_{i=1}^n, \ell : M_S \to M'_S)$$

From log twisted curves to twisted curves

Now we fix a log twisted curve $(C/S, \{\sigma_i, a_i\}_{i=1}^n, \ell : M_S \to M'_S)$. In particular we have the log smooth curve

$$(C, M_C) \to (S, M_S)$$

which is the coarse moduli space of a putative twisted curve. We want to describe $\mathcal{C}/C$ as the stack parametrizing natural objects over $T \to C$. Here it is! If we denote the relevant maps as follows

$$\begin{array}{ccc} T & \xrightarrow{s} & C \\ & \searrow^{h} & \downarrow \\ & & S, \end{array}$$

then $\mathcal{C}$ is the groupoid of diagrams

$$\begin{array}{ccc} h^*M_S & \xrightarrow{\ell} & H^*M'_S \\ \downarrow & & \downarrow \\ s^*M_C & \xrightarrow{k} & M'_C, \end{array}$$

where

(1) k is simple and for every geometric point $\bar{t} \to T$ the map $\overline{M}'_{S,\bar{t}} \to \overline{M}'_{C,\bar{t}}$ is either an isomorphism (at a general point), or of the form

$$\mathbb{N}^r \xhookrightarrow{(\mathrm{id},0)} \mathbb{N}^r \oplus \mathbb{N}$$

(at a marked point), or

$$\mathbb{N}^{r-1} \oplus \mathbb{N} \xhookrightarrow{(\mathrm{id},\Delta)} \mathbb{N}^{r-1} \oplus \mathbb{N}^2$$

(at a node).

(2) for all i and geometric point $\bar{t} \to T$ with $s(\bar{t}) \subset \sigma_i(S) \subset C$ the group

$$\mathrm{Coker}(\overline{M}^{gp}_{C,\bar{t}} \to \overline{M'}^{gp}_{C,\bar{t}})$$

is cyclic of order a_i.

12. Log stable maps

From curves to maps and expansions

Curves In Section 4 we discussed how prestable curves can be encoded as log smooth curves, and how in particular the stack of Deligne–Mumford stable curves can be interpreted as a logarithmic stack, representing log smooth stable curves over the category of fine and saturated log schemes. Stability in this situation just means that $\Omega^1_{(C,\mathcal{M}_C)}$ is an ample line bundle. This is the same as saying that $\omega_C(D)$ is ample, where D is the divisor of markings.

Maps Kontsevich [35] introduced the moduli of *stable maps* of prestable curves into a projective target variety X. This is a proper Deligne–Mumford stack $\overline{\mathcal{M}}_\Gamma(X)$ having projective coarse moduli space, where $\Gamma = (g, n, \beta)$ is the relevant numerical data: genus, number of markings and homology class of the image curve. It parametrizes maps $f : C \to X$, where this time stability means that ω_C is f-ample.

Kontsevich's moduli space has the property that it carries a *perfect relative obstruction theory* giving rise to a *virtual fundamental class* $[\overline{\mathcal{M}}_\Gamma(X)]^{vir}$, see [40, 4]. This is a key ingredient in defining Gromov–Witten invariants, with their applications in enumerative geometry and theoretical physics. The simplest GW invariants are

$$\langle \gamma_1 \cdots \gamma_n \rangle := \int_{[\overline{\mathcal{M}}_\Gamma(X)]^{vir}} e_1^*\gamma_1 \cdots e_n^*\gamma_n,$$

where $\gamma_i \in H^*(X, \mathbb{Q})$ and $e_i : \overline{\mathcal{M}}_\Gamma(X) \to X$ are the natural *evaluation maps* at the n markings.

So far no logarithmic structures are necessary.

Degenerations Among the methods of computing GW invariants, the *degeneration formula* is among the most powerful ones. It was introduced by A.-M. Li and Y. Ruan in symplectic geometry [37], see also Ionel-Parker [21, 22]. For our purposes, Jun Li's treatment in algebraic geometry [38, 39] is most relevant.

We are interested in the invariants of the smooth projective variety X. Since these are deformation invariant, it is natural to consider a degeneration of X, with smooth total space, into a union $X_0 = Y_1 \sqcup_D Y_2$ of two smooth varieties Y_i meeting transversally along a smooth divisor D. One wants to show that invariants of X

coincide with suitably defined invariants of X_0, and these in turn can be computed in terms of appropriately defined invariants of Y_i relative to D.

This is where logarithmic structures begin to show up, but there is still some way to go.

Perfect obstruction and Li's approach The difficulty with the degeneration is precisely the fact that the variety X_0 is singular, and therefore the natural obstruction theory on the moduli space of stable maps is not perfect in general. A similar situation occurs when considering the pair (Y_i, D), but we will not get into this discussion.

The problem occurs when a component of the source curve C maps entirely into D.

Jun Li's approach uses *expanded degenerations*. There are similar ideas in [37], but the symplectic approach builds in deformations of Cauchy-Riemann equations and has, at least on the surface, a significantly different flavor.

The idea is, that just as in stable pointed curves, if a marking travels towards a node one sprouts a new component of the curve, Jun Li says that when a component of C travels into D we can let X_0 sprout a new component.

Expansions Here is what the new component looks like. Denote by $N_{D\subset Y_i}$ the normal bundle of D in Y_i. Since the total space is smooth, we have $N_{D\subset Y_1} \cong N^\vee_{D\subset Y_2}$. Let $\mathbb{P} = \mathbb{P}\mathrm{roj}_D(1_D \oplus N_{D\subset Y_1})$. We have $\mathbb{P} \cong \mathbb{P}\mathrm{roj}_D(N_{D\subset Y_2} \oplus 1_D)$ so we can denote by D^+ and D^- the smooth divisors in $\mathbb{P}$ which correspond to the normal bundle $N_{D\subset Y_1}$ and $N_{D\subset Y_2}$ respectively. Note that the divisor D^+ and D^- are canonically isomorphic to D.

We now glue things up. Let $\mathbb{P}_i$ for $i \in \mathbb{N}$ be copies of $\mathbb{P}$. We can glue Y_1 and $\mathbb{P}_1$ along D and D^- respectively, $\mathbb{P}_i$ and $\mathbb{P}_{i+1}$ along D^+ and D^- respectively, and $\mathbb{P}_n$ and Y_2 along D^+ and D respectively. We denote the resulting gluing by

$$X_0[n] = Y_1 \bigsqcup_{D_1} \mathbb{P}_1 \bigsqcup_{D_2} \cdots \bigsqcup_{D_n} \mathbb{P}_n \bigsqcup_{D_{n+1}} Y_2, \tag{12.1}$$

where $D_1, \cdots, D_{n+1}$ are the disjoint singular loci of $X_0[n]$.

Such a beast is known as an *expanded degeneration*, or by the more folksy name *an n-accordion*.

This led Jun Li to define *degeneration stable maps with target* X_0 as *nondegenerate maps* $C \to X_0[n]$*:* a map is nondegenerate if no component of C maps into any of the D_i.

Predeformability But here Jun Li meets another phenomenon, already present in the space of admissible covers of Harris–Mumford [17]: nondegenerate maps to $C \to X_0[n]$ have many redundant components which have nothing to do with maps to the generic fiber X. Near a singular point of $X_0[n]$ which looks like $\{xy = 0\}$, a curve map locally deforms to a smoothing of X_0 if and only if the curve looks like $\{uv = 0\}$, and the map given by $x = u^m, y = v^m$. Such nice maps are called *predeformable*. But predeformable maps are clearly not open in the space of maps - they are actually closed among nondegenerate maps. This means that the restricted obstruction theory on them is "wrong" - definitely not perfect

Of course the virtual fundamental class of Gromov–Witten theory has no problem dealing with the total moduli space with its extra components, but these extra components do get in the way of decomposing invariants of X_0 in terms of (Y_i, D). So we really do want to stick to predeformable maps.

Logarithmic methods: from Jun Li to Bumsig Kim

Predeformable deformations At this point Jun Li's reasoning arrives at a point where a new obstruction theory on predeformable maps is needed. Having read this chapter, the reader will immediately recognize that

(1) nodal curves are log-smooth,
(2) n-accordions are d-semistable and admit a canonical log-smooth structure, and
(3) predeformable maps can be viewed as log maps from the log-smooth curve to the d-semistable target.

Jun Li also recognized this fact, as did Shin Mochizuki before him [42] when he in his turn revisited the space of admissible covers of Harris–Mumford. What he lacked at the time was a formalism for logarithmic deformation theory of singular spaces, such as the moduli space itself: a perfect obstruction theory is a two-term complex mapping to the cotangent complex of the moduli space, but the moduli space is, as usual, highly singular, even taking its logarithmic structure into account.

So Jun Li resorted to an ad-hoc construction of his perfect obstruction theory. This is the most difficult part of his work.

In Section 7 we saw how log deformation theory works in the necessary generality. This is where Bumsig Kim's paper [33] comes in: he provides a correct formalism for nondegenerate logarithmic stable maps into expanded degenerations, and shows that it carries a perfect obstruction theory. The degeneration formula in this formalism has been worked out by one of us (Q. Chen), and should appear as part of a larger project indicated below.

One aspect that deserves mention is Kim's notion of *minimal log structures* on maps. Recall that the log stack $\overline{\mathcal{M}}_{g,n}$ can be constructed as a stack over the category of fine saturated log schemes, whose objects over $S = (\underline{S}, M)$ are log smooth stable pointed curves over S. But in order to exhibit the underlying stack, one needs to use the *canonical* log structure on $\underline{S}$, which is initial among all possible ones carrying the log smooth curve.

Kim describes his stack similarly - given a predeformable map $\underline{C} \to \underline{X}$ of underlying schemes over $\underline{S}$, it amounts to describing what he calls *minimal* log structure S on $\underline{S}$ carrying a log map $C \to X$. Kim does this by explicitly describing the combinatorial structure of such log structures.

Unexpanded log maps: from Siebert into the future

A very different approach was proposed in a 2001 lecture by Bernd Siebert, but lay dormant for almost a decade.

The point is this. If one embraces logarithmic structures, and logarithmic maps from log smooth curves to some logarithmic scheme, then expansions are no longer necessary. Defined correctly, the space of log maps automatically has a perfect log-obstruction theory, which in view of sections 6 and 7 can be viewed as an obstruction theory relative to the stack LOG. This automatically results in a virtual fundamental class.

With this way of thinking, one can consider much more general logarithmic stable maps, gaining access to invariants of much more general degenerations of varieties. This has been a desired goal for a number of years.

So what is the correct definition? Consider a fine and saturated log scheme X. Following the work of F. Kato [26] as discussed in section 4, one comes up with a definition of a category $\overline{\mathcal{M}}_\Gamma(X)$ fibered over the category LSch_{fs} of fine and saturated log schemes: an object over a fine saturated log scheme S is a log smooth curve $C \to S$ and a log map $f : C \to X$. We further require it to be *stable*: the line bundle of logarithmic differentials $\Omega^1_{C/S}$ is required to be f-ample. This is tantamount to requiring the map of underlying schemes to be a stable map.

So the main claim is: $\overline{\mathcal{M}}_\Gamma(X)$ is represented by a logarithmic Deligne–Mumford stack with projective coarse moduli space. In fact this stack is proper and quasi-finite over the usual stack of stable maps $\overline{\mathcal{M}}_\Gamma(\underline{X})$ of the underlying scheme. As in the discussion of Kim's work, the underlying stack can be viewed as a moduli of log maps with *minimal* log structure.

This is the subject of current work - of Gross and Siebert on the one hand and of two of us (mainly Chen, and to a lesser extent Abramovich) on the other, so it would not be appropriate to get into details until definite results appear.

Let us instead put this in a larger context. Consider logarithmic schemes $Z \to B$ and X, and assume we are given a morphism of underlying schemes $\underline{f} : \underline{Z} \to \underline{X}$. We can define a category $\mathrm{Lift}_{\underline{f}}$ fibered over LSch_{fs} whose objects over a fine saturated log B-scheme $S \to B$ are lifts $f_S : Z_S \to X$ of the morphism of underlying schemes $\underline{f}_{\underline{S}} : \underline{Z}_{\underline{S}} \to \underline{X}$.

One can ask the following general questions:

Question 12.2. (1) Under what conditions is $\mathrm{Lift}_{\underline{f}}$ a log stack locally of finite type over B?
(2) What natural numerical data cut out a substack of finite type?
(3) Under what conditions is the result proper?

We want to stress our belief that this question is natural, important and quite tractable. For instance, the case where $B = \operatorname{Spec} \mathbb{C}$ with trivial structure and $Z = X = \operatorname{Spec}(\mathbb{N} \to \mathbb{C})$ the result is a countable union of components. It is similar in nature to an inertia stack. The more general case where $B = X$, $Z = X \times \operatorname{Spec}(\mathbb{N} \to \mathbb{C})$ and $\underline{f}$ is the diagonal, is the relevant analogue of the inertia stack of X. It is important for Gromov–Witten theory - up to $\mathbb{C}^*$ action the result is the natural target for evaluation maps associated to log smooth curves. Its components account for the contact orders of relative stable maps. Further examples of a similar nature govern gluing of nodes of log-smooth curves.

References

[1] Dan Abramovich, Martin Olsson, and Angelo Vistoli. Twisted stable maps to tame artin stacks. *J. Algebraic Geom.* **20**(3) 399–477, 2011. ← 49, 50

[2] Dan Abramovich and Angelo Vistoli. Compactifying the space of stable maps. *J. Amer. Math. Soc.*, **15**(1), 27–75 (electronic), 2002. ← 49

[3] Valery Alexeev. Complete moduli in the presence of semiabelian group action. *Ann. of Math. (2)*, **155**(3), 611–708, 2002. ← 46

[4] K. Behrend and B. Fantechi. The intrinsic normal cone. *Invent. Math.*, **128**(1), 45–88, 1997. ← 54

[5] Weimin Chen and Yongbin Ruan. Orbifold Gromov-Witten theory, in: Orbifolds in mathematics and physics (Madison, WI, 2001), 25–85. *Contemp. Math.*,**310**, Amer. Math. Soc., Providence, RI, 2002. ← 49

[6] Olga V. Chuvashova. The main component of the toric Hilbert scheme. *Tohoku Math. J. (2)*, **60**(3), 365–382, 2008. ← 47

[7] Pierre Deligne. Équations différentielles à points singuliers réguliers. *Lecture Notes in Mathematics*, vol. **163**. Springer-Verlag, Berlin, 1970. ← 44

[8] Pierre Deligne. Travaux de Griffiths, in: Séminaire Bourbaki. Vol. 1969/70: Exposés 364–381. *Lecture Notes in Mathematics*, vol. **180**, 213–237: Exposé 376. Springer-Verlag, Berlin, 1971. ← 39, 40

[9] P. Deligne and D. Mumford. The irreducibility of the space of curves of given genus. *Inst. Hautes Études Sci. Publ. Math.*, (36), 75–109, 1969. ← 18

[10] Robert Friedman. Global smoothings of varieties with normal crossings. *Ann. of Math. (2)*, **118**(1), 75–114, 1983. ← 3, 19, 21

[11] Robert Friedman and Francesco Scattone. Type III degenerations of K3 surfaces. *Invent. Math.*, **83**(1), 1–39, 1985. ← 48

[12] Ofer Gabber and Lorenzo Ramero. Foundations of p-adic hodge theory, 2009. `http://arxiv.org/abs/0409584`. ← 4

[13] Phillip Griffiths. Logarithmic hodge structures: Report on the work of kato-usui. Notes from a talk given at MSRI on March 18, 2009. `http://math.brown.edu/~abrmovic/Griffiths-log.pdf`. ← 38

[14] A. Grothendieck. Éléments de géométrie algébrique. IV. Étude locale des schémas et des morphismes de schémas. I. *Inst. Hautes Études Sci. Publ. Math.*, (20), 259, 1964. ← 8

[15] A. Grothendieck and M. Raynaud. *Revêtements étales et groupe fondamental.* Springer-Verlag, Berlin, 1971. Séminaire de Géométrie Algébrique du Bois Marie 1960–1961 (SGA 1), Dirigé par Alexandre Grothendieck. Augmenté de deux exposés de M. Raynaud, Lecture Notes in Mathematics, vol. **224**. ← 13

[16] Mark Haiman and Bernd Sturmfels. Multigraded Hilbert schemes. *J. Algebraic Geom.*, **13**(4), 725–769, 2004. ← 47

[17] Joe Harris and David Mumford. On the Kodaira dimension of the moduli space of curves. *Invent. Math.*, **67**(1), 23–88, 1982. With an appendix by William Fulton. ← 55

[18] W. J. Harvey. Boundary structure of the modular group, in: Riemann surfaces and related topics: Proceedings of the 1978 Stony Brook Conference (State Univ. New York, Stony Brook, N.Y., 1978), 245–251. *Ann. of Math. Stud.*, **97**, Princeton Univ. Press, Princeton, N.J., 1981. ← 34

[19] Luc Illusie. Complexe cotangent et déformations. I. *Lecture Notes in Mathematics*, vol.**239**. Springer-Verlag, Berlin, 1971. ← 29

[20] Luc Illusie. Complexe cotangent et déformations. II. *Lecture Notes in Mathematics*, vol.**283**. Springer-Verlag, Berlin, 1972. ← 29

[21] Eleny-Nicoleta Ionel and Thomas H. Parker. Relative Gromov-Witten invariants. *Ann. of Math. (2)*, **157**(1), 45–96, 2003. ← 54

[22] Eleny-Nicoleta Ionel and Thomas H. Parker. The symplectic sum formula for Gromov-Witten invariants. *Ann. of Math. (2)*, **159**(3), 935–1025, 2004. ← 54

[23] Takeshi Kajiwara and Chikara Nakayama. Higher direct images of local systems in log Betti cohomology. *J. Math. Sci. Univ. Tokyo*, **15**(2), 291–323, 2008. ← 34, 43

[24] Fumiharu Kato. Log smooth deformation theory. *Tohoku Math. J. (2)*, **48**(3), 317–354, 1996. ← 13, 19, 20, 22

[25] Fumiharu Kato. The relative log Poincaré lemma and relative log de Rham theory. *Duke Math. J.*, **93**(1), 179–206, 1998. ← 42

[26] Fumiharu Kato. Log smooth deformation and moduli of log smooth curves. *Internat. J. Math.*, **11**(2), 215–232, 2000. ← 13, 14, 15, 16, 18, 46, 52, 53, 57

[27] Kazuya Kato. Logarithmic structures of Fontaine-Illusie, in: Algebraic analysis, geometry, and number theory (Baltimore, MD, 1988), 191–224. Johns Hopkins Univ. Press, Baltimore, MD, 1989. ← 1, 4, 8, 11, 13, 14, 19, 24, 37, 42

[28] Kazuya Kato. Toric singularities. *Amer. J. Math.*, **116**(5), 1073–1099, 1994. ← 11

[29] Kazuya Kato, Toshiharu Matsubara, and Chikara Nakayama. Log C^∞-functions and degenerations of Hodge structures, in: Algebraic geometry 2000, Azumino (Hotaka), 269–320. *Adv. Stud. Pure Math.*, **36**, Math. Soc. Japan, Tokyo, 2002. ← 38, 43

[30] Kazuya Kato and Chikara Nakayama. Log Betti cohomology, log étale cohomology, and log de Rham cohomology of log schemes over C. *Kodai Math. J.*, **22**(2), 161–186, 1999. ← 33, 34, 35, 37, 38, 41, 42

[31] Kazuya Kato and Sampei Usui. Classifying spaces of degenerating polarized Hodge structures. *Annals of Mathematics Studies*, **169**. Princeton University Press, Princeton, NJ, 2009. ← 38, 39, 41, 43, 45

[32] G. Kempf, Finn Faye Knudsen, D. Mumford, and B. Saint-Donat. Toroidal embeddings. I. *Lecture Notes in Mathematics*, vol. **339**, Springer-Verlag, Berlin, 1973. ← 11

[33] Bumsig Kim. Logarithmic stable maps, in: New developments in algebraic geometry, integrable systems and mirror symmetry (RIMS, Kyoto, 2008), 167–200, *Adv. Stud. Pure Math.*, **59**, Math. Soc. Japan, Tokyo, 2010. ← 56

[34] János Kollár. Algebraic geometry, in: Timothy Gowers, June Barrow-Green, and Imre Leader, editors, *The Princeton companion to mathematics*, 363–372, Princeton University Press, Princeton, NJ, 2008. ← 3

[35] Maxim Kontsevich. Enumeration of rational curves via torus actions, in: The moduli space of curves (Texel Island, 1994), 335–368. *Progr. Math.*, **129**, Birkhäuser Boston, Boston, MA, 1995. ← 54

[36] Gérard Laumon and Laurent Moret-Bailly. Champs algébriques. *Ergebnisse der Mathematik und ihrer Grenzgebiete. 3. Folge. A Series of Modern Surveys in Mathematics [Results in Mathematics and Related Areas. 3rd Series. A Series of Modern Surveys in Mathematics]*, **39**, Springer-Verlag, Berlin, 2000. ← 26

[37] An-Min Li and Yongbin Ruan. Symplectic surgery and Gromov-Witten invariants of Calabi-Yau 3-folds. *Invent. Math.*, **145**(1), 151–218, 2001. ← 54, 55

[38] Jun Li. Stable morphisms to singular schemes and relative stable morphisms. *J. Differential Geom.*, **57**(3), 509–578, 2001. ← 54

[39] Jun Li. A degeneration formula of GW-invariants. *J. Differential Geom.*, **60**(2), 199–293, 2002. ← 54

[40] Jun Li and Gang Tian. Virtual moduli cycles and Gromov-Witten invariants of algebraic varieties. *J. Amer. Math. Soc.*, **11**(1), 119–174, 1998. ← 54

[41] Toshiharu Matsubara. On log Hodge structures of higher direct images. *Kodai Math. J.*, **21**(2), 81–101, 1998. ← 43, 44

[42] Shinichi Mochizuki. The geometry of the compactification of the Hurwitz scheme. *Publ. Res. Inst. Math. Sci.*, **31**(3), 355–441, 1995. ← 56

[43] Chikara Nakayama and Arthur Ogus. Relative rounding in toric and logarithmic geometry. *Geom. Topol.* **14**(4), 2189–2241, 2010. ← 33, 34, 37

[44] Arthur Ogus. F-crystals, Griffiths transversality, and the Hodge decomposition. *Astérisque*, (221), ii+183, 1994. ← 9

[45] Arthur Ogus. Lectures on logarithmic algebraic geometry. TeXed notes, 2001. `http://math.berkeley.edu/~ogus/preprints/log_book/logbook.pdf`. ← 4, 5, 7, 9, 10, 12

[46] Martin C. Olsson. Logarithmic geometry and algebraic stacks. *Ann. Sci. École Norm. Sup. (4)*, **36**(5), 747–791, 2003. ← 23, 25, 26, 27, 28, 29

[47] Martin C. Olsson. Universal log structures on semi-stable varieties. *Tohoku Math. J. (2)*, **55**(3), 397–438, 2003. ← 19, 21, 22, 23

[48] Martin C. Olsson. Semistable degenerations and period spaces for polarized K3 surfaces. *Duke Math. J.*, **125**(1), 121–203, 2004. ← 46, 48, 49

[49] Martin C. Olsson. The logarithmic cotangent complex. *Math. Ann.*, **333**(4), 859–931, 2005. ← 13, 27, 28, 29, 30, 32, 33

[50] Martin C. Olsson. Deformation theory of representable morphisms of algebraic stacks. *Math. Z.*, **253**(1), 25–62, 2006. ← 29, 32

[51] Martin C. Olsson. (Log) twisted curves. *Compos. Math.*, **143**(2), 476–494, 2007. ← 50, 52, 53

[52] Martin C. Olsson. Compactifying moduli spaces for abelian varieties. *Lecture Notes in Mathematics*, **1958** Springer-Verlag, Berlin, 2008. ← 46

[53] Martin C. Olsson. Logarithmic interpretation of the main component in toric Hilbert schemes, in: Curves and abelian varieties, 231–252. *Contemp. Math.*, **465** , Amer. Math. Soc., Providence, RI, 2008. ← 46, 47, 48

[54] Ulf Persson and Henry Pinkham. Some examples of nonsmoothable varieties with normal crossings. *Duke Math. J.*, **50**(2), 477–486, 1983. ← 21

(Abramovich) Department of Mathematics, Brown University, Providence, RI 02912, U.S.A.
E-mail address: `abrmovic@math.brown.edu`

(Chen) Department of Mathematics, Brown University, Providence, RI 02912, U.S.A.
Current address, Chen: Department of Mathematics, Columbia University, Rm 628, MC 4421, 2990 Broadway, New York, NY 10027, USA
E-mail address: `q_chen@math.columbia.edu`

(Gillam) Department of Mathematics, Brown University, Providence, RI 02912, U.S.A.
Current address: Departement Mathematik, Universitatät Zürich, Rämistrasse 101, 8092 Zürich, Switzerland
E-mail address: `wdgillam@gmail.com`

(Huang) Department of Mathematics, University of California, Berkeley, CA 94720, U.S.A.
E-mail address: `huang@math.berkeley.edu`

(Olsson) Department of Mathematics, University of California, Berkeley, CA 94720, U.S.A.
E-mail address: `molsson@math.berkeley.edu`

(Satriano) Department of Mathematics, University of California, Berkeley, CA 94720, U.S.A.
Current address: Department of Mathematics, University of Michigan, 2074 East Hall, Ann Arbor, MI 48109, USA
E-mail address: `satriano@umich.edu`

(Sun) Department of Mathematics, University of California, Berkeley, CA 94720, U.S.A.
Current address: Mathematical Sciences Center, Room 131, Jin Chun Yuan West Building, Tsinghua University, Haidian District, Beijing 100084, China
E-mail address: `shsun@math.tsinghua.edu.cn`

Invariant Hilbert schemes

Michel Brion

Abstract. This is a survey article on moduli of affine schemes equipped with an action of a reductive group. The emphasis is on examples and applications to the classification of spherical varieties.

Contents

1. Introduction

The Hilbert scheme is a fundamental object of projective algebraic geometry; it parametrizes those closed subschemes of the projective space $\mathbb{P}^N$ over a field k, that have a prescribed Hilbert polynomial. Many moduli schemes (for example, the moduli space of curves of a fixed genus) are derived from the Hilbert scheme by taking locally closed subschemes and geometric invariant theory quotients.

In recent years, several versions of the Hilbert scheme have been constructed in the setting of algebraic group actions on affine varieties. One of them, the *G-Hilbert scheme* G–Hilb(V), is associated to a linear action of a finite group G on a finite-dimensional complex vector space V; it classifies those G-stable subschemes $Z \subset V$ such that the representation of G in the coordinate ring $\mathcal{O}(Z)$ is the regular representation. The G-Hilbert scheme comes with a morphism to the quotient variety V/G, that associates with Z the point Z/G. This *Hilbert-Chow morphism* has an inverse over the open subset of V/G consisting of orbits with trivial isotropy group, as every such orbit Z is a point of G–Hilb(V). In favorable cases (e.g. in dimension 2), the Hilbert-Chow morphism is a desingularization of V/G; this construction plays an essential role in the McKay correspondence (see e.g. [27, 28, 17]).

Another avatar of the Hilbert scheme is the *multigraded Hilbert scheme* introduced by Haiman and Sturmfels in [26]; it parametrizes those homogeneous ideals I of a polynomial ring $k[t_1, \dots, t_N]$, graded by an abelian group Γ, such that each homogeneous component of the quotient $k[t_1, \dots, t_N]/I$ has a prescribed (finite) dimension. In contrast to the construction (due to Grothendieck) of the Hilbert scheme which relies on homological methods, that of Haiman and Sturmfels is based on commutative algebra and algebraic combinatorics only; it is valid over any base ring k. Examples of multigraded Hilbert schemes include the Grothendieck-Hilbert scheme (as follows from a result of Gotzmann, see [24]) and the *toric Hilbert scheme* (defined by Peeva and Sturmfels in [49]) where the homogeneous components of I have codimension 0 or 1.

The invariant Hilbert scheme may be viewed as a common generalization of G-Hilbert schemes and multigraded Hilbert schemes; given a complex reductive group G and a finite-dimensional G-module V, it classifies those closed G-subschemes $Z \subset V$ such that the G-module $\mathcal{O}(Z)$ has prescribed (finite) multiplicities. If G is diagonalizable with character group Λ, then Z corresponds to a homogeneous ideal of the polynomial ring $\mathcal{O}(V)$ for the Λ-grading defined by the G-action; we thus recover the multigraded Hilbert scheme. But actually, the construction of the invariant Hilbert scheme in [3] relies on a reduction to the multigraded case via highest weight theory.

The Hilbert scheme of $\mathbb{P}^N$ is known to be projective and connected; invariant Hilbert schemes are quasi-projective (in particular, of finite type) but possibly non-projective. Also, they may be disconnected, already for certain toric Hilbert schemes (see [53]). One may wonder how such moduli schemes can exist in the setting of affine algebraic geometry, since for example any curve in the affine plane is a flat limit of curves of higher degree. In fact, the condition for the considered subschemes $Z \subset X$ to have a coordinate ring with finite multiplicities is quite strong; for example, it yields a proper morphism to a punctual Hilbert scheme, that associates with Z the categorical quotient $Z/\!/G \subset X/\!/G$.

In this article, we present the construction and fundamental properties of the invariant Hilbert scheme, and survey some of its applications to varieties with algebraic group actions. The prerequisites are (hopefully) quite modest: basic notions of algebraic geometry; the definitions and results that we need about actions and representations of algebraic groups are reviewed at the beginning of Section 2. Then we introduce flat families of closed G-stable subschemes of a fixed affine G-scheme X, where G is a fixed reductive group, and define the Hilbert function which encodes the G-module structure of coordinate rings. Given such a function h, our main result asserts the existence of a universal family with base a quasi-projective scheme: the invariant Hilbert scheme $\mathrm{Hilb}_h^G(X)$.

Section 3 presents a number of basic properties of invariant Hilbert schemes. We first reduce the construction of $\mathrm{Hilb}_h^G(X)$ to the case (treated by Haiman and Sturmfels) where G is diagonalizable and X is a G-module. For this, we slightly generalize the approach of [3], where G was assumed to be connected. Then we describe the Zariski tangent space at any closed point Z, with special attention to the case where Z is a G-orbit closure in view of its many applications. Here also, we adapt the results of [3]. More original are the next developments on the action of the equivariant automorphism group and on a generalization of the Hilbert-Chow morphism, which aim at filling some gaps in the literature.

In Section 4, we first give a brief overview of invariant Hilbert schemes for finite groups and their applications to resolving quotient singularities; here the reader may consult [5] for a detailed exposition. Then we survey very different applications of invariant Hilbert schemes, namely, to the classification of *spherical varieties*. These form a remarkable class of varieties equipped with an action of a connected reductive group G, that includes toric varieties, flag varieties and symmetric homogeneous spaces. A normal affine G-variety Z is spherical if and only if the G-module $\mathcal{O}(Z)$ is multiplicity-free; then Z admits an equivariant degeneration to an affine spherical variety Z_0 with a simpler structure, e.g., the decomposition of $\mathcal{O}(Z_0)$ into simple G-modules is a grading of that algebra. Thus, Z_0 is uniquely determined by the highest weights of these simple modules, which form a finitely generated monoid Γ. We show (after [3]) that the affine spherical G-varieties with weight monoid Γ are parametrized by an affine scheme of finite type M_Γ, a locally closed subscheme of some invariant Hilbert scheme.

Each subsection ends with examples which illustrate its main notions and results; some of these examples have independent interest and are developed along the whole text. In particular, we present results of Jansou (see [29]) that completely describe invariant Hilbert schemes associated with the "simplest" data: G is semi-simple, X is a simple G-module, and the Hilbert function h is that of the cone of highest weight vectors (the affine cone over the closed G-orbit Y in the projective space $\mathbb{P}(X)$). Quite remarkably, $\mathrm{Hilb}_h^G(X)$ turns out to be either a (reduced) point or an affine line $\mathbb{A}^1$; moreover, the latter case occurs precisely when Y can be embedded into another projective variety $\tilde{Y}$ as an ample divisor, where $\tilde{Y}$ is homogeneous under a semi-simple group $\tilde{G} \supset G$, and G acts transitively on the complement $\tilde{Y} \setminus Y$. Then the universal family is just the affine cone over $\tilde{Y}$ embedded via the sections of Y.

This relation between invariant Hilbert schemes and projective geometry has been further developed in recent works on the classification of arbitrary spherical G-varieties. In loose terms, one reduces to classifying *wonderful G-varieties* which are smooth projective G-varieties having the simplest orbit structure; examples include the just considered G-varieties $\tilde{Y}$. Taking an appropriate affine multi-cone, one associates to each wonderful variety a family of affine spherical varieties over an affine space $\mathbb{A}^r$, which turns out to be the universal family. This approach is presented in more details in the final Subsections 4.4 and 4.5.

Throughout this text, the emphasis is on geometric methods and very little space is devoted to the combinatorial and Lie-theoretical aspects of the domain, which are however quite important. The reader may consult [7, 22, 44, 53, 55] for the combinatorics of toric Hilbert schemes, [51] for the classification of spherical embeddings, [9] for that of wonderful varieties, and [39] for uniqueness properties of spherical varieties. Also, we do not present the close relations between certain invariant Hilbert schemes and moduli of polarized projective schemes with algebraic group action; see [2] for the toric case (and, more generally, semiabelic varieties), and [4] for the spherical case. These relations would deserve further investigation.

Also, it would be interesting to obtain a modular interpretation of the *main component* of certain invariant Hilbert schemes, that contains the irreducible varieties. For toric Hilbert schemes, such an interpretation is due to Olsson (see [46]) in terms of logarithmic structures.

Finally, another interesting line of investigation concerns the moduli scheme M_Γ of affine spherical varieties with weight monoid Γ. In all known examples, the irreducible components of M_Γ are just affine spaces, and it is tempting to conjecture that this always holds. A positive answer would yield insight into the classification of spherical varieties and the multiplication in their coordinate rings.

Acknowledgements. I thank Stéphanie Cupit-Foutou, Bart Van Steirteghem, Ronan Terpereau, and the referee for their careful reading and valuable comments.

2. Families of affine schemes with reductive group action

2.1. Algebraic group actions

In this subsection, we briefly review some basic notions about algebraic groups and their actions; details and proofs of the stated results may be found e.g. in the notes [20]. We begin by fixing notation and conventions.

Throughout this article, we consider algebraic varieties, algebraic groups, and schemes over a fixed algebraically closed field k of characteristic zero. Unless explicitly mentioned, schemes are assumed to be *noetherian*.

A **variety** is a reduced separated scheme of finite type; thus, varieties need not be irreducible. By a point of such a variety X, we mean a closed point, or equivalently a k-rational point.

An **algebraic group** is a variety G equipped with morphisms $\mu_G : G \times G \to G$ (the multiplication), $\iota_G : G \to G$ (the inverse) and with a point e_G (the neutral element) that satisfy the axioms of a group; this translates into the commutativity of certain diagrams.

Examples of algebraic groups include closed subgroups of the general linear group GL_n consisting of all $n \times n$ invertible matrices, where n is a positive integer; such algebraic groups are called **linear**. We will only consider linear algebraic groups in the sequel.

An (algebraic) **action** of an algebraic group G on a scheme X is a morphism

$$\alpha : G \times X \longrightarrow X, \quad (g, x) \longmapsto g \cdot x$$

such that the composite morphism

$$X \xrightarrow{e_G \times \mathrm{id}_X} G \times X \xrightarrow{\alpha} X$$

is the identity (i.e., e_G acts on X via the identity id_X), and the square

$$\begin{array}{ccc} G \times G \times X & \xrightarrow{\mathrm{id}_G \times \alpha} & G \times X \\ {\scriptstyle \mu_G \times \mathrm{id}_X}\downarrow & & \downarrow{\scriptstyle \alpha} \\ G \times X & \xrightarrow{\alpha} & X \end{array} \tag{2.1}$$

commutes (the associativity property of an action).

A scheme equipped with a G-action is called a G-**scheme**. Given two G-schemes X, Y with action morphisms α, β, a morphism $f : X \to Y$ is called **equivariant** (or a G-**morphism**) if the square

$$\begin{array}{ccc} G \times X & \xrightarrow{\alpha} & X \\ {\scriptstyle \mathrm{id}_G \times f}\downarrow & & \downarrow{\scriptstyle f} \\ G \times Y & \xrightarrow{\beta} & Y \end{array}$$

commutes. If β is the trivial action, i.e., the projection $G \times Y \to Y$, then we say that f is G-**invariant**.

An (algebraic, or rational) G-**module** is a vector space V over k, equipped with a linear action of G which is algebraic in the following sense: every $v \in V$ is contained in a G-stable finite-dimensional subspace $V_v \subset V$ on which G acts algebraically. (We do not assume that V itself is finite-dimensional). Equivalently, G acts on V via a representation $\rho : G \to GL(V)$ which is a union of finite-dimensional algebraic subrepresentations.

A G-stable subspace W of a G-module V is called a G-**submodule**; V is **simple** if it has no non-zero proper submodule. Note that simple modules are finite-dimensional, and correspond to irreducible representations.

A G-**algebra** is an algebra A over k, equipped with an action of G by algebra automorphisms which also makes it a G-module.

Given a G-scheme X, the algebra $\mathcal{O}(X)$ of global sections of the structure sheaf is equipped with a linear action of G via

$$(g \cdot f)(x) = f(g^{-1} \cdot x).$$

In fact, this action is algebraic and hence $\mathcal{O}(X)$ *is a* G-*algebra*. The assignment $X \mapsto \mathcal{O}(X)$ defines a bijective correspondence between *affine* G-schemes and G-algebras.

Each G-algebra of finite type A is generated by a finite-dimensional G-submodule V. Hence A is a quotient of the symmetric algebra $\mathrm{Sym}(V) \cong \mathcal{O}(V^*)$ by a G-stable ideal. It follows that *every affine* G-*scheme of finite type is isomorphic to a closed* G-*subscheme of a finite-dimensional* G-*module.*

Examples 2.2. (i) Let $G := GL_1$ be the **multiplicative group**, denoted by $\mathbb{G}_m$. Then $\mathcal{O}(G) \cong k[t, t^{-1}]$ and the G-modules are exactly the graded vector spaces

$$V = \bigoplus_{n \in \mathbb{Z}} V_n \tag{2.3}$$

where $\mathbb{G}_m$ acts via $t \cdot \sum_n v_n = \sum_n t^n v_n$. In particular, the G-algebras are just the $\mathbb{Z}$-graded algebras.

(ii) More generally, consider a **diagonalizable group** G, i.e., a closed subgroup of a **torus** $(\mathbb{G}_m)^n$. Then G is uniquely determined by its **character group** $\mathcal{X}(G)$, the set of homomorphisms $\chi : G \to \mathbb{G}_m$ equipped with pointwise multiplication. Moreover, the abelian group $\mathcal{X}(G)$ is finitely generated, and the assignment $G \mapsto \mathcal{X}(G)$ defines a bijective correspondence between diagonalizable groups and finitely generated abelian groups. This correspondence is contravariant, and G-modules correspond to $\mathcal{X}(G)$-graded vector spaces.

Any character of $\mathbb{G}_m$ is of the form $t \mapsto t^n$ for some integer n; this identifies $\mathcal{X}(\mathbb{G}_m)$ with $\mathbb{Z}$.

2.2. Reductive groups

In this subsection, we present some basic results on reductive groups, their representations and invariants; again, we refer to the notes [20] for details and proofs.

A linear algebraic group G is called **reductive** if every G-module is semi-simple, i.e., isomorphic to a direct sum of simple G-modules. In view of the characteristic-0 assumption, this is equivalent to the condition that G has no non-trivial closed normal subgroup isomorphic to the additive group of a finite-dimensional vector space; this is the group-theoretical notion of reductivity.

Examples of reductive groups include finite groups, diagonalizable groups, and the classical groups such as GL_n and the special linear group SL_n (consisting of $n \times n$ matrices of determinant 1).

Given a reductive group G, we denote by $\mathrm{Irr}(G)$ the set of isomorphism classes of simple G-modules (or of irreducible G-representations). The class of the trivial G-module k is denoted by 0.

For any G-module V, the map

$$\bigoplus_{M\in\mathrm{Irr}(G)} \mathrm{Hom}^G(M,V)\otimes_k M \longrightarrow V, \quad \sum_M f_M\otimes x_M \longmapsto \sum_M f_M(x_M)$$

is an isomorphism of G-modules, where $\mathrm{Hom}^G(M,V)$ denotes the vector space of morphisms of G-modules from M to V, and G acts on the left-hand side via

$$g\cdot\sum_M f_M\otimes x_M = \sum_M f_M\otimes g\cdot x_M.$$

Thus, the dimension of $\mathrm{Hom}^G(M,V)$ is the multiplicity of M in V (which may be infinite). This yields the **isotypical decomposition**

$$V\cong\bigoplus_{M\in\mathrm{Irr}(G)} V_M\otimes_k M \quad \text{where} \quad V_M:=\mathrm{Hom}^G(M,V). \tag{2.4}$$

In particular, V_0 is the subspace of G-invariants in V, denoted by V^G.

For a G-algebra A, the invariant subspace A^G is a subalgebra. Moreover, each A_M is an A^G-module called the **module of covariants of type** M. Denoting by $X=\mathrm{Spec}(A)$ the associated G-scheme, we also have

$$A_M\cong\mathrm{Mor}^G(X,M^*)$$

(the set of G-morphisms from X to the dual module M^*; note that M^* is simple). Also, we have an isomorphism of A^G-G-modules in an obvious sense

$$A\cong\bigoplus_{M\in\mathrm{Irr}(G)} A_M\otimes_k M. \tag{2.5}$$

Example 2.6. Let G be a diagonalizable group with character group Λ. Then G is reductive and its simple modules are exactly the lines k where G acts via a character $\lambda\in\Lambda$. The decompositions (2.4) and (2.5) give back the Λ-gradings of G-modules and G-algebras.

Returning to an arbitrary reductive group G, the modules of covariants satisfy an important finiteness property:

Lemma 2.7. *Let A be a G-algebra, finitely generated over a subalgebra $R\subset A^G$. Then the algebra A^G is also finitely generated over R. Moreover, the A^G-module A_M is finitely generated for any $M\in\mathrm{Irr}(G)$.*

In the case where $R=k$, this statement is a version of the classical Hilbert-Nagata theorem, see e.g. [20, Theorem 1.25 and Lemma 2.1]. The proof given there adapts readily to our setting, which will be generalized to the setting of families in Subsection 2.3.

In particular, for an affine G-scheme of finite type $X = \mathrm{Spec}(A)$, the subalgebra A^G is finitely generated and hence corresponds to an affine scheme of finite type $X//G$, equipped with a G-invariant morphism

$$\pi : X \longrightarrow X//G.$$

In fact, π is universal among all invariant morphisms with source X; thus, π is called the **categorical quotient**. Also, given a closed G-subscheme $Y \subset X$, the restriction $\pi_{|Y}$ is the categorical quotient of Y, and the image of $\pi_{|Y}$ is closed in $X//G$. As a consequence, π is surjective and each fiber contains a unique closed G-orbit.

We now assume G to be *connected*. Then the simple G-modules may be explicitly described via highest weight theory that we briefly review. Choose a **Borel subgroup** $B \subset G$, i.e., a maximal connected solvable subgroup. Every simple G-module V contains a unique line of eigenvectors of B, say kv; then v is called a **highest weight vector**. The corresponding character $\lambda : B \to \mathbb{G}_m$, such that $b \cdot v = \lambda(b)v$ for all $b \in B$, is called the **highest weight of** V; it uniquely determines the G-module V up to isomorphism. We thus write $V = V(\lambda)$, $v = v_\lambda$, and identify $\mathrm{Irr}(G)$ with a subset of the character group $\Lambda := \mathcal{X}(B)$: the set Λ^+ of **dominant weights**, which turns out to be a finitely generated submonoid of the **weight lattice** Λ. Moreover, $V(0)$ is the trivial G-module k.

In this setting, the modules of covariants admit an alternative description in terms of highest weights. To state it, denote by U the unipotent part of B; this is a closed connected normal subgroup of B, and a maximal unipotent subgroup of G. Moreover, B is the semi-direct product UT where $T \subset B$ is a maximal torus, and Λ is identified with the character group $\mathcal{X}(T)$ via the restriction. Given a G-module V, the subspace of U-invariants V^U is a T-module and hence a direct sum of eigenspaces V^U_λ where $\lambda \in \Lambda$. Moreover, the map

$$\mathrm{Hom}^G(V(\lambda), V) \longrightarrow V^U_\lambda, \quad f \longmapsto f(v_\lambda)$$

is an isomorphism for any $\lambda \in \Lambda^+$, and $V^U_\lambda = 0$ for any $\lambda \notin \Lambda^+$. As a consequence, given a G-algebra A, we have isomorphisms

$$A_{V(\lambda)} \cong A^U_\lambda$$

of modules of covariants over

$$A^G = A^B = A^U_0.$$

The algebra of U-invariants also satisfies a useful finiteness property (see e.g. [20, Theorem 2.7]):

Lemma 2.8. *With the notation and assumptions of Lemma 2.7, the algebra A^U is finitely generated over* R.

This in turn yields a categorical quotient

$$\psi : X \longrightarrow X//U$$

for an affine G-scheme of finite type X, where $X/\!/U := \operatorname{Spec} \mathcal{O}(X)^U$ is an affine T-scheme of finite type. Moreover, for any closed G-subscheme $Y \subset X$, the restriction $\psi_{|Y}$ is again the categorical quotient of Y (but ψ is generally not surjective).

Also, many properties of X may be read off $X/\!/U$. For example, *an affine G-scheme X is of finite type (resp. reduced, irreducible, normal) if and only if so is* $X/\!/U$ (see [25, Chapter 18]).

Examples 2.9. (i) If G is a torus, then $G = B = T$ and $\Lambda^+ = \Lambda$; each $V(\lambda)$ is just the line k where T acts via $t \cdot z = \lambda(t)z$.

(ii) Let $G = SL_2$. We may take for B the subgroup of upper triangular matrices with determinant 1. Then U is the subgroup of upper triangular matrices with diagonal entries 1, isomorphic to the additive group $\mathbb{G}_a$. Moreover, we may take for T the subgroup of diagonal matrices with determinant 1, isomorphic to the multiplicative group $\mathbb{G}_m$. Thus, $\Lambda \cong \mathbb{Z}$.

In fact, the simple G-modules are exactly the spaces $V(n)$ of homogeneous polynomials of degree n in two variables x, y where G acts by linear change of variables; here $n \in \mathbb{N} \cong \Lambda^+$. A highest weight vector in $V(n)$ is the monomial y^n. Moreover, $V(n)$ is isomorphic to its dual module, and hence to the n-th symmetric power $\operatorname{Sym}^n(k^2)$ where $k^2 \cong V(1)^* \cong V(1)$ denotes the standard G-module.

Since G acts transitively on $V(1) \setminus \{0\}$, the categorical quotient $V(1)/\!/G$ is just a point. One easily shows that the quotient by U is the map

$$V(1) \longrightarrow \mathbb{A}^1, \quad ax + by \longmapsto a.$$

Also, the categorical quotient of $V(2)$ by G is given by the discriminant

$$\Delta : V(2) \longrightarrow \mathbb{A}^1, \quad ax^2 + 2bxy + cy^2 \longmapsto ac - b^2$$

and the categorical quotient by U is the map

$$V(2) \longrightarrow \mathbb{A}^2, \quad ax^2 + 2bxy + cy^2 \longmapsto (a, ac - b^2).$$

For large n, no explicit description of the categorical quotients $V(n)/\!/G$ and $V(n)/\!/U$ is known, although the corresponding algebras $\operatorname{Sym}(V(n))^G$ and $\operatorname{Sym}(V(n))^U$ (the invariants and covariants of binary forms of degree n) have been extensively studied since the mid-19th century.

2.3. Families

In this subsection, we fix a reductive group G and an affine G-scheme of finite type, $X = \operatorname{Spec}(A)$. We now introduce our main objects of interest.

Definition 2.10. **A family of closed G-subschemes of X over a scheme** S is a closed subscheme $\mathcal{Z} \subset X \times S$, stable by the action of G on $X \times S$ via $g \cdot (x, s) = (g \cdot x, s)$.

The family $\mathcal{Z}$ is **flat**, if so is the morphism $p : \mathcal{Z} \to S$ induced by the projection $p_2 : X \times S \to S$.

Given a family $\mathcal{Z}$ as above, the morphism p is G-invariant; thus, for any k-rational point s of S, the scheme-theoretic fiber $\mathcal{Z}_s$ of p at s is a closed G-subscheme of X. More generally, an arbitrary point $s \in S$ yields a closed subscheme $\mathcal{Z}_{k(\bar{s})} \subset X_{k(\bar{s})}$ where $k(s)$ denotes the residue field of s, and $k(\bar{s})$ is an algebraic closure of that field. The scheme $\mathcal{Z}_{k(\bar{s})}$ is the geometric fiber of $\mathcal{Z}$ at s, also denoted by $\mathcal{Z}_{\bar{s}}$; this is a closed subscheme of $X_{\bar{s}}$ equipped with an action of G (viewed as a constant group scheme) or equivalently of $G_{\bar{s}}$ (the group of $k(\bar{s})$-rational points of G).

Since X is affine, the data of the family $\mathcal{Z}$ is equivalent to that of the sheaf $p_*\mathcal{O}_{\mathcal{Z}}$ as a quotient of $(p_2)_*\mathcal{O}_{X\times S} = A \otimes_k \mathcal{O}_S$, where both are viewed as sheaves of $\mathcal{O}_S$-G-algebras. Moreover, as G acts trivially on S, we have a canonical decomposition

$$p_*\mathcal{O}_{\mathcal{Z}} \cong \bigoplus_{M\in \mathrm{Irr}(G)} \mathcal{F}_M \otimes_k M, \tag{2.11}$$

where each **sheaf of covariants**

$$\mathcal{F}_M := \mathrm{Hom}^G(M, p_*\mathcal{O}_{\mathcal{Z}}) = (p_*\mathcal{O}_{\mathcal{Z}} \otimes_k M^*)^G \tag{2.12}$$

is a sheaf of $\mathcal{O}_S$-modules, and (2.11) is an isomorphism of sheaves of $\mathcal{O}_S$-G-modules.

Also, $p_*\mathcal{O}_{\mathcal{Z}}$ is a sheaf of finitely generated $\mathcal{O}_S$-algebras, since X is of finite type. In view of Lemma 2.7, it follows that

$$\mathcal{F}_0 = (p_*\mathcal{O}_{\mathcal{Z}})^G$$

is a sheaf of finitely generated $\mathcal{O}_S$-algebras as well; moreover, $\mathcal{F}_M$ is a coherent sheaf of $\mathcal{F}_0$-modules, for any $M \in \mathrm{Irr}(G)$. By (2.11), the family $\mathcal{Z}$ is flat if and only if each sheaf of covariants $\mathcal{F}_M$ is flat.

Definition 2.13. With the preceding notation, the family $\mathcal{Z}$ is **multiplicity-finite** if the sheaf of $\mathcal{O}_S$-modules $\mathcal{F}_M$ is coherent for any $M \in \mathrm{Irr}(G)$; equivalently, $\mathcal{F}_0$ is coherent.

We say that $\mathcal{Z}$ is **multiplicity-free** if each $\mathcal{F}_M$ is zero or invertible.

Since flatness is equivalent to local freeness for a finitely generated module over a Noetherian ring, we see that *a multiplicity-finite family is flat iff each sheaf of covariants is locally free of finite rank.* When the base S is connected, the ranks of these sheaves are well-defined and yield a numerical invariant of the family: the **Hilbert function**

$$h = h_{\mathcal{Z}} : \mathrm{Irr}(G) \longrightarrow \mathbb{N}, \quad M \longmapsto \mathrm{rk}_{\mathcal{O}_S}(\mathcal{F}_M).$$

This motivates the following:

Definition 2.14. Given a function $h : \mathrm{Irr}(G) \to \mathbb{N}$, a **flat family of closed subschemes of X with Hilbert function** h is a closed subscheme $\mathcal{Z} \subset X \times S$ such that each sheaf of covariants $\mathcal{F}_M$ is locally free of rank $h(M)$.

Remarks 2.15. (i) In the case where $S = \mathrm{Spec}(k)$, a family is just a closed G-subscheme $Z \subset X$. Then Z is multiplicity-finite if and only if the quotient $Z/\!/G$ is finite; equivalently, Z contains only finitely many closed G-orbits. For example, *any G-orbit closure is multiplicity-finite.*

Also, Z is multiplicity-free if and only if the G-module $\mathcal{O}(Z)$ has multiplicities 0 or 1. If Z is an irreducible variety, this is equivalent to the condition that Z contains an open orbit of a Borel subgroup $B \subset G$ (see e.g. [20, Lemma 2.12]).

In particular, when G is a torus, say T, an affine irreducible T-variety Z is multiplicity-free if and only if it contains an open T-orbit. Then each non-zero eigenspace $\mathcal{O}(Z)_\lambda$ is a line, and the set of those λ such that $\mathcal{O}(Z)_\lambda \neq 0$ consists of those linear combinations $n_1\lambda_1 + \cdots + n_r\lambda_r$, where $n_1, \dots, n_r$ are non-negative integers, and $\lambda_1, \dots, \lambda_r$ are the weights of homogeneous generators of the algebra $\mathcal{O}(Z)$. Thus, this set is a finitely generated submonoid of Λ that we denote by $\Lambda^+(Z)$ and call the **weight monoid of** Z.

Each affine irreducible multiplicity-free T-variety Z is uniquely determined by its weight monoid Γ: in fact, *the Λ-graded algebra $\mathcal{O}(Z)$ is isomorphic to* $k[\Gamma]$, the algebra of the monoid Γ over k. Moreover, Z is normal if and only if Γ is **saturated**, i.e., equals the intersection of the group that it generates in Λ, with the convex cone that it generates in $\Lambda \otimes_{\mathbb{Z}} \mathbb{R}$. Under that assumption, Z is called an (affine) **toric variety**.

Returning to an affine irreducible G-variety Z, note that Z is multiplicity-free if and only if so is $Z//U$. In that case, we have an isomorphism of G-modules

$$\mathcal{O}(Z) \cong \bigoplus_{\lambda \in \Lambda^+(Z)} V(\lambda) \tag{2.16}$$

where $\Lambda^+(Z) := \Lambda^+(Z//U)$ is again called the **weight monoid of** Z. In other words, the Hilbert function of Z is given by

$$h(\lambda) = \begin{cases} 1 & \text{if } \lambda \in \Lambda^+(Z), \\ 0 & \text{otherwise.} \end{cases} \tag{2.17}$$

Also, Z *is normal if and only if* $\Lambda^+(Z)$ *is saturated;* then Z is called an (affine) **spherical G-variety**. In contrast to the toric case, spherical varieties are not uniquely determined by their weight monoid, see e.g. Example 2.19(ii).

(ii) A flat family $\mathcal{Z}$ over a *connected* scheme S is multiplicity-finite (resp. is multiplicity-free, or has a prescribed Hilbert function h) if and only if so does some geometric fiber $\mathcal{Z}_{\bar{s}}$. In particular, if some geometric fiber is a spherical variety, then the family is multiplicity-free.

(iii) Any family of closed G-subschemes $\mathcal{Z} \subset X \times S$ yields a family of closed T-subschemes $\mathcal{Z}//U \subset X//U \times S$; moreover, the sheaves of covariants of $\mathcal{Z}$ and $\mathcal{Z}//U$ are isomorphic. Thus, $\mathcal{Z}$ is flat (multiplicity-finite, multiplicity-free) if and only if so is $\mathcal{Z}//U$. Also, $\mathcal{Z}$ has Hilbert function h if and only if $\mathcal{Z}//U$ has Hilbert function $\tilde{h}$

such that

$$\bar{h}(\lambda) = \begin{cases} h(\lambda) & \text{if } \lambda \in \Lambda^+ \\ 0 & \text{otherwise.} \end{cases} \tag{2.18}$$

Examples 2.19. (i) The surface $\mathcal{Z}$ of equation $xy - z = 0$ in $\mathbb{A}^3$ is stable under the action of $\mathbb{G}_m$ via $t \cdot (x, y, z) = (tx, t^{-1}y, z)$. The morphism $z : \mathcal{Z} \to \mathbb{A}^1$ is a flat family of closed $\mathbb{G}_m$-subschemes of $\mathbb{A}^2$ (the affine plane with coordinates x, y). The fibers over non-zero points of $\mathbb{A}^1$ are all isomorphic to $\mathbb{G}_m$; they are exactly the orbits of points of $\mathbb{A}^2$ minus the coordinate axes. In particular, the family has Hilbert function the constant function 1. The fiber at 0 is the (reduced) union of the coordinate axes.

For the $\mathbb{G}_m$-action on $\mathbb{A}^3$ via $t \cdot (x, y, z) = (t^2x, t^{-1}y, z)$, the surface $\mathcal{W}$ of equation $xy^2 = z$ yields a family with the same fibers at non-zero points, but the fiber at 0 is non-reduced.

More generally, consider a torus T acting linearly on the affine space $V = \mathbb{A}^N$ with pairwise distinct weights. Denote by $\lambda_1, \ldots, \lambda_N$ the opposites of these weights, i.e., the weights of the coordinate functions. Also, let $v \in V$ be a **general point** in the sense that all its coordinates are non-zero. Then the orbit closure

$$Z := \overline{T \cdot v} \subset V$$

is an irreducible multiplicity-free variety, and different choices of v yield isomorphic T-varieties; moreover, all irreducible multiplicity-free T-varieties may be obtained in this way. The weight monoid of Z is generated by $\lambda_1, \ldots, \lambda_N$.

We construct flat families over $\mathbb{A}^1$ with general fiber Z as follows. Let the torus $T \times \mathbb{G}_m$ act linearly on $V \times \mathbb{A}^1 = \mathbb{A}^{N+1}$ such that the coordinate functions have weights

$$(\lambda_1, a_1), \ldots, (\lambda_N, a_N), (0,1)$$

where $a_1, \ldots, a_N$ are integers, viewed as characters of $\mathbb{G}_m$. Then the orbit closure

$$\mathcal{Z} := \overline{(T \times \mathbb{G}_m) \cdot (v,1)} \subset V \times \mathbb{A}^1$$

may be viewed as a T-variety. The projection $p : \mathcal{Z} \to \mathbb{A}^1$ is T-invariant, and flat since $\mathcal{Z}$ is an irreducible variety. Moreover, p is trivial over $\mathbb{A}^1 \setminus \{0\}$; specifically, the map

$$p^{-1}(\mathbb{A}^1 \setminus \{0\}) \longrightarrow Z \times (\mathbb{A}^1 \setminus \{0\}), \quad (v, s) \longmapsto (s^{-1}v, s)$$

is a T-equivariant isomorphism of families over $\mathbb{A}^1 \setminus \{0\}$. In particular, the fibers of p at non-zero points are all isomorphic to Z, and p is multiplicity-free with Hilbert function being 1 on the monoid generated by $\lambda_1, \ldots, \lambda_N$, and 0 elsewhere. On the other hand, the special fiber $\mathcal{Z}_0$ is non-empty if and only if p (viewed as a regular function on $\mathcal{Z}$) is not invertible; this translates into the condition that the convex cone generated by $(\lambda_1, a_1), \ldots, (\lambda_N, a_N)$ does not contain $(0, -1)$.

One can show that the preceding construction yields all one-parameter families with generic fiber a multiplicity-free T-variety. Also, one can show that the special fiber is reducible unless the whole family is trivial; this contrasts with our next example, where all fibers are irreducible varieties and the special fiber is singular while all others are smooth.

(ii) Let $G = SL_2$ and $\Delta : V(2) \to \mathbb{A}^1$ the discriminant as in Example 2.9(ii). Then the graph

$$\mathcal{Z} = \{(f, s) \in V(2) \times \mathbb{A}^1 \mid \Delta(f) = s\}$$

is a flat family of G-stable closed subschemes of $V(2)$. The fibers at non-zero closed points are exactly the G-orbits of non-degenerate quadratic forms, while the fiber at 0 consists of two orbits: the squares of non-zero linear forms, and the origin. Since $\mathcal{Z} \cong V(2)$ as G-varieties, the Hilbert function of $\mathcal{Z}$ is given by

$$h_2(n) = \begin{cases} 1 & \text{if } n \text{ is even,} \\ 0 & \text{if } n \text{ is odd.} \end{cases}$$

Thus, $\mathcal{Z}$ is multiplicity-free. Moreover, the family $\mathcal{Z}//U$ is trivial with fiber $\mathbb{A}^1$, as follows from the description of $V(2)//U$ in Example 2.9(ii). In particular, each fiber $\mathcal{Z}_s$ is a spherical variety.

Next, consider the quotient $\mathcal{W}$ of $V(2)$ by the involution $\sigma : f \mapsto -f$. Then $\mathcal{W}$ is the affine G-variety associated with the subalgebra of $\mathcal{O}(V(2)) \cong \mathrm{Sym}(V(2))$ consisting of even polynomial functions, i.e., the subalgebra generated by $\mathrm{Sym}^2(V(2)) \cong V(4) \oplus V(0)$. In other words,

$$\mathcal{W} \subset V(4) \times \mathbb{A}^1$$

and the resulting projection $q : \mathcal{W} \to \mathbb{A}^1$ may be identified with the σ-invariant map $V(2) \to \mathbb{A}^1$ given by the discriminant. It follows that q is a flat family of G-stable closed subschemes of $V(4)$, with Hilbert function given by

$$h_4(n) = \begin{cases} 1 & \text{if } n \text{ is a multiple of 4,} \\ 0 & \text{otherwise.} \end{cases}$$

In particular, $\mathcal{W}$ is multiplicity-free. Moreover, its fibers at non-zero closed points are exactly the orbits $G \cdot f^2 \subset V(4)$, where f is a non-degenerate quadratic form, while the fiber at 0 consists of two orbits: the fourth powers of non-zero linear forms, and the origin. The family $\mathcal{W}//U$ is again trivial with fiber $\mathbb{A}^1$, so that the fibers of $\mathcal{W}$ are spherical varieties.

We will show that both families just constructed are universal (in the sense of the next subsection), and that no family with similar properties exists in $V(n)$ for $n \neq 2,4$. For this, we will apply the various techniques that we successively introduce; see Examples 2.23(ii), 3.12(ii), 3.18(ii) and 3.24(ii).

2.4. The universal family

In the setting of the previous subsection, there is a natural construction of **pull-back** for families of G-stable subschemes of X: given such a family $\mathcal{Z} \subset X \times S$

and a morphism of schemes $f : S' \to S$, we can form the cartesian square

$$\begin{array}{ccc} \mathcal{Z}' & \longrightarrow & X \times S' \\ \downarrow & & \downarrow{\scriptstyle \mathrm{id}_X \times f} \\ \mathcal{Z} & \longrightarrow & X \times S \end{array}$$

where the horizontal arrows are inclusions. This yields a family of closed G-subschemes of X over S': the pull-back of $\mathcal{Z}$ under f, which may also be defined via the cartesian square

$$\begin{array}{ccc} \mathcal{Z}' & \longrightarrow & S' \\ \downarrow & & \downarrow{\scriptstyle f} \\ \mathcal{Z} & \xrightarrow{p} & S. \end{array}$$

Note that $\mathcal{Z}'$ is flat over S' whenever $\mathcal{Z}$ is flat over S; moreover, multiplicity-finiteness and -freeness are preserved under pull-back, as well as the Hilbert function.

We may now state our main result, which asserts the existence of a universal family:

Theorem 2.20. *Given a reductive group* G*, an affine* G*-scheme of finite type* X *and a function* $h : \mathrm{Irr}(G) \to \mathbb{N}$*, there exists a family of closed* G*-subschemes with Hilbert function* h*,*

$$\mathrm{Univ}_h^G(X) \subset X \times \mathrm{Hilb}_h^G(X), \tag{2.21}$$

such that any family $\mathcal{Z} \subset X \times S$ *of closed* G*-subschemes with Hilbert function* h *is obtained from (2.21) by pull-back under a unique morphism* $f : S \to \mathrm{Hilb}_h^G(X)$*. Moreover, the scheme* $\mathrm{Hilb}_h^G(X)$ *is quasi-projective (in particular, of finite type).*

The family (2.21) is of course uniquely determined up to a unique isomorphism by its universal property. The scheme $\mathrm{Hilb}_h^G(X)$ is called the **invariant Hilbert scheme** associated with the affine G-scheme X and the function h.

Theorem 2.20 may be reformulated as asserting that *the* **Hilbert functor** $\mathcal{H}ilb_h^G(X)$ *that associates with any scheme* S*, the set of flat families* $\mathcal{Z} \subset X \times S$ *with Hilbert function* h*, is represented by the quasi-projective scheme* $\mathrm{Hilb}_h^G(X)$.

By taking $S = \mathrm{Spec}(R)$ where R is an arbitrary algebra, this yields an algebraic description of the R-points of the invariant Hilbert scheme: these are those G-stable ideals $I \subset A \otimes_k R$ such that each R-module of covariants

$$\big((R \otimes_k A)/I\big)_M = (R \otimes_k A)_M / I_M$$

is locally free of rank $h(M)$.

In particular, the k-rational points of $\mathrm{Hilb}_h^G(X)$ (which are the same as its closed points, since this scheme is of finite type) are those G-stable ideals I of $A = \mathcal{O}(X)$ such that each simple G-module M has multiplicity $h(M)$ in the quotient A/I. These points may also be identified with the closed G-stable subschemes $Z \subset X$ with Hilbert function h.

Remarks 2.22. (i) The case of the **trivial group** G is already quite substantial. There, X is just an affine scheme of finite type, and a family with finite multiplicities is exactly a closed subscheme $\mathcal{Z} \subset X \times S$ such that the projection $p : \mathcal{Z} \to S$ is finite. Moreover, a Hilbert function is just a non-negative integer n. In that case, $\mathrm{Hilb}_n(X)$ is the **punctual Hilbert scheme** (also known as the **Hilbert scheme of points**) that parametrizes the closed subschemes of length n of X. In fact, $\mathrm{Hilb}_n(X)$ exists more generally for any quasi-projective scheme X over a field of arbitrary characteristic.

(ii) If G is the multiplicative group $\mathbb{G}_m$, we know that X corresponds to a $\mathbb{Z}$-graded algebra of finite type A. For a Hilbert function $h : \mathbb{Z} \to \mathbb{N}$, the scheme $\mathrm{Hilb}_h^{\mathbb{G}_m}(X)$ parametrizes those graded ideals $I \subset A$ such that the vector space $(A/I)_n$ has dimension $h(n)$ for all $n \in \mathbb{Z}$.

Of special interest is the case where X is the affine space $\mathbb{A}^N$ on which $\mathbb{G}_m$ acts via scalar multiplication, i.e., A is a polynomial ring in N indeterminates of weight 1. Then a necessary condition for the existence of such ideals I is that $h(n) = P(n)$ for all $n \gg 0$, where $P(t)$ is a (uniquely determined) polynomial: the **Hilbert polynomial** of the graded algebra A/I. In that case, we also have the **Hilbert scheme** $\mathrm{Hilb}_P(\mathbb{P}^{N-1})$ that parametrizes closed subschemes of the projective $(N-1)$-space with Hilbert polynomial P, or equivalently, graded ideals $I \subset A$ such that $\dim(A/I)_n = P(n)$ for all $n \gg 0$. This yields a morphism

$$\mathrm{Hilb}_h^{\mathbb{G}_m}(\mathbb{A}^N) \longrightarrow \mathrm{Hilb}_P(\mathbb{P}^{N-1})$$

which is in fact an isomorphism for an appropriate choice of the Hilbert function h, associated to a given Hilbert polynomial P (see [26, Lemma 4.1]).

(iii) More generally, if G is diagonalizable with character group Λ, and X is a G-module of finite dimension N, then A is a polynomial ring on homogeneous generators with weights $\lambda_1, \ldots, \lambda_N \in \Lambda$. Moreover, $\mathrm{Hilb}_h^G(X)$ parametrizes those Λ-graded ideals $I \subset A$ such that each vector space $(A/I)_\lambda$ has a prescribed dimension $h(\lambda)$. In that case, $\mathrm{Hilb}_h^G(X)$ is the **multigraded Hilbert scheme** of [26]. As shown there, that scheme exists over any base ring, and no Noetherian assumption is needed in the definition of the corresponding functor.

Examples 2.23. (i) Consider a torus T acting linearly on the affine space $\mathbb{A}^N$ via pairwise distinct weights and take for h the Hilbert function of a general T-orbit closure. Denoting by $\lambda_1, \ldots, \lambda_N$ the weights of the coordinate functions on $\mathbb{A}^N$ and by Γ the submonoid of Λ generated by these weights, we have

$$h(\lambda) = \begin{cases} 1 & \text{if } \lambda \in \Gamma, \\ 0 & \text{otherwise.} \end{cases} \tag{2.24}$$

The associated invariant Hilbert scheme $\mathrm{Hilb}_h^T(\mathbb{A}^N)$ is called the **toric Hilbert scheme**; it was constructed by Peeva and Stillman (see [49]) prior to the more general construction of multigraded Hilbert schemes. Since $\mathrm{Hilb}_h^T(\mathbb{A}^N)$ only depends on T and $\underline{\lambda} = (\lambda_1, \ldots, \lambda_N)$, we will denote it by $\mathrm{Hilb}^T(\underline{\lambda})$.

(ii) Let $G = SL_2$ and take for X the simple G-module $V(n)$. Then X contains a distinguished closed G-stable subvariety Z, consisting of the n-th powers of linear forms. In other words, Z is the affine cone over the image of the n-uple embedding of $\mathbb{P}^1$ in $\mathbb{P}^n = \mathbb{P}(V(n))$. Since that image is the unique closed G-orbit, Z is the smallest non-zero closed G-stable subcone of $V(n)$. Also, Z is a normal surface with singular locus the origin if $n \geqslant 2$, while $Z = V(1)$ if $n = 1$. Moreover, the Hilbert function of Z is given by

$$h_n(m) = \begin{cases} 1 & \text{if } m \text{ is a multiple of } n, \\ 0 & \text{otherwise.} \end{cases} \tag{2.25}$$

As we will show, the corresponding invariant Hilbert scheme is the affine line if $n = 2$ or 4; in both cases, the universal family is that constructed in Example 2.19. For all other values of n, the invariant Hilbert scheme consists of the (reduced) point Z.

(iii) More generally, let G be an arbitrary connected reductive group, $V = V(\lambda)$ a non-trivial simple G-module, and $v = v_\lambda$ a highest weight vector. Then the corresponding point $[v]$ of the projective space $\mathbb{P}(V)$ is the unique B-fixed point. Hence $G \cdot [v]$ is the unique closed G-orbit in $\mathbb{P}(V)$, by Borel's fixed point theorem. Thus,

$$Z := \overline{G \cdot v} = G \cdot v \cup \{0\}$$

is the smallest non-zero G-stable cone in V: the **cone of highest weight vectors**. Moreover, we have an isomorphism of graded G-modules

$$\mathcal{O}(Z) \cong \bigoplus_{n=0}^{\infty} V(n\lambda)^* \tag{2.26}$$

where $V(n\lambda)^*$ has degree n. Thus, denoting by λ^* the highest weight of the simple G-module $V(\lambda)^*$, we see that the T-algebra $\mathcal{O}(Z)^U$ is a polynomial ring in one variable of weight λ^*. In particular, $Z/\!/U$ is normal, and hence Z is a spherical variety. Its Hilbert function is given by

$$h_\lambda(\mu) = \begin{cases} 1 & \text{if } \mu \text{ is a multiple of } \lambda^*, \\ 0 & \text{otherwise.} \end{cases}$$

Again, it turns out that the corresponding invariant Hilbert scheme is the affine line for certain dominant weights λ, and is trivial (i.e., consists of the reduced point Z) for all other weights. This result is due to Jansou (see [29, Théorème 1.1]), who also constructed the universal family in the non-trivial cases, as follows.

Assume that the G-module $V(\lambda) \oplus V(0) \cong V(\lambda) \times \mathbb{A}^1$ carries a linear action of a connected reductive group $\tilde{G} \supset G$. Assume moreover that this $\tilde{G}$-module is simple, say $V(\tilde{\lambda})$, and that the associated cone of highest vectors $\tilde{Z} \subset V(\tilde{\lambda})$ satisfies

$Z = \tilde{Z} \cap V(\lambda)$ as schemes. Then the projection $p : \tilde{Z} \to \mathbb{A}^1$ is a flat family of G-subschemes of $V(\lambda)$ with fiber Z at 0, and hence has Hilbert function h_λ. By [29, Section 2.2], this is in fact the universal family; moreover, all non-trivial cases are obtained from this construction.

One easily shows that the projectivization $\tilde{Y} := \tilde{G} \cdot [v]$, the closed $\tilde{G}$-orbit in $\mathbb{P}(V(\tilde{\lambda}))$, consists of two G-orbits: the closed orbit $Y := G \cdot [v]$, a hyperplane section of $\tilde{Y}$, and its (open affine) complement. Moreover, the projective data $Y \subset \tilde{Y}$ uniquely determine the affine data $Z \subset \tilde{Z} \subset V(\tilde{\lambda})$, since the space of global sections of the ample divisor Y on $\tilde{Y}$ is $V(\tilde{\lambda})^*$.

In fact, the non-trivial cases correspond bijectively to the smooth projective varieties where a connected algebraic group acts with two orbits, the closed one being an ample divisor (see [loc. cit., Section 2], based on Akhiezer's classification of certain varieties with two orbits in [1]).

Returning to the case where $G = SL_2$, the universal family for $n = 2$ is obtained by taking $\tilde{G} = SL_2 \times SL_2$ where SL_2 is embedded as the diagonal. Moreover,

$$V(\tilde{\lambda}) = V(1,1) = V(1) \otimes_k V(1) \cong V(2) \oplus V(0)$$

where the latter isomorphism is as SL_2-modules; also, $Y = \mathbb{P}^1$ is the diagonal in $\tilde{Y} = \mathbb{P}^1 \times \mathbb{P}^1$.

For $n = 4$, one replaces SL_2 with its quotient $PSL_2 = PGL_2$ (that we will keep denoting by G for simplicity), and takes $\tilde{G} = SL_3$ where G is embedded via its representation in the 3-dimensional space $V(2)$. Moreover, $V(\tilde{\lambda})$ is the symmetric square of the standard representation k^3 of $\tilde{G}$, so that

$$V(\tilde{\lambda}) \cong \mathrm{Sym}^2(V(2)) \cong V(4) \oplus V(0)$$

as G-modules. Here $Y = \mathbb{P}^1$ is embedded in $\tilde{Y} \cong \mathbb{P}^2$ as a conic.

(iv) As another generalization of (ii) above, take again $G = SL_2$ and $X = V(n)$. Assume that $n = 2m$ is even and consider the function $h = h_4$ if m is even, and $h = h_2$ if m is odd. The invariant Hilbert scheme $\mathrm{Hilb}_h^G(X)$ has been studied in detail by Budmiger in his thesis [21]. A closed point of that scheme is the (closed) orbit $G \cdot x^m y^m$, which in fact lies in an irreducible component whose underlying reduced scheme is isomorphic to $\mathbb{A}^1$. But $\mathrm{Hilb}_h^G(X)$ turns out to be non-reduced for $m = 6$, and reducible for $m = 8$; see [21, Section III.1].

3. Basic properties

3.1. Existence

In this subsection, we show how to deduce the existence of the invariant Hilbert scheme (Theorem 2.20) from that of the multigraded Hilbert scheme, proved in [26]. We begin with three intermediate results which are of some independent interest. The first one will allow us to enlarge the acting group G. To state it, we need some preliminaries on *induced schemes*.

Consider an inclusion of reductive groups $G \subset \tilde{G}$; then the homogeneous space $\tilde{G}/G$ is an affine $\tilde{G}$-variety equipped with a base point, the image of $e_{\tilde{G}}$. Let X be an affine G-scheme of finite type. Then there exists an affine $\tilde{G}$-scheme of finite type $\tilde{X}$, equipped with a $\tilde{G}$-morphism

$$f : \tilde{X} \longrightarrow \tilde{G}/G$$

such that the fiber of f at the base point is isomorphic to X as a G-scheme. Moreover, $\tilde{X}$ is the quotient of $\tilde{G} \times X$ by the action of G via

$$g \cdot (\tilde{g}, x) := (\tilde{g}g^{-1}, g \cdot x)$$

and this identifies f with the morphism obtained from the projection $\tilde{G} \times X \to \tilde{G}$. (These assertions follow e.g. from descent theory; see [45, Proposition 7.1] for a more general result).

The scheme $\tilde{X}$ satisfies the following property, analogous to *Frobenius reciprocity* relating induction and restriction in representation theory: For any $\tilde{G}$-scheme $\tilde{Y}$, we have an isomorphism

$$\mathrm{Mor}^{\tilde{G}}(\tilde{X}, \tilde{Y}) \cong \mathrm{Mor}^{G}(X, \tilde{Y})$$

that assigns to any $f : \tilde{X} \to \tilde{Y}$ its restriction to X. The inverse isomorphism assigns to any $\varphi : X \to \tilde{Y}$ the morphism $\tilde{G} \times X \to \tilde{Y}$, $(\tilde{g}, x) \mapsto \varphi(\tilde{g} \cdot x)$ which is G-invariant and hence descends to a morphism $\tilde{X} \to \tilde{Y}$. Thus, $\tilde{X}$ is called an **induced scheme**; we denote it by $\tilde{G} \times^G X$.

Taking for $\tilde{Y}$ a $\tilde{G}$-module $\tilde{V}$, we obtain an isomorphism

$$\mathrm{Hom}^{\tilde{G}}(\tilde{V}, \mathcal{O}(\tilde{X})) \cong \mathrm{Hom}^{G}(\tilde{V}, \mathcal{O}(X)). \tag{3.1}$$

Also, we have isomorphisms of $\tilde{G}$-modules

$$\mathcal{O}(\tilde{X}) \cong \mathcal{O}(\tilde{G} \times X)^G \cong \left(\mathcal{O}(\tilde{G}) \otimes_k \mathcal{O}(X)\right)^G \cong \mathrm{Ind}_G^{\tilde{G}}(\mathcal{O}(X))$$

where $\mathrm{Ind}_G^{\tilde{G}}$ denotes the induction functor from G-modules to $\tilde{G}$-modules.

We may now state our first reduction result:

Lemma 3.2. *Let $G \subset \tilde{G}$ be an inclusion of reductive groups, X an affine G-scheme of finite type, and $\tilde{X} := \tilde{G} \times^G X$. Let $\mathcal{Z} \subset X \times S$ be a flat family of closed G-stable subschemes with Hilbert function h. Then*

$$\tilde{\mathcal{Z}} := \tilde{G} \times^G \mathcal{Z} \subset \tilde{X} \times S$$

is a flat family of closed $\tilde{G}$-stable subschemes, having a Hilbert function $\tilde{h}$ that depends only on h. Moreover, if $\mathrm{Hilb}_{\tilde{h}}^{\tilde{G}}(\tilde{X})$ is represented by a scheme $\mathcal{H}$, then $\mathrm{Hilb}_h^G(X)$ is represented by a union of connected components of $\mathcal{H}$.

Proof. Consider a simple $\tilde{G}$-module $\tilde{M}$ and the associated sheaf of covariants

$$\mathcal{F}_{\tilde{M}} = \mathrm{Hom}^{\tilde{G}}(\tilde{M}, \tilde{p}_* \mathcal{O}_{\tilde{G} \times^G \mathcal{Z}})$$

where $\tilde{p} : \tilde{G} \times^G \mathcal{Z} \to S$ denotes the projection. By using (3.1), this yields

$$\mathcal{F}_{\tilde{M}} \cong \mathrm{Hom}^G(\tilde{M}, p_*\mathcal{O}_{\mathcal{Z}}).$$

Thus, the sheaf of $\mathcal{O}_S$-modules $\mathcal{F}_{\tilde{M}}$ is locally free of rank

$$\sum_{M \in \mathrm{Irr}(G)} \dim \mathrm{Hom}^G(M, \tilde{M})\, h(M) =: \tilde{h}(\tilde{M}).$$

It follows that $\tilde{G} \times^G \mathcal{Z}$ is flat with Hilbert function $\tilde{h}$ just defined. This shows the first assertion, and defines a morphism of functors $\mathrm{Hilb}^G_h(X) \to \mathrm{Hilb}^{\tilde{G}}_{\tilde{h}}(\tilde{X})$, $\mathcal{Z} \mapsto \tilde{\mathcal{Z}} := \tilde{G} \times^G \mathcal{Z}$.

Next, consider a family of closed $\tilde{G}$-stable subschemes $\tilde{\mathcal{Z}} \subset \tilde{X} \times S$. Then the composite morphism

$$\tilde{\mathcal{Z}} \xrightarrow{\ \tilde{q}\ } \tilde{X} \xrightarrow{\ f\ } \tilde{G}/G$$

is $\tilde{G}$-equivariant. It follows that $\tilde{\mathcal{Z}} = \tilde{G} \times^G \mathcal{Z}$ for some G-stable subscheme $\mathcal{Z} \subset X \times S$. If $\tilde{\mathcal{Z}}$ is flat over S, then by the preceding step, the sheaf of $\mathcal{O}_S$-modules $\mathrm{Hom}^G(\tilde{M}, p_*\mathcal{O}_{\mathcal{Z}})$ is locally free for any simple $\tilde{G}$-module $\tilde{M}$. But every simple G-module M is a submodule of some simple $\tilde{G}$-module $\tilde{M}$ (indeed, M is a quotient of $\mathrm{Ind}_G^{\tilde{G}}(M)$, and hence a quotient of a simple $\tilde{G}$-submodule). It follows that the sheaf of $\mathcal{O}_S$-modules $\mathrm{Hom}^G(M, p_*\mathcal{O}_{\mathcal{Z}})$ is a direct factor of $\mathrm{Hom}^G(\tilde{M}, p_*\mathcal{O}_{\mathcal{Z}})$ and hence is locally free of finite rank. Thus, $\mathcal{Z}$ is flat and multiplicity-finite over S; hence $\mathcal{Z}$ has a Hilbert function h' such that $\widetilde{h'} = \tilde{h}$, if S is connected. When $h' = h$, the assignments $\mathcal{Z} \mapsto \tilde{\mathcal{Z}}$ and $\tilde{\mathcal{Z}} \mapsto \mathcal{Z}$ are mutually inverse. Taking for S a connected component of $\mathcal{H}$ and for $\tilde{\mathcal{Z}}$ the pull-back of the universal family, we obtain the final assertion. □

By the preceding lemma, we may replace G with GL_n; in particular, we may assume that G is *connected*. Our second result, a variant of [3, Theorem 1.7], will allow us to replace G with a torus. As in Subsection 2.2, we choose a Borel subgroup $B \subset G$ with unipotent part U, and a maximal torus $T \subset B$. We consider an affine G-scheme $X = \mathrm{Spec}(A)$ and a function $h : \Lambda^+ \to \mathbb{N}$; we extend h to a function $\bar{h} : \Lambda \to \mathbb{N}$ with values 0 outside Λ^+, as in (2.18).

Lemma 3.3. *With the preceding notation, assume that* $\mathrm{Hilb}^T_{\bar{h}}(X/\!/U)$ *is represented by a scheme* $\mathcal{H}$. *Then* $\mathrm{Hilb}^G_h(X)$ *is represented by a closed subscheme of* $\mathcal{H}$.

Proof. We closely follow the argument of [3, Theorem 1.7]. Given a scheme S and a flat family $\mathcal{Z} \subset X \times S$ of closed G-stable subschemes with Hilbert function h, we obtain a family $\mathcal{Z}/\!/U \subset X/\!/U \times S$ of closed T-stable subschemes which is again flat and has Hilbert function $\bar{h}$, by Remark 2.15(iii). Observe that $\mathcal{Z}/\!/U$ uniquely determines $\mathcal{Z}$: indeed, $\mathcal{Z}$ corresponds to a G-stable sheaf of ideals

$$\mathcal{I} \subset A \otimes_k \mathcal{O}_S$$

such that each quotient $(A \otimes_k \mathcal{O}_S)^U_\lambda / \mathcal{I}^U_\lambda$ is locally free of rank $\bar{h}(\lambda)$. Moreover, $\mathcal{Z}//U$ corresponds to the T-stable sheaf of ideals

$$\mathcal{I}^U \subset A^U \otimes_k \mathcal{O}_S$$

which generates $\mathcal{I}$ as a sheaf of $\mathcal{O}_S$-G-modules.

We now express the condition for a given T-stable sheaf of ideals

$$\mathcal{J} \subset A^U \otimes_k \mathcal{O}_S$$

such that each quotient $(A \otimes_k \mathcal{O}_S)^U_\lambda / \mathcal{J}_\lambda$ is locally free of rank $\bar{h}(\lambda)$, to equal $\mathcal{I}^U$ for some G-stable sheaf of ideals $\mathcal{I}$ as above. This is equivalent to the condition that the $\mathcal{O}_S$-G-module

$$\mathcal{I} := \langle G \cdot \mathcal{J} \rangle \subset A \otimes_k \mathcal{O}_S$$

generated by $\mathcal{J}$, is a sheaf of ideals, i.e., $\mathcal{I}$ is stable under multiplication by A. By highest weight theory, this means that

$$(\mathcal{I} \cdot A)^U \subset \mathcal{J}.$$

We will translate the latter condition into the vanishing of certain morphisms of locally free sheaves over S, arising from the universal family of $\mathcal{H}$ via the classifying morphism

$$f : S \longrightarrow \mathcal{H}.$$

For this, consider three dominant weights λ, μ, ν and a copy of the simple G-module $V(\nu)$ in $V(\lambda) \otimes_k V(\mu)$, with highest weight vector

$$v \in \big(V(\lambda) \otimes_k V(\mu)\big)^U_\nu.$$

We may write

$$v = \sum_i c_i (g_i \cdot v_\lambda) \otimes (h_i \cdot v_\mu),$$

a finite sum where $c_i \in k$ and $g_i, h_i \in G$. This defines a morphism of sheaves of $\mathcal{O}_S$-modules

$$A^U_\lambda \otimes_k \mathcal{J}_\mu \longrightarrow A^U_\nu \otimes_k \mathcal{O}_S, \quad a \otimes b \longmapsto \sum_i c_i (g_i \cdot a)(h_i \cdot b).$$

Composing with the quotient by $\mathcal{J}_\nu$ yields a morphism of sheaves of $\mathcal{O}_S$-modules,

$$F_v : A^U_\lambda \otimes_k \mathcal{J}_\mu \longrightarrow (A^U_\nu \otimes_k \mathcal{O}_S)/\mathcal{J}_\nu.$$

Our condition is the vanishing of these morphisms F_v for all triples (λ, μ, ν) and all v as above. Now $(A^U_\nu \otimes_k \mathcal{O}_S)/\mathcal{J}_\nu$ and $\mathcal{J}_\mu$ are the pull-backs under f of the analogous locally free sheaves on $\mathcal{H}$. This shows that the Hilbert functor $\mathrm{Hilb}^G_h(X)$ is represented by the closed subscheme of $\mathcal{H}$ obtained as the intersection of the zero loci of the F_v. □

Our final reduction step will allow us to enlarge X:

Lemma 3.4. *Let X be a closed G-subscheme of an affine G-scheme Y of finite type. If $\mathrm{Hilb}_h^G(Y)$ is represented by a scheme $\mathcal{H}$, then $\mathrm{Hilb}_h^G(X)$ is represented by a closed subscheme of $\mathcal{H}$.*

Proof. Let $X = \mathrm{Spec}(A)$ and $Y = \mathrm{Spec}(B)$, so that we have an exact sequence

$$0 \longrightarrow I \longrightarrow B \longrightarrow A \longrightarrow 0$$

where I is a G-stable ideal of B. For any $M \in \mathrm{Irr}(G)$, this yields an exact sequence for modules of covariants over B^G:

$$0 \longrightarrow I_M \longrightarrow B_M \longrightarrow A_M \longrightarrow 0.$$

Next, consider a scheme S and a flat family $p : \mathcal{Z} \to S$ of closed G-stable subschemes of Y, with Hilbert function h. Then each associated sheaf of covariants $\mathcal{F}_M$ is a locally free quotient of $B_M \otimes_k \mathcal{O}_S$, of rank $h(M)$; this defines a linear map $q_M : B_M \to H^0(S, \mathcal{F}_M)$. Moreover, $\mathcal{Z}$ is contained in $X \times S$ if and only if the image of $I \otimes_k \mathcal{O}_S$ in $p_*\mathcal{O}_{\mathcal{Z}}$ is zero; equivalently, $q_M(I_M) = 0$ for all $M \in \mathrm{Irr}(G)$. Taking for p the universal family of $\mathcal{H}$, it follows that the invariant Hilbert functor $\mathrm{Hilb}_h^G(X)$ is represented by the closed subscheme of $\mathcal{H}$, intersection of the zero loci of the subspaces $q_M(I_M) \subset \Gamma(\mathrm{Hilb}_h^G(Y), \mathcal{F}_M)$ for all $M \in \mathrm{Irr}(G)$. □

Summarizing, we may reduce to the case where G is a maximal torus of GL_n by combining Lemmas 3.2 and 3.3, and then to the case where X is a finite-dimensional G-module by Lemma 3.4. Then the invariant Hilbert scheme is exactly the multigraded one, as noted in Remark 2.22(iii).

Remarks 3.5. (i) The proof of Lemma 3.3 actually shows that the invariant Hilbert functor $\mathrm{Hilb}_h^G(X)$ is a closed subfunctor of $\mathrm{Hilb}_h^T(X/\!/U)$. Likewise, in the setting of Lemma 3.2 (resp. of Lemma 3.4), $\mathrm{Hilb}_h^G(X)$ is a closed subfunctor of $\mathrm{Hilb}_h^{\tilde{G}}(\tilde{X})$ (resp. of $\mathrm{Hilb}_h^G(Y)$).

(ii) The arguments of this subsection establish the existence of the invariant Hilbert scheme over any field of characteristic 0. Indeed, highest weight theory holds for GL_n in that setting, whereas it fails for non-split reductive groups.

3.2. Zariski tangent space

In this subsection, we consider a reductive group G, an affine G-scheme of finite type $X = \mathrm{Spec}(A)$, and a function $h : \mathrm{Irr}(G) \to \mathbb{N}$. We study the Zariski tangent space $T_Z\mathrm{Hilb}_h^G(X)$ to the invariant Hilbert scheme at an arbitrary closed point Z, i.e., at a closed G-stable subscheme of X with Hilbert function h. As a first step, we obtain:

Proposition 3.6. *With the preceding notation, we have*

$$T_Z\mathrm{Hilb}_h^G(X) = \mathrm{Hom}_A^G(I, A/I) = \mathrm{Hom}_{\mathcal{O}(Z)}^G(I/I^2, \mathcal{O}(Z)) \tag{3.7}$$

where $I \subset A$ *denotes the ideal of* Z, *and* Hom_A^G *stands for the space of* A*-linear,* G*-equivariant maps.*

Indeed, $\mathrm{Hom}_A(I, A/I)$ parametrizes the **first-order deformations of** Z **in** X, i.e., those closed subschemes

$$\mathcal{Z} \subset X \times \operatorname{Spec} k[\varepsilon]$$

(where $\varepsilon^2 = 0$) that are flat over $\operatorname{Spec} k[\varepsilon]$ and satisfy $\mathcal{Z}_s = Z$ where s denotes the closed point of $\operatorname{Spec} k[\varepsilon]$; see e.g. [54, Section 3.2]). The subspace $\mathrm{Hom}_A^G(I, A/I)$ parametrizes the G-stable deformations.

Example 3.8. Let $G = SL_2$ and $X := rV(2)$ (the direct sum of r copies of $V(2)$), where r is a positive integer. We consider the invariant Hilbert scheme $\mathrm{Hilb}_h^G(X)$, where $h = h_2$ is as defined in (2.25), and show that the Zariski tangent space at any closed point Z has dimension r.

Indeed, the r projections $p_1, \ldots, p_r : Z \to V(2)$ are all proportional, since $h_2(2) = 1$. Thus, we may assume that Z is contained in the first copy of $V(2)$, for an appropriate choice of projections. Then the condition that $h_2(0) = 1$ implies that Z is contained in the scheme-theoretic fiber of the discriminant Δ at some scalar t. Since that fiber has also Hilbert function h_2, we see that equality holds: the ideal of Z satisfies

$$I = \big(\Delta(p_1) - t, p_2, \cdots, p_r\big).$$

In particular, Z is a complete intersection in X, and the $\mathcal{O}(Z)$-G-module I/I^2 is freely generated by the images of $\Delta(p_1) - t, p_2, \ldots, p_r$. This yields an isomorphism of $\mathcal{O}(Z)$-G-modules

$$I/I^2 \cong \mathcal{O}(Z) \otimes_k \big(V(0) \oplus (r-1)V(2)\big).$$

As a consequence,

$$\mathrm{Hom}^G_{\mathcal{O}(Z)}\big(I/I^2, \mathcal{O}(Z)\big) \cong \mathrm{Hom}^G\big(V(0) \oplus (r-1)V(2), \mathcal{O}(Z)\big)$$

has dimension $h_2(0) + (r-1)h_2(2) = r$. Together with (3.7), this implies the statement.

In fact, $\mathrm{Hilb}_h^G(X)$ is a smooth irreducible variety of dimension r, as we will see in Example 3.18(iv). Specifically, $\mathrm{Hilb}_h^G(X)$ is the total space of the line bundle of degree -2 on $\mathbb{P}^{r-1}$, see Example 3.24(iv).

The isomorphism (3.7) is the starting point of a local analysis of the invariant Hilbert scheme, in relation to deformation theory (for the latter, see [54]). We will present a basic and very useful result in that direction; it relies on the following:

Lemma 3.9. *Let* $\mathcal{M}$ *be a coherent sheaf on an affine scheme* Z, *and* $M = H^0(Z, \mathcal{M})$ *the associated finitely generated module over* $R := \mathcal{O}(Z)$. *Let* $Z_0 \subset Z$ *be a dense open subscheme and denote by* $\iota : Z_0 \to Z$ *the inclusion map. Then the pull-back*

$$\iota^* : \mathrm{Hom}_R(M, R) = \mathrm{Hom}_Z(\mathcal{M}, \mathcal{O}_Z) \longrightarrow \mathrm{Hom}_{Z_0}(\mathcal{M}_{|Z_0}, \mathcal{O}_{Z_0})$$

is injective.

If Z *is a normal irreducible variety and the complement* $Z \setminus Z_0$ *has codimension* $\geqslant 2$ *in* Z*, then* ι^* *is an isomorphism.*

Proof. Choose a presentation of the R-module M,

$$R^m \xrightarrow{A} R^n \longrightarrow M \longrightarrow 0,$$

where A is a matrix with entries in R. This yields an exact sequence of R-modules

$$0 \longrightarrow \mathrm{Hom}_R(M, R) \longrightarrow R^n \xrightarrow{B} R^m$$

where B denotes the transpose of A. In other words, $\mathrm{Hom}_R(M, R)$ consists of those $(f_1, \dots, f_n) \in R^n$ that are killed by B. This implies both assertions, since $\iota^* : R = \mathcal{O}(Z) \to \mathcal{O}(Z_0)$ is injective, and is an isomorphism under the additional assumptions. □

We may now obtain a more concrete description of the Zariski tangent space at a G-orbit closure:

Proposition 3.10. *Let* G *be a reductive group,* V *a finite-dimensional* G*-module, and* v *a point of* V*. Denote by* $Z \subset V$ *the closure of the orbit* $G \cdot v$ *and by* h *the Hilbert function of* $\mathcal{O}(Z)$*. Let* $G_v \subset G$ *be the isotropy group of* v*, and* $\mathfrak{g}$ *the Lie algebra of* G*. Then*

$$T_Z \mathrm{Hilb}_h^G(V) \hookrightarrow (V/\mathfrak{g}\cdot v)^{G_v} \qquad (3.11)$$

where G_v *acts on* $V/\mathfrak{g}\cdot v$ *via its linear action on* V *which stabilizes the subspace* $\mathfrak{g}\cdot v = T_v(G\cdot v)$*.*

Moreover, equality holds in (3.11) if Z *is normal and the boundary* $Z \setminus G\cdot v$ *has codimension* $\geqslant 2$ *in* Z*.*

Proof. We apply Proposition 3.6 and Lemma 3.9 by taking $M = I/I^2$ and $Z_0 = G \cdot v$. This yields an injection of $T_Z \mathrm{Hilb}_h^G(V)$ into

$$W := \mathrm{Hom}_{Z_0}^G\big((\mathcal{I}/\mathcal{I}^2)_{|Z_0}, \mathcal{O}_{Z_0}\big),$$

where $\mathcal{I}$ denotes the ideal sheaf of Z in V. Moreover, $T_Z \mathrm{Hilb}_h^G(V) = W$ under the additional assumptions. Since Z_0 is a smooth subvariety of the smooth irreducible variety V, the conormal sheaf $(\mathcal{I}/\mathcal{I}^2)_{|Z_0}$ is locally free. Denoting the dual (normal) sheaf by $\mathcal{N}_{Z_0/V}$, we have

$$W = H^0(Z_0, \mathcal{N}_{Z_0/V})^G.$$

But $\mathcal{N}_{Z_0/V}$ is the sheaf of local sections of the normal bundle, and the total space of that bundle is the G-variety $G \times^{G_v} \mathcal{N}_{Z_0/V,v}$ equipped with the projection to $G/G_v = G\cdot v$. Moreover, we have isomorphisms of G_v-modules

$$\mathcal{N}_{Z_0/V,v} \cong T_v(V)/T_v(G\cdot v) \cong V/\mathfrak{g}\cdot v.$$

It follows that

$$H^0(Z_0, \mathcal{N}_{Z_0/V})^G \cong \big(\mathcal{O}(G/G_v) \otimes_k \mathcal{N}_{Z_0/V,v}\big)^G \cong \mathrm{Hom}^G(G/G_v, \mathcal{N}_{Z_0/V,v}) \cong \mathcal{N}_{Z_0/V,v}^{G_v}.$$

This implies our assertions. □

We refer to [3, Section 1.4] for further developments along these lines, including a relation to (non-embedded) first-order deformations, and to [48, Section 3] for generalizations where the boundary may have irreducible components of codimension 1. The obstruction space for G-invariant deformations is considered in [23, Section 3.5].

Examples 3.12. (i) Let $\underline{\lambda} = (\lambda_1, \ldots, \lambda_N)$ be a list of pairwise distinct weights of a torus T, and $\mathrm{Hilb}^T(\underline{\lambda})$ the associated *toric Hilbert scheme* as in Example 2.23(i). Let $Z = \overline{T \cdot v}$ where v is a general point of $V = \mathbb{A}^N$ (i.e., all of its coordinates are non-zero). Then the stabilizer T_v is the kernel of the homomorphism

$$\underline{\lambda} : T \longrightarrow (\mathbb{G}_m)^N, \quad t \longmapsto (\lambda_1(t), \ldots, \lambda_N(t)) \tag{3.13}$$

and hence acts trivially on V. Thus, the preceding proposition just yields an inclusion

$$\iota : T_Z \mathrm{Hilb}^T(\underline{\lambda}) \hookrightarrow V/\mathfrak{t} \cdot v$$

where $\mathfrak{t}$ denotes the Lie algebra of T.

In fact, ι *is an isomorphism*. Indeed, moving v among the general points defines a family $p : \mathcal{Z} \to (\mathbb{G}_m)^N$ of T-orbit closures in V, and hence a morphism $f : (\mathbb{G}_m)^N \to \mathrm{Hilb}^T(\underline{\lambda})$. Moreover, the differential of f at v composed with ι yields the quotient map $V \to V/\mathfrak{t} \cdot v$; hence ι is surjective. (See Example 3.18(i) for another version of this argument, based on the natural action of $(\mathbb{G}_m)^N$ on $\mathrm{Hilb}^T(\underline{\lambda})$.)

(ii) As in Example 2.23(ii), let $G = SL_2$, $X = V(n)$ and Z the variety of n-th powers of linear maps. Then $Z = \overline{G \cdot v} = G \cdot v \cup \{0\}$, where $v := y^n$ is a highest weight vector; moreover, Z is a normal surface. Thus, we may apply the preceding proposition to determine $T_Z \mathrm{Hilb}_h^G(V(n))$, where $h = h_n$ is the function (2.25).

The stabilizer G_{y^n} is the semi-direct product of the additive group U (acting via $x \mapsto x + ty$, $y \mapsto y$) with the group μ_n of n-th roots of unity acting via $x \mapsto \zeta x$, $y \mapsto \zeta^{-1} y$. Also, $\mathfrak{g} \cdot v$ is spanned by the monomials y^n and xy^{n-1}, and $V/\mathfrak{g} \cdot v$ has basis the images of the remaining monomials $x^n, x^{n-1}y, \ldots, x^2 y^{n-2}$. It follows that $(V/\mathfrak{g} \cdot v)^U$ is spanned by the image of $x^2 y^{n-2}$; the latter is fixed by μ_n if and only if $n = 2$ or $n = 4$. We thus obtain:

$$T_Z \mathrm{Hilb}_h^G(V) = \begin{cases} k & \text{if } n = 2 \text{ or } 4, \\ 0 & \text{otherwise.} \end{cases}$$

(iii) More generally, let G be an arbitrary connected reductive group, $V = V(\lambda)$ a simple G-module of dimension $\geqslant 2$, $v = v_\lambda$ a highest weight vector, and $Z = \overline{G \cdot v}$ as in Example 2.23(iii). Then the stabilizer of the highest weight line $[v]$ is a parabolic subgroup $P \supset B$, and the character λ of B extends to P; moreover, G_v is the kernel of that extended character. Also, Z is normal and its boundary (the origin) has codimension $\geqslant 2$. Thus, Proposition 3.10 still applies to this situation. Combined with arguments of combinatorial representation theory, it yields that $T_Z\mathrm{Hilb}_h^G(V) = 0$ unless λ belongs to an explicit (and rather small) list of dominant weights; in that case, $T_Z\mathrm{Hilb}_h^G(V) = k$ (see [29, Section 1.3]).

3.3. Action of equivariant automorphism groups

As in the previous subsection, we fix an affine G-scheme of finite type X and a function $h : \mathrm{Irr}(G) \to \mathbb{N}$. We obtain a natural equivariance property of the corresponding invariant Hilbert scheme.

Proposition 3.14. *Let* H *be an algebraic group, and* $\beta : H \times X \to X$ *an action by* G*-automorphisms. Then* β *induces an* H*-action on* $\mathrm{Hilb}_h^G(X)$ *that stabilizes the universal family* $\mathrm{Univ}_h^G(X) \subset X \times \mathrm{Hilb}_h^G(X)$.

Proof. Given a flat family of G-stable subschemes $\mathcal{Z} \subset X \times S$ with Hilbert function h, we construct a flat family of G-stable subschemes

$$\mathcal{W} \subset H \times X \times S$$

with the same Hilbert function, where G acts on $H \times X \times S$ via $g \cdot (h, x, s) = (h, g \cdot x, s)$. For this, form the cartesian square

$$\begin{array}{ccc} \mathcal{W} & \longrightarrow & H \times X \times S \\ \downarrow & & \downarrow{\scriptstyle \beta \times \mathrm{id}_S} \\ \mathcal{Z} & \longrightarrow & X \times S. \end{array} \tag{3.15}$$

Then $\mathcal{W}$ is a closed subscheme of $H \times X \times S$, stable under the given G-action since β is G-equivariant. Moreover, the map

$$H \times X \longrightarrow H \times X, \quad (h, x) \longmapsto (h, h \cdot x)$$

is an isomorphism; thus, the square

$$\begin{array}{ccc} H \times X \times S & \longrightarrow & H \times S \\ {\scriptstyle \beta \times \mathrm{id}_S}\downarrow & & \downarrow \\ X \times S & \longrightarrow & S \end{array} \tag{3.16}$$

(where the non-labeled arrows are the projections) is cartesian. By composing (3.15) and (3.16), we obtain a cartesian square

$$\begin{array}{ccc} \mathcal{W} & \longrightarrow & H \times S \\ \downarrow & & \downarrow \\ \mathcal{Z} & \longrightarrow & S \end{array}$$

i.e., an isomorphism

$$\mathcal{W} \cong (H \times S) \times_S \mathcal{Z} \cong H \times \mathcal{Z}$$

where the morphism $H \times \mathcal{Z} \to \mathcal{W}$ is given by $(h, z) \mapsto \beta(h^{-1}, z)$. It follows that $\mathcal{W}$ is flat over $H \times S$ with Hilbert function h.

Applying this construction to $\mathcal{Z} = \mathrm{Univ}_h^G(X)$ and $S = \mathrm{Hilb}_h^G(X)$ yields a flat family $\mathcal{W}$ with Hilbert function h and base $H \times \mathrm{Hilb}_h^G(X)$, and hence a morphism of schemes

$$\varphi : H \times \mathrm{Hilb}_h^G(X) \longrightarrow \mathrm{Hilb}_h^G(X)$$

such that $\mathcal{W}$ is the pull-back of the universal family. Since the composite morphism

$$X \xrightarrow{e_H \times \mathrm{id}_X} H \times X \xrightarrow{\beta} X$$

is the identity, it follows that the same holds for the composite

$$\mathrm{Hilb}_h^G(X) \xrightarrow{e_H \times \mathrm{id}_{\mathrm{Hilb}_h^G(X)}} H \times \mathrm{Hilb}_h^G(X) \xrightarrow{\varphi} \mathrm{Hilb}_h^G(X).$$

Likewise, φ satisfies the associativity property (2.1). Thus, φ is an H-action on $\mathrm{Hilb}_h^G(X)$.

To show that the induced H-action on $X \times \mathrm{Hilb}_h^G(X)$ stabilizes the closed subscheme $\mathrm{Univ}_h^G(X)$, note that $\mathcal{W} \subset H \times X \times \mathrm{Hilb}_h^G(X)$ is the closed subscheme defined by $((\beta(h^{-1}, x), s) \in \mathrm{Univ}_h^G(X)$ with an obvious notation (as follows from the cartesian square (3.15)). But $\mathcal{W}$ is also defined by $(x, \varphi(h, s)) \in \mathrm{Univ}_h^G(X)$ (since it is the pull-back of $\mathrm{Univ}_h^G(X)$). This yields the desired statement. □

In the case where X is a finite-dimensional G-module, say V, a natural choice for H is the automorphism group of the G-module V, i.e., the centralizer of G in GL(V); we denote that group by $\mathrm{GL}(V)^G$. To describe it, consider the isotypical decomposition

$$V \cong m_1 V(\lambda_1) \oplus \cdots \oplus m_r V(\lambda_r), \tag{3.17}$$

where $\lambda_1, \ldots, \lambda_r$ are pairwise distinct dominant weights, and the multiplicities $m_1, \ldots, m_r$ are positive integers. By Schur's lemma, $\mathrm{GL}(V)^G$ preserves each isotypical component $m_i V(\lambda_i) \cong k^{m_i} \otimes_k V(\lambda_i)$ and acts there via a homomorphism to GL_{m_i}. It follows that

$$H \cong \mathrm{GL}_{m_1} \times \cdots \times \mathrm{GL}_{m_r}.$$

In particular, the center of $\mathrm{GL}(V)^G$ is a torus $(\mathbb{G}_m)^r$ acting on V via

$$(t_1, \ldots, t_r) \cdot (v_1, \ldots, v_r) = (t_1 v_1, \ldots, t_r v_r) \quad \text{where} \quad v_i \in m_i V(\lambda_i).$$

Examples 3.18. (i) For a torus T acting linearly on $V = \mathbb{A}^N$ via pairwise distinct weights $\lambda_1, \ldots, \lambda_N$, the group $\mathrm{GL}(V)^T$ is just the diagonal torus $(\mathbb{G}_m)^N$. In particular, this yields an action of $(\mathbb{G}_m)^N$ on the toric Hilbert scheme $\mathrm{Hilb}^T(\underline{\lambda})$ where $\underline{\lambda} = (\lambda_1, \ldots, \lambda_N)$. The stabilizer of a general orbit closure $Z = \overline{T \cdot v}$ is the image of the homomorphism $\underline{\lambda} : T \to (\mathbb{G}_m)^N$ (3.13) with kernel T_v. Thus, the orbit $(\mathbb{G}_m)^N \cdot Z$ is a smooth subvariety of $\mathrm{Hilb}^T(\underline{\lambda})$, of dimension

$$N - \dim \underline{\lambda}(T) = N - \dim(T) + \dim(T_v) = \dim(V/t \cdot v).$$

Since $\dim T_Z \mathrm{Hilb}^T(\underline{\lambda}) \leqslant \dim(V/t \cdot v)$ by Example 3.12(i), it follows that equality holds. We conclude that *the orbit* $(\mathbb{G}_m)^N \cdot Z$ *is open in* $\mathrm{Hilb}^T(\underline{\lambda})$.

As a consequence, the closure of that orbit is an irreducible component of the toric Hilbert scheme, equipped with its reduced subscheme structure: the **main component**, also called the **coherent component**. Its points are the general T-orbit closures and their flat limits as closed subschemes of $\mathbb{A}^N$. The normalization of the main component is a quasi-projective toric variety under the quotient torus $(\mathbb{G}_m)^N/\underline{\lambda}(T)$.

In particular, the main component is a point if and only if the homomorphism $\underline{\lambda}$ is surjective, i.e., the weights $\lambda_1, \dots, \lambda_N$ are linearly independent. In that case, one easily sees that the whole toric Hilbert scheme is a (reduced) point.

(ii) The automorphism group of the simple SL_2-module $V(n)$ is just $\mathbb{G}_m$ acting by scalar multiplication. For the induced action on the invariant Hilbert scheme, a closed point is fixed if and only if its ideal is homogeneous. If the Hilbert function is the function h_n defined in (2.25), then there is a unique such ideal I: that of the variety of n-th powers. Indeed, we have an isomorphism of SL_2-modules

$$\mathcal{O}(V(n))/I \cong \mathrm{Sym}(V(n))/I \cong \bigoplus_{m=0}^{\infty} V(mn).$$

Moreover, the graded G-algebra $\mathrm{Sym}(V(n))/I$ is generated by (the image of) $V(n)$, its component of degree 1. By an easy induction on m, it follows that the component of an arbitrary degree m of that algebra is isomorphic to $V(mn)$. But the G-submodule $V(mn) \subset \mathrm{Sym}^m(V(n))$ has a unique G-stable complement: the sum of all other simple submodules. This determines each homogeneous component of the graded ideal I.

(iii) The preceding argument adapts to show that the cone of highest weight vectors is the unique fixed point for the $\mathbb{G}_m$-action on $\mathrm{Hilb}^G_{h_\lambda}(V(\lambda))$, with the notation of Example 2.23(iii).

(iv) As in Example 3.8, let $G = SL_2$, $V = rV(2)$ and $h = h_2$. Then $H = GL(V)^G$ is the general linear group GL_r acting on $V \cong k^r \otimes_k V(2)$ via its standard action on k^r. For the induced action of H on $\mathrm{Hilb}^G_h(V)$, the closed points form two orbits Ω_1, Ω_0, with representatives Z_1, Z_0 defined by $\Delta(p_1) = 1$ ($\Delta(p_1) = 0$) and $p_2 = \dots = p_r = 0$. One easily checks that the isotropy group H_{Z_0} is the parabolic subgroup of GL_r that stabilizes the first coordinate line ke_1; as a consequence, $\Omega_0 \cong \mathbb{P}^{r-1}$. Also, H_{Z_1} is the stabilizer of $\pm e_1$, and hence $\Omega_1 \cong (\mathbb{A}^r \setminus \{0\})/\pm 1$ where GL_r acts linearly on $\mathbb{A}^r$. Since the family $\mathcal{Z}$ of Example 2.19(ii) has general fibers in Ω_1 and special fiber in Ω_0, we see that Ω_0 is contained in the closure of Ω_1. As a consequence, $\mathrm{Hilb}^G_h(V)$ *is irreducible of dimension* r. Since its Zariski tangent space has dimension r at each closed point, it follows that $\mathrm{Hilb}^G_h(V)$ *is a smooth variety.*

3.4. The quotient-scheme map

We keep the notation of the previous subsection, and consider a family of G-stable closed subschemes $\mathcal{Z} \subset X \times S$ over some scheme S. Recall that the sheaf of

$\mathcal{O}_S$-algebras $\mathcal{F}_0 = (p_*\mathcal{O}_\mathcal{Z})^G$ is a quotient of $A^G \otimes_k \mathcal{O}_S$ where $A = \mathcal{O}(X)$. This defines a family of closed subschemes

$$\mathcal{Z}//G \subset X//G \times S$$

where we recall that $X//G = \mathrm{Spec}(A^G)$. If $\mathcal{Z}$ is flat over S with Hilbert function h, then $\mathcal{F}_0$ is locally free over S, of rank

$$n := h(0).$$

Thus, $\mathcal{Z}//G$ defines a morphism to the punctual Hilbert scheme of the quotient,

$$f : S \longrightarrow \mathrm{Hilb}_n(X//G).$$

Applying this construction to the universal family yields a morphism

$$\gamma : \mathrm{Hilb}_h^G(X) \longrightarrow \mathrm{Hilb}_n(X//G) \tag{3.19}$$

that we call the **quotient-scheme map**.

Proposition 3.20. *With the preceding notation, the morphism (3.19) is projective. In particular, if the scheme $X//G$ is finite, or equivalently, if X contains only finitely many closed G-orbits, then $\mathrm{Hilb}_h^G(X)$ is projective.*

Proof. Since $\mathrm{Hilb}_h^G(X)$ is quasi-projective, it suffices to show that γ is proper. For this, we use the valuative criterion of properness for schemes of finite type: let R be a discrete valuation ring containing k and denote by K the fraction field of R. Let $\mathcal{Z}_K \in \mathrm{Hilb}_h^G(X)(K)$ and assume that $\gamma(\mathcal{Z}_K) \in \mathrm{Hilb}_n(X//G)(K)$ admits a lift to $\mathrm{Hilb}_n(X//G)(R)$. Then we have to show that $\mathcal{Z}_K$ admits a lift to $\mathrm{Hilb}_h^G(X)(R)$.

In other words, we have a family

$$\mathcal{Z}_K \subset X \times \mathrm{Spec}(K)$$

of closed G-stable subschemes with Hilbert function h, such that the family

$$\mathcal{Z}_K//G \subset X//G \times \mathrm{Spec}(K)$$

extends to a family in $X//G \times \mathrm{Spec}(R)$. These data correspond to a G-stable ideal

$$I_K \subset A \otimes_k K$$

such that $I_K^G \subset A^G \otimes_k K$ equals $J \otimes_R K$, where

$$J \subset A^G \otimes_k R$$

is an ideal such that the R-module $(A^G \otimes_k R)/J$ is free of rank $h(0)$. Then

$$J \subset (A^G \otimes_k R) \cap (J \otimes_R K),$$

where the intersection is considered in $A \otimes_R K$. Moreover, the quotient R-module $(A^G \otimes_k R) \cap (J \otimes_R K)/J$ is torsion; on the other hand, this quotient is contained in the free R-module $(A^G \otimes_k R)/J$. Thus,

$$J = (A^G \otimes_k R) \cap (J \otimes_R K) = (A^G \otimes_k R) \cap I_K.$$

We now consider

$$I := (A \otimes_k R) \cap I_K.$$

Clearly, I is an ideal of $A \otimes_k R$ satisfying $I^G = J$ and $I \otimes_R K = I_K$. Thus, the R-module

$$\big((A \otimes_k R)/I\big)^G = (A^G \otimes_k R)/I^G$$

is free of rank n. Moreover, each module of covariants $\big((A \otimes_k R)/I\big)_M$ is finitely generated over $\big((A \otimes_k R)/I\big)^G$, and torsion-free by construction. Hence the R-module $\big((A \otimes_k R)/I\big)_M$ is free; tensoring with K, we see that its rank is $h(M)$. Thus, I corresponds to an R-point of $\mathrm{Hilb}_h^G(X)$, which is the desired lift. □

Remarks 3.21. (i) The morphism (3.19) is analogous to the **Hilbert-Chow morphism**, or **cycle map**, that associates with any closed subscheme Z of the projective space $\mathbb{P}^N$, the support of Z (with multiplicities) viewed as a point of the Chow variety of $\mathbb{P}^N$.

The cycle map may be refined in the setting of punctual Hilbert schemes: given a quasi-projective scheme X and a positive integer n, there is a natural morphism

$$\varphi_n : \mathrm{Hilb}_n(X) \longrightarrow X^{(n)} \tag{3.22}$$

where $X^{(n)}$ denotes the n**-th symmetric product of** X, i.e., the quotient of the product $X \times \cdots \times X$ (n factors) by the action of the symmetric group S_n that permutes the factors; this is a quasi-projective scheme with closed points the effective 0-cycles of degree n on X. Moreover, φ_n induces the cycle map on closed points, and is a projective morphism (for these results, see [5, Theorem 2.16]).

In the setting of Proposition 3.20, let

$$\delta : \mathrm{Hilb}_h^G(X) \longrightarrow (X/\!/G)^{(n)} \tag{3.23}$$

denote the composite of (3.19) with (3.22). Then δ is a projective morphism, and $(X/\!/G)^{(n)}$ is affine. As a consequence, *the invariant Hilbert scheme is projective over an affine scheme.*

(ii) The quotient-scheme map satisfies a natural equivariance property: with the notation and assumptions of Proposition 3.14, the H-action on X induces an action on $X/\!/G$ so that the quotient morphism π is equivariant. This yields in turn an H-action on the punctual Hilbert scheme $\mathrm{Hilb}_n(X/\!/G)$; moreover, *the morphism (3.19) is equivariant* (as may be checked along the lines of that proposition).

Also, H acts on the symmetric product $(X/\!/G)^{(n)}$ and the morphism (3.23) is H-equivariant.

(iii) In the case where X is a finite-dimensional G-module V, and $H = (\mathbb{G}_m)^r$ is the center of $GL(V)^G$, the closed H-fixed points of $\mathrm{Hilb}_h^G(X)$ may be identified with those G-stable ideals $I \subset \mathcal{O}(V)$ that are homogeneous with respect to the isotypical decomposition (3.17). Moreover, *the closure of each* H*-orbit in* $\mathrm{Hilb}_h^G(V)$ *contains a fixed point*: indeed, the closure of each H-orbit in V contains the origin, and hence the same holds for the induced action of H on the symmetric product $(V//G)^{(n)}$ (where the origin is the image of $(0,\ldots,0)$). Together with the properness of the morphism δ and Borel's fixed point theorem, this implies the assertion.

Examples 3.24. (i) For the toric Hilbert scheme $\mathrm{Hilb}^T(\lambda_1,\ldots,\lambda_N) = \mathrm{Hilb}^T(\underline{\lambda})$ of Example 2.23(i), the quotient-scheme map may be refined as follows. Given λ in the monoid Γ generated by $\lambda_1,\ldots,\lambda_N$, consider the graded subalgebra

$$\mathcal{O}(\mathbb{A}^N)^{(\lambda)} := \bigoplus_{m=0}^{\infty} \mathcal{O}(\mathbb{A}^N)_{m\lambda} \subset \mathcal{O}(\mathbb{A}^N)$$

with degree-0 component $\mathcal{O}(\mathbb{A}^N)_0 = \mathcal{O}(\mathbb{A}^N//T)$. Replacing λ with a positive integer multiple, we may assume that the algebra $\mathcal{O}(\mathbb{A}^N)^{(\lambda)}$ is finite over its subalgebra generated by its components of degrees 0 and 1. Then

$$\mathbb{A}^N//_\lambda T := \mathrm{Proj}\big(\mathcal{O}(\mathbb{A}^N)^{(\lambda)}\big)$$

is a projective variety over $\mathbb{A}^N//_\lambda T$, and the twisting sheaf $\mathcal{O}(1)$ is an ample invertible sheaf on $\mathbb{A}^N//_\lambda T$, generated by its subspace $\mathcal{O}(\mathbb{A}^N)_\lambda$ of global sections. Moreover, one can define a morphism

$$\gamma_\lambda : \mathrm{Hilb}^T(\underline{\lambda}) \longrightarrow \mathbb{A}^N//_\lambda T$$

which lifts the quotient-scheme morphism γ. The collection of these morphisms forms a finite inverse system; its inverse limit is called the **toric Chow quotient** and denoted by $\mathbb{A}^N//_C T$. This construction yields the **toric Chow morphism**

$$\mathrm{Hilb}^T(\underline{\lambda}) \longrightarrow \mathbb{A}^N//_C T$$

which induces an isomorphism on the associated reduced schemes, under some additional assumptions (see [26, Section 5]).

(ii) With the notation of Example 2.23(ii), we may now show that $\mathrm{Hilb}^{SL_2}_{h_n}(V(n))$ is either an affine line if $n = 2$ or 4, or a reduced point for all other values of n.

Indeed, for the natural action of $\mathbb{G}_m$, each orbit closure contains the unique fixed point Z. If $n \neq 2$ and $n \neq 4$, then it follows that $\mathrm{Hilb}^{SL_2}_{h_n}(V(n))$ is just Z, since its Zariski tangent space at that point is trivial.

On the other hand, we have constructed a family $\mathcal{Z} \subset V(2) \times \mathbb{A}^1$ (Example 2.19) and hence a morphism

$$f : \mathbb{A}^1 \to \mathrm{Hilb}^{SL_2}_{h_2}(V(2)).$$

Moreover, f is injective (on closed points), since the fibers of $\mathcal{Z}$ are pairwise distinct subschemes of $V(2)$. Also, $\mathcal{Z}$ is stable under the action of $\mathbb{G}_m$ on $V(2) \times \mathbb{A}^1$ via $t \cdot (x, y) = (tx, t^2 y)$ and hence f is $\mathbb{G}_m$-equivariant for the action on $\mathbb{A}^1$ via $t \cdot y = t^2 y$. In particular, $\mathrm{Hilb}^{SL_2}_{h_2}(V(2))$ has dimension $\geqslant 1$ at Z. Since its Zariski tangent space has dimension 1, it follows that $\mathrm{Hilb}^{SL_2}_{h_2}(V(2))$ is smooth and 1-dimensional at Z. Using the $\mathbb{G}_m$-action, it follows in turn that f is an isomorphism; hence the natural map $\mathcal{Z} \to \mathrm{Univ}_{h_2}^{SL_2}(V(2))$ is an isomorphism as well. The quotient-scheme map is also an isomorphism in that case.

Likewise, the family $\mathcal{W}$ of Example 2.19(ii) yields isomorphisms

$$\mathbb{A}^1 \longrightarrow \mathrm{Hilb}_{h_4}^{SL_2}(V(4)), \quad \mathcal{W} \longrightarrow \mathrm{Univ}_{h_4}^{SL_2}(V(4)).$$

Moreover, the quotient-scheme map is a closed immersion

$$\mathbb{A}^1 \hookrightarrow V(4)//SL_2 \cong \mathbb{A}^2.$$

(iii) The preceding argument adapts to show that $\mathrm{Hilb}_{h_\lambda}^G(V(\lambda))$ is either an affine line or a reduced point, with the notation of Example 2.23(iii). The quotient-scheme map is again a closed immersion.

(iv) With the notation of Example 3.8, we describe the quotient-scheme map γ; it takes values in $V//G$ since $h(0) = 1$. Observe that the image of G in $GL(V(2)) \cong GL_3$ is the special orthogonal group associated with the non-degenerate quadratic form Δ (the discriminant). By classical invariant theory, it follows that the algebra of invariants of $rV(2)$ is generated by the maps

$$(A_1, \dots, A_r) \longmapsto \Delta(A_i), \quad \delta(A_i, A_j), \quad \det(A_i, A_j, A_k),$$

where δ denotes the bilinear form associated with Δ. But $\det(A_i, A_j, A_k)$ vanishes identically on the image of γ, which may thus be identified with the variety of symmetric $r \times r$ matrices of rank $\leqslant 1$. In other words, $\gamma(\mathrm{Hilb}_h^G(V))$ *is the affine cone over* $\mathbb{P}^{r-1}$ *embedded via* $\mathcal{O}_{\mathbb{P}^{r-1}}(2)$. Moreover, γ is a GL_r-equivariant desingularization of that cone, with exceptional locus the homogeneous divisor Ω_0. This easily yields an isomorphism of GL_r-varieties

$$\mathrm{Hilb}_h^G(V) \cong O_{\mathbb{P}^{r-1}}(-2)$$

where $O_{\mathbb{P}^{r-1}}(-2)$ denotes the total space of the line bundle of degree -2 over $\mathbb{P}^{r-1}$, i.e., the blow-up at the origin of the image of γ.

We now apply the construction of the quotient-scheme map to obtain a "flattening" of the categorical quotient $\pi : X \to X//G$, where X *is an irreducible variety*. Then $X//G$ is an irreducible variety as well, and there exists a largest open subset $Y_0 \subset Y := X//G$ such that the pull-back

$$\pi_0 : X_0 := \pi^{-1}(Y_0) \longrightarrow Y_0$$

is flat. It follows that the (scheme-theoretic) fibers of π at all closed points of its **flat locus** Y_0 have the same Hilbert function, say $h = h_X$. Since $F//G$ is a (reduced) point for each such fiber F, we have $h(0) = 1$. Thus, the invariant-scheme map associated with this special Hilbert function h is just a morphism

$$\gamma : \mathrm{Hilb}_h^G(X) \longrightarrow X//G.$$

Proposition 3.25. *With the preceding notation and assumptions, the diagram*

$$
\begin{array}{ccc}
\mathrm{Univ}_h^G(X) & \xrightarrow{q} & X \\
{\scriptstyle p}\downarrow & & \downarrow{\scriptstyle \pi} \\
\mathrm{Hilb}_h^G(X) & \xrightarrow{\gamma} & X/\!/G
\end{array}
\tag{3.26}
$$

commutes, where the morphisms from $\mathrm{Univ}_h^G(X)$ *are the projections. Moreover, the pull-back of* γ *to the flat locus of* π *is an isomorphism.*

Proof. Set for simplicity $\mathcal{Z} := \mathrm{Univ}_h^G(X)$ and $S := \mathrm{Hilb}_h^G(X)$. Then the natural map $\mathcal{O}_S \to (p_*\mathcal{O}_{\mathcal{Z}})^G$ is an isomorphism, by the definition of the Hilbert scheme and the assumption that $h(0) = 1$. Thus, p factors as the quotient morphism $\mathcal{Z} \to \mathcal{Z}/\!/G$ followed by an isomorphism $\mathcal{Z}/\!/G \to S$. In view of the definition of γ, this yields the first assertion.

For the second assertion, let $S_0 := \gamma_0^{-1}(Y_0)$ and $\mathcal{Z}_0 := p^{-1}(S_0)$. By the preceding step, we have a closed immersion $\iota : \mathcal{Z}_0 \subset X_0 \times_{Y_0} S_0$ of G-schemes. But both are flat over S_0 with the same Hilbert function h. Thus, the associated sheaves of covariants $\mathcal{F}_M$ (for $\mathcal{Z}_0$) and $\mathcal{G}_M$ (for $X_0 \times_{Y_0} S_0$) are locally free sheaves of $\mathcal{O}_S$-modules of the same rank, and come with a surjective morphism $\iota_M^* : \mathcal{G}_M \to \mathcal{F}_M$. It follows that ι_M^* is an isomorphism, and in turn that so is ι. □

This construction is of special interest in the case where G is a finite group, see [5, Section 4] and also Subsection 4.1. It also deserves further study in the setting of connected algebraic groups.

Example 3.27. Let G be a semi-simple algebraic group acting on its Lie algebra $\mathfrak{g}$ via the adjoint representation. By a classical result of Kostant, the categorical quotient $\mathfrak{g}/\!/G$ is an affine space of dimension equal to the rank r of G (the dimension of a maximal torus $T \subset G$). Moreover, the quotient morphism π is flat; its fibers are exactly the orbit closures of regular elements of $\mathfrak{g}$ (i.e., those with centralizer of minimal dimension r), and the corresponding Hilbert function $h = h_{\mathfrak{g}}$ is given by

$$h(\lambda) = \dim V(\lambda)^T. \tag{3.28}$$

Thus, the invariant-scheme map yields an isomorphism $\mathrm{Hilb}_h^G(\mathfrak{g}) \cong \mathfrak{g}/\!/G$.

If $G = SL_2$, then $\mathfrak{g} \cong V(2)$ and we get back that $\mathrm{Hilb}_{h_2}^{SL_2}(V(2)) \cong \mathbb{A}^1$. More generally, when applied to $G = SL_n$, we recover a result of Jansou and Ressayre, see [30, Theorem 2.5]. They also show that $\mathrm{Hilb}_h^{SL_n}(\mathfrak{sl}_n)$ is an affine space whenever h is the Hilbert function of an *arbitrary* orbit closure, and they explicitly describe the universal family in these cases; see [loc. cit., Theorem 3.6].

4. Some further developments and applications

4.1. Resolution of certain quotient singularities

In this subsection, we assume that the group G is *finite*. We discuss a direct construction of the invariant Hilbert scheme in that setting, as well as some applications; the reader may consult the notes [5] for further background, details, and developments.

Recall that Irr(G) denotes the set of isomorphism classes of simple G-modules. Since this set is finite, functions $h : \mathrm{Irr}(G) \to \mathbb{N}$ may be identified with isomorphism classes of arbitrary finite-dimensional modules, by assigning to h the G-module $\bigoplus_{M \in \mathrm{Irr}(G)} h(M)M$. For example, the function $M \mapsto \dim(M)$ corresponds to the regular representation, i.e., $\mathcal{O}(G)$ where G acts via left multiplication.

Given such a function h and a G-scheme X, we may consider the invariant Hilbert functor $\mathrm{Hilb}_h^G(X)$ as in Subsection 2.3: it associates with any scheme S the set of closed G-stable subschemes $\mathcal{Z} \subset X \times S$ such that the projection $p : \mathcal{Z} \to S$ is flat and the module of covariants $\mathrm{Hom}^G(M, p_*\mathcal{O}_\mathcal{Z})$ is locally free of rank $h(M)$ for any $M \in \mathrm{Irr}(G)$.

For such a family, the sheaf $p_*\mathcal{O}_\mathcal{Z}$ is locally free of rank

$$n = n(h) := \sum_{M \in \mathrm{Irr}(G)} h(M) \dim(M),$$

in view of the isotypical decomposition (2.11). In other words, $\mathcal{Z}$ is finite and flat over S of constant degree m, the dimension of the representation associated with h. If X is *quasi-projective*, then the punctual Hilbert scheme $\mathrm{Hilb}_n(X)$ exists and is equipped with an action of G (see Proposition 3.14). Thus, we have a morphism $f : S \to \mathrm{Hilb}_n(X)$ which is readily seen to be G-invariant. In other words, f factors through a morphism to the **fixed point subscheme** $\mathrm{Hilb}_n(X)^G \subset \mathrm{Hilb}_n(X)$, i.e., the largest closed G-stable subscheme on which G acts trivially. Moreover, the pull-back of the universal family $\mathrm{Univ}_n(X)$ to $\mathrm{Hilb}_n^G(X)$ is a finite flat family of G-stable subschemes of X, and has a well-defined Hilbert function on each connected component (see Remark 2.15(iii)).

This easily implies the following version of Theorem 2.20 for finite groups:

Proposition 4.1. *With the preceding notation and assumptions, the Hilbert functor* $\mathrm{Hilb}_h^G(X)$ *is represented by a union of connected components of the fixed point subscheme* $\mathrm{Hilb}_n(X)^G$.

Also, the quotient $\pi : X \to X/G$ exists, where the underlying topological space to X/G is just the orbit space, and the structure sheaf $\mathcal{O}_{X/G}$ equals $(\pi_*\mathcal{O}_X)^G$. Moreover, the set-theoretical fibers of π are exactly the G-orbits. As in Subsection 3.4, this yields a quotient-scheme map in this setting,

$$\gamma : \mathrm{Hilb}_h^G(X) \longrightarrow \mathrm{Hilb}_n(X/G).$$

(In fact, the assignment $Z \mapsto Z//G$ yields a morphism from $\mathrm{Hilb}_n^G(X)$ to the disjoint union of the punctual Hilbert schemes $\mathrm{Hilb}_m(X/G)$ for $m \leqslant n$).

We now assume that X *is an irreducible variety on which* G *acts faithfully*. Then X contains G-**regular points**, i.e., points with trivial isotropy groups, and they form an open G-stable subset $X_{\mathrm{reg}} \subset X$. Moreover, the (scheme-theoretic) fiber of π at a given $x \in X(k)$ equals the orbit $G \cdot x$ if and only if x is G-regular. In other words, the **regular locus** X_{reg}/G is the largest open subset of X/G over which π induces a Galois covering with group G; it is contained in the flat locus. Thus, the Hilbert function h_X associated with the general fibers of π (as in Subsection 3.4) is just that of the regular representation. The corresponding invariant Hilbert scheme is called the G-**Hilbert scheme** and denoted by $G\text{–}\mathrm{Hilb}(X)$. It is a union of connected components of $\mathrm{Hilb}_n(X)^G$, where n is the order of G, and comes with a projective morphism

$$\gamma : G - -\mathrm{Hilb}(X) \longrightarrow X/G \tag{4.2}$$

which induces an isomorphism above the regular locus X_{reg}/G. Moreover, γ fits into a commutative square

$$\begin{array}{ccc} G - -\mathrm{Univ}(X) & \xrightarrow{q} & X \\ p \downarrow & & \pi \downarrow \\ G - -\mathrm{Hilb}(X) & \xrightarrow{\gamma} & X/G \end{array}$$

by Proposition 3.25. In other words, $G\text{–}\mathrm{Univ}(X)$ is a closed G-stable subscheme of the fibered product $X \times_{X/G} G - -\mathrm{Hilb}(X)$.

We denote by $G\text{–}\mathcal{H}_X$ the closure of $\gamma^{-1}(X_{\mathrm{reg}}/G)$ in $G\text{–}\mathrm{Hilb}(X)$, equipped with the reduced subscheme structure.. This is an irreducible component of $G\text{–}\mathrm{Hilb}(X)$: the **main component**, also called the **orbit component**. The points of $G\text{–}\mathcal{H}_X$ are the regular G-orbits and their flat limits as closed subschemes of X; also, the quotient-scheme map restricts to a projective *birational* morphism $G\text{–}\mathcal{H}_X \to X/G$.

Examples 4.3. (i) If X and X/G are smooth, then *the quotient-scheme map (4.2) is an isomorphism*. (Indeed, π is flat in that case, and the assertion follows from Proposition 3.25).

In particular, if V is a finite-dimensional vector space and $G \subset GL(V)$ a finite subgroup generated by pseudo-reflections, then $\mathcal{O}(V)^G$ is a polynomial algebra by a theorem of Chevalley and Shepherd-Todd. Thus, (4.2) is an isomorphism under that assumption.

(ii) If X is a smooth surface, then every punctual Hilbert scheme $\mathrm{Hilb}_n(X)$ is smooth. Since smoothness is preserved by taking fixed points under finite (or, more generally, reductive) group actions, it follows that $\mathrm{Hilb}_h^G(X)$ *is smooth for any function* h. Thus, the quotient-scheme map $\gamma : G - -\mathcal{H}_X \to X/G$ is a resolution of singularities.

In particular, if G is a finite subgroup of GL_2 which is not generated by pseudo-reflections, then the quotient $\mathbb{A}^2/G$ is a normal surface with singular locus (the image of) the origin, and $\gamma : G - -\mathcal{H}_{\mathbb{A}^2} \to \mathbb{A}^2/G$ is a canonical desingularization. If in addition $G \subset SL_2$, then G contains no pseudo-reflection, and the resulting singularity is a rational double point. In that case, γ yields the *minimal* desingularization (this result is due to Ito and Nakamura, see [27, 28]; a self-contained proof is provided in [5, Section 5]).

(iii) The preceding argument does not extend to smooth varieties X of dimension $\geqslant 3$, since $\mathrm{Hilb}_n(X)$ is generally singular in that setting. Yet it was shown by Bridgeland, King and Reid via homological methods that G–Hilb(X) *is irreducible and has trivial canonical sheaf, if* $\dim(X) \leqslant 3$ *and the canonical sheaf of* X *is equivariant trivial* (see [17, Theorem 1.2]). As a consequence, if $G \subset SL_n$ with $n \leqslant 3$, then $\gamma : G - \mathrm{Hilb}(\mathbb{A}^n) \to \mathbb{A}^n/G$ is a crepant resolution; in particular, $G - \mathrm{Hilb}(\mathbb{A}^n)$ is irreducible. This result fails in dimension 4 as the G-Hilbert scheme may be reducible; this is the case for the binary tetrahedral group $G \subset SL_2$ acting on k^4 via the direct sum of two copies of its natural representation, see [35].

4.2. The horospherical family

In this subsection, G denotes a *connected* reductive group. We present a classical algebraic construction that associates with each affine G-scheme Z, a "simpler" affine G-scheme Z_0 called its horospherical degeneration (see [52] or [25, §15]). Then we interpret this construction in terms of families, after [3].

We freely use the notation and conventions from highest weight theory (Subsection 2.2) and denote by $\alpha_1, \ldots, \alpha_r$ the simple roots of (G, T) associated with the Borel subgroup B (i.e., the corresponding positive roots are those of (B, T)).

Given a G-algebra A, recall the isotypical decomposition

$$A = \bigoplus_{\lambda \in \Lambda^+} A_{(\lambda)} \quad \text{where} \quad A_{(\lambda)} \cong A_\lambda^U \otimes_k V(\lambda). \tag{4.4}$$

Also recall that when G is a torus, $A_{(\lambda)}$ is just the weight space A_λ and (4.4) is a *grading* of A, i.e., $A_\lambda \cdot A_\mu \subset A_{\lambda+\mu}$ for all λ, μ. For an arbitrary group G, (4.4) is no longer a grading, but gives rise to a *filtration* of A. To see this, we study the multiplicative properties of the isotypical decomposition.

Given $\lambda, \mu \in \Lambda^+$, there is an isomorphism of G-modules

$$V(\lambda) \otimes_k V(\mu) \cong \bigoplus_{\nu \in \Lambda^+} c_{\lambda,\mu}^\nu V(\nu) \tag{4.5}$$

where the $c_{\lambda,\mu}^\nu$'s are non-negative integers, called the **Littlewood-Richardson coefficients**. Moreover, if $c_{\lambda,\mu}^\nu \neq 0$ then $\nu \leqslant \lambda + \mu$ where $\leqslant$ is the partial ordering on Λ defined by:

$$\mu \leqslant \lambda \Leftrightarrow \lambda - \mu = \sum_{i=1}^r n_i \alpha_i \text{ for some non-negative integers } n_1, \ldots, n_r.$$

Finally, $c_{\lambda,\mu}^{\lambda+\mu} = 1$, i.e., the simple module with the largest weight $\lambda + \mu$ occurs in the tensor product $V(\lambda) \otimes_k V(\mu)$ with multiplicity 1. This special component is called the **Cartan component** of $V(\lambda) \otimes_k V(\mu)$.

We set

$$A_{(\leqslant\lambda)} := \bigoplus_{\mu\in\Lambda^+,\, \mu\leqslant\lambda} A_{(\mu)}.$$

In view of (4.5), we have

$$A_{(\leqslant\lambda)} \cdot A_{(\leqslant\mu)} \subset A_{(\leqslant\lambda+\mu)} \tag{4.6}$$

for all dominant weights λ, μ. In other words, the G-submodules $A_{(\leqslant\lambda)}$ form an increasing filtration of the G-algebra A, indexed by the partially ordered group Λ. The associated graded algebra gr(A) is a G-algebra, isomorphic to A as a G-module but where the product of any two isotypical components $A_{(\lambda)}$ and $A_{(\mu)}$ is obtained from their product in A by projecting on the Cartan component $A_{(\lambda+\mu)}$. Thus, the product of any two simple submodules of gr(A) is either their Cartan product, or zero. Also, note that $\mathrm{gr}(A)^U \cong A^U$ as T-algebras, since $A^U_{(\lambda)} = A^U_\lambda$ for all λ.

Now consider the **Rees algebra** associated to this filtration:

$$\mathcal{R}(A) := \bigoplus_{\mu\in\Lambda} A_{(\leqslant\mu)}\, e^{\mu} = \bigoplus_{\lambda\in\Lambda^+,\, \mu\in\Lambda,\, \lambda\leqslant\mu} A_{(\lambda)}\, e^{\mu} \tag{4.7}$$

where e^{μ} denotes the character μ viewed as a regular function on T (so that $e^{\mu}e^{\nu} = e^{\mu+\nu}$ for all μ and ν). Thus, $\mathcal{R}(A)$ is a subspace of

$$A \otimes_k \mathcal{O}(T) = \bigoplus_{\lambda\in\Lambda,\, \mu\in\Lambda} A_{(\lambda)}\, e^{\mu}.$$

In fact, $A \otimes_k \mathcal{O}(T)$ is a $G \times T$-algebra, and $\mathcal{R}(A)$ is a $G \times T$-subalgebra by the multiplicative property (4.6). Also, note that $\mathcal{R}(A)$ contains variables

$$t_1 := e^{\alpha_1}, \ldots, t_r := e^{\alpha_r} \tag{4.8}$$

associated with the simple roots; the monomials in these variables are just the $e^{\mu-\lambda}$ where $\lambda \leqslant \mu$. By (4.7), we have

$$\mathcal{R}(A) \cong \bigoplus_{\lambda\in\Lambda^+} A_{(\lambda)}\, e^{\lambda} \otimes_k k[t_1, \ldots, t_r] \cong A[t_1, \ldots, t_r]$$

as G–$k[t_1, \ldots, t_r]$-modules. In particular, $\mathcal{R}(A)$ is a free module over the polynomial ring $k[t_1, \ldots, t_r] \subset \mathcal{O}(T)$. Moreover, we have an isomorphism of T–$k[t_1, \ldots, t_r]$-algebras

$$\mathcal{R}(A)^U \cong A^U[t_1, \ldots, t_r]$$

that maps each $f \in \mathcal{R}(A)^U_\lambda$ to fe^{λ}. Also, the ideal $(t_1, \ldots, t_r) \subset \mathcal{R}(A)$ is G-stable, and the quotient by that ideal is just the G-module A where the product of any two components $A_{(\lambda)}, A_{(\mu)}$ is the same as in gr(A). In other words,

$$\mathcal{R}(A)/(t_1, \ldots, t_r) \cong \mathrm{gr}(A).$$

On the other hand, when inverting $t_1, \dots, t_r$, we obtain

$$\mathcal{R}(A)[t_1^{-1}, \dots, t_r^{-1}] \cong \bigoplus A_{(\lambda)}\, e^{\mu}$$

where the sum is over those $\lambda \in \Lambda^+$ and $\mu \in \Lambda$ such that $\lambda - \mu = n_1\alpha_1 + \cdots + n_r\alpha_r$ for some integers $n_1, \dots n_r$ (of arbitrary signs). In other words, $\lambda - \mu$ is in the **root lattice**, i.e., the sublattice of the weight lattice Λ generated by the roots. The torus associated with the root lattice is the **adjoint torus** T_{ad}, isomorphic to $(\mathbb{G}_m)^r$ via the homomorphism

$$\underline{\alpha} : T \longrightarrow (\mathbb{G}_m)^r, \quad t \longmapsto \big(\alpha_1(t), \dots, \alpha_r(t)\big).$$

Moreover, the kernel of $\underline{\alpha}$ is the center of G, that we denote by $Z(G)$. This identifies T_{ad} with $T/Z(G)$, a maximal torus of the adjoint group $G/Z(G)$ (whence the name). Now $Z(G)$ acts on $A \otimes_k \mathcal{O}(T)$ via its action on A as a subgroup of G (then each isotypical component $A_{(\lambda)}$ is an eigenspace of weight $\lambda_{|Z(G)}$) and its action on $\mathcal{O}(T)$ as a subgroup of T (then each e^{μ} is an eigenspace of weight $-\mu$). Moreover, the invariant ring satisfies

$$(4.9) \qquad \big(A \otimes_k \mathcal{O}(T)\big)^{Z(G)} \cong \mathcal{R}(A)[t_1^{-1}, \dots, t_r^{-1}]$$

as $G \times T$-algebras over $\mathcal{O}(T)^{Z(G)} \cong k[t_1, t_1^{-1}, \dots, t_r, t_r^{-1}]$.

Translating these algebraic constructions into geometric terms yields the following statement:

Proposition 4.10. *Let* $Z = \mathrm{Spec}(A)$ *be an affine* G*-scheme and* $p : \mathcal{Z} \to \mathbb{A}^r$ *the morphism associated with the inclusion* $k[t_1, \dots, t_r] \subset \mathcal{R}(A)$*, where* $\mathcal{R}(A)$ *denotes the Rees algebra (4.7), and* $t_1, \dots, t_r$ *the variables (4.8).*

Then p *is a flat family of affine* G*-schemes, and the induced family of* T*-schemes* $\mathcal{Z}/\!/U \to \mathbb{A}^r$ *is trivial with fiber* $Z/\!/U$*. The fiber of* p *at* 0 *is* $\mathrm{Spec}\big(\mathrm{gr}(A)\big)$*.*

Moreover, the pull-back of p *to* $T_{ad} \subset \mathbb{A}^r$ *is isomorphic to the projection* $Z \times^{Z(G)} T \to T/Z(G) = T_{ad}$*. In particular, all fibers of* p *at general points of* $\mathbb{A}^r$ *are isomorphic to* Z*.*

This *special fiber* Z_0 is an affine G-scheme such that the isotypical decomposition of $\mathcal{O}(Z_0)$ is a grading; such schemes are called **horospherical**, and

$$Z_0 := \mathrm{Spec}\big(\mathrm{gr}(A)\big)$$

is called the **horospherical degeneration** of the affine G-scheme Z. We say that $p : \mathcal{Z} \to \mathbb{A}^r$ is the **horospherical family** (this terminology originates in hyperbolic geometry; note that horospherical varieties need not be spherical).

For example, the cone of highest weight vectors $Z = \overline{G \cdot v_\lambda} = G \cdot v_\lambda \cup \{0\}$ is a horospherical G-variety, in view of the isomorphism (2.26). In that case, the fixed point subscheme Z^U is the highest weight line $kv_\lambda = V(\lambda)^U$, and thus $Z = G \cdot Z^U$. In fact, the latter property characterizes horospherical G-schemes:

Proposition 4.11. *An affine* G*-scheme* Z *is horospherical if and only if* $Z = G \cdot Z^U$ *(as schemes).*

Proof. First, note that the closed subscheme $Z^U \subset Z$ is stable under the Borel subgroup B; it follows that $G \cdot Z^U$ is closed in Z for an arbitrary G-scheme Z. Indeed, the morphism

$$\varphi : G \times Z^U \longrightarrow Z, \quad (g, z) \longmapsto g \cdot z \tag{4.12}$$

factors as the morphism

$$\psi : G \times Z^U \longrightarrow G/B \times Z, \quad (g, z) \longmapsto (gB, g \cdot z)$$

followed by the projection $G/B \times Z \to Z$. The latter morphism is proper, since G/B is complete; moreover, ψ is easily seen to be a closed immersion.

Also, note that the ideal of $G \cdot Z^U$ in $A = \mathcal{O}(Z)$ is the intersection of the G-translates of the ideal I of Z^U. Thus, $Z = G \cdot Z^U$ if and only if I contains no non-zero simple G-submodule of A. Moreover, the ideal I is generated by the $g \cdot f - f$ where $g \in U$ and $f \in A$.

We now assume that Z is horospherical. Consider a simple G-submodule $V(\lambda) \subset A$. Then $V(\lambda)$ admits a unique lowest weight (with respect to the partial ordering $\leqslant$), equal to $-\lambda^*$, and the corresponding eigenspace is a line. Moreover, the span of the $g \cdot v - v$, where $g \in U$ and $v \in V(\lambda)$, is just the sum of all other T-eigenspaces; we denote that span by $V(\lambda)_+$. Since the product $V(\lambda) \cdot V(\mu)$, where $V(\mu)$ is some other simple submodule of A, is either 0 or the Cartan product $V(\lambda+\mu)$, we see that

$$V(\lambda)_+ \cdot V(\mu) \subset V(\lambda + \mu)_+.$$

Thus, the sum of the $V(\lambda)_+$ over all simple submodules is an ideal of A, and hence equals I. In particular, I contains no non-zero simple G-submodule of A.

Conversely, assume that $Z = G \cdot Z^U$, i.e., the morphism (4.12) is surjective. Note that φ is invariant under the action of U via $u \cdot (g, z) = (gu^{-1}, z)$, and also equivariant for the action of G on $G \times Z^U$ via left multiplication on G, and for the given action on Z. Thus, φ yields an inclusion of G-algebras

$$\mathcal{O}(Z) \hookrightarrow \mathcal{O}(G)^U \otimes_k \mathcal{O}(Z^U)$$

where G acts on the right-hand side through its action on $\mathcal{O}(G)^U$ via left multiplication. But $\mathcal{O}(G)^U$ is also a T-algebra via right multiplication; this action commutes with that of G, and we have an isomorphism of $G \times T$-modules

$$\mathcal{O}(G)^U \cong \bigoplus_{\lambda \in \Lambda^+} V(\lambda)^*$$

where G acts naturally on each $V(\lambda)^*$, and T acts via its character λ (see e.g. [20, Section 2.1]). In particular, the isotypical decomposition of $\mathcal{O}(G)^U$ is a grading; thus, the same holds for $\mathcal{O}(G)^U \otimes_k \mathcal{O}(Z^U)$ and for $\mathcal{O}(Z)$. □

Next, we relate the preceding constructions to the invariant Hilbert scheme of a finite-dimensional G-module V. Here it should be noted that the full horospherical family of a closed G-subscheme $Z \subset V$ need not be a family of closed G-subschemes of $V \times \mathbb{A}^r$ (see e.g. Example 4.13). Yet *the pull-back of the horospherical family to* T_{ad} *may be identified with a family of closed* G*-subschemes of* $V \times T_{ad}$ as follows.

By 4.9, we have an isomorphism of $G \times T$-algebras

$$\mathcal{O}(V \times T)^{Z(G)} \cong \mathcal{R}(\mathcal{O}(V))[t_1^{-1}, \dots, t_r^{-1}]$$

over $\mathcal{O}(T_{ad}) \cong k[t_1^{\pm 1}, \dots, t_r^{\pm 1}]$. Also, each isotypical component $V_{(\lambda)} = m_\lambda V(\lambda)$ yields a subspace

$$m_\lambda V(\lambda)^* e^{\lambda^*} \subset \mathcal{O}(V \times T)^{Z(G)},$$

stable under the action of $G \times T$. Moreover, all these subspaces generate a $G \times T$-subalgebra of $\mathcal{O}(V \times T)^{Z(G)}$, equivariantly isomorphic to $\mathcal{O}(V)$ where V is a G-module via the given action, and T acts linearly on V so that each $V_{(\lambda)}$ is an eigenspace of weight $-\lambda^*$. Finally, we have an isomorphism

$$\mathcal{O}(V \times T)^{Z(G)} \cong \mathcal{O}(V) \otimes_k \mathcal{O}(T_{ad})$$

of $G \times T$-algebras over $\mathcal{O}(T_{ad})$, which translates into an isomorphism

$$p^{-1}(T_{ad}) \cong V \times T_{ad}$$

of families of G-schemes over T_{ad}. (In geometric terms, we have trivialized the homogeneous vector bundle $V \times^{Z(G)} T \to T/Z(G)$ by extending the $Z(G)$-action on V to a T-action commuting with that of G).

This construction extends readily to the setting of families, i.e., given a family of closed G-subschemes $\mathcal{Z} \subset V \times S$, we obtain a family of closed G-subschemes $\mathcal{W} \subset V \times T_{ad} \times S$. By arguing as in the proof of Proposition 3.14, this defines an *action of* T_{ad} *on the invariant Hilbert scheme* $\mathrm{Hilb}_h^G(V)$.

In fact, *this action arises from the linear* T*-action on* V *for which each* $V_{(\lambda)}$ *has weight* $-\lambda^*$: since $\lambda + \lambda^*$ is in the root lattice for any $\lambda \in \Lambda$, the induced action of the center $Z(G) \subset T$ coincides with its action as a subgroup of G, so that $Z(G)$ acts trivially on $\mathrm{Hilb}_h^G(V)$.

Example 4.13. Let $G = SL_2$ and $Z = G \cdot xy \subset V(2)$. Then $Z = (\Delta = 1)$ is a closed G-subvariety of $V(2)$ with Hilbert function h_2. One checks that the G-submodules $\mathcal{O}(Z)_{\leqslant 2n}$ are just the restrictions to Z of the spaces of polynomial functions on $V(2)$ with degree $\leqslant n$. Moreover, $Z_0 = \overline{G \cdot y^2}$ and the horospherical family is that of Example 2.19(ii).

Likewise, if $Z = G \cdot x^2y^2 \subset V(4)$ then $Z_0 = \overline{G \cdot y^4}$ and the horospherical family is again that of Example 2.19(ii).

Also, $V(1)$ is its own horospherical degeneration, but the horospherical degeneration of $V(2)$ is the singular hypersurface $\{(z,t) \in V(2) \oplus V(0) \mid \Delta(z) = 0\}$.

4.3. Moduli of multiplicity-free varieties with prescribed weight monoid

In this subsection, we still consider a connected reductive group G, and fix a finitely generated submonoid $\Gamma \subset \Lambda^+$. We will construct a moduli space for irreducible multiplicity-free G-varieties Z with weight monoid Γ or equivalently, with Hilbert function $h = h_\Gamma$ (2.24). Recall that $Z//U$ is an irreducible multiplicity-free T-variety with weight monoid Γ, and hence is isomorphic to $Y := \mathrm{Spec}(k[\Gamma])$.

We begin by constructing a special such variety Z_0. Choose generators $\lambda_1, \dots, \lambda_N$ of the monoid Γ. Consider the G-module

$$V := V(\lambda_1)^* \oplus \cdots \oplus V(\lambda_N)^*,$$

the sum of highest weight vectors

$$v := v_{\lambda_1^*} + \cdots + v_{\lambda_N^*},$$

and define

$$Z_0 := \overline{G \cdot v} \subset V.$$

Since v is fixed by U, the irreducible variety Z_0 is horospherical in view of Proposition 4.11. The Λ-graded algebra $\mathcal{O}(Z_0)$ is generated by $V(\lambda_1), \dots, V(\lambda_N)$; thus, $\Lambda^+(Z_0) = \Gamma$. This yields a special algebra structure on the G-module

$$V(\Gamma) := \bigoplus_{\lambda \in \Gamma} V(\lambda)$$

such that the subalgebra $V(\Gamma)^U$ is isomorphic to $\mathcal{O}(Y) = k[\Gamma]$.

Each irreducible multiplicity-free variety Z with weight monoid Γ satisfies $\mathcal{O}(Z) \cong V(\Gamma)$ as G-modules and $\mathcal{O}(Z)^U \cong V(\Gamma)^U$ as T-algebras. This motivates the following:

Definition 4.14. A family of algebra structures of type Γ over a scheme S consists of the structure of an $\mathcal{O}_S$-G-algebra on $V(\Gamma) \otimes_k \mathcal{O}_S$ that extends the given T-algebra structure on $V(\Gamma)^U \cong k[\Gamma]$.

In other words, a family of algebra structures of type Γ over S is a multiplication law m on $V(\Gamma) \otimes_k \mathcal{O}_S$ which makes it an $\mathcal{O}_S$-G-algebra and restricts to the multiplication of the $\mathcal{O}_S$-T-algebra $V(\Gamma)^U \otimes_k \mathcal{O}_S$. We may write

$$m = \sum_{\lambda,\mu,\nu \in \Gamma} m_{\lambda,\mu}^{\nu} \qquad (4.15)$$

where each component

$$m_{\lambda,\mu}^{\nu} : (V(\lambda) \otimes_k \mathcal{O}_S) \otimes_{\mathcal{O}_S} (V(\mu) \otimes_k \mathcal{O}_S) \longrightarrow V(\nu) \otimes_k \mathcal{O}_S$$

is an $\mathcal{O}_S$-G-morphism. Moreover, the commutativity of m and its compatibility with the multiplication on $V(\Gamma)^U \otimes_k \mathcal{O}_S$ translate into linear relations between the $m_{\lambda,\mu}^{\nu}$'s, while the associativity translates into quadratic relations. Also, each $m_{\lambda,\mu}^{\nu}$ may be viewed as a linear map

$$\mathrm{Hom}^G(V(\lambda) \otimes_k V(\mu), V(\nu)) \longrightarrow H^0(S, \mathcal{O}_S)$$

or equivalently, as a morphism of schemes

$$S \longrightarrow \mathrm{Hom}^G(V(\nu), V(\lambda) \otimes_k V(\mu)),$$

and the polynomial relations between the $m_{\lambda,\mu}^{\nu}$'s are equivalent to polynomial relations between these morphisms. It follows that *the functor* M_Γ *that associates with any scheme* S, *the set of families of algebra structures of type* Γ *over* S, *is represented by an affine scheme* M_Γ, a closed subscheme of an infinite-dimensional affine space. Also, note that $m_{\lambda,\mu}^{\nu} = 0$ unless $\nu \leqslant \lambda + \mu$.

In fact, *the scheme* M_Γ *is of finite type*; yet one does not know how to obtain this directly from the preceding algebraic description. This finiteness property will rather be derived from a relation of M_Γ to the invariant Hilbert scheme $\mathrm{Hilb}_h^G(V)$ that we now present.

Given a family of algebra structures of type Γ over S, the inclusion of G-modules $V^* \subset V(\Gamma)$ yields a homomorphism of $\mathcal{O}_S$-G-algebras

$$\varphi : \mathrm{Sym}(V^*) \otimes_k \mathcal{O}_S \longrightarrow V(\Gamma) \otimes_k \mathcal{O}_S.$$

Moreover, the $\mathcal{O}_S$-T-algebra $V(\Gamma)^U \otimes_k \mathcal{O}_S$ is generated by the images of the highest weight lines $V(\lambda_1)^U, \ldots, V(\lambda_N)^U \subset (V^*)^U$. In particular, the restriction

$$\mathrm{Sym}\big((V^*)^U\big) \otimes_k \mathcal{O}_S \longrightarrow V(\Gamma)^U \otimes_k \mathcal{O}_S$$

is surjective; thus, φ is surjective as well. This defines a family of closed G-subschemes $\mathcal{Z} \subset V \times S$ with Hilbert function h, such that the sheaf of $\mathcal{O}_S$-algebras $(p_*\mathcal{O}_\mathcal{Z})^U$ is generated by the preceding highest weight lines. Choosing highest weight vectors $v_{\lambda_1}, \ldots, v_{\lambda_N}$, we obtain a surjective homomorphism of $\mathcal{O}_S$-T-algebras $\mathcal{O}_S[t_1, \ldots, t_N] \to (p_*\mathcal{O}_\mathcal{Z})^U$ that maps each t_i to v_{λ_i}. Equivalently, we obtain a closed immersion $\mathcal{Z}/\!/U \hookrightarrow \mathbb{A}^N \times S$ of families of closed T-subschemes with Hilbert function h, where T acts linearly on $\mathbb{A}^N$ with weights $-\lambda_1, \ldots, -\lambda_N$. This is also equivalent to a morphism

$$f : S \to \mathrm{Hilb}^T(\underline{\lambda})$$

where the target is the toric Hilbert scheme of Example 2.23(i). Now the condition that our family of algebra structures extends the given algebra structure on $V(\Gamma)^U$ means that f maps S to the closed point $Z_0/\!/U$, viewed as a general T-orbit closure in $\mathbb{A}^N$.

Conversely, given a family of closed G-subschemes $\mathcal{Z} \subset V \times S$ with Hilbert function h such that $(p_*\mathcal{O}_\mathcal{Z})^U$ is generated by $v_{\lambda_1}, \ldots, v_{\lambda_N}$ and the resulting morphism f maps S to the point $Z_0/\!/U$, we obtain an isomorphism of $\mathcal{O}_S$-G-modules $\mathcal{O}_S \otimes_k V(\Gamma) \cong p_*\mathcal{O}_\mathcal{Z}$ which restricts to an isomorphism of $\mathcal{O}_S$-T-algebras $\mathcal{O}_S \otimes_k V(\Gamma)^U \cong (p_*\mathcal{O}_\mathcal{Z})^U$. This translates into a family of algebra structures of type Γ over S.

Summarizing, we have the following link between algebra structures and invariant Hilbert schemes:

Theorem 4.16. *With the preceding notation and assumptions, there exists an open subscheme*

$$\mathrm{Hilb}_h^G(V)_0 \subset \mathrm{Hilb}_h^G(V)$$

that parametrizes those families $\mathcal{Z}$ such that the $\mathcal{O}_S$-algebra $(p_*\mathcal{O}_\mathcal{Z})^U$ is generated by the image of $(V^*)^U$ under φ. Moreover, there exists a morphism

$$f : \mathrm{Hilb}_h^G(V)_0 \longrightarrow \mathrm{Hilb}^T(\underline{\lambda})$$

that sends $\mathcal{Z}$ to $\mathcal{Z}//U$. The fiber of f at the closed point $Z_0//U \in \mathrm{Hilb}^T(\underline{\lambda})$ represents the functor $\mathcal{M}_\Gamma$.

We denote the fiber of f at $Z_0//U$ by M_Γ and call it the **moduli scheme of multiplicity-free varieties with weight monoid** Γ. Since it represents the functor $\mathcal{M}_\Gamma$, the scheme M_Γ is independent of the choices of generators of Γ and of highest weight vectors. It comes with a special point Z_0, the common horospherical degeneration to all of its closed points.

We may also consider the preimage in $\mathrm{Hilb}_h^G(V)_0$ of the open $(\mathbb{G}_m)^N$-orbit $\mathrm{Hilb}_{\underline{\lambda}}^T \subset \mathrm{Hilb}^T(\underline{\lambda})$ that consists of general T-orbit closures. This preimage is an open subscheme of $\mathrm{Hilb}_h^G(V)_0$, that we denote by $\mathrm{Hilb}_{\underline{\lambda}}^G$. Its closed points are exactly the irreducible multiplicity-free varieties $Z \subset V$ having weight monoid Γ and such that the projections $Z \to V(\lambda_1)^*, \ldots, V(\lambda_N)^*$ are all non-zero; equivalently, the horospherical degeneration Z_0 is contained in V. Such varieties Z are called **non-degenerate**.

Next, we relate these constructions to the action of the adjoint torus T_{ad} on $\mathrm{Hilb}_h^G(V)$, defined in the previous subsection. The torus $(\mathbb{G}_m)^N$ acts on $\mathrm{Hilb}_h^G(V)$ as the equivariant automorphism group of the G-module V. This action stabilizes the open subschemes $\mathrm{Hilb}_h^G(V)_0$ and $\mathrm{Hilb}_{\underline{\lambda}}^G$; moreover, f is equivariant for the natural action of $(\mathbb{G}_m)^N$ on the toric Hilbert scheme. Also, note that the $(\mathbb{G}_m)^N$-orbit $\mathrm{Hilb}_{\underline{\lambda}}^T$ is isomorphic to $(\mathbb{G}_m)^N/\underline{\lambda}(T)$ where $\underline{\lambda}$ denotes the homomorphism (3.13). This yields an action of T on M_Γ and one checks that the center $Z(G)$ acts trivially. Thus, *T_{ad} acts on M_Γ and each $(\mathbb{G}_m)^N$-orbit in $\mathrm{Hilb}_{\underline{\lambda}}^G$ intersects M_Γ along a unique T_{ad}-orbit.*

Given a family $\mathcal{Z}$ of non-degenerate subvarieties of V, one shows that the associated horospherical family $\mathcal{X}$ is a family of non-degenerate subvarieties of V as well. It follows that *the T_{ad}-action on M_Γ extends to an $\mathbb{A}^r$-action such that the origin of $\mathbb{A}^r$ acts via the constant map to Z_0*. In particular, Z_0 is the unique T_{ad}-fixed point and is contained in each T_{ad}-orbit closure; thus, *the scheme M_Γ is connected.*

The T_{ad}-action on M_Γ may also be seen in terms of multiplication laws (4.15): by [3, Proposition 2.11], *each non-zero component $m_{\lambda,\mu}^\nu$ is a T_{ad}-eigenvector of weight* $\lambda+\mu-\nu$ (note that $\lambda+\mu-\nu$ lies in the root lattice, and hence is indeed a character of T_{ad}).

As a consequence, given an irreducible multiplicity-free variety Z with weight monoid Γ, the T_{ad}-orbit closure of Z (viewed as a closed point of M_Γ) has for weight monoid the submonoid $\Sigma_Z \subset \Lambda$ generated by the weights $\lambda+\mu-\nu$ where $V(\nu)$ is contained in the product $V(\lambda) \cdot V(\mu) \subset \mathcal{O}(Z)$. In particular, *the monoid Σ_Z is finitely generated.* Again, it is not known how to obtain this result directly from the algebraic definition of the **root monoid** Σ_Z.

In view of a deep theorem of Knop (see [33, Theorem 1.3]), the saturation of the monoid Σ_Z is free, i.e., generated (as a monoid) by linearly independent elements. Equivalently, *the normalization of each* T_{ad}*-orbit closure in* M_Γ *is equivariantly isomorphic to an affine space on which* T_{ad} *acts linearly.*

We also mention a simple relation between the Zariski tangent space to M_Γ at a closed point Z and the space $T^1(Z)$ parametrizing first-order deformations of Z: namely, *the normal space to the orbit* $T_{ad} \cdot Z$ *in* M_Γ *is isomorphic to the* G*-invariant subspace* $T^1(Z)^G$ (see [3, Proposition 1.13]). In particular,

$$T_{Z_0}M_\Gamma = T^1(Z_0)^G$$

as Z_0 is fixed by T_{ad}.

In fact, many results of this subsection hold in the more general setting where the algebra $k[\Gamma]$ is replaced with an arbitrary T-algebra of finite type; see [3]. The multiplicity-free case presents remarkable special features; namely, finiteness properties that will be surveyed in the next subsection.

Example 4.17. If the monoid Γ is free, then of course we choose $\lambda_1, \dots, \lambda_N$ to be its minimal generators. Since they are linearly independent, $\mathrm{Hilb}^T(\underline{\lambda})$ is a (reduced) point and hence

$$\mathrm{Hilb}^G_h(V)_0 = \mathrm{Hilb}^G_{\underline{\lambda}} = M_\Gamma.$$

Also, since the homomorphism $\underline{\lambda}$ is surjective, each $(\mathbb{G}_m)^N$-orbit in $\mathrm{Hilb}^G_{\underline{\lambda}}$ is a unique T_{ad}-orbit. The pull-back $\pi : \mathrm{Univ}_\Gamma \to M_\Gamma$ of the universal family of $\mathrm{Hilb}^G_h(V)$ may be viewed as the universal family of non-degenerate spherical subvarieties of V.

4.4. Finiteness properties of spherical varieties

In this subsection, we survey finiteness and uniqueness results relative to the structure and classification of spherical varieties. We still denote by G a connected reductive group; we fix a Borel subgroup $B \subset G$ and a maximal torus $T \subset B$.

Recall that a (possibly non-affine) G-variety X is spherical, if X is normal and contains an open B-orbit; in particular, X contains an open G-orbit X_0. Choosing a base point $x \in X_0$ and denoting by H its isotropy group, we may identify X_0 with the homogeneous space G/H. We say that H is a **spherical subgroup** of G, and the pair (X, x) is an **equivariant embedding** of G/H; the complement $X \setminus X_0$ is called the **boundary**. Morphisms of embeddings are defined as those equivariant morphisms that preserve base points. If the variety X is complete, then X is called an **equivariant completion** (or equivariant compactification) of G/H.

One can show that *any spherical* G*-variety contains only finitely many* B*-orbits*; as a consequence, *any equivariant embedding of a spherical* G*-homogeneous space contains only finitely many* G*-orbits*. Conversely, if a G-homogeneous space X_0 satisfies the property that all of its equivariant embeddings contain finitely many orbits, then X_0 is spherical.

Spherical homogeneous spaces admit a further remarkable characterization, in terms of the existence of equivariant completions with nice geometric properties. Specifically, consider an embedding X of a homogeneous space $X_0 = G/H$. Assume that X is smooth and $X \setminus X_0$ is a union of smooth prime divisors that intersect transversally; in other words, the boundary is a **divisor with simple normal crossings**. Then we may consider the associated **sheaf of logarithmic vector fields**, consisting of those derivations of $\mathcal{O}_X$ that preserve the ideal sheaf of $D := X \setminus X_0$. This subsheaf, denoted by $T_X(-\log D)$, is a locally free subsheaf of the tangent sheaf T_X of all derivations of $\mathcal{O}_X$; both sheaves coincide along X_0. The **logarithmic tangent bundle** is the vector bundle on X associated with $T_X(-\log D)$. The G-action on (X, D) yields an action of the Lie algebra $\mathfrak{g}$ by derivations that preserve D, i.e., a homomorphism of Lie algebras

$$\mathfrak{g} \longrightarrow H^0(X, T_X(-\log D)).$$

We say that the pair (X, D) is **log homogeneous under** G if $\mathfrak{g}$ generates the sheaf of logarithmic vector fields. Now *any complete log homogeneous* G*-variety is spherical; moreover, any spherical* G*-homogeneous space admits a log homogeneous equivariant completion* (as follows from [6, Sections 2.2, 2.5]; see [19] for further developments on log homogeneous varieties and their relation to spherical varieties). We will need a stronger version of part of this result:

Lemma 4.18. *Let* X *be a smooth spherical* G*-variety with boundary a divisor* D *with simple normal crossings and denote by* S_X *the subsheaf of* $T_X(-\log D)$ *generated by* $\mathfrak{g}$. *If* S_X *is locally free, then* $S_X = T_X(-\log D)$. *In particular,* X *is log homogeneous.*

Proof. Clearly, S_X and $T_X(-\log D)$ coincide along the open G-orbit. Since these sheaves are locally free, the support of the quotient $T_X(-\log D)/S_X$ has pure codimension 1 in X. But this support is G-stable, and contains no G-orbit of codimension 1 by [6, Section 2]. Thus, this support is empty; this yields our assertion. □

Log homogeneous pairs satisfy an important rigidity property, namely,

$$H^1(X, T_X(-\log D)) = 0 \qquad (4.19)$$

whenever X is complete (as follows from a vanishing theorem due to Knop, see [32, Theorem 4.1]). This is the main ingredient for proving the following finiteness result ([3, Theorem 3.1]):

Theorem 4.20. *For any* G*-variety* X*, only finitely many conjugacy classes of spherical subgroups of* G *occur as isotropy groups of points of* X.

In other words, *any* G*-variety contains only finitely many isomorphism classes of spherical* G*-orbits*.

For the proof, one reduces by general arguments of algebraic transformation groups to the case where X is irreducible and admits a geometric quotient

$$p : X \longrightarrow S$$

where the fibers of p are exactly the G-orbits. Arguing by induction on the dimension, we may replace S with an open subset; thus, we may further assume that X and the morphism p are smooth. Then the spherical G-orbits form an open subset of X, since the same holds for the B-orbits of dimension $\dim(X) - \dim(S)$. So we may assume that all fibers are spherical. Now, by general arguments of algebraic transformation groups again, there exists an equivariant fiberwise completion of X, i.e., a G-variety $\bar{X}$ equipped with a proper G-invariant morphism

$$\bar{p} : \bar{X} \longrightarrow S$$

such that $\bar{X}$ contains X as a G-stable open subset, and $\bar{p}$ extends p. We may further perform equivariant blow-ups and hence assume that $\bar{X}$ is smooth, the boundary $\bar{X} \setminus X$ is a divisor with simple normal crossings, and the subsheaf $S_{\bar{X}} \subset T_{\bar{X}}$ generated by $\mathfrak{g}$ is locally free. By Lemma 4.18, it follows that $\bar{X}$ is a family of log homogeneous varieties (possibly after shrinking S again). Now the desired statement is a consequence of rigidity (4.19) together with arguments of deformation theory; see [3, pp. 113–115] for details.

As a direct consequence of Theorem 4.20, *any finite-dimensional* G*-module* V *contains only finitely many closures of spherical* G*-orbits, up to the action of* $GL(V)^G$ (see [3, p. 116]). In view of the results of Subsection 4.3, it follows that *every moduli scheme* M_Γ *contains only finitely many* T_{ad}*-orbits*. In particular, *up to equivariant isomorphism, there are only finitely many affine spherical varieties having a prescribed weight monoid.*

This suggests that spherical varieties may be classified by combinatorial invariants. Before presenting a number of results in this direction, we associate three such invariants to a spherical homogeneous space $X_0 = G/H$.

The first invariant is the set of weights of B-eigenvectors in the field of rational functions $k(X_0) = k(G)^H$; this is a subgroup of Λ, denoted by $\Lambda(X_0)$ and called the **weight lattice** of X_0. The rank of this lattice is called the **rank** of X_0 and denoted by $\mathrm{rk}(X_0)$. Note that any B-eigenfunction is determined by its weight up to a non-zero scalar, since $k(X_0)^B = k$ as X_0 contains an open B-orbit.

The second invariant is the set $\mathcal{V}(X_0)$ of those discrete valuations of the field $k(X_0)$, with values in the field $\mathbb{Q}$ of rational numbers, that are invariant under the natural G-action. One can show that any such valuation is uniquely determined by its restriction to B-eigenvectors; moreover, this identifies $\mathcal{V}(X_0)$ to a convex polyhedral cone in the $\mathbb{Q}$-vector space $\mathrm{Hom}(\Lambda(X_0), \mathbb{Q})$. Thus, $\mathcal{V}(X_0)$ is called the **valuation cone**.

The third invariant is the set $\mathcal{D}(X_0)$ of B-stable prime divisors in X_0, called **colors**; these are exactly the irreducible components of the complement of the open B-orbit. Any $D \in \mathcal{D}(X_0)$ defines a discrete valuation of $k(X_0)$, and hence (by restriction to B-eigenvectors) a point $\varphi_D \in \mathrm{Hom}(\Lambda(X_0), \mathbb{Q})$. Moreover, the stabilizer of D in G is a parabolic subgroup G_D containing B, and hence corresponds to a set of simple roots. Thus, $\mathcal{D}(X_0)$ may be viewed as an abstract finite set equipped with maps $D \mapsto \varphi_D$ to $\mathrm{Hom}(\Lambda(X_0), \mathbb{Q})$ and $D \mapsto G_D$ to subsets of simple roots.

The invariants $\Lambda(X_0)$, $\mathcal{V}(X_0)$, $\mathcal{D}(X_0)$ are the main ingredients of a *classification of all equivariant embeddings of* X_0, due to Luna and Vust (see [51] for an overview, and [31] for an exposition; the original article [43] addresses embeddings of arbitrary homogeneous spaces, see also [56]). This classification is formulated in terms of **colored fans**, combinatorial objects that generalize the fans of toric geometry. Indeed, toric varieties are exactly the equivariant embeddings of a torus T viewed as a homogeneous space under itself. In that case, $\Lambda(T)$ is just the character lattice, and the set $\mathcal{D}(T)$ is empty; one shows that $\mathcal{V}(T)$ is the whole space $\mathrm{Hom}(\Lambda(T), \mathbb{Q})$.

Another important result, due to Losev (see [36, Theorem 1]), asserts that *any spherical homogeneous space is uniquely determined by its weight lattice, valuation cone and colors, up to equivariant isomorphism*. The proof combines many methods, partial classifications, and earlier results, including the Luna-Vust classification and the finiteness theorem 4.20.

Returning to an affine spherical variety Z, one can show that the valuation cone of the open G-orbit Z_0 is dual (in the sense of convex geometry) to the cone generated by the root monoid Σ_Z. Also, recall that the saturation of Σ_Z is a free monoid; its generators are called the **spherical roots** of Z. By another uniqueness result of Losev (see [37, Theorem 1.2]), *any affine spherical G-variety is uniquely determined by its weight monoid and spherical roots, up to equivariant isomorphism*. Moreover, *any smooth affine spherical G-variety is uniquely determined by its weight monoid*, again up to equivariant isomorphism ([37, Theorem 1.3]).

Note that all smooth affine spherical varieties are classified in [34]; yet one does not know how to deduce the preceding uniqueness result (a former conjecture of Knop) from that classification.

Example 4.21. The spherical subgroups of $G = SL_2$ are exactly the closed subgroups of positive dimension. Here is the list of these subgroups up to conjugation in G:

(i) $H = B$ (the Borel subgroup of upper triangular matrices of determinant 1). Then $G/H \cong \mathbb{P}^1$ has rank 0 and a unique color, the B-fixed point ∞.

(ii) $H = U\mu_n$ where U denotes the unipotent part of B, and μ_n the group of diagonal matrices with eigenvalues ζ, ζ^{-1} where $\zeta^n = 1$; here n is a positive integer. Then $H \subset B$ and via the resulting map $G/H \to G/B$, we see that G/H is the total space of the line bundle $\mathcal{O}_{\mathbb{P}^1}(n)$ minus the zero section. Moreover, G/H has rank 1 and a unique color, the fiber at ∞ of the projection to $\mathbb{P}^1$. We have $\Lambda(G/H) = n\mathbb{Z} \subset \mathbb{Z} = \Lambda$, and the valuation cone is the whole space $\mathrm{Hom}(\Lambda(G/H), \mathbb{Q}) \cong \mathbb{Q}$.

A smooth equivariant completion of G/H is $\mathbb{P}(\mathcal{O}_{\mathbb{P}^1} \oplus \mathcal{O}_{\mathbb{P}^1}(n))$, the rational ruled surface of index n. By contracting the unique curve of negative self-intersection, i.e., the section of self-intersection $-n$, we obtain another equivariant completion which is singular if $n \neq 1$, and isomorphic to $\mathbb{P}(V(1) \oplus V(0)) \cong \mathbb{P}^2$ if $n = 1$.

(iii) $H = T$ (the torus of diagonal matrices of determinant 1). Then $G/H \cong G \cdot xy \subset V(2)$ is the affine quadric ($\Delta = 1$); it has rank 1 and weight lattice $2\mathbb{Z}$. Note that H is the intersection of B with the opposite Borel subgroup B^- (of lower triangular matrices). Thus, G/H admits $\mathbb{P}^1 \times \mathbb{P}^1$ as an equivariant completion via the natural morphism $G/H \to G/B \times G/B^-$; this is in fact the unique non-trivial equivariant embedding. Also, G/H has two colors D^+, D^- (the fibers at ∞ of the two morphisms to $\mathbb{P}^1$). The valuation cone is the negative half-line in $\mathrm{Hom}(\Lambda(G/H), \mathbb{Q}) \cong \mathbb{Q}$, and D^+, D^- are mapped to the same point of the positive half-line.

(iv) $H = N_G(T)$ (the normalizer of T in G). Then $G/H \cong G \cdot [xy] \subset \mathbb{P}(V(2))$ is the open affine subset ($\Delta = 0$) in the projective plane $\mathbb{P}(V(2))$, which is the unique non-trivial embedding. Moreover, G/H has rank 1 and weight lattice $4\mathbb{Z}$. There is a unique color D, with closure the projective line $\mathbb{P}(yV(1)) \subset \mathbb{P}(V(2))$. The valuation cone is again the negative half-line in $\mathrm{Hom}(\Lambda(G/H), \mathbb{Q}) \cong \mathbb{Q}$, and D is mapped to a point of the positive half-line.

4.5. Towards a classification of wonderful varieties

In this subsection, we introduce the class of wonderful varieties, which play an essential role in the structure of spherical varieties. Then we present a number of recent works that classify wonderful varieties (possibly with additional assumptions) via Lie-theoretical or geometric methods.

Definition 4.22. A variety X is called **wonderful** if it satisfies the following properties:

(1) X is smooth and projective.
(2) X is equipped with an action of a connected reductive group G having an open orbit X_0.
(3) The boundary $D := X \setminus X_0$ is a divisor with simple normal crossings, and its irreducible components $D_1, \dots, D_r$ meet.
(4) The G-orbit closures are exactly the partial intersections of $D_1, \dots, D_r$.

Then $D_1, \dots, D_r$ are called the *boundary components*; their number r is the *rank* of X. By (4), X has a unique closed orbit

$$Y := D_1 \cap \cdots \cap D_r.$$

The wonderful G-varieties of rank 0 are just the complete G-homogeneous varieties, i.e., the homogeneous spaces G/P where $P \subset G$ is a parabolic subgroup containing B. Those of rank 1 are exactly the smooth equivariant completions of a homogeneous space by a homogeneous divisor; they have been classified by Akhiezer (see [1]) and they turn out to be spherical. The latter property extends to all ranks: in fact, *the wonderful varieties are exactly the complete log homogeneous varieties having a unique closed orbit*, as follows from [40] combined with [6, Propositions 2.2.1 and

2.5]. Moreover, the rank of a wonderful variety coincides with the rank of its open G-orbit.

The combinatorial invariants $\Lambda(X_0)$, $\mathcal{V}(X_0)$, $\mathcal{D}(X_0)$ associated with the open G-orbit of a wonderful variety X admit simple geometric interpretations. To state them, let X_1 denote the complement in X of the union of the closures of the colors. Then X_1 is an open B-stable subset of X. One shows that X_1 is isomorphic to an affine space and meets each G-orbit in X along its open B-orbit; it easily follows that *the closure in* X *of each color* $D \in \mathcal{D}(X_0)$ *is a base-point-free divisor, and these divisors form a basis of the Picard group of* X. In particular, we have equalities in Pic(X):

$$D_i = \sum_{D \in \mathcal{D}} c_{D,i} D \quad (i = 1, \ldots, r) \tag{4.23}$$

where the $c_{D,i}$ are uniquely determined integers. Also, $X_1 \cap Y$ is the open Bruhat cell in Y, and hence equals $B \cdot y$ for a unique T-fixed point $y \in Y$. Thus, T acts in the normal space $T_y(X)/T_y(Y)$ of dimension $r = \mathrm{rk}(X)$; one shows that the corresponding weights $\sigma_1, \ldots, \sigma_r$ (called the **spherical roots** of the wonderful variety X) are linearly independent. Now *the spherical roots form a basis of* $\Lambda(X_0)$, *and generate the dual cone to* $\mathcal{V}(X_0)$. The dual basis of $\mathrm{Hom}(\Lambda(X_0), \mathbb{Q})$ consists of the opposites of the valuations $\nu_1, \ldots, \nu_r$ associated with the boundary divisors. Moreover, (4.23) implies that the map $\varphi : \mathcal{D} \to \mathrm{Hom}(\Lambda(X_0), \mathbb{Q})$ is given by

$$\varphi(\nu_D) = \sum_{i=1}^{r} c_{D,i} \nu_i \quad (D \in \mathcal{D}).$$

To each spherical homogeneous space $X_0 = G/H$, one associates a wonderful variety as follows. Denote by $N_G(H)$ the normaliser of H in G, so that the quotient group $N_G(H)/H$ is isomorphic to the equivariant automorphism group $\mathrm{Aut}^G(X_0)$. Since X_0 is spherical, the algebraic group $N_G(H)/H$ is diagonalizable; moreover, $N_G(H)$ equals the normalizer of the Lie algebra $\mathfrak{h}$. Thus, the homogeneous space $G/N_G(H)$ is the G-orbit of $\mathfrak{h}$ viewed as a point of the Grassmannian variety $\mathrm{Gr}(\mathfrak{g})$ of subspaces of $\mathfrak{g}$ (or alternatively, of the scheme of Lie subalgebras of $\mathfrak{g}$). The orbit closure

$$X := \overline{G \cdot \mathfrak{h}} \subset \mathrm{Gr}(\mathfrak{g})$$

is a projective equivariant completion of $G/N_G(H)$, called the *Demazure embedding* of that homogeneous space. In fact, *the variety* X *is wonderful* by a result of Losev (see [38]) based on earlier results of several mathematicians, including Demazure and Knop (see [33, Corollary 7.2]). Moreover, by embedding theory of spherical homogeneous spaces, *the log homogeneous embeddings of* G/H *are exactly those smooth equivariant embeddings that admit a morphism to* X; *then the logarithmic tangent bundle is the pull-back of the tautological quotient bundle on* $\mathrm{Gr}(\mathfrak{g})$. Also, by embedding theory again, *a complete log homogeneous variety* X' *is wonderful if and only if the morphism* $X' \to X$ *is finite.*

It follows that every spherical homogeneous space G/H such that $H = N_G(H)$ admits a wonderful equivariant completion; in the converse direction, if G/H admits such a completion X, then X is unique, and the quotient $N_G(H)/H$ is finite. In particular, the center of G acts on X via a finite quotient; thus, one can assume that G is semi-simple when considering wonderful G-varieties.

Since the G-variety $\mathrm{Gr}(\mathfrak{g})$ contains only finitely many isomorphism classes of spherical G-orbits, and any G-homogeneous space admits only finitely many finite equivariant coverings, we see that *the number of isomorphism classes of wonderful G-varieties is finite* (for a given group G). Also, note that the wonderful varieties are exactly those log homogeneous varieties that are **log Fano**, i.e., the determinant of the logarithmic tangent sheaf is ample.

To classify wonderful G-varieties, it suffices to characterize those triples $(\Lambda, \mathcal{V}, \mathcal{D})$ that occur as combinatorial invariants of their open G-orbits, in view of Losev's uniqueness result. In fact, part of the information contained in such triples is more conveniently encoded by abstract combinatorial objects called **spherical systems**. These were introduced by Luna, who obtained a complete classification of wonderful G-varieties for those groups G of type A, in terms of spherical systems only. For an arbitrary group G, Luna also showed how to reduce the classification of spherical G-homogeneous spaces to that of wonderful G-varieties, and he conjectured that *wonderful varieties are classified by spherical systems* (see [41]).

Luna's Lie theoretic methods were further developed by Bravi and Pezzini to classify wonderful varieties in classical types B, C, D (see [13, 14, 15, 16]); the case of exceptional types E was treated by Bravi in [8]. Thus, Luna's conjecture has been checked in almost all cases. The exceptional type F_4 is considered by Bravi and Luna in [12]; they listed the 266 spherical systems in that case, and they constructed many examples of associated wonderful varieties.

Luna's conjecture has also been confirmed for those wonderful G-varieties that arise as orbit closures in projectivizations of simple G-modules, via a classification due to Bravi and Cupit-Foutou (see [11]). These wonderful varieties are called **strict**; they are characterized by the property that the isotropy group of each point equals its normalizer, as shown by Pezzini (see [50, Theorem 2]). In [10], Bravi and Cupit-Foutou applied that classification to explicitly describe certain moduli schemes of spherical varieties with a prescribed weight monoid; we now survey their results.

We say that a submonoid $\Gamma \subset \Lambda^+$ is G-**saturated**, if

$$\Gamma = \Lambda^+ \cap \mathbb{Z}\Gamma$$

where $\mathbb{Z}\Gamma$ denotes the subgroup of Λ generated by Γ. Then Γ is finitely generated, and saturated in the sense arising from toric geometry. By [47], the G-saturated submonoids of Λ^+ are exactly the weight monoids of those affine horospherical G-varieties Z_0 such that Z_0 is normal and contains an open G-orbit with boundary of codimension $\geqslant 2$.

Next, fix a G-saturated submonoid $\Gamma \subset \Lambda^+$ that is freely generated, with basis $\lambda_1, \ldots, \lambda_N$. Consider the associated moduli scheme M_Γ equipped with the action of the adjoint torus T_{ad} (Subsection 4.3). Then M_Γ is an open subscheme of $\mathrm{Hilb}_h^G(V)$ by Example 4.17, where $V = V(\lambda_1)^* \oplus \cdots \oplus V(\lambda_N)^*$ and $h = h_\Gamma$.

By the results of [10, Section 2.3], M_Γ *is isomorphic to a* T_{ad}*-module with linearly independent weights, say* $\sigma_1, \ldots, \sigma_r$. *Moreover, the union of all non-degenerate* G*-subvarieties*

$$Z \subset V(\lambda_1)^* \oplus \cdots \oplus V(\lambda_N)^*$$

with weight monoid Γ *is the affine multi-cone over a wonderful* G*-variety*

$$X \subset \mathbb{P}(V(\lambda_1)^*) \times \cdots \times \mathbb{P}(V(\lambda_N)^*)$$

which is strict and has spherical roots $\sigma_1, \ldots, \sigma_r$. The closures of the colors of X are exactly the pull-backs of the B-stable hyperplanes in $\mathbb{P}(V(\lambda_i)^*)$ (defined by the highest weight vectors of $V(\lambda_i)$) under the projections $X \to \mathbb{P}(V(\lambda_i)^*)$.

To prove these results, one first studies the tangent space to M_Γ at the horospherical degeneration Z_0, based on 3.10. This T_{ad}-module turns out to be multiplicity-free with weights $\sigma_1, \ldots, \sigma_r$ among an explicit list of spherical roots. Then one shows that the data of $\lambda_1, \ldots, \lambda_N$ and $\sigma_1, \ldots, \sigma_r$ define a spherical system; finally, by the classification of strict wonderful varieties, this spherical system corresponds to a unique such variety X.

Yet several G-saturated monoids may well yield the same strict wonderful variety, for instance in the case where Γ has basis a dominant weight λ (any such monoid is G-saturated); see the final example below.

Another natural example of a G-saturated monoid is the whole monoid Λ^+ of dominant weights. The affine spherical varieties Z having that weight monoid are called a **model G-variety**, as every simple G-module occurs exactly once in $\mathcal{O}(Z)$; then the horospherical degeneration of Z is $Z_0 = G//U$. The strict wonderful varieties associated with model varieties have been described in detail by Luna (see [42]).

More recently, Cupit-Foutou generalizes the approach of [10] in view of a geometric classification of wonderful varieties and of a proof of Luna's conjecture in full generality (see [23]). For this, she associates with any wonderful variety of rank r a family of (affine) spherical varieties over the affine space $\mathbb{A}^r$, having a prescribed free monoid. Then she shows that this family is the universal family.

The first step is based on the construction of the **total coordinate ring** (also called the **Cox ring**) of a wonderful variety X. Recall that the set $\mathcal{D}$ of (closures of) colors freely generates the Picard group of X, and consider the $\mathbb{Z}^{\mathcal{D}}$-graded ring

$$R(X) := \bigoplus_{(n_D) \in \mathbb{Z}^{\mathcal{D}}} H^0(X, \mathcal{O}_X(\sum_{D \in \mathcal{D}} n_D D))$$

relative to the multiplication of sections. We may assume that G is semi-simple and simply connected; then each space $H^0(X, \mathcal{O}_X(\sum_{D \in \mathcal{D}} n_D D))$ has a unique structure of a G-module, and the total coordinate ring $R(X)$ is a $\mathbb{Z}^{\mathcal{D}}$-graded G-algebra. It is also a finitely generated unique factorization domain. Moreover, the invariant subring $R(X)^U$ is freely generated by the canonical sections s_D of the colors and by those $s_1, \ldots, s_r$ of the boundary divisors; each s_i is homogeneous of weight $(c_{D,i})_{D \in \mathcal{D}}$. Each s_D is a B-eigenvector of weight (say) ω_D, and hence generates a simple submodule

$$V_D \cong V(\omega_D) \subset H^0(X, \mathcal{O}_X(D)).$$

As a consequence, the graded ring $R(X)$ is generated by $s_1, \ldots, s_r$ and by the V_D where $D \in \mathcal{D}$. Moreover, $s_1, \ldots, s_r$ form a regular sequence in $R(X)$ (see [18, Section 3.1]).

In geometric terms, the affine variety $\tilde{X} := \operatorname{Spec}(R(X))$ is equipped with an action of the connected reductive group $\tilde{G} := G \times (\mathbb{G}_m)^{\mathcal{D}}$ and with a flat, G-invariant morphism

$$\pi = (s_1, \ldots, s_r) : \tilde{X} \longrightarrow \mathbb{A}^r \tag{4.24}$$

which is also $(\mathbb{G}_m)^{\mathcal{D}}$-equivariant for the linear action of that torus on $\mathbb{A}^r$ with weights $\sum_{D \in \mathcal{D}} c_{D,i} \varepsilon_D$ where $i = 1, \ldots, r$ and $\varepsilon_D : (\mathbb{G}_m)^{\mathcal{D}} \to \mathbb{G}_m$ denotes the coefficient on D. Moreover, the $\tilde{G}$-variety $\tilde{X}$ is spherical and equipped with a closed immersion into the $\tilde{G}$-module $\left(\mathbb{A}^r \times \prod_{D \in \mathcal{D}} V_D\right)^*$ that identifies π with the projection to $(\mathbb{A}^r)^* \cong \mathbb{A}^r$. Here $(\mathbb{G}_m)^{\mathcal{D}}$ acts on $\prod_{D \in \mathcal{D}} V_D^*$ via multiplication by $-\varepsilon_D$ on V_D^*.

It follows that π may be viewed as a family of non-degenerate spherical $G \times C$-subvarieties of

$$V := \bigoplus_{D \in \mathcal{D}} V_D^*$$

where C denotes the neutral component of the kernel of the homomorphism

$$(\mathbb{G}_m)^{\mathcal{D}} \longrightarrow (\mathbb{G}_m)^r, \quad (t_D)_{D \in \mathcal{D}} \longmapsto \Big(\prod_{D \in \mathcal{D}} t_D^{c_{D,i}} \Big)_{i=1,\ldots,r}.$$

Thus, C is a torus, and $G \times C$ a connected reductive group with maximal torus $T \times C$ and adjoint torus T_{ad}. The weight monoid Γ is freely generated by the restrictions to $T \times C$ of the weights $(\omega_D, \varepsilon_D)$ of $T \times (\mathbb{G}_m)^{\mathcal{D}}$, where $D \in \mathcal{D}$.

Now the main results of [23] assert that the moduli scheme M_Γ is isomorphic to $\mathbb{A}^r$, and $\tilde{X}$ to the universal family. Moreover, X is the quotient by $(\mathbb{G}_m)^{\mathcal{D}}$ of the union of non-degenerate orbits (an open subset of $\tilde{X}$, stable under $\tilde{G}$). In particular, the wonderful G-variety X is uniquely determined by the monoid Γ.

As in [10], the first step in the proof is the determination of $T_{Z_0}(M_\Gamma)$. Then a new ingredient is the vanishing of $T^2(X_0)^G$, an obstruction space for the functor of invariant infinitesimal deformations of X_0. This yields the smoothness of M_Γ at Z_0, which implies easily the desired isomorphism $M_\Gamma \cong \mathbb{A}^r$.

Example 4.25. Let $G = SL_2$ as in Example 4.21. Then the wonderful G-varieties X are those of the following list, up to G-isomorphism:

(i) $\mathbb{P}^1 = G/B$. Then X is strict of rank 0: it has no spherical root. Moreover, $R(X) = \mathrm{Sym}(V(1))$, $\tilde{G} \cong \mathbb{G}_m \times G$, and $\tilde{X} = V(1)$ where $\mathbb{G}_m$ acts via scalar multiplication; the map (4.24) is constant.

(ii) $\mathbb{P}^1 \times \mathbb{P}^1$, of rank 1 with open orbit G/T and closed orbit $Y = G/B$ embedded as the diagonal. Then X is not strict, of rank 1, and its spherical root is the simple root α; we have $Y = D^+ + D^-$ in Pic(X). Moreover, $R(X) \cong \mathrm{Sym}(V(1) \oplus V(1))$, $\tilde{G} = G \times (\mathbb{G}_m)^2$, and $\tilde{X} = V(1) \oplus V(1)$ where $(\mathbb{G}_m)^2$ acts via componentwise multiplication, and G acts diagonally. The map (4.24) is the determinant. The torus C is one-dimensional, and the monoid Γ has basis $(1,1)$ and $(1,-1)$.

(iii) $\mathbb{P}^2 = \mathbb{P}(V(2))$, of rank 1 with open orbit $G/N_G(T)$ and closed orbit $Y = G/B$ embedded as the conic $(\Delta = 0)$. Here X is strict, of rank 1, and its spherical root is 2α; we have $Y = 2D$ in Pic(X). Moreover, $R(X) = \mathrm{Sym}(V(2))$, $\tilde{G} \cong G \times \mathbb{G}_m$, and $\tilde{X} = V(2)$ where $\mathbb{G}_m$ acts via scalar multiplication. The map (4.24) is the discriminant Δ. The torus C is trivial, and Γ is generated by 2. We have $M_\Gamma = \mathrm{Hilb}^G_{h_2}(V(2)) \cong \mathbb{A}^1$. Note that the monoid generated by 4 yields the same wonderful variety.

References

[1] D. Akhiezer. Equivariant completion of homogeneous algebraic varieties by homogeneous divisors. *Ann. Glob. Anal. Geom.*, **1**, 49–78, 1983. ← 79, 109

[2] V. Alexeev. Complete moduli in the presence of semiabelian group actions. *Ann. of Math. (2)*, **155**(3), 611–708, 2002. ← 66

[3] V. Alexeev and M. Brion. Moduli of affine schemes with reductive group action. *J. Algebraic Geom.*, **14**, 83–117, 2005. ← 64, 65, 81, 86, 97, 104, 105, 106, 107

[4] V. Alexeev and M. Brion. Stable spherical varieties and their moduli. *Int. Math. Res. Pap.*, 57 pp, 2006. Article ID 46293. ← 66

[5] J. Bertin. The punctual Hilbert scheme: an introduction, in: Geometric Methods in Representation Theory, , 1–100, *Sémin. Congr.*, **24-I**, Soc. Math. France, Paris, 2012. ← 65, 91, 94, 95, 97

[6] F. Bien and M. Brion. Automorphisms and local rigidity of regular varieties. *Compositio Math.*, **104**, 1–26, 1996. ← 106, 110

[7] R. Birkner. Non-coherent components of the toric Hilbert scheme. arXiv: 1009:4321. ← 66

[8] P. Bravi. Wonderful varieties of type E. *Represent. Theory*, **11**, 174–191, 2007. ← 111

[9] P. Bravi. Classification of spherical varieties, in: Notes of the course "Hamiltonian actions : invariants and classification" (Luminy, 2009). Available at http://ccirm.cedram.org/ccirm-bin/feuilleter ← 66

[10] P. Bravi and S. Cupit-Foutou. Equivariant deformations of the affine multicone over a flag variety. *Adv. Math.*, **217**(6), 2800–2821, 2008. ← 111, 112, 113

[11] P. Bravi and S. Cupit-Foutou. Classification of strict wonderful varieties. *Ann. Inst. Fourier (Grenoble)*, **60**(2), 641–681, 2010. ← 111

[12] P. Bravi and D. Luna. An introduction to wonderful varieties with many examples of type F_4. *J. Algebra*, **329**, 4–51, 2011. ← 111

[13] P. Bravi and G. Pezzini. Wonderful varieties of type D. *Represent. Theory*, **9**, 578–637, 2005. ← 111

[14] P. Bravi and G. Pezzini. A constructive approach to the classification of wonderful varieties. `arXiv:1103.0380`. ← 111

[15] P. Bravi and G. Pezzini. Primitive wonderful varieties. `arXiv:1106.3187`. ← 111

[16] P. Bravi and G. Pezzini. The spherical systems of the wonderful reductive subgroups. `arXiv:1109.6777`. ← 111

[17] T. Bridgeland, A. King, and M. Reid. The McKay correspondence as an equivalence of derived categories. *J. Amer. Math. Soc.*, **14**(3), 535–554, 2001. ← 64, 97

[18] M. Brion. The total coordinate ring of a wonderful variety. *J. Algebra*, **313**(1), 61–99, 2007. ← 113

[19] M. Brion. Log homogeneous varieties, in: Actas del XVI Coloquio Latinoamericano de Álgebra, 1–39, Revista Matemática Iberoamericana, Madrid, 2007. ← 106

[20] M. Brion. Introduction to actions of algebraic groups, in: Notes of the course "Hamiltonian actions : invariants and classification" (Luminy, 2009). Available at `http://ccirm.cedram.org/ccirm-bin/feuilleter` ← 66, 68, 69, 70, 73, 100

[21] J. Budmiger. Deformation of orbits in minimal sheets. Available at `http://tel.archives-ouvertes.fr/docs/00/49/25/15/PDF/Budmiger.pdf` ← 79

[22] O. Chuvashova. The main component of the toric Hilbert scheme. *Tohoku Math. J.* (2), **60**(3), 365-382, 2008. ← 66

[23] S. Cupit-Foutou. Wonderful varieties: A geometrical realization. `arXiv:0907.2852`. ← 86, 112, 113

[24] G. Gotzmann. Eine Bedingung für die Flachheit und das Hilbertpolynom eines graduierten Ringes. *Math. Z.*, **158**(1), 61–70, 1978. ← 64

[25] F. D. Grosshans. Algebraic homogeneous spaces and invariant theory. *Lecture Notes in Math.*, **1673**, Springer, New York, 1997. ← 71, 97

[26] M. Haiman and B. Sturmfels. Multigraded Hilbert schemes. *J. Algebraic Geom.*, **13**, 725–769, 2004. ← 64, 77, 79, 92

[27] Y. Ito and I. Nakamura. McKay correspondence and Hilbert schemes. *Proc. Japan Acad. Ser. A Math. Sci.* **72**(7), 135–138, 1996. ← 64, 97

[28] Y. Ito and I. Nakamura. Hilbert schemes and simple singularities, in: New trends in algebraic geometry (Warwick, 1996), 151–233. *London Math. Soc. Lecture Notre Ser.* **264**, Cambridge Univ. Press, Cambridge, 1999. ← 64, 97

[29] S. Jansou. Déformations des cônes de vecteurs primitifs. *Math. Ann.*, **338**, 627–667, 2007. ← 65, 78, 79, 87

[30] S. Jansou and N. Ressayre. Invariant deformations of orbit closures in $\mathfrak{sl}(n)$, *Represent. Theory*, **13**, 50–62, 2009. ← 94

[31] F. Knop. The Luna-Vust theory of spherical embeddings, in: Proceedings of the Hyderabad Conference on Algebraic Groups (Hyderabad, 1989), 225–249, Manoj Prakashan, Madras, 1991. ← 108

[32] F. Knop. A Harish-Chandra homomorphism for reductive group actions. *Ann. of Math. (2)*, **140**(2), 253–288, 1994. ← 106

[33] F. Knop. Automorphisms, root systems, and compactifications of homogeneous varieties. *J. Amer. Math. Soc.*, **9**(1), 153–174, 1996. ← 105, 110

[34] F. Knop and B. Van Steirteghem. Classification of smooth affine spherical varieties. *Transform. Groups* , **11**(3), 495–516, 2006. ← 108

[35] M. Lehn and C. Sorger. A symplectic resolution for the binary tetrahedral group, in: Geometric Methods in Representation Theory, 427–433 *Sémin. Congr.*, **24-II**, Soc. Math. France, Paris. 2012. ← 97

[36] I. Losev. Uniqueness property for spherical homogeneous spaces. *Duke Math. J.*, **147**(2), 315–343, 2009. ← 108

[37] I. Losev. Proof of the Knop conjecture. *Ann. Inst. Fourier (Grenoble)*, **59**(3), 1105–1134, 2009. ← 108

[38] I. Losev. Demazure embeddings are smooth. *Int. Math. Res. Not.*, **14**, 2588–2596, 2009. ← 110

[39] I. Losev. Uniqueness properties for spherical varieties, in: Notes of the course "Hamiltonian actions : invariants and classification" (Luminy, 2009). Available at `http://ccirm.cedram.org/ccirm-bin/feuilleter` ← 66

[40] D. Luna. Toute variété magnifique est sphérique. *Transform. Groups*, **1**(3), 249–258, 1996. ← 109

[41] D. Luna. Variétés sphériques de type A. *Publ. Math. Inst. Hautes Études Sci.*, **94**, 161–226, 2001. ← 111

[42] D. Luna. La variété magnifique modèle. *J. Algebra*, **313**(1), 292–319, 2007. ← 112

[43] D. Luna and Th. Vust. Plongements d'espaces homogènes. *Comment. Math. Helv.*, **58**(2), 186–245, 1983. ← 108

[44] D. Maclagan and R. R. Thomas. Combinatorics of the toric Hilbert scheme. *Discrete Comput. Geom.*, **27**(2), 249–272, 2002. ← 66

[45] D. Mumford, J. Fogarty, and F. Kirwan. Geometric Invariant Theory. Third Edition. *Ergeb. der Math.*, Springer, 1994. ← 80

[46] M. Olsson. Logarithmic interpretation of the main component in toric Hilbert schemes, in: Curves and abelian varieties, 231–252. *Contemp. Math.*, **465**, Amer. Math. Soc., Providence, RI, 2008. ← 66

[47] D. Panyushev. On deformation method in invariant theory. *Ann. Inst. Fourier (Grenoble)*, **47**(4), 985–1012, 1997. ← 111

[48] S. A. Papadakis and B. Van Steirteghem. Equivariant degenerations of spherical modules for groups of type A. `arXiv:1008.0911`, to appear at *Ann. Inst. Fourier (Grenoble)*. ← 86

[49] I. Peeva and M.Stillman. Toric Hilbert schemes. *Duke Math. J.*, **111**, 419–449, 2002. ← 64, 77

[50] G. Pezzini. Simple immersions of wonderful varieties. *Math. Z.*, **255**(4), 793–812, 2007. ← 111

[51] G. Pezzini. Lectures on spherical and wonderful varieties. In: *Notes of the course "Hamiltonian actions : invariants and classification" (Luminy, 2009)*. Available at `http://ccirm.cedram.org/ccirm-bin/feuilleter` ← 66, 108

[52] V. L. Popov. Contraction of the actions of reductive algebraic groups. *Mat. Sb. (N.S.)*, **130**(3), 310–334, 1986. ← 97

[53] F. Santos. Non-connected toric Hilbert schemes. *Math. Ann.*, **332**(3), 645–665, 2005. ← 64, 66

[54] E. Sernesi. Deformations of algebraic schemes. *Grundlehren der mathematischen Wissenschaften*, **334**, Springer, New York, 2006. ← 84

[55] B. Sturmfels. Gröbner bases and convex polytopes. *University Lecture Series*, **8**, Amer. Math. Soc., Providence, 1996. ← 66

[56] D. A. Timashev. Equivariant embeddings of homogeneous spaces. In: *Surveys in geometry and number theory: reports on contemporary Russian mathematics*, 226–278. *London Math. Soc. Lecture Note Ser.*, **338**, Cambridge Univ. Press, Cambridge, 2007. ← 108

Algebraic and tropical curves: comparing their moduli spaces

Lucia Caporaso

Abstract. We construct the moduli space for equivalence classes of n-pointed tropical curves of genus g, together with its bordification given by weighted tropical curves, and its compactification by extended (weighted) tropical curves. We compare it to the moduli spaces of smooth and stable, n-pointed algebraic curves, from the combinatorial, the topological, and the Teichmüller point of view.

Contents

2000 *Mathematics Subject Classification.* Primary 14H10, 14TXX; Secondary 05CXX.

Key words and phrases. smooth curve, stable curve, tropical curve, combinatorial graph, metric graph, dual graph, moduli space, Teichmüller theory.

1. Introduction

1.1. Overview of the paper

This is a largely expository article whose main goal is to construct the moduli space of tropical curves with marked points, and to compare it to the moduli space of algebraic curves, highlighting the symmetries and the analogies which occur at various places.

The moduli space of nonsingular algebraic curves is a well known object that has been thoroughly studied, together with its compactifications, over the last five decades (to say the least). There are several expository papers and monographs attesting the depth and richness of the subject. In this paper we shall give the basic definitions and state some famous results in a way that should be accessible to the non expert reader. We will provide references where details can be found.

On the other hand tropical geometry is a rather young branch of mathematics, which has seen a flourish of diverse activities over the recent years. Thus, quite naturally, the field still lacks exhaustive bibliographical sources, and even solid foundations. Hence we shall treat tropical curves and their moduli quite thoroughly, including technical details and proofs, although some of the results we shall describe are known, if maybe only as folklore.

Let us illustrate the content of the paper, section by section. We begin with algebraic nodal curves of genus g with n marked points, define the notion of stably equivalent curves, and introduce the moduli spaces for these equivalence classes, the well known Deligne-Mumford spaces $\overline{M}_{g,n}$.

Next, in Section 2, after some preliminaries about graphs, we give the original definition of abstract tropical curves and of tropically equivalent curves. Tropical equivalence should be viewed as the analogue of stable equivalence. Tropical curves up to tropical equivalence are parametrized by metric graphs with no vertex of valence less than 3. We extend the set-up to curves with marked points.

As it turns out, tropical curves do not behave well in families, in the sense that the genus may drop under specialization. One solution to this problem, introduced in [5] for unpointed curves, is to add a weight function on the vertices of their corresponding graph, thus generalizing the original notion (when the weight function is zero we get back the original definition). We do that in the last part of Section 2. Of course, this may not be the only, or the best, solution. But it is one that can be worked out, and whose analogies with moduli of stable algebraic curves are strong and interesting. To distinguish, we refer to the original definition as "pure" tropical curve, and to the generalized one as "weighted" tropical curve.

In Section 3 we explicitly construct, as topological spaces, the moduli space for n-pointed weighted tropical curves of genus g, denoted by $M^{\text{trop}}_{g,n}$, and the analog for pure tropical curves, denoted by $M^{\text{pure}}_{g,n}$. It is clear from the construction that these two spaces are almost never topological manifolds; but they turn out to be connected, Hausdorff, of pure dimension $3g - 3 + n$. Also, $M_{g,n}^{pure}$ is open and dense in $M_{g,n}^{trop}$; see Theorem 3.21.

Furthermore, $M_{g,n}^{pure}$ and $M_{g,n}^{trop}$ are not tropical varieties in general. The case $g = 0$, which has been extensively studied (see for example [22], [23], [16], [15]), is an exception, because of the absence of automorphisms. Indeed $M_{0,n}^{trop}$ (which coincides with $M_{0,n}^{pure}$) is a tropical variety for every $n \geqslant 3$.

In the literature there are a few constructions giving these moduli spaces some special structure, resembling that of a tropical variety; see [15], [5] and [9] for example; nevertheless we prefer to treat them just as topological spaces, as the categorical picture does not look clear at the moment.

Essentially by definition, $M_{g,n}^{trop}$ is closed under specialization, but it is not compact. A natural compactification for it, by what we call "extended tropical curves", is studied in Subsection 3.3.

In Section 4 we compare the space $M_{g,n}^{trop}$ to the Deligne-Mumford space $\overline{M}_{g,n}$. Both spaces admit a natural partition indexed by stable graphs, with a poset structure defined by the inclusion of closures. The properties of the two partitions, and the correspondence between them, is described in Theorem 4.7.

Section 5 considers the Teichmüller approach to the moduli space of smooth curves M_g, and to the moduli space of metric graphs. While the Teichmüller point of view constitutes one of the principal chapters of the complex algebraic theory, its analog for metric graphs, by means of the Culler-Vogtmann space (see [11]), is less known in tropical geometry, and awaits to be further studied. We conclude the paper with a table summarizing these analogies, and with a list of open problems and research directions.

Acknowledgements. I wish to thank M. Melo, G. Mikhalkin and F. Viviani for several useful comments, and a referee for a very detailed report.

1.2. Algebraic curves

Details about the material of this and the next subsection may be found, for example, in [18, Chapt 1-4] or in [2, Chapt. 10 and 12].

By an "algebraic curve", or simply "curve" when no confusion is possible, in this paper we mean a reduced, connected projective variety (not necessarily irreducible) of dimension one, defined over an algebraically closed field. The arithmetic genus of a curve X is $g_X = h^1(X, \mathcal{O}_X)$.

We denote by X_{sing} the set of singular points of X.

We will only be interested in nodal curves, i.e. curves having at most ordinary double points as singularities. The reason is that the class of nodal curves, up to an important equivalence relation which we will define in Subsection 1.3, has a moduli space with very nice properties: it is a projective, irreducible (reduced and normal) variety containing as a dense open subset the moduli space of nonsingular curves.

The need for an equivalence relation is quite fundamental, as the set of all nodal curves is too big to admit a separated moduli space. It is in fact easy to produce families of nodal, or even nonsingular, curves admitting (infinitely many) non isomorphic nodal curves as limits.

Before being more specific, we want to extend the present discussion to so-called n-pointed curves.

Let g and n be nonnegative integers. A *nodal n-pointed curve* of genus g, denoted by $(X; p_1, \dots, p_n)$, is a nodal curve X of arithmetic genus g, together with n distinct nonsingular points $p_i \in X$, $i = 1, \dots, n$.

The nodal n-pointed curve $(X; p_1, \dots, p_n)$ is *stable* (in the sense of Deligne and Mumford) if the set of automorphisms of X fixing $\{p_1, \dots, p_n\}$ is finite; or, equivalently, if the line bundle $\omega_X(p_1 + \dots + p_n)$ has positive degree on every subcurve of X (where ω_X is the dualizing line bundle[1] of X).

Finally, since for any irreducible component $E \subset X$ we have

$$\deg_E \omega_X\Big(\sum_1^n p_i\Big) = 2g_E - 2 + |E \cap \{p_1, \dots, p_n\}| + |E \cap \overline{X \smallsetminus E}|, \tag{1.1}$$

we have that our nodal n-pointed curve is stable if and only if for every irreducible component E of arithmetic genus at most 1 we have

$$|E \cap \{p_1, \dots, p_n\}| + |E \cap \overline{X \smallsetminus E}| \geqslant \begin{cases} 3 & \text{if } g_E = 0 \\ 1 & \text{if } g_E = 1. \end{cases}$$

Note that the requirement for $g_E = 1$ is saying that there exist no stable curves for $g = 1$ and $n = 0$. More generally it is easy to check that stable curves exist if and only if $2g - 2 + n \geqslant 1$.

Fact 1.2. *Assume $2g-2+n \geqslant 1$. There exists an irreducible projective scheme of dimension $3g - 3 + n$, denoted by $\overline{M}_{g,n}$, which is the (coarse) moduli space of n-pointed stable curves of genus g. The moduli space of nonsingular n-pointed curves of genus g is an open (dense) subset $M_{g,n} \subset \overline{M}_{g,n}$.*

Remark 1.3. What do we mean by "moduli space"? In algebraic geometry there are various categories (schemes, stacks, algebraic spaces) in which moduli spaces can be axiomatically defined and their properties rigorously proved.

The basic axiom in the schemes category is, of course, the fact that closed points[2] be in bijection with the isomorphism classes of the objects under investigation (stable pointed curves, in our case). Other axioms require that to a family of objects parametrized by a scheme B there is a unique associated morphism from B to the moduli space, and ensure that a moduli space is unique when it exists.

---

[1] For smooth curves ω_X coincides with the canonical bundle.

[2] I.e. points defined over the base field, versus points defined over some field extension of it.

The word "coarse" in the statement indicates that n-pointed stable curves may have nontrivial automorphisms, so that to a morphism from B to $\overline{M}_{g,n}$ there may correspond more than one family of n-pointed stable curves, or no family at all. We refer to [18] for details about this topic.

By contrast, as of this writing, in tropical geometry there is no clear understanding of what a good categorical framework for moduli theory could be. Therefore we shall simply view tropical moduli spaces of curves as geometric objects (specifically: topological spaces) whose points are in bijection with isomorphism classes of pointed tropical curves up to tropical equivalence. Essentially by construction, the topological structure of our tropical moduli spaces will reflect the most basic notion of a continuously varying family of tropical curves.

The results stated in Fact 1.2 may be considered a great achievement of twentieth century algebraic geometry. They are due to P. Deligne - D. Mumford [12], F. Knudsen [20], and D. Gieseker [17], with fundamental contributions from other mathematicians (among whom A. Grothendieck and A. Mayer). Some details about the construction of the moduli space of stable curves will be given at the beginning of Section 5.

Remark 1.4. It is important to keep in mind that the points $p_1, \dots, p_n$ of a stable n-pointed curves are always meant to be labeled (or ordered), as the following example illustrates.

Example 1.5. Let $g = 0$. Then there exist stable curves only if $n \geqslant 3$, and if $n = 3$ there exists a unique one (up to isomorphisms): $\mathbb{P}^1$ with 3 distinct points, indeed it is well known that the automorphisms of $\mathbb{P}^1$ act transitively on the set of triples of distinct points. Notice also that there exist no nontrivial automorphisms of $\mathbb{P}^1$ fixing 3 points, therefore *for every $n \geqslant 3$ a smooth n-pointed curve of genus 0 has no nontrivial automorphisms.*

The next case, $n = 4$, is quite interesting. We may set the three first points to be $p_1 = 0, p_2 = 1, p_3 = \infty$ and let the fourth point p_4 vary in $\mathbb{P}^1 \smallsetminus \{0, 1, \infty\}$. So we have a simple description of $M_{0,4}$ as a nonsingular, rational, affine curve. Its completion, $\overline{M}_{0,4}$, the moduli space of stable curves, is isomorphic to $\mathbb{P}^1$, and the three boundary points are the three singular 4-pointed stable rational curves such that $X = C_1 \cup C_2$ with $|C_1 \cap C_2| = 1$ and, of course, $C_i \cong \mathbb{P}^1$.

What distinguishes the three curves from one another is the distribution of the four points among the two components. Since each of them must contain two of the four points, and since the two components can be interchanged, we have three different non-isomorphic curves corresponding to the three different partitions $(p_1, p_2 | p_3, p_4)$, $(p_1, p_3 | p_2, p_4)$, $(p_1, p_4 | p_2, p_3)$.

Example 1.6. If $g = 1$ there exist stable curves only if $n \geqslant 1$, and for $n = 1$ we have a 1-dimensional family of smooth ones, and a unique singular one: an irreducible curve of arithmetic genus 1 with one node and one marked point.

1.3. Stably equivalent nodal curves

We now describe an explicit procedure to construct the *stabilization*, or *stable model* of a nodal pointed curve. As we mentioned, this is needed to have a separated moduli space for equivalence classes of nodal curves. We will do this by a sequence of steps, which will be useful later on.

Let $(X; p_1, \dots, p_n)$ be an n-pointed nodal curve of genus g, with $2g-2+n \geqslant 1$. Suppose X is not stable. Then X has components where the degree of $\omega_X(\sum_1^n p_i)$ is at most 0.

• *Step 1.* Suppose there exists a component E of X where this degree is negative. Then, by (1.1), E is a smooth rational component containing none of the p_i and such that $|E \cap X_{\text{sing}}| = 1$ (so, $\deg_E \omega_X(\sum_1^n p_i) = -1$). These components are called unpointed rational tails. We remove E from X; this operation does not alter the genus of X, its connectedness (as $|E \cap \overline{X \smallsetminus E}| = 1$), the nature of its singularities, or the number of marked points (which remain nonsingular and distinct). On the other hand it may create one new unpointed rational tail, in which case we repeat the operation. So, the first step consists in iterating this operation until there are no unpointed rational tails remaining. It is clear that this process terminates after finitely many steps. We denote by $(X'; p'_1, \dots, p'_n)$ the resulting n-pointed curve, which is unique up to isomorphism.

• *Step 2.* Now the degree of $\omega_{X'}(\sum_1^n p'_i)$ on every component of X' is at least 0, and if $(X'; p'_1, \dots, p'_n)$ is not stable there exists a component E on which this degree is 0. By (1.1) we have $E \cong \mathbb{P}^1$ and E contains a total of 2 among marked and singular points of X'. There are two cases according to whether E contains a marked point or not (E necessarily contains at least one singular point of X', for $E \subsetneq X'$).

• • *Step 2.a.* Suppose $p'_i \notin E$ for all $i = 1, \dots, n$. Then E intersects $\overline{X' \smallsetminus E}$ in 2 points (for X' contains no unpointed rational tail). Such an E is are called exceptional component. This step consists in contracting each of these exceptional components to a node. Notice that this step does not touch the marked points, nor does it add any new rational tail. The resulting curve $(X''; p''_1, \dots, p''_n)$ is again an n-pointed genus g curve free from exceptional components or unpointed rational tails.

• • *Step 2.b.* Suppose E contains some of the marked points. Since E must intersect $\overline{X'' \smallsetminus E}$, we have $|E \cap \overline{X'' \smallsetminus E}| = 1$ and E contains only one marked point, p''_1 say. These components are called uni-pointed rational tails. Now we remove E as in the first step, but we need to keep track of p''_1. To do that we mark the attaching point $E \cap \overline{X'' \smallsetminus E}$, which, after we remove E, becomes a nonsingular point replacing the lost point p''_1. By iterating this process until there are no uni-pointed rational tails left, we arrive at a genus g stable curve $(X'''; p'''_1, \dots, p'''_n)$. It is easy to check that this curve is unique up to isomorphism.

This curve $(X'''; p'''_1, \dots, p'''_n)$ is called the stabilization, or stable model, of the given one.

Remark 1.7. Following [18], we say that two n-pointed nodal curves are *stably equivalent* if they have the same stabilization (always assuming $2g-2+n \geqslant 1$). It is easy to check that two stable curves are stably equivalent only if they are isomorphic. So, in every stable equivalence class of n-pointed curves there is a unique stable representative.

By Fact 1.2, there exists a projective variety, $\overline{M}_{g,n}$, parametrizing stable equivalence classes of n-pointed nodal curves of genus g.

Remark 1.8. There are many good reasons for extending our field of interest from curves to pointed curves. Here is a basic and useful one:

Stability of pointed curves is preserved under normalization.

More precisely, let $(X; p_1, \ldots, p_n)$ be an n-pointed stable curve of genus g. Pick a node $q \in X_{\text{sing}}$, let $\nu_q : X_q^\nu \to X$ be the normalization at q; let $q_1, q_2 \in X_q^\nu$ be the two branches of q, and abuse notation by setting $p_i = \nu_q^{-1}(p_i)$ for $i = 1, \ldots, n$. Then one easily checks that the $(n+2)$-pointed curve $(X_q^\nu; p_1, \ldots, p_n, q_1, q_2)$ is either stable of genus $g-1$ (if q is not a separating node of X), or the disjoint union of two pointed stable curves of genera summing to g (if q is a separating node).

2. Graphs and pure tropical curves

2.1. Graphs

By a *topological graph*, or simply a graph, we mean a one dimensional finite simplicial (or CW) complex Γ; we denote by $V(\Gamma)$ the set of its vertices (or 0-cells) and by $E(\Gamma)$ the set of its edges (or 1-cells). To every $e \in E(\Gamma)$ one associates the pair $\{v, v'\}$ of possibly equal vertices which form the boundary of e; we call v and v' the *endpoints* of e. If $v = v'$ we say that e is a *loop-edge* based at v.

The *valence* of a vertex v is the number of edges having v as endpoint, with the convention that a loop based at v be counted twice.

For a fixed integer p, we say that a vertex of valence p is p-*valent*, and we say that a graph is p-*regular* if all of its vertices are p-valent.

The genus $g(\Gamma)$ of Γ is its first Betti number

$$g(\Gamma) = b_1(\Gamma) := \mathrm{rk}_{\mathbb{Z}} H_1(\Gamma, \mathbb{Z}) = |E(\Gamma)| - |V(\Gamma)| + c,$$

where c is the number of connected components of Γ. Morphisms between topological graphs are, as usual, cellular maps.

We also need to define graphs in purely combinatorial terms, for which there are various possibilities. Our definition, almost the same as in [2], is most convenient to simultaneously treat tropical and algebraic curves.

Definition 2.1. A *combinatorial graph* Γ with n *legs* is the following set of data:

(1) A finite non-empty set $V(\Gamma)$, the set of *vertices*.
(2) A finite set $H(\Gamma)$, the set of *half-edges*.

(3) An involution

$$\iota : H(\Gamma) \longrightarrow H(\Gamma) \qquad h \mapsto \overline{h}$$

with n fixed points, called *legs*, whose set is denoted by $L(\Gamma)$.

(4) An *endpoint* map $\epsilon : H(\Gamma) \to V(\Gamma)$.

A pair $e = \{h, \overline{h}\}$ of distinct elements in $H(\Gamma)$ interchanged by the involution is called an *edge* of the graph; the set of edges is denoted by $E(\Gamma)$. If $\epsilon(h) = v$ we say that h, or e, is *adjacent* to v.

The *valence* of a vertex v is the number $|\epsilon^{-1}(v)|$ of half-edges adjacent to v.

An edge adjacent to a vertex of valence 1 is called a *leaf*.

An edge whose endpoints coincide is called a *loop-edge*.

Two legs are called *disjoint* if their endpoints are distinct.

Definition 2.2. A morphism α between combinatorial graphs Γ and Γ' is a map $\alpha : V(\Gamma) \cup H(\Gamma) \to V(\Gamma') \cup H(\Gamma')$ such that $\alpha(L(\Gamma)) \subset L(\Gamma')$, and such that the two diagrams below are commutative.

$$\begin{array}{ccc} V(\Gamma) \cup H(\Gamma) & \xrightarrow{\alpha} & V(\Gamma') \cup H(\Gamma') \\ \downarrow{\scriptstyle (id_V, \epsilon)} & & \downarrow{\scriptstyle (id_{V'}, \epsilon')} \\ V(\Gamma) \cup H(\Gamma) & \xrightarrow{\alpha} & V(\Gamma') \cup H(\Gamma') \end{array} \tag{2.3}$$

In particular, $\alpha(V(\Gamma)) \subset V(\Gamma')$. Next

$$\begin{array}{ccc} V(\Gamma) \cup H(\Gamma) & \xrightarrow{\alpha} & V(\Gamma') \cup H(\Gamma') \\ \downarrow{\scriptstyle (id_V, \iota)} & & \downarrow{\scriptstyle (id_{V'}, \iota')} \\ V(\Gamma) \cup H(\Gamma) & \xrightarrow{\alpha} & V(\Gamma') \cup H(\Gamma') \end{array} \tag{2.4}$$

We say that a morphism α as above is an *isomorphism* if α induces, by restriction, three bijections $\alpha_V : V(\Gamma) \to V(\Gamma')$, $\alpha_E : E(\Gamma) \to E(\Gamma')$ and $\alpha_L : L(\Gamma) \to L(\Gamma')$.

An automorphism of Γ is an isomorphism of Γ with itself.

Remark 2.5. The image of an edge $e \in E(\Gamma)$ is either an edge, or a vertex v' of Γ'; in the latter case the endpoints of e are also also mapped to v', and we say that e is contracted by α.

A trivial example of morphism is the map forgetting some legs.

Example 2.6. Let $e = \{h, \overline{h}\}$ be a loop-edge of Γ. Then Γ has a "loop-inversion" automorphism, exchanging h and $\overline{h}$ and fixing everything else.

It is clear that to every topological graph we can associate a unique combinatorial graph with no legs.

Conversely, to every combinatorial graph with vertex set V and edge set E, we can associate a unique topological graph, as follows. We take V as the set of 0-cells;

then we add a 1-cell for every $e = \{h, \overline{h}\} \in E$, in such a way that the boundary of this 1-cell is $\{\epsilon(h), \epsilon(\overline{h})\}$.

Now, if the combinatorial graph has a non empty set of legs L, we add to the topological graph associated to it a 1-cell for every $h \in L$ in such a way that one extreme of the 1-cell contains $\epsilon(h)$ in its closure. The topological space we obtain in this way will be called a topological graph with n legs. So, legs are 1-cells only one end of which is adjacent to a vertex (which will be called the endpoint of the leg). From now on we shall freely switch between the combinatorial and topological structure on graphs, intermixing the two points of view without mention. The notions of valence, genus, endpoints, p-regularity and so on, can be given in each setting and coincide.

Remark 2.7. From now on, we shall assume our graphs to be connected, unless we specify otherwise.

We now describe a type of morphism between graphs, named *(edge) contraction*, which will play an important role. We do that for topological graphs, leaving the translation for combinatorial graphs to the reader.

Let Γ be a topological graph and $e \in E(\Gamma)$ be an edge. Let $\Gamma_{/e}$ be the graph obtained by contracting e to a point and leaving everything else unchanged [13, sect I.1.7]. Then there is a natural continuous surjective map $\Gamma \to \Gamma_{/e}$, called the *contraction* of e. More generally, if $S \subset E(\Gamma)$ is a set of edges, we denote by $\Gamma_{/S}$ the contraction of every edge in S and denote by $\sigma : \Gamma \to \Gamma_{/S}$ the associated map. Let $T := E(\Gamma) \smallsetminus S$. Then there is a natural identification between $E(\Gamma_{/S})$ and T. Moreover the map σ induces a bijection between the sets of legs $L(\Gamma)$ and $L(\Gamma_{/S})$, and a surjection

$$\sigma_V : V(\Gamma) \longrightarrow V(\Gamma_{/S}); \quad v \mapsto \sigma(v).$$

Let us disregard the legs, as they play essentially no role in what we are going to describe. Notice that every connected component of $\Gamma - T$ (the graph obtained from Γ by removing every edge in T) gets contracted to a vertex of $\Gamma_{/S}$; conversely, for every vertex $\overline{v}$ of $\Gamma_{/S}$ its preimage $\sigma^{-1}(\overline{v}) \subset \Gamma$ is a connected component of $\Gamma - T$. In particular, we have

$$b_1(\Gamma - T) = \sum_{\overline{v} \in V(\Gamma_{/S})} b_1(\sigma^{-1}(\overline{v})). \tag{2.8}$$

Let $\sigma : \Gamma \to \Gamma_{/S}$ be the contraction of S as above. Then

$$b_1(\Gamma) = b_1(\Gamma_{/S}) + b_1(\Gamma - T). \tag{2.9}$$

2.2. Tropical curves

A *metric graph* is a leg-free graph Γ endowed with the structure of a metric space, so that every edge is locally isometric to an interval in $\mathbb{R}$, and where the distance between two points is the shortest length of an edge-path joining them. In particular, on a metric graph we have a length function

$$\ell : E(\Gamma) \to \mathbb{R}_{>0}$$

mapping an edge to the distance between its endpoints. It is clear, conversely, that the datum of such an ℓ determines on Γ the structure of a metric space with every edge e having length $\ell(e)$.

If Γ has legs, it is convenient to extend the function ℓ to the legs of Γ, by setting $\ell(x) = \infty$ for every $x \in L(\Gamma)$.

The genus of a metric graph is the genus of the underlying topological graph.

We shall now define tropical curves following G. Mikhalkin. An abstract (pure[3]) tropical curve is almost the same as a metric graph; see [22, Prop. 5.1] or [24, Prop. 3.6]. The difference is in the length of those edges adjacent to vertices of valence 1, i.e. the leaves; the length of a leaf is set to be equal to ∞ for a tropical curve, whereas for a metric graph is finite.

Definition 2.10. A *(pure) tropical curve* of genus g is a pair (Γ, ℓ) where Γ is a leg-free graph of genus g and ℓ a length function on the edges $\ell : E(\Gamma) \to \mathbb{R}_{>0} \cup \{\infty\}$ such that $\ell(e) = \infty$ if and only if e is a leaf.

Two tropical curves are *(tropically) equivalent* if they can be obtained from one another by adding or removing vertices of valence 2, or vertices of valence 1 together with their adjacent leaf.

Remark 2.11. A tropical curve is not a metric space, as the distance from a 1-valent vertex to another point is not defined. Of course, if all 1-valent vertices are removed what remains is a metric space. We shall see in 2.16 that equivalence classes of tropical curves are bijectively parametrized by certain metric graphs.

Let us illustrate tropical equivalence in details with some pictures.

(1) *Addition/removal of a vertex of valence 1 and of of its adjacent edge (a leaf).* The next picture illustrates the removal of the one-valent vertex u_0 and of the leaf e_0 adjacent to it. The opposite move is the addition of a leaf, where the length of

Figure 1. Removal of the 1-valent vertex u_0 and of its adjacent edge e_0.

the added edge e_0 is set equal to ∞.

[3] A explained in the introduction, "pure" is added for reasons that will be clear later.

(2) *Addition/removal of a vertex of valence* 2. Pick an edge $e \in E(\Gamma)$ and denote by $v, w \in V(\Gamma)$ its endpoints. We can add a vertex u in the interior of e. This move replaces the edge e of length $\ell(e)$ by two edges e_v (with endpoints v, u) and e_w (with endpoints w, u), whose lengths satisfy $\ell(e) = \ell(e_v) + \ell(e_w)$. If one of the two endpoints of e, say v, has valence 1 we set the length of e_v equal to ∞, whereas the length of e_w can be arbitrary. The opposite procedure, which should be clear, is represented in the figure 2 below:

Figure 2. Removal of a vertex of valence 2.

2.3. Pointed tropical curves

Points on a tropical curve are conveniently represented as legs on the corresponding graph. Indeed, let C be a tropical curve and $p \in C$; if p is a vertex we add a leg based at p, if p is in the interior of an edge, we add a vertex at p and a leg based at it; since the added vertex is 2-valent, this operation does not change the equivalence class of C.

Definition 2.12. An n-*pointed (pure) tropical curve* of genus g is a pair (Γ, ℓ) where Γ is a combinatorial graph of genus g with a set $L(\Gamma) = \{x_1, \ldots, x_n\}$ of legs, and ℓ a length function

$$\ell : E(\Gamma) \cup L(\Gamma) \longrightarrow \mathbb{R}_{>0} \cup \infty$$

such that $\ell(x) = \infty$ if and only if x is a leaf or a leg.

The legs $\{x_1, \ldots, x_n\}$ are called *(marked) points* of the curve.

We say that two pointed tropical curves are *(tropically) equivalent* if they can be obtained one from the other by adding or removing vertices of valence 2 or vertices of valence 1 with their adjacent leaves.

We say that two tropical curves (Γ, ℓ) and (Γ', ℓ') with n marked points $L(\Gamma) = \{x_1, \ldots, x_n\}$ and $L(\Gamma') = \{x'_1, \ldots, x'_n\}$ are *isomorphic* if there exists an isomorphism α from Γ to Γ' as defined in 2.2, such that $\ell(e) = \ell'(\alpha(e))$ for every $e \in E(\Gamma)$, and $\alpha(x_i) = x'_i$ for every $i = 1, \ldots, n$.

Remark 2.13. Just as for stable curves, marked points are always labeled.

Two pointed tropical curves are equivalent if they can be obtained from one another by a finite sequence of the two moves described in the pictures of Subsection 2.2. Notice that now in move (2) we allow adding 2-valent vertices in the interior of a leg, as well as removing the endpoint v of a leg if v has valence 2, but we do not allow adding or removing legs.

Remark 2.14. Tropical equivalence preserves the number of marked points, but not their being disjoint, as the next picture shows.

Figure 3. After removing the 2-valent vertex v the two legs are no longer disjoint

Example 2.15. Let Γ have one vertex and no edges or legs. Then Γ is an unpointed tropical curve of genus 0, equivalent to any tropical curve of genus 0. From the moduli point of view, its equivalence class is viewed as a trivial one, and will be excluded in future considerations.

Let now Γ be a graph with one vertex, v, and one loop attached to it; so Γ is an unpointed tropical curve of genus 1. Now v is 2-valent, hence can be removed, leaving us with something which is not a tropical curve. For this reason, these curves are viewed as degenerate, and will also be excluded. Hence if $g = 1$ we shall always assume $n \geqslant 1$.

By a similar reasoning, if $g = 0$ we shall always assume $n \geqslant 3$. This motivates the future assumption $2g - 2 + n \geqslant 1$.

Proposition 2.16. *Assume* $2g - 2 + n \geqslant 1$.

(1) *Every equivalence class of n-pointed pure tropical curve contains a representative whose n marked points are disjoint.*
(2) *The set of equivalence classes of n-pointed pure tropical curves of genus g is in bijection with the set of metric graphs of genus g with n legs having no vertex of valence less than 3.*

Proof. Let us prove (1). Pick a representative (Γ, ℓ) such that the legs of Γ are not disjoint. Hence there is a vertex $v \in V(\Gamma)$ which is the endpoint of $m \geqslant 2$ legs, $x_1, \ldots, x_m$. We add a vertex v_i' (of valence 2) in the interior of x_i for all $i = 2, \ldots, m$; this operation does not change the equivalence class. This creates, for $i = 1, \ldots, m$, a new edge e_i' and a leg x_i' adjacent to v_i'. In this way we obtain a new graph Γ' whose legs $x_1, x_2', \ldots, x_m'$ are disjoint. Of course Γ and Γ' are tropically equivalent. After repeating this process finitely many times we arrive at a graph with disjoint legs, tropically equivalent to the original one.

Now part (2). Let (Γ, ℓ) be a graph with n legs, i.e. an n-pointed tropical curve. We shall describe a construction to produce, in the same equivalence class of the given curve, a metric graph with n legs having no vertex of valence less than 3. This construction is entirely similar to the construction described in subsection 1.3 for algebraic curves. We itemize it so as to highlight the analogies.

$*$ *Step 1.* Suppose Γ has some 1-valent vertex. Then we remove it, together with its leaf; by definition this operation does not change the equivalence class. We thus get a new graph Γ^*, with a natural inclusion $E(\Gamma^*) \subset E(\Gamma)$ and a natural identification $L(\Gamma^*) = L(\Gamma)$. We define the length function ℓ^* on $E(\Gamma^*) \cup L(\Gamma^*)$ by restricting ℓ. Hence (Γ^*, ℓ^*) is an n-pointed curve equivalent to the given one. We can obviously iterate the above procedure until there are no 1-valent vertices left. We denote by (Γ', ℓ') the n-pointed tropical curve obtained at the end.

$*$ *Step 2.* If Γ' has no vertex of valence 2 we are done. So, let v be a vertex of valence 2 of Γ'.

$**$ *Step 2.a.* If v has no leg based at it, then we remove v as described in subsection 2.2. (i.e. in such a way that the two edges adjacent to v are merged into an edge of length equal to the sum of their lengths). It is clear that this operation does not add any new leaf or leg, and diminishes the number of these 2-valent vertices. So we can repeat it finitely many times until there are no such 2-valent vertices left. The resulting tropical curve, (Γ'', ℓ''), has n points and it is tropically equivalent to the given one.

$**$ *Step 2b.* If v has a (necessarily unique) leg l based at it, let e be the (also unique) edge adjacent to v. Now by removing v the edge e changes into a leg, merging with l (see the figure in Remark 2.14). This operation preserves the equivalence class, the number of legs, and does not add any new leaf; it clearly diminishes the number of 2-valent vertices. So after finitely many iterations we arrive at a tropical curve, (Γ''', ℓ''') with n points and such that the graph has no vertex of valence less than 3.

We have thus shown that every tropical equivalence class of n-pointed tropical curves has a representative all of whose vertices have valence at least 3. The uniqueness of this representative is trivial to prove. □

Remark 2.17. Recall from Section 1.3 that two nodal algebraic curves are stably equivalent if they have isomorphic stabilizations. *To say that two n-pointed tropical curves are tropically equivalent is analogous to say that two n-pointed algebraic curves are stably equivalent.* This should be clear by comparing the construction of 1.3 with the proof of Proposition 2.16. Indeed: (Step 1) the removal of a 1-valent vertex and its adjacent edge corresponds to removing an unpointed rational tail. Next (Step 2.a) the removal of a 2-valent vertex adjacent to no legs corresponds to contracting an exceptional component to a node. Finally (Step 2.b) the removal of a 2-valent vertex having an adjacent leg corresponds to removing a uni-pointed rational tail.

Example 2.18. Let $g = 0$ and $n = 4$. A graph with 4 legs and no vertex of valence $\leqslant 2$ can have at most 1 edge. The graph with 0 edges is unique. On the other hand there are three non-isomorphic combinatorial graphs with one edge, according to how the 4 legs are distributed. They are drawn in the following picture. Each of these graphs supports a one dimensional family of metric graphs, as the length of their unique edge varies in $\mathbb{R}_{>0}$.

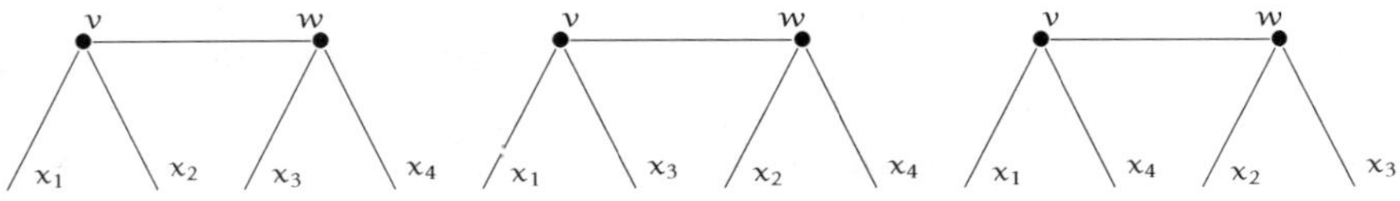

Figure 4. The three genus 0, 3-regular graphs with 4 legs.

Example 2.19. *Rational tropical curves.* A detailed description of the case $g = 0$ and $n \geqslant 4$, with nice pictures, can be found in [23].

This case is special for two reasons. First: it is easy to see that an n-pointed rational pointed curve free from vertices of valence $\leqslant 2$ has no nontrivial automorphisms. This is the key reason why the moduli space for these curves is a tropical variety. As we saw Example 1.5, this happens also for algebraic curves and, as a consequence, the moduli spaces $\overline{M}_{0,n}$ are smooth varieties (see [20]), whereas $\overline{M}_{g,n}$ is singular in general. As we said in the introduction, this discrepancy seems to have its tropical analogue in the fact that moduli spaces of tropical curves of higher genus are not tropical varieties in general.

The second reason why the genus zero case is easier to handle is that a family of genus zero tropical curves specializes to a genus zero curve. This is not the case if $g \geqslant 1$ as we are going to explain.

2.4. Adding a weight function on tropical curves

The definition of pure tropical curve presents a problem when studying families. To explain why, let us first identify tropical curves with metric graphs having no vertex of valence less than 3 (which, up to tropical equivalence, we can do). Now, by varying the lengths of the edges of a metric graph we obtain a family. Let us make this precise; fix a graph Γ as above, write $E(\Gamma) = \{e_1, \ldots, e_{|E(\Gamma)|}\}$, and consider the space of all metric graphs, i.e. of all tropical curves, supported on it. This space is easily identified with $\mathbb{R}_{>0}^{E(\Gamma)}$, indeed, to a vector

$$(l_1, \ldots, l_{|E(\Gamma)|}) \in \mathbb{R}_{>0}^{E(\Gamma)}$$

there corresponds the tropical curve (Γ, ℓ) such that $\ell(e_i) = l_i$, $\forall i$. It is then natural to ask what happens when some of the lengths go to zero. So, with the above notation, let l_1 say, tend to 0. How do we give the limit an interpretation in tropical, or geometric, language? There is a simple candidate: as l_1 tends to zero, (Γ, ℓ) specializes to a metric graph $(\overline{\Gamma}, \overline{\ell})$ where $\overline{\Gamma} = \Gamma_{/e_1}$ is obtained by contracting e_1 to a point, and $\overline{\ell}(e_i) = \ell(e_i)$, $\forall i \geqslant 2$ (as $E(\overline{\Gamma}) = E(\Gamma) \smallsetminus \{e_1\} = \{e_2, \ldots, e_{|E(\Gamma)|}\}$). But there is a drawback with this limit: its genus may be smaller than $g(\Gamma)$. Indeed we have

$$g(\overline{\Gamma}) = \begin{cases} g(\Gamma) - 1 & \text{if } e_1 \text{ is a loop} \\ g(\Gamma) & \text{otherwise.} \end{cases}$$

From a geometric perspective this is quite unpleasant. We like the genus to remain constant under specialization. A solution to this problem is provided by S. Brannetti, M. Melo and F. Viviani in [5]. The idea is to extend the definition of a tropical curve by adding a weight function on the vertices.

Definition 2.20. A *weighted graph* with n legs is a pair (Γ, w) where Γ is a graph with n legs, and $w : V(\Gamma) \to \mathbb{Z}_{\geqslant 0}$ a *weight* function on the vertices.

The genus $g(\Gamma, w)$ is defined as follows:

$$g(\Gamma, w) = b_1(\Gamma) + \sum_{v \in V(\Gamma)} w(v) = b_1(\Gamma) + |w|. \tag{2.21}$$

A *weighted metric graph* is defined exactly in the same way, assuming that Γ is a metric graph to start with.

Definition 2.22. Let $S \subset E(\Gamma)$ be a set of edges of a weighted graph (Γ, w). Using the notations of subsection 2.1 we define the *weighted contraction* of S as the weighted graph $(\Gamma_{/S}, w_{/S})$ where $\Gamma_{/S}$ is the contraction of S. The weight function $w_{/S}$ is defined by setting, for every $\overline{v} \in V(\Gamma_{/S})$,

$$w_{/S}(\overline{v}) = b_1(\sigma^{-1}(\overline{v})) + \sum_{v \in \sigma_V^{-1}(\overline{v})} w(v). \tag{2.23}$$

Example 2.24. Let $S = \{e\}$. If e is a loop based at a vertex v_0, let $\overline{v_0} \in V(\Gamma_{/S})$ be the image of v_0, hence the image of the contracted loop e. Then $w_{/S}(\overline{v_0}) = w(v_0) + 1$. Whereas for $\overline{v} \in V(\Gamma_{/S})$ with $\overline{v} \neq \overline{v_0}$ we have $w_{/S}(\overline{v}) = w(v)$. If e is not a loop, then $w_{/S}(\overline{v}) = \sum_{v \in \sigma_V^{-1}(\overline{v})} w(v)$ for every $\overline{v} \in V(\Gamma_{/S})$.

Example 2.25. In the next picture we have two weighted contractions; the starting graph (Γ, w) has all vertices of weight zero, represented by a "$\circ$", so that Γ has genus 3. We first contract the non-loop edge e_1, so that the weighted contraction has again weight function equal to zero. Then we contract a loop edge, so that the weighted contraction has one vertex of weight 1, represented by a "$\bullet$".

Remark 2.26. By the identities (2.8) and (2.9), we have

$$g(\Gamma_{/S}, w_{/S}) = g(\Gamma, w).$$

Let (Γ', w') be a weighted graph. We denote

$$(\Gamma, w) \geqslant (\Gamma', w') \text{ if } (\Gamma', w') \text{ is a weighted contraction of } (\Gamma, w). \tag{2.27}$$

The next definition generalizes [5, Def 3.1.3] (with changes in terminology).

Definition 2.28. An *n-pointed (weighted) tropical curve* of genus g is a triple (Γ, w, ℓ) where (Γ, w) is a weighted graph of genus g with legs $L(\Gamma) = \{x_1, \ldots, x_n\}$, ℓ a function $\ell : E(\Gamma) \cup L(\Gamma) \to \mathbb{R}_{>0} \cup \{\infty\}$ such that $\ell(x) = \infty$ if and only if x is a leg or an edge adjacent to 1-valent vertex of weight 0.

(As in Definition 2.12, the legs $\{x_1, \ldots, x_n\}$ are the marked points.)

If $w(v) = 0$ for every $v \in V(\Gamma)$, we write $w = \underline{0}$ and $(\Gamma, \underline{0}, \ell)$, and say that the tropical curve is *pure*, consistently with Definition 2.12.

A tropical curve is called *regular* if it is pure and if Γ is a 3-regular graph (every vertex has valence 3).

Two n-pointed tropical curves are tropically equivalent if they can be obtained from one another by adding or removing 2-valent vertices of weight 0, or 1-valent vertices of weight 0 together with their adjacent edge.

The terminology "pure" tropical curve is useful to keep track of the original definition, to connect to the rest of the literature. The notion of regular tropical curve, together with referring to the 3-regularity of its graph, suggests that, in our view, pure tropical curves with a 3-regular graph play the role of regular (i.e. smooth) curves in the moduli theory of algebraic curves; in fact, as we shall see, regular tropical curves have the following property.

If a tropical curve C *specializes to a regular one, then* C *itself is regular.*

The same holds for nonsingular algebraic curves: if an algebraic curve X specializes to a nonsingular one, then X is nonsingular. See Theorem 4.7 for more on this point.

Observe, however, that in Section 5 we shall describe a scenario in which the role of smooth algebraic curves is played by pure tropical curves.

Example 2.29. (This generalizes Example 2.15.) A graph made of one vertex of weight g and no edges nor legs is a weighted tropical curve of genus g. If $g \leqslant 1$ such a curve is the specialization of the degenerate cases described in Example 2.15, and will thus be excluded by our future numerical assumption, namely $2g - 2 + n > 0$. If $g \geqslant 2$ these curves play a role and must be considered.

To study moduli of tropical curves we need to generalize Proposition 2.16, which is straightforward, once we provide the correct replacement for graphs having no vertex of valence less than 3. Here is how to do that:

Definition 2.30. A weighted graph, or a metric weighted graph, is called *stable*[4] if any vertex of weight 0 has valence at least 3, and any vertex of weight 1 has valence at least 1.

For instance, $(\Gamma, \underline{0})$ is stable if and only if Γ has no vertex of valence $\leqslant 2$. All graphs in Example 2.18 are stable.

[4]The terminology is motivated by the subsequent Remark 4.2.

Remark 2.31. If (Γ, w) is stable and $(\Gamma', w') \leqslant (\Gamma, w)$, then (Γ', w') is stable.

Now we can generalize Proposition 2.16. The proof is the same, provided that "1-valent vertices" are replaced by "1-valent vertices of weight 0", and hence "leaves" are replaced by "edges adjacent to a 1-valent vertex of weight 0".

Proposition 2.32. *Assume* $2g - 2 + n \geqslant 1$.

(1) Every tropical equivalence class of n*-pointed, weighted tropical curves contains a representative whose* n *marked points are disjoint.*

(2) The set of tropical equivalence classes of n*-pointed, weighted tropical curves of genus* g *is in bijection with the set of stable metric weighted graphs of genus* g *with* n *legs.*

The following well known, elementary fact will be useful.

Lemma 2.33. *Let* (Γ, w) *be a genus* g *stable graph with* n *legs. Then* $|E(\Gamma)| \leqslant 3g - 3 + n$ *and equality holds if and only if* Γ *is a 3-regular graph with* $b_1(\Gamma) = g$*. Moreover, in such a case* $w = \underline{0}$.

Proof. We use induction on n. Let us begin with the base case, $n = 0$. We have $g = b_1(\Gamma) + |w| = |E(\Gamma)| - |V(\Gamma)| + 1 + |w|$; hence, as $|w| \geqslant 0$,

$$|E(\Gamma)| = g - 1 - |w| + |V(\Gamma)| \leqslant g - 1 + |V(\Gamma)|$$

and the maximum is achieved for $|w| = 0$. In this case $g = b_1(\Gamma)$ and every vertex of Γ must have valence at least 3 (as (Γ, w) is stable), therefore

$$g = |E(\Gamma)| - |V(\Gamma)| + 1 \geqslant 3|V(\Gamma)|/2 - |V(\Gamma)| + 1 = |V(\Gamma)|/2 + 1$$

hence $|V(\Gamma)| \leqslant 2g - 2$ and equality holds if and only if Γ is 3-regular. In such a case we have $|E(\Gamma)| = 3g - 3$ and $b_1(\Gamma) = g$.

Let us assume $n > 0$. Let (Γ, w) be a stable graph of genus g for which $|E(\Gamma)|$ is maximum. We can reduce to assume that $\{g, n\} \neq \{0,3\}$ and $\{g, n\} \neq \{1,1\}$ by treating these trivial cases separately. Denote by $(l_1, \ldots, l_n)$ the legs of Γ. Let $\Gamma' := \Gamma - l_n$, hence Γ' has $n - 1$ legs, and same vertices and edges as Γ. We claim that Γ' is not stable. Indeed, if Γ' is stable, we can construct a new stable graph Γ'', of genus g as follows. Pick an edge $e' \in E(\Gamma')$, insert a weight-0 vertex u in its interior, and add a leg adjacent to u. This gives a graph, stable of genus g and having a number of edges equal to $|E(\Gamma)| + 1$. This contradicts the maximality of $|E(\Gamma)|$.

So, Γ' is not a stable graph; this means that the vertex $v \in V(\Gamma)$ adjacent to l_n has weight 0 and valence 3. Now look at v as a vertex of Γ'; it has valence 2 and, by our reduction, Γ' has either two distinct edges, or an edge and a leg, adjacent to v. We consider the weighted contraction of Γ' given by contracting an edge adjacent to v, a stable graph (Γ^*, w^*) of genus g with $n - 1$ legs. By induction, we have $|E(\Gamma^*)| = 3g - 3 + n - 1$, moreover Γ^* is 3-regular with weight function constantly zero. But then $|E(\Gamma)| = |E(\Gamma^*)| + 1 = 3g - 3 + n - 1 + 1 = 3g - 3 + n$ and Γ is 3-regular by construction. The proof is complete. □

3. The moduli space of tropical curves

From now on, we shall assume $2g-2+n \geqslant 1$ and we shall consider weighted tropical curves up to tropical equivalence. Hence, by Proposition 2.32 we can identify our n-pointed tropical curves with stable metric graphs.

Everything we shall say about weighted tropical curves holds for pure tropical curves, with obvious modifications.

3.1. Tropical curves with fixed combinatorial type

Definition 3.1. We say that two tropical curves (Γ, w, ℓ) and (Γ', w', ℓ') with n marked points $L(\Gamma)=\{x_1,\dots,x_n\}$ and $L(\Gamma')=\{x'_1,\dots,x'_n\}$ are isomorphic, and write $(\Gamma, w, \ell)\cong(\Gamma', w', \ell')$, if there is an isomorphism α of the underlying unweighted n-pointed curves (Γ,ℓ) and (Γ',ℓ') as in Definition 2.12, such that $\forall v\in V(\Gamma)$ we have $w(v)=w'(\alpha(v))$.

We write $\mathrm{Aut}(\Gamma, w, \ell)$ for the set of automorphisms of a tropical curve (Γ, w, ℓ).

Let us fix a stable graph (Γ, w) of genus g with labeled legs $L(\Gamma)=\{x_1,\dots,x_n\}$, and let us introduce the set $M^{\mathrm{trop}}(\Gamma, w)$ of isomorphism classes of n-pointed tropical curves supported on (Γ, w). In order to study $M^{\mathrm{trop}}(\Gamma, w)$, we introduce the open cone

$$R(\Gamma, w) := \mathbb{R}_{>0}^{E(\Gamma)}.$$

Any element in $R(\Gamma, w)$ defines a unique metric graph supported on (Γ, w). Therefore there is a natural surjection

$$\pi : R(\Gamma, w) \longrightarrow M^{\mathrm{trop}}(\Gamma, w); \qquad (3.2)$$

it is clear that $\pi(\Gamma, w, \ell)=\pi(\Gamma, w, \ell')$ if and only if $(\Gamma, w, \ell)\cong(\Gamma, w, \ell')$.

The closure of $R(\Gamma, w)$ is, of course, the closed cone

$$\overline{R(\Gamma, w)} = \mathbb{R}_{\geqslant 0}^{E(\Gamma)} \subset \mathbb{R}^{E(\Gamma)}.$$

Let $p\in\overline{R(\Gamma, w)}\smallsetminus R(\Gamma, w)$ and, to simplify the notation, suppose that

$$p=(t_1,\dots,t_m,0,\dots,0)$$

with $t_i>0$ for all $i=1,\dots,m$, for some $0\leqslant m<|E(\Gamma)|$. Let us show that there is a unique n-pointed tropical curve (Γ_p, w_p, ℓ_p) of genus g associated to p. Let

$$S=\{e_{m+1},\dots,e_{|E(\Gamma)|}\}\subset E(\Gamma)=\{e_1,\dots,e_{|E(\Gamma)|}\};$$

then $(\Gamma_p, w_p)=(\Gamma_{/S}, w_{/S})$, i.e. (Γ_p, w_p) is the weighted contraction of (Γ, w) obtained by contracting S (defined in 2.22). We thus have a natural identification

$$E(\Gamma_p)=E(\Gamma)\smallsetminus S=\{e_1,\dots,e_m\}.$$

The length function ℓ_p is defined by setting $\ell_p(e_i)=t_i$ for all $e_i\in E(\Gamma_p)$. As we noticed in Remark 2.26, we have $g=g(\Gamma_p, w_p)$.

Summarizing: we showed that the boundary points of $\overline{R(\Gamma, w)}$ parametrize n-pointed tropical curves of genus g whose underlying weighted graph is a contraction of (Γ, w). More precisely, for any $I \subset \{1,2,\dots,|E(\Gamma)|\}$, denote by $\overline{R(\Gamma,w)}_I \subset \overline{R(\Gamma,w)}$ the open face

$$\overline{R(\Gamma,w)}_I := \{(t_1,\dots,t_{|E(\Gamma)|}) \in \overline{R(\Gamma,w)} : \ t_i = 0 \ \forall i \in I, \ t_i > 0 \ \forall i \notin I\}.$$

Next, write $S_I := \{e_i, \forall i \in I\} \subset E(\Gamma)$. We have proved the following.

Lemma 3.3. *With the above notation, the partition*

$$\overline{R(\Gamma,w)} = \bigsqcup_{I \subset \{1,2,\dots,|E(\Gamma)|\}} \overline{R(\Gamma,w)}_I$$

is such that for every I *there is a natural isomorphism* $\overline{R(\Gamma,w)}_I \cong R(\Gamma',w')$ *where* $(\Gamma',w') = (\Gamma_{/S_I}, w_{/S_I})$.

Example 3.4. If $I = E(\Gamma)$ then $\overline{R(\Gamma,w)}_I = \{0\}$, corresponding to the graph with no edges, one vertex of weight g, and n legs. It is clear that this graph can be obtained as weighted contraction from every genus g weighted graph.

Consider two points $p_1, p_2 \in \overline{R(\Gamma,w)}$. By what we said there exist two metric weighted graphs associated to them, denoted (Γ_1, w_1, ℓ_1) and (Γ_2, w_2, ℓ_2). We have an equivalence on $\overline{R(\Gamma,w)}$:

$$p_1 \sim p_2 \ \text{ if } \ (\Gamma_1, w_1, \ell_1) \cong (\Gamma_2, w_2, \ell_2). \tag{3.5}$$

The quotient with respect to this equivalence relation will be denoted

$$\overline{\pi} : \overline{R(\Gamma,w)} \longrightarrow \overline{M^{\text{trop}}(\Gamma,w)} := \overline{R(\Gamma,w)}/\sim \tag{3.6}$$

and $\overline{M^{\text{trop}}(\Gamma,w)}$ is a topological space, with the quotient topology of the Euclidean topology on $\overline{R(\Gamma,w)} = \mathbb{R}_{\geqslant 0}^{E(\Gamma)}$. A precise description of the fibers of $\overline{\pi}$ is given in Lemma 3.12 below.

Now we must study the automorphisms of (Γ, w), and how they act $\overline{R(\Gamma,w)}$.

Definition 3.7. Let (Γ, w) be a weighted graph with labeled legs $L(\Gamma) = \{x_1,\dots,x_n\}$. An automorphism of (Γ, w) is an isomorphism, α, of Γ with itself as defined in 2.2, such that $\alpha(x_i) = x_i$ for every $i = 1,\dots,n$, and $w(v) = w(\alpha(v))$ for every $v \in V(\Gamma)$. We denote by $\text{Aut}(\Gamma, w)$ the group of automorphisms of (Γ, w).

Remark 3.8. If (Γ, w, ℓ) is a pointed curve, then $\text{Aut}(\Gamma, w, \ell) \subset \text{Aut}(\Gamma, w)$. The automorphism group of a weighted graph, and of a tropical curve, is finite.

Of course, $\text{Aut}(\Gamma, w)$ is made of pairs $\alpha = (\alpha_V, \alpha_E)$ of permutations on the vertices and the edges satisfying some compatibility conditions; in fact, by definition, α has to fix the legs (i.e. with the notation of Definition 2.2, $\alpha_L = id_{L(\Gamma)}$). Therefore $\text{Aut}(\Gamma, w)$ acts on $\overline{R(\Gamma,w)}$ by permuting the coordinates according to α_E, in particular, $\text{Aut}(\Gamma, w)$ acts as group of isometries.

$\text{Aut}(\Gamma, w)$ may contain non-trivial elements acting trivially on $\overline{R(\Gamma,w)}$.

Example 3.9. The loop-inversion automorphism described in Example 2.6 acts trivially on $\overline{R(\Gamma, w)}$.

Example 3.10. Assume $L(\Gamma) = \emptyset$ and $V(\Gamma) = \{v_1, v_2\}$ with $w(v_1) = w(v_2)$; suppose $E(\Gamma) = \{e_1, \dots, e_n\}$ and let Γ have no loops, as in the picture below. So (Γ, w) has genus $n - 1 + |w|$.

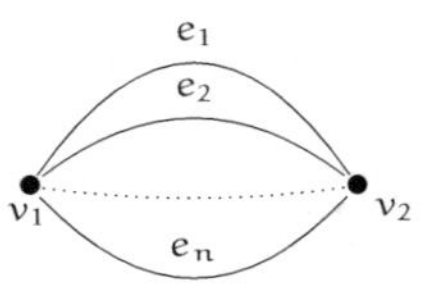

Now, Γ has an involution swapping v_1 and v_2, and every conjugate pair of half-edges $\{h, \overline{h}\}$ (so that the edges are all fixed). This automorphism acts trivially on $\overline{R(\Gamma, w)}$.

Denote by $\mathcal{S}_n$ the symmetric group, then

$$\mathrm{Aut}(\Gamma, w) \cong \mathcal{S}_n \times \mathbb{Z}/2\mathbb{Z}$$

where the $\mathcal{S}_n$ factor accounts for automorphisms permuting the edges, which clearly act non-trivially on $\overline{R(\Gamma, w)}$.

Observe that if $L(\Gamma) \neq \emptyset$, or if $w(v_1) \neq w(v_2)$, then $\mathrm{Aut}(\Gamma, w) \cong \mathcal{S}_n$.

In the sequel we simplify the notation and set, for $I \subset (1, \dots, |E(\Gamma)|)$,

$$F_I := \overline{R(\Gamma, w)}_I, \quad \text{and} \quad G_I := \mathrm{Aut}(\Gamma_{/S_I}, w_{/S_I}) \tag{3.11}$$

so that G_I acts on $\overline{F_I}$ (closure of F_I in $R(\Gamma, w)$) by permuting the coordinates. As F_I varies among the (open) faces of $\overline{R(\Gamma, w)}$ it may very well happen that different faces correspond to isomorphic weighted graphs. Let us introduce some notation to keep track of this fact. For a fixed $I \subset \{1, \dots, |E(\Gamma)|\}$, we denote

$$\mathcal{I}so(I) := \{J \subset (1, \dots, |E(\Gamma)|) : \ (\Gamma_{/S_J}, w_{/S_J}) \cong (\Gamma_{/S_I}, w_{/S_I})\}.$$

Next, for every $J \in \mathcal{I}so(I)$ we fix an isomorphism $\Phi_J : (\Gamma_{/S_I}, w_{/S_I}) \to (\Gamma_{/S_J}, w_{/S_J})$, and the isometry

$$\phi_J : \overline{F_I} \xrightarrow{\cong} \overline{F_J}$$

induced by Φ_J. Notice that ϕ_J is induced by a bijection between the natural coordinates of F_I and F_J. If $I = J$ we shall assume that Φ_I is the identity. For every point $p \in F_I$ we denote $\mathcal{I}so(p) := \mathcal{I}so(I)$.

Lemma 3.12. *Let* $p \in \overline{R(\Gamma, w)}$. *Then, with the notation* (3.6),

$$\overline{\pi}^{-1}(\overline{\pi}(p)) = \{g\phi_J(p), \ \forall J \in \mathcal{I}so(p), \ \forall g \in G_J\}.$$

Proof. The inclusion $\overline{\pi}^{-1}(\overline{\pi}(p)) \supset \{g\phi_J(p), \forall J \in \mathcal{I}so(p), \forall g \in G_J\}$ is obvious, as the set on the right parametrizes isomorphic metric weighted graphs.

Let F_I be the face containing p. Let $r \in \overline{R(\Gamma, w)}$ be such that $\overline{\pi}(p) = \overline{\pi}(r)$. We have an isomorphism $(\Gamma_p, w_p, \ell_p) \cong (\Gamma_r, w_r, \ell_r)$, hence $F_I \cong R(\Gamma_p, w_p) \cong R(\Gamma_r, w_r)$. Suppose first that $r \in F_I$. Then the underlying weighted graph of p and r is the same, namely $(\Gamma_{/S_I}, w_{/S_I})$, and an isomorphism between them is an element of $\mathrm{Aut}(\Gamma_{/S_I}, w_{/S_I}) = G_I$ preserving the lengths of the edges. In other words, there exists $g \in G_I$ such that $r = gp$ (recall that $\phi_I = \mathrm{Id}_{F_I}$).

Now let $F_J \cong R(\Gamma_r, w_r)$ be the face containing r, with $J \neq I$. Of course, $J \in \mathcal{I}so(p)$, hence we have an isometry $\phi_J : F_I \to F_J$, induced by an isomorphism between the underlying graphs. It is clear that $r' := \phi_J(p)$ parametrizes a metric weighted graph isomorphic to the one parametrized by p. Therefore $\overline{\pi}(r') = \overline{\pi}(r)$, hence, by the previous part, there exists $g \in G_J$ such that $r = gr' = g\phi_J(p)$. □

For any group G acting on $\overline{R(\Gamma, w)}$ and any subset $Z \subset \overline{R(\Gamma, w)}$, we denote by $Z^G \subset \overline{R(\Gamma, w)}$ the union of the G-orbits of the elements in Z.

Proposition 3.13. *Let (Γ, w) be a stable graph.*

(1) There is a canonical decomposition (notation in (2.27))

$$\overline{M^{trop}(\Gamma, w)} = \bigsqcup_{(\Gamma', w') \leqslant (\Gamma, w)} M^{trop}(\Gamma', w'),$$

where $M^{trop}(\Gamma, w)$ is open and dense in $\overline{M^{trop}(\Gamma, w)}$.

(2) The quotient map $\overline{\pi} : \overline{R(\Gamma, w)} \to \overline{M^{trop}(\Gamma, w)}$ factors as follows:

$$\overline{\pi} : \overline{R(\Gamma, w)} \xrightarrow{\tau} \overline{R(\Gamma, w)}/\mathrm{Aut}(\Gamma, w) \xrightarrow{\gamma} \overline{M^{trop}(\Gamma, w)}.$$

Moreover τ is open and $\overline{\pi}$ has finite fibers.

(3) $\overline{M^{trop}(\Gamma, w)}$ is a Hausdorff topological space.

Proof. The existence of the decomposition is a straightforward consequence of the definition of $\overline{M^{trop}(\Gamma, w)}$ and of Lemma 3.3.

Recall from (3.2) that $M^{trop}(\Gamma, w) = R(\Gamma, w)/\sim$. Now $R(\Gamma, w)$ is open and dense in $\overline{R(\Gamma, w)}$ and we have $\overline{\pi}^{-1}(M^{trop}(\Gamma, w)) = R(\Gamma, w)$. Therefore $M^{trop}(\Gamma, w)$ is open and dense in $\overline{M^{trop}(\Gamma, w)}$.

Now let $G := \mathrm{Aut}(\Gamma, w)$. The restriction of $\overline{\pi}$ to $R(\Gamma, w)$ is the quotient $\pi : R(\Gamma, w) \to M^{trop}(\Gamma, w) = R(\Gamma, w)/G$, so in this case part (2) is proved.

Let now $p \in \overline{R(\Gamma, w)}_I$ with $I \neq \emptyset$ (cf. Lemma 3.3). Then $p = (t_1, \dots, t_{|E(\Gamma)|})$ with $t_i = 0$ for every $i \in I$, and $t_i > 0$ otherwise. Let $\alpha \in G$, then α acts as a permutation on $\{1, \dots, |E(\Gamma)|\}$, hence

$$\alpha(p) = (t_{\alpha^{-1}(1)}, \dots, t_{\alpha^{-1}(|E(\Gamma)|)}) \in \overline{R(\Gamma, w)}_{\alpha(I)},$$

where $\alpha(I) = \{\alpha(i), \forall i \in I\} \subset \{1, 2, \dots, |E(\Gamma)|\}$. Let $S = S_I \subset E(\Gamma)$ and $\alpha(S) \subset E(\Gamma)$ be the set of edges corresponding, respectively, to I and $\alpha(I)$. Lemma 3.3 and the discussion preceding it yield

$$\overline{R(\Gamma,w)_I} = R(\Gamma_{/S}, w_{/S}), \quad \overline{R(\Gamma,w)_{\alpha(I)}} = R(\Gamma_{/\alpha(S)}, w_{/\alpha(S)}).$$

The automorphism α maps S into $\alpha(S)$, hence it induces an isomorphism

$$\overline{\alpha} : \Gamma_{/S} \xrightarrow{\cong} \Gamma/\alpha(S).$$

Moreover, α (as any automorphism of any graph) maps bijectively cycles to cycles, therefore $\overline{\alpha}$ induces an isomorphism $(\Gamma_{/S}, w_{/S}) \cong (\Gamma/\alpha(S), w/\alpha(S))$. Finally, the point p corresponds to a metric graph on $(\Gamma_{/S}, w_{/S})$ with length function $\ell(e_i) = t_i$ for every $i \notin I$ (we have $E(\Gamma_{/S}) = E(\Gamma) \smallsetminus S$); it is clear that $\alpha(p)$ corresponds to a metric graph on $(\Gamma_{/S}, w_{/S})$ with length $\alpha(\ell)$ given by $\alpha(\ell)(e_{\alpha(i)}) = t_{\alpha^{-1}(\alpha(i))} = t_i$ for every $i \notin I$. But then p and $\alpha(p)$ parametrize isomorphic metric weighted graphs. Therefore $\overline{\pi}(p) = \overline{\pi}(\alpha(p))$ and the factorization of part (2) is proved.

The fact that τ is open follows easily from G being a finite group of isometries of $\overline{R(\Gamma,w)}$. Similarly, we easily get that $\overline{R(\Gamma,w)}/\mathrm{Aut}(\Gamma,w)$ is Hausdorff.

The fibers of $\overline{\pi}$ are finite by Lemma 3.12.

It remains to show that $\overline{M^{trop}(\Gamma,w)}$ is Hausdorff; to do that we will use the notation (3.11). Let $\overline{p}$ and $\overline{q}$ be two distinct points in $\overline{M^{trop}(\Gamma,w)}$. Set $\overline{\pi}^{-1}(\overline{p}) = \{p_1,\dots,p_m\}$ and $\overline{\pi}^{-1}(\overline{q}) = \{q_1,\dots,q_n\}$. There exists an $\epsilon > 0$ small enough that the following holds.

(1) For every i,j the open balls $B_{p_i}(\epsilon)$ and $B_{q_j}(\epsilon)$ do not intersect each other.

(2) If $\overline{F_I} \cap B_{p_i}(\epsilon) \neq \emptyset$ then $p_i \in \overline{F_I}$. Similarly, if $\overline{F_I} \cap B_{q_j}(\epsilon) \neq \emptyset$ then $q_j \in \overline{F_I}$.

We now set

$$U := \cup_{i=1}^m B_{p_i}(\epsilon) \cap \overline{R(\Gamma,w)} \quad \text{and} \quad V := \cup_{j=1}^n B_{q_j}(\epsilon) \cap \overline{R(\Gamma,w)}.$$

It is clear that U and V are open subsets of $\overline{R(\Gamma,w)}$; moreover $U \cap V = \emptyset$, by (1) above. We claim that

$$\pi^{-1}(\pi(U)) = U \quad \text{and} \quad \pi^{-1}(\pi(V)) = V. \tag{3.14}$$

To prove it we set some simplifying conventions. Pick $I \subset \{1,\dots,|E(\Gamma)|\}$ and consider the group G_I, which acts on $\overline{F_I}$. Consider also, for every $J \in \mathcal{I}so(I)$, the isometry $\phi_J : \overline{F_I} \to \overline{F_J}$ (cf. Lemma 3.12). We extend to $\overline{R(\Gamma,w)}$ the action of G_I and the map ϕ_J as follows. For every $p \in \overline{R(\Gamma,w)} \smallsetminus \overline{F_I}$, every $g \in G_I$ and every ϕ_J as above, we set $gp = p$ and $\phi_J(p) = p$. Now, by Lemma 3.12 to prove the claim it suffices to prove the following two facts.

(a) For every I and every $g \in G_I$ we have $U^g \subset U$.

(b) For every $u \in U$ and every $J \in \mathcal{I}so(u)$ we have $\phi_J(U) \subset U$.

Pick I and $g \in G_I$; as g acts as the identity away from $\overline{F_I}$ we can assume that $\overline{F_I} \cap U \neq \emptyset$, i.e. that there exists p_i such that $\overline{F_I} \cap B_{p_i}(\epsilon) \neq \emptyset$. By (2) above, this implies that $p_i \in \overline{F_i}$; hence $gp_i \in \pi^{-1}(\overline{p})$ and

$$(\overline{F_I} \cap B_{p_i}(\epsilon))^g = \overline{F_I} \cap B_{gp_i}(\epsilon) \subset U$$

(as G_I preserves the metric of $\overline{F_i}$). This proves (a). To prove (b), let $u \in U \cap F_I$, let $J \in \mathcal{I}so(u)$ and let $\phi_J : \overline{F_I} \to \overline{F_J}$. As before, there exists p_i such that $u \in \overline{F_I} \cap B_{p_i}(\epsilon)$. Therefore

$$\phi_J(\overline{F_I} \cap B_{p_i}(\epsilon)) = \overline{F_J} \cap B_{\phi_J(p_i)}(\epsilon).$$

By what we proved before, $\overline{\pi}(p_i) = \overline{\pi}(\phi_J(p_i))$, hence $\overline{F_J} \cap B_{\phi_J(p_i)}(\epsilon) \subset U$. Now (b) is proved, and claim (3.14) with it. This yields that $\overline{\pi}(U)$ and $\overline{\pi}(V)$ are open and disjoint in $\overline{M^{trop}(\Gamma, w)}$. Since obviously $\overline{p} \in \overline{\pi}(U)$ and $\overline{q} \in \overline{\pi}(V)$ we are done. □

Remark 3.15. The map γ in Proposition 3.13 (2) identifies isomorphic curves which are not identified by the automorphisms of (Γ, w). The point is: in general, $\mathrm{Aut}(\Gamma, w)$ does not induce every automorphism of its weighted contractions, nor does it induces all the isomorphisms between its weighted contractions.

For instance, consider the graph (Γ, w) in Example 2.25. Then the graph in the middle of the picture, $(\Gamma_{/e_1}, w_{/e_1})$, has an $\mathcal{S}_3$ of automorphisms permuting its three loops; these automorphisms act non trivially on $\overline{R(\Gamma, w)}$, and are not all induced by automorphisms of (Γ, w).

Next, contracting any one of the three loops of $(\Gamma_{/e_1}, w_{/e_1})$ gives three isomorphic graphs (one of which is drawn at the right of that picture). It is clear that $\mathrm{Aut}(\Gamma, w)$ does not act transitively on them.

3.2. Construction and properties of $M_{g,n}^{trop}$

We shall now construct the moduli space of n-pointed tropical curves of genus g, $M_{g,n}^{trop}$, as a topological space. Recall that we always assume $2g - 2 + n \geqslant 1$. We set (denoting by "$\cong$" isomorphism of tropical curves)

$$M_{g,n}^{trop} := \Big(\bigsqcup_{\substack{\Gamma \text{ 3-regular} \\ b_1(\Gamma)=g, |L(\Gamma)|=n}} \overline{M^{trop}(\Gamma, \underline{0})} \Big) / \cong . \tag{3.16}$$

Remark 3.17. Consider the quotient map

$$\pi_g : \bigsqcup_{\substack{\Gamma \text{ 3-regular} \\ b_1(\Gamma)=g, |L(\Gamma)|=n}} \overline{M^{trop}(\Gamma, \underline{0})} \longrightarrow M_{g,n}^{trop}. \tag{3.18}$$

If Γ is 3-regular with $b_1(\Gamma) = g$, then $\pi_g^{-1}(\pi_g([(\Gamma, \underline{0}, \ell)])) = [(\Gamma, \underline{0}, \ell)]$.

Comparing with (3.6) we have another description of $M_{g,n}^{trop}$:

$$M_{g,n}^{trop} = \Big(\bigsqcup_{\substack{\Gamma \text{ 3-regular} \\ b_1(\Gamma)=g, |L(\Gamma)|=n}} \overline{R(\Gamma, \underline{0})} \Big) / \sim . \tag{3.19}$$

$M_{g,n}^{trop}$ is defined as the topological quotient space of a topological space; the two expressions of $M_{g,n}^{trop}$ as a quotient clearly yield the same topology.

Remark 3.20. $M_{g,n}^{trop}$ *is connected.* Indeed, every $\overline{M^{trop}(\Gamma, \underline{0})}$ appearing in (3.16) is connected and contains the point parametrizing the metric weighted graph with no edges and one vertex of weight g; see Example 3.4. We shall prove later, in Proposition 4.11, that $M_{g,n}^{\mathrm{trop}}$ is actually connected through codimension one.

Theorem 3.21. *Assume* $2g-2+n \geqslant 1$.

(1) *The points of* $M_{g,n}^{\mathrm{trop}}$ *bijectively parametrize isomorphism classes of* n*-pointed tropical curves of genus* g *(up to tropical equivalence).*

(2) *Let* $M_{g,n}^{\mathrm{reg}} \subset M_{g,n}^{\mathrm{trop}}$ *be the subset parametrizing regular curves, i.e.*

$$M_{g,n}^{\mathrm{reg}} = \bigsqcup_{\substack{\Gamma\ 3-\mathit{regular} \\ b_1(\Gamma)=g, |L(\Gamma)|=n}} M^{\mathrm{trop}}(\Gamma, w) \subset M_{g,n}^{\mathrm{trop}},$$

and $M_{g,n}^{\mathrm{reg}}$ *is open and dense.*

(3) *Let* $M_{g,n}^{\mathrm{pure}}$ *be the subset parametrizing pure tropical curves. Then* $M_{g,n}^{\mathrm{pure}}$ *is open and dense.*

(4) $M_{g,n}^{\mathrm{trop}}$ *is Hausdorff.*

Proof. Let (Γ, w, ℓ) be a stable metric graph of genus g with n legs, and let us prove that its isomorphism class corresponds to a point in $M_{g,n}^{\mathrm{trop}}$. This amounts to showing that (Γ, w, ℓ) is in the closure of $R(\Gamma_0, \underline{0})$ for some 3-regular graph Γ_0 with $b_1(\Gamma_0) = g$. In other words, by Lemma 3.3, we must prove that (Γ, w) is a weighted contraction of $(\Gamma_0, \underline{0})$, with Γ_0 a 3-regular graph.

Suppose first that $w = \underline{0}$. A proof of this fact is the proof of [8, Prop. A.2.4](that Proposition is concerned with 3-edge-connected curves, but the proof is easily seen to work in this case).

We continue by induction on $|w|$, the basis being the case $w = \underline{0}$, which we just did. Let us now assume that $w(v_1) \geqslant 1$ for some $v_1 \in V(\Gamma)$. Let (Γ', w') be the genus g stable graph defined as follows: Γ' is obtained from Γ by just adding a loop, e_0, based at v_1. Therefore we have $V(\Gamma') = V(\Gamma)$ and $E(\Gamma') = E(\Gamma) \cup \{e_0\}$. Now let

$$w'(v) = \begin{cases} w(v) & \text{if } v \neq v_1 \\ w(v_1) - 1 & \text{otherwise.} \end{cases}$$

It is clear that (Γ, w) is a weighted contraction of (Γ', w'), indeed

$$(\Gamma, w) = (\Gamma'/e_0, w'/e_0).$$

We can apply induction, as $|w'| = |w| - 1$; hence (Γ', w') is a weighted contraction of $(\Gamma_0, \underline{0})$ for some 3-regular graph Γ_0, i.e $(\Gamma', w') = ((\Gamma_0)_{/S}, \underline{0}_{/S})$ for some $S \subset E(\Gamma_0)$. Now, $e_0 \in E(\Gamma') \subset E(\Gamma_0)$, hence, denoting $S_0 = \{e_0\} \cup S$, we have $(\Gamma, w) = (\Gamma_{0\ /S_0}, \underline{0}_{/S_0})$. The first part is proved.

Let us prove (2); the description of $M_{g,n}^{\mathrm{reg}}$ follows from Lemma 2.33. Now fix a 3-regular graph Γ; by Proposition 3.13 we know that $M^{\mathrm{trop}}(\Gamma, \underline{0})$ is open and dense in $\overline{M^{\mathrm{trop}}(\Gamma, \underline{0})}$. Consider the definition of $M_{g,n}^{\mathrm{trop}}$ given in (3.16). Pick a point $[(\Gamma, \underline{0}, \ell)] \in M^{\mathrm{trop}}(\Gamma, \underline{0})$; by Remark 3.17, the map π_g induces a homeomorphism of

$M^{trop}(\Gamma,\underline{0})$ with its image such that

$$\pi_g^{-1}(\pi_g(M^{trop}(\Gamma,\underline{0})) = M^{trop}(\Gamma,\underline{0}).$$

Therefore $M^{trop}(\Gamma,\underline{0})$ is open in $M_{g,n}^{trop}$. Moreover, the union of the $M^{trop}(\Gamma,\underline{0})$ as Γ runs through all 3-regular graphs with $b_1(\Gamma) = g$ is obviously dense in $M_{g,n}^{trop}$. Therefore the above union is open and dense.

Now (3). We have, of course

$$M_{g,n}^{reg} \subset M_{g,n}^{pure} \subset M_{g,n}^{trop}.$$

Hence $M_{g,n}^{pure}$ is dense by part (2). Now recall that

$$M_{g,n}^{pure} = \{[(\Gamma,w,\ell)] : b_1(\Gamma) = g\} = \{[(\Gamma,w,\ell)] : |w| = 0\}.$$

Let us denote by $R(\Gamma,w)^+ \subset \overline{R(\Gamma,w)}$ the union of all loci corresponding to weighted graphs (Γ',w') such that $b_1(\Gamma') < g$ (cf. Lemma 3.3). The set $R(\Gamma,w)^+$ is closed, as the first Betti number does not grow under edge contraction. Hence its complement, the locus parametrizing pure tropical curve, is open. Hence the locus in $\overline{M^{trop}(\Gamma,\underline{0})}$ corresponding to pure tropical curves is also open.

Let now $\overline{p} \in M_{g,n}^{pure} \subset M_{g,n}^{trop}$ and let $(\Gamma_p,\underline{0})$ be its supporting graph. For every 3-regular graph Γ with $b_1(\Gamma) = g$ let $p_\Gamma \in \overline{M^{trop}(\Gamma,\underline{0})}$ be the preimage of p. By what we just said p_Γ admits an open neighborhood $U_\Gamma \subset \overline{M^{trop}(\Gamma,\underline{0})}$ such that U_Γ parametrizes only pure tropical curves. Up to shrinking each U_Γ around p_Γ we can assume that $\pi_g(U_\Gamma) = \pi_g(U_{\Gamma'})$ for all such Γ and Γ'. But then

$$\pi_g(\cup U_\Gamma) \subset M_{g,n}^{pure}, \text{ and } \pi_g^{-1}(\pi_g(\cup U_\Gamma)) = \cup U_\Gamma$$

where the union is over all 3-regular graphs Γ with $b_1(\Gamma) = g$. This implies that $\pi_g(\cup U_\Gamma)$ is an open neighborhood of $\overline{p}$ all contained in $M_{g,n}^{pure}$. Part (3) is proved.

Finally, let us show that $M_{g,n}^{trop}$ is Hausdorff. The quotient map π_g (3.18) induces a bijection of $\overline{M^{trop}(\Gamma,\underline{0})}$ with its image. Therefore, by Proposition 3.13, $M_{g,n}^{trop}$ is obtained by gluing together finitely many Hausdorff spaces. Let $\overline{p},\overline{q} \in M_{g,n}^{trop}$; for every 3-regular Γ with $b_1(\Gamma) = g$ let $p_\Gamma = \overline{M^{trop}(\Gamma,\underline{0})} \cap \pi_g^{-1}(\overline{p})$ and $q_\Gamma = \overline{M^{trop}(\Gamma,\underline{0})} \cap \pi_g^{-1}(\overline{q})$ and pick, in the Hausdorff space $\overline{M^{trop}(\Gamma,\underline{0})}$, disjoint open neighborhoods of the two points: $p_\Gamma \in U_\Gamma$, $q_\Gamma \in V_\Gamma$. If $\pi_g^{-1}(\overline{p})$ doesn't intersect $\overline{M^{trop}(\Gamma,\underline{0})}$ we don't do anything, similarly for $\overline{q}$. Arguing as before, we can assume that $\pi_g(U_\Gamma) = \pi_g(U_{\Gamma'})$ for all Γ and Γ' as above, hence $\pi_g(U_\Gamma)$ is open; similarly for V_Γ. Then, $\pi_g(U_\Gamma)$ and $\pi_g(V_\Gamma)$ are open disjoint neighborhoods of $\overline{p}$ and $\overline{q}$ in $M_{g,n}^{trop}$. □

Remark 3.22. For every stable graph (Γ,w) as above, the space $M^{trop}(\Gamma,w)$ is the quotient of the topological manifold $R(\Gamma,w)$ by the finite group $\mathrm{Aut}(\Gamma,w)$. Its dimension can thus be defined as follows:

$$\dim M^{trop}(\Gamma,w) := \dim R(\Gamma,w) = |E(\Gamma)|.$$

More generally, let X be a topological space containing a dense open subset which is an orbifold of dimension n (locally the quotient of an n-dimensional topological manifold by a finite group); then we say that X *has pure dimension* n.

Remark 3.23. $M_{g,n}^{trop}$ *has pure dimension equal to* $3g - 3 + n$.

Indeed, by Theorem 3.21 the moduli space of regular curves, $M_{g,n}^{reg}$, is open and dense in $M_{g,n}^{trop}$. Now, $M_{g,n}^{reg}$ is the disjoint union of finitely many spaces of type $M^{trop}(\Gamma, w)$. By what we just observed, each of these spaces has dimension $|E(\Gamma)|$; by Lemma 2.33 we have $|E(\Gamma)| = 3g - 3 + n$.

Example 3.24. Suppose $g = 1$, hence $n \geqslant 1$. The case $n = 1$ is very simple: $M_{1,1}^{trop}$ is homeomorphic to $\mathbb{R}_{\geqslant 0}$ with the point 0 identified with the curve with no edges and a vertex of weight 1 (see Figure 5). In the next pictures we list all the combinatorial types in cases $n = 1$ and $n = 3$; the 0-weight vertices are pictured as "∘", whereas the 1-weight vertices are pictured as "•", together with a "+1" next to them. The dashed arrows represent specializations, i.e. weighted contractions.

Figure 5. The case $g = 1$, $n = 1$

Next is the case $n = 3$. It is clear that $M_{1,3}^{trop}$ is not a manifold. In Figure 6, the notation "×2", or "×3", next to a specialization arrow means that the specialization is obtained in two, or three, different ways, i.e. by contracting two, or three, different edges.

3.3. Compactification of the moduli space of tropical curves

In this subsection we describe a natural compactification of the moduli space of tropical curves, following an indication of G.Mikhalkin (see [23] for example).

Definition 3.25. An *extended, n-pointed tropical curve* (up to tropical equivalence) of genus g will be a triple (Γ, w, ℓ) where (Γ, w) is a stable graph of genus g with n labeled legs, and

$$\ell : E(\Gamma) \cup L(\Gamma) \to \mathbb{R}_{>0} \cup \{\infty\}$$

is a length function on the edges such that $\ell(x) = \infty$ for every $x \in L(\Gamma)$.

We are thus generalizing the previous definition by allowing the length of any edge to be infinite.

Remark 3.26. In the sequel, we view $\mathbb{R} \cup \{\infty\}$ as a topological space with the Alexandroff one-point topology (i.e. the open sets are all the usual open subsets of $\mathbb{R}$, and

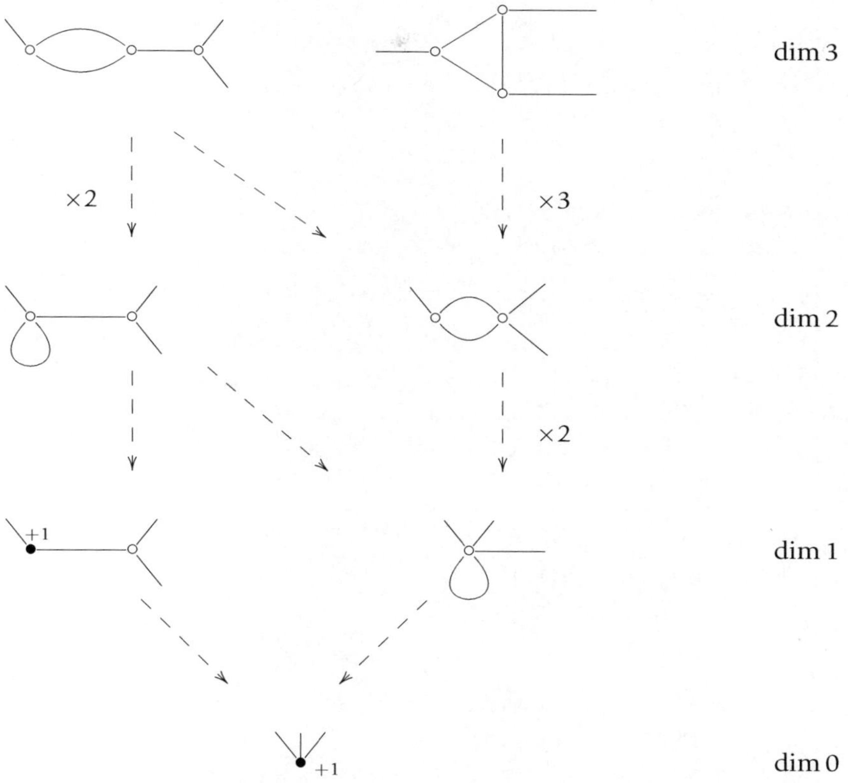

Figure 6. The case $g = 1$, $n = 3$

all the complements of compact subsets of $\mathbb{R}$). Therefore, for every n the spaces $(\mathbb{R} \cup \{\infty\})^n$ are compact, Hausdorff spaces. Hence also the spaces $(\mathbb{R}_{\geqslant 0} \cup \{\infty\})^n$ are are compact and Hausdorff.

The definitions of isomorphism, or automorphism, of extended tropical curves are given exactly in the same way as for tropical curves. In particular, the automorphism group of an extended tropical curve is finite. For a stable graph (Γ, w) we denote

$$R_\infty(\Gamma, w) := (\mathbb{R}_{>0} \cup \{\infty\})^{E(\Gamma)};$$

every element in $R_\infty(\Gamma, w)$ corresponds to an extended tropical curve whose underlying graph is (Γ, w). We extend the notation for tropical curves as follows. We write $M_\infty^{\mathrm{trop}}(\Gamma, w)$ for the set of isomorphism classes of extended tropical curves having (Γ, w) as underlying graph; hence we have a surjection

$$R_\infty(\Gamma, w) \longrightarrow M_\infty^{\mathrm{trop}}(\Gamma, w).$$

The closure of $R_\infty(\Gamma, w)$ inside $(\mathbb{R}\cup\{\infty\})^{E(\Gamma)}$ is denoted by

$$\overline{R_\infty(\Gamma, w)} := (\mathbb{R}_{\geqslant 0}\cup\{\infty\})^{E(\Gamma)}.$$

Lemma 3.3 trivially extends, so that we have a decomposition

$$\overline{R_\infty(\Gamma, w)} = \bigsqcup_{I\subset\{1,2,\ldots,E(\Gamma)\}} \overline{R_\infty(\Gamma, w)}_I$$

where for every I we have $\overline{R_\infty(\Gamma, w)}_I \cong R_\infty(\Gamma_{/S_I}, w_{/S_I})$.

We continue by introducing the quotient space

$$\overline{\pi_\infty} : \overline{R_\infty(\Gamma, w)} \longrightarrow \overline{M_\infty^{\text{trop}}(\Gamma, w)} := \overline{R_\infty(\Gamma, w)}/\sim$$

where $p_1 \sim p_2$ if and only if the extended tropical curves parametrized by p_1 and p_2 are isomorphic. Naturally, $\overline{M_\infty^{\text{trop}}(\Gamma, w)}$ is endowed with the quotient topology induced by $\overline{R_\infty(\Gamma, w)}$. The finite group $\text{Aut}(\Gamma, w)$ acts on $\overline{R_\infty(\Gamma, w)}$ by permuting the coordinates, and hence it acts as group of homeomorphisms. Proposition 3.13 extends:

Proposition 3.27. *Let (Γ, w) be a stable graph.*

(1) There is a decomposition

$$\overline{M_\infty^{\text{trop}}(\Gamma, w)} = \bigsqcup_{(\Gamma', w')\leqslant(\Gamma, w)} M_\infty^{\text{trop}}(\Gamma', w'),$$

where $M_\infty^{\text{trop}}(\Gamma, w)$ is open and dense in $\overline{M_\infty^{\text{trop}}(\Gamma, w)}$.

(2) The map $\overline{\pi_\infty}$ has finite fibers.

(3) $\overline{M_\infty^{\text{trop}}(\Gamma, w)}$ is a compact Hausdorff topological space.

Proof. The only new statement with respect to Proposition 3.13 is the compactness of $\overline{M_\infty(\Gamma, w)}$, which follows from its being a quotient of the compact topological space $\overline{R_\infty(\Gamma, w)}$.

For the rest, it suffices to explain what needs to be modified in the proof of Proposition 3.13, whose notation we continue to use here. One easily checks that the only part where some changes are needed is the proof of Hausdorffness. Indeed, we need to replace the open neighborhoods $B_p(\epsilon)$ with neighborhoods of a different type. To do that, pick a point $p = (t_1, \ldots, t_{|E(\Gamma)|}) \in \overline{R_\infty(\Gamma, w)}$, and any pair of positive real numbers ϵ, η; we define the following open neighborhood of p

$$A_p(\epsilon, \eta) = \prod_{j=1}^{|E(\Gamma)|} A_p(\epsilon, \eta)_j \subset (\mathbb{R}_{\geqslant 0}\cup\{\infty\})^{E(\Gamma)}$$

where $A_p(\epsilon, \eta)_j \subset \mathbb{R}_{\geqslant 0}\cup\{\infty\}$ is the following open subset

$$A_p(\epsilon, \eta)_j = \begin{cases} (t_j - \epsilon, t_j + \epsilon)\cap\mathbb{R}_{\geqslant 0} & \text{if } t_j \in \mathbb{R} \\ \mathbb{R}_{\geqslant 0}\cup\{\infty\}\smallsetminus[-\eta, +\eta] & \text{if } t_j = \infty. \end{cases}$$

It is clear that $A_p(\epsilon, \eta)$ is open for all p, ϵ and η as above.

Now we show that $\overline{M_\infty^{trop}(\Gamma,w)}$ is Hausdorff, by small modifications of the proof of Proposition 3.13. Let $\overline{p}$ and $\overline{q}$ be two distinct points in $\overline{M_\infty(\Gamma,w)}$. Let $\pi_\infty^{-1}(p)=\{p_1,\dots,p_n\}$ and $\pi_\infty^{-1}(q)=\{q_1,\dots,q_m\}$. It is easy to check that there exist $\epsilon\in\mathbb{R}_{>0}$ small enough and $\eta\in\mathbb{R}$ big enough so that
(1) $A_{p_i}(\epsilon,\eta)\cap A_{q_j}(\epsilon,\eta)=\emptyset$ for every $i=1,\dots,n$ and $j=1,\dots,m$.
(2) If $\overline{F_{\infty,I}}\cap A_{p_i}(\epsilon,\eta)\neq\emptyset$ then $p_i\in\overline{F_{\infty,I}}$. If $\overline{F_{\infty,I}}\cap A_{q_j}(\epsilon,\eta)\neq\emptyset$ then $q_j\in\overline{F_{\infty,I}}$, where $F_{\infty,I}:=R_\infty(\Gamma,w)_I$.

Set $U:=\cup_{i=1}^n A_{p_i}(\epsilon,\eta)$ and $V=\cup_{j=1}^m A_{q_j}(\epsilon,\eta)$. It is clear that U and V are open and disjoint. We claim that

$$\pi_\infty^{-1}(\pi_\infty(U))=U \quad\text{and}\quad \pi_\infty^{-1}(\pi_\infty(V))=V. \tag{3.28}$$

The proof is identical to the proof of the analogous claim used for Proposition 3.13. And (3.28) implies the statement, as in Proposition 3.13. □

Similarly to what we did in (3.21) we set

$$\overline{M_{g,n}^{trop}}:=\Big(\bigsqcup_{\substack{\Gamma\ 3-\text{regular}\\ b_1(\Gamma)=g}}\overline{M_\infty^{trop}(\Gamma,\underline{0})}\Big)/\cong \tag{3.29}$$

where $\cong$ is isomorphism of extended tropical curves. We obviously have a quotient map extending the map (3.18)

$$\pi_g:\bigsqcup_{\substack{\Gamma\ 3-\text{regular}\\ b_1(\Gamma)=g}}\overline{M_\infty^{trop}(\Gamma,\underline{0})}\longrightarrow\overline{M_{g,n}^{trop}}\supset M_{g,n}^{trop}$$

so that $\overline{M_{g,n}^{trop}}$ has the quotient topology induced by π_g. If Γ is a 3-regular graph with $b_1(\Gamma)=g$, then $\pi_g^{-1}(\pi_g([(\Gamma,\underline{0},l)]))=[(\Gamma,\underline{0},l])$. Extending Theorem 3.21 we have

Theorem 3.30. *Let $g\geqslant 2$. The points of $\overline{M_{g,n}^{trop}}$ bijectively parametrize isomorphism classes of extended n-pointed tropical curves of genus g. The topological space $\overline{M_{g,n}^{trop}}$ is compact and Hausdorff. Furthermore, it has pure dimension $3g-3+n$ and it is connected through codimension one*[5].

Proof. By Proposition 3.27 the spaces $\overline{M_\infty^{trop}(\Gamma,\underline{0})}$ are all compact. Therefore $\overline{M_{g,n}^{trop}}$ is the quotient of a compact space (the union in (3.29) is finite) and hence it is compact. The proof of its being connected through codimension one is the same as for $M_{g,n}^{trop}$, given in 4.11; indeed that proof is concerned with the underlying weighted graphs, and completely ignores the length function. The proof of its being Hausdorff is a simple modification of the proof of Theorem 3.21, along the same lines as the modifications used to prove Proposition 3.27 □

[5]See Definition 4.10

4. Comparing moduli spaces

4.1. Dual graph of a curve and combinatorial partition of $\overline{M}_{g,n}$

We recall the definition of a useful graph associated to a pointed nodal curve.

Definition 4.1. If $(X,\underline{p}) = (X; p_1, \dots p_n)$ is a pointed nodal curve, its *(weighted) dual graph* is the weighted graph $(\Gamma_{(X,\underline{p})}, w_{(X,\underline{p})})$ with n legs such that $\Gamma_{(X,\underline{p})}$ has a vertex for every irreducible component of X and an edge for every node of X joining the two (possibly equal) vertices corresponding to the components on which the node lies; for every point p_i there is a leg of $\Gamma_{(X,\underline{p})}$ adjacent to the vertex corresponding to the component containing p_i. The weight function $w_{(X,\underline{p})}$ assigns to every vertex the geometric genus of the corresponding component.

So the (arithmetic) genus g of X is the same as that of its dual graph

$$g = b_1(\Gamma_{(X,\underline{p})}) + \sum_{v \in V(\Gamma_{(X,\underline{p})})} w_{(X,\underline{p})}(v) = g(\Gamma_{(X,\underline{p})}, w_{(X,\underline{p})}).$$

Remark 4.2. The n-pointed curve $(X;\underline{p})$ is stable if and only if its weighted dual graph $(\Gamma_{(X,\underline{p})}, w_{(X,\underline{p})})$ is stable.

Example 4.3. Assume $g = 0$ and $n = 4$. We described all stable 4-pointed rational curves in Example 1.5. Their dual graphs are pictured in Example 2.18.

Now, for any stable graph (Γ, w) of genus g with n labeled legs, we denote by $M^{alg}(\Gamma, w) \subset \overline{M}_{g,n}$ the locus of n-pointed curves with (Γ, w) as dual graph:

$$M^{alg}(\Gamma, w) := \{(X;\underline{p}) \in \overline{M}_{g,n} : (\Gamma_{(X,\underline{p})}, w_{(X,\underline{p})}) = (\Gamma, w)\}^6.$$

The following is the *combinatorial partition* of $\overline{M}_{g,n}$:

$$\overline{M}_{g,n} = \bigsqcup_{(\Gamma,w) \text{ stable, } n \text{ legs, genus } g} M^{alg}(\Gamma, w). \tag{4.4}$$

The next well known fact is an easy consequence of Fact 1.2.

Lemma 4.5. *Assume $2g - 2 + n > 0$. Then for any stable graph of genus g with n legs, (Γ, w), we have that $M^{alg}(\Gamma, w)$ is an irreducible quasiprojective variety and its codimension in $\overline{M}_{g,n}$ is equal to $|E(\Gamma)|$.*

Proof. Set $\delta := |E(\Gamma)| > 0$ (otherwise we have a special case of 1.2). Pick a curve $(X;\underline{p}) \in M^{alg}(\Gamma, w)$, denote by $C_1, \dots, C_\gamma$ its irreducible components, and let n_i be the number of marked points contained in C_i, so that $\sum_1^\gamma n_i = n$.

Our $(X;\underline{p})$ determines, for every $i = 1, \dots, \gamma$, a nonsingular curve of genus g_i with $n_i + \delta_i$ marked points in it, $(C_i^\nu; \underline{p}^{(i)})$, where $\nu : \sqcup_1^\gamma C_i^\nu \to X$ denotes the normalization of X and

$$\delta_i := |\nu^{-1}(X_{sing}) \cap C_i^\nu|.$$

---

[6]The superscript "alg" stands for "algebraic" and is used to keep the distinction with $M^{trop}(\Gamma, w)$.

Moreover, as we observed in Remark 1.8, we have that $(C_i^\gamma; \underline{p}^{(i)})$ is a stable curve, i.e. $(C_i^\gamma; \underline{p}^{(i)}) \in M_{g_i, n_i+\delta_i}$. Once we have such γ pointed curves, the gluing data of $\nu^{-1}(X_{\text{sing}})$ are uniquely determined by the graph Γ. We conclude that there is a surjective morphism

$$M_{g_1,n_1+\delta_1} \times \ldots \times M_{g_\gamma,n_\gamma+\delta_\gamma} \longrightarrow M^{\text{alg}}(\Gamma, w).$$

The above morphism extends to $\overline{M}_{g_1,n_1+\delta_1} \times \ldots \times \overline{M}_{g_\gamma,n_\gamma+\delta_\gamma} \to \overline{M}_{g,n}$ and $M^{\text{alg}}(\Gamma, w)$ is open in its image; hence $M^{\text{alg}}(\Gamma, w)$ is quasiprojective. Now for every i, Fact 1.2 applies, hence $M_{g_i,n_i+\delta_i}$ is irreducible of dimension $3g_i - 3 + n_i + \delta_i$ for all $i = 1, \ldots, \gamma$. Hence $M^{\text{alg}}(\Gamma, w)$ is irreducible. Since the above surjection has clearly finite fibers we get

$$\dim M^{\text{alg}}(\Gamma, w) = \sum_{i=1}^{\gamma}(3g_i - 3 + n_i + \delta_i) = 3\sum_{i=1}^{\gamma} g_i - 3\gamma + n + 2\delta$$

(since $\sum_{i=1}^{\gamma} \delta_i = 2\delta$). Now $g = \sum_{i=1}^{\gamma} g_i + \delta - \gamma + 1$ hence

$$\dim M^{\text{alg}}(\Gamma, w) = 3g - 3\delta + 3\gamma - 3 - 3\gamma + n + 2\delta = 3g - 3 + n - \delta.$$

□

4.2. Partition analogies

By Theorem 3.21 and Proposition 3.13 we have on $M_{g,n}^{\text{trop}}$ a partition similar to (4.4)

$$M_{g,n}^{\text{trop}} = \bigsqcup_{(\Gamma,w) \text{ stable, } n \text{ legs, genus } g} M^{\text{trop}}(\Gamma, w). \tag{4.6}$$

In the next statement we highlight some topological and combinatorial analogies which, undoubtedly, are known to the experts. We include a (partial) proof, as we believe it is quite instructive. By $\dim M^{\text{alg}}(\Gamma, w)$ we mean the dimension as an algebraic variety, while by $\operatorname{codim} M^{\text{trop}}(\Gamma, w)$ we mean the orbifold codimension in $M_{g,n}^{\text{trop}}$.

Theorem 4.7. *Assume $2g - 2 + n \geqslant 1$. Consider the partitions (4.4) and (4.6), and the bijection*

$$M^{\text{alg}}(\Gamma, w) \mapsto M^{\text{trop}}(\Gamma, w)$$

where (Γ, w) varies among stable graphs of genus g with n legs. Then the following properties hold.

(1) $\dim M^{\text{alg}}(\Gamma, w) = \operatorname{codim} M^{\text{trop}}(\Gamma, w) = 3g - 3 + n - |E(\Gamma)|$.

(2) With the notation (2.27),

$$M^{\text{alg}}(\Gamma, w) \subset \overline{M^{\text{alg}}(\Gamma', w')} \Leftrightarrow M^{\text{trop}}(\Gamma', w') \subset \overline{M^{\text{trop}}(\Gamma, w)} \Leftrightarrow (\Gamma, w) \geqslant (\Gamma', w').$$

(3) Let (Γ, w) be such that $E(\Gamma) \neq \emptyset$. Then there exists a stable graph (Γ', w') with $|E(\Gamma')| = |E(\Gamma)| - 1$ such that $M^{\text{alg}}(\Gamma, w) \subset \overline{M^{\text{alg}}(\Gamma', w')}$ and $M^{\text{trop}}(\Gamma', w') \subset \overline{M^{\text{trop}}(\Gamma, w)}$.

Proof. For (1), the statement about $M^{\mathrm{alg}}(\Gamma, w)$ follows from Lemma 4.5. On the other hand, we observed above that $\dim M^{\mathrm{trop}}(\Gamma, w) = |E(\Gamma)|$; hence we get $\operatorname{codim} M^{\mathrm{trop}}(\Gamma, w) = \dim M_{g,n}^{\mathrm{trop}} - |E(\Gamma)| = 3g - 3 + n - |E(\Gamma)|$.

Now (2), whose second double implication follows from Proposition 3.13(1). We give the proof of the remaining part only in case $n = 0$, leaving the (straightforward) generalization to the reader.

Let $X \in M^{\mathrm{alg}}(\Gamma, w)$ be a stable curve; if X lies in the closure of $M^{\mathrm{alg}}(\Gamma', w')$ there exist families of curves with dual graph (Γ', w') specializing to X. We pick one of these families, $f : \mathcal{X} \to B$ with a one dimensional base B with a marked point b_0, so that the fiber of f over every $b \neq b_0$ is a stable curve $X_b \in M^{\mathrm{alg}}(\Gamma, w)$, while the fiber over b_0 is isomorphic to X.

Under such a specialization every node of X_b specializes to a node of X, and distinct nodes specialize to distinct nodes. This singles out a set $T \subset E(\Gamma)$ of nodes of X, namely T is the set of nodes that are specializations of nodes of X_b. Let $S = E(\Gamma) \smallsetminus T$, so S parametrizes the nodes of X that do not come from nodes of X_b. Consider the graph $(\Gamma_{/S}, w_{/S})$; we claim that $(\Gamma_{/S}, w_{/S}) = (\Gamma', w')$. To prove it we shall use the notation of Definition 2.22.

By construction we have a bijection between $E(\Gamma')$ and $E(\Gamma_{/S}) = T$, mapping an edge of Γ', i.e. a node of X', to the node in X to which it specializes.

The total space $\mathcal{X}$ of our family of curves is singular along the nodes of the fibers X_b, for $b \neq b_0$. Let us desingularize $\mathcal{X}$ at such loci (exactly $|T|$ of them); we thus obtain a new family $\mathcal{Y} \to B$ whose fiber over $b \neq b_0$ is the normalization of X_b. The fiber over b_0 is the partial normalization of X at T, which we denote by Y; so its dual graph satisfies $\Gamma_Y = \Gamma - T$.

Notice that $\mathcal{Y} \to B$ is a union of families parametrized by the irreducible components of X_b, i.e. by the vertices of Γ'. Let us denote these families by $\mathcal{Y}(v') \to B$. So if $b \neq b_0$ the fiber of $\mathcal{Y}(v')$ over b is the smooth irreducible component corresponding to $v' \in V(\Gamma')$. The fiber over b_0 of $\mathcal{Y}(v') \to B$ is a connected component of Y, which we denote $Y(v')$. Of course $Y(v')$ determines a set of vertices of Γ (those corresponding to its components). Now notice that two different vertices of Γ' determine in this way disjoint sets of vertices of Γ. Therefore we have a surjection $\phi : V(\Gamma) \to V(\Gamma')$ mapping each vertex v to the vertex v' such that the component corresponding to v lies in $Y(v')$. It is clear that $\phi(v_1) = \phi(v_2)$ if and only if v_1 and v_2 belong to the same connected component of $\Gamma \smallsetminus T$. Therefore ϕ is the same map as the map $\sigma_V : V(\Gamma) \to V(\Gamma_{/S})$. This shows that $V(\Gamma')$ and $V(\Gamma_{/S})$ are in natural bijection, and hence that $\Gamma' \cong \Gamma_{/S}$. Finally, since the arithmetic genus of a family of algebraic curves is constant, we have, for any $v' \in V(\Gamma')$, that the genus of the component corresponding to v', i.e. the weight $w'(v')$, is equal to the arithmetic genus of the limit curve $Y(v')$. Therefore

$$w'(v') = b_1(\Gamma_{Y(v')}) + \sum_{v \in \sigma_V^{-1}(v')} w(v) = b_1(\sigma^{-1}(v')) + \sum_{v \in \sigma_V^{-1}(v')} w(v).$$

By (2.23) the weight function w' coincides with $w_{/S}$; so we are done.

Conversely, suppose that $(\Gamma, w) \geqslant (\Gamma', w')$ with $(\Gamma', w') = (\Gamma_{/S}, w_{/S})$ for some $S \subset E(\Gamma)$; let $T := E(\Gamma) \setminus S$. We shall show how to reverse the procedure we just described. Let $X \in M^{alg}(\Gamma, w)$ and let $Y \to X$ be the normalization of X at T, so that $\Gamma_Y = \Gamma - T$. Notice that Y is endowed with $|T|$ pairs of smooth distinguished points, namely the branches over the nodes in T, and it is thus a disjoint union of stable pointed curves. Therefore, by Remark 1.8, there exists a family of curves $\mathcal{Y} \to B$ with $|T|$ pairs of disjoint sections (with $\dim B = 1$, $b_0 \in B$, and $\mathcal{Y}$ a possibly disconnected surface) whose fiber over b_0 is Y (as a pointed curve) and whose fiber over $b \neq b_0$ is a disjoint union of smooth curves with $2|T|$ distinct points. We let $\mathcal{X}$ be the surface obtained by gluing together the $|T|$ pairs of sections. It is clear that $\mathcal{X}$ is a family of nodal curves over B, whose fiber over b_0 is X and whose fiber over $b \neq b_0$ lies in $M^{alg}(\Gamma', w')$.

Now (3). Let $e \in E(\Gamma)$ and set $(\Gamma', w') = (\Gamma_{/e}, w_{/e})$. Then (Γ', w') is stable and $|E(\Gamma')| = |E(\Gamma)| - 1$. By (2) we are done. □

Corollary - Definition 4.8. *The* k-dimensional stratum *of the combinatorial partition of $\overline{M}_{g,n}$ given in (4.4) is defined as the following closed subscheme of $\overline{M}_{g,n}$:*

$$C_g^{(k)} := \overline{\bigsqcup_{\dim M^{alg}(\Gamma,w)=k} M^{alg}(\Gamma, w)} = \bigsqcup_{\dim M^{alg}(\Gamma,w)\leqslant k} M^{alg}(\Gamma, w). \tag{4.9}$$

4.3. Connectedness properties

The next definition is adapted from [21, Definition 3.3.2].

Definition 4.10. Let X be a topological space of pure dimension d; see Remark 3.22. Assume that X is endowed with a decomposition $X = \sqcup_{i\in I} X_i$, where every X_i is a connected orbifold. We say that X is *connected through codimension one* if the subset

$$\bigsqcup_{i\in I:\dim X_i \geqslant d-1} X_i \subset X$$

is connected.

Connectedness through codimension one is a strong form of connectedness (if X is connected through codimension one, it is also connected). Tropical varieties (associated to prime ideals) have this property, which in fact is a fundamental one; see the Structure Theorem in [21, Ch. 3].

Although $M_{g,n}^{trop}$ is not a tropical variety in general, we may still ask whether $M_{g,n}^{trop}$ is connected through codimension one. The answer is yes (see [5] for the unpointed case).

The second part of the next proposition, stating that $C_g^{(k)}$ is connected, may be known, although we don't have a reference; our proof, exhibiting as a a simple consequence of the first part, is a good illustration of the interplay between the combinatorial and the algebro-geometric point of view.

Proposition 4.11. *(1) $M^{trop}_{g,n}$ is connected through codimension one.*

(2) If $k \geqslant 1$ the k-dimensional stratum $C_g^{(k)}$ of the combinatorial partition of $\overline{M}_{g,n}$ is connected.

Proof. By Theorem 4.7, the second part in case $k = 1$ is an immediate consequence of the first, and the case $k \geqslant 2$ follows from it.

So, let us concentrate on the tropical moduli space. We saw in 3.23 that $M^{trop}_{g,n}$ is of pure dimension $3g-3+n$. We know that (4.6) is a decomposition of $M^{trop}_{g,n}$ as disjoint union of orbifolds of known dimensions. We must therefore show that its subset

$$\bigsqcup_{\text{codim } M^{trop}(\Gamma,w)\leqslant 1} M^{trop}(\Gamma,w) = \bigsqcup_{|E(\Gamma)|\geqslant 3g-4+n} M^{trop}(\Gamma,w)$$

is connected. Recall that in the above decomposition the top dimensional orbifolds, of dimension $3g-3+n$, are precisely the ones for which Γ is a 3-regular graph. To prove the statement we apply the following result of Hatcher-Thurston [19, Prop. page 236] (a purely combinatorial proof, with other applications to moduli of tropical curves, may be found in [6]):

Fact 4.12. *Let Γ and Γ' be two 3-regular connected graphs of the same genus, free from legs. Then there exists a finite sequence*

$$\begin{array}{ccccccccc} \Gamma=\Gamma_1 & & & & \Gamma_3 & \dots & \dots & & \Gamma_{2h+1}=\Gamma' \\ & \searrow & & \swarrow & & \searrow & & \swarrow & \\ & & \Gamma_2 & & & & \dots & \Gamma_{2h} & \end{array} \tag{4.13}$$

where every arrow is the map contracting precisely one edge which is not a loop. Moreover, every odd-indexed graph in the diagram above is 3-regular.

More precisely, every even-indexed graph above satisfies

$$\Gamma_{2i} = (\Gamma_{2i-1})_{/e} = (\Gamma_{2i+1})_{/e}$$

with $e \in E(\Gamma_{2i-1})$ and $e' \in E(\Gamma_{2i+1})$.

To use this for our statement, let us first suppose that $n = 0$. Then we may apply the above diagram to two weighted graphs $(\Gamma, \underline{0})$ and $(\Gamma', \underline{0})$. Since no loop gets contracted, we may view the arrows of the above diagram as weighted contractions, where the weight function is always the zero function. By hypothesis the odd-indexed sets $M^{trop}(\Gamma_{2i-1}, \underline{0})$ are codimension zero sets of the decomposition (4.6), hence every $M^{trop}(\Gamma_{2i}, \underline{0})$ has codimension one.

The above fact says that the closures of two consecutive odd-indexed sets $M^{trop}(\Gamma_{2i-1}, \underline{0})$ and $M^{trop}(\Gamma_{2i+1}, \underline{0})$ intersect in $M^{trop}(\Gamma_{2i}, \underline{0})$. Therefore, if $n = 0$, we are done.

If $n > 0$ we may assume $3g-3+n \geqslant 2$, for otherwise the result follows from what we know already (namely, that $M^{trop}_{g,n}$ is connected). Now, we need to show that Fact 4.12 holds for graphs with n legs, in such a way that no leg gets contracted by the maps in the diagram. This is easily proved by induction on n, noticing that if we remove from our 3-regular graph Γ a leg together with its base vertex, we obtain a 3-regular graph of the same genus as Γ, and $n-1$ legs (as $3g-3+n \geqslant 2$). Conversely, if we add a leg in the interior of any edge of a 3-regular graph with $n-1$ legs, we get a 3-regular graph with n legs, and all 3-regular graphs with n legs can be obtained in this way (more details can be found in [6, Prop. 3.3.1]). □

5. Moduli spaces via Teichmüller theory

From now on, we restrict to the unpointed case $n = 0$. Up to now we considered the moduli space of stable algebraic curves $\overline{M}_g$, or, which is the same, the moduli space of stable equivalence classes of nodal curves. The construction of $\overline{M}_g$ as a projective variety, as given in [17], can be summarized as follows (we refer also to [18, Chapt. 4] for details and references).

By definition, stable curves have an ample dualizing bundle, so they can all be embedded in some projective space using a suitable power of it. Furthermore, this projective space can be chosen to be the same $\mathbb{P}^r$ for all stable curves of fixed genus g. Since the dualizing bundle is preserved by isomorphisms, two such projective curves are abstractly isomorphic if and only if they are projectively equivalent, i.e. conjugate by an element of the group $G = \mathrm{Aut}(\mathbb{P}^r) = PGL(r+1)$.

Let us denote by H the set of all these projective models of our curves. By what we said there is an obvious bijection between the quotient set H/G and the set of isomorphism classes of genus g stable curves.

To give the set H/G an algebraic structure, one proceeds by giving H (via Grothendieck's theory of Hilbert schemes), and then H/G (via Mumford's Geometric Invariant Theory) the structure of an algebraic variety.

The approach we just sketched has many advantages; we wish to mention a few of them. One is the fact that stable curves are treated at the same time as smooth curves, so that the same construction yields a projective algebraic variety $\overline{M}_g = H/G$, the moduli space of stable curves, and an open dense subset $M_g \subset \overline{M}_g$, the moduli space of smooth curves.

Another consequence of this construction is that, being purely algebraic, it works in any characteristic. Since the space H turns out to be irreducible and smooth, one obtains that $\overline{M}_g$, and hence M_g, is irreducible (an important fact that was not known in positive characteristic) and normal, i.e. mildly singular.

Finally, the quotient $H \to H/G$ can be rather explicitly described, locally at every point. By what we said, the stabilizers of the action of G are the automorphism groups of our abstract stable curves, which are finite by definition. In particular, locally at curves having no nontrivial automorphisms (and there is a dense open subset of them as soon as $g \geqslant 3$) the space $\overline{M}_g$ is nonsingular.

A somewhat more sophisticated construction of $\overline{M}_g$ can be given, by first constructing the related algebraic stack, $\overline{\mathcal{M}_g}$, and then showing that this stack admits a projective moduli scheme $\overline{M}_g$. The stack $\overline{\mathcal{M}_g}$ is preferable from the point of view of moduli theory, as it retains more information about the moduli problem than the moduli variety $\overline{M}_g$. Stacks form a larger category than algebraic varieties or schemes, and may be viewed, loosely speaking, as the algebraic counterparts of topological orbifolds. We have mentioned this more general point of view as it relates to some of the open problems that we shall list at the end of the paper. More details can be found in [12] or [2, Chapt 12 and 14].

We will end the section by speaking about another approach to the moduli theory of curves that can be used both in the algebraic and tropical situation.

5.1. The Teichmüller approach to moduli of complex curves

In complex geometry, a different, quite natural, point of view is the so-called Teichmüller approach, which puts emphasis on the topological aspects. We shall now give a very short overview of profound and fundamental results; for details and references we refer to [2, Chapt. XV].

The starting observation is that every smooth complex curve C of genus g has the same underlying topological manifold, up to homeomorphism; namely a compact, connected, orientable surface of genus g, which we shall fix and denote by S_g from now on.

These topological 2-manifolds have been deeply studied in the past, and their classification dates back to the nineteenth century. Moreover, the group of isotopy classes of orientation-preserving homeomorphisms[7] of S_g with itself, known as the *mapping class group* and denoted by Γ_g, has also been heavily studied. Among its properties, we need to recall that, denoting by $\Pi_g := \pi_1(S_g)$, we have

$$\Gamma_g \cong \mathrm{Out}^+(\Pi_g) \subset \mathrm{Out}(\Pi_g) := \mathrm{Aut}(\Pi_g)/\mathrm{Inn}(\Pi_g)$$

where the superscript "+" indicates restriction to orientation preserving automorphisms (see loc. cit. for the precise definition). Let us go back to our moduli problem. To parametrize complex curves one proceeds by "marking them", i.e. one fixes a homeomorphism, called *marking*, $\mu : S_g \to C$, up to isotopy, and considers the set of isomorphism classes of marked complex curves (C, μ). So, (C, μ) and (C', μ') are in the same class if there exists an algebraic isomorphism $\iota : C \to C'$ such that the diagram below commutes

(5.1)

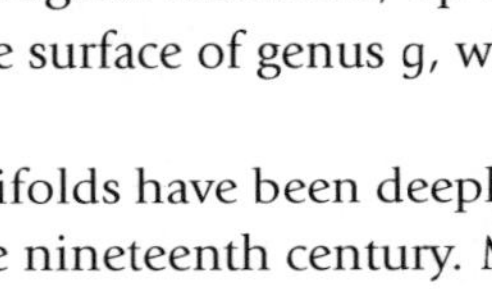

[7]Two homeomorphisms are isotopic if they can be deformed to one another by a continuous family of homeomorphisms.

The *Teichmüller space* T_g is defined as the set of all such classes

$$T_g := \{[(C, \mu)], \ C \ \text{of genus } g\}.$$

Using the basic deformation theory of curves, in particular the properties of their so-called Kuranishi family, the Teichmüller space T_g is endowed with a natural structure of complex manifold of dimension $3g - 3$. T_g is known to be a contractible space.

Now, it is clear that there is a surjection

$$T_g \longrightarrow M_g; \quad [(C, \mu)] \mapsto [C].$$

In fact, this surjection is the quotient of T_g by the natural action of the mapping class group Γ_g given by, for $\gamma \in \Gamma_g$

$$[(C, \mu)] \mapsto [(C, \gamma \circ \mu)].$$

Let us conclude this brief description by mentioning that the Teichmüller approach has had remarkable applications in establishing some geometric properties of M_g and of $M_{g,n}$; see [18] or [2].

5.2. The Teichmüller approach to moduli of metric graphs

We shall now briefly overview some work of M. Culler and K. Vogtmann ([11]) on moduli of metric graphs. Their approach, inspired by the previously described Teichmüller theory, presents strong analogies with the algebraic situation. By contrast, their main goal was not the study of the moduli space of metric graphs, which appears more as a tool, however important, than a goal. Rather, the principal scope of [11] was to study an important group: the automorphism group of the free group of rank g, F_g.

Following an extremely successful line of research, groups may be studied through their actions on some geometric spaces. To study the automorphism group of F_g, or rather, its quotient by inner automorphisms, $\mathrm{Out}(F_g)$, in [11] the authors introduce a space, called the *outer space* and denoted here by O_g (as in [26]) on which $\mathrm{Out}(F_g)$ acts.

The starting point is that F_g is the fundamental group, $\pi_1(R_g)$, of the connected graph of genus g with one vertex and g loops, often called the "rose with g petals", and written R_g.

Elements in $\mathrm{Out}(F_g)$ are thus geometrically represented by homotopy equivalences of R_g with itself.

Now, a connected graph Γ of genus g is homotopy equivalent to R_g. By adding a metric structure ℓ to Γ, and letting ℓ vary, one obtains a new interesting space on which $\mathrm{Out}(F_g)$ acts, provided Γ is "marked", i.e. provided a homotopy equivalence $\mu : R_g \to \Gamma$ is fixed. Now, to ensure that the space of all marked genus g metric graphs has finite dimension one needs to assume that graphs have no vertices of valence 1 or 2 (sounds familiar!). An extra, normalization, assumption made in [11] is that the sum of the lengths of the edges, i.e. the *volume* of the graph, $vol(\Gamma, \ell) = \sum_{e \in E(\Gamma)} \ell(e)$, be equal to 1. The outer space O_g is defined as follows

$$O_g := \{[(\Gamma, \ell, \mu)] \text{ of genus } g, \text{ volume } 1, \text{ having no vertex of valence } \leqslant 2\}$$

where (Γ, ℓ, μ) and (Γ', ℓ', μ') are in the same class if there exists an isometry between (Γ, ℓ) and (Γ', ℓ') commuting with the markings up to homotopy.

So, O_g plays the same role for metric graphs as the Teichmüller space T_g for complex curves. The group $Out(F_g)$ plays the role of the mapping class group Γ_g.

ALGEBRAIC CURVES	METRIC GRAPHS
S_g topological surface of genus g compact connected orientable	R_g connected graph with one vertex and g loops.
$\Pi_g := \pi_1(S_g)$	$F_g := \pi_1(R_g)$
C algebraic curve, genus g smooth projective over $\mathbb{C}$	(Γ, ℓ) metric graph, genus g, volume 1 connected, no vertex of valence $\leqslant 2$
a *marking* of C: a homeomorphism up to isotopy $\mu : S_g \longrightarrow C$	a *marking* of Γ: a homotopy equivalence $\mu : R_g \longrightarrow \Gamma$
The *Teichmüller Space*: $T_g := \{(C, \mu)\}/\sim$ $(C, \mu) \sim (C', \mu')$ if $\exists\, C \xrightarrow{\iota} C'$ ι an algebraic isomorphism such that $\iota\mu = \mu'$	The *Outer Space*: $O_g := \{(\Gamma, \ell, \mu)\}/\sim$ $(\Gamma, \ell, \mu) \sim (\Gamma', \ell', \mu')$ if $\exists\, \Gamma \xrightarrow{\iota} \Gamma''$ ι an isometry such that $\iota\mu = \mu'$ up to homotopy
T_g is a contractible $\mathbb{C}$-manifold $\dim T_g = 3g - 3$	O_g is a contractible cell complex not a manifold $\dim O_g = 3g - 4$
The *mapping class group*: $\Gamma_g = \{\phi : S_g \to S_g$ homeomorphism$\}$ $\Gamma_g = Out^+(\Pi_g)$	$\{\phi : R_g \to R_g$ homotopy equivalence$\}$ $Out(F_g)$
Γ_g acts on T_g with finite stabilizers	$Out(F_g)$ acts on O_g with finite stabilizers
$M_g = T_g/\Gamma_g$ moduli space of smooth genus g algebraic curves	$O_g/Out(F_g)$ moduli space for genus g metric graphs of volume 1

Table 1. Analogies between algebraic curves and metric graphs

In [11] the outer space O_g is given the structure of a cell complex, it is shown not to be a manifold (not hard), and to be a contractible space (a theorem, also attributed to Gersten).

The quotient space $O_g/\mathrm{Out}(F_g)$ is thus a moduli space for metric graphs of fixed volume and genus, having no vertex of valence $\leqslant 2$. This quotient is thus closely related to the moduli space of tropical curves.

This connection has not yet been thoroughly investigated, perhaps because of the relative youth of tropical geometry, or the complexity of the group $\mathrm{Out}(F_g)$. Also because, due to the recent, fast development of tropical geometry, interest in moduli of metric graphs is probably stronger nowadays than it was twenty years ago. We believe it deserves to be studied as it offers an entirely new perspective. For example, using what is known for the outer space O_g (see the survey [25]) one may construct bordifications of the moduli space of pure tropical curves other than the one we described in the present paper (via weighted, or extended tropical curves). This may lead to new, profound insights, just like it has been the case for the moduli space of smooth algebraic curves, for which different compactifications exist and have been used in different ways.

5.3. Analogies

The analogies between the Teichmüller construction of M_g and the Culler-Vogtmann space, are summarized in Table 1, where the word "homeomorphism" stands for "orientation preserving homeomorphism".

6. Open problems

We conclude with some interesting lines of research closely related to the topics treated in this paper.

(1) *Compactifications of* $M_{g,n}^{\mathrm{pure}}$. We constructed the bordification of the moduli space of pure pointed tropical curves using weighted tropical curves (as done in [5] for the unpointed case, using a different set up) and its compactification, by extended weighted tropical curves.

Question 1.

(A) What geometric interpretations can these compactifications be given?

(B) Do there exist other compactifications, or bordifications? If so, how do they relate among each other?

As we already mentioned, it should be especially interesting to study these issues in connection with the set up described in Section 5.

(2) *Divisor theory.* In analogy with the classical theory, for pure tropical curves there is a good notion of divisors, linear equivalence and linear systems. In particular, the theorem of Riemann-Roch is known to hold for pure tropical curves, by work of A. Gathmann and M. Kerber [14], building upon work of M. Baker and

S. Norine (in [4]) for combinatorial graphs; see also the work on G. Mikhalkin and I. Zharkov in [24].

One appealing research direction is to explore the connection with the divisor theory for algebraic curves, smooth or singular. There is a clear relation between the combinatorial and the algebraic divisor theory. In fact, let X be a nodal curve and Γ its dual graph; then for any divisor D on X its multidegree $\underline{\deg} D$ is naturally a divisor on Γ, and on any tropical curve supported on Γ. The interaction between the two theories is not trivial, and its investigation may bring new fertile perspectives also on the classical theory. This in fact has already happened, as shown by the new proof of the famous Brill-Noether theorem for algebraic curves given in [10], using the groundwork developed in [3].

Observe that the "Specialization Lemma" of [3], a powerful tool to relate the combinatorial with the algebro-geometric divisor theory, can be refined to be applicable in the classical algebro-geometric set-up (see [7]) and extended to weighted graphs (see [1]).

A related problem is the extension of the Riemann-Roch theorem, and of other classical theorems from Riemann surfaces, to weighted tropical curves, More specifically:

Question 2. Does a Riemann-Roch theorem hold for line bundles on weighted tropical curves? What about Clifford's theorem?

A natural approach to these last problems would require, first of all, an answer to Question 1.A, raised in part (1). A proof of the Riemann-Roch formula for a definition of rank extending the one used in [14] is given in [1]. Other definitions of rank are used, for which the Riemann-Roch formula is not known.

(3) *Tropical orbifolds or stacks.* Establish a categorical framework for tropical moduli theory. This was also the leading theme of a school in tropical geometry, organized by E. Brugallé, I. Itenberg and G. Mikhalkin, which took place in March 2010. As we emphasized a few times in the paper, the spaces $M_{g,n}^{pure}$ and $M_{g,n}^{trop}$ are not manifolds, and seem too complicated to be tropical varieties (with a few known exceptions).

Question 3. What is a good category to use in tropical geometry? Can this category be defined so as to include tropical varieties?

Comparing with the algebraic setting, as we mentioned in Remark 1.3, moduli spaces for algebraic curves can be studied within the category of (singular) algebraic schemes, or, more generally, within that of Deligne-Mumford stacks (but also, that of algebraic spaces, or that of orbifolds).

A promising, seemingly unexplored, line of research could be, again, to study the connection with the work in [11] and the later developments.

References

[1] O. Amini and L. Caporaso. *Riemann-Roch theory for weighted graphs and tropical curves.* Preprint. ← 158

[2] E. Arbarello, M. Cornalba, and P. A. Griffiths. Geometry of algebraic curves, vol. II. With a contribution by J.D. Harris. *Grundlehren der Mathematischen Wissenschaften*, **268**, Springer-Verlag, 2011. ← 121, 125, 154, 155

[3] M. Baker. Specialization of linear systems from curves to graphs. *Algebra and Number Theory*, **2**(6), 613–653, 2008. ← 158

[4] M. Baker and S. Norine. Riemann Roch and Abel-Jacobi theory on a finite graph. *Adv. of Math.*, **215**(2), 766–788, 2007. ← 158

[5] S. Brannetti, M. Melo, and F. Viviani. On the tropical Torelli map. *Adv. in Math.*, **226**, 2546–2586, 2011. `arXiv:0907.3324 v3` ← 120, 121, 133, 151, 157

[6] L. Caporaso. Geometry of tropical moduli spaces and linkage of graphs. *J. of Comb. Theory, Ser. A.*, **119**, 579–598, 2012. `arXiv: 1001.2815` ← 152, 153

[7] L. Caporaso. Algebraic and combinatorial Brill-Noether theory. To appear in: Compact moduli spaces and vector bundles. *The Contemporary Mathematics Volume*, Edited by Valery Alexeev, Elham Izadi, Angela Gibney, Jànos Kollár, Eduard Loojenga. Amer. Math. Soc. Bookstore. `arXiv:1106.1140` ← 158

[8] L. Caporaso and F. Viviani. Torelli theorem for graphs and tropical curves. *Duke Math. Jour. Vol.*, **153** (1), 129–171, 2010. `arXiv:0901.1389` ← 142

[9] M. Chan. Combinatorics of the tropical Torelli map. To appear in *Algebra and Number Theory*. `arXiv:1012.4539` ← 121

[10] F. Cools, J. Draisma, S. Payne, and E. Robeva. A tropical proof of the Brill-Noether Theorem. Preprint. `arXiv:1001.2774` ← 158

[11] M. Culler and K. Vogtmann. Moduli of graphs and automorphisms of free groups. *Invent. Math.*, **84** (1), 91–119, 1986. ← 121, 155, 157, 158

[12] P. Deligne and D. Mumford. The irreducibility of the space of curves of given genus. *Inst. Hautes Études Sci. Publ. Math.*, (36), 75–120, 1969. ← 123, 154

[13] R. Diestel. Graph theory. *Graduate Text in Math.*, **173**, Springer-Verlag, Berlin, 1997. ← 127

[14] A. Gathmann and M. Kerber. A Riemann-Roch theorem in tropical geometry. *Math. Z.*, **259**(1), 217–230, 2008. ← 157, 158

[15] A. Gathmann, M. Kerber, and H. Markwig. Tropical fans and the moduli spaces of tropical curves. *Compos. Math.*, **145**, 173–195, 2009. ← 121

[16] A. Gathmann and H. Markwig. Kontsevich's formula and the WDVV equations in tropical geometry. *Adv. in Math.*, **217**, 537–560, 2008. ← 121

[17] D. Geseker. Lectures on moduli of curves. *Tata Institute of Fundamental Research Lectures on Mathematics and Physics*, **69**, Springer-Verlag, Berlin-New York, 1982. ← 123, 153

[18] J. Harris and I. Morrison. Moduli of curves. *Graduate Texts in Math.*, **187**, Springer-Verlag, New York, 1998. ← 121, 123, 125, 153, 155

[19] A. Hatcher and W. Thurston. A presentation of the mapping class group of a closed orientable surface. *Topology*, **19**, 221–237, 1980. ← 152

[20] F. Knudsen. The projectivity of the moduli space of stable curves. II and III. *Math. Scand.*, **52** (2), 161–199 and 200–212, 1983. ← 123, 132

[21] D. Maclagan and B. Sturmfels. *Introduction to Tropical Geometry.* Book in preparation. ← 151

[22] G. Mikhalkin. Tropical geometry and its applications. *International Congress of Mathematicians. Vol. II*, 827–852, Eur. Math. Soc., Zürich, 2006. ← 121, 128

[23] G. Mikhalkin. Moduli spaces of rational tropical curves. *Proceedings of Gökova Geometry-Topology Conference 2006*, 39–51, Gökova Geometry/Topology Conference (GGT) 2007. arXiv:0704.0839 ← 121, 132, 144

[24] G. Mikhalkin and I. Zharkov. Tropical curves, their Jacobians and Theta functions. *Contemporary Mathematics 465: Proceedings of the International Conference on Curves and Abelian Varieties in honor of Roy Smith's 65th birthday*, 203–231. ← 128, 158

[25] K. Vogtmann. Automorphisms of free groups and outer space, in: Proceedings of the Conference on Geometric and Combinatorial Group Theory, Part 1 (Haifa, 2000), 1–31. *Geom. Dedicata*, **94**, 2002. ← 157

[26] K. Vogtmann. The cohomology of automorphism groups of free groups. In *Proceedings of the International Congress of Mathematicians, Madrid 2006, vol. II*, 1101–1119, European Mathematical Society, Zurich, Switzerland, 2006. ← 155

Dipartimento di Matematica - Università Roma Tre - Roma - ITALY
E-mail address: caporaso@mat.uniroma3.it

A superficial working guide to deformations and moduli

F. Catanese

Dedicated to David Mumford with admiration.

Contents

The present work took place in the realm of the DFG Forschergruppe 790 "Classification of algebraic surfaces and compact complex manifolds". A major part of the article was written, and the article was completed, when the author was a visiting research scholar at KIAS. I owe to David Buchsbaum the joke that an expert on algebraic surfaces is a 'superficial' mathematician.

Introduction

There are several ways to look at moduli theory, indeed the same name can at a first glance disguise completely different approaches to mathematical thinking; yet there is a substantial unity since, although often with different languages and purposes, the problems treated are substantially the same.

The most classical approach and motivation is to consider moduli theory as the fine part of classification theory: the big quest is not just to prove that certain moduli spaces exist, but to use the study of their structure in order to obtain geometrical information about the varieties one wants to classify; and using each time the most convenient incarnation of 'moduli'.

For instance, as a slogan, we might think of moduli theory and deformation theory as analogues of the global study of an algebraic variety versus a local study of its singularities, done using power series methods. On the other hand, the shape of an algebraic variety is easily recognized when it has singularities!

In this article most of our attention will be cast on the case of complex algebraic surfaces, which is already sufficiently intricate to defy many attempts of investigation. But we shall try, as much as possible, to treat the higher dimensional and more general cases as well. We shall also stick to the world of complex manifolds and complex projective varieties, which allows us to find so many beautiful connections to related fields of mathematics, such as topology, differential geometry and symplectic geometry.

David Mumford clarified the concept of biregular moduli through a functorial definition, which is extremely useful when we want a precise answer to questions concerning a certain class of algebraic varieties.

The underlying elementary concepts are the concept of normal forms, and of quotients of parameter spaces by a suitable equivalence relation, often given by the action of an appropriate group. To give an idea through an elementary geometric problem: how many are the projective equivalence classes of smooth plane curves of degree 4 admitting 4 distinct collinear hyperflexes?

A birational approach to moduli existed before, since, by the work of Cayley, Bertini, Chow and van der Waerden, varieties $X^n_d \subset \mathbb{P}^N$ in a fixed projective space, having a fixed dimension n and a fixed degree d are parametrized by the so called Chow variety $\mathcal{C}h(n; d; N)$, over which the projective group $G := \mathbb{P}GL(N+1, \mathbb{C})$ acts. And, if Z is an irreducible component of $\mathcal{C}h(n; d; N)$, the transcendence degree of the field of invariant rational functions $\mathbb{C}(Z)^G$ was classically called the number of polarized moduli for the class of varieties parametrized by Z. This topic: 'embedded varieties' is treated in the article by Joe Harris in this Handbook.

A typical example leading to the concept of stability was: take the fourfold symmetric product Z of $\mathbb{P}^2$, parametrizing 4-tuples of points in the plane. Then Z has dimension 8 and the field of invariants has transcendence degree 0. This is not a surprise, since 4 points in linear general position are a projective basis, hence they are projectively equivalent; but, if one takes 4 point to lie on a line, then there is a modulus, namely, the cross ratio. This example, plus the other basic example given by the theory of Jordan normal forms of square matrices (explained in [111] in detail) guide our understanding of the basic problem of Geometric Invariant Theory: in which sense may we consider the quotient of a variety by the action of an algebraic group. In my opinion geometric invariant theory, in spite of its beauty and its conceptual simplicity, but in view of its difficulty, is a foundational but not a fundamental tool in classification theory. Indeed one of the most difficult results, due to Gieseker, is the asymptotic stability of pluricanonical images of surfaces of general type; it has as an important corollary the existence of a moduli space for the canonical models of surfaces of general type, but the methods of proof do not shed light on the classification of such surfaces (indeed boundedness for the families of surfaces with given invariants had followed earlier by the results of Moishezon, Kodaira and Bombieri).

We use in our title the name 'working': this may mean many things, but in particular here our goal is to show how to use the methods of deformation theory in order to classify surfaces with given invariants.

The order in our exposition is more guided by historical development and by our education than by a stringent logical nesting.

The first guiding concepts are the concepts of Teichmüller space and moduli space associated to an oriented compact differentiable manifold M of even dimension. These however are only defined as topological spaces, and one needs the Kodaira-Spencer-Kuranishi theory in order to try to give the structure of a complex space to them.

A first question which we investigate, and about which we give some new results (proposition 1.15 and theorem 4.11), is: when is Teichmüller space locally homeomorphic to Kuranishi space?

This equality has been often taken for granted, of course under the assumption of the validity of the so called Wavrik condition (see theorem 1.5), which requires the dimension of the space of holomorphic vector fields to be locally constant under deformation .

An important role plays the example of Atiyah about surfaces acquiring a node: we interpret it here as showing that Teichmüller space is non separated (theorem 2.4). In section 4 we see that it also underlies some recent pathological behaviour of automorphisms of surfaces, recently discovered together with Ingrid Bauer: even if deformations of canonical and minimal models are essentially the same, up to finite base change, the same does not occur for deformations of automorphisms (theorems 4.6 and 4.7). The connected components for deformation of automorphisms of canonical models (X, G, α) are bigger than the connected components for deformation of automorphisms of minimal models (S, G, α'), the latter yielding locally closed sets of the moduli spaces which are locally closed but not closed.

To describe these results we explain first the Gieseker coarse moduli space for canonical models of surfaces of general type, which has the same underlying reduced space as the coarse moduli stack for minimal models of surfaces of general type. We do not essentially talk about stacks (for which an elementary presentation can be found in [67]), but we clarify how moduli spaces are obtained by gluing together Kuranishi spaces, and we show the fundamental difference for the étale equivalence relation in the two respective cases of canonical and minimal models: we exhibit examples showing that the relation is not finite (proper) in the case of minimal models (a fact which underlies the definition of Artin stacks given in [5]).

We cannot completely settle here the question whether Teichmüller space is locally homeomorphic to Kuranishi space for all surfaces of general type, as this question is related to a fundamental question about the non existence of complex automorphisms which are isotopic to the identity, but different from the identity (see however the already mentioned theorem 4.11).

Chapter five is dedicated to the connected components of moduli spaces, and to the action of the absolute Galois group on the set of irreducible components of the moduli space, and surveys many recent results.

We end by discussing concrete issues showing how one can determine a connected component of the moduli space by resorting to topological or differential arguments; we overview several results, without proofs but citing the references, and finally we prove a new result, theorem 5.6, obtained in collaboration with Ingrid Bauer.

There would have been many other interesting topics to treat, but these should probably better belong to a 'part 2' of the working guide.

1. Analytic moduli spaces and local moduli spaces: Teichmüller and Kuranishi space

1.1. Teichmüller space

Consider, throughout this subsection, an oriented real differentiable manifold M of real dimension $2n$ (without loss of generality we may a posteriori assume M and all the rest to be $\mathcal{C}^\infty$ or even $\mathcal{C}^\omega$, i.e., real-analytic).

At a later point it will be convenient to assume that M is compact.

Ehresmann ([63]) defined an **almost complex structure** on M as the structure of a complex vector bundle on the real tangent bundle $TM_{\mathbb{R}}$: namely, the action of $\sqrt{-1}$ on $TM_{\mathbb{R}}$ is provided by an endomorphism

$$J : TM_{\mathbb{R}} \to TM_{\mathbb{R}}, \text{ with } J^2 = -\mathrm{Id}.$$

It is completely equivalent to give the decomposition of the complexified tangent bundle $TM_{\mathbb{C}} := TM_{\mathbb{R}} \otimes_{\mathbb{R}} \mathbb{C}$ as the direct sum of the i, respectively $-i$ eigenbundles:

$$TM_{\mathbb{C}} = TM^{1,0} \oplus TM^{0,1} \text{ where } TM^{0,1} = \overline{TM^{1,0}}.$$

In view of the second condition, it suffices to give the subbundle $TM^{1,0}$, or, equivalently, a section of the associated Grassmannian bundle $\mathcal{G}(n, TM_{\mathbb{C}})$ whose fibre at a point $x \in M$ is the variety of n-dimensional vector subspaces of the complex tangent space at x, $TM_{\mathbb{C},x}$ (note that the section must take values in the open set $\mathcal{T}_n$ of subspaces V such that V and $\bar{V}$ generate).

The space $\mathcal{AC}(M)$ of almost complex structures, once $TM_{\mathbb{R}}$ (hence all associated bundles) is endowed with a Riemannian metric, has a countable number of seminorms (locally, the sup norm on a compact K of all the derivatives of the endomorphism J), and is therefore a Fréchet space. One may for instance assume that M is embedded in some $\mathbb{R}^N$.

Assuming that M is compact, one can also consider the Sobolev k-norms (i.e., for derivatives up order k).

A closed subspace of $\mathcal{AC}(M)$ consists of the set $\mathcal{C}(M)$ of complex structures: these are the almost complex structures for which there are at each point x local holomorphic coordinates, i.e., functions $z_1, \dots, z_n$ whose differentials span the dual $(TM_y^{1,0})^\vee$ of $TM_y^{1,0}$ for each point y in a neighbourhood of x.

In general, the splitting

$$TM_{\mathbb{C}}^\vee = (TM^{1,0})^\vee \oplus (TM^{0,1})^\vee$$

yields a decomposition of exterior differentiation of functions as $df = \partial f + \bar{\partial} f$, and a function is said to be holomorphic if its differential is complex linear, i.e., $\bar{\partial} f = 0$.

This decomposition $d = \partial + \bar{\partial}$ extends to higher degree differential forms.

The theorem of Newlander-Nirenberg ([112]), first proven by Eckmann and Frölicher in the real analytic case ([61], see also [34] for a simple proof) characterizes the complex structures through an explicit equation:

Theorem 1.1. (Newlander-Nirenberg) *An almost complex structure J yields the structure of a complex manifold if and only if it is integrable, which means $\bar{\partial}^2 = 0$.*

Obviously the group of oriented diffeomorphisms of M acts on the space of complex structures, hence one can define in few words some basic concepts.

Definition 1.2. *Let $\mathcal{D}iff^+(M)$ be the group of orientation preserving diffeomorphisms of M, and let $\mathcal{C}(M)$ the space of complex structures on M. Let $\mathcal{D}iff^0(M) \subset \mathcal{D}iff^+(M)$ be the connected component of the identity, the so called subgroup of diffeomorphisms which are isotopic to the identity.*

Then Dehn ([54]) defined the mapping class group of M as

$$\mathcal{M}ap(M) := \mathcal{D}iff^+(M)/\mathcal{D}iff^0(M),$$

while the Teichmüller space of M, respectively the moduli space of complex structures on M are defined as

$$\mathcal{T}(M) := \mathcal{C}(M)/\mathcal{D}iff^0(M), \ \mathfrak{M}(M) := \mathcal{C}(M)/\mathcal{D}iff^+(M).$$

These definitions are very clear, however, they only show that these objects are topological spaces, and that

$$(*)\ \mathfrak{M}(M) = \mathcal{T}(M)/\mathcal{M}ap(M).$$

The simplest examples here are two: complex tori and compact complex curves.

The example of complex tori sheds light on the important question concerning the determination of the connected components of $\mathcal{C}(M)$, which are called the deformation classes in the large of the complex structures on M (cf. [39], [41]).

Complex tori are parametrized by an open set $\mathcal{T}_n$ of the complex Grassmann Manifold $Gr(n,2n)$, image of the open set of matrices

$$\{\Omega \in Mat(2n, n; \mathbb{C}) \mid (i)^n \det(\Omega\overline{\Omega}) > 0\}.$$

This parametrization is very explicit: if we consider a fixed lattice $\Gamma \cong \mathbb{Z}^{2n}$, to each matrix Ω as above we associate the subspace

$$V = (\Omega)(\mathbb{C}^n),$$

so that $V \in Gr(n,2n)$ and $\Gamma \otimes \mathbb{C} \cong V \oplus \bar{V}$.

Finally, to Ω we associate the torus $Y_V := V/p_V(\Gamma)$, $p_V : V \oplus \bar{V} \to V$ being the projection onto the first addendum.

Not only we obtain in this way a connected open set inducing all the small deformations (cf. [87]), but indeed, as it was shown in [39] (cf. also [41]) $\mathcal{T}_n$ is a connected component of Teichmüller space (as the letter $\mathcal{T}$ suggests).

It was observed however by Kodaira and Spencer already in their first article ([88], and volume II of Kodaira's collected works) that for $n \geqslant 2$ the mapping class group $SL(2n, \mathbb{Z})$ does not act properly discontinuously on $\mathcal{T}_n$. More precisely, they show that for every non empty open set $U \subset \mathcal{T}_n$ there is a point t such that the orbit $SL(2n, \mathbb{Z}) \cdot t$ intersects U in an infinite set.

This shows that the quotient is not Hausdorff at each point, probably it is not even a non separated complex space.

Hence the moral is that for compact complex manifolds it is better to consider, rather than the Moduli space, the Teichmüller space.

Moreover, after some initial constructions by Blanchard and Calabi (cf. [16], [17], , [18], [28]) of non Kähler complex structures X on manifolds diffeomorphic to a product $C \times T$, where C is a compact complex curve and T is a 2-dimensional complex torus, Sommese generalized their constructions, obtaining ([132]) that the space of complex structures on a six dimensional real torus is not connected.

These examples were then generalized in [39] [41] under the name of **Blanchard-Calabi manifolds** showing (corollary 7.8 of [41]) that also the space of complex structures on the product of a curve C of genus $g \geqslant 2$ with a four dimensional real torus is not connected, and that there is no upper bound for the dimension of Teichmüller space (even when M is fixed).

The case of compact complex curves C is instead the one which was originally considered by Teichmüller.

In this case, if the genus g is at least 2, the Teichmüller space $\mathcal{T}_g$ is a bounded domain, diffeomorphic to a ball, contained in the vector space of quadratic differentials $H^0(C, \mathcal{O}_C(2K_C))$ on a fixed such curve C.

In fact, for each other complex structure on the oriented 2-manifold M underlying C we obtain a complex curve C', and there is a unique extremal quasi-conformal map $f : C \to C'$, i.e., a map such that the Beltrami distortion $\mu_f := \bar{\partial} f / \partial f$ has minimal norm (see for instance [79] or [2]).

The fact that the Teichmüller space $\mathcal{T}_g$ is homeomorphic to a ball (see [138] for a simple proof) is responsible for the fact that the moduli space of curves $\mathfrak{M}_g$ is close to be a classifying space for the mapping class group (see [110] and the articles by Edidin and Wahl in this Handbook).

1.2. Kuranishi space

Interpreting the Beltrami distortion as a closed (0,1)- form with values in the dual $(TC^{1,0})$ of the cotangent bundle $(TC^{1,0})^{\vee}$, we obtain a particular case of the Kodaira-Spencer-Kuranishi theory of local deformations.

In fact, by Dolbeault 's theorem, such a closed form determines a cohomology class in $H^1(\Theta_C)$, where Θ_C is the sheaf of holomorphic sections of the holomorphic tangent bundle $(TC^{1,0})$: these cohomology classes are interpreted, in the Kodaira-Spencer-Kuranishi theory, as infinitesimal deformations (or derivatives of a family of deformations) of a complex structure: let us review briefly how.

Local deformation theory addresses precisely the study of the **small deformations** of a complex manifold $Y = (M, J_0)$.

We shall use here unambiguously the double notation $TM^{0,1} = TY^{0,1}$, $TM^{1,0} = TY^{1,0}$ to refer to the splitting determined by the complex structure J_0.

J_0 is a point in $\mathcal{C}(M)$, and a neighbourhood in the space of almost complex structures corresponds to a distribution of subspaces which are globally defined as graphs of an endomorphism

$$\phi : TM^{0,1} \to TM^{1,0},$$

called a **small variation of complex structure**, since one then defines

$$TM^{0,1}_{\phi} := \{(u, \phi(u)) | \, u \in TM^{0,1}\} \subset TM^{0,1} \oplus TM^{1,0}.$$

In terms of the new $\bar{\partial}$ operator, the new one is simply obtained by considering

$$\bar{\partial}_{\phi} := \bar{\partial} + \phi,$$

and the integrability condition is given by the Maurer-Cartan equation

$$(MC) \;\; \bar{\partial}(\phi) + \frac{1}{2}[\phi, \phi] = 0,$$

where $[\phi, \phi]$ denotes the Schouten bracket, which is the composition of exterior product of forms followed by Lie bracket of vector fields, and which is graded commutative.

Observe for later use that the form $F(\phi) := (\bar{\partial}(\phi) + \frac{1}{2}[\phi, \phi])$ is $\bar{\partial}$ closed, if $\bar{\partial}(\phi) = 0$, since then

$$\bar{\partial}F(\phi) = \frac{1}{2}\bar{\partial}[\phi, \phi] = \frac{1}{2}([\bar{\partial}\phi, \phi] + [\phi, \bar{\partial}\phi]) = 0.$$

Recall also the theorem of Dolbeault: if Θ_Y is the sheaf of holomorphic sections of $TM^{1,0}$, then $H^j(\Theta_Y)$ is isomorphic to the quotient space $\frac{\mathrm{Ker}(\bar{\partial})}{\mathrm{Im}(\bar{\partial})}$ of the space of $\bar{\partial}$ closed $(0, j)$-forms with values in $TM^{1,0}$ modulo the space of $\bar{\partial}$-exact $(0, j)$-forms with values in $TM^{1,0}$.

Our F is a map of degree 2 between two infinite dimensional spaces, the space of $(0,1)$-forms with values in the bundle $TM^{1,0}$, and the space of $(0,2)$-forms with values in $TM^{1,0}$.

Observe that, since our original complex structure J_0 corresponds to $\phi = 0$, the derivative DF of the above equation F at $\phi = 0$ is simply

$$\bar{\partial}(\phi) = 0,$$

hence the tangent space to the space of complex structures consists of the space of $\bar{\partial}$-closed forms of type $(0,1)$ and with values in the bundle $TM^{1,0}$. Moreover the derivative of F surjects onto the space of $\bar{\partial}$-exact $(0,2)$-forms with values in $TM^{1,0}$.

We are now going to show why we can restrict our consideration only to the class of such forms ϕ in the Dolbeault cohomology group

$$H^1(\Theta_Y) := \ker(\bar{\partial})/\mathrm{Im}(\bar{\partial}).$$

This is done by answering the question: how does the group of diffeomorphisms act on an almost complex structure J?

This is in general difficult to specify, but we can consider the infinitesimal action of a 1-parameter group of diffeomorphisms

$$\{\psi_t := \exp(t(\theta + \bar{\theta})|t \in \mathbb{R}\},$$

corresponding to a differentiable vector field θ with values in $TM^{1,0}$; from now on, we shall assume that M is compact, hence the diffeomorphism ψ_t is defined $\forall t \in \mathbb{R}$.

We refer to [92] and [81], lemma 6.1.4 , page 260, for the following calculation of the Lie derivative:

Lemma 1.3. *Given a 1-parameter group of diffeomorphisms* $\{\psi_t := \exp(t(\theta + \bar{\theta})|t \in \mathbb{R}\}$, $(\frac{d}{dt})_{t=0}(\psi_t^*(J_0))$ *corresponds to the small variation* $\bar{\partial}(\theta)$.

The lemma says, roughly speaking, that locally at each point J the orbit for the group of diffeomorphisms in $\mathfrak{Diff}^0(M)$ contains a submanifold, having as tangent space the forms in the same Dolbeault cohomology class of 0, which has finite codimension inside another submanifold with tangent space the space of $\bar{\partial}$-closed forms ϕ. Hence the tangent space to the orbit space is the space of such Dolbeault cohomology classes.

Even if we 'heuristically' assume $\bar{\partial}(\phi) = 0$, it looks like we are still left with another equation with values in an infinite dimensional space. However, the derivative DF surjects onto the space of exact forms, while the restriction of F to the subspace of $\bar{\partial}$-closed forms ($\{\bar{\partial}(\phi) = 0\}$ takes values in the space of $\bar{\partial}$-closed forms: this is the moral reason why indeed one can reduce the above equation $F = 0$, associated to a map between infinite dimensional spaces, to an equation $k = 0$ for a map $k : H^1(\Theta_Y) \to H^2(\Theta_Y)$, called the Kuranishi map.

This is done explicitly via a miraculous equation (see [87], [86],[93] and [34] for details) set up by Kuranishi in order to reduce the problem to a finite dimensional one (here Kuranishi, see [92], uses the Sobolev r- norm in order to be able to use the implicit function theorem for Banach spaces).

Here is how the Kuranishi equation is set up.

Let $\eta_1, \dots, \eta_m \in H^1(\Theta_Y)$ be a basis for the space of harmonic (0,1)-forms with values in $TM^{1,0}$, and set $t := (t_1, \dots, t_m) \in \mathbb{C}^m$, so that $t \mapsto \sum_i t_i \eta_i$ establishes an isomorphism $\mathbb{C}^m \cong H^1(\Theta_Y)$.

Then the *Kuranishi slice* (see [114] for a general theory of slices) is obtained by associating to t the unique power series solution of the following equation:

$$\phi(t) = \sum_i t_i \eta_i + \frac{1}{2} \bar{\partial}^* G[\phi(t), \phi(t)],$$

satisfying moreover $\phi(t) = \sum_i t_i \eta_i +$ higher order terms (G denotes here the Green operator).

The upshot is that for these forms the integrability equation simplifies drastically; the result is summarized in the following definition.

Definition 1.4. *The Kuranishi space $\mathfrak{B}(Y)$ is defined as the germ of complex subspace of $H^1(\Theta_Y)$ defined by $\{t \in \mathbb{C}^m | \ H[\phi(t), \phi(t)] = 0\}$, where H is the harmonic projector onto the space $H^2(\Theta_Y)$ of harmonic forms of type (0,2) and with values in $TM^{1,0}$.*

Kuranishi space $\mathfrak{B}(Y)$ parametrizes exactly the set of small variations of complex structure $\phi(t)$ which are integrable. Hence over $\mathfrak{B}(Y)$ we have a family of complex structures which deform the complex structure of Y.

It follows from the above arguments that the Kuranishi space $\mathfrak{B}(Y)$ surjects onto the germ of the Teichmüller space at the point corresponding to the given complex structure $Y = (M, J_0)$.

It fails badly to be a homeomorphism, and my favorite example for this is (see [30]) the one of the Segre ruled surfaces $\mathbb{F}_n$, obtained as the blow up at the origin of the projective cone over a rational normal curve of degree n, and described by Hirzebruch biregularly as $\mathbb{P}(\mathcal{O}_{\mathbb{P}^1} \oplus \mathcal{O}_{\mathbb{P}^1}(n)), n \geqslant 0$.

Kuranishi space is here the vector space

$$H^1(\Theta_{\mathbb{F}_n}) \cong \mathrm{Ext}^1(\mathcal{O}_{\mathbb{P}^1}(n), \mathcal{O}_{\mathbb{P}^1})$$

parametrizing projectivizations $\mathbb{P}(E)$, where the rank 2 bundle E occurs as an extension

$$0 \to \mathcal{O}_{\mathbb{P}^1} \to E \to \mathcal{O}_{\mathbb{P}^1}(n) \to 0.$$

By Grothendieck's theorem, however, E is a direct sum of two line bundles, hence we get as a possible surface only a surface $\mathbb{F}_{n-2k}$, for each $k \leqslant \frac{n}{2}$. Indeed Teichmüller space, in a neighbourhood of the point corresponding to $\mathbb{F}_n$ consists just of a finite number of points corresponding to each $\mathbb{F}_{n-2k}$, and where $\mathbb{F}_{n-2k}$ is in the closure of $\mathbb{F}_{n-2h}$ if and only if $k \leqslant h$.

The reason for this phenomenon is the following. Recall that the form ϕ can be infinitesimally changed by adding $\bar\partial(\theta)$; now, for $\phi = 0$, nothing is changed if $\bar\partial(\theta) = 0$. i.e., if $\theta \in H^0(\Theta_Y)$ is a holomorphic vector field. But the exponentials of these vector fields, which are holomorphic on $Y = \mathbb{F}_n$, but not necessarily for $\mathbb{F}_{n-2k}$, act transitively on each stratum of the stratification of $\mathrm{Ext}^1(\mathcal{O}_{\mathbb{P}^1}(n), \mathcal{O}_{\mathbb{P}^1})$ given by isomorphism type (each stratum is thus the set of surfaces isomorphic to $\mathbb{F}_{n-2k}$).

In other words, the jumping of the dimension of $H^0(\Theta_{Y_t})$ for $t \in \mathfrak{B}(Y)$ is responsible for the phenomenon.

Indeed Kuranishi, improving on a result of Wavrik ([145]) obtained in [92] the following result.

Theorem 1.5. (Kuranishi's third theorem) *Assume that the dimension of* $H^0(\Theta_{Y_t})$ *for* $t \in \mathfrak{B}(Y)$ *is a constant function in a neighbourhood of* 0.

Then there is $k >> 0$ *and a neighbourhood* $\mathfrak{U}$ *of the identity map in the group* $\mathcal{D}\mathrm{iff}(M)$*, with respect to the* k*-th Sobolev norm, and a neighbourhood* U *of* 0 *in* $\mathfrak{B}(Y)$ *such that, for each* $f \in \mathfrak{U}$*, and* $t \neq t' \in U$*,* f *cannot yield a holomorphic map between* Y_t *and* $Y_{t'}$.

Kuranishi's theorem ([90],[91]) shows that Teichmüller space can be viewed as being locally dominated by a complex space of locally finite dimension (its dimension, as we already observed, may however be unbounded, cf. cor. 7.7 of [41]).

A first consequence is that Teichmüller space is locally connected by holomorphic arcs, hence the determination of the connected components of $\mathcal{C}(M)$, respectively of $\mathcal{T}(M)$, can be done using the original definition of deformation equivalence, given by Kodaira and Spencer in [88].

Corollary 1.6. *Let* $Y = (M, J)$*,* $Y' = (M, J')$ *be two different complex structures on* M.

Define deformation equivalence as the equivalence relation generated by direct deformation equivalence, where Y*,* Y' *are said to be* **direct disk deformation equivalent** *if and only if there is a proper holomorphic submersion with connected fibres* $f\colon \mathcal{Y} \to \Delta$, *where $\mathcal{Y}$ is a complex manifold, $\Delta \subset \mathbb{C}$ is the unit disk, and moreover there are two fibres of* f *biholomorphic to* Y*, respectively* Y'.

Then two complex structures on M *yield points in the same connected component of* $\mathcal{T}(M)$ *if and only if they are in the same deformation equivalence class.*

In the next subsections we shall illustrate the meaning of the condition that the vector spaces $H^0(\Theta_{Y_t})$ have locally constant dimension, in terms of deformation theory. Moreover, we shall give criteria implying that Kuranishi and Teichmüller space do locally coincide.

1.3. Deformation theory and how it is used

One can define deformations not only for complex manifolds, but also for complex spaces. The technical assumption of flatness replaces then the condition that π be a submersion.

Definition 1.7. *1) A* **deformation** *of a compact complex space* X *is a pair consisting of*

1.1) a flat proper morphism $\pi : \mathcal{X} \to T$ *between connected complex spaces (i.e.,* $\pi^* : \mathcal{O}_{T,t} \to \mathcal{O}_{\mathcal{X},x}$ *is a flat ring extension for each* x *with* $\pi(x) = t$*)*

1.2) an isomorphism $\psi : X \cong \pi^{-1}(t_0) := X_0$ *of* X *with a fibre* X_0 *of* π.

2.1) A **small deformation** *is the germ* $\pi : (\mathcal{X}, X_0) \to (T, t_0)$ *of a deformation.*

2.2) Given a deformation $\pi : \mathcal{X} \to T$ *and a morphism* $f : T' \to T$ *with* $f(t'_0) = t_0$*, the* **pull-back** $f^*(\mathcal{X})$ *is the fibre product* $\mathcal{X}' := \mathcal{X} \times_T T'$ *endowed with the projection onto the second factor* T' *(then* $X \cong X'_0$*).*

3.1) A small deformation $\pi : \mathcal{X} \to T$ *is said to be* **versal** *or* **complete** *if every other small deformation* $\pi : \mathcal{X}' \to T'$ *is obtained from it via pull back; it is said to be* **semi-universal** *if the differential of* $f : T' \to T$ *at* t'_0 *is uniquely determined, and* **universal** *if the morphism* f *is uniquely determined.*

4) Two compact complex manifolds X, Y *are said to be* **direct deformation equivalent** *if there are a deformation* $\pi : \mathcal{X} \to T$ *of* X *with* T *irreducible and where all the fibres are smooth, and an isomorphism* $\psi' : Y \cong \pi^{-1}(t_1) := X_1$ *of* Y *with a fibre* X_1 *of* π.

Let's however come back to the case of complex manifolds, observing that in a small deformation of a compact complex manifold one can shrink the base T and assume that all the fibres are smooth.

We can now state the results of Kuranishi and Wavrik (([90], [91], [145]) in the language of deformation theory.

Theorem 1.8. (Kuranishi). *Let* Y *be a compact complex manifold: then*

I) the Kuranishi family $\pi : (\mathcal{Y}, Y_0) \to (\mathfrak{B}(Y), 0)$ *of* Y *is semiuniversal.*

II) $(\mathfrak{B}(Y), 0)$ *is unique up to (non canonical) isomorphism, and is a germ of analytic subspace of the vector space* $H^1(Y, \Theta_Y)$*, inverse image of the origin under a local holomorphic map (called Kuranishi map and denoted by* k*)* $k : H^1(Y, \Theta_Y) \to H^2(Y, \Theta_Y)$ *whose differential vanishes at the origin.*

Moreover the quadratic term in the Taylor development of the Kuranishi map k *is given by the bilinear map* $H^1(Y, \Theta_Y) \times H^1(Y, \Theta_Y) \to H^2(Y, \Theta_Y)$*, called Schouten bracket, which is the composition of cup product followed by Lie bracket of vector fields.*

III) The Kuranishi family is a versal deformation of Y_t *for* $t \in \mathfrak{B}(Y)$.

IV) The Kuranishi family is universal if $H^0(Y, \Theta_Y) = 0$.

V) **(Wavrik)** *The Kuranishi family is universal if* $\mathfrak{B}(Y)$ *is reduced and* $h^0(Y_t, \Theta_{Y_t}) := \dim H^0(Y_t, \Theta_{Y_t})$ *is constant for* $t \in \mathfrak{B}(Y)$ *in a suitable neighbourhood of* 0.

In fact Wavrik in his article ([145]) gives a more general result than V); as pointed out by a referee, the same criterion has also been proven by Schlessinger (prop. 3.10 of [127]).

Wavrik says that the Kuranishi space is a local moduli space under the assumption that $h^0(Y_t, \Theta_{Y_t})$ is locally constant. This terminology can however be confusing, as we shall show, since in no way the Kuranishi space is like the moduli space locally, even if one divides out by the action of the group $Aut(Y)$ of biholomorphisms of Y.

The first most concrete question is how one can calculate the Kuranishi space and the Kuranishi family. In this regard, the first resource is to try to use the implicit functions theorem.

For this purpose one needs to calculate the Kodaira Spencer map of a family $\pi : (\mathcal{Y}, Y_0) \to (T, t_0)$ of complex manifolds having a smooth base T. This is defined as follows: consider the cotangent bundle sequence of the fibration

$$0 \to \pi^*(\Omega^1_T) \to \Omega^1_{\mathcal{Y}} \to \Omega^1_{\mathcal{Y}|T} \to 0,$$

and the direct image sequence of the dual sequence of bundles,

$$0 \to \pi_*(\Theta_{\mathcal{Y}|T}) \to \pi_*(\Theta_{\mathcal{Y}}) \to \Theta_T \to \mathcal{R}^1\pi_*(\Theta_{\mathcal{Y}|T}).$$

Evaluation at the point t_0 yields a map ρ of the tangent space to T at t_0 into $H^1(Y_0, \Theta_{Y_0})$, which is the derivative of the variation of complex structure (see [87] for a more concrete description, but beware that the definition given above is the most effective for calculations).

Corollary 1.9. *Let* Y *be a compact complex manifold and assume that we have a family* $\pi : (\mathcal{Y}, Y_0) \to (T, t_0)$ *with smooth base* T*, such that* $Y \cong Y_0$*, and such that the Kodaira Spencer map* ρ_{t_0} *surjects onto* $H^1(Y, \Theta_Y)$.

Then the Kuranishi space $\mathfrak{B}(Y)$ *is smooth and there is a submanifold* $T' \subset T$ *which maps isomorphically to* $\mathfrak{B}(Y)$*; hence the Kuranishi family is the restriction of* π *to* T'.

The key point is that, by versality of the Kuranishi family, there is a morphism $f : T \to \mathfrak{B}(Y)$ inducing π as a pull back, and ρ is the derivative of f.

This approach clearly works only if Y is **unobstructed**, which simply means that $\mathfrak{B}(Y)$ is smooth. In general it is difficult to describe the Kuranishi map, and even calculating the quadratic term is nontrivial (see [78] for an interesting example).

In general, even if it is difficult to calculate the Kuranishi map, Kuranishi theory gives a lower bound for the 'number of moduli' of Y, since it shows that $\mathfrak{B}(Y)$ has dimension $\geqslant h^1(Y,\Theta_Y) - h^2(Y,\Theta_Y)$. In the case of curves $H^2(Y,\Theta_Y) = 0$, hence curves are unobstructed; in the case of a surface S

$$\dim\mathfrak{B}(S) \geqslant h^1(\Theta_S) - h^2(\Theta_S) = -\chi(\Theta_S) + h^0(\Theta_S) = 10\chi(\mathcal{O}_S) - 2K_S^2 + h^0(\Theta_S).$$

The above is the Enriques' inequality ([66], observe that Max Noether postulated equality), proved by Kuranishi in all cases and also for non algebraic surfaces.

There have been recently two examples where resorting to the Kuranishi theorem in the obstructed case has been useful.

The first one appeared in a preprint by Clemens ([52]), who then published the proof in [53]; it shows that if a manifold is Kählerian, then there are fewer obstructions than foreseen, since a small deformation Y_t of a Kähler manifold is again Kähler, hence the Hodge decomposition still holds for Y_t.

Another independent proof was given by Manetti in [98].

Theorem 1.10. (Clemens-Manetti) *Let Y be a compact complex Kähler manifold.*

Then there exists an analytic automorphism of $H^2(Y,\Theta_Y)$ with linear part equal to the identity, such that the Kuranishi map $k : H^1(Y,\Theta_Y) \to H^2(Y,\Theta_Y)$ takes indeed values in the intersection of the subspaces

$$\mathrm{Ker}(H^2(Y,\Theta_Y) \to \mathrm{Hom}(H^q(\Omega^p_Y), H^{q+2}(\Omega^{p-1}_Y))$$

(the linear map is induced by cohomology cup product and tensor contraction).

Clemens' proof uses directly the Kuranishi equation, and a similar method was used by Sönke Rollenske in [123], [124] in order to consider the deformation theory of complex manifolds yielding left invariant complex structures on nilmanifolds. Rollenske proved, among other results, the following

Theorem 1.11. (Rollenske) *Let Y be a compact complex manifold corresponding to a left invariant complex structure on a real nilmanifold. Assume that the following condition is verified:*

(*) the inclusion of the complex of left invariant forms of pure antiholomorphic type in the Dolbeault complex

$$(\bigoplus_p H^0(\mathcal{A}^{(0,p)}(Y)), \bar\partial)$$

yields an isomorphism of cohomology groups.

Then every small deformation of the complex structure of Y consists of left invariant complex structures.

The main idea, in spite of the technical complications, is to look at Kuranishi's equation, and to see that everything is then left invariant.

Rollenske went over in [126] and showed that for the complex structures on nilmanifolds which are complex parallelizable Kuranishi space is defined by explicit polynomial equations, and most of the time singular.

There have been several attempts to have a more direct approach to the understanding of the Kuranishi map, namely to do things more algebraically and giving up to consider the Kuranishi slice. This approach has been pursued for instance in [128] and effectively applied by Manetti. For instance, as already mentioned, Manetti ([98]) gave a nice elegant proof of the above theorem 1.10 using the notion of differential graded Lie algebras, abbreviated by the acronym DGLA 's.

The typical example of such a DGLA is provided by the Dolbeault complex

$$(\bigoplus_p H^0(\mathcal{A}^{(0,p)}(TM_Y^{1,0})), \bar{\partial})$$

further endowed with the operation of Schouten bracket (here: the composition of exterior product followed by Lie bracket of vector fields), which is graded commutative.

The main thrust is to look at solutions of the Maurer Cartan equation $\bar{\partial}(\phi) + \frac{1}{2}[\phi, \phi] = 0$ modulo gauge transformations, that is, exponentials of sections in $H^0(\mathcal{A}^{(0,0)}(TM_Y^{1,0}))$.

The deformation theory concepts generalize from the case of deformations of compact complex manifolds to the more general setting of DGLA's, which seem to govern almost all of the deformation type problems (see for instance [99]).

1.4. Kuranishi and Teichmüller

Returning to our setting where we considered the closed subspace $\mathcal{C}(M)$ of $\mathcal{AC}(M)$ consisting of the set of complex structures on M, it is clear that there is a universal tautological family of complex structures parametrized by $\mathcal{C}(M)$, and with total space

$$\mathfrak{U}_{\mathcal{C}(M)} := M \times \mathcal{C}(M),$$

on which the group $\mathcal{D}iff^+(M)$ naturally acts, in particular $\mathcal{D}iff^0(M)$.

A rather simple observation is that $\mathcal{D}iff^0(M)$ acts freely on $\mathcal{C}(M)$ if and only if for each complex structure Y on M the group of biholomorphisms $\mathrm{Aut}(Y)$ contains no automorphism which is differentiably isotopic to the identity (other than the identity).

Definition 1.12. *A compact complex manifold Y is said to be* **rigidified** *if* $\mathrm{Aut}(Y) \cap \mathcal{D}iff^0(Y) = \{Id_Y\}$. *A compact complex manifold Y is said to be cohomologically rigidified if* $\mathrm{Aut}(Y) \to \mathrm{Aut}(H^*(Y, \mathbb{Z}))$ *is injective, and rationally cohomologically rigidified if* $\mathrm{Aut}(Y) \to \mathrm{Aut}(H^*(Y, \mathbb{Q}))$ *is injective.*

The condition of being rigidified is obviously stronger than the condition $H^0(\Theta_Y) = 0$, which is necessary, else there is a positive dimensional Lie group of biholomorphic self maps, and is weaker than the condition of being cohomologically rigidified.

Compact curves of genus $g \geqslant 2$ are rationally cohomologically rigidified since if $\tau : C \to C$ is an automorphism acting trivially on cohomology, then in the product

$C \times C$ the intersection number of the diagonal Δ_C with the graph Γ_τ equals the self intersection of the diagonal, which is the Euler number $e(C) = 2 - 2g < 0$. But, if τ is not the identity, Γ_τ and Δ_C are irreducible and distinct, and their intersection number is a non negative number, equal to the number of fixed points of τ, counted with multiplicity: a contradiction.

It is an interesting question whether compact complex manifolds of general type are rigidified. It is known that already for surfaces of general type there are examples which are not rationally cohomologically rigidified (see a partial classification done by Jin Xing Cai in [27]), while examples which are not cohomologically rigidified might exist among surfaces isogenous to a product (potential candidates have been proposed by Wenfei Liu).

Jin Xing Cai pointed out to us that, for simply connected (compact) surfaces, by a result of Quinn ([118]), every automorphism acting trivially in rational cohomology is isotopic to the identity, and that he conjectures that simply connected surfaces of general type are rigidified (equivalently, rationally cohomologically rigidified).

Remark 1.13. Assume that the complex manifold Y has $H^0(\Theta_Y) = 0$, or satisfies Wavrik's condition, but is not rigidified: then by Kuranishi' s third theorem, there is an automorphism $f \in Aut(Y) \cap \mathcal{D}iff^0(Y)$ which lies outside of a fixed neighbourhood of the identity. f acts therefore on the Kuranishi space, hence, in order that the natural map from Kuranishi space to Teichmüller space be injective, f must act trivially on $\mathfrak{B}(Y)$, which means that f remains biholomorphic for all small deformations of Y.

At any case, the condition of being rigidified implies that the tautological family of complex structures descends to a universal family of complex structures on Teichmüller space:

$$\mathfrak{U}_{\mathcal{T}(M)} := (M \times \mathcal{C}(M))/\mathcal{D}iff^0(M) \to \mathcal{C}(M))/\mathcal{D}iff^0(M) = \mathcal{T}(M).$$

on which the mapping class group acts.

Fix now a complex structure yielding a compact complex manifold Y, and compare with the Kuranishi family

$$\mathcal{Y} \to \mathfrak{B}(Y).$$

Now, we already remarked that there is a locally surjective continuous map of $\mathfrak{B}(Y)$ to the germ $\mathcal{T}(M)_Y$ of $\mathcal{T}(M)$ at the point corresponding to the complex structure yielding Y. For curves this map is a local homeomorphism, and this fact provides a complex structure on Teichmüller space.

Remark 1.14. Indeed we observe that more generally, if

1) the Kuranishi family is universal at any point

2) $\mathfrak{B}(Y) \to \mathcal{T}(M)_Y$ is a local homeomorphism at every point, then

Teichmüller space has a natural structure of complex space.

Moreover

3) since $\mathfrak{B}(Y) \to \mathcal{T}(M)_Y$ is surjective, it is a local homeomorphism iff it is injective; in fact, since $\mathcal{T}(M)$ has the quotient topology and it is the quotient by a group action, and $\mathfrak{B}(Y)$ is a local slice for a subgroup of $\mathcal{D}iff^0(M)$, the projection $\mathfrak{B}(Y) \to \mathcal{T}(M)_Y$ is open.

The simple idea used by Arbarello and Cornalba ([2]) to reprove the result for curves is to establish the universality of the Kuranishi family for continuous families of complex structures.

In fact, if any family is locally induced by the Kuranishi family, and we have rigidified manifolds only, then there is a continuous inverse to the map $\mathfrak{B}(Y) \to \mathcal{T}(M)_Y$, and we have the desired local homeomorphism between Kuranishi space and Teichmüller space.

Since there are many cases (for instance, complex tori) where Kuranishi and Teichmüller space coincide, yet the manifolds are not rigidified, we give a simple criterion.

Proposition 1.15. *1) The continuous map $\pi\colon \mathfrak{B}(Y) \to \mathcal{T}(M)_Y$ is a local homeomorphism between Kuranishi space and Teichmüller space if there is an injective continuous map $f\colon \mathfrak{B}(Y) \to Z$, where Z is Hausdorff, which factors through π.*

2) Assume that Y is a compact Kähler manifold and that the local period map f is injective: then $\pi\colon \mathfrak{B}(Y) \to \mathcal{T}(M)_Y$ is a local homeomorphism.

3) In particular, this holds if Y is Kähler with trivial canonical divisor [1].

Proof. 1) : observe that, since $\mathfrak{B}(Y)$ is locally compact and Z is Hausdorff, it follows that f is a homeomorphism with its image $Z' := \operatorname{Im} f \subset Z$. Given the factorization $f = F \circ \pi$, then the inverse of π is the composition $f^{-1} \circ F$, hence π is a homeomorphism.

2) : if Y is Kähler, then every small deformation Y_t of Y is still Kähler, as it is well known (see [87]).

Therefore one has the Hodge decomposition

$$H^*(M, \mathbb{C}) = H^*(Y_t, \mathbb{C}) = \bigoplus_{p,q} H^{p,q}(Y_t)$$

and the corresponding period map $f\colon \mathfrak{B}(Y) \to \mathfrak{D}$, where $\mathfrak{D}$ is the period domain classifying Hodge structures of type $\{(h_{p,q}) | 0 \leqslant p, q, p+q \leqslant 2n\}$.

As shown by Griffiths in [75], see also [76] and [144], the period map is indeed holomorphic, in particular continuous, and $\mathfrak{D}$ is a separated complex manifold, hence 1) applies.

3) the previous criterion applies in several situations, for instance, when Y is a compact Kähler manifold with trivial canonical bundle.

[1] As observed by a referee, the same proof works when Y is Kähler with torsion canonical divisor, since one can consider the local period map of the canonical cover of Y

In this case the Kuranishi space is smooth (this is the so called Bogomolov-Tian-Todorov theorem, compare [19], [134], [137], and see also [119] and [83] for more general results) and the local period map for the period of holomorphic n-forms is an embedding, since the derivative of the period map, according to [75] is given by cup product

$$\mu\colon H^1(Y,\Theta_Y) \to \oplus_{p,q} \mathrm{Hom}(H^q(\Omega^p_Y), H^{q+1}(\Omega^{p-1}_Y))$$

$$= \oplus_{p,q} \mathrm{Hom}(H^{p,q}(Y), H^{p-1,q+1}(Y)).$$

If we apply it for $q = 0, p = n$, we get that μ is injective, since by Serre duality $H^1(Y,\Theta_Y) = H^{n-1}(Y,\Omega^1_Y \otimes \Omega^n_Y)^\vee$ and cup product with $H^0(\Omega^n_Y)$ yields an isomorphism with $H^{n-1}(Y,\Omega^1_Y)^\vee$ which is by Serre duality exactly isomorphic to $H^1(\Omega^{n-1}_Y)$.

□

As we shall see later, a similar criterion applies to show 'Kuranishi= Teichmüller' for most minimal models of surfaces of general type.

For more general complex manifolds, such that the Wavrik condition holds, then the Kuranishi family is universal at any point, so a program which has been in the air for a quite long time has been the one to glue together these Kuranishi families, by a sort of analytic continuation giving another variant of Teichmüller space.

We hope to be able to return on this point in the future.

2. The role of singularities

2.1. Deformation of singularities and singular spaces

The basic analytic result is the generalization due to Grauert of Kuranishi's theorem ([73], see also [130] for the algebraic analogue)

Theorem 2.1. Grauert's Kuranishi type theorem for complex spaces. *Let X be a compact complex space: then*

I) there is a semiuniversal deformation $\pi : (\mathcal{X}, X_0) \to (T, t_0)$ of X, i.e., a deformation such that every other small deformation $\pi' : (\mathcal{X}', X'_0) \to (T', t'_0)$ is the pull-back of π for an appropriate morphism $f : (T', t'_0) \to (T, t_0)$ whose differential at t'_0 is uniquely determined.

II) (T, t_0) is unique up to isomorphism, and is a germ of analytic subspace of the vector space $\mathbb{T}^1$ of first order deformations.

(T, t_0) is the inverse image of the origin under a local holomorphic map (called Kuranishi map and denoted by k)

$$k : \mathbb{T}^1 \to \mathbb{T}^2$$

to the finite dimensional vector space $\mathbb{T}^2$ (called **obstruction space***), and whose differential vanishes at the origin (the point corresponding to the point t_0).*

If X *is reduced, or if the singularities of* X *are local complete intersection singularities, then* $\mathbb{T}^1 = \mathrm{Ext}^1(\Omega^1_X, \mathcal{O}_X)$.

If the singularities of X *are local complete intersection singularities, then* $\mathbb{T}^2 = \mathrm{Ext}^2(\Omega^1_X, \mathcal{O}_X)$.

Recall once more that this result reproves the theorem of Kuranishi ([90], [91]), which dealt with the case of compact complex manifolds, where $\mathbb{T}^j = \mathrm{Ext}^j(\Omega^1_X, \mathcal{O}_X) \cong H^j(X, \Theta_X)$, $\Theta_X := \mathcal{H}om(\Omega^1_X, \mathcal{O}_X)$ being the sheaf of holomorphic vector fields.

There is also the local variant, concerning isolated singularities, which was obtained by Grauert in [72] extending the earlier result by Tyurina in the unobstructed case where $\mathcal{E}xt^2(\Omega^1_X, \mathcal{O}_X)_{x_0} = 0$ ([135]).

Theorem 2.2. Grauert's theorem for deformations of isolated singularities.. *Let* (X, x_0) *be the germ of an isolated singularity of a reduced complex space: then*

I) there is a semiuniversal deformation $\pi : (\mathcal{X}, X_0, x_0) \to (\mathbb{C}^n, 0) \times (T, t_0)$ *of* (X, x_0), *i.e., a deformation such that every other small deformation* $\pi' : (\mathcal{X}', X'_0, x'_0) \to (\mathbb{C}^n, 0) \times (T', t'_0)$ *is the pull-back of* π *for an appropriate morphism* $f : (T', t'_0) \to (T, t_0)$ *whose differential at* t'_0 *is uniquely determined.*

II) (T, t_0) *is unique up to isomorphism, and is a germ of analytic subspace of the vector space* $\mathbb{T}^1_{x_0} := \mathcal{E}xt^1(\Omega^1_X, \mathcal{O}_X)_{x_0}$, *inverse image of the origin under a local holomorphic map (called Kuranishi map and denoted by* k*)*

$$k : \mathbb{T}^1_{x_0} = \mathcal{E}xt^1(\Omega^1_X, \mathcal{O}_X)_{x_0} \to \mathbb{T}^2_{x_0}$$

to the finite dimensional vector space $\mathbb{T}^2_{x_0}$ *(called* **obstruction space***), and whose differential vanishes at the origin (the point corresponding to the point* t_0*).*

The obstruction space $\mathbb{T}^2_{x_0}$ *equals* $\mathcal{E}xt^2(\Omega^1_X, \mathcal{O}_X)_{x_0}$ *if the singularity of* X *is normal.*

For the last assertion, see [130], prop. 3.1.14, page 114.

The case of complete intersection singularities was shown quite generally to be unobstructed by Tyurina in the hypersurface case ([135]), and then by Kas-Schlessinger in [82].

This case lends itself to a very explicit description.

Let $(X,0) \subset \mathbb{C}^n$ be the complete intersection $f^{-1}(0)$, where

$$f = (f_1, \dots, f_p) \colon (\mathbb{C}^n, 0) \to (\mathbb{C}^p, 0).$$

Then the ideal sheaf $\mathcal{I}_X$ of X is generated by $(f_1, \dots, f_p)$ and the conormal sheaf $\mathcal{N}^\vee_X := \mathcal{I}_X / \mathcal{I}^2_X$ is locally free of rank p on X.

Dualizing the exact sequence

$$0 \to \mathcal{N}^\vee_X \cong \mathcal{O}^p_X \to \Omega^1_{\mathbb{C}^n} \otimes \mathcal{O}_X \cong \mathcal{O}^n_X \to \Omega^1_X \to 0$$

we obtain (as $\Theta_X := \mathcal{H}om(\Omega^1_X, \mathcal{O}_X)$)

$$0 \to \Theta_X \to \Theta_{\mathbb{C}^n} \otimes \mathcal{O}_X \cong \mathcal{O}^n_X \to \mathcal{N}_X \cong \mathcal{O}^p_X \to \mathcal{E}xt^1(\Omega^1_X, \mathcal{O}_X) \to 0$$

which represents $\mathbb{T}^1_0 := \mathcal{E}xt^1(\Omega^1_X, \mathcal{O}_X)_0$ as a quotient of $\mathcal{O}^p_{X,0}$, and as a finite dimensional vector space (whose dimension will be denoted as usual by τ, which is the so called Tyurina number).

Let $(g^1, \dots, g^\tau) \in \mathcal{O}^p_{X,0}$, $g^i = (g^i_1, \dots, g^i_p)$ represent a basis of $\mathbb{T}^1_0$.

Consider now the complete intersection

$$(\mathfrak{X}, 0) := V(F_1, \dots, F_p) \subset (\mathbb{C}^n \times \mathbb{C}^\tau, 0)$$

where

$$F_j(x,t) := f_j(x) + \sum_{i=1}^{\tau} t_i g^i_j(x).$$

Then

$$(X,0) \stackrel{i}{\hookrightarrow} (\mathfrak{X}, 0) \stackrel{\phi}{\longrightarrow} (\mathbb{C}^\tau, 0)$$

where i is the inclusion and ϕ is the projection, yields the semiuniversal deformation of $(X,0)$.

In the case $p = 1$ of hypersurfaces, this representation of $\mathbb{T}^1_0 := \mathcal{E}xt^1(\Omega^1_X, \mathcal{O}_X)_0$ as a quotient of $\mathcal{O}_{X,0}$ yields the well known formula:

$$\mathbb{T}^1_0 = \mathcal{O}_{\mathbb{C}^n,0}/(f, f_{x_1}, \dots, f_{x_n}),$$

where $f_{x_i} := \frac{\partial f}{\partial x_i}$.

The easiest example is then the one of an ordinary quadratic singularity, or node, where we have $p = 1$, and $f = \sum_{i=1,\dots n} x_i^2$.

Then our module $\mathbb{T}^1_0 = \mathcal{O}_{\mathbb{C}^n,0}/(x_i)$ and the deformation is

$$f + t = \sum_{i=1}^{n} x_i^2 + t = 0.$$

2.2. Atiyah's example and three of its implications

Around 1958 Atiyah ([6]) made a very important discovery concerning families of surfaces acquiring ordinary double points. His result was later extended by Brieskorn and Tyurina ([136], [23], [24]) to the more general case of rational double points, which are the rational hypersurface singularities, and which are referred to as RDP's or as Du Val singularities (Patrick Du Val classified them as the surface singularities which do not impose adjunction conditions, see[59], [3], [120], [121])) or as Kleinian singularities (they are analytically isomorphic to a quotient $\mathbb{C}^2/G$, with $G \subset SL(2, \mathbb{C})$).

The crucial part of the story takes place at the local level, i.e., when one deforms the ordinary double point singularity

$$X = \{(u,v,w) \in \mathbb{C}^3 | w^2 = uv\}.$$

In this case the semiuniversal deformation is, as we saw, the family

$$\mathfrak{X} = \{(u,v,w,t) \in \mathbb{C}^4 | w^2 - t = uv\}$$

mapping to $\mathbb{C}$ via the projection over the variable t; and one observes here that $\mathfrak{X} \cong \mathbb{C}^3$.

The minimal resolution of X is obtained blowing up the origin, but we cannot put the minimal resolutions of the X_t together.

One can give two reasons for this fact. The first is algebro geometrical, in that any normal modification of $\mathfrak{X}$ which is an isomorphism outside the origin, and is such that the fibre over the origin has dimension at most 1, must be necessarily an isomorphism.

The second reason is that the restriction of the family (of manifolds with boundary) to the punctured disk $\{t \neq 0\}$ is not topologically trivial, its monodromy being given by a Dehn twist around the vanishing two dimensional sphere (see[102]).

As a matter of fact the square of the Dehn twist is differentiably isotopic to the identity, as it is shown by the fact that the family X_t admits a simultaneous resolution after that we perform a base change

$$t = \tau^2 \Rightarrow w^2 - \tau^2 = uv.$$

Definition 2.3. *Let $\mathfrak{X} \to T'$ be the family where*

$$\mathfrak{X} = \{(u, v, w, \tau) | w^2 - \tau^2 = uv\}$$

and T' is the affine line with coordinate τ.

$\mathfrak{X}$ has an isolated ordinary quadratic singularity which can be resolved either by blowing up the origin (in this way we get an exceptional divisor $\cong \mathbb{P}^1 \times \mathbb{P}^1$) or by taking the closure of one of two distinct rational maps to $\mathbb{P}^1$. The two latter resolutions are called the **small resolutions**.

One defines $\mathcal{S} \subset \mathfrak{X} \times \mathbb{P}^1$ to be one of the small resolutions of $\mathfrak{X}$, and $\mathcal{S}'$ to be the other one, namely:

$$\mathcal{S} : \{(u, v, w, \tau)(\xi) \in \mathfrak{X} \times \mathbb{P}^1 | \frac{w - \tau}{u} = \frac{v}{w + \tau} = \xi\}$$

$$\mathcal{S}' : \{(u, v, w, \tau)(\eta) \in \mathfrak{X} \times \mathbb{P}^1 | \frac{w + \tau}{u} = \frac{v}{w - \tau} = \eta\}.$$

Now, the two families on the disk $\{\tau \in \mathbb{C} || \tau| < \epsilon\}$ are clearly isomorphic by the automorphism σ_4 such that $\sigma_4(u, v, w, \tau) = (u, v, w, -\tau)$,

On the other hand, the restrictions of the two families to the punctured disk $\{\tau \neq 0\}$ are clearly isomorphic by the automorphism acting as the identity on the variables (u, v, w, τ), since over the punctured disk these two families coincide with the family $\mathfrak{X}$.

This automorphism yields a birational map $\iota : \mathcal{S} \dashrightarrow \mathcal{S}'$ which however does not extend biregularly, since $\xi u = v\eta^{-1}$.

The automorphism $\sigma := \sigma_4 \circ \iota$ acts on the restriction $\mathcal{S}^*$ of the family $\mathcal{S}$ to the punctured disk, and it acts on the given differentiably trivialized family $\mathcal{S}^*$ of manifolds with boundary via the Dehn twist on the vanishing 2-sphere.

For $\tau = 0$ the Dehn twist cannot yield a holomorphic map $\phi\colon S_0 \to S_0$, since every biholomorphism ϕ sends the (-2)-curve E to itself (E is the only holomorphic curve in its homology class), hence it acts on the normal bundle of E by scalar multiplication, therefore by an action which is homotopic to the identity in a neighbourhood of E: a contradiction.

From the above observations, one can derive several 'moral' consequences, when one globalizes the procedure.

Assume now that we have a family of compact algebraic surfaces X_t such that X_t is smooth for $t \neq 0$, and, for $t = 0$, it acquires a node.

We can then take the corresponding families S_τ and S'_τ of smooth surfaces.

We can view the family S_τ as the image of a 1 dimensional complex disk in the Teichmüller space $\mathcal{T}(S_0)$ of S_0, and then the Dehn twist σ yields a self map

$$\sigma^*\colon \mathcal{T}(S_0) \to \mathcal{T}(S_0).$$

It has the property that $\sigma^*(S_\tau) = S_{-\tau}$ for $\tau \neq 0$, but for $\tau = 0$, we have that $\sigma^*(S_0) \neq S_0$, since a map homotopically equivalent to the Dehn twist cannot yield a biholomorphic map.

Hence we get two different points of $\mathcal{T}(S_0)$, namely, $\sigma^*(S_0) \neq S_0$, which are both limits $\lim_{\tau\to 0}\sigma^*(S_\tau) = \lim_{\tau\to 0} S_{-\tau}$ and the conclusion is the following theorem, which is a slightly different version of a result of Burns and Rapoport ([25]).

Theorem 2.4. *Let S_0 be a compact complex surface which contains a (-2)-curve E, i.e., a smooth rational curve with self intersection equal to -2, obtained from the resolution of a normal surface X_0 with exactly one singular point , which is an ordinary quadratic singularity.*

Assume further that X_0 admits a smoothing deformation.

Then the Teichmüller space $\mathcal{T}(S_0)$ is not separated.

That such a surface exists is obvious: it suffices, for each degree $d \geqslant 2$, to consider a surface X_0 in $\mathbb{P}^3$, with equation $f_0(x_0, x_1, x_2, x_3) = 0$, and such that there is no monomial divisible by x_0^{d-1} appearing in f_0 with non zero coefficient. The required smoothing is gotten by setting $X_t := \{f_t := f_0 + t x_0^d = 0\}$.

This example can of course be interpreted in a second way, and with a completely different wording (non separatedness of some Artin moduli stack), which I will try to briefly explain in a concrete way.

It is clear that $\sigma^*(S_0) \neq S_0$ in Teichmüller space, but $\sigma^*(S_0)$ and S_0 yield the same point in the moduli space.

Think of the family S_τ as a 1 dimensional complex disk in the Kuranishi space of S_0: then when we map this disk to the moduli space we have two isomorphic surfaces, namely, since $\sigma^*(S_\tau) = S_{-\tau}$ for $\tau \neq 0$, we identify the point τ with the point $-\tau$.

If we consider a disk Δ, then we get an equivalence relation in $\Delta \times \Delta$ which identifies τ with the point $-\tau$. We do not need to say that $\tau = 0$ is equivalent to itself, because this is self evident. However, we have seen that we cannot extend the self map σ of the family $\mathcal{S}^*$ to the full family $\mathcal{S}$. Therefore, if we require that equivalences come from families, or, in other words, when we glue Kuranishi families, we obtain the following.

The equivalence relation in $\Delta \times \Delta$ is the image of two complex curves, one being the disk Δ, the other being the punctured disk Δ^*.

Δ maps to the diagonal $\Delta \times \Delta$, i.e., $\tau \mapsto (\tau, \tau)$, while the punctured disk Δ^* maps to the antidiagonal, deprived of the origin, that is, $\tau \neq 0, \tau \mapsto (\tau, -\tau)$.

The quotient in the category of complex spaces is indifferent to the fact that we cannot have a family extending the isomorphism ι given previously across $\tau = 0$, and the quotient is the disk Δ_t with coordinate $t := \tau^2$.

But over the disk Δ_t there will not be, as already remarked, a family of smooth surfaces.

This example by Atiyah motivated Artin in [5] to introduce his theory of Artin stacks, where one takes quotients by maps which are étale on both factors, but not proper (as the map of Δ^* into $\Delta \times \Delta$).

A third implication of Atiyah's example will show up in the section on automorphisms.

3. Moduli spaces for surfaces of general type

3.1. Canonical models of surfaces of general type

In the birational class of a non ruled surface there is, by the theorem of Castelnuovo (see e.g. [21]), a unique (up to isomorphism) minimal model S.

We shall assume from now on that S is a smooth minimal (projective) surface of general type: this is equivalent (see [21]) to the two conditions:

(*) $K_S^2 > 0$ and K_S is nef

(we recall that a divisor D is said to be **nef** if, for each irreducible curve C, we have $D \cdot C \geqslant 0$).

It is very important that, as shown by Kodaira in [85], the class of non minimal surfaces is stable by small deformation; on the other hand, a small deformation of a minimal algebraic surface of general type is again minimal (see prop. 5.5 of [8]). Therefore, the class of minimal algebraic surfaces of general type is stable by deformation in the large.

Even if the canonical divisor K_S is nef, it does not however need to be an ample divisor, indeed

The canonical divisor K_S *of a minimal surface of general type* S *is ample iff there does not exist an irreducible curve* C $(\neq 0)$ *on* S *with* $K \cdot C = 0 \Leftrightarrow$ *there is no (-2)-curve* C *on* S, *i.e., a curve such that* $C \cong \mathbb{P}^1$, *and* $C^2 = -2$.

The number of (-2)-curves is bounded by the rank of the Neron Severi lattice NS(S) of S, and these curves can be contracted by a contraction $\pi\colon S \to X$, where X is a normal surface which is called the **canonical model** of S.

The singularities of X are exactly Rational Double Points (in the terminology of [3]), also called Du Val or Kleinian singularities, and X is Gorenstein with canonical divisor K_X such that $\pi^*(K_X) = K_S$.

The canonical model is directly obtained from the 5-th pluricanonical map of S, but it is abstractly defined as the Projective Spectrum (set of homogeneous prime ideals) of the canonical ring

$$\mathcal{R}(S) := (\mathcal{R}(S, K_S)) := \bigoplus_{m \geqslant 0} H^0(\mathcal{O}_S(mK_S).$$

In fact if S is a surface of general type the canonical ring $\mathcal{R}(S)$ is a graded $\mathbb{C}$-algebra of finite type (as first proven by Mumford in [106]), and then the canonical model is $X = \mathrm{Proj}(\mathcal{R}(S, K_S)) = \mathrm{Proj}(\mathcal{R}(X, K_X))$.

By choosing a minimal homogeneous set of generators of $\mathcal{R}(S)$ of degrees $d_1, \ldots, d_r$ one obtains a natural embedding of the canonical model X into a weighted projective space (see[56]). This is however not convenient in order to apply Geometric Invariant Theory, since one has then to divide by non reductive groups, unlike the case of pluricanonical maps, which we now discuss.

In this context the following is the content of the theorem of Bombieri ([20]), which shows with a very effective estimate the boundedness of the family of surfaces of general type with fixed invariants K_S^2 and $\chi(S) := \chi(\mathcal{O}_S)$.

Theorem 3.1. (Bombieri) *Let S be a minimal surface of general type, and consider the linear system $|mK_S|$ for $m \geqslant 5$, or for $m = 4$ when $K_S^2 \geqslant 2$.*

Then $|mK_S|$ yields a birational morphism φ_m onto its image, called the m-th pluricanonical map of S, which factors through the canonical model X as $\varphi_m = \psi_m \circ \pi$, and where ψ_m is the m-th pluricanonical map of X, associated to the linear system $|mK_X|$, and gives an embedding of the canonical model

$$\psi_m\colon X \to^{\cong} X_m \subset \mathbb{P}H^0(\mathcal{O}_X(mK_X))^\vee = \mathbb{P}H^0(\mathcal{O}_S(mK_S))^\vee.$$

3.2. The Gieseker moduli space

The theory of deformations of complex spaces is conceptually simple but technically involved because Kodaira, Spencer, Kuranishi, Grauert et al. had to prove the convergence of the power series solutions which they produced.

It is a matter of life that tori and algebraic K3 surfaces have small deformations which are not algebraic. But there are cases, like the case of curves and of surfaces of general type, where all small deformations are still projective, and then life simplifies incredibly, since one can deal only with projective varieties or projective subschemes.

For these, the most natural parametrization, from the point of view of deformation theory, is given by the Hilbert scheme, introduced by Grothendieck ([77]). Let us illustrate this concept through the case of surfaces of general type.

For these, as we already wrote, the first important consequence of the theorem on pluricanonical embeddings is the finiteness, up to deformation, of the minimal surfaces S of general type with fixed invariants $\chi(S) = a$ and $K_S^2 = b$.

In fact, their 5-canonical models X_5 are surfaces with Rational Double Points as singularities and of degree $25b$ in a fixed projective space $\mathbb{P}^N$, where $N + 1 = P_5 := h^0(5K_S) = \chi(S) + 10K_S^2 = a + 10b$.

The Hilbert polynomial of X_5 equals

$$P(m) := h^0(5mK_S) = a + \frac{1}{2}(5m - 1)5mb.$$

Grothendieck ([77]) showed that there is

i) an integer d and

ii) a subscheme $\mathcal{H} = \mathcal{H}_P$ of the Grassmannian of codimension $P(d)$- subspaces of $H^0(\mathbb{P}^N, \mathcal{O}_{\mathbb{P}^N}(d))$, called Hilbert scheme, such that

iii) $\mathcal{H}$ parametrizes the degree d graded pieces $H^0(\mathcal{I}_\Sigma(d))$ of the homogeneous ideals of all the subschemes $\Sigma \subset \mathbb{P}^N$ having the given Hilbert polynomial P.

We can then talk about the Hilbert point of Σ as the Plücker point

$$\Lambda^{P(d)}(r_\Sigma^\vee)$$

$$r_\Sigma : H^0(\mathbb{P}^N, \mathcal{O}_{\mathbb{P}^N}(d)) \to H^0(\Sigma, \mathcal{O}_\Sigma(d))$$

being the restriction homomorphism (surjective for d large).

Inside $\mathcal{H}$ one has the open set

$$\mathcal{H}^0 := \{\Sigma | \Sigma \text{ is reduced with only R.D.P.'s as singularities}\}.$$

This is plausible, since rational double points are hypersurface singularities, and first of all the dimension of the Zariski tangent space is upper semicontinuous, as well as the multiplicity: some more work is needed to show that the further property of being a 'rational' double point is open. The result has been extended in greater generality by Elkik in [65].

One can use the following terminology (based on results of Tankeev in [133]).

Definition 3.2. *The 5-pseudo moduli space of surfaces of general type with given invariants* K^2, χ *is the closed subscheme* $\mathcal{H}_0 \subset \mathcal{H}^0$ *(defined by fitting ideals of the direct image of* $\omega_\Sigma^{\otimes 5} \otimes \mathcal{O}_\Sigma(-1)$*),*

$$\mathcal{H}_0(\chi, K^2) := \{\Sigma \in \mathcal{H}^0 | \omega_\Sigma^{\otimes 5} \cong \mathcal{O}_\Sigma(1)\}$$

Since $\mathcal{H}_0$ is a quasi-projective scheme, it has a finite number of irreducible components, called the **deformation types** of the surfaces of general type with given invariants K^2, χ.

As we shall see, the above deformation types of canonical models coincide with the equivalence classes for the relation of deformation equivalence between minimal surfaces of general type.

Remark 3.3. The group $\mathbb{P}GL(N+1,\mathbb{C})$ acts on $\mathcal{H}_0$ with finite stabilizers (corresponding to the groups of automorphisms of each surface) and the orbits correspond to the isomorphism classes of minimal surfaces of general type with invariants K^2, χ.

Tankeev in [133] showed that a quotient by this action exists not only as a complex analytic space, but also as a Deligne Mumford stack ([55]).

Saying that the quotient is a stack is a way to remedy the fact that, over the locus of surfaces with automorphisms, there does not exist a universal family, so we have only, in Mumford's original terminology, a coarse and not a fine moduli space.

In a technically very involved paper ([71]) Gieseker showed that, if one replaces the 5-canonical embedding by an m-canonical embedding with much higher m, then the Hilbert point $\Lambda^{P(d)}(r_\Sigma^\vee)$ is a stable point; this means that, beyond the already mentioned property that the stabilizer is finite, that there are polynomial functions which are invariant for the action of $SL(N+1,\mathbb{C})$ and which do not vanish at the point, so that the Hilbert point maps to a point of the Projective spectrum of the ring of $SL(N+1,\mathbb{C})$-invariants.

The result of Gieseker leads then to the following

Theorem 3.4. **(Gieseker)** *For m very large, the quotient*

$$\mathfrak{M}^{can}_{\chi,K^2} := \mathcal{H}_0(\chi,K^2)/SL(N+1,\mathbb{C})$$

exists as a quasi-projective scheme. It is independent of m *and called the* **Gieseker moduli space** *of canonical models of surfaces of general type with invariants* χ, K^2.

It should be noted that at that time Gieseker only established the result for a field of characteristic zero; as he remarks in the paper, the only thing which was missing then in characteristic p was the boundedness of the surfaces of general type with given invariants χ, K^2. This result was provided by Ekedahl's extension of Bombieri's theorem to characteristic p ([64], see also [45] and [46] for a simpler proof).

3.3. Minimal models versus canonical models

Let us go back to the assertion that deformation equivalence classes of minimal surfaces of general type are the same thing as deformation types of canonical models (a fact which is no longer true in higher dimension).

We have more precisely the following theorem.

Theorem 3.5. *Given two minimal surfaces of general type* S, S' *and their respective canonical models* X, X'*, then*

S *and* S' *are deformation equivalent* $\Leftrightarrow$ X *and* X' *are deformation equivalent.*

The idea of the proof can be simplified by the elementary observation that, in order to analyse deformation equivalence, one may restrict oneself to the case of families parametrized by a base T with $\dim(T) = 1$: since two points in a complex space $T \subset \mathbb{C}^n$ (or in an algebraic variety) belong to the same irreducible component of T if and only if they belong to an irreducible curve $T' \subset T$. And one may further reduce to the case where T is smooth simply by taking the normalization $T^0 \to T_{red} \to T$ of the reduction T_{red} of T, and taking the pull-back of the family to T^0.

But the crucial point underlying the result is the theorem on the so-called simultaneous resolution of singularities (cf. [136],[22], [23], [24])

Theorem 3.6. (Simultaneous resolution according to Brieskorn and Tjurina). *Let $T := \mathbb{C}^\tau$ be the basis of the semiuniversal deformation of a Rational Double Point* (X,0). *Then there exists a ramified Galois cover $T' \to T$, with T' smooth $T' \cong \mathbb{C}^\tau$ such that the pull-back $\mathcal{X}' := \mathcal{X} \times_T T'$ admits a simultaneous resolution of singularities $p : \mathcal{S}' \to \mathcal{X}'$ (i.e., p is bimeromorphic and all the fibres of the composition $\mathcal{S}' \to \mathcal{X}' \to T'$ are smooth and equal, for t'_0, to the minimal resolution of singularities of* (X,0).

We reproduce Tjurina' s proof for the case of A_n-singularities, observing that the case of the node was already described in the previous section.

Proof. Assume that we have the A_n-singularity

$$\{(x, y, z) \in \mathbb{C}^3 | xy = z^{n+1}\}.$$

Then the semiuniversal deformation is given by

$$\mathcal{X} := \{((x, y, z), (a_2, \dots a_{n+1})) \in \mathbb{C}^3 \times \mathbb{C}^n | xy = z^{n+1} + a_2 z^{n-1} + \dots a_{n+1}\},$$

the family corresponding to the natural deformations of the simple cyclic covering.

We take a ramified Galois covering with group $\mathcal{S}_{n+1}$ corresponding to the splitting polynomial of the deformed degree $n+1$ polynomial

$$\mathcal{X}' := \{((x, y, z), (\alpha_1, \dots \alpha_{n+1})) \in \mathbb{C}^3 \times \mathbb{C}^{n+1} | \sum_j \alpha_j = 0,\ xy = \prod_j (z - \alpha_j)\}.$$

One resolves the new family $\mathcal{X}'$ by defining $\phi_i : \mathcal{X}' \dashrightarrow \mathbb{P}^1$ as

$$\phi_i := (x, \prod_{j=1}^{i} (z - \alpha_j))$$

and then taking the closure of the graph of $\Phi := (\phi_1, \dots \phi_n) : \mathcal{X}' \dashrightarrow (\mathbb{P}^1)^n$.

□

Here the Galois group G of the covering $T' \to T$ in the above theorem is the Weyl group corresponding to the Dynkin diagram of the singularity (whose vertices are the (-2) curves in the minimal resolution, and whose edges correspond to the intersection points).

I.e., if $\mathcal{G}$ is the simple algebraic group corresponding to the Dynkin diagram (see [80]), and H is a Cartan subgroup, N_H its normalizer, then the Weyl group is the factor group $W := N_H/H$. For example, A_n corresponds to the group $SL(n+1, \mathbb{C})$, its Cartan subgroup is the subgroup of diagonal matrices, which is normalized by the symmetric group $\mathcal{S}_{n+1}$, and N_H is here a semidirect product of H with $\mathcal{S}_{n+1}$. E. Brieskorn ([24]) found later a direct explanation of this phenomenon.

The Weyl group W and the quotient $T = T'/W$ play a crucial role in the understanding of the relations between the deformations of the minimal model S and the canonical model X, which is a nice discovery by Burns and Wahl ([26]).

But, before we do that, let us make the following important observation, saying that the local analytic structure of the Gieseker moduli space is determined by the action of the group of automorphisms of X on the Kuranishi space of X.

Remark 3.7. Let X be the canonical model of a minimal surface of general type S with invariants χ, K^2. The isomorphism class of X defines a point $[X] \in \mathfrak{M}^{can}_{\chi,K^2}$.

Then the germ of complex space $(\mathfrak{M}^{can}_{\chi,K^2}, [X])$ is analytically isomorphic to the quotient $\mathfrak{B}(X)/Aut(X)$ of the Kuranishi space of X by the finite group $Aut(X) = Aut(S)$.

Forgetting for the time being about automorphisms, and concentrating on families, we want to explain the 'local contributions to global deformations of surfaces', in the words of Burns and Wahl ([26]).

Let S be a minimal surface of general type and let X be its canonical model. To avoid confusion between the corresponding Kuranishi spaces, denote by Def(S) the Kuranishi space for S, respectively Def(X) the Kuranishi space of X.

Their result explains the relation holding between Def(S) and Def(X).

Theorem 3.8. (Burns - Wahl) *Assume that K_S is not ample and let $\pi : S \to X$ be the canonical morphism.*

Denote by $\mathcal{L}_X$ the space of local deformations of the singularities of X (Cartesian product of the corresponding Kuranishi spaces) and by $\mathcal{L}_S$ the space of deformations of a neighbourhood of the exceptional locus of π. Then Def(S) is realized as the fibre product associated to the Cartesian diagram

$$\begin{array}{ccc} \mathrm{Def}(S) & \longrightarrow & \mathrm{Def}(S_{\mathrm{Exc}(\pi)}) =: \mathcal{L}_S \cong \mathbb{C}^\nu, \\ \downarrow & & \downarrow \lambda \\ \mathrm{L:}\ \mathrm{Def}(X) & \longrightarrow & \mathrm{Def}(X_{\mathrm{Sing}X}) =: \mathcal{L}_X \cong \mathbb{C}^\nu, \end{array}$$

where ν is the number of rational (-2)-curves in S, and λ is a Galois covering with Galois group $W := \oplus_{i=1}^r W_i$, the direct sum of the Weyl groups W_i of the singular points of X (these are generated by reflections, hence yield a smooth quotient, see [50]).

An immediate consequence is the following

Corollary 3.9. (Burns - Wahl) *1) ψ : Def(S) $\to$ Def(X) is a finite morphism, in particular, ψ is surjective.*

2) If the derivative of Def(X) $\to \mathcal{L}_X$ *is not surjective (i.e., the singularities of* X *cannot be independently smoothen by the first order infinitesimal deformations of* X*), then* Def(S) *is singular.*

Moreover one has a further corollary

Corollary 3.10. [35]
If the morphism L *is constant, then* Def(S) *is everywhere non reduced,*

$$\mathrm{Def}(S) \cong \mathrm{Def}(X) \times \lambda^{-1}(0).$$

In [35] several examples were exhibited, extending two previous examples by Horikawa and Miranda. In these examples the canonical model X is a hypersurface of degree d in a weighted projective space:

$$X_d \subset \mathbb{P}(1,1,p,q),\, d > 2+p+q,$$

where

- $X_d \subset \mathbb{P}(1,1,2,3)$, $d = 1+6k$, X has one singularity of type A_1 and one of type A_2, or
- $X_d \subset \mathbb{P}(1,1,p,p+1)$, $d = p(k(p+1)-1)$, X has one singularity of type A_p, or
- $X_d \subset \mathbb{P}(1,1,p,rp-1)$, $d = (kp-1)(rp-1)$, $r > p-2$, X has one singularity of type A_{p-1}.

The philosophy in these examples (some hard calculations are however needed) is that all the deformations of X remain hypersurfaces in the same projective space, and this forces X to preserve, in view of the special arithmetic properties of the weights and of the degree, its singularities.

3.4. Number of moduli done right

The interesting part of the discovery of Burns and Wahl is that they completely clarified the background of an old dispute going on in the late 1940's between Francesco Severi and Beniamino Segre. The (still open) question was: given a degree d, which is the maximum number $\mu(d)$ of nodes that a normal surface $X \subset \mathbb{P}^3$ of degree d can have ?

The answer is known only for small degree $d \leqslant 6$: $\mu(2) = 1$, $\mu(3) = 4$ (Cayley's cubic), $\mu(4) = 16$ (Kummer surfaces), $\mu(5) = 31$ (Togliatti quintics), $\mu(6) = 65$ (Barth's sextic), and Severi made the following bold assertion: an upper bound is clearly given by the 'number of moduli', i.e., the dimension of the moduli space of the surfaces of degree d in $\mathbb{P}^3$; this number equals the difference between the dimension of the underlying projective space $\frac{(d+3)(d+2)(d+1)}{6} - 1$ and the dimension of the group of projectivities, at least for $d \geqslant 4$ when the general surface of degree d has only a finite group of projective automorphisms.

One should then have $\mu(d) \leqslant \nu(d) := \frac{(d+3)(d+2)(d+1)}{6} - 16$, but Segre ([129]) found some easy examples contradicting this inequality, the easiest of which are some surfaces of the form

$$L_1(x) \cdot \dots \cdot L_d(x) - M(x)^2,$$

where d is even, the $L_i(x)$ are linear forms, and $M(x)$ is a homogeneous polynomial of degree $\frac{d}{2}$.

Whence the easiest Segre surfaces have $\frac{1}{4}d^2(d-1)$ nodes, corresponding to the points where $L_i(x) = L_j(x) = M(x) = 0$, and this number grows asymptotically as $\frac{1}{4}d^3$, versus Severi's upper bound, which grows like $\frac{1}{6}d^3$ (in fact we know nowadays, by Chmutov in [51], resp. Miyaoka in [103], that $\frac{5}{12}d^3 \leqslant \mu(d) \leqslant \frac{4}{9}d^3$).

The problem with Severi's claim is simply that the nodes impose independent conditions infinitesimally, but only for the smooth model S: in other words, if X has δ nodes, and S is its desingularization, then $\mathrm{Def}(S)$ has Zariski tangent dimension at least δ, while it is not true that $\mathrm{Def}(S)$ has dimension at least δ. Burns and Wahl, while philosophically rescuing Severi' s intuition, showed in this way that there are a lot of examples of obstructed surfaces S, thereby killing Kodaira and Spencer's dream that their cohomology dimension $h^1(\Theta_S)$ would be the expected number of moduli.

3.5. The moduli space for minimal models of surfaces of general type

In this section we shall derive some further moral consequences from the result of Burns and Wahl.

For simplicity, consider the case where the canonical model X has only one double point, and recall the notation introduced previously, concerning the local deformation of the node, given by the family

$$uv = w^2 - t,$$

the pull back family

$$\mathfrak{X} = \{(u, v, w, \tau) | w^2 - \tau^2 = uv\}$$

and the two families

$$\mathcal{S} : \{(u, v, w, \tau)(\xi) \in \mathfrak{X} \times \mathbb{P}^1 | \frac{w - \tau}{u} = \frac{v}{w + \tau} = \xi\}$$

$$\mathcal{S}' : \{(u, v, w, \tau)(\eta) \in \mathfrak{X} \times \mathbb{P}^1 | \frac{w + \tau}{u} = \frac{v}{w - \tau} = \eta\}.$$

There are two cases to be considered in the following oversimplified example:

1) $t \in \Delta$ is a coordinate of an effective smoothing of the node, hence we have a family $\mathcal{S}$ parametrized by $\tau \in \Delta$

2) we have no first order smoothing of the node, hence the Spectrum of the ring $\mathbb{C}[\tau]/(\tau^2)$ replaces Δ.

In case 1), we have two families $\mathcal{S}, \mathcal{S}'$ on a disk Δ with coordinate τ, which are isomorphic over the punctured disk. This induces for the punctured disk Δ with coordinate τ, base of the first family, an equivalence relation induced by the isomorphism with itself where τ is exchanged with $-\tau$; in this case the ring of germs of holomorphic functions at the origin which are invariant for the resulting equivalence relation is just the ring of power series $\mathbb{C}\{t\}$, and we have a smooth 'moduli space'.

In case 2), there is no room for identifying τ with $-\tau$, since if we do this on $\mathbb{C}[\tau]/(\tau^2)$, which has only one point, we glue the families without inducing the identity on the base, and this is not allowed. In this latter case we are left with the non reduced scheme $\mathrm{Spec}(\mathbb{C}[\tau]/(\tau^2))$ as a 'moduli space'.

Recall now Mumford's definition of a coarse moduli space ([107], page 99, definition 5.6 , and page 129, definition 7.4) for a functor of type $\mathcal{S}urf$, such as $\mathcal{S}urf^{min}$, associating to a scheme T the set $\mathcal{S}urf^{min}(T)$ of isomorphism classes of families of smooth minimal surfaces of general type over T, or as $\mathcal{S}urf^{can}$, associating to a scheme T the set $\mathcal{S}urf^{can}(T)$ of isomorphism classes of families of canonical models of surfaces of general type over T.

It should be a scheme A, given together with a morphism Φ from the functor $\mathcal{S}urf$ to $h_A := \mathcal{H}om(-, A)$, such that

(1) for all algebraically closed fields k, $\Phi(\mathrm{Spec}\ k)\colon \mathcal{S}urf(\mathrm{Spec}\ k) \to h_A(\mathrm{Spec}\ k)$ is an isomorphism
(2) any other morphism Ψ from the functor $\mathcal{S}urf$ to a functor h_B factors uniquely through $\chi\colon h_A \to h_B$.

Since any family of canonical models $p\colon \mathcal{X} \to T$ induces, once we restrict T and we choose a local frame for the direct image sheaf $p_*(\omega^m_{\mathcal{X}|T})$ a family of pluricanonical models embedded in a fixed $\mathbb{P}^{P_m - 1}$, follows

Theorem 3.11. *The Gieseker moduli space $\mathfrak{M}^{can}_{\chi,K^2}$ is the coarse moduli space for the functor $\mathcal{S}urf^{can}_{\chi,K^2}$, i.e., for canonical models of surfaces S of general type with given invariants χ, K^2. Hence it gives a natural complex structure on the topological space $\mathfrak{M}(S)$, for S as above.*

As for the case of algebraic curves, we do not have a fine moduli space, i.e., the functor is not representable by this scheme. Here, automorphisms are the main obstruction to the existence of a fine moduli space: dividing the universal family over the Hilbert scheme by the linear group we obtain a family over the quotient coarse moduli space such that the fibre over the isomorphism class of a canonical model X, in the case where the group of automorphisms $\mathrm{Aut}(X)$ is non trivial, is the quotient $X/\mathrm{Aut}(X)$. And $X/\mathrm{Aut}(X)$ is then not isomorphic to X.

Instead, in the case of the functor $\mathcal{S}urf^{min}(T)$, there is a further problem: that the equivalence relation (of isomorphism of families) is not proper on the parameter space, as we already mentioned.

While for curves we have a Deligne-Mumford stack, which amounts roughly speaking to take more general functors than functors which are set valued, this no longer holds for surfaces of general type. Therefore Artin in [5] had to change the definition, allowing more general equivalence relations. The result is ([5], Example 5.5 page 182)

Theorem 3.12. (Artin) *There exists a moduli space* $\mathfrak{M}^{min}_{\chi,K^2}$ *which is an algebraic Artin stack for minimal surfaces of general type with given invariants* χ, K^2.

The beginning step for Artin was to show that there is a finite number of algebraic families parametrizing all the minimal models with given invariants: this step is achieved by Artin in [4] showing that the simultaneous resolution of a family of canonical models can be done not only in the holomorphic category, but also over schemes, provided that one allows a base change process producing locally quasi-separated algebraic spaces.

After that, one can consider the equivalence relation given by isomorphisms of families.

We shall illustrate the local picture by considering the restriction of this equivalence relation to the base $\mathrm{Def}(S)$ of the Kuranishi family.

(I) First of all we have an action of the group $\mathrm{Aut}(S)$ on $\mathrm{Def}(S)$, and we must take the quotient by this action.

In fact, if $g \in \mathrm{Aut}(S)$, then g acts on $\mathrm{Def}(S)$, and if $\mathcal{S} \to \mathrm{Def}(S)$ is the universal family, we can take the pull back family $g^*\mathcal{S}$. By the universality of the Kuranishi family, we have an isomorphism $g^*\mathcal{S} \cong \mathcal{S}$ lying over the identity of $\mathrm{Def}(S)$, and by property (2) we must take the quotient of $\mathrm{Def}(S)$ by this action of $\mathrm{Aut}(S)$.

(II) Let now $w \in W$ be an element of the Weyl group which acts on $\mathrm{Def}(S)$ via the Burns-Wahl fibre product. We let U_w be the open set of $\mathcal{L}_S$ where the transformation w acts freely (equivalently, w being a pseudo reflection, U_w is the complement of the hyperplane of fixed points of w), and we let $\mathrm{Def}(S)_w$ be equal to the open set inverse image of U_w.

Since the action of w is free on $\mathrm{Def}(S)_w$, we obtain that w induces an isomorphism of the family $\mathcal{S}_w \to \mathrm{Def}(S)_w$ with its pull back under w, inducing the identity on the base: hence we have to take the graph of w on $\mathrm{Def}(S)_w$, and divide $\mathrm{Def}(S)_w$ by the action of w.

(III) The 'equivalence relation' on $\mathrm{Def}(S)$ is thus generated by (I) and (II), but it is not really a proper equivalence relation.

The complex space underlying $\mathfrak{M}^{min}_{\chi,K^2}$ is obtained taking the subsheaf of $\mathcal{O}_{\mathrm{Def}(S)}$ consisting of the functions which are invariant for this equivalence relation (i.e., in case (II) , their restriction to $\mathrm{Def}(S)_w$ should be w- invariant).

$\mathfrak{M}^{min}_{\chi,K^2}$ has the same associated reduced complex space as $\mathfrak{M}^{can}_{\chi,K^2}$, but a different ringed space structure, as the examples of [35] mentioned after corollary 3.10 show, see the next subsection.

In fact, the main difference is that $\mathfrak{M}^{can}_{\chi,K^2}$ is locally, by Burns-Wahl's fibre product theorem, the quotient of $\mathrm{Def}(S)$ by the group G' which is the semidirect product of the Weyl group W by $\mathrm{Aut}(S) = \mathrm{Aut}(X)$, as a ringed space (the group G' will make its appearance again in the concrete situation of Lemma 4.9).

Whereas, for $\mathfrak{M}^{min}_{X,K^2}$ the action on the set is the same, but the action of an element w of the Weyl group on the sheaf of regular functions is only there on an open set ($\mathrm{Def}(S)_w$) and this set $\mathrm{Def}(S)_w$ may be empty if $\mathrm{Def}(X)$ maps to a branch divisor of the quotient map $\mathcal{L}_S \to \mathcal{L}_X$.

A general question for which we have not yet found the time to provide an answer is whether there is a quasi-projective scheme whose underlying complex space is $\mathfrak{M}^{min}_{X,K^2}$: we suspect that the answer should be positive.

3.6. Singularities of moduli spaces

In general one can define

Definition 3.13. *The local moduli space* $(\mathfrak{M}^{min,loc}_{X,K^2}, [S])$ *of a smooth minimal surface of general type* S *is the quotient* $\mathrm{Def}(S)/\mathrm{Aut}(S)$ *of the Kuranishi space of* S *by the action of the finite group of automorphisms of* S.

Caveat: whereas, for the canonical model X, $\mathrm{Def}(X)/\mathrm{Aut}(X)$ is just the analytic germ of the Gieseker moduli space at the point corresponding to the isomorphism class of X, the local moduli space $(\mathfrak{M}^{min,loc}_{X,K^2}, [S]) := \mathrm{Def}(S)/\mathrm{Aut}(S)$ is different in general from the analytic germ of the moduli space $(\mathfrak{M}^{min}_{X,K^2}, [S])$, though it surjects onto the latter. But it is certainly equal to it in the special case where the surface S has ample canonical divisor K_S.

The Cartesian diagram by Burns and Wahl was used in [35] to construct everywhere non reduced moduli spaces $(\mathfrak{M}^{min,loc}_{X,K^2}, [S]) := \mathrm{Def}(S)/\mathrm{Aut}(S)$ for minimal models of surfaces of general type.

In this case the basic theorem is

Theorem 3.14. **([35])** *There are (generically smooth) connected components of Gieseker moduli spaces* $\mathfrak{M}^{can}_{X,K^2}$ *such that all the canonical models in it are singular.*

Hence the local moduli spaces $(\mathfrak{M}^{min,loc}_{X,K^2}, [S])$ *for the corresponding minimal models are everywhere non reduced, and the same occurs for the germs* $(\mathfrak{M}^{min}_{X,K^2}, [S])$.

The reason is simple: we already mentioned that if we take the fibre product

$$\begin{array}{ccc} \mathrm{Def}(S) & \longrightarrow & \mathrm{Def}(S_{\mathrm{Exc}(\pi)}) =: \mathcal{L}_S \cong \mathbb{C}^\nu, \\ \downarrow & & \downarrow \lambda \\ \mathrm{Def}(X) & \longrightarrow & \mathrm{Def}(X_{\mathrm{Sing}X}) =: \mathcal{L}_X \cong \mathbb{C}^\nu, \end{array}$$

the lower horizontal arrow maps to a reduced point, hence $\mathrm{Def}(S)$ is just the product $\mathrm{Def}(X) \times \lambda^{-1}(0)$, and $\lambda^{-1}(0)$ is a non reduced point, spectrum of an Artin local ring of length equal to the cardinality of the Galois group $W := \oplus_{i=1}^r W_i$.

Hence Def(S) is everywhere non reduced. Moreover, one can show that, in the examples considered, the general surface has no automorphisms, i.e., there is an open set for which the analytic germ of the Gieseker moduli space coincides with the Kuranishi family Def(X), and the family of canonical models just obtained is equisingular.

Hence, once we consider the equivalence relation on Def(S) induced by isomorphisms, the Weyl group acts trivially (because of equisingularity, we get $\mathrm{Def}(S)_w = \emptyset$ $\forall w \in W$). Moreover, by our choice of a general open set, $\mathrm{Aut}(S)$ is trivial.

The conclusion is that $(\mathfrak{M}^{min}_{\chi,K^2}, [S])$ is locally isomorphic to the Kuranishi family Def(S), hence everywhere non reduced. □

Using an interesting result of M'nev about line configurations, Vakil ([141]) was able to show that 'any type of singularity' can occur for the Gieseker moduli space (his examples are such that $S = X$ and S has no automorphisms, hence they produce the desired singularities also for the local moduli space, for the Kuranishi families; they also produce singularities for the Hilbert schemes, because his surfaces have $q(S) := h^1(\mathcal{O}_S) = 0$).

Theorem 3.15. (Vakil's 'Murphy's law') *Given any singularity germ of finite type over the integers, there is a Gieseker moduli space $\mathfrak{M}^{can}_{\chi,K^2}$ and a surface S with ample canonical divisor K_S (hence $S = X$) such that $(\mathfrak{M}^{can}_{\chi,K^2}, [X])$ realizes the given singularity germ.*

In the next section we shall see more instances where automorphisms play an important role.

4. Automorphisms and moduli

4.1. Automorphisms and canonical models

The good thing about differential forms is that any group action on a complex manifold leads to a group action on the vector spaces of differential forms.

Assume that G is a group acting on a surface S of general type, or more generally on a Kähler manifold Y: then G acts linearly on the Hodge vector spaces $H^{p,q}(Y) \cong H^q(\Omega^p_Y)$, and also on the vector spaces $H^0((\Omega^n_Y)^{\otimes m}) = H^0(\mathcal{O}_Y(mK_Y))$, hence on the canonical ring

$$\mathcal{R}(Y) := (\mathcal{R}(Y, K_Y)) := \bigoplus_{m \geqslant 0} H^0(\mathcal{O}_Y(mK_Y)).$$

If Y is a variety of general type, then the group G acts linearly on the vector space $H^0(\mathcal{O}_Y(mK_Y))$, hence linearly on the m-th pluricanonical image Y_m, which is an algebraic variety bimeromorphic to Y. Hence G is contained in the algebraic group $\mathrm{Aut}(Y_m)$ and, if G were infinite, as observed by Matsumura ([101]), $\mathrm{Aut}(Y_m)$ would contain a non trivial Cartan subgroup (hence $\mathbb{C}$ or $\mathbb{C}^*$) and Y would be uniruled, a contradiction. This was the main argument of the following

Theorem 4.1. (Matsumura) *The automorphism group of a variety Y of general type is finite.*

Let us specialize now to S a surface of general type, even if most of what we say shall remain valid also in higher dimension.

Take an m-pseudo moduli space $\mathcal{H}_0(\chi, K^2)$ with m so large that the corresponding Hilbert points of the varieties X_m are stable, and let G be a finite group acting on a minimal surface of general type S whose m-th canonical image is in $\mathcal{H}_0(\chi, K^2)$.

Since G acts on the vector space $V_m := H^0(\mathcal{O}_S(mK_S))$, the vector space splits uniquely, up to permutation of the summands, as a direct sum of irreducible representations

$$(**)\ V_m = \bigoplus_{\rho \in Irr(G)} W_\rho^{n(\rho)}.$$

We come now to the basic notion of a family of G-automorphisms

Definition 4.2. *A family of* G*-automorphisms is a triple*

$$((p\colon \mathcal{S} \to T), G, \alpha)$$

where:

(1) $(p\colon \mathcal{S} \to T)$ *is a family in a given category (a smooth family for the case of minimal models of general type)*
(2) G *is a (finite) group*
(3) $\alpha\colon G \times \mathcal{S} \to \mathcal{S}$ *yields a biregular action* $G \to Aut(\mathcal{S})$, *which is compatible with the projection* p *and with the trivial action of* G *on the base* T *(i.e.,* $p(\alpha(g, x)) = p(x),\ \forall g \in G, x \in \mathcal{S}$*).*

As a shorthand notation, one may also write $g(x)$ *instead of* $\alpha(g, x)$, *and by abuse of notation say that the family of automorphisms is a deformation of the pair* (S_t, G) *instead of the triple* (S_t, G, α_t).

Proposition 4.3. *1) A family of automorphisms of surfaces of general type (not necessarily minimal models) induces a family of automorphisms of canonical models.*

2) A family of automorphisms of canonical models induces, if the basis T *is connected, a constant decomposition type* (∗∗) *for* $V_m(t)$.

3) A family of automorphisms of surfaces of general type admits a differentiable trivialization, i.e., in a neighbourhood of $t_0 \in T$, *a diffeomorphism as a family with* $(S_0 \times T, p_T, \alpha_0 \times Id_T)$*; in other words, with the trivial family for which* $g(y, t) = (g(y), t)$.

Proof.

We sketch only the main ideas.

1) follows since one can take the relative canonical divisor $K := K_{\mathcal{S}|T}$, the sheaf of graded algebras

$$\mathcal{R}(p) := \oplus_m p_*(\mathcal{O}_{\mathcal{S}}(mK))$$

and take the relative Proj, yielding $\mathfrak{X} := Proj(\mathcal{R}(p))$, whose fibres are the canonical models.

2) follows since for a representation space (V, ρ') the multiplicity with which an irreducible representation W occurs in V is the dimension of $Hom(W, V)^G$, which

in turn is calculated by the integral on G of the trace of $\rho''(g)$, where $\rho''(g)$ is the representation $\mathrm{Hom}(W, V)$. If we have a family, we get a continuous integer valued function, hence a constant function.

3) Since G acts trivially on the base T, it follows that for each $g \in G$ the fixed locus $\mathrm{Fix}(g)$ is a relative submanifold with a submersion onto T. By the use of stratified spaces (see [100]), and control data, one finds then a differentiable trivialization for the quotient analytic space $\mathcal{S}/G$, hence a trivialization of the action. □

Let us then consider the case of a family of canonical models: by 2) above, and shrinking the base in order to make the addendum $\mathcal{R}(p)_m = p_*(\mathcal{O}_\mathcal{S}(mK))$ free, we get an embedding of the family

$$(\mathfrak{X}, G) \hookrightarrow T \times (\mathbb{P}(V_m = \bigoplus_{\rho \in \mathrm{Irr}(G)} W_\rho^{n(\rho)}), G).$$

In other words, all the canonical models X_t are contained in a fixed projective space, where also the action of G is fixed.

Now, the canonical model X_t is left invariant by the action of G if and only if its Hilbert point is fixed by G. Hence, we get a closed set

$\mathcal{H}_0(\chi, K^2)^G \subset \mathcal{H}_0(\chi, K^2)$

of the pseudomoduli space, and a corresponding closed subset of the moduli space. Hence we get the following theorem.

Theorem 4.4. *The surfaces of general type which admit an action of a given pluricanonical type* $(**)$ *i.e., with a fixed irreducible G- decomposition of their canonical ring, form a closed subvariety* $(\mathfrak{M}^{can}_{\chi,K^2})^{G,(**)}$ *of the moduli space* $\mathfrak{M}^{can}_{\chi,K^2}$.

We shall see that the situation for the minimal models is different, because then the subset of the moduli space where one has a fixed differentiable type is not closed.

4.2. Kuranishi subspaces for automorphisms of a fixed type

Proposition 4.3 is quite useful when one analyses the deformations of a given G-action.

In the case of the canonical models, we just have to look at the fixed points for the action on a subscheme of the Hilbert scheme; whereas, for the case of the deformations of the minimal model, we have to look at the complex structures for which the given differentiable action is biholomorphic. Hence we derive

Proposition 4.5. *Consider a fixed action of a finite group G on a minimal surface of general type S, and let X be its canonical model. Then we obtain closed subsets of the respective Kuranishi spaces, corresponding to deformations which preserve the given action, and yielding a maximal family of deformations of the G-action.*

These subspaces are $\mathfrak{B}(S) \cap H^1(\Theta_S)^G = \mathrm{Def}(S) \cap H^1(\Theta_S)^G$, *respectively* $\mathfrak{B}(X) \cap \mathrm{Ext}^1(\Omega^1_X, \mathcal{O}_X)^G = \mathrm{Def}(X) \cap \mathrm{Ext}^1(\Omega^1_X, \mathcal{O}_X)^G$.

We refer to [34] for a proof of the first fact, while for the second the proof is based again on Cartan's lemma ([29]), that the action of a finite group in an analytic neighbourhood of a fixed point can be linearized.

Just a comment about the contents of the proposition: it says that in each of the two cases, the locus where a group action of a fixed type is preserved is a locally closed set of the moduli space. We shall see more clearly the distinction in the next subsection.

4.3. Deformations of automorphisms differ for canonical and for minimal models

The scope of this subsection is to illustrate the main principles of a rather puzzling phenomenon which we discovered in my joint work with Ingrid Bauer ([10], [11]) on the moduli spaces of Burniat surfaces.

Before dwelling on the geometry of these surfaces, I want to explain clearly what happens, and it suffices to take the example of nodal secondary Burniat surfaces, which I will denote by BUNS in order to abbreviate the name.

For BUNS one has $K_S^2 = 4, p_g(S) := h^0(K_S) = 0$, and the bicanonical map is a Galois cover of the Del Pezzo surface Y of degree 4 with just one node as singularity (the resolution of Y is the blow up Y' of the plane in 5 points, of which exactly 3 are collinear). The Galois group is $G = (\mathbb{Z}/2)^2$, and over the node of Y lies a node of the canonical model X of S, which does not have other singularities.

Then we have BUES, which means extended secondary Burniat surfaces, whose bicanonical map is again a finite $(\mathbb{Z}/2)^2$ - Galois cover of the 1-nodal Del Pezzo surface Y of degree 4 (and for these $S = X$, i.e., the canonical divisor K_S is ample).

All these actions on the canonical models fit together into a single family, but, if we pass to the minimal models, then the topological type of the action changes in a discontinuous way when we pass from the closed set of BUNS to the open set of BUES, and we have precisely two families.

We have , more precisely, the following theorems ([11]):

Theorem 4.6. (Bauer-Catanese) *An irreducible connected component, normal, of dimension 3 of the moduliunirational space of surfaces of general type $\mathfrak{M}_{1,4}^{can}$ is given by the subset $\mathcal{NEB}_4$, formed by the disjoint union of the open set corresponding to BUES (*extended *secondary Burniat surfaces), with the irreducible closed set parametrizing BUNS (nodal secondary Burniat surfaces).*

For all surfaces S in $\mathcal{NEB}_4$ the bicanonical map of the canonical model X is a finite cover of degree 4, with Galois group $G = (\mathbb{Z}/2)^2$, of the 1-nodal Del Pezzo surface Y of degree 4 in $\mathbb{P}^4$.

Moreover the Kuranishi space $\mathfrak{B}(S)$ of any such a minimal model S is smooth.

Theorem 4.7. (Bauer-Catanese) *The deformations of nodal secondary Burniat surfaces (secondary means that $K_S^2 = 4$) to extended secondary Burniat surfaces yield examples where* $\mathrm{Def}(S, (\mathbb{Z}/2\mathbb{Z})^2) \to \mathrm{Def}(X, (\mathbb{Z}/2\mathbb{Z})^2)$ *is not surjective.*

Indeed the pairs (X, G), *where* $G := (\mathbb{Z}/2\mathbb{Z})^2$ *and X is the canonical model of an extended or nodal secondary Burniat surface, where the action of G on X is induced by the bicanonical map of X, belong to only one deformation type.*

If S is a BUNS, then $\mathrm{Def}(S, (\mathbb{Z}/2\mathbb{Z})^2) \subsetneq \mathrm{Def}(S)$, *and* $\mathrm{Def}(S, (\mathbb{Z}/2\mathbb{Z})^2)$ *consists exactly of all the BUNS '; while for the canonical model X of S we have:* $\mathrm{Def}(X, (\mathbb{Z}/2\mathbb{Z})^2) = \mathrm{Def}(X)$.

Indeed for the pairs (S, G), *where S is the minimal model of an extended or nodal Burniat surface,* $G := (\mathbb{Z}/2\mathbb{Z})^2$ *and the action is induced by the bicanonical map (it is unique up to automorphisms of G), they belong to exactly two distinct deformation types, one given by BUNS, and the other given by BUES.*

The discovery of BUES came later as a byproduct of the investigation of tertiary (3-nodal) Burniat surfaces, where we knew by the Enriques-Kuranishi inequality that tertiary Burniat surfaces cannot form a component of the moduli space: and knowing that there are other deformations helped us to find them eventually.

For BUNS, we first erroneously thought (see [10]) that they form a connected component of the moduli space, because $G = (\mathbb{Z}/2\mathbb{Z})^2 \subset \mathrm{Aut}(S) = \mathrm{Aut}(X)$ for a BUNS, and BUNS are exactly the surfaces S for which the action deforms, while we proved that for all deformations of the canonical model X the action deforms.

The description of BUNS and especially of BUES is complicated, so I refer simply to [11]; but the essence of the pathological behaviour can be understood from the local picture around the node of the Del Pezzo surface Y.

We already described most of this local picture in a previous section.

We make here a first additional observation:

Proposition 4.8. *Let* $t \in \mathbb{C}$ *, and consider the action of* $G := (\mathbb{Z}/2\mathbb{Z})^2$ *on* $\mathbb{C}^3$ *generated by* $\sigma_1(u, v, w) = (u, v, -w)$, $\sigma_2(u, v, w) = (-u, -v, w)$. *Then the hypersurfaces* $X_t = \{(u, v, w) | w^2 = uv + t\}$ *are G-invariant, and the quotient* X_t/G *is the hypersurface*

$$Y_t = Y_0 = Y := \{(x, y, z) | z^2 = xy\},$$

which has a nodal singularity at the point $x = y = z = 0$.

$X_t \to Y$ *is a family of finite bidouble coverings (Galois coverings with group* $G := (\mathbb{Z}/2\mathbb{Z})^2$*).*

We get in this way a flat family of (non flat) bidouble covers.

Proof. The invariants for the action of G on $\mathbb{C}^3 \times \mathbb{C}$ are:

$$x := u^2, y := v^2, z := uv, s := w^2, t.$$

Hence the family $\mathfrak{X}$ of the hypersurfaces X_t is the inverse image of the family of hypersurfaces $s = z + t$ on the product

$$Y \times \mathbb{C}^2 = \{(x, y, z, s, t) | xy = z^2\}.$$

Hence the quotient of X_t is isomorphic to Y.

□

The following is instead a rephrasing and a generalization of the discovery of Atiyah in the context of automorphisms, which is the main local content of the above theorem. It says that the family of automorphisms of the canonical models X_t, i.e., the automorphism group of the family $\mathfrak{X}$, does not lift, even after base change, to the family $\mathcal{S}$ of minimal surfaces S_τ.

Lemma 4.9. *Let* G *be the group* $G \cong (\mathbb{Z}/2\mathbb{Z})^2$ *acting on* $\mathfrak{X}$ *trivially on the variable* τ, *and else as follows on* $\mathfrak{X}$: *the action of* $G := (\mathbb{Z}/2\mathbb{Z})^2$ *on* $\mathbb{C}^3$ *is generated by* $\sigma_1(u, v, w) = (u, v, -w)$, $\sigma_2(u, v, w) = (-u, -v, w)$ *(we set then* $\sigma_3 := \sigma_1\sigma_2$, *so that* $\sigma_3(u, v, w) = (-u, -v, -w)$*).*

The invariants for the action of G *on* $\mathbb{C}^3 \times \mathbb{C}$ *are:*

$$x := u^2, y := v^2, z := uv, s := w^2, t.$$

Observe that the hypersurfaces $X_t = \{(u, v, w) | w^2 = uv + t\}$ *are* G*-invariant, and the quotient* X_t/G *is the hypersurface*

$$Y_t \cong Y_0 = \{(x, y, z) | z^2 = xy\},$$

which has a nodal singularity at the point $x = y = z = 0$.

Let further σ_4 *act by* $\sigma_4(u, v, w, \tau) = (u, v, w, -\tau)$, *let* $G' \cong (\mathbb{Z}/2\mathbb{Z})^3$ *be the group generated by* G *and* σ_4, *and let* $H \cong (\mathbb{Z}/2\mathbb{Z})^2$ *be the subgroup* $\{\mathrm{Id}, \sigma_2, \sigma_1\sigma_4, \sigma_3\sigma_4\}$.

The biregular action of G' *on* $\mathfrak{X}$ *lifts only to a birational action on* $\mathcal{S}$, *respectively* $\mathcal{S}'$. *The subgroup* H *acts on* $\mathcal{S}$, *respectively* $\mathcal{S}'$, *as a group of biregular automorphisms.*

The elements of $G' \setminus H = \{\sigma_1, \sigma_3, \sigma_4, \sigma_2\sigma_4\}$ *yield isomorphisms between* $\mathcal{S}$ *and* $\mathcal{S}'$.

The group G *acts on the punctured family* $\mathcal{S} \setminus S_0$, *in particular it acts on each fibre* S_τ.

Since σ_4 *acts trivially on* S_0, *the group* G' *acts on* S_0 *through its direct summand* G.

The biregular actions of G *on* $\mathcal{S} \setminus S_0$ *and on* S_0 *do not patch together to a biregular action on* $\mathcal{S}$, *in particular* σ_1 *and* σ_3 *yield birational maps which are not biregular: they are called Atiyah flops (cf. [6]).*

Another more geometrical way to see that there is no G-action on the family $\mathcal{S}$ is the following: if G would act on $\mathcal{S}$, and trivially on the base, then the fixed loci for the group elements would be submanifolds with a smooth map onto the parameter space $\mathbb{C}$ with parameter τ. Hence all the quotients S_τ/G would be homeomorphic.

But for BUNS the quotient of S_0 by G is the blow up Y' of Y at the origin, while for $\tau \neq 0$, S_τ/G is just Y! [2] In fact, if one wants to construct the family of smooth models as a family of bidouble covers of a smooth surface, one has to take the blown up surface Y' and its exceptional divisor N (N is called the nodal curve).

Remark 4.10. i) The simplest way to view X_t is to see $\mathbb{C}^2$ as a double cover of Y branched only at the origin, and then X_t as a family of double covers of $\mathbb{C}^2$ branched on the curve $uv + t = 0$, which acquires a double point for $t = 0$.

[2]In the case of BUNS, Y is a nodal Del Pezzo surface of degree 4, whereas in the local analysis we use the same notation Y for the quadric cone, which is the germ of the nodal Del Pezzo surface at the nodal singular point.

ii) If we pull back the bidouble cover X_t to Y', and we normalize it, we can see that the three branch divisors, corresponding to the fixed points for the three non trivial elements of the group G, are as follows:

- D_3 is, for $t = 0$, the nodal curve N, and is the empty divisor for $t \neq 0$;
- D_1 is, for $t \neq 0$, the inverse image of the curve $z + t = 0$; while, for $t = 0$, it is only its strict transform, i.e. a divisor made up of F_1, F_2, the proper transforms of the two branch lines ($\{x=z=0\}$, resp. $\{y=z=0\}$) on the quadric cone Y
- D_2 is an empty divisor for $t = 0$, and the nodal curve N for $t \neq 0$.

The above remark shows then that in order to construct the smooth models, one has first of all to take a discontinuous family of branch divisors; and, moreover, for $t \neq 0$, we obtain then a non minimal surface which contains two (-1)-curves ($S_t = X_t$ is then gotten by contracting these two (-1)-curves).

4.4. Teichmüller space for surfaces of general type

Recall the fibre product considered by Burns and Wahl:

$$\begin{array}{ccc} \mathrm{Def}(S) & \longrightarrow & \mathrm{Def}(S_{\mathrm{Exc}(\pi)}) =: \mathcal{L}_S \cong \mathbb{C}^\nu, \\ \downarrow & & \downarrow \lambda \\ \mathrm{Def}(X) & \longrightarrow & \mathrm{Def}(X_{\mathrm{Sing} X}) =: \mathcal{L}_X \cong \mathbb{C}^\nu, \end{array}$$

This gives a map $f\colon \mathrm{Def}(S) \to \mathrm{Def}(X)/\mathrm{Aut}(X)$ of the Kuranishi space of S into an open set of a quasiprojective variety, which factors through Teichmüller space.

Theorem 4.11. *Let S be the minimal model of a surface of general type.*

Then the continuous map π: $\mathrm{Def}(S) \to \mathcal{T}(M)_S$ is a local homeomorphism between Kuranishi space and Teichmüller space if

1) $\mathrm{Aut}(S)$ is a trivial group, or

2) K_S is ample and S is rigidified.

Proof. We need only to show that π is injective. Assume the contrary: then there are two points $t_1, t_2 \in \mathrm{Def}(S)$ yielding surfaces S_1 and S_2 isomorphic through a diffeomorphism Ψ isotopic to the identity.

By the previous remark, the images of t_1, t_2 inside $\mathrm{Def}(X)/\mathrm{Aut}(X)$ must be the same.

Case 1): there exists then an element w of the Weyl group of λ carrying t_1 to t_2, hence the composition of w and Ψ yields an automorphism of S_1. Since $\mathrm{Aut}(S_1) = \mathrm{Aut}(X_1)$ and the locus of canonical models with non trivial automorphisms is closed, we conclude that, taking $\mathrm{Def}(S)$ as a suitably small germ, then this automorphism is the identity. This is however a contradiction, since w acts non trivially on the cohomology of the exceptional divisor, while Ψ acts trivially.

Case 2) : In this case there is an automorphism g of S carrying t_1 to t_2, and again the composition of g and Ψ yields an automorphism of S_1. We apply the same argument, since g is not isotopic to the identity by our assumption.

□

Remark 4.12. With more work one should be able to treat the more general case where we assume that $\mathrm{Aut}(S)$ is non trivial, but S is rigidified. In this case one should show that a composition $g \circ w$ as above is not isotopic to the identity.

The most interesting question is however whether every surface of general type is rigidified.

5. Connected components and arithmetic of moduli spaces for surfaces

5.1. Gieseker's moduli space and the analytic moduli spaces

As we saw, all 5-canonical models of surfaces of general type with invariants K^2, χ occur in a big family parametrized by an open set of the Hilbert scheme $\mathcal{H}^0$ parametrizing subschemes with Hilbert polynomial $P(m) = \chi + \frac{1}{2}(5m-1)5mK^2$, namely the open set

$$\mathcal{H}^0(\chi, K^2) := \{\Sigma | \Sigma \text{ is reduced with only R.D.P.'s as singularities }\}.$$

Indeed, it is not necessary to consider the 5-pseudo moduli space of surfaces of general type with given invariants K^2, χ, which was defined as the closed subset $\mathcal{H}_0 \subset \mathcal{H}^0$,

$$\mathcal{H}_0(\chi, K^2) := \{\Sigma \in \mathcal{H}^0 | \omega_\Sigma^{\otimes 5} \cong \mathcal{O}_\Sigma(1)\}.$$

At least, if we are only interested about having a family which contains all surfaces of general type, and are not interested about taking the quotient by the projective group.

Observe however that if $\Sigma \in \mathcal{H}^0(\chi, K^2)$, then Σ is the canonical model X of a surface of general type , embedded by a linear system $|D|$, where D is numerically equivalent to $5K_S$, i.e., $D = 5K_S + \eta$, where η is numerically equivalent to 0.

Therefore the connected components $\mathcal{N}$, respectively the irreducible components $\mathcal{Z}$ of the Gieseker moduli space correspond to the connected , resp. irreducible, components of $\mathcal{H}_0(\chi, K^2)$, and in turn to the connected , resp. irreducible, components of $\mathcal{H}^0(\chi, K^2)$ which intersect $\mathcal{H}_0(\chi, K^2)$.

We shall however, for the sake of brevity, talk about connected components $\mathcal{N}$ of the Gieseker moduli space $\mathfrak{M}^{can}_{a,b}$ even if these do not really parametrize families of canonical models.

We refer to [44] for a more ample discussion of the basic ideas which we are going to sketch here.

$\mathfrak{M}^{can}_{a,b}$ has a finite number of connected components, and these parametrize the deformation classes of surfaces of general type. By the classical theorem of Ehresmann

([62]), deformation equivalent varieties are diffeomorphic, and moreover, by a diffeomorphism carrying the canonical class to the canonical class.

Hence, fixed the two numerical invariants $\chi(S) = a$, $K_S^2 = b$, which are determined by the topology of S (indeed, by the Betti numbers of S), we have a finite number of differentiable types.

It is clear that the analytic moduli space $\mathfrak{M}(S)$ that we defined at the onset is then the union of a finite number of connected components of $\mathfrak{M}^{can}_{a,b}$. But how many, and how?

A very optimistic guess was: one.

A basic question was really whether a moduli space $\mathfrak{M}(S)$ would correspond to a unique connected component of the Gieseker moduli space, and this question was abbreviated as the DEF = DIFF question.

I.e., the question whether differentiable equivalence and deformation equivalence would coincide for surfaces.

I conjectured (in [140]) that the answer should be negative, on the basis of some families of simply connected surfaces of general type constructed in [31]: these were then homeomorphic by the results of Freedman (see [68], and [69]), and it was then relatively easy to show then ([32]) that there were many connected components of the moduli space corresponding to homeomorphic but non diffeomorphic surfaces. It looked like the situation should be similar even if one would fix the diffeomorphism type.

Friedman and Morgan instead made the 'speculation' that the answer to the DEF= DIFF question should be positive (1987) (see [70]), motivated by the new examples of homeomorphic but not diffeomorphic surfaces discovered by Donaldson (see [57] for a survey on this topic).

The question was finally answered in the negative, and in every possible way ([97],[84],[40],[48],[13].

Theorem 5.1. *(Manetti '98, Kharlamov -Kulikov 2001, C. 2001, C. - Wajnryb 2004, Bauer- C. - Grunewald 2005)*

The Friedman- Morgan speculation does not hold true and the DEF= DIFF question has a negative answer.

In my joint work with Bronek Wajnryb ([48]) the question was also shown to have a negative answer even for simply connected surfaces.

I showed later ([39]) that each minimal surface of general type S has a natural symplectic structure with class of the symplectic form equal to $c_1(K_S)$, and in such a way that to each connected component $\mathcal{N}$ of the moduli space one can associate the pair of a differentiable manifold with a symplectic structure, unique up to symplectomorphism.

Would this further datum determine a unique connected component, so that DEF = SIMPL ?

This also turned out to have a negative answer ([43]).

Theorem 5.2. *Manetti surfaces provide counterexamples to the DEF = SIMPL question.*

I refer to [42] for a rather comprehensive treatment of the above questions.

Let me just observe that the Manetti surfaces are not simply connected, so that the DEF=SYMPL question is still open for the case of simply connected surfaces. Concerning the question of canonical symplectomorphism of algebraic surfaces, Auroux and Katzarkov ([7]) defined asymptotic braid monodromy invariants of a symplectic manifold, extending old ideas of Moishezon (see [104]).

Quite recent work, not covered in [42], is my joint work with Lönne and Wajnryb ([47]), which investigates in this direction the braid monodromy invariants (especially the 'stable' ones) for the surfaces introduced in [31].

5.2. Arithmetic of moduli spaces

A basic remark is that all these schemes are defined by equations involving only $\mathbb{Z}$ coefficients, since the defining equation of the Hilbert scheme is a rank condition on a multiplication map (see for instance [74]), and similarly the condition $\omega_\Sigma^{\otimes 5} \cong \mathcal{O}_\Sigma(1)$ is also closed (see [108]) and defined over $\mathbb{Z}$..

It follows that the absolute Galois group $Gal(\overline{\mathbb{Q}}, \mathbb{Q})$ acts on the Gieseker moduli space $\mathfrak{M}^{can}_{a,b}$.

To explain how it concretely acts, it suffices to recall the notion of a conjugate variety.

Remark 5.3. 1) $\phi \in Aut(\mathbb{C})$ acts on $\mathbb{C}[z_0, \dots z_n]$, by sending $P(z) = \sum_{i=0}^n a_i z^i \mapsto \phi(P)(z) := \sum_{i=0}^n \phi(a_i) z^i$.

2) Let X be as above a projective variety

$$X \subset \mathbb{P}^n_{\mathbb{C}}, X := \{z | f_i(z) = 0 \ \forall i\}.$$

The action of ϕ extends coordinatewise to $\mathbb{P}^n_{\mathbb{C}}$, and carries X to another variety, denoted X^ϕ, and called the **conjugate variety**. Since $f_i(z) = 0$ implies $\phi(f_i)(\phi(z)) = 0$, we see that

$$X^\varphi = \{w | \phi(f_i)(w) = 0 \ \forall i\}.$$

If ϕ is complex conjugation, then it is clear that the variety X^ϕ that we obtain is diffeomorphic to X, but in general, what happens when ϕ is not continuous ?

Observe that, by the theorem of Steiniz, one has a surjection $Aut(\mathbb{C}) \to Gal(\bar{\mathbb{Q}}/\mathbb{Q})$, and by specialization the heart of the question concerns the action of $Gal(\bar{\mathbb{Q}}/\mathbb{Q})$ on varieties X defined over $\bar{\mathbb{Q}}$.

For curves, since in general the dimensions of spaces of differential forms of a fixed degree and without poles are the same for X^ϕ and X, we shall obtain a curve of the same genus, hence X^ϕ and X are diffeomorphic.

But for higher dimensional varieties this breaks down, as discovered by Jean Pierre Serre in the 60's ([131]), who proved the existence of a field automorphism $\phi \in \mathrm{Gal}(\bar{\mathbb{Q}}/\mathbb{Q})$, and a variety X defined over $\bar{\mathbb{Q}}$ such that X and the Galois conjugate variety X^{ϕ} have non isomorphic fundamental groups.

In work in collaboration with Ingrid Bauer and Fritz Grunewald ([14], [15]) we discovered wide classes of algebraic surfaces for which the same phenomenon holds.

A striking result in a similar direction was obtained by Easton and Vakil ([60])

Theorem 5.4. *The absolute Galois group* $\mathrm{Gal}(\bar{\mathbb{Q}}/\mathbb{Q})$ *acts faithfully on the set of irreducible components of the (coarse) moduli space of canonical surfaces of general type,*

$$\mathfrak{M}^{can} := \cup_{a,b\geqslant 1}\mathfrak{M}^{can}_{a,b}.$$

5.3. Topology sometimes determines connected components

There are cases where the presence of a big fundamental group implies that a connected component of the moduli space is determined by some topological invariants.

A typical case is the one of surfaces isogenous to a product ([38]), where a surface is said to be isogenous to a (higher) product if and only if it is a quotient

$$(C_1 \times C_2)/G,$$

where C_1, C_2 are curves of genera $g_1, g_2 \geqslant 2$, and G is a finite group acting freely on $(C_1 \times C_2)$.

Theorem 5.5. *(see* [38]*).*

a) A projective smooth surface is isogenous to a higher product if and only if the following two conditions are satisfied:

1) there is an exact sequence

$$1 \to \Pi_{g_1} \times \Pi_{g_2} \to \pi = \pi_1(S) \to G \to 1,$$

where G is a finite group and where Π_{g_i} denotes the fundamental group of a compact curve of genus $g_i \geqslant 2$;

2) $e(S)(= c_2(S)) = \frac{4}{|G|}(g_1 - 1)(g_2 - 1)$.

b) Any surface X with the same topological Euler number and the same fundamental group as S is diffeomorphic to S. The corresponding subset of the moduli space, $\mathfrak{M}^{top}_S = \mathfrak{M}^{diff}_S$, *corresponding to surfaces orientedly homeomorphic, resp. orientedly diffeomorphic to S, is either irreducible and connected or it contains two connected components which are exchanged by complex conjugation.*

In particular, if S′ is orientedly diffeomorphic to S, then S′ is deformation equivalent to S or to $\bar{S}$.

Other non trivial examples are the cases of Keum-Naie surfaces, Burniat surfaces and Kulikov surfaces ([9] , [12], [49]): for these classes of surfaces the main result is that any surface homotopically equivalent to a surface in the given class belongs to a unique irreducible connected component of the moduli space.

Just to give a flavour of some of the arguments used, let us consider a simple example which I worked out together with Ingrid Bauer.

Let S be a minimal surface of general type with $q(S) \geqslant 2$. Then we have the Albanese map

$$\alpha\colon S \to A := \mathrm{Alb}(S),$$

and S is said to be of Albanese general type if $\alpha(S) := Z$ is a surface. This property is a topological property (see [36]), since α induces a homomorphism of cohomology algebras

$$\alpha^*\colon H^*(A, \mathbb{Z}) \to H^*(S, \mathbb{Z})$$

and $H^*(A, \mathbb{Z})$ is the full exterior algebra $\Lambda^*(H^1(A, \mathbb{Z})) \cong \Lambda^*(H^1(S, \mathbb{Z}))$ over $H^1(S, \mathbb{Z})$.

In particular, in the case where $q(S) = 2$, the degree d of the Albanese map equals the index of the image of $\Lambda^4 H^1(S, \mathbb{Z})$ inside $H^4(S, \mathbb{Z}) = \mathbb{Z}[S]$.

The easiest case is the one when $d = 2$, because then $K_S = R$, R being the ramification divisor. Observe that the Albanese morphism factors through the canonical model X of S, and a morphism $a\colon X \to A$.

Assume now that a is a finite morphism, so that $2K_X = a^*(a_*(K_X))$. In particular, if we set $D := a_*(K_X)$, then $D^2 = 2K_X^2 = 2K_S^2$, and this number is also a topological invariant.

By the standard formula for double covers we have that $p_g(S) = h^0(L) + 1$, where D is linearly equivalent to $2L$; hence, if L is a polarization of type (d_1, d_2), then $p_g(S) = d_1 d_2 + 1$, D is a polarization of type $(2d_1, 2d_2)$, and $4d_1 d_2 = 2L^2 = K_S^2$, hence in particular we have

$$K_S^2 = 4(p_g - 1) = 4\chi(S),$$

since $q(S) = 2$.

I can moreover recover the polarization type (d_1, d_2) (where d_1 divides d_2) using the fact that $2d_1$ is exactly the divisibility index of D. This is in turn the divisibility of K_S, since K_S gives a linear form on $H^2(A, \mathbb{Z})$ simply by the composition of pushforward and cup product, and this linear form is represented by the class of D. Finally, the canonical class K_S is a differentiable invariant of S (see [58] or [105]).

The final argument is that, by formulae due to Horikawa ([78]), necessarily if $K_S^2 = 4\chi(S)$ the branch locus has only negligible singularities (see [9]), which means that the normal finite cover branched over D has rational double points as singularities.

Theorem 5.6. (Bauer-Catanese) *Let S be a minimal surface of general type whose canonical model X is a finite double cover of an Abelian surface A, branched on a divisor D of type $(2d_1, 2d_2)$. Then S belongs to an irreducible connected component $\mathcal{N}$ of the moduli space of dimension $4d_1 d_2 + 2 = 4\chi(S) + 2$.*

Moreover,

1) any other surface which is diffeomorphic to such a surface S belongs to the component $\mathcal{N}$.

2) The Kuranishi space $\mathrm{Def}(X)$ *is always smooth.*

The assumption that X is a finite double cover is a necessary one.

For instance, Penegini and Polizzi ([117]) construct surfaces with $p_g(S) = q(S) = 2$ and $K_S^2 = 6$ such that for the general surface the canonical divisor is ample (whence $S = X$), while the Albanese map, which is generically finite of degree 2, contracts an elliptic curve Z with $Z^2 = -2$ to a point. The authors show then that the corresponding subset of the moduli space consists of three irreducible connected components.

Other very interesting examples with degree $d = 3$ have been considered by Penegini and Polizzi in [116].

6. Smoothings and surgeries

Lack of time prevents me to develop this section.

I refer the reader to [42] for a general discussion, and to the articles [97] and [94] for interesting applications of the $\mathbb{Q}$-Gorenstein smoothings technique (for the study of components of moduli spaces, respectively for the construction of new interesting surfaces).

There is here a relation with the topic of compactifications of moduli spaces. Arguments to show that certain subsets of the moduli spaces are closed involved taking limits of canonical models and studying certain singularities (see [33], [96], see also [122] for the relevant results on deformations of singularities); in [89] a more general study was pursued of the singularities allowed on the boundary of the moduli space of surfaces. I refer to the article by Kollár in this Handbook for the general problem of compactifying moduli spaces (also Viehweg devoted a big effort into this enterprise, see [142], [143], another important reference is [139]).

An explicit study of compactifications of the moduli spaces of surfaces of general type was pursued in [113], [1], [95], [125].

There is here another relation, namely to the article by Abramovich and others in this Handbook, since the deformation of pairs (Y, D) where Y is a smooth complex manifold and $D = \cup_{i=1,\dots,h} D_i$ is a normal crossing divisor, are governed by the cohomology groups

$$H^i(\Theta_Y(-\log D_1, \dots, -\log D_h)),$$

for $i = 1,2$, and where the sheaf $\Theta_Y(-\log D_1, \dots, -\log D_h)$ is the Serre dual of the sheaf $\Omega^1_Y(\log D_1, \dots, \log D_h)(K_Y)$, with its residue sequence

$$0 \to \Omega^1_Y(K_Y) \to \Omega^1_Y(\log D_1, \dots, \log D_h)(K_Y) \to \bigoplus_{i=1}^{3} \mathcal{O}_{D_i}(K_Y) \to 0.$$

These sheaves are the appropriate ones to calculate the deformations of ramified coverings, see for instance [31]),[115], [37], [10], and especially [11]).

I was taught about these by David Mumford back in 1977, when he had just been working on the Hirzebruch proportionality principle in the non compact case ([109]).

Acknowledgements.

I would like to thank Ingrid Bauer, Barbara Fantechi and Keiji Oguiso for interesting conversations, Wenfei Liu and JinXing Cai for email correspondence, Sönke Rollenske, Antonio J. Di Scala and two referees for reading the manuscript with care and pointing out corrections.

I am grateful to Jun Muk Hwang, Jongnam Lee and Miles Reid for organizing the winter school 'Algebraic surfaces and related topics', which took place at the KIAS in Seoul in March 2010: some themes treated in those lectures are reproduced in this text.

A substantial part of the paper was written when I was a visiting KIAS scholar in March-April 2011: I am very grateful for the excellent atmosphere I found at the KIAS and for the very nice hospitality I received from Jun Muk Hwang and Jong Hae Keum.

References

[1] V. Alexeev and R. Pardini. Explicit compactifications of moduli spaces of Campedelli and Burniat surfaces. arXiv:0901.4431 ← 205

[2] E. Arbarello and M. Cornalba. Teichmüller space via Kuranishi families. *Ann. Sc. Norm. Super. Pisa Cl. Sci. (5)*, **8** (1), 89–116, 2009. ← 167, 176

[3] M. Artin. On isolated rational singularities of surfaces. *Amer. J. Math.*, **88**, 129–136, 1966. ← 179, 183

[4] M. Artin. Algebraic construction of Brieskorn's resolutions. *J. of Algebra*, **29**, 330–348, 1974. ← 191

[5] M. Artin. Versal deformations and algebraic stacks. *Invent. Math.*, **27**, 165–189, 1974. ← 164, 182, 190, 191

[6] M. F. Atiyah. On analytic surfaces with double points. *Proc. Roy. Soc. London. Ser. A*, **247**, 237–244, 1958. ← 179, 198

[7] D. Auroux and L. Katzarkov. Branched coverings of $\mathbb{CP}^2$ and invariants of symplectic 4-manifolds. *Inv. Math.*, **142**, 631–673, 2000. ← 202

[8] W. Barth, C. Peters, and A. Van de Ven. Compact complex surfaces. *Ergebnisse der Mathematik und ihrer Grenzgebiete (3)*, **4**, Springer-Verlag, Berlin, 1984; second edition by W. Barth, K. Hulek, C. Peters, A. Van de Ven, *Ergebnisse der Mathematik und ihrer Grenzgebiete.*, **3**, Folge. A , 4, Springer-Verlag, Berlin, 2004. ← 182

[9] I. Bauer and F. Catanese. The moduli space of Keum-Naie surfaces. arXiv:0909.1733. *Groups Geom. Dyn.*, **5**(2), 231–250, 2011. ← 203, 204

[10] I. Bauer and F. Catanese. Burniat surfaces. II. Secondary Burniat surfaces form three connected components of the moduli space. *Invent. Math.*, **180**(3), 559–588, 2010. ← 196, 197, 205

[11] I. Bauer and F. Catanese. Burniat surfaces III: deformations of automorphisms and extended Burniat surfaces. `arXiv:1012.3770` ← 196, 197, 205

[12] I. Bauer and F. Catanese. Burniat surfaces I: fundamental groups and moduli of primary Burniat surfaces. *Classification of algebraic varieties-Schiermonnikoog 2009*, C. Faber, G. van der Geer, E. Looijenga editors, EMS Congress Reports, 1–31, 2011. ← 203

[13] I. Bauer, F. Catanese, and F. Grunewald. Beauville surfaces without real structures, in: Geometric methods in algebra and number theory, 1–42. *Progr. Math.*, **235**, Birkhäuser, 2005. ← 201

[14] I. Bauer, F. Catanese, and F. Grunewald. Chebycheff and Belyi polynomials, dessins d'enfants, Beauville surfaces and group theory. *Mediterranean J. Math.*, **3**(2), 119–143, 2006. ← 203

[15] I. Bauer, F. Catanese, and F. Grunewald. *The absolute Galois group acts faithfully on the connected components of the moduli spaces of surfaces of general type?* preliminary version 2007. `arXiv:0706.1466` ← 203

[16] A. Blanchard. Recherche de structures analytiques complexes sur certaines variétés. *C. R. Acad. Sci. Paris*, **236**, 657–659, 1953. ← 166

[17] A. Blanchard. Espaces fibrés kählériens compacts. *C. R. Acad. Sci. Paris*, **238**, 2281–2283, 1954. ← 166

[18] A. Blanchard. Sur les variétés analytiques complexes. *Ann. Sci. Ecole Norm. Sup.(3)*, **73**, 157–202, 1956. ← 166

[19] F. A. Bogomolov. Hamiltonian Kählerian manifolds. *Dokl. Akad. Nauk SSSR*, **243** (5), 1101–1104, 1978. ← 177

[20] E. Bombieri. Canonical models of surfaces of general type. *Publ. Math. I.H.E.S.*, **42**, 173–219, 1973. ← 183

[21] E. Bombieri and D. Husemoller. Classification and embeddings of surfaces. In: Algebraic geometry, Humboldt State Univ., Arcata, Calif., 1974, *Proc. Sympos. Pure Math.*, **29**, 329–420, 1975, Amer. Math. Soc., Providence, R.I. ← 182

[22] E. Brieskorn. Rationale Singularitäten komplexer Flächen. *Invent. Math.*, **4** , 336–358, 1967/1968. ← 186

[23] E. Brieskorn. Die Auflösung der rationalen Singularitäten holomorpher Abbildungen. *Math. Ann.*, **178**, 255–270, 1968. ← 179, 186

[24] E. Brieskorn. Singular elements of semi-simple algebraic groups. 'Actes du Congrés International des Mathématiciens' (Nice, 1970), Tome 2, *Gauthier-Villars*, 279–284, 1971, Paris. ← 179, 186, 187

[25] D. Burns and M. Rapoport. On the Torelli problem for kählerian K? 3 surfaces. *Ann. Sci. École Norm. Sup. (4)*, **8**(2), 235–273, 1975. ← 181

[26] D. Burns and J. Wahl. Local contributions to global deformations of surfaces. *Invent. Math.*, **26**, 67–88, 1974. ← 187

[27] J. X. Cai. Classification of fiber surfaces of genus 2 with automorphisms acting trivially in cohomology. *Pacific J. Math.*, **232** (1), 43–59, 2007. ← 175

[28] E. Calabi. Construction and properties of some 6-dimensional almost complex manifolds. *Trans. Amer. Math. Soc.*, **87**, 407–438, 1958. ← 166

[29] H. Cartan. Quotient d'un espace analytique par un groupe d'automorphismes. In *A symposium in honor of S. Lefschetz, Algebraic geometry and topology*, 90–102, Princeton University Press , Princeton, N. J., 1957. ← 196

[30] F. Catanese. Moduli of surfaces of general type. In 'Algebraic geometry—open problems' (Ravello, 1982). *Lecture Notes in Math.*, **997**, 90–112, Springer, Berlin-New York, 1983. ← 169

[31] F. Catanese. On the Moduli Spaces of Surfaces of General Type. *J. Differential Geom.*, **19**, 483–515, 1984. ← 201, 202, 205

[32] F. Catanese. Connected Components of Moduli Spaces. *J. Differential Geom.*, **24**, 395–399, 1986. ← 201

[33] F. Catanese. Automorphisms of Rational Double Points and Moduli Spaces of Surfaces of General Type. *Comp. Math.*, **61**, 81–102, 1987. ← 205

[34] F. Catanese. Moduli of algebraic surfaces. Theory of moduli (Montecatini Terme, 1985), *Lecture Notes in Math.*, **1337**, 1–83, Springer, Berlin, 1988. ← 165, 169, 196

[35] F. Catanese. Everywhere non reduced Moduli Spaces. *Inv. Math.*, **98**, 293–310, 1989. ← 188, 191, 192

[36] F. Catanese. Moduli and classification of irregular Kähler manifolds (and algebraic varieties) with Albanese general type fibrations. *Invent. math.*, **104**, 263–289, 1991. ← 204

[37] F. Catanese. Singular bidouble covers and the construction of interesting algebraic surfaces. *Amer. Math. Soc. Cont. Math.*, **241**, 97–120, 1999. ← 205

[38] F. Catanese. Fibred surfaces, varieties isogenous to a product and related moduli spaces. *Amer. J. Math.*, **122** (1), 1–44, 2000. ← 203

[39] F. Catanese. *Symplectic structures of algebraic surfaces and deformation*. 14 pages. math.AG/0207254 ← 166, 201

[40] F. Catanese. Moduli Spaces of Surfaces and Real Structures. *Ann. Math.(2)*, **158** (2), 577–592, 2003. ← 201

[41] F. Catanese. Deformation in the large of some complex manifolds, I. Ann. *Mat. Pura Appl. (4)*, **183** (3), 261–289, Volume in Memory of Fabio Bardelli, 2004. ← 166, 170

[42] F. Catanese. Differentiable and deformation type of algebraic surfaces, real and symplectic structures. *Symplectic 4-manifolds and algebraic surfaces*, 55–167, Lecture Notes in Math., 1938; Springer, Berlin, 2008. ← 202, 205

[43] F. Catanese. Canonical symplectic structures and deformations of algebraic surfaces. *Comm. in Contemp. Math.*, **11** (3), 481–493, 2009. ← 202

[44] F. Catanese. Algebraic surfaces and their moduli spaces: real, differentiable and symplectic structures. *Boll. Unione Mat. Ital. (9)*, **2**(3), 537–558, 2009. ← 200

[45] F. Catanese and M. Franciosi. Divisors of small genus on algebraic surfaces and projective embeddings, in: Proceedings of the Hirzebruch 65 Conference on Algebraic Geometry (Ramat Gan, 1993). *Israel Math. Conf. Proc.*, *9*, 109–140, Bar-Ilan Univ., Ramat Gan, 1996. ← 185

[46] F. Catanese, M. Franciosi, K. Hulek, and M. Reid. Embeddings of curves and surfaces. *Nagoya Math. J.*, **154**, 185–220, 1999. ← 185

[47] F. Catanese, M. Lönne, and B. Wajnryb. Moduli spaces and braid monodromy types of bidouble covers of the quadric. *Geometry – Topology*, **15**, 351–396, 2011. URL: `http://www.msp.warwick.ac.uk/gt/2011/15-01/p010.xhtml`, DOI: `10.2140/gt.2011.15.351` ← 202

[48] F. Catanese and B. Wajnryb. Diffeomorphism of simply connected algebraic surfaces. *J. Differential Geom.*, **76** (2), 177–213, 2007. ← 201

[49] M. Chan and S. Coughlan. Kulikov surfaces form a connected component of the moduli space. `arXiv:1011.5574` ← 203

[50] C. Chevalley. Invariants of finite groups generated by reflections. *Amer. J. Math.*, **77**, 778–782, 1955. ← 187

[51] S. V. Chmutov. Examples of projective surfaces with many singularities. *J. Algebraic Geom.*, **1**(2), 191–196, 1992. ← 189

[52] H. Clemens. Cohomology and obstructions I: geometry of formal Kuranishi theory. `arxiv:math/9901084` ← 173

[53] H. Clemens. Geometry of formal Kuranishi theory. *Adv. Math.* **198**(1), 311-365, 2005. ← 173

[54] M. Dehn. Die Gruppe der Abbildungsklassen(Das arithmetische Feld auf Flächen). *Acta Math.*, **69**, 135–206, 1938. ← 165

[55] P. Deligne and D. Mumford. The irreducibility of the space of curves of given genus. *Inst. Hautes Études Sci. Publ. Math.*, **36**, 75–109, 1969. ← 185

[56] I. Dolgachev. Weighted projective varieties, in: Group actions and vector fields (Vancouver, B.C., 1981), 34–71. *Lecture Notes in Math.*, **956**, Springer, Berlin, 1982. ← 183

[57] S. K. Donaldson. Gauge theory and four-manifold topology. [CA] Joseph, A. (ed.) et al., First European congress of mathematics (ECM), Paris, France, July 6–10, 1992. Volume I: Invited lectures (Part 1). *Basel: Birkhäuser, Prog. Math.*, **119**, 121–151, 1994. ← 201

[58] S. K. Donaldson. The Seiberg-Witten Equations and 4-manifold topology. *Bull. Am. Math. Soc., (N S)*, **33** (1), 45–70, 1996. ← 204

[59] P. Du Val. On singularities of surfaces which do not impose adjunction conditions. *Proc. Cambridge Philos. Soc.*, **30**, 483–491, 1934. ← 179

[60] R. W. Easton and R. Vakil. Absolute Galois acts faithfully on the components of the moduli space of surfaces: a Belyi-type theorem in higher dimension. *Int. Math. Res. Not. IMRN*, (20), 10 pp, 2007, Art. ID `rnm080` ← 203

[61] B. Eckmann and A. Frölicher. Sur l'intégrabilité des structures presque complexes. *C. R. Acad. Sci. Paris*, **232**, 2284–2286, 1951. ← 165

[62] C. Ehresmann. Sur les espaces fibrés différentiables. *C. R. Acad. Sci. Paris*, 1611-1612, 224, 1947. ← 201

[63] C. Ehresmann. Sur la théorie des espaces fibrés. In: Topologie algébrique. *Colloques Internationaux du Centre National de la Recherche Scientifique*, **12**, 3–15, Centre de la Recherche Scientifique, Paris, 1949. ← 164

[64] T. Ekedahl. Canonical models of surfaces of general type in positive characteristic. *Inst. Hautes Études Sci. Publ. Math.*, (67), 97–144, 1988. ← 185

[65] R. Elkik. Singularités rationnelles et déformations. *Invent. Math.*, **47**(2), 139 –147, 1978. ← 184

[66] F. Enriques, *Le Superficie Algebriche*. Zanichelli, Bologna , 1949. ← 173

[67] B. Fantechi. Stacks for everybody, *European Congress of Mathematics*, vol. **I**, 349–359, (Barcelona, 2000); *Progr. Math.*, **201**, Birkhäuser, Basel, 2001. ← 164

[68] M. Freedman. The topology of four-dimensional manifolds. *J. Differential Geom.*, **17**(3), 357–454, 1982. ← 201

[69] M. Freedman and F. Quinn. Topology of 4- manifolds. *Princeton Math. Series*, **39**, 1990. ← 201

[70] R. Friedman and J. W. Morgan. Algebraic surfaces and four-manifolds: some conjectures and speculations. *Bull. Amer. Math.Soc.*, **18**, 1–19, 1988. ← 201

[71] D. Gieseker. Global moduli for surfaces of general type. *Invent. Math.*, **43** (3), 233–282, 1977. ← 185

[72] H. Grauert. Über die Deformation isolierter Singularitäten analytischer Mengen. *Invent. Math.*, **15**, 171–198, 1972. ← 178

[73] H. Grauert. Der Satz von Kuranishi für kompakte komplexe Räume. *Invent. Math.*, **25**, 107–142, 1974. ← 177

[74] M. Green. Restrictions of linear series to hyperplanes, and some results of Macaulay and Gotzmann, in: Algebraic curves and projective geometry (Trento, 1988), 76–86. *Lecture Notes in Math.*, **1389**, Springer, Berlin, 1989. ← 202

[75] P. Griffiths. Periods of integrals on algebraic manifolds.I. Construction and properties of the modular varieties. -II. Local study of the period mapping. *Amer. J. Math.*, **90**, 568–626 and 805–865, 1968. ← 176, 177

[76] P. Griffiths. Periods of integrals on algebraic manifolds. III. Some global differential-geometric properties of the period mapping. *Inst. Hautes Études Sci. Publ. Math.*, **38**, 125–180, 1970. ← 176

[77] A. Grothendieck. Techniques de construction et théoremes d'existence en géométrie algébrique. IV, Les schemas de Hilbert. *Sem. Bourbaki*, **13**, 1–28, 1960/1961. ← 184

[78] E. Horikawa. On deformations of quintic surfaces. *Invent. Math.*, **31**, 43–85, 1975. ← 172, 204

[79] J. H. Hubbard. *Teichmüller theory and applications to geometry, topology, and dynamics. Vol. 1. Teichmüller theory.* Matrix Editions, Ithaca, NY, 2006. ← 167

[80] J. E. Humphreys. Linear algebraic groups. *Graduate Texts in Mathematics*, **21**, Springer-Verlag, New York-Heidelberg-Berlin, 1975. ← 186

[81] D. Huybrechts. Complex geometry. An introduction. *Universitext*, Springer-Verlag, Berlin, 2005. ← 168

[82] A. Kas and M. Schlessinger. On the versal deformation of a complex space with an isolated singularity. *Math. Ann.*, **196**, 23–29, 1972. ← 178

[83] Y. Kawamata. Unobstructed deformations. A remark on a paper of Z. Ran: Deformations of manifolds with torsion or negative canonical bundle. *J. Algebraic Geom.*, **1** (2), 183–190, 1992. ← 177

[84] V. M. Kharlamov and V. S. Kulikov. On real structures of real surfaces. *Izv. Ross. Akad. Nauk Ser. Mat.*, **66** (1), 133–152, 2000; translation in *Izv. Math.*, **66**(1), 133–150, 2002. ← 201

[85] K. Kodaira. On stability of compact submanifolds of complex manifolds. *Am. J. Math.*, **85**, 79–94, 1963. ← 182

[86] K. Kodaira. Complex manifolds and deformation of complex structures. Translated from the Japanese by Kazuo Akao. With an appendix by Daisuke Fujiwara. *Grundlehren der Mathematischen Wissenschaften*, **283**, Springer-Verlag, New York, 1986. ← 169

[87] K. Kodaira and J. Morrow. *Complex manifolds.* Holt, Rinehart and Winston, Inc., New York-Montreal, Que.-London, 1971. ← 166, 169, 172, 176

[88] K. Kodaira and D. C. Spencer. On deformations of complex analytic structures. I, II. *Ann. of Math. (2)*, **67**, 328–466, 1958. ← 166, 170

[89] J. Kollár and N. I. Shepherd-Barron. Threefolds and deformations of surface singularities. *Inv. Math.*, **91**, 299–338, 1988. ← 205

[90] M. Kuranishi. On the locally complete families of complex analytic structures. *Ann. Math. (2)*, **75**, 536–577, 1962. ← 170, 171, 178

[91] M. Kuranishi. New proof for the existence of locally complete families of complex structures. *Proc. Conf. Complex Analysis, Minneapolis*, **1964**, 142–154, 1965. ← 170, 171, 178

[92] M. Kuranishi. A note on families of complex structures, in:*Global Analysis (Papers in Honor of K. Kodaira), (1969)*, 309–313. Princeton Univ. Press and Univ. Tokyo Press, Tokyo, 1969. ← 168, 169, 170

[93] M. Kuranishi. Deformations of compact complex manifolds. *Séminaire de Mathématiques Supérieures*, **39**, (Été 1969. Les Presses de l'Université de Montréal, Montreal, Que., 1971. ← 169

[94] Y. Lee and J. Park. A simply connected surface of general type with $p_g = 0$ and $K^2 = 2$. *Invent. Math.*, **170** (3), 483–505, 2007. ← 205

[95] Wenfei Liu. Stable Degenerations of Surfaces Isogenous to a Product, II. *Trans. A.M.S.*, **362**(5), 2411–2427, 2012. ← 205

[96] M. Manetti. Iterated Double Covers and Connected Components of Moduli Spaces. *Topology*, **36** (3), 1997, 745–764. ← 205

[97] M. Manetti. On the Moduli Space of diffeomorphic algebraic surfaces. *Inv. Math.*, **143**, 29–76, 2001. ← 201, 205

[98] M. Manetti. Cohomological constraint on deformations of compact Kähler manifolds. *Adv. Math.*, **186**(1), 125–142, 2004. ← 173, 174

[99] M. Manetti. Differential graded Lie algebras and formal deformation theory, in: Algebraic geometry-Seattle 2005. Part 2, 785–810, *Proc. Sympos. Pure Math.*, **80**, Part 2, Amer. Math. Soc., Providence, RI, 2009. ← 174

[100] J. Mather. Notes on topological stability. *Harvard University Math. Notes* , Boston, 1970. ← 195

[101] H. Matsumura. On algebraic groups of birational transformations. *Atti Accad. Naz. Lincei Rend. Cl. Sci. Fis. Mat. Natur. (8)*, **34**, 151–155, 1963. ← 193

[102] J. W. Milnor. Singular points of complex hypersurfaces. *Annals of Mathematics Studies*, **61**, Princeton University Press and the University of Tokyo Press, 1968. ← 180

[103] Y. Miyaoka. The maximal number of quotient singularities on surfaces with given numerical invariants. *Math. Ann.*, **268** (2), 159–171, 1984. ← 189

[104] B. Moishezon. The arithmetics of braids and a statement of Chisini, in: Geometric Topology, Haifa 1992, 151–175. *Contemp. Math.*, **164**, Amer. Math. Soc., Providence, RI, 1994. ← 202

[105] J. W. Morgan. The Seiberg-Witten equations and applications to the topology of smooth four-manifolds. *Mathematical Notes*, **44**, Princeton Univ. Press, Princeton, 1996. ← 204

[106] D. Mumford. The canonical ring of an algebraic surface. *Ann. of Math. (2)*, **76**, 612–615, 1962. Appendix to *The theorem of Riemann-Roch for high multiples of an effective divisor on an algebraic surface*, by O. Zariski, ibid. 560–611. ← 183

[107] D. Mumford. Geometric invariant theory. *Ergebnisse der Mathematik und ihrer Grenzgebiete, Neue Folge, Band*, **34**, Springer-Verlag, Berlin-New York, 1965. ← 190

[108] D. Mumford. Abelian varieties. *Tata Institute of Fundamental Research Studies in Mathematics*, **5**, Published for the Tata Institute of Fundamental Research, Bombay; Oxford University Press, London, 1970. ← 202

[109] D. Mumford. Hirzebruch's proportionality theorem in the noncompact case. *Invent. Math.*, **42**, 239–272, 1977. ← 206

[110] D. Mumford. Towards an enumerative geometry of the moduli space of curves, in: Arithmetic and geometry, Pap. dedic. I. R. Shafarevich, Vol. II: Geometry, 271–328. *Prog. Math.*, **36**, Birkhäuser, 1983. ← 167

[111] D. Mumford and K. Suominen. Introduction to the theory of moduli, in : Algebraic geometry, Oslo 1970, *Proc. Fifth Nordic Summer-School in Math.*, 171–222. Wolters-Noordhoff, Groningen, 1972. ← 163

[112] A. Newlander and L. Nirenberg. Complex analytic coordinates in almost complex manifolds. *Ann. of Math. (2)*, **65**, 391–404, 1957. ← 165

[113] M. van Opstall. Stable degenerations of surfaces isogenous to a product of curves. *Proc. Amer. Math. Soc.*, **134** (10), 2801–2806, 2006. ← 205

[114] R. S. Palais. On the existence of slices for actions of non-compact Lie groups. *Ann. of Math.(2)*, **73**, 295–323, 1961. ← 169

[115] R. Pardini. Abelian covers of algebraic varieties. *J. Reine Angew. Mathematik*, **417**, 191–213, 1991. ← 205

[116] M. Penegini and F. Polizzi. On surfaces with $p_g = q = 2, K^2 = 5$ and Albanese map of degree 3. `arXiv:1011.4388` ← 205

[117] M. Penegini and F. Polizzi. Surfaces with $p_g = q = 2, K^2 = 6$ and Albanese map of degree 2. To appear in *Can. J. Math.*, `doi:10.4153/CJM-2012-007-0`, 2012. ← 205

[118] F. Quinn. Isotopy of 4-manifolds, *J. Differential Geom.*, **24** (3), 343–372, 1986. ← 175

[119] Z. Ran. Deformations of manifolds with torsion or negative canonical bundle. *J. Algebraic Geom.*, **1** (2), 279–291, 1992. ← 177

[120] M. Reid. Canonical 3-folds. In: Journées de géometrie algébrique, Angers, France 1979, *Sijthoff and Nordhoff*, 273–310, 1980. ← 179

[121] M. Reid. Young person's guide to canonical singularities, in: Algebraic geometry, Proc. Summer Res. Inst., Brunswick/Maine 1985, part 1, *Proc. Symp. Pure Math.*, **46**, 345–414, 1987. ← 179

[122] O. Riemenschneider. Deformationen von Quotientensingularitäten (nach zyklischen Gruppen). *Math. Ann.*, **209**, 211–248, 1974. ← 205

[123] S. Rollenske. Lie-algebra Dolbeault-cohomology and small deformations of nilmanifolds. *J. Lond. Math. Soc. (2)*, **79**(2), 346–362, 2009. ← 173

[124] S. Rollenske. Geometry of nilmanifolds with left-invariant complex structure and deformations in the large. *Proc. Lond. Math. Soc. (3)*, **99** (2), 425–460, 2009. ← 173

[125] S. Rollenske. Compact moduli for certain Kodaira fibrations. *Ann. Sc. Norm. Super. Pisa Cl. Sci. IX (4)*, 851–874, 2010. ← 205

[126] S. Rollenske. The Kuranishi-space of complex parallelisable nilmanifolds. *Journal of the European Mathematical Society*, **13**(3), 513–531, 2011. ← 173

[127] M. Schlessinger. Functors of Artin rings. *Trans. Amer. Math. Soc.*, **130**, 208–222, 1968. ← 172

[128] M. Schlessinger and J. Stasheff. The Lie algebra structure of tangent cohomology and deformation theory. *J. Pure Appl. Algebra*, **38**(2–3), 313–322, 1985. ← 174

[129] B. Segre. Sul massimo numero di nodi delle superficie di dato ordine. *Boll. Un. Mat. Ital. (3)*, **2**, 204–212, 1947. ← 189

[130] E. Sernesi. Deformations of algebraic schemes. *Grundlehren der Mathematischen Wissenschaften*, **334**, Springer-Verlag, Berlin, 2006. ← 177, 178

[131] J. P. Serre. Exemples de variétés projectives conjuguées non homéomorphes. *C. R. Acad. Sci. Paris*, **258**, 4194–4196, 1964. ← 203

[132] A. J. Sommese. Quaternionic manifolds. *Math. Ann.*, **212**, 191–214, 1974/75. ← 166

[133] S. G. Tankeev. A global theory of moduli for algebraic surfaces of general type. *Izv. Akad. Nauk SSSR Ser. Mat.*, **36**, 1220–1236, 1972. ← 184, 185

[134] G. Tian. Smoothness of the universal deformation space of compact Calabi-Yau manifolds and its Petersson-Weil metric, in: Yau, S. T. *Mathematical Aspects of String theory*, 629–646, World Scientific, Singapore, 1987. ← 177

[135] G. N. Tjurina. Locally semi-universal flat deformations of isolated singularities of complex spaces. *Izv. Akad. Nauk SSSR Ser. Mat.*, **33**, 1026–1058, 1969. ← 178

[136] G. N. Tjurina. Resolution of singularities of plane (= flat) deformations of double rational points. *Funk.Anal. i Prilozen* , **4**, 77–83, 1970. ← 179, 186

[137] A. N. Todorov. The Weil-Petersson geometry of the moduli space of SU($n \geqslant 3$) (Calabi-Yau) manifolds. I. *Comm. Math. Phys.*, **126** (2), 325–346, 1989. ← 177

[138] A. Tromba. Dirichlet's energy on Teichmüller 's moduli space and the Nielsen realization problem. *Math. Z.*, **222**, 451–464, 1996. ← 167

[139] N. Tziolas. Q-Gorenstein smoothings of nonnormal surfaces. *Amer. J. Math.*, **131** (1), 171–193, 2009. ← 205

[140] Kenji Ueno. (Ed.) Open problems: Classification of algebraic and analytic manifolds. Classification of algebraic and analytic manifolds, Proc. Symp. Katata/Jap. 1982. *Progress in Mathematics*, **39**, 591–630, Birkhäuser, Boston, Mass. 1983. ← 201

[141] R. Vakil. Murphy's law in algebraic geometry: badly-behaved deformation spaces. *Invent. Math.*, **164**(3), 569–590, 2006. ← 193

[142] E. Viehweg. Quasi-projective moduli for polarized manifolds. *Ergebnisse der Mathematik und ihrer Grenzgebiete*, **3**, Folge, 30, Springer-Verlag, Berlin, 1995. ← 205

[143] E. Viehweg. Compactifications of smooth families and of moduli spaces of polarized manifolds. *Annals of Mathematics*, **172**, 809–910, 2010. ← 205

[144] C. Voisin. Théorie de Hodge et géométrie algébrique complexe. *Cours Spécialisés,* **10**, Société Mathématique de France, Paris, 2002. ← 176

[145] J. Wavrik. Obstructions to the existence of a space of moduli, in: Global Analysis (Papers in Honor of K. Kodaira), 403–414, Princeton University and Univ. Tokyo Press, Tokyo, 1969. ← 170, 171, 172

Lehrstuhl Mathematik VIII, Mathematisches Institut der Universität Bayreuth, NW II, Universitätsstr. 30, 95447 Bayreuth

E-mail address: `fabrizio.catanese@uni-bayreuth.de`

Moduli spaces of hyperbolic surfaces and their Weil–Petersson volumes

Norman Do

Abstract. Moduli spaces of hyperbolic surfaces may be endowed with a symplectic structure via the Weil–Petersson form. Mirzakhani proved that Weil–Petersson volumes exhibit polynomial behaviour and that their coefficients store intersection numbers on moduli spaces of curves. In this survey article, we discuss these results as well as some consequences and applications.

Contents

2000 *Mathematics Subject Classification.* Primary 32G15; Secondary 14H15, 53D30.
Key words and phrases. moduli space, hyperbolic surface, Weil–Petersson symplectic form.

1. Introduction

Over the last few decades, moduli spaces of curves have become increasingly important objects of study in mathematics. In fact, they now lie at the centre of a rich confluence of seemingly disparate areas such as geometry, topology, combinatorics, integrable systems, matrix models and theoretical physics.

Smooth curves with labelled points are equivalent to Riemann surfaces with punctures, and these are in turn equivalent to punctured surfaces with complete constant curvature Riemannian metrics.[1] For all but finitely many pairs (g, n), a genus g surface with n punctures admits a hyperbolic metric. Thus, we are led to the study of hyperbolic surfaces and their corresponding moduli spaces. In this article, we adopt such a hyperbolic geometric perspective, which allows for lines of thought that have no natural analogue in the realms of algebraic geometry and complex analysis.

For an n-tuple $\mathbf{L} = (L_1, L_2, \dots, L_n)$ of positive real numbers, let $\mathcal{M}_{g,n}(\mathbf{L})$ denote the set of genus g hyperbolic surfaces with n geodesic boundary components whose lengths are prescribed by $\mathbf{L}$. We require the boundary components to be labelled from 1 up to n and consider hyperbolic surfaces up to isometries which preserve these labels. The Teichmüller theory construction of this space endows it with an orbifold structure and local coordinates known as Fenchel–Nielsen coordinates. These may be used to define the Weil–Petersson symplectic form ω, thus providing the moduli space of hyperbolic surfaces $\mathcal{M}_{g,n}(\mathbf{L})$ with a natural symplectic structure. Our primary focus will be on the Weil–Petersson volume

$$V_{g,n}(\mathbf{L}) = \int_{\mathcal{M}_{g,n}(\mathbf{L})} \frac{\omega^{3g-3+n}}{(3g-3+n)!}.$$

The foundation of this article is a result due to Mirzakhani [30] which states that the Weil–Petersson volume is given by the following polynomial.

$$V_{g,n}(\mathbf{L}) = \sum_{|\boldsymbol{\alpha}|+m=3g-3+n} \frac{(2\pi^2)^m \int_{\overline{\mathcal{M}}_{g,n}} \psi_1^{\alpha_1}\psi_2^{\alpha_2}\cdots\psi_n^{\alpha_n}\kappa_1^m}{2^{|\boldsymbol{\alpha}|}\alpha_1!\alpha_2!\cdots\alpha_n!m!} L_1^{2\alpha_1}L_2^{2\alpha_2}\cdots L_n^{2\alpha_n}.$$

Here, $\psi_1, \psi_2, \dots, \psi_n \in H^2(\overline{\mathcal{M}}_{g,n}; \mathbb{Q})$ are the psi-classes while $\kappa_1 \in H^2(\overline{\mathcal{M}}_{g,n}; \mathbb{Q})$ is the first Mumford–Morita–Miller class on the Deligne–Mumford compactification of the moduli space of curves. We use $\boldsymbol{\alpha} = (\alpha_1, \alpha_2, \dots, \alpha_n)$ to denote an n-tuple of non-negative integers and $|\boldsymbol{\alpha}|$ to denote the sum $\alpha_1+\alpha_2+\cdots+\alpha_n$. For a concise exposition of intersection theory on moduli spaces of curves, including the definitions of the psi-classes and the Mumford–Morita–Miller classes, see Appendix A.

Mirzakhani originally demonstrated the polynomiality of Weil–Petersson volumes using symplectic reduction [30]. She subsequently provided an alternative proof where the main idea is to unfold the integral on $\mathcal{M}_{g,n}(\mathbf{L})$ to a more tractable

[1] We decree that all surfaces referred to in this article are to be connected and oriented. We also decree that all algebraic curves referred to in this article are to be complex, connected and complete.

cover over the moduli space [29]. One of the tools required is a generalization of McShane's identity which, in its original form, states that a certain sum over the simple closed geodesics on a hyperbolic once-punctured torus is constant. The end result is a recursive formula which can be used to calculate all Weil–Petersson volumes. This in turn yields a recursion for intersection numbers on $\overline{\mathcal{M}}_{g,n}$. As a consequence, Mirzakhani was able to deduce the celebrated Witten–Kontsevich theorem.

Some of the recent work on moduli spaces of hyperbolic surfaces has focused on the behaviour of Weil–Petersson volumes under various limits. For example, the Weil–Petersson volume polynomials exhibit interesting behaviour when one of the lengths formally approaches $2\pi i$. This manifests as non-trivial relations between $V_{g,n+1}(\mathbf{L}, L_{n+1})$ and $V_{g,n}(\mathbf{L})$, which generalise the string and dilaton equations [14]. These can be proven using arguments from algebraic geometry. However, the form of the relations also suggests that a proof may entail the geometry of hyperbolic cone surfaces. In particular, one can interpret the evaluation $L = \theta i$ as the degeneration of the corresponding geodesic boundary component to a cone point with angle θ.

A direct implementation of Mirzakhani's recursive formula produces a rather slow algorithm for the calculation of Weil–Petersson volumes. However, Zograf has provided an alternative algorithm which is empirically much faster and calculated enough numerical data to produce interesting conjectures concerning Weil–Petersson volumes in the large g limit [52]. Some progress on these conjectures has recently been made by Mirzakhani [32].

It is natural to consider the asymptotic behaviour of $V_{g,n}(N\mathbf{x})$ for a fixed $\mathbf{x} = (x_1, x_2, \ldots, x_n)$ as N approaches infinity. We are thus motivated to analyse a certain normalisation of the Weil–Petersson form on $\mathcal{M}_{g,n}(N\mathbf{x})$ as N approaches infinity [13]. In the limit, we obtain a 2-form originally defined by Kontsevich in his proof of Witten's conjecture [22]. In this way, we obtain yet another proof of the Witten–Kontsevich theorem which makes explicit the connection between the work of Kontsevich and Mirzakhani.

The structure of the article is as follows.

- In Section 2, we introduce hyperbolic surfaces and their moduli spaces via Teichmüller theory. We discuss the Weil–Petersson symplectic structure of the space $\mathcal{M}_{g,n}(\mathbf{L})$ and its Deligne–Mumford compactification. We conclude by showing how the hyperbolic geometry leads to a cell decomposition of the moduli space based on the combinatorial notion of a ribbon graph.
- In Section 3, we begin the study of Weil–Petersson volumes. Some early results in the area are presented, followed by some preparatory remarks on symplectic reduction. We then show how Mirzakhani applies this technique to prove the polynomiality of Weil–Petersson volumes [30].
- In Section 4, we discuss Mirzakhani's recursion for Weil–Petersson volumes [29]. We calculate the volume of $\mathcal{M}_{1,1}(0)$ as a motivating example. The generalised McShane identity is presented and then used as one of

the main ingredients in the proof of Mirzakhani's recursive formula. We consider some applications of the recursion, including a proof of the Witten–Kontsevich theorem.

- In Section 5, we analyse Weil–Petersson volumes under various limits. In particular, we examine the recent results on the behaviour of $V_{g,n}(\mathbf{L})$ as one of the lengths approaches $2\pi i$ [14], in the large g limit [32], and as the lengths approach infinity [13].

2. Moduli spaces of hyperbolic surfaces

2.1. Hyperbolic surfaces

Consider the smooth surface $\Sigma_{g,n}$ with genus g and n boundary components. A *hyperbolic surface of type* (g, n) is the surface $\Sigma_{g,n}$ equipped with a complete Riemannian metric of constant curvature -1. We restrict our attention to the case when every boundary component is smooth and totally geodesic. A mild restriction on the pair of non-negative integers (g, n) is required due to the Gauss-Bonnet theorem. In its simplest form, it states that the integral of the Gaussian curvature over a surface with Riemannian metric and totally geodesic boundary is equal to 2π multiplied by the Euler characteristic of the surface.

$$\int_S K\, dA = 2\pi\chi(S).$$

So for a metric of constant curvature -1 to exist, the Euler characteristic must be negative — in other words, $2g - 2 + n > 0$. This is a rather mild restriction since it only prohibits the pairs $(0, 0)$, $(0, 1)$, $(0,2)$ and $(1,0)$. Note that these exceptional cases are precisely the pairs (g, n) for which a Riemann surface of genus g with n punctures possesses infinitely many automorphisms.

An alternative definition of a hyperbolic surface uses the notion of an atlas. Thus, we define a hyperbolic surface to be a smooth surface covered by charts $\phi_U : U \to \mathbb{H}^2$ which map open subsets of the surface homeomorphically onto their image in the hyperbolic plane. We require that if $U \cap V \neq \emptyset$, then the two charts are compatible in the sense that the transition function $\phi_V \circ \phi_U^{-1} : \phi_U(U \cap V) \to \phi_V(U \cap V)$ is an isometry. As usual, a hyperbolic surface is defined by a maximal atlas — that is, a maximal collection of compatible charts. The two definitions provided coincide since any two-dimensional domain with a Riemannian metric of constant curvature -1 is locally isometric to a subset of the hyperbolic plane. Which to use as a working definition is largely a matter of taste, though it is advantageous to keep both viewpoints in mind.

Yet another way to define a hyperbolic surface is via its universal cover. Every hyperbolic surface S with geodesic boundary has a universal cover isometric to a convex domain in $\mathbb{H}^2$ with geodesic boundary. Therefore, hyperbolic surfaces with geodesic boundary arise from taking the quotient of a convex domain in $\mathbb{H}^2$ with geodesic boundary by a discrete subgroup of $PSL(2, \mathbb{R})$, the group of orientation-preserving isometries in the hyperbolic plane. The following result states an important geometric property of hyperbolic surfaces. For a proof, see Buser's *Geometry and spectra of compact Riemann surfaces* [10], a valuable reference on the hyperbolic geometry of surfaces.

Proposition 2.1. *On a hyperbolic surface, non-trivial homotopy classes of closed curves have unique geodesic representatives. Furthermore, such geodesic representatives realise minimal intersection and self-intersection numbers. In particular, every simple closed curve is homotopic to a simple closed geodesic.*

For an n-tuple $\mathbf{L} = (L_1, L_2, \ldots, L_n)$ of positive real numbers, we define the *moduli space of hyperbolic surfaces* as follows.

$$\mathcal{M}_{g,n}(\mathbf{L}) = \left\{ (S, \beta_1, \beta_2, \ldots, \beta_n) \;\middle|\; \begin{array}{l} S \text{ is a hyperbolic surface of type} \\ (g, n) \text{ with boundary components} \\ \beta_1, \beta_2, \ldots, \beta_n \text{ of lengths } L_1, L_2, \ldots, L_n \end{array} \right\} \Big/ \sim$$

Here, $(S, \beta_1, \beta_2, \ldots, \beta_n) \sim (T, \gamma_1, \gamma_2, \ldots, \gamma_n)$ if and only if there exists an isometry from S to T which sends β_k to γ_k for all k.

Note that when the length of a boundary component approaches zero, we obtain a hyperbolic cusp in the limit. By a *hyperbolic cusp*, we mean a subset of the surface isometric to $\mathbb{R}/\mathbb{Z} \times [2, \infty)$, in the Poincaré upper half-plane model for $\mathbb{H}^2$. In particular, we denote the moduli space of genus g hyperbolic surfaces with n labelled cusps by $\mathcal{M}_{g,n}(\mathbf{0})$.

One of the goals of this article is to explain how the study of moduli spaces of hyperbolic surfaces can lead to results concerning moduli spaces of curves. This is possible through the interplay between hyperbolic surfaces, Riemann surfaces, and algebraic curves. It is well-known that the category of smooth algebraic curves is equivalent to the category of compact Riemann surfaces. Due to this equivalence, the boundary between these two fields is rather porous, with techniques from complex analysis flowing into algebraic geometry and vice versa. The uniformisation theorem then provides a one-to-one correspondence between Riemann surfaces and hyperbolic surfaces.

Theorem 2.2 (Uniformisation theorem). *Every Riemannian metric on a surface is conformally equivalent to a complete constant curvature metric. If the Euler characteristic of the surface is negative and we require that the curvature is -1, then the metric is unique.*

From the previous discussion, a smooth genus g algebraic curve with n labelled points corresponds to a genus g Riemann surface with n labelled points, which we usually think of as punctures.

$$\left\{ \begin{array}{c} \text{smooth algebraic curves with} \\ \text{genus } g \text{ and } n \text{ labelled points} \end{array} \right\} \longleftrightarrow \left\{ \begin{array}{c} \text{Riemann surfaces with} \\ \text{genus } g \text{ and } n \text{ punctures} \end{array} \right\}$$

The complex structure on a Riemann surface defines a conformal class of metrics which, by the uniformisation theorem, contains a hyperbolic metric as long as $2g - 2 + n > 0$. Furthermore, if we demand that the resulting surface is complete, then this hyperbolic metric is unique and gives each puncture the structure of a hyperbolic cusp. So we have the following one-to-one correspondence.

$$\left\{ \begin{array}{c} \text{Riemann surfaces with} \\ \text{genus } g \text{ and } n \text{ punctures} \end{array} \right\} \longleftrightarrow \left\{ \begin{array}{c} \text{hyperbolic surfaces with} \\ \text{genus } g \text{ and } n \text{ cusps} \end{array} \right\}$$

At the moment, the moduli space of hyperbolic surfaces $\mathcal{M}_{g,n}(\mathbf{L})$ has only been defined as a set. In Section 2.2, we will see that it possesses not only a topology, but also an orbifold structure. The equivalence above sets up a bijection between the moduli space of curves $\mathcal{M}_{g,n}$ and the moduli space of hyperbolic surfaces $\mathcal{M}_{g,n}(\mathbf{0})$. This map respects the topology of both spaces as well as the structure-preserving automorphism groups. As a result, $\mathcal{M}_{g,n}$ and $\mathcal{M}_{g,n}(\mathbf{0})$ are homeomorphic as orbifolds.

2.2. Teichmüller theory

Teichmüller theory will enable us to construct the moduli space of hyperbolic surfaces $\mathcal{M}_{g,n}(\mathbf{L})$ and endow it with a natural symplectic structure. Begin by fixing a smooth surface $\Sigma_{g,n}$ with genus g and n boundary components labelled from 1 up to n, where $2g - 2 + n > 0$. Define a *marked hyperbolic surface of type* (g, n) to be a pair (S, f) where S is a hyperbolic surface and $f : \Sigma_{g,n} \to S$ is a diffeomorphism. We call f the marking of the hyperbolic surface and define the *Teichmüller space* as follows.

$$\mathcal{T}_{g,n}(\mathbf{L}) = \left\{ (S, f) \;\middle|\; \begin{array}{l} (S, f) \text{ is a marked hyperbolic surface of} \\ \text{type } (g, n) \text{ with boundary components} \\ \text{of lengths } L_1, L_2, \ldots, L_n \end{array} \right\} \Big/ \sim$$

Here, $(S, f) \sim (T, g)$ if and only if there exists an isometry $\phi : S \to T$ such that $\phi \circ f$ is isotopic to g.

One can informally think of Teichmüller space as the space of deformations of the hyperbolic structure on a given hyperbolic surface. For example, consider applying a hyperbolic Dehn twist to a marked hyperbolic surface. By this, we mean cutting along a simple closed geodesic, twisting the two sides relative to each other, and gluing the two sides back together. This gives a one parameter family of deformations of the hyperbolic structure. Once a full twist has been applied, the end result is a hyperbolic surface isometric to the original and hence, corresponds to the same point in the moduli space. On the other hand, the end result has a different marking to the original and hence, corresponds to a different point in the Teichmüller space.

The geometry and topology of Teichmüller space will become much more apparent once we define global coordinates, known as Fenchel–Nielsen coordinates. We use the idea that pairs of pants — spheres with three boundary components — can be used as building blocks to create surfaces with negative Euler characteristic. Start by considering a pants decomposition of the surface $\Sigma_{g,n}$, which is a collection of disjoint simple closed curves whose complement is a disjoint union of pairs of pants. Alternatively, a pants decomposition is a maximal collection of disjoint simple closed curves such that no curve is parallel to the boundary and no two are homotopic. Since the Euler characteristic is additive over surfaces glued along circles, the number of pairs of pants in any such decomposition must be $-\chi(\Sigma_{g,n}) = 2g - 2 + n$. A simple combinatorial argument can be used to show that every pants decomposition of $\Sigma_{g,n}$ consists of precisely $3g - 3 + n$ simple closed curves.

A marking $f : \Sigma_{g,n} \to S$ maps a fixed pants decomposition to a collection of simple closed curves on S, each of which is homotopic to a unique simple closed geodesic by Proposition 2.1. Denote these simple closed geodesics by $\gamma_1, \gamma_2, \ldots, \gamma_{3g-3+n}$ and let their lengths be $\ell_1, \ell_2, \ldots, \ell_{3g-3+n}$, respectively. Cutting S along these geodesics leaves a disjoint union of $2g-2+n$ hyperbolic pairs of pants. The following elementary result guarantees that the lengths $\ell_1, \ell_2, \ldots, \ell_{3g-3+n}$ are sufficient to determine the hyperbolic structure on each pair of pants.

Lemma 2.3. *There exists a unique hyperbolic pair of pants up to isometry with geodesic boundary components of prescribed non-negative length. As usual, we interpret a geodesic boundary component with length zero as a hyperbolic cusp. The three simple geodesic arcs perpendicular to the boundary components and joining them in pairs are referred to as seams. Cutting along the seams decomposes a hyperbolic pair of pants into two congruent right-angled hexagons.*

Note that the lengths $\ell_1, \ell_2, \ldots, \ell_{3g-3+n}$ provide insufficient information to reconstruct the hyperbolic structure on all of S, since there are infinitely many ways to glue together the pairs of pants. This extra gluing information is stored in the twist parameters, which we denote by $\tau_1, \tau_2, \ldots, \tau_{3g-3+n}$. To construct them, fix a collection C of disjoint smooth curves on $\Sigma_{g,n}$ which are either closed or have endpoints on the boundary. We require that C meets the pants decomposition transversely, such that its restriction to any pair of pants consists of three disjoint arcs, connecting the boundary components pairwise. Now to construct the twist parameter τ_k, take a curve $\gamma \in C$ such that $f(\gamma)$ meets γ_k. Homotopic to $f(\gamma)$, relative to the boundary of S, is a unique length-minimising piecewise geodesic curve which is entirely contained in the seams of the pairs of pants and the curves $\gamma_1, \gamma_2, \ldots, \gamma_{3g-3+n}$. The twist parameter τ_k is the signed distance that this curve travels along γ_k, according to the following sign convention. Lemma 2.3 guarantees that the twist parameter is independent of the choice of curve $\gamma \in C$.

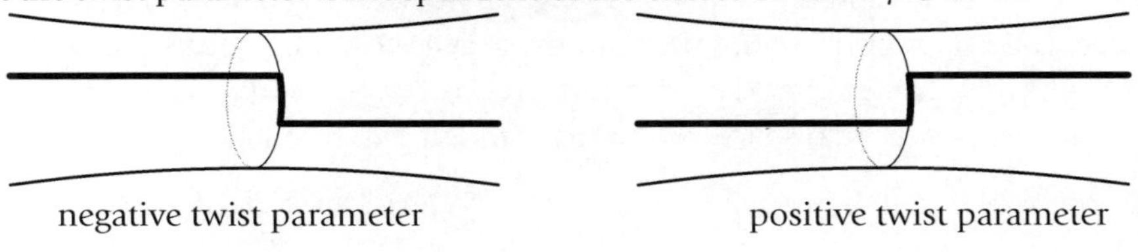

negative twist parameter   positive twist parameter

For further details, one can consult Thurston's book [45], in which he writes the following.

> "That a twist parameter takes values in $\mathbb{R}$, rather than S^1, tends to be a confusing issue ... But, remember, to determine a point in Teichmüller space we need to consider how many times the leg of the pajama suit is twisted before it fits onto the baby's foot."

More prosaically, the length parameters and the twist parameters modulo the length parameters are sufficient to reconstruct the hyperbolic structure on S. However, to recover the marking as well, it is necessary to consider the twist parameters as elements of $\mathbb{R}$. Despite the fact that the length and twist parameters — collectively known as *Fenchel–Nielsen coordinates* — depend on the choice of pants decomposition and the construction of twist parameters, we have the following result.

Theorem 2.4. *The map $\mathcal{T}_{g,n}(\mathbf{L}) \to \mathbb{R}_+^{3g-3+n} \times \mathbb{R}^{3g-3+n}$, which associates to a marked hyperbolic surface its length and twist parameters, is a bijection. In fact, if Teichmüller space is considered with its natural topology, then the map is a homeomorphism.*

Clearly, there is a projection map $\mathcal{T}_{g,n}(\mathbf{L}) \to \mathcal{M}_{g,n}(\mathbf{L})$ which simply forgets the marking. In fact, the moduli space is obtained as a quotient of Teichmüller space by a group action. Define the *mapping class group* as

$$\mathrm{Mod}_{g,n} = \mathrm{Diff}^+(\Sigma_{g,n})/\mathrm{Diff}_0^+(\Sigma_{g,n}),$$

where Diff^+ denotes the group of orientation preserving diffeomorphisms fixing the boundary components and Diff_0^+ denotes the normal subgroup consisting of those diffeomorphisms isotopic to the identity. There is a natural action of the mapping class group on Teichmüller space such that $[\phi] \in \mathrm{Mod}_{g,n}$ sends the marked hyperbolic surface (X, f) to the marked hyperbolic surface $(X, f \circ \phi)$. The moduli space $\mathcal{M}_{g,n}(\mathbf{L})$ is obtained by taking the quotient of the Teichmüller space $\mathcal{T}_{g,n}(\mathbf{L})$ by the action of the mapping class group $\mathrm{Mod}_{g,n}$.

Proposition 2.5. *The action of $\mathrm{Mod}_{g,n}$ on $\mathcal{T}_{g,n}(\mathbf{L})$ is properly discontinuous, though not necessarily free. Therefore, the quotient $\mathcal{M}_{g,n}(\mathbf{L}) = \mathcal{T}_{g,n}(\mathbf{L})/\mathrm{Mod}_{g,n}$ is an orbifold of dimension $6g - 6 + 2n$.*

So the moduli space of hyperbolic surfaces $\mathcal{M}_{g,n}(\mathbf{L})$ has not only a topology, but also an orbifold structure. The orbifold group at a point is canonically isomorphic to the automorphism group of the corresponding hyperbolic surface. However, the situation is not so bad, since the following theorem — which follows from results of Boggi and Pikaart [6] — allows one to make sense of calculations on the orbifold by lifting to a finite cover. For this reason, it is convenient for us to consider the cohomology of moduli spaces with rational, rather than integral, coefficients.

Theorem 2.6. *A finite cover $\widetilde{\mathcal{M}}_{g,n}(\mathbf{L}) \to \mathcal{M}_{g,n}(\mathbf{L})$ exists such that $\widetilde{\mathcal{M}}_{g,n}(\mathbf{L})$ is a smooth manifold.*

We have only very briefly touched upon the vast area that is Teichmüller theory. For more information, see the *Handbook of Teichmüller theory* [37, 38].

2.3. Symplectification and compactification

The Teichmüller space $\mathcal{T}_{g,n}(\mathbf{L})$ can be endowed with the canonical symplectic form

$$\omega = \sum_{k=1}^{3g-3+n} d\ell_k \wedge d\tau_k$$

using the Fenchel–Nielsen coordinates. Although this is a rather trivial statement, it is a deep fact that this form is invariant under the action of the mapping class group. Therefore, ω descends to a symplectic form on the quotient, namely the moduli space $\mathcal{M}_{g,n}(\mathbf{L})$. This is referred to as the *Weil–Petersson symplectic form* and we will also denote it by ω. Its existence allows us to introduce the techniques of symplectic geometry to the study of moduli spaces. For all values of $\mathbf{L}$, the spaces $\mathcal{M}_{g,n}(\mathbf{L})$ are diffeomorphic to each other, but not necessarily symplectomorphic to each other. It is therefore natural to ask how the symplectic structure varies as $\mathbf{L}$ varies, a question which we will pursue in Section 3.3.

The uniformisation theorem allows us to deduce that the moduli spaces $\mathcal{M}_{g,n}(\mathbf{L})$ are not only diffeomorphic to each other, but also diffeomorphic to the moduli space of curves $\mathcal{M}_{g,n}$. It is often more natural to work with the Deligne–Mumford compactification $\overline{\mathcal{M}}_{g,n}$, obtained by introducing the notion of a stable algebraic curve — see Appendix A for the relevant definitions. There is an analogous construction in the hyperbolic setting, where a node of an algebraic curve corresponds to degenerating the length of a simple closed curve on a hyperbolic surface to zero. This intuition leads to the following construction of $\overline{\mathcal{T}}_{g,n}(\mathbf{L})$, the Teichmüller space of marked stable hyperbolic surfaces. Define a *stable hyperbolic surface of type* (g, n) to be a pair (S, M) where S is a surface of genus g with n labelled boundary components and M is a collection of disjoint simple closed curves on S, none of which are homotopic to the boundary. We require that $S \setminus M$ be endowed with a finite area hyperbolic metric such that the boundary components are geodesic. It is useful to think of a stable hyperbolic surface as a collection of hyperbolic surfaces whose cusps have been formally identified in pairs. As usual, we refer to a diffeomorphism $f : \Sigma_{g,n} \to S$ as a marking and define the *compactified Teichmüller space* as follows.

$$\overline{\mathcal{T}}_{g,n}(\mathbf{L}) = \left\{ (S, M, f) \;\middle|\; \begin{array}{l} (S, M, f) \text{ is a marked stable hyperbolic} \\ \text{surface of type } (g, n) \text{ with boundary} \\ \text{components of lengths } L_1, L_2, \ldots, L_n \end{array} \right\} \Big/ \sim$$

Here, $(S, M, f) \sim (T, N, g)$ if and only if there exists a homeomorphism $\phi : S \to T$ such that $\phi(M) = N$, ϕ restricted to $S \setminus M$ is an isometry, and $\phi \circ f$ is isotopic to g on each connected component of $\Sigma_{g,n} \setminus f^{-1}(M)$. Once again, the mapping class group acts on the compactified Teichmüller space and one may define the *Deligne–Mumford compactification* of the moduli space of hyperbolic surfaces as

$$\overline{\mathcal{M}}_{g,n}(\mathbf{L}) = \overline{\mathcal{T}}_{g,n}(\mathbf{L})/\mathrm{Mod}_{g,n}.$$

Let us make some remarks on the Deligne–Mumford compactification. First, the compactification locus $\overline{\mathcal{M}}_{g,n}(\mathbf{L}) \setminus \mathcal{M}_{g,n}(\mathbf{L})$ is a union of submanifolds of positive codimension. Second, $\overline{\mathcal{M}}_{g,n}(\mathbf{0})$ can be canonically identified with the Deligne–Mumford compactification of the moduli space of curves $\overline{\mathcal{M}}_{g,n}$ via the uniformisation theorem. Hence, $\overline{\mathcal{M}}_{g,n}(\mathbf{0})$ possesses a natural complex structure, whereas the moduli space $\overline{\mathcal{M}}_{g,n}(\mathbf{L})$ does not, for $\mathbf{L} \neq \mathbf{0}$. However, by the work of Wolpert, the Fenchel–Nielsen coordinates do induce a real analytic structure [1, 51].

Wolpert [51] used the real analytic structure on $\mathcal{M}_{g,n}(\mathbf{L})$ to show that the Weil–Petersson form extends smoothly to a closed non-degenerate form on the Deligne–Mumford compactification $\overline{\mathcal{M}}_{g,n}(\mathbf{L})$. In the particular case $\mathbf{L} = \mathbf{0}$, he showed that this extension defines a cohomology class $[\omega] \in H^2(\overline{\mathcal{M}}_{g,n}, \mathbb{R})$ which satisfies the following.

Theorem 2.7. *The de Rham cohomology class of the Weil–Petersson symplectic form on $\overline{\mathcal{M}}_{g,n}(\mathbf{0})$ satisfies $[\omega] = 2\pi^2\kappa_1 \in H^2(\overline{\mathcal{M}}_{g,n}, \mathbb{R})$, where κ_1 denotes the first Mumford–Morita–Miller class.*

2.4. Combinatorial moduli space

An important notion in the study of moduli spaces of curves is the combinatorial structure known in the literature as a ribbon graph or fatgraph. A *ribbon graph of type* (g, n) is essentially the 1-skeleton of a cell decomposition of a genus g surface with n faces. We require the vertices to have degree at least three and the faces to be labelled from 1 up to n. Note that such a graph may possibly have loops or multiple edges. The orientation of the surface gives a cyclic ordering to the oriented edges pointing toward each vertex. Conversely, given the underlying graph and the cyclic ordering of the oriented edges pointing toward each vertex, the genus of the surface and its cell decomposition may be recovered. This is accomplished by using the extra structure to thicken the graph into a surface with boundaries. These boundaries may then be filled in with disks to produce a closed surface with an associated cell decomposition.

One usually draws ribbon graphs with the convention that the cyclic ordering of the oriented edges pointing toward each vertex is induced by the orientation of the page. For example, the following diagram portrays a ribbon graph of type (1,1) as well as the surface obtained by thickening the graph.

It is often useful to think of a ribbon graph in the following more precise way. Given a cell decomposition Γ of a surface, let X denote the set of its oriented edges and let s_0 be the permutation on X which cyclically permutes all oriented edges pointing toward the same vertex in an anticlockwise manner. Also, let s_1 be the permutation on X which interchanges each pair of oriented edges which correspond to the same underlying edge. The set $X/\langle s_0\rangle$ is canonically equivalent to the set of vertices of Γ while the set $X/\langle s_1\rangle$ is canonically equivalent to the set of edges of Γ. Furthermore, if we let $s_2 = s_1 s_0^{-1}$, then the set $X/\langle s_2\rangle$ is canonically equivalent to the set of faces of Γ. Therefore, one can alternatively define a ribbon graph to be a triple (X, s_0, s_1) where X is a finite set, s_0 is a permutation on X without fixed points or transpositions, and s_1 is an involution on X without fixed points. We also require a labelling in the form of a bijection from $X/\langle s_2\rangle$ to $\{1, 2, \ldots, n\}$. Define two ribbon graphs (X, s_0, s_1) and $(\overline{X}, \overline{s}_0, \overline{s}_1)$ to be isomorphic if and only if there exists a bijection $f : X \to \overline{X}$ such that $f \circ s_0 = \overline{s}_0 \circ f$ and $f \circ s_1 = \overline{s}_1 \circ f$. We also impose the condition that f must preserve the labelling of the boundary components. A ribbon graph automorphism is, of course, an isomorphism from a ribbon graph to itself. The set of automorphisms of a ribbon graph Γ forms a group which is denoted by $\mathrm{Aut}(\Gamma)$.

A ribbon graph with a positive real number assigned to each edge is referred to as a *metric ribbon graph*. The metric associates to each face in the cell decomposition a perimeter, which is simply the sum of the numbers appearing around the boundary of the face. We define the *combinatorial moduli space* as follows.

$$\mathcal{M}RG_{g,n}(\mathbf{L}) = \left\{ \begin{array}{c} \text{metric ribbon graphs of type } (g, n) \\ \text{with perimeters } L_1, L_2, \ldots, L_n \end{array} \right\} \Big/ \sim$$

Here, two metric ribbon graphs are equivalent if and only if there exists an isometry between them which corresponds to a ribbon graph isomorphism.

For a ribbon graph Γ of type (g, n), consider $\mathcal{M}RG_\Gamma(\mathbf{L}) \subseteq \mathcal{M}RG_{g,n}(\mathbf{L})$, the subset consisting of those metric ribbon graphs whose underlying ribbon graph is Γ. Note that $\mathcal{M}RG_\Gamma(\mathbf{L})$ can be naturally identified with the following quotient of a possibly empty polytope by a finite group.

$$\mathcal{M}RG_\Gamma(\mathbf{L}) \cong \left\{ \mathbf{e} \in \mathbb{R}_+^{E(\Gamma)} \;\middle|\; A_\Gamma \mathbf{e} = \mathbf{L} \right\} \Big/ \mathrm{Aut}(\Gamma)$$

Here, $\mathbf{e}$ represents the lengths of the edges in the metric ribbon graph, $E(\Gamma)$ denotes the edge set of Γ, and A_Γ is the linear map which represents the adjacency between faces and edges in the cell decomposition corresponding to Γ. Thus, $\mathcal{M}RG_\Gamma(\mathbf{L})$ is an orbifold cell and these naturally glue together via edge degenerations — in other words, when an edge length goes to zero, the edge contracts to give a ribbon graph with fewer edges. So this cell decomposition for $\mathcal{M}RG_{g,n}(\mathbf{L})$ equips it with not only a topology, but also an orbifold structure. The main reason for considering $\mathcal{M}RG_{g,n}(\mathbf{L})$ is the following result.

Theorem 2.8. *The moduli spaces* $\mathcal{M}_{g,n}(\mathbf{L})$ *and* $\mathcal{MRG}_{g,n}(\mathbf{L})$ *are homeomorphic as orbifolds.*

One can prove this fact by generalising the work of Bowditch and Epstein [8], who consider the case of cusped hyperbolic surfaces. The main idea is to associate to a hyperbolic surface S with geodesic boundary its spine $\Gamma(S)$. For every point $p \in S$, let $n(p)$ denote the number of shortest paths from p to the boundary. Generically, we have $n(p) = 1$ and we define the *spine* as

$$\Gamma(S) = \{p \in S \mid n(p) \geqslant 2\}.$$

The locus of points with $n(p) = 2$ consists of a disjoint union of open geodesic segments. These correspond precisely to the edges of a graph embedded in S. The locus of points with $n(p) \geqslant 3$ forms a finite set which corresponds to the set of vertices of the aforementioned graph. In fact, if $n(p) \geqslant 3$, then the corresponding vertex will have degree $n(p)$. In this way, $\Gamma(S)$ has the structure of a ribbon graph. Furthermore, it is a deformation retract of the original hyperbolic surface, so if S is a hyperbolic surface of type (g, n), then $\Gamma(S)$ will be a ribbon graph of type (g, n).

Now for each vertex p of $\Gamma(S)$, consider the $n(p)$ shortest paths from p to the boundary. We refer to these geodesic segments as *ribs* and note that they are perpendicular to the boundary of S. The diagram below shows part of a hyperbolic surface, along with its spine and ribs. Cutting S along its ribs leaves a collection of hexagons, each with four right angles and a reflective axis of symmetry along one of the diagonals. In fact, this diagonal is one of the edges of $\Gamma(S)$ and we assign to it the length of the side of the hexagon which lies along the boundary of S. Of course, there are two such sides — however, the reflective symmetry guarantees that they are equal in length. In this way, $\Gamma(S)$ becomes a metric ribbon graph of type (g, n). By construction, the perimeters of $\Gamma(S)$ correspond precisely with the lengths of the boundary components of S, so we have a map $\Gamma : \mathcal{M}_{g,n}(\mathbf{L}) \to \mathcal{MRG}_{g,n}(\mathbf{L})$. It is possible, though more difficult, to construct the inverse map $S : \mathcal{MRG}_{g,n}(\mathbf{L}) \to \mathcal{M}_{g,n}(\mathbf{L})$ and show that it preserves the orbifold structure of the moduli spaces. The omitted details may be found elsewhere in the literature, such as Section 4 of Bowditch and Epstein's paper [8, 12].

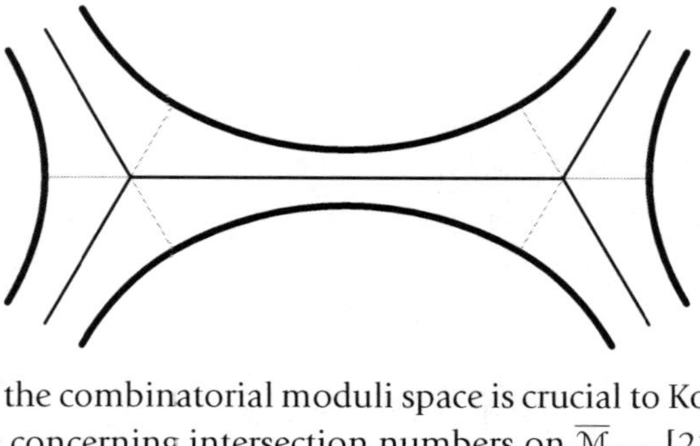

The notion of the combinatorial moduli space is crucial to Kontsevich's proof of Witten's conjecture concerning intersection numbers on $\overline{\mathcal{M}}_{g,n}$ [22]. We remark that

Kontsevich uses a different construction of the combinatorial moduli space which begins with punctured Riemann surfaces. Metric ribbon graphs arise via the existence of Jenkins–Strebel quadratic differentials on Riemann surfaces, an observation which Kontsevich attributes to Harer, Mumford, Penner and Thurston.

3. Weil–Petersson volumes

3.1. Early results

By raising the Weil–Petersson symplectic form to the appropriate exterior power, one obtains the following volume form on the Teichmüller space $\mathcal{T}_{g,n}(\mathbf{L})$.

$$\frac{\omega^{3g-3+n}}{(3g-3+n)!} = d\ell_1 \wedge d\tau_1 \wedge d\ell_2 \wedge d\tau_2 \wedge \cdots \wedge d\ell_{3g-3+n} \wedge d\tau_{3g-3+n}.$$

Of course, $\mathcal{T}_{g,n}(\mathbf{L})$ has infinite volume with respect to this form. However, the action of the mapping class group is such that the volume of $\mathcal{M}_{g,n}(\mathbf{L})$ is finite. One way to see this is via Wolpert's observation [51] that the Weil–Petersson symplectic form extends smoothly to a closed non-degenerate form on the Deligne–Mumford compactification $\overline{\mathcal{M}}_{g,n}(\mathbf{L})$. Another way is via Bers' observation [4] that every hyperbolic surface possesses a pants decomposition where the length of each curve is bound by a constant depending only on the topology of the surface and the maximum boundary length. Thus, let us define the *Weil–Petersson volume*

$$V_{g,n}(\mathbf{L}) = \int_{\mathcal{M}_{g,n}(\mathbf{L})} \frac{\omega^{3g-3+n}}{(3g-3+n)!}.$$

Below we present a brief selection of some early results concerning Weil–Petersson volumes. When comparing these results with those in the literature, there may be some discrepancy due to two issues. First, there are distinct normalisations of the Weil–Petersson symplectic form which differ by a factor of two. We have scaled the results, where appropriate, to correspond to the Weil–Petersson symplectic form defined in Section 2.3. Second, one must treat the special cases of $V_{1,1}(L_1)$ and $V_{2,0}$ with some care. This is due to the fact that every hyperbolic surface of type $(1,1)$ or $(2,0)$ possesses a twofold symmetry. Correspondingly, every point in $\mathcal{M}_{1,1}(L)$ and $\mathcal{M}_{2,0}$ is an orbifold point, generically with orbifold group equal to $\mathbb{Z}/2\mathbb{Z}$. Cleaner statements of results are obtained if one considers $V_{1,1}(L_1)$ and $V_{2,0}$ as orbifold volumes — in this case, half of the true volumes. The upshot is that one should not be alarmed if results concerning Weil–Petersson volumes from distinct sources differ by a factor which is a power of two.

- Wolpert [49, 50] proved that $V_{0,4}(0,0,0,0) = 2\pi^2$, $V_{1,1}(0) = \frac{\pi^2}{12}$ and $V_{g,n}(\mathbf{0}) = q(2\pi^2)^{3g-3+n}$ for some rational number q. This last fact is a corollary of Theorem 2.7, from which it follows that $q = \int_{\overline{\mathcal{M}}_{g,n}} \kappa_1^{3g-3+n}$.
- Penner [39] proved that $V_{1,2}(0,0) = \frac{\pi^4}{4}$.

- Zograf [52] proved that $V_{0,n}(\mathbf{0}) = \frac{(2\pi^2)^{n-3}}{(n-3)!} a_n$, where $a_3 = 1$ and

$$a_n = \frac{1}{2}\sum_{k=1}^{n-3} \frac{k(n-k-2)}{n-1}\binom{n-4}{k-1}\binom{n}{k+1} a_{k+2}a_{n-k} \qquad \text{for } n \geqslant 4.$$

- Nakanishi and Näätänen [35] calculated the Weil–Petersson volumes of the two-dimensional moduli spaces.

$$V_{0,4}(L_1, L_2, L_3, L_4) = \frac{1}{2}(L_1^2 + L_2^2 + L_3^2 + L_4^2 + 4\pi^2),$$
$$V_{1,1}(L_1) = \frac{1}{48}(L_1^2 + 4\pi^2).$$

Nakanishi and Näätänen's result shows that $V_{1,1}(L_1)$ and $V_{0,4}(L_1, L_2, L_3, L_4)$ are both polynomials in the squares of the boundary lengths. That this is the case for all Weil–Petersson volumes $V_{g,n}(\mathbf{L})$ was proven by Mirzakhani in two distinct ways. In Section 3.3, we discuss the first of Mirzakhani's proofs, which uses symplectic reduction in a fundamental way [30].

3.2. Symplectic reduction

Symplectic geometry has its origins in the mathematical formulation and generalisation of the phase space of a classical mechanical system. Physicists have often taken advantage of the fact that when a symmetry group of dimension n acts on a system, then the number of degrees of freedom for the positions and momenta can be reduced by $2n$. The analogous mathematical phenomenon is known as *symplectic reduction*.

More precisely, take a symplectic manifold (M, ω) of dimension $2d$ with a

$$T^n = \underbrace{S^1 \times S^1 \times \cdots \times S^1}_{n \text{ times}}$$

action that preserves the symplectic form. Suppose that there exists a moment map $\mu : M \to \mathbb{R}^n$. In other words, μ is invariant under the T^n action and, for each basis vector of $\mathbb{R}^n$, we have the equation

$$d\mu^X(\cdot) = \omega(X^\#, \cdot).$$

Here, $\mu^X : M \to \mathbb{R}$ is the component of μ along X and $X^\#$ is the vector field generated by the 1-parameter subgroup $\{\exp tX \mid t \in \mathbb{R}\} \subseteq T^n$. If $\mathbf{0}$ is a regular value of μ, we can define the manifold $M_\mathbf{0} = \mu^{-1}(\mathbf{0})/T^n$, since T^n must act on the level sets of μ.

Theorem 3.1 (Marsden–Weinstein theorem). *The orbit space $M_\mathbf{0} = \mu^{-1}(\mathbf{0})/T^n$ is a symplectic manifold of dimension $2d - 2n$ with respect to the unique 2-form $\omega_\mathbf{0}$ which satisfies $i^*\omega = \pi^*\omega_\mathbf{0}$. Here, $\pi : \mu^{-1}(\mathbf{0}) \to M_\mathbf{0}$ and $i : \mu^{-1}(\mathbf{0}) \to M$ are the natural projection and inclusion maps.*

Since $\mathbf{0}$ is a regular value, there exists an $\varepsilon > 0$ such that all $\mathbf{a} \in \mathbb{R}^n$ satisfying $|\mathbf{a}| < \varepsilon$ are also regular values. So it is possible to define symplectic manifolds

(M_a, ω_a) for all such a. If we think of the T^n action as n commuting circle actions, then the kth copy of S^1 induces a circle bundle $\mathcal{S}_k$ on M_0. The variation of the symplectic form ω_a can be described in terms of the first Chern classes $\phi_k = c_1(\mathcal{S}_k)$.

Theorem 3.2. *For $a = (a_1, a_2, \ldots, a_n)$ sufficiently close to $\mathbf{0}$, (M_a, ω_a) is symplectomorphic to M_0 equipped with a symplectic form whose cohomology class is equal to $[\omega_0] + a_1\phi_1 + a_2\phi_2 + \cdots + a_n\phi_n$.*

From this theorem, one obtains as a direct corollary an expression for the variation of the volume.

Corollary 3.3. *For $a = (a_1, a_2, \ldots, a_n)$ sufficiently close to $\mathbf{0}$, the volume of (M_a, ω_a) is a polynomial in $a_1, a_2, \ldots, a_n$ of degree $d = \frac{1}{2}\dim(M_a)$ given by the formula*

$$\sum_{|\alpha|+m=d} \frac{\int_{M_0} \phi_1^{\alpha_1}\phi_2^{\alpha_2}\cdots\phi_n^{\alpha_n}\omega^m}{\alpha_1!\alpha_2!\cdots\alpha_n!m!} a_1^{\alpha_1}a_2^{\alpha_2}\cdots a_n^{\alpha_n}.$$

There are several good introductions to symplectic geometry, such as the book by Cannas da Silva [41]. The rich subject of T^n actions on symplectic manifolds is discussed at length by Guillemin [17].

3.3. Polynomiality of Weil–Petersson volumes

We now consider Mirzakhani's construction of a setup in which Corollary 3.3 may be used to produce Weil–Petersson volumes [30]. This allows us to prove that $V_{g,n}(\mathbf{L})$ is a polynomial and, furthermore, that its coefficients store intersection numbers on the moduli space of curves $\overline{\mathcal{M}}_{g,n}$. We start by considering the space

$$\widehat{\mathcal{M}}_{g,n} = \left\{ (X, p_1, p_2, \ldots, p_n) \;\middle|\; \begin{array}{l} X \text{ is a genus } g \text{ hyperbolic surface with} \\ n \text{ geodesic boundary components} \\ \beta_1, \beta_2, \ldots, \beta_n \text{ and } p_k \in \beta_k \text{ for all } k \end{array} \right\}.$$

There is a T^n action on this space, where the kth copy of S^1 moves the point p_k along the boundary β_k at a constant speed in the direction given by the orientation of the surface.

We now show that $\widehat{\mathcal{M}}_{g,n}$ has a T^n invariant symplectic structure. Fix a tuple $\gamma = (\gamma_1, \gamma_2, \ldots, \gamma_n)$ of homotopy classes of disjoint simple closed curves on the surface $\Sigma_{g,2n}$ with genus g and $2n$ labelled boundary components such that γ_k bounds a pair of pants with the boundaries labelled $2k-1$ and $2k$. Since mapping classes act on homotopy classes of curves, elements of $\mathrm{Mod}_{g,2n}$ act on γ componentwise. Now define

$$\mathcal{M}^*_{g,2n} = \{(X, \eta_1, \eta_2, \ldots, \eta_n) \mid X \in \mathcal{M}_{g,2n}(\mathbf{0}) \text{ and } (\eta_1, \eta_2, \ldots, \eta_n) \in \mathrm{Mod}_{g,2n} \cdot \gamma\}.$$

Equivalently, we can use the definition $\mathcal{M}^*_{g,2n} = \mathcal{T}_{g,2n}(\mathbf{0})/\mathrm{Stab}(\gamma)$, where the stabiliser

$$\mathrm{Stab}(\gamma) = \{[\phi] \in \mathrm{Mod}_{g,2n} \mid \phi(\gamma_k) \text{ is homotopic to } \gamma_k \text{ for all } k\} \leqslant \mathrm{Mod}_{g,2n}$$

acts on the Teichmüller space in the usual way. Since the Weil–Petersson symplectic form on the Teichmüller space $\mathcal{T}_{g,2n}(\mathbf{0})$ is invariant under the action of the mapping class group, it must also be invariant under $\mathrm{Stab}(\gamma)$. Therefore, it descends to a symplectic form on $\mathcal{M}^*_{g,2n}$.

There is a natural map $f : \widehat{\mathcal{M}}_{g,n} \to \mathcal{M}^*_{g,2n}$ which is easy to describe. Simply take $(X, p_1, p_2, \ldots, p_n)$ where $X \in \mathcal{M}_{g,n}(\mathbf{L})$ and, to the kth boundary component, glue in a pair of pants with two cusps labelled $2k-1$ and $2k$ and a boundary component of length L_k. Of course, this can be done in infinitely many ways and we choose the unique way such that the seam from the cusp labelled $2k$ meets the point p_k. The map f can be used to pull back the symplectic form from $\mathcal{M}^*_{g,2n}$ to $\widehat{\mathcal{M}}_{g,n}$, where it is invariant under the T^n action. Furthermore, by the definition of the Weil–Petersson symplectic form, the canonical map $\ell^{-1}(\mathbf{L})/T^n \to \mathcal{M}_{g,n}(\mathbf{L})$ is a symplectomorphism, where $\ell : \widehat{\mathcal{M}}_{g,n} \to \mathbb{R}^n$ sends a hyperbolic surface to the n-tuple of its boundary lengths. One may check that the T^n action is the Hamiltonian flow for the moment map $\mu : \widehat{\mathcal{M}}_{g,n} \to \mathbb{R}^n$ defined by $\mu(X, p_1, p_2, \ldots, p_n) = (\frac{1}{2}L_1^2, \frac{1}{2}L_2^2, \ldots, \frac{1}{2}L_n^2)$, where L_k denotes the length of the geodesic boundary component β_k.

By construction, the symplectic quotient $\mu^{-1}(\frac{1}{2}L_1^2, \frac{1}{2}L_2^2, \ldots, \frac{1}{2}L_n^2)/T^n$ is the moduli space $\mathcal{M}_{g,n}(\mathbf{L})$. As usual, the T^n action gives rise to n circle bundles on the symplectic quotient. Although the moment map is only regular away from $\mathbf{0}$, one obtains circle bundles $\mathcal{S}_1, \mathcal{S}_2, \ldots, \mathcal{S}_n$ on $\mathcal{M}_{g,n}(\mathbf{0})$ by taking the limit as $\mathbf{L} \to \mathbf{0}$. Mirzakhani proved the following fact concerning the Chern classes of these circle bundles.

Proposition 3.4. *For* $k = 1, 2, \ldots, n$, $c_1(\mathcal{S}_k) = \psi_k \in H^2(\overline{\mathcal{M}}_{g,n}; \mathbb{Q})$.

This proposition states that $c_1(\mathcal{S}_k)$ is an element of $H^2(\overline{\mathcal{M}}_{g,n}; \mathbb{Q})$, even though it is apparent that $\mathcal{S}_k$ is a circle bundle over the uncompactified space $\mathcal{M}_{g,n}(\mathbf{0})$. However, with a little more care, all of the previous discussion generalises to the Deligne–Mumford compactifications of the moduli spaces involved. We are now ready to state and prove one of the most important results underlying this article.

Theorem 3.5 (Mirzakhani's theorem). *The Weil–Petersson volume* $V_{g,n}(\mathbf{L})$ *is given by the formula*

$$\sum_{|\alpha|+m=3g-3+n} \frac{(2\pi^2)^m \int_{\overline{\mathcal{M}}_{g,n}} \psi_1^{\alpha_1}\psi_2^{\alpha_2}\cdots\psi_n^{\alpha_n}\kappa_1^m}{2^{|\alpha|}\alpha_1!\alpha_2!\cdots\alpha_n!m!} L_1^{2\alpha_1}L_2^{2\alpha_2}\cdots L_n^{2\alpha_n}.$$

Proof. We simply apply Corollary 3.3 to the symplectic manifold $\widehat{\mathcal{M}}_{g,n}$ with the moment map μ defined above. This implies that the Weil–Petersson volume of $\mathcal{M}_{g,n}(\mathbf{L})$ for $\mathbf{L} \neq \mathbf{0}$ is a polynomial in $\frac{1}{2}L_1^2, \frac{1}{2}L_2^2, \ldots, \frac{1}{2}L_n^2$. The coefficients are given by integrating products of Chern classes of certain circle bundles alongside powers of the reduced symplectic form. In the $\mathbf{L} \to \mathbf{0}$ limit, Proposition 3.4 states that these Chern classes are precisely the psi-classes on $\overline{\mathcal{M}}_{g,n}$. Furthermore, the reduced symplectic form converges to the usual Weil–Petersson symplectic form in the limit.

All that is required now is to invoke Corollary 3.3 and substitute $\omega = 2\pi^2\kappa_1$, which is true in cohomology by Theorem 2.7.[2] □

We remark that Theorem 3.2 applied to this setup yields a generalisation of Theorem 2.7. The generalisation states that the de Rham cohomology class of the Weil–Petersson symplectic form on $\overline{\mathcal{M}}_{g,n}(\mathbf{L})$ satisfies

$$[\omega] = 2\pi^2\kappa_1 + \frac{1}{2}L_1^2\psi_1 + \frac{1}{2}L_2^2\psi_2 + \cdots + \frac{1}{2}L_n^2\psi_n \in H^2(\overline{\mathcal{M}}_{g,n};\mathbb{R}).$$

Mirzakhani's theorem shows that the Weil–Petersson volume $V_{g,n}(\mathbf{L})$ is a polynomial whose coefficients store intersection numbers on $\overline{\mathcal{M}}_{g,n}$. One of the consequences is that a meaningful statement about the volume $V_{g,n}(\mathbf{L})$ often yields a meaningful statement about the intersection theory on $\overline{\mathcal{M}}_{g,n}$, and vice versa. The strength of such an approach lies in the fact that one need not consider the geometry of the Deligne–Mumford compactification, which is rather subtle. This is because the compactification locus is the union of submanifolds of positive codimension, which do contribute to the Weil–Petersson volume .

In this section, we have only outlined the proof of Mirzakhani's theorem, whereas the technical details can be found in Mirzakhani's original paper [30]. The reader may also like to compare Mirzakhani's theorem with the earlier results of Kaufmann, Manin and Zagier [19] on higher Weil–Petersson volumes and their relationship with intersection theory on $\overline{\mathcal{M}}_{g,n}$.

4. A recursion for Weil–Petersson volumes

4.1. The volume of $\mathcal{M}_{1,1}(0)$

One of the main obstacles in the calculation of Weil–Petersson volumes is the fact that the Fenchel–Nielsen coordinates for Teichmüller space are not well-behaved under the action of the mapping class group. In particular, there is no concrete description for a fundamental domain of $\mathcal{M}_{g,n}(\mathbf{L})$ in $\mathcal{T}_{g,n}(\mathbf{L})$ in the general case. The workaround successfully applied by Mirzakhani [29] is to *unfold the integral* in the following way. Let $\pi : X_1 \to X_2$ be a covering map, dv_2 a volume form on X_2, and $dv_1 = \pi^* dv_2$ the pullback volume form on X_1. If π is a finite covering, then for any function $f : X_1 \to \mathbb{R}$, one can construct the pushforward function $\pi_* f : X_2 \to \mathbb{R}$ defined by

$$(\pi_* f)(y) = \sum_{x \in \pi^{-1}(y)} f(x).$$

In fact, even if π is an infinite covering, then the pushforward function may still exist, provided f is sufficiently well-behaved. The main reason for considering this setup

[2]The literature on symplectic reduction generally does not discuss the case of symplectic orbifolds. However, one can get around such problems by lifting to a manifold cover, which is possible by Theorem 2.6. This takes a little extra care, but essentially causes no problems.

is the fact that, under mild integrability assumptions, we have

$$\int_{X_1} f \, dv_1 = \int_{X_2} (\pi_* f) \, dv_2.$$

We will use this strategy to calculate the volume of $\mathcal{M}_{1,1}(0)$, which will serve as a motivating example for the general case of $\mathcal{M}_{g,n}(\mathbf{L})$. For this, set $X_2 = \mathcal{M}_{1,1}(0)$ and

$$X_1 = \mathcal{M}^*_{1,1}(0) = \{(X, \gamma) \mid X \in \mathcal{M}_{1,1}(0) \text{ and } \gamma \text{ a simple closed geodesic on } X\}.$$

Equivalently, we can use the definition $\mathcal{M}^*_{1,1}(0) = \mathcal{T}_{1,1}(0)/\mathrm{Stab}(\alpha)$, where α is a simple closed curve on the interior of the once-punctured torus. The stabiliser

$$\mathrm{Stab}(\alpha) = \{[\phi] \in \mathrm{Mod}_{1,1} \mid \phi(\alpha) \text{ is homotopic to } \alpha\} \leqslant \mathrm{Mod}_{1,1}$$

acts on the Teichmüller space in the usual way. Using Fenchel–Nielsen coordinates, we see that each $(X, \gamma) \in \mathcal{M}^*_{1,1}(0)$ can be described by the pair (ℓ, τ), where ℓ denotes the length of γ and τ the corresponding twist parameter. The only redundancy in this description comes from the fact that the pair $(\ell, \tau + \ell)$ may also be used to describe the same point in $\mathcal{M}^*_{1,1}(0)$. Hence, we can write

$$\mathcal{M}^*_{1,1}(0) \cong \{(\ell, \tau) \mid \ell \in \mathbb{R}_+ \text{ and } 0 \leqslant \tau \leqslant \ell\}/\sim,$$

where $(\ell, 0) \sim (\ell, \ell)$ for all $\ell \in \mathbb{R}_+$.

The map $\pi : \mathcal{M}^*_{1,1}(0) \to \mathcal{M}_{1,1}(0)$ is the obvious projection map defined by $\pi(X, \gamma) = X$. Through the tower of coverings $\mathcal{T}_{1,1}(0) \to \mathcal{M}^*_{1,1}(0) \to \mathcal{M}_{1,1}(0)$, we see that the Weil–Petersson form pulls back to $\pi^*\omega = d\ell \wedge d\tau$ on the intermediate cover $\mathcal{M}^*_{1,1}(0)$. Let $\ell : \mathcal{M}^*_{1,1}(0) \to \mathbb{R}$ be the geodesic length function so that $\ell(X, \gamma)$ equals the length of γ on X. Unfolding the integral and using the description above for $\mathcal{M}^*_{1,1}(0)$ yields the following equalities.

$$\int_{\mathcal{M}_{1,1}(0)} \sum_{\pi(Y)=X} f(\ell(Y)) \, dX = \int_{\mathcal{M}^*_{1,1}(0)} f(\ell(Y)) \, dY = \int_0^\infty \int_0^\ell f(\ell) \, d\tau \, d\ell.$$

Therefore, in order to obtain the volume of $\mathcal{M}_{1,1}(0)$, we would like an identity of the form

$$\sum_{\pi(Y)=X} f(\ell(Y)) = 1,$$

valid for all $X \in \mathcal{M}_{1,1}(0)$. Note that the summation is over the set of simple closed geodesics on X. Such an identity had been discovered by McShane [27] prior to Mirzakhani's work on Weil–Petersson volumes.

Theorem 4.1 (McShane identity). *If X is a hyperbolic torus with one cusp, then*

$$\sum_{\gamma} \frac{2}{1 + \exp \ell(\gamma)} = 1.$$

Here, the summation is over the set of simple closed geodesics on X and $\ell(\gamma)$ denotes the length of γ.

So to complete the calculation of the volume of $\mathcal{M}_{1,1}(0)$, take $f(\ell) = \frac{2}{1+\exp \ell}$.

$$\int_{\mathcal{M}_{1,1}(0)} 1 \, dX = \int_0^\infty \int_0^\ell \frac{2}{1+\exp \ell} \, d\tau \, d\ell = \int_0^\infty \frac{2\ell}{1+\exp \ell} \, d\ell = \frac{\pi^2}{6}.$$

However, recall that the case of $\mathcal{M}_{1,1}(0)$ is exceptional in the sense that a generic point of the moduli space is an orbifold point with orbifold group $\mathbb{Z}/2\mathbb{Z}$. This fact arises from the existence of the elliptic involution on every hyperbolic torus with one cusp. Since we consider orbifold volumes in this article, it is necessary to divide the integral calculation above by two. Therefore, we finally have the result

$$V_{1,1}(0) = \frac{\pi^2}{12}.$$

4.2. McShane identities

In order to unfold the integral required to calculate $V_{g,n}(\mathbf{L})$, it is necessary to obtain a more general version of McShane's identity.

Theorem 4.2 (Generalised McShane identity). *On a hyperbolic surface with geodesic boundary components $\beta_1, \beta_2, \ldots, \beta_n$ of lengths $L_1, L_2, \ldots, L_n$, respectively,*

$$\sum_{(\alpha_1,\alpha_2)} D(L_1, \ell(\alpha_1), \ell(\alpha_2)) + \sum_{k=2}^{n} \sum_{\gamma} R(L_1, L_k, \ell(\gamma)) = L_1.$$

Here, the first summation is over unordered pairs (α_1, α_2) of simple closed geodesics which bound a pair of pants with β_1, while the second summation is over simple closed geodesics γ which bound a pair of pants with β_1 and β_k. The functions $D : \mathbb{R}^3 \to \mathbb{R}$ and $R : \mathbb{R}^3 \to \mathbb{R}$ are given by the equations

$$D(x, y, z) = 2 \log \left(\frac{e^{\frac{x}{2}} + e^{\frac{y+z}{2}}}{e^{-\frac{x}{2}} + e^{\frac{y+z}{2}}} \right) \quad \text{and} \quad R(x, y, z) = x - \log \left(\frac{\cosh \frac{y}{2} + \cosh \frac{x+z}{2}}{\cosh \frac{y}{2} + \cosh \frac{x-z}{2}} \right).$$

The proof supplied by Mirzakhani [29] is a modification of the argument originally applied by McShane to the case of the once-punctured torus [27]. The main idea is to consider, for each point $x \in \beta_1$, the geodesic γ_x which meets β_1 orthogonally at x. If we start at x and walk along γ_x, then one of the following situations must arise.

(1) The geodesic γ_x intersects itself.
(2) The geodesic γ_x intersects β_1 without intersecting itself.
(3) The geodesic γ_x intersects β_k for $2 \leqslant k \leqslant n$ without intersecting itself.
(4) The geodesic γ_x never intersects itself or a boundary component.

We now use this observation to construct a map from a subset $\beta_1^* \subseteq \beta_1$ to the set

$$\mathcal{P}_1 = \left\{ \begin{array}{c} \text{embedded hyperbolic pairs of pants, one of} \\ \text{whose geodesic boundary components is } \beta_1 \end{array} \right\}.$$

Note that the generalised McShane identity is not a summation over simple closed geodesics, but over $\mathcal{P}_1$. It just so happens that on a once-punctured torus, the two notions coincide. In fact, McShane's original identity — see Theorem 4.1 — may be recovered by using $(g, n) = (1,1)$, dividing both sides of the identity by L_1, and taking the $L_1 \to 0$ limit.

In cases (1) and (2), consider the union of β_1 and the geodesic path γ_x from x to the intersection point. For $\epsilon > 0$ sufficiently small, the ϵ-neighbourhood of this embedded graph is topologically a pair of pants. By taking geodesic representatives in the homotopy classes of the boundary components, we obtain an embedded hyperbolic pair of pants, one of whose geodesic boundary components is β_1. Let $f(x) \in \mathcal{P}_1$ denote this pair of pants.

In case (3), consider the union of β_1, β_k and the geodesic path γ_x from x to the intersection point. Again, for $\epsilon > 0$ sufficiently small, the ϵ-neighbourhood of this embedded graph is topologically a pair of pants. By taking geodesic representatives in the homotopy classes of the boundary components, we obtain an embedded hyperbolic pair of pants, one of whose geodesic boundary components is β_1. Let $f(x) \in \mathcal{P}_1$ denote this pair of pants.

Thus, we have defined a function $f : \beta_1^* \to \mathcal{P}_1$, where β_1^* is the set consisting of those points in β_1 for which cases (1), (2) or (3) occur. The points in β_1 for which f is undefined are those for which case (4) occurs. A result due to Birman and Series [5] states that the union of all complete simple geodesics on a closed hyperbolic surface has Hausdorff dimension one. By doubling the surface along its boundary, we can generalise the statement to hyperbolic surfaces with boundary and complete simple geodesics perpendicular to the boundary. Hence, we may deduce that $\beta \setminus \beta_1^*$ is a subset of zero measure with respect to the hyperbolic line element μ on β_1. In fact, Mirzakhani [29] shows that it is homeomorphic to the union of a Cantor set and countably many isolated points. The upshot of this discussion is the equation

$$\sum_{P \in \mathcal{P}_1} \mu(f^{-1}(P)) = L_1.$$

Theorem 4.2 now follows from a couple of simple facts.

Lemma 4.3.

- *If* $P \in \mathcal{P}_1$ *is bound by* β_1 *and two simple closed geodesics* α_1 *and* α_2*, then*

$$\mu(f^{-1}(P)) = D(L_1, \ell(\alpha_1), \ell(\alpha_2)).$$

- *If* $P \in \mathcal{P}_1$ *is bound by* β_1*,* β_k *and a simple closed geodesic* γ*, then*

$$\mu(f^{-1}(P)) = R(L_1, L_k, \ell(\gamma)).$$

Note that the calculation of $\mu(f^{-1}(P))$ is local in the sense that it depends only on the geometry of P and not on the geometry of the entire surface. To a hyperbolic pair of pants with geodesic boundary components α, β, γ, we associate four distinguished points on α. Such a pair of pants necessarily contains exactly four complete simple geodesics which meet α orthogonally exactly once and are disjoint from β and γ. One of them intersects α at B_1 and spirals around β one way, while another intersects α at B_2 and spirals around β the other way. Similarly, one of them intersects α at C_1 and spirals around γ one way, while another intersects α at C_2 and spirals around γ the other way. Note that the orientation reversing isometry which reflects the pair of pants through its seams interchanges B_1 with B_2 and C_1 with C_2. This is encapsulated in the schematic diagram below.

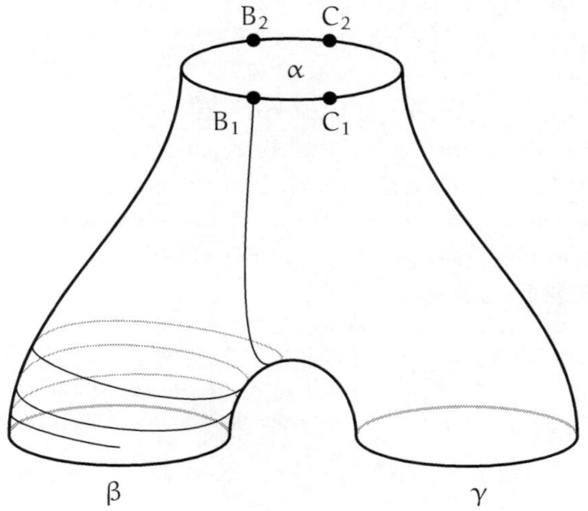

Suppose that $x \in \alpha$ lies on the interval B_1C_1 which does not include B_2 and C_2 or on the interval B_2C_2 which does not include B_1 and C_1. Then the geodesic γ_x will intersect itself or α, so case (1) or (2) occurs. Therefore, we define $D(\ell(\alpha), \ell(\beta), \ell(\gamma))$ to be twice the length of the interval B_1C_1 which does not include B_2 and C_2. Now suppose that $x \in \alpha$ lies on the interval B_1B_2 which does not include C_1 and C_2. Then the geodesic γ_x will intersect β, so case (3) occurs. Therefore, we define $R(\ell(\alpha), \ell(\beta), \ell(\gamma))$ to be the length of the interval C_1C_2 which includes B_1 and B_2. The proof of Lemma 4.3 now follows from these definitions.

All that remains is to explicitly compute the functions D and R. In the universal cover, the value of $D(\ell(\alpha), \ell(\beta), \ell(\gamma))$ is twice the distance between the projection of β and γ on α and the value of $R(\ell(\alpha), \ell(\beta), \ell(\gamma))$ is $\ell(\alpha)$ minus the length of the projection of γ on α. We do not complete the calculation here but remark that it can be carried out by applying some elementary results from hyperbolic trigonometry [10].

There are now many variations on the McShane theme.

- Bowditch used the notion of Markoff triples to give an alternative proof of McShane's identity for the once-punctured torus [7].
- Akiyoshi, Miyachi and Sakuma produced variants of McShane's identity for quasi-Fuchsian punctured surface groups and hyperbolic punctured surface bundles over the circle [2].

- McShane determined an identity for simple geodesics on a closed hyperbolic surface of genus two [28]. This work capitalises on the existence of the hyperelliptic involution and Weierstrass points on a genus two surface.
- Tan, Wong and Zhang gave a generalisation of McShane's identity to hyperbolic cone surfaces, where all cone points have angles bounded above by π [42]. They also found variations concerning representations of punctured torus groups to $SL(2, \mathbb{C})$ [43] and also classical Schottky groups [44].
- Luo and Tan have recently found a McShane identity for all closed hyperbolic surfaces [25]. Their proof draws some inspiration from Calegari's elegant and unified treatment [11] of the following two results — namely, the identities of Basmajian [3] and Bridgeman [9]. Despite the fact that these share a similar flavour with McShane identities, they concern lengths of orthogeodesics[3] in hyperbolic manifolds of arbitrary dimension and do not provide any explicit information about lengths of simple closed geodesics in hyperbolic surfaces.

Theorem 4.4. *Let M be a compact hyperbolic n-manifold with totally geodesic boundary ∂M and let $(\ell_1, \ell_2, \ell_3, \ldots)$ denote the lengths of the ortho geodesics of M, listed with multiplicity. There exist functions A_n and V_n depending only on n such that*

$$\text{area } \partial M = \sum A_n(\ell_i) \quad \textit{and} \quad \text{volume } M = \sum V_n(\ell_i).$$

4.3. Mirzakhani's recursion

In this section, we prove a formula for Weil–Petersson volumes, originally due to Mirzakhani [29]. We use the convention that $V_{0,1}(L_1) = 0$, $V_{0,2}(L_1, L_2) = 0$ and $V_{0,3}(L_1, L_2, L_3) = 1$.

Theorem 4.5 (Mirzakhani's recursion). *For $2g + n > 3$, the Weil–Petersson volumes satisfy the following relation.*

$$2\frac{\partial}{\partial L_1} L_1 V_{g,n}(\mathbf{L}) =$$

$$\int_0^\infty \int_0^\infty xy\, H(x+y, L_1)\, V_{g-1,n+1}(x, y, \widehat{\mathbf{L}})\, dx\, dy$$

$$+ \sum_{\substack{g_1+g_2=g \\ I \sqcup J = [2,n]}} \int_0^\infty \int_0^\infty xy\, H(x+y, L_1)\, V_{g_1,|I|+1}(x, \mathbf{L}_I)\, V_{g_2,|J|+1}(y, \mathbf{L}_J)\, dx\, dy$$

$$+ \sum_{k=2}^{n} \int_0^\infty x\, [H(x, L_1 + L_k) + H(x, L_1 - L_k)]\, V_{g,n-1}(x, \widehat{\mathbf{L}}_k)\, dx$$

---

[3] An *orthogeodesic* of M is a geodesic arc which is perpendicular to ∂M at its endpoints.

Here, we have used the notation $\widehat{\mathbf{L}} = (L_2, L_3, \ldots, L_n)$, $\mathbf{L}_I = (L_{i_1}, L_{i_2}, \ldots, L_{i_m})$ *for* $I = \{i_1, i_2, \ldots, i_m\}$, *and* $\widehat{\mathbf{L}}_k = (L_2, \ldots, \widehat{L}_k, \ldots, L_n)$ *where the hat denotes omission. The function* $H : \mathbb{R}^2 \to \mathbb{R}$ *is defined by*

$$H(x, y) = \frac{1}{1 + \exp\frac{x+y}{2}} + \frac{1}{1 + \exp\frac{x-y}{2}}.$$

The proof uses the calculation of $V_{1,1}(0)$ from Section 4.1 as a model. Our point of departure is the generalised McShane identity — see Theorem 4.2 — which we rewrite in the following way.

$$\mathcal{D}_{\text{con}}(X) + \sum_{\substack{g_1+g_2=g \\ I \sqcup J = [2,n]}} \mathcal{D}_{g_1,I}(X) + \sum_{k=2}^{n} \mathcal{R}_k(X) = L_1.$$

We have grouped the left hand side into terms of three distinct types.

- The first type is

  $$\mathcal{D}_{\text{con}}(X) = \sum_{(\alpha_1,\alpha_2)} D(L_1, \ell(\alpha_1), \ell(\alpha_2)),$$

  where the summation is over unordered pairs (α_1, α_2) of simple closed geodesics which bound a pair of pants with β_1, whose complement is a connected surface.
- The second type is

  $$\mathcal{D}_{g_1,I}(X) = \sum_{(\alpha_1,\alpha_2)} D(L_1, \ell(\alpha_1), \ell(\alpha_2)),$$

  where the summation is over unordered pairs (α_1, α_2) of simple closed geodesics which bound a pair of pants with β_1, whose complement is a disconnected surface. We require that one component of this disconnected surface has genus g_1 and includes only those boundary components from the original surface labelled by elements of I.
- The third type is

  $$\mathcal{R}_k(X) = \sum_{\gamma} R(L_1, L_k, \ell(\gamma)),$$

  where the summation is over simple closed geodesics γ which bound a pair of pants with β_1 and β_k.

The rationale for expressing the generalised McShane identity in this way is that each term is now a summation over a mapping class group orbit. This is due to the fact that two sets of disjoint simple closed curves on a surface are in the same mapping class group orbit if and only if their complements have the same topological type and labelling of boundary components. Sums over mapping class group orbits can be expressed as pushforwards of functions on appropriate covers of the moduli space. And these are the functions which we are able to integrate over the moduli space itself.

Now take the generalised McShane identity and integrate both sides over the moduli space $\mathcal{M}_{g,n}(\mathbf{L})$.

$$L_1 V_{g,n}(\mathbf{L}) = \int_{\mathcal{M}_{g,n}(\mathbf{L})} \mathcal{D}_{\text{con}}(X)\, dX + \sum_{\substack{g_1+g_2=g \\ I \sqcup J = [2,n]}} \int_{\mathcal{M}_{g,n}(\mathbf{L})} \mathcal{D}_{g_1,I}(X)\, dX$$
$$+ \sum_{k=2}^{n} \int_{\mathcal{M}_{g,n}(\mathbf{L})} \mathcal{R}_k(X)\, dX.$$

From the previous discussion, we know that it is possible to unfold each of the integrals using the strategy employed in Section 4.1. For example, let us concentrate on the term

$$\int_{\mathcal{M}_{g,n}(\mathbf{L})} \mathcal{R}_k(X)\, dX.$$

In order to unfold the integral, recall that we require a covering map $\pi : X_1 \to X_2$, a volume form dv_2 on X_2, and the pullback volume form $dv_1 = \pi^* dv_2$ on X_1. For this, set $X_2 = \mathcal{M}_{g,n}(\mathbf{L})$ and

$$X_1 = \mathcal{M}^*_{g,n}(\mathbf{L}) = \left\{ (X, \gamma) \;\middle|\; \begin{array}{l} X \in M_{g,n}(\mathbf{L}) \text{ and } \gamma \text{ a simple closed geodesic on} \\ X \text{ which bounds a pair of pants with } \beta_1 \text{ and } \beta_k \end{array} \right\}.$$

Equivalently, we can use the definition $\mathcal{M}^*_{g,n}(\mathbf{L}) = \mathcal{T}_{g,n}(\mathbf{L})/\text{Stab}(\alpha)$, where α is a simple closed curve on the surface $\Sigma_{g,n}$ which bounds a pair of pants with the boundary components labelled 1 and k. The stabiliser

$$\text{Stab}(\alpha) = \{[\phi] \in \text{Mod}_{g,n} \mid \phi(\alpha) \text{ is homotopic to } \alpha\} \leqslant \text{Mod}_{g,n}$$

acts on the Teichmüller space in the usual way. Using Fenchel–Nielsen coordinates, we see that each $(X, \gamma) \in \mathcal{M}^*_{g,n}(\mathbf{L})$ can be described by the triple $(\ell, \tau, \widehat{X})$, where ℓ denotes the length of γ and τ the corresponding twist parameter. The surface $\widehat{X} \in \mathcal{M}_{g,n-1}(\ell, \widehat{\mathbf{L}}_k)$ is simply the complement of the pair of pants bound by β_1, β_k and γ. The only redundancy in this description comes from the fact that the triple $(\ell, \tau + \ell, \widehat{X})$ may also be used to describe the same point in $\mathcal{M}^*_{g,n}(\mathbf{L})$. Hence, we can write

$$\mathcal{M}^*_{g,n}(\mathbf{L}) \cong \{(\ell, \tau, \widehat{X}) \mid \ell \in \mathbb{R}_+, 0 \leqslant \tau \leqslant \ell \text{ and } \widehat{X} \in \mathcal{M}_{g,n-1}(\ell, \widehat{\mathbf{L}}_k)\}/\sim,$$

where $(\ell, \tau, \widehat{X}) \sim (\ell, \tau + \ell, \widehat{X})$.

The map $\pi : \mathcal{M}^*_{g,n}(\mathbf{L}) \to \mathcal{M}_{g,n}(\mathbf{L})$ is the obvious projection map defined by $\pi(X, \gamma) = X$. Through the tower of coverings $\mathcal{T}_{g,n}(\mathbf{L}) \to \mathcal{M}^*_{g,n}(\mathbf{L}) \to \mathcal{M}_{g,n}(\mathbf{L})$, we see that the Weil–Petersson form pulls back to $\pi^*\omega = d\ell \wedge d\tau \wedge \widehat{\omega}$ on the intermediate cover $\mathcal{M}^*_{g,n}(\mathbf{L})$, where $\widehat{\omega}$ is the Weil–Petersson form on the lower dimensional moduli space $\mathcal{M}_{g,n-1}(\ell, \widehat{\mathbf{L}}_k)$. Let $\ell : \mathcal{M}^*_{g,n}(\mathbf{L}) \to \mathbb{R}$ be the geodesic length function so that $\ell(X, \gamma)$ equals the length of γ on X. Unfolding the integral and using the description above for $\mathcal{M}^*_{g,n}(\mathbf{L})$ yields the following equalities.

$$\begin{aligned}\int_{\mathcal{M}_{g,n}(\mathbf{L})} \mathcal{R}_k(X)\, dX &= \int_{\mathcal{M}_{g,n}(\mathbf{L})} \sum_{\pi(Y)=X} R(L_1, L_k, \ell(Y))\, dX \\ &= \int_{\mathcal{M}^*_{g,n}(\mathbf{L})} R(L_1, L_k, \ell(Y))\, dY \\ &= \int_0^\infty \int_0^\ell \int_{\mathcal{M}_{g,n-1}(\ell, \widehat{\mathbf{L}}_k)} R(L_1, L_k, \ell)\, \widehat{\omega}\, d\tau\, d\ell \\ &= \int_0^\infty x\, R(L_1, L_k, x)\, V_{g,n-1}(x, \widehat{\mathbf{L}}_k)\, dx.\end{aligned}$$

For the other terms in the generalised McShane identity, although the details may be different, the argument remains the same. After unfolding each of the integrals and summing, the end result is the following formula.

$$\begin{aligned}L_1 V_{g,n}(\mathbf{L}) =& \frac{1}{2} \int_0^\infty \int_0^\infty xy\, D(L_1, x, y)\, V_{g-1,n+1}(x, y, \widehat{\mathbf{L}})\, dx\, dy \\ &+ \frac{1}{2} \sum_{\substack{g_1+g_2=g \\ I \sqcup J = [2,n]}} \int_0^\infty \int_0^\infty xy\, D(L_1, x, y)\, V_{g_1,|I|+1}(x, \mathbf{L}_I)\, V_{g_2,|J|+1}(y, \mathbf{L}_J)\, dx\, dy \\ &+ \sum_{k=2}^{n} \int_0^\infty x\, R(L_1, L_k, x)\, V_{g,n-1}(x, \widehat{\mathbf{L}}_k)\, dx.\end{aligned}$$

Note that the factor of $\frac{1}{2}$ in front of the first two terms of the right hand side is to account for the twofold symmetry between x and y. Theorem 4.5 expresses the recursion in a more useful form, which is obtained by applying $2\frac{\partial}{\partial L_1}$ to both sides of this equation. For this, we use the straightforward calculations

$$\frac{\partial}{\partial x} D(x, y, z) = H(y + z, x) \quad \text{and} \quad \frac{\partial}{\partial x} R(x, y, z) = \frac{1}{2} \left[H(z, x + y) + H(z, x - y) \right].$$

4.4. Applications of Mirzakhani's recursion

The mechanism behind Mirzakhani's recursion is based on removing pairs of pants from the surface $\Sigma_{g,n}$ which contain at least one boundary component. Therefore, the calculation of any Weil–Petersson volume can be reduced to the base cases $V_{0,3}(L_1, L_2, L_3) = 1$ and $V_{1,1}(L_1) = \frac{1}{48}(L_1^2 + 4\pi^2)$. In the practical application of Mirzakhani's recursion, we require the following explicit integral calculations.

$$\int_0^\infty x^{2k-1} H(x, t)\, dx = F_{2k-1}(t),$$

$$\int_0^\infty \int_0^\infty x^{2a-1} y^{2b-1} H(x + y, t)\, dx\, dy = \frac{(2a-1)!(2b-1)!}{(2a+2b-1)!} F_{2a+2b-1}(t).$$

Here, $F_{2k-1}(t)$ is the following even polynomial of degree $2k$ in t, where the coefficient of t^{2m} is a rational multiple of π^{2k-2m}.

$$F_{2k-1}(t) = (2k-1)! \sum_{i=0}^{k} \frac{\zeta(2i)(2^{2i+1}-4)}{(2k-2i)!} t^{2k-2i}.$$

We list below the polynomials $F_{2k-1}(t)$ for the first few values of k.

$$\begin{aligned}
F_1(t) &= \frac{t^2}{2} + \frac{2\pi^2}{3}, \\
F_3(t) &= \frac{t^4}{4} + 2\pi^2 t^2 + \frac{28\pi^4}{15}, \\
F_5(t) &= \frac{t^6}{6} + \frac{10\pi^2 t^4}{3} + \frac{56\pi^4 t^2}{3} + \frac{992\pi^6}{63}, \\
F_7(t) &= \frac{t^8}{8} + \frac{14\pi^2 t^6}{3} + \frac{196\pi^4 t^4}{3} + \frac{992\pi^6 t^2}{3} + \frac{4064\pi^8}{15}.
\end{aligned}$$

As an example of Mirzakhani's recursion, consider the following computation of $V_{1,2}(L_1, L_2)$.

Example 4.6. *For $(g, n) = (1,2)$, we obtain a contribution from only two terms on the right hand side of Mirzakhani's recursion — one involving $V_{0,3}$ and the other involving $V_{1,1}$. This corresponds to the fact that removing a pair of pants from the surface $\Sigma_{1,2}$ which contains at least one boundary component must leave $\Sigma_{0,3}$ or $\Sigma_{1,1}$.*

$$\begin{aligned}
&2\frac{\partial}{\partial L_1} L_1 V_{1,2}(L_1, L_2) \\
&= \int_0^\infty \int_0^\infty xy\, H(x+y, L_1)\, V_{0,3}(x, y, L_2)\, dx\, dy \\
&\quad + \int_0^\infty x[H(x, L_1 + L_2) + H(x, L_1 - L_2)]\, V_{1,1}(x)\, dx \\
&= \int_0^\infty \int_0^\infty xy\, H(x+y, L_1)\, dx\, dy \\
&\quad + \int_0^\infty x[H(x, L_1 + L_2) + H(x, L_1 - L_2)] \left(\frac{x^2 + 4\pi^2}{48}\right) dx \\
&= \frac{F_3(L_1)}{6} + \frac{F_3(L_1 + L_2) + F_3(L_1 - L_2)}{48} + \frac{\pi^2 F_1(L_1 + L_2) + \pi^2 F_1(L_1 - L_2)}{12} \\
&= \frac{5L_1^4}{96} + \frac{L_1^2 L_2^2}{16} + \frac{L_2^4}{96} + \frac{\pi^2 L_1^2}{2} + \frac{\pi^2 L_2^2}{6} + \frac{\pi^4}{2}.
\end{aligned}$$

Now integrate with respect to L_1 and divide by $2L_1$ to obtain the desired result. Observe that no constant of integration appears, since $V_{1,2}(0, 0)$ is finite, as noted in Section 3.1.

$$V_{1,2}(L_1, L_2) = \frac{L_1^4}{192} + \frac{L_1^2 L_2^2}{96} + \frac{L_2^4}{192} + \frac{\pi^2 L_1^2}{12} + \frac{\pi^2 L_2^2}{12} + \frac{\pi^4}{4}.$$

Mirzakhani's recursion can be used to provide an alternative proof of the following fact, which is a direct corollary of Theorem 3.5.

Corollary 4.7. *The Weil–Petersson volume* $V_{g,n}(\mathbf{L})$ *is an even symmetric polynomial in* $L_1, L_2, \ldots, L_n$ *of degree* $6g-6+2n$. *Furthermore, the coefficient of* $L_1^{2\alpha_1} L_2^{2\alpha_2} \cdots L_n^{2\alpha_n}$ *is a rational multiple of* $\pi^{6g-6+2n-2|\alpha|}$.

The symmetry of $V_{g,n}(\mathbf{L})$ is a consequence of the symmetry of the boundary labels. However, the symmetry is not present in Mirzakhani's recursion, which treats one of the boundary components as distinguished. The remainder of Corollary 4.7 can be proven with a straightforward application of induction on the value of $2g-2+n$.

Mirzakhani's theorem — see Theorem 3.5 — shows that $V_{g,n}(\mathbf{L})$ is a polynomial whose coefficients store information about the intersection theory on $\overline{\mathcal{M}}_{g,n}$. In fact, all psi-class intersection numbers on $\overline{\mathcal{M}}_{g,n}$ can be recovered from the top degree part of $V_{g,n}(\mathbf{L})$ alone. On the other hand, Mirzakhani's recursion — see Theorem 4.5 — shows that the Weil–Petersson volume $V_{g,n}(\mathbf{L})$ can be calculated in an explicit manner. So the conjunction of these two results provides an algorithm to compute all psi-class intersection numbers on $\overline{\mathcal{M}}_{g,n}$. Thus, Mirzakhani was able to give a new proof of the Witten–Kontsevich theorem [30]. Although several proofs of the Witten–Kontsevich theorem now exist in the literature, there are three novel features of Mirzakhani's proof. First, she proved it by directly verifying the Virasoro constraints. Second, her proof was the first to appear which did not make explicit use of a matrix model. Third, her work uses hyperbolic geometry in a fundamental way.

Following Mirzakhani's work, Liu and Xu have demonstrated that Mirzakhani's recursion is in fact equivalent to the Witten–Kontsevich theorem [24]. Mulase and Safnuk have derived a differential version of Mirzakhani's recursion and shown that it is a Virasoro constraint for Weil–Petersson volumes [34]. In this way, they prove that a certain generating function for intersection numbers on $\overline{\mathcal{M}}_{g,n}$ involving psi-classes and the class κ_1 yields a 1-parameter family of solutions to the KdV hierarchy.

Further mileage can be obtained from integration over moduli spaces of hyperbolic surfaces. For example, Mirzakhani has applied this technique to obtain the following result concerning the number of simple closed geodesics of bounded length on a hyperbolic surface [31].

Theorem 4.8. *For* γ *a simple closed curve on* $X \in \mathcal{M}_{g,n}(\mathbf{0})$, *let* $s(X,\gamma,N)$ *denote the number of simple closed geodesics in the mapping class group orbit of* γ *whose length is at most* N. *Then*

$$\lim_{N\to\infty} \frac{s(X,\gamma,N)}{N^{6g-6+2n}} = \frac{c(\gamma)B(X)}{\int_{\mathcal{M}_{g,n}(\mathbf{0})} B(X)}.$$

Here, $c(\gamma) \in \mathbb{Q}$ *depends only on the topological type of* γ *and* $B(X)$ *is the volume of the unit ball centred at* X *in the space of measured geodesic laminations.*

5. Limits of Weil–Petersson volumes

5.1. Hyperbolic cone surfaces

Mirzakhani's recursion — see Theorem 4.5 — allows us to calculate Weil–Petersson volumes explicitly. The table in Appendix B contains $V_{g,n}(\mathbf{L})$ for various values of g and n. These data suggest the striking observation that $V_{g,1}(2\pi i) = 0$. This statement does indeed hold true for all positive integers g, which indicates that the Weil–Petersson volume polynomials display interesting behaviour when the lengths are formally set to $2\pi i$. Further investigation into the matter yields the following results [14].

Theorem 5.1 (String and dilaton equations for Weil–Petersson volumes). *For $2g - 2 + n > 0$, the Weil–Petersson volumes satisfy the following relations.*

$$V_{g,n+1}(\mathbf{L}, 2\pi i) = \sum_{k=1}^{n} \int_0^{L_k} L_k \, V_{g,n}(\mathbf{L}) \, dL_k,$$

$$\frac{\partial V_{g,n+1}}{\partial L_{n+1}}(\mathbf{L}, 2\pi i) = 2\pi i \, (2g - 2 + n) \, V_{g,n}(\mathbf{L}).$$

These equations must follow from Mirzakhani's recursion since it uniquely determines all Weil–Petersson volumes. The proof based on this observation is rather unwieldy and not so transparent [12]. An alternative proof expresses the string and dilaton equations as relations between the coefficients of $V_{g,n+1}(\mathbf{L}, L_{n+1})$ and of $V_{g,n}(\mathbf{L})$. By Mirzakhani's theorem — see Theorem 3.5 — this translates to relations between intersection numbers on $\overline{\mathcal{M}}_{g,n+1}$ and on $\overline{\mathcal{M}}_{g,n}$. Thus, we may equivalently write Theorem 5.1 in the following way.

$$\sum_{j=0}^{m} (-1)^j \binom{m}{j} \int_{\overline{\mathcal{M}}_{g,n+1}} \psi_1^{\alpha_1} \cdots \psi_n^{\alpha_n} \psi_{n+1}^{j} \kappa_1^{m-j} = \sum_{k=1}^{n} \int_{\overline{\mathcal{M}}_{g,n}} \psi_1^{\alpha_1} \cdots \psi_k^{\alpha_k - 1} \cdots \psi_n^{\alpha_n} \kappa_1^{m},$$

$$\sum_{j=0}^{m} (-1)^j \binom{m}{j} \int_{\overline{\mathcal{M}}_{g,n+1}} \psi_1^{\alpha_1} \cdots \psi_n^{\alpha_n} \psi_{n+1}^{j+1} \kappa_1^{m-j} = (2g - 2 + n) \int_{\overline{\mathcal{M}}_{g,n}} \psi_1^{\alpha_1} \cdots \psi_n^{\alpha_n} \kappa_1^{m}.$$

Observe that ψ_k on the left hand side refers to the psi-class on $\overline{\mathcal{M}}_{g,n+1}$ while on the right hand side it refers to the psi-class on $\overline{\mathcal{M}}_{g,n}$. These are generalisations of the string and dilaton equations — see Theorem A.1 — which correspond to the case $m = 0$. They may be proven using arguments from algebraic geometry [14] and were also obtained by Liu and Xu in their work on higher Weil–Petersson volumes [23]. The succinct statement of Theorem 5.1 indicates that the Weil–Petersson volume polynomial $V_{g,n}(\mathbf{L})$ provides a useful way to package intersection numbers on $\overline{\mathcal{M}}_{g,n}$.

One predicts yet another approach to the string and dilaton equations, which may prove to be the most interesting. A phenomenon often occurring in hyperbolic geometry is the fact that a purely imaginary length can be interpreted as an angle. As an example, consider the work of Tan, Wong and Zhang [42], in which they show that the generalised McShane identity — see Theorem 4.2 — holds for hyperbolic cone surfaces. In fact, one need only substitute $i\theta$ into the formula to represent a cone point with angle θ. It follows that one can extend the definition of the Weil–Petersson volume polynomials to the case of moduli spaces of hyperbolic cone surfaces. Thus, it is tempting to think of the string and dilaton equations as describing the Weil–Petersson volume and its derivative as one of the boundary components degenerates to a cone point with angle 2π and hence, is removable. Unfortunately, a proof of the string and dilaton equations following this intuition is yet to be formalised. One of the main obstacles is the fact that the Teichmüller theory for hyperbolic cone surfaces breaks down when cone points have angles larger than π. Indeed, on such surfaces, it ceases to be true that every homotopy class of closed curves contains a geodesic representative.

Note that Mirzakhani's recursion does not produce Weil–Petersson volumes of moduli spaces of closed hyperbolic surfaces. One application of Theorem 5.1 is the computation of these numbers. The following result is a direct corollary of the dilaton equation in the $n = 0$ case [14].

Corollary 5.2. *The Weil–Petersson volumes of moduli spaces of closed hyperbolic surfaces satisfy the formula*

$$V_{g,0} = \frac{V'_{g,1}(2\pi i)}{2\pi i(2g-2)}.$$

Another application of the string and dilaton equations is the computation of small genus Weil–Petersson volumes [14].

Proposition 5.3. *The string equation alone uniquely determines* $V_{0,n+1}(\mathbf{L}, L_{n+1})$ *from* $V_{0,n}(\mathbf{L})$ *and the string and dilaton equations together uniquely determine* $V_{1,n+1}(\mathbf{L}, L_{n+1})$ *from* $V_{1,n}(\mathbf{L})$.

The proof of Proposition 5.3 is elementary and can be converted to algorithms for the computation of $V_{0,n}(\mathbf{L})$ and $V_{1,n}(\mathbf{L})$. These are empirically more efficient than a direct implementation of Mirzakhani's recursion, which requires $V_{0,k}$ for $3 \leqslant k \leqslant n$ for the computation of $V_{0,n+1}$.

The string and dilaton equations for Weil–Petersson volumes relate the value and derivative of $V_{g,n+1}(\mathbf{L}, L_{n+1})$ evaluated at $L_{n+1} = 2\pi i$ to $V_{g,n}(\mathbf{L})$. Therefore, one might wonder whether there are similar expressions for higher derivatives. In fact, we have the following equation involving the second derivative, although it is in some sense equivalent to the string equation.

$$\frac{\partial^2 V_{g,n+1}}{\partial L_{n+1}^2}(\mathbf{L}, 2\pi i) = \sum_{k=1}^{n} L_k \frac{\partial V_{g,n}(\mathbf{L})}{\partial L_k} - (4g - 4 + 2n)\, V_{g,n}(\mathbf{L}).$$

There is reason to believe that such equations for higher derivatives simply do not exist. For example, see the work of Eynard and Orantin [15], which considers Weil–Petersson volumes as analogous to correlation functions arising from matrix models.

They predict string and dilaton equations for functions which emerge from a vast generalisation of Mirzakhani's recursion.

5.2. The large g limit

Mirzakhani's recursion can in theory be used to calculate all Weil–Petersson volumes. However, a direct implementation of the recursion yields a computer program which is practical only for small genus. Zograf has provided an alternative algorithm which is empirically much faster [53]. In particular, he has managed to gather enough numerical evidence to suggest two interesting conjectures involving Weil–Petersson volumes in the large g limit. In order to state the first, we use the following notation introduced by Mirzakhani [32] to express a certain normalisation of the coefficients of the polynomial $V_{g,n}(\mathbf{L})$.

$$[\tau_{\alpha_1}\tau_{\alpha_2}\cdots\tau_{\alpha_n}]_{g,n} = \frac{\prod 2^{2\alpha_k}(2\alpha_k+1)!!}{(3g-3+n-|\boldsymbol{\alpha}|)!}\int_{\overline{\mathcal{M}}_{g,n}} \psi_1^{\alpha_1}\psi_2^{\alpha_2}\cdots\psi_n^{\alpha_n}\,\omega^{3g-3+n-|\boldsymbol{\alpha}|}.$$

Proposition 5.4. *For a fixed tuple $(\alpha_1, \alpha_2, \ldots, \alpha_n)$ of non-negative integers,*

$$\lim_{g\to\infty} \frac{[\tau_{\alpha_1}\tau_{\alpha_2}\cdots\tau_{\alpha_n}]_{g,n}}{V_{g,n}(\mathbf{0})} = 1.$$

Although stated as a conjecture by Zograf [53], the result follows from certain Weil–Petersson volume estimates due to Mirzakhani [32]. These are obtained by rewriting her recursion — see Theorem 4.5 — in the following form.

$$\begin{aligned}
&[\tau_{\alpha_1}\tau_{\alpha_2}\cdots\tau_{\alpha_n}]_{g,n}\\
&=\frac{1}{2}\sum_{m=0}^{3g-3+n-|\alpha|}\ \sum_{i+j=\alpha_1+m-2} b_m\,[\tau_i\tau_j\tau_{\alpha_2}\tau_{\alpha_3}\cdots\tau_{\alpha_n}]_{g-1,n+1}\\
&+\frac{1}{2}\sum_{\substack{g_1+g_2=g\\ I\sqcup J=\{2,n\}}}\ \sum_{m=0}^{3g-3+n-|\alpha|}\ \sum_{i+j=\alpha_1+m-2} b_m\,[\tau_i\tau_{\alpha_I}]_{g_1,|I|+1}\,[\tau_j\tau_{\alpha_J}]_{g_2,|J|+1}\\
&+\sum_{k=2}^{n}\ \sum_{m=0}^{3g-3+n-|\alpha|} (2\alpha_k+1)\,b_m\,[\tau_{\alpha_2}\cdots\tau_{\alpha_k+\alpha_1+m-1}\cdots\tau_{\alpha_n}]_{g,n-1}.
\end{aligned}$$

Here, we have used the notation $\tau_{\alpha_I} = \tau_{\alpha_{i_1}}\tau_{\alpha_{i_2}}\cdots\tau_{\alpha_{i_m}}$ for $I = \{i_1, i_2, \ldots, i_m\}$. It is particularly useful to observe that the sequence $b_n = \zeta(2n)(1 - \frac{1}{2^{2n-1}})$ consists only of positive terms, is strictly increasing, and limits to the value of 1.

The second of Zograf's conjectures gives the asymptotic behaviour of $V_{g,n}(\mathbf{0})$.

Conjecture 5.5. *For a fixed non-negative integer n,*

$$V_{g,n}(\mathbf{0}) = \frac{1}{\sqrt{g\pi}}(4\pi^2)^{2g-3+n}(2g-3+n)!\left[1 + c_n g^{-1} + O(g^{-2})\right] \quad \textit{as} \quad g\to\infty.$$

This is consistent with the previously obtained bounds for Weil–Petersson volumes [16, 39, 40] which show the existence of positive constants C_1 and C_2 for a fixed non-negative integer n such that

$$C_1^g(2g)! < V_{g,n}(\mathbf{0}) < C_2^g(2g)!.$$

Although this conjecture is yet to be proven, Mirzakhani [32] has offered some further supporting evidence.

Theorem 5.6. *For a fixed non-negative integer* n,

$$\frac{V_{g,n+1}(\mathbf{0})}{2gV_{g,n}(\mathbf{0})} = 4\pi^2 + O(g^{-1}) \quad \textit{and} \quad \frac{V_{g,n}(\mathbf{0})}{V_{g-1,n+2}(\mathbf{0})} = 1 + O(g^{-1}) \quad \textit{as} \quad g \to \infty.$$

Mirzakhani has used these and other Weil–Petersson volume estimates in the large g limit to investigate the geometric properties of random hyperbolic surfaces [32]. In particular, she has obtained results concerning the length of the shortest simple closed geodesic, the diameter, and the Cheeger constant of a random surface with large genus, chosen with respect to the Weil–Petersson measure.

It is worth remarking that the $g \to \infty$ limit for fixed n is rather distinct from the $n \to \infty$ limit for fixed g. The latter case was investigated by Manin and Zograf, who obtain generating functions for the Weil–Petersson volumes $V_{g,n}(\mathbf{0})$ and prove the following result [26].

Theorem 5.7. *There exist constants* $a_0, a_1, a_2, \ldots$ *and* C *such that, for a fixed non-negative integer* g,

$$V_{g,n}(\mathbf{0}) = n!C^n n^{(5g-7)/2}\left[a_g + O(n^{-1})\right] \quad \textit{as} \quad n \to \infty.$$

5.3. The asymptotic Weil–Petersson form

If we are only interested in psi-class intersection numbers on $\overline{\mathcal{M}}_{g,n}$, then we need only look at the top degree part of the polynomial $V_{g,n}(\mathbf{L})$. This observation leads us to consider the following asymptotics of the Weil–Petersson volume for a fixed value of $\mathbf{x} = (x_1, x_2, \ldots, x_n)$.

$$(5.8) \qquad \lim_{N\to\infty} \frac{V_{g,n}(N\mathbf{x})}{N^{6g-6+2n}} = \sum_{|\alpha|=3g-3+n} \frac{\int_{\overline{\mathcal{M}}_{g,n}} \psi_1^{\alpha_1}\psi_2^{\alpha_2}\cdots\psi_n^{\alpha_n}}{2^{3g-3+n}\alpha_1!\alpha_2!\cdots\alpha_n!} x_1^{2\alpha_1}x_2^{2\alpha_2}\cdots x_n^{2\alpha_n}.$$

One way to access the asymptotics of the Weil–Petersson volume is via the following map on the combinatorial moduli space.

$$f : \mathcal{MRG}_{g,n}(\mathbf{x}) \to \mathcal{MRG}_{g,n}(N\mathbf{x}) \to \mathcal{M}_{g,n}(N\mathbf{x}).$$

This homeomorphism of orbifolds is the composition of two maps — the first scales the ribbon graph metric by N while the second uses the Bowditch–Epstein construction described in Section 2.4. In one direction, this construction associates to a hyperbolic surface with boundary its spine — in other words, the set of points which have at least two equal shortest paths to the boundary. The inverse of this construction produces a hyperbolic surface $S(\Gamma) \in \mathcal{M}_{g,n}(\mathbf{L})$ for every metric ribbon graph $\Gamma \in \mathcal{MRG}_{g,n}(\mathbf{L})$.

The normalised Weil–Petersson form $\frac{\omega}{N^2}$ on $\mathcal{M}_{g,n}(N\mathbf{x})$ pulls back via f to a symplectic form on the combinatorial moduli space. We will be interested in the limiting behaviour of this symplectic form since we may alternatively express the asymptotics of the Weil–Peterson volume in the following way.

$$\begin{aligned}\lim_{N\to\infty} \frac{V_{g,n}(N\mathbf{x})}{N^{6g-6+2n}} &= \frac{1}{(3g-3+n)!} \lim_{N\to\infty} \int_{\mathcal{M}_{g,n}(N\mathbf{x})} \left(\frac{\omega}{N^2}\right)^{3g-3+n} \\ &= \frac{1}{(3g-3+n)!} \int_{\mathcal{MRG}_{g,n}(\mathbf{x})} \left(\lim_{N\to\infty} \frac{f^*\omega}{N^2}\right)^{3g-3+n} \\ &= \frac{1}{(3g-3+n)!} \sum_{\Gamma} \int_{\mathcal{MRG}_\Gamma(\mathbf{x})} \left(\lim_{N\to\infty} \frac{f^*\omega}{N^2}\right)^{3g-3+n}.\end{aligned}$$

To obtain the second line from the first, we pull back the integral to the combinatorial moduli space and invoke the Lebesgue dominated convergence theorem to move the limit inside the integral. To obtain the third line from the second, we use the orbifold cell decomposition of the combinatorial moduli space described in Section 2.4. Recall that the combinatorial moduli space $\mathcal{MRG}_{g,n}(\mathbf{x})$ is the disjoint union of orbifold cells $\mathcal{MRG}_\Gamma(\mathbf{x})$, where Γ ranges over the ribbon graphs of type (g,n). Here, the sum is only over the set of trivalent ribbon graphs of type (g,n), since these correspond precisely to the open cells of this decomposition.

The previous discussion suggests that we should study the asymptotic behaviour of the Weil–Petersson form. In order to do this, fix a trivalent ribbon graph Γ of type (g,n) and label its edges from 1 up to $6g-6+3n$. As noted in Section 2.4, the lengths of these edges $e_1, e_2, \ldots, e_{6g-6+3n}$ provide a set of natural coordinates on $\mathcal{MRG}_\Gamma(\mathbf{x})$ and we can write

$$\mathcal{MRG}_\Gamma(\mathbf{x}) \cong \left\{\mathbf{e} \in \mathbb{R}_+^{6g-6+3n} \;\middle|\; A_\Gamma \mathbf{e} = \mathbf{x}\right\} \Big/ \operatorname{Aut}(\Gamma).$$

Here, A_Γ is the linear map which represents the adjacency between faces and edges in the cell decomposition corresponding to Γ.

Theorem 5.9. *In the $N \to \infty$ limit, the symplectic form $\frac{f^*\omega}{N^2}$ converges pointwise to a 2-form Ω on $\mathcal{MRG}_\Gamma(\mathbf{x})$. There exists a $(6g-6+2n) \times (6g-6+2n)$ skew-symmetric matrix B_Γ such that, after an appropriate permutation of the edge labels,*

$$\Omega = \sum_{1 \leqslant i < j \leqslant 6g-6+2n} [B_\Gamma]_{ij} \, de_i \wedge de_j.$$

We can use this theorem to write the asymptotics of the Weil–Petersson volume in the following way, where $\operatorname{pf}(B_\Gamma)$ denotes the Pfaffian of B_Γ.

$$\lim_{N\to\infty} \frac{V_{g,n}(N\mathbf{x})}{N^{6g-6+2n}} = \sum_{\Gamma} \frac{\operatorname{pf}(B_\Gamma)}{|\operatorname{Aut}(\Gamma)|} \int_{A_\Gamma \mathbf{e} = \mathbf{x}} de_1 \wedge de_2 \wedge \cdots \wedge de_{6g-6+2n}. \tag{5.10}$$

Now we can equate the expressions appearing in Equations (5.8) and (5.10). The Pfaffian and integral in Equation (5.10) can be calculated explicitly in terms of the combinatorics of Γ. Upon doing so and taking the Laplace transform of both sides, we recover the following identity.

Theorem 5.11 (Kontsevich's combinatorial formula). *For $2g - 2 + n > 0$, we have the following equality of rational polynomials in $s_1, s_2, \ldots, s_n$.*

$$\sum_{|\alpha|=3g-3+n} \int_{\overline{\mathcal{M}}_{g,n}} \psi_1^{\alpha_1} \psi_2^{\alpha_2} \cdots \psi_n^{\alpha_n} \prod_{k=1}^{n} \frac{(2\alpha_k - 1)!!}{s_k^{2\alpha_k+1}} = \sum_{\Gamma} \frac{2^{2g-2+n}}{|\mathrm{Aut}(\Gamma)|} \prod_{e \in E(\Gamma)} \frac{1}{s_{\ell(e)} + s_{r(e)}}.$$

The sum on the right hand side is over trivalent ribbon graphs of type (g, n). For an edge e, the expressions $\ell(e)$ and $r(e)$ denote the labels of the faces on its left and right.[4]

This is the main identity used by Kontsevich in his proof of the Witten–Kontsevich theorem. At first blush, the equality seems rather surprising, since the left hand side is manifestly polynomial in $\frac{1}{s_1}, \frac{1}{s_2}, \ldots, \frac{1}{s_n}$ while the right hand side is not. Our proof of this result highlights the close relationship between the combinatorial methods pioneered by Kontsevich [22] and the hyperbolic geometry used by Mirzakhani [29, 30].

It is worth making a few remarks on the proof of Theorem 5.9. The result appears implicitly in the work of Mondello [33] although we will discuss an alternative proof which is less computational in nature [13]. Underlying this work is the observation that a hyperbolic surface with large boundary lengths resembles a ribbon graph after appropriately scaling the hyperbolic metric. In order to make this statement precise, take a metric ribbon graph Γ and consider the surface $\frac{1}{N}S(N\Gamma)$ for large values of N. Here, we use the notation λX to denote the result of scaling the metric on X by a positive real number λ. By the Gauss–Bonnet theorem, the area of the surface $\frac{1}{N}S(N\Gamma)$ goes to zero as N increases to infinity. On the other hand, the length of the boundaries remains fixed. So in the $N \to \infty$ limit, one expects the entire surface to collapse onto the spine Γ. The following result formalises this intuition in a precise way [13].

Theorem 5.12. *In the Gromov–Hausdorff topology, for every metric ribbon graph Γ, we have*

$$\lim_{N \to \infty} \frac{1}{N} S(N\Gamma) = \Gamma.$$

The intuitive observation behind this result suggests a great deal about the geometry of a hyperbolic surface with large boundary lengths. For example, one expects the length of a closed geodesic on $\frac{1}{N}S(N\Gamma)$ to converge to a sum of lengths of edges in Γ. Furthermore, one expects the acute angle at which two closed geodesics meet to converge to 0. We recover the limiting behaviour of the Weil–Petersson form

---

[4] Although the left and right of an edge are not well-defined, the expression $s_{\ell(e)} + s_{r(e)}$ certainly is.

from the limiting behaviour of such lengths and angles via the following result of Wolpert [51].

Proposition 5.13. *Let $C_1, C_2, \ldots, C_{6g-6+2n}$ be simple closed geodesics with lengths $\ell_1, \ell_2, \ldots, \ell_{6g-6+2n}$ in a hyperbolic surface $S \in \mathcal{M}_{g,n}(\mathbf{L})$. If C_i and C_j meet at a point p for $i < j$, denote by θ_p the angle between the curves, measured anticlockwise from C_i to C_j. Define the $(6g-6+2n) \times (6g-6+2n)$ skew-symmetric matrix X by the formula*

$$X_{ij} = \sum_{p \in C_i \cap C_j} \cos\theta_p \qquad \text{for } i < j.$$

If X is invertible, then $\ell_1, \ell_2, \ldots, \ell_{6g-6+2n}$ are local coordinates at $S \in \mathcal{M}_{g,n}(\mathbf{L})$ and the Weil–Petersson form is given by

$$\omega = -\sum_{i<j} [X^{-1}]_{ij}\, d\ell_i \wedge d\ell_j.$$

A judicious choice of curves allows us to use this result to obtain Theorem 5.9, including a concrete description of the matrix B_Γ in terms of the combinatorics of the ribbon graph Γ. Furthermore, one finds that the asymptotic Weil–Petersson form Ω coincides with the 2-form on the combinatorial moduli space introduced by Kontsevich in his proof of the Witten–Kontsevich theorem [22].

A. Intersection theory on moduli spaces of curves

The reader will find a wealth of information concerning moduli spaces of curves and their intersection theory elsewhere in the literature [18, 46, 47]. The aim of this appendix is to provide a concise exposition of the topic in order to keep this article reasonably self-contained.

For non-negative integers g and n satisfying the Euler characteristic condition $2g-2+n>0$, define the *moduli space of curves* as follows.

$$\mathcal{M}_{g,n} = \left\{ (C, p_1, p_2, \ldots, p_n) \;\middle|\; \begin{array}{l} C \text{ is a smooth algebraic curve with genus } g \\ \text{and } n \text{ distinct points } p_1, p_2, \ldots, p_n \end{array} \right\} \Big/ \sim$$

Here, $(C, p_1, p_2, \ldots, p_n) \sim (D, q_1, q_2, \ldots, q_n)$ if and only if there exists an isomorphism from C to D which sends p_k to q_k for all k. It is often more natural to work with the *Deligne–Mumford compactification* of the moduli space of curves.

$$\overline{\mathcal{M}}_{g,n} = \left\{ (C, p_1, p_2, \ldots, p_n) \;\middle|\; \begin{array}{l} C \text{ is a stable algebraic curve with genus } g \\ \text{and } n \text{ distinct smooth points } p_1, p_2, \ldots, p_n \end{array} \right\} \Big/ \sim$$

Again, $(C, p_1, p_2, \ldots, p_n) \sim (D, q_1, q_2, \ldots, q_n)$ if and only if there exists an isomorphism from C to D which sends p_k to q_k for all k. An algebraic curve is called *stable* if it has at worst nodal singularities and a finite automorphism group. The practical interpretation of this latter condition is that the normalisation of every rational component must have at least three distinguished points which are nodes or labelled points. One of the virtues of the Deligne–Mumford compactification amongst the various competing options is the fact that it is modular — in other words, each point in $\overline{\mathcal{M}}_{g,n}$ represents an algebraic curve. The set $\overline{\mathcal{M}}_{g,n}$ possesses a rich geometric structure and is an example of a Deligne–Mumford stack, although one will not go too far wrong thinking of it as a complex orbifold.

A natural approach to understanding the structure of geometric spaces is through algebraic invariants, such as homology and cohomology. And so it is with moduli spaces of curves, but for the fact that its full cohomology ring is notoriously intractable in general. However, a great deal of progress can be made by calculating intersection numbers with respect to certain characteristic classes. The classes that we consider live in the cohomology ring $H^*(\overline{\mathcal{M}}_{g,n};\mathbb{Q})$ and arise from taking Chern classes of natural complex vector bundles.[5] One obtains a more natural theory using rational, rather than integral, coefficients for cohomology due to the orbifold nature of $\overline{\mathcal{M}}_{g,n}$.

Given a stable genus g curve with $n+1$ labelled points, one can forget the point labelled $n+1$ to obtain a genus g curve with n labelled points. The resulting curve may not be stable, but gives rise to a well-defined stable curve after contracting all unstable rational components to a point. This yields a map $\pi : \overline{\mathcal{M}}_{g,n+1} \to \overline{\mathcal{M}}_{g,n}$ known as the *forgetful morphism*, which can be interpreted as the universal family over $\overline{\mathcal{M}}_{g,n}$. Thus, given a pair (C, p) consisting of a stable curve $C \in \overline{\mathcal{M}}_{g,n}$ and a point p on the curve, it is possible to associate to it a unique stable curve $D \in \overline{\mathcal{M}}_{g,n+1}$ such that $\pi(D) = C$. In particular, the fibre over $C \in \overline{\mathcal{M}}_{g,n}$ is essentially the stable curve corresponding to C. So the point labelled k defines a section $\sigma_k : \overline{\mathcal{M}}_{g,n} \to \overline{\mathcal{M}}_{g,n+1}$ for $k = 1, 2, \ldots, n$. The forgetful morphism can be used to pull back cohomology classes, but it will also be useful to push them forward. This is possible via the Gysin map $\pi_* : H^*(\overline{\mathcal{M}}_{g,n+1};\mathbb{Q}) \to H^*(\overline{\mathcal{M}}_{g,n};\mathbb{Q})$, the homomorphism of graded rings with grading -2 which represents integration along fibres.

Consider the vertical cotangent bundle on $\overline{\mathcal{M}}_{g,n+1}$ whose fibre at the point associated to the pair (C, p) is equal to the cotangent line T_p^*C. Unfortunately, this definition is nonsensical when p is a singular point of C. Therefore, it is necessary to consider the relative dualising sheaf, the unique line bundle on $\overline{\mathcal{M}}_{g,n+1}$ which extends the vertical cotangent bundle. More precisely, it can be defined as $\mathcal{L} = \mathcal{K}_X \otimes \pi^*\mathcal{K}_B^{-1}$, where $\mathcal{K}_X$ denotes the canonical line bundle on $\overline{\mathcal{M}}_{g,n+1}$ and $\mathcal{K}_B$ denotes the canonical line bundle on $\overline{\mathcal{M}}_{g,n}$. Sections of $\mathcal{L}$ correspond to meromorphic 1-forms with at worst simple poles allowed at the nodes which also satisfy the condition that the two residues at the preimages of each node under normalisation must sum to zero.

The tautological line bundles on $\overline{\mathcal{M}}_{g,n}$ are formed by pulling back $\mathcal{L}$ along the sections σ_k for $k = 1, 2, \ldots, n$. Taking Chern classes of these line bundles, we obtain

[5] Readers with a more algebraic predilection may prefer to think of these classes as living in the Chow ring $A^*(\overline{\mathcal{M}}_{g,n};\mathbb{Q})$.

the *psi-classes*

$$\psi_k = c_1(\sigma_k^* \mathcal{L}) \in H^2(\overline{\mathcal{M}}_{g,n}; \mathbb{Q}) \qquad \text{for } k = 1, 2, \ldots, n.$$

Define the twisted Euler class by $e = c_1(\mathcal{L}(D_1 + D_2 + \cdots + D_n))$, where D_k is the divisor on $\overline{\mathcal{M}}_{g,n+1}$ representing the image of the section σ_k. Taking the pushforwards of its powers, we obtain the *Mumford–Morita–Miller classes*

$$\kappa_m = \pi_*(e^{m+1}) \in H^{2m}(\overline{\mathcal{M}}_{g,n}; \mathbb{Q}) \qquad \text{for } m = 0, 1, 2, \ldots, 3g - 3 + n.$$

A great deal of attention has been paid to the subring of $H^*(\overline{\mathcal{M}}_{g,n}; \mathbb{Q})$ known as the *tautological ring*. It has the benefit of being more tractable than the full cohomology ring and possessing a rich combinatorial structure, while still containing all known classes of geometric interest. Any top intersections in the tautological ring can be determined from the top intersections of psi-classes alone. Thus, we are motivated to study intersection numbers of the form

$$\langle \tau_{\alpha_1} \tau_{\alpha_2} \cdots \tau_{\alpha_n} \rangle = \int_{\overline{\mathcal{M}}_{g,n}} \psi_1^{\alpha_1} \psi_2^{\alpha_2} \cdots \psi_n^{\alpha_n} \in \mathbb{Q},$$

where $|\boldsymbol{\alpha}| = 3g - 3 + n$ or equivalently, $g = \frac{1}{3}(|\boldsymbol{\alpha}| - n + 3)$. The angle bracket notation above — originally introduced by Witten — suppresses the genus and encodes the symmetry between the psi-classes. We treat the τ variables as commuting, so that we can write intersection numbers in the form $\langle \tau_0^{d_0} \tau_1^{d_1} \tau_2^{d_2} \cdots \rangle$ and we set $\langle \tau_{\alpha_1} \tau_{\alpha_2} \cdots \tau_{\alpha_n} \rangle = 0$ if $n = 0$ or if the genus $g = \frac{1}{3}(|\boldsymbol{\alpha}| - n + 3)$ is non-integral or negative. In this way, we have defined a linear functional $\langle \cdot \rangle : \mathbb{Q}[\tau_0, \tau_1, \tau_2, \ldots] \to \mathbb{Q}$. The psi-class intersection numbers contain a great deal of structure, as evidenced by the following result [48].

Proposition A.1 (String and dilaton equations). *For $2g - 2 + n > 0$, the psi-class intersection numbers satisfy the following relations.*

$$\langle \tau_0 \tau_{\alpha_1} \tau_{\alpha_2} \cdots \tau_{\alpha_n} \rangle = \sum_{k=1}^{n} \langle \tau_{\alpha_1} \cdots \tau_{\alpha_k - 1} \cdots \tau_{\alpha_n} \rangle,$$

$$\langle \tau_1 \tau_{\alpha_1} \tau_{\alpha_2} \cdots \tau_{\alpha_n} \rangle = (2g - 2 + n) \langle \tau_{\alpha_1} \tau_{\alpha_2} \cdots \tau_{\alpha_n} \rangle.$$

One of the landmark results concerning intersection theory on moduli spaces of curves is Witten's conjecture, now Kontsevich's theorem. In conjunction with the string equation, it allows us to calculate any psi-class intersection number from the base case $\langle \tau_0^3 \rangle = 1$. In order to precisely describe the result, we let $\mathbf{t} = (t_0, t_1, t_2, \ldots)$ and $\boldsymbol{\tau} = (\tau_0, \tau_1, \tau_2, \ldots)$ and consider the generating function $F(\mathbf{t}) = \langle \exp(\mathbf{t} \cdot \boldsymbol{\tau}) \rangle$. Here, the expression is to be expanded using multilinearity in the variables $t_0, t_1, t_2, \ldots$. Equivalently, we may define

$$F(t_0, t_1, t_2, \ldots) = \sum_{\mathbf{d}} \prod_{k=0}^{\infty} \frac{t_k^{d_k}}{d_k!} \langle \tau_0^{d_0} \tau_1^{d_1} \tau_2^{d_2} \cdots \rangle,$$

where the summation is over all sequences $\mathbf{d} = (d_0, d_1, d_2, \ldots)$ of non-negative integers with finitely many non-zero terms. In his foundational paper [48], Witten argued on physical grounds that the formal series $U = \frac{\partial^2 F}{\partial t_0^2}$ satisfies the KdV hierarchy of partial differential equations. This is the prototypical example of an exactly solvable model, whose soliton solutions have attracted tremendous mathematical interest over the past few decades. More explicitly, the Witten–Kontsevich theorem can be stated in the following way.

Theorem A.2 (Witten–Kontsevich theorem). *The generating function F satisfies the following partial differential equation for every non-negative integer n.*

$$(2n+1)\frac{\partial^3 F}{\partial t_n \partial t_0^2} = \left(\frac{\partial^2 F}{\partial t_{n-1}\partial t_0}\right)\left(\frac{\partial^3 F}{\partial t_0^3}\right) + 2\left(\frac{\partial^3 F}{\partial t_{n-1}\partial t_0^2}\right)\left(\frac{\partial^2 F}{\partial t_0^2}\right) + \frac{1}{4}\frac{\partial^5 F}{\partial t_{n-1}\partial t_0^4}.$$

An equivalent formulation of the Witten–Kontsevich theorem states that the Virasoro operators annihilate the generating function $\exp F$. These operators span the Virasoro Lie algebra and are defined by

$$L_{-1} = -\frac{\partial}{\partial t_0} + \frac{1}{2}t_0^2 + \sum_{k=0}^{\infty} t_{k+1}\frac{\partial}{\partial t_k}, \qquad L_0 = -\frac{3}{2}\frac{\partial}{\partial t_1} + \frac{1}{2}\sum_{k=0}^{\infty}(2k+1)t_k\frac{\partial}{\partial t_k} + \frac{1}{16},$$

and for positive integers n,

$$L_n = -\frac{(2n+3)!!}{2^{n+1}}\frac{\partial}{\partial t_{n+1}} + \frac{1}{2^{n+1}}\sum_{k=0}^{\infty}\frac{(2k+2n+1)!!}{(2k-1)!!}t_k\frac{\partial}{\partial t_{n+k}}$$
$$+ \frac{1}{2^{n+2}}\sum_{i+j=n-1}(2i+1)!!(2j+1)!!\frac{\partial^2}{\partial t_i \partial t_j}.$$

There now exist several proofs of the Witten–Kontsevich theorem, due to Kontsevich [22], Okounkov and Pandharipande [36], Kim and Liu [21], Kazarian and Lando [20], and Mirzakhani [30]. That there are so many proofs, each with their own distinct flavour, is testament to the importance and richness of the result.

B. Table of Weil–Petersson volumes

The following table shows some examples of Weil–Petersson volumes. We use the notation $m_{(\alpha_1,\alpha_2,\ldots,\alpha_k)}$ to denote the monomial symmetric polynomial

$$\sum_{(\beta_1,\beta_2,\ldots,\beta_n)} L_1^{2\beta_1}L_2^{2\beta_2}\cdots L_n^{2\beta_n},$$

where the summation ranges over all permutations of $(\alpha_1, \alpha_2, \ldots, \alpha_k, 0, 0, \ldots, 0)$. For example, we have the following when $n = 3$.

$$m_{(3,2,1)} = L_1^6L_2^4L_3^2 + L_1^6L_2^2L_3^4 + L_1^4L_2^6L_3^2 + L_1^4L_2^2L_3^6 + L_1^2L_2^6L_3^4 + L_1^2L_2^4L_3^6,$$
$$m_{(2)} = L_1^4 + L_2^4 + L_3^4,$$
$$m_{(1,1,1)} = L_1^2L_2^2L_3^2.$$

g	n	$V_{g,n}(L_1, L_2, \ldots, L_n)$
0	3	1
	4	$\frac{1}{2}m_{(1)} + 2\pi^2$
	5	$\frac{1}{8}m_{(2)} + \frac{1}{2}m_{(1,1)} + 3\pi^2 m_{(1)} + 10\pi^4$
	6	$\frac{1}{48}m_{(3)} + \frac{3}{16}m_{(2,1)} + \frac{3}{4}m_{(1,1,1)} + \frac{3\pi^2}{2}m_{(2)} + 6\pi^2 m_{(1,1)} + 26\pi^4 m_{(1)} + \frac{244\pi^6}{3}$
	7	$\frac{1}{384}m_{(4)} + \frac{1}{24}m_{(3,1)} + \frac{3}{32}m_{(2,2)} + \frac{3}{8}m_{(2,1,1)} + \frac{3}{2}m_{(1,1,1,1)} + \frac{5\pi^2}{12}m_{(3)}$
		$+\frac{15\pi^2}{4}m_{(2,1)} + 15\pi^2 m_{(1,1,1)} + 20\pi^4 m_{(2)} + 80\pi^4 m_{(1,1)} + \frac{910\pi^6}{3}m_{(1)} + \frac{2758\pi^8}{3}$
1	1	$\frac{1}{48}m_{(1)} + \frac{\pi^2}{12}$
	2	$\frac{1}{192}m_{(2)} + \frac{1}{96}m_{(1,1)} + \frac{\pi^2}{12}m_{(1)} + \frac{\pi^4}{4}$
	3	$\frac{1}{1152}m_{(3)} + \frac{1}{192}m_{(2,1)} + \frac{1}{96}m_{(1,1,1)} + \frac{\pi^2}{24}m_{(2)} + \frac{\pi^2}{8}m_{(1,1)} + \frac{13\pi^4}{24}m_{(1)} + \frac{14\pi^6}{9}$
	4	$\frac{1}{9216}m_{(4)} + \frac{1}{768}m_{(3,1)} + \frac{1}{384}m_{(2,2)} + \frac{1}{128}m_{(2,1,1)} + \frac{1}{64}m_{(1,1,1,1)} + \frac{7\pi^2}{576}m_{(3)}$
		$+\frac{\pi^2}{12}m_{(2,1)} + \frac{\pi^2}{4}m_{(1,1,1)} + \frac{41\pi^4}{96}m_{(2)} + \frac{17\pi^4}{12}m_{(1,1)} + \frac{187\pi^6}{36}m_{(1)} + \frac{529\pi^8}{36}$
	5	$\frac{1}{92160}m_{(5)} + \frac{1}{4608}m_{(4,1)} + \frac{7}{9216}m_{(3,2)} + \frac{1}{384}m_{(3,1,1)} + \frac{1}{192}m_{(2,2,1)}$
		$+\frac{1}{64}m_{(2,1,1,1)} + \frac{1}{32}m_{(1,1,1,1,1)} + \frac{11\pi^2}{4608}m_{(4)} + \frac{35\pi^2}{1152}m_{(3,1)} + \frac{\pi^2}{16}m_{(2,2)}$
		$+\frac{5\pi^2}{24}m_{(2,1,1)} + \frac{5\pi^2}{8}m_{(1,1,1,1)} + \frac{13\pi^4}{72}m_{(3)} + \frac{253\pi^4}{192}m_{(2,1)} + \frac{35\pi^4}{8}m_{(1,1,1)}$
		$+\frac{809\pi^6}{144}m_{(2)} + \frac{703\pi^6}{36}m_{(1,1)} + \frac{4771\pi^8}{72}m_{(1)} + \frac{16751\pi^{10}}{90}$
2	0	$\frac{43\pi^6}{2160}$
	1	$\frac{1}{442368}m_{(4)} + \frac{29\pi^2}{138240}m_{(3)} + \frac{139\pi^4}{23040}m_{(2)} + \frac{169\pi^6}{2880}m_{(1)} + \frac{29\pi^8}{192}$
	2	$\frac{1}{4423680}m_{(5)} + \frac{1}{294912}m_{(4,1)} + \frac{29}{2211840}m_{(3,2)} + \frac{11\pi^2}{276480}m_{(4)} + \frac{29\pi^2}{69120}m_{(3,1)}$
		$+\frac{7\pi^2}{7680}m_{(2,2)} + \frac{19\pi^4}{7680}m_{(3)} + \frac{181\pi^4}{11520}m_{(2,1)} + \frac{551\pi^6}{8640}m_{(2)} + \frac{7\pi^6}{36}m_{(1,1)}$
		$+\frac{1085\pi^8}{1728}m_{(1)} + \frac{787\pi^{10}}{480}$
	3	$\frac{1}{53084160}m_{(6)} + \frac{1}{2211840}m_{(5,1)} + \frac{11}{4423680}m_{(4,2)} + \frac{1}{147456}m_{(4,1,1)}$
		$+\frac{29}{6635520}m_{(3,3)} + \frac{29}{1105920}m_{(3,2,1)} + \frac{7}{122880}m_{(2,2,2)} + \frac{\pi^2}{172800}m_{(5)}$
		$+\frac{11\pi^2}{110592}m_{(4,1)} + \frac{5\pi^2}{13824}m_{(3,2)} + \frac{29\pi^2}{27648}m_{(3,1,1)} + \frac{7\pi^2}{3072}m_{(2,2,1)} + \frac{41\pi^4}{61440}m_{(4)}$
		$+\frac{211\pi^4}{27648}m_{(3,1)} + \frac{37\pi^4}{2304}m_{(2,2)} + \frac{223\pi^4}{4608}m_{(2,1,1)} + \frac{77\pi^6}{2160}m_{(3)} + \frac{827\pi^6}{3456}m_{(2,1)}$
		$+\frac{419\pi^6}{576}m_{(1,1,1)} + \frac{30403\pi^8}{34560}m_{(2)} + \frac{611\pi^8}{216}m_{(1,1)} + \frac{75767\pi^{10}}{8640}m_{(1)} + \frac{1498069\pi^{12}}{64800}$
3	0	$\frac{176557\pi^{12}}{1209600}$
	1	$\frac{1}{53508833280}m_{(7)} + \frac{77\pi^2}{9555148800}m_{(6)} + \frac{3781\pi^4}{2786918400}m_{(5)} + \frac{47209\pi^6}{418037760}m_{(4)}$
		$+\frac{127189\pi^8}{26127360}m_{(3)} + \frac{8983379\pi^{10}}{87091200}m_{(2)} + \frac{8497697\pi^{12}}{9331200}m_{(1)} + \frac{9292841\pi^{14}}{4082400}$
	2	$\frac{1}{856141332480}m_{(8)} + \frac{1}{21403533312}m_{(7,1)} + \frac{77}{152882380800}m_{(6,2)}$
		$+\frac{503}{267544166400}m_{(5,3)} + \frac{607}{214035333120}m_{(4,4)} + \frac{17\pi^2}{22295347200}m_{(7)}$
		$+\frac{77\pi^2}{3185049600}m_{(6,1)} + \frac{17\pi^2}{88473600}m_{(5,2)} + \frac{1121\pi^2}{2229534720}m_{(4,3)} + \frac{1499\pi^4}{7431782400}m_{(6)}$
		$+\frac{899\pi^4}{185794560}m_{(5,1)} + \frac{10009\pi^4}{371589120}m_{(4,2)} + \frac{191\pi^4}{4128768}m_{(3,3)} + \frac{3859\pi^6}{139345920}m_{(5)}$
		$+\frac{33053\pi^6}{69672960}m_{(4,1)} + \frac{120191\pi^6}{69672960}m_{(3,2)} + \frac{195697\pi^8}{92897280}m_{(4)} + \frac{110903\pi^8}{4644864}m_{(3,1)}$
		$+\frac{6977\pi^8}{138240}m_{(2,2)} + \frac{37817\pi^{10}}{430080}m_{(3)} + \frac{2428117\pi^{10}}{4147200}m_{(2,1)} + \frac{5803333\pi^{12}}{3110400}m_{(2)}$
		$+\frac{18444319\pi^{12}}{3110400}m_{(1,1)} + \frac{20444023\pi^{14}}{1209600}m_{(1)} + \frac{2800144027\pi^{16}}{65318400}$

g	n	$V_{g,n}(L_1, L_2, \ldots, L_n)$
4	0	$\frac{1959225867017\pi^{18}}{493807104000}$
	1	$\frac{1}{2958824450508800}\mathfrak{m}_{(10)} + \frac{149\pi^2}{3698530556313600}\mathfrak{m}_{(9)} + \frac{48689\pi^4}{2397195730944000}\mathfrak{m}_{(8)} + \frac{50713\pi^6}{8989483991040}\mathfrak{m}_{(7)} + \frac{30279589\pi^8}{32105299968000}\mathfrak{m}_{(6)} + \frac{43440449\pi^{10}}{445906944000}\mathfrak{m}_{(5)} + \frac{274101371\pi^{12}}{44590694400}\mathfrak{m}_{(4)} + \frac{66210015481\pi^{14}}{292626432000}\mathfrak{m}_{(3)} + \frac{221508280867\pi^{16}}{50164531200}\mathfrak{m}_{(2)} + \frac{74706907467169\pi^{18}}{1975228416000}\mathfrak{m}_{(1)} + \frac{92480712720869\pi^{20}}{987614208000}$
5	0	$\frac{84374265930915479\pi^{24}}{355541114880000}$
	1	$\frac{1}{487424903779901767680000}\mathfrak{m}_{(13)} + \frac{7\pi^2}{1339079405988741120000}\mathfrak{m}_{(12)} + \frac{1823\pi^4}{3106765667303424000}\mathfrak{m}_{(11)} + \frac{296531\pi^6}{7766914168258560000}\mathfrak{m}_{(10)} + \frac{68114707\pi^8}{4271802792542208000}\mathfrak{m}_{(9)} + \frac{2123300941\pi^{10}}{4746447547269120000}\mathfrak{m}_{(8)} + \frac{42408901133\pi^{12}}{494421619507200000}\mathfrak{m}_{(7)} + \frac{19817320001\pi^{14}}{176579149824000}\mathfrak{m}_{(6)} + \frac{11171220559409\pi^{16}}{113515167744000}\mathfrak{m}_{(5)} + \frac{62028372646367\pi^{18}}{111244864389120}\mathfrak{m}_{(4)} + \frac{202087901261599\pi^{20}}{10534551552000}\mathfrak{m}_{(3)} + \frac{626693680890100121\pi^{22}}{1738201006080000}\mathfrak{m}_{(2)} + \frac{881728936440038779\pi^{24}}{289700167680000}\mathfrak{m}_{(1)} + \frac{2118524149898372944 1\pi^{26}}{2824576634880000}$

References

[1] W. Abikoff. The real analytic theory of Teichmüller space. *Lecture Notes in Mathematics*, **820**, Springer, Berlin, 1980. ← 226

[2] H. Akiyoshi, H. Miyachi, and M. Sakuma. Variations of McShane's identity for punctured surface groups. In: Spaces of Kleinian groups, 151–185. *London Math. Soc. Lecture Note Ser.*, **329**, Cambridge Univ. Press, Cambridge, 2006. ← 237

[3] A. Basmajian. The orthogonal spectrum of a hyperbolic manifold. *Amer. J. Math.*, **115** (5), 1139–1159, 1993. ← 238

[4] L. Bers. An inequality for Riemann surfaces. In: *Differential geometry and complex analysis*, 87–93, Springer, Berlin, 1985. ← 229

[5] J. S. Birman and C. Series. Geodesics with bounded intersection number on surfaces are sparsely distributed. *Topology*, **24** (2), 217–225, 1985. ← 236

[6] M. Boggi and M. Pikaart. Galois covers of moduli of curves. *Compositio Math.*, **120** (2), 171–191, 2000. ← 224

[7] B. H. Bowditch. A proof of McShane's identity via Markoff triples. *Bull. London Math. Soc.*, **28**(1), 73–78, 1996. ← 237

[8] B. H. Bowditch and D. B. A. Epstein. Natural triangulations associated to a surface. *Topology*, **27** (1), 91–117, 1988. ← 228

[9] M. Bridgeman and J. Kahn. Hyperbolic volume of manifolds with geodesic boundary and orthospectra. *Geom. Funct. Anal.*, **20** (5), 1210–1230, 2010. ← 238

[10] P. Buser. Geometry and spectra of compact Riemann surfaces. *Progress in Mathematics*, **106**, Birkhäuser Boston Inc., Boston, MA, 1992. ← 221, 237

[11] D. Calegari. Chimneys, leopard spots and the identities of Basmajian and Bridgeman. *Algebr. Geom. Topol.*, **10** (3), 1857–1863, 2010. ← 238

[12] N. Do. *Intersection theory on moduli spaces of curves via hyperbolic geometry*. PhD thesis, The University of Melbourne, 2008. ← 228, 244

[13] N. Do. The asymptotic Weil–Petersson form and intersection theory on $\overline{\mathcal{M}}_{g,n}$. `arXiv:1010.4126v1 [math.GT]` ← 219, 220, 249

[14] N. Do and P. Norbury. Weil-Petersson volumes and cone surfaces. *Geom. Dedicata*, **141**, 93–107, 2009. ← 219, 220, 244, 245

[15] B. Eynard and N. Orantin. Topological recursion in enumerative geometry and random matrices. *J. Phys. A*, **42** (29), 117, 2009. ← 245

[16] S. Grushevsky. An explicit upper bound for Weil-Petersson volumes of the moduli spaces of punctured Riemann surfaces. *Math. Ann.*, **321** (1), 1–13, 2001. ← 247

[17] V. Guillemin. Moment maps and combinatorial invariants of Hamiltonian T^n-spaces. *Progress in Mathematics*, **122**, Birkhäuser Boston Inc., Boston, MA, 1994. ← 231

[18] J. Harris and I. Morrison. Moduli of curves. *Graduate Texts in Mathematics*, **187**, Springer-Verlag, New York, 1998. ← 250

[19] R. Kaufmann, Y. Manin, and D. Zagier. Higher Weil-Petersson volumes of moduli spaces of stable n-pointed curves. *Comm. Math. Phys.*, **181** (3), 763–787, 1996. ← 233

[20] M. E. Kazarian and S. K. Lando. An algebro-geometric proof of Witten's conjecture. *J. Amer. Math. Soc.*, **20** (4), 1079–1089, 2007. ← 253

[21] Y.-S. Kim and K. Liu. A simple proof of Witten conjecture through localization. `arXiv:math/0508384v2 [math.AG]` ← 253

[22] M. Kontsevich. Intersection theory on the moduli space of curves and the matrix Airy function. *Comm. Math. Phys.*, **147**(1), 1–23, 1992. ← 219, 228, 249, 250, 253

[23] K. Liu and H. Xu. Recursion formulae of higher Weil-Petersson volumes. *Int. Math. Res. Not. IMRN*, **5**, 835–859, 2009. ← 244

[24] K. Liu and H. Xu. Mirzakharni's recursion formula is equivalent to the Witten-Kontsevich theorem. *Astérisque*, 328, 2009; 223–235, 2010. ← 243

[25] F. Luo and S.-P. Tan. A dilogarithm identity on moduli spaces of curves. `arXiv:1102.2133v2 [math.GT]` ← 238

[26] Y. I. Manin and P. Zograf. Invertible cohomological field theories and Weil-Petersson volumes. *Ann. Inst. Fourier (Grenoble)*, **50** (2), 519–535, 2000. ← 247

[27] G. McShane. Simple geodesics and a series constant over Teichmuller space. *Invent. Math.*, **132** (3), 607–632, 1998. ← 234, 235

[28] G. McShane. Simple geodesics on surfaces of genus 2. *Ann. Acad. Sci. Fenn. Math.*, **31**(1), 31–38, 2006. ← 238

[29] M. Mirzakhani. Simple geodesics and Weil-Petersson volumes of moduli spaces of bordered Riemann surfaces. *Invent. Math.*, **167** (1), 179–222, 2007. ← 219, 233, 235, 236, 238, 249

[30] M. Mirzakhani. Weil-Petersson volumes and intersection theory on the moduli space of curves. *J. Amer. Math. Soc.*, **20** (1), 1–23 (electronic), 2007. ← 218, 219, 230, 231, 233, 243, 249, 253

[31] M. Mirzakhani. Growth of the number of simple closed geodesics on hyperbolic surfaces. *Ann. of Math. (2)*, **168** (1), 97–125, 2008. ← 243

[32] M. Mirzakhani. Growth of Weil–Petersson volumes and random hyperbolic surfaces of large genus.
`arXiv:1012.2167v1 [math.GN]`. ← 219, 220, 246, 247

[33] G. Mondello. Triangulated Riemann surfaces with boundary and the Weil-Petersson Poisson structure. *J. Differential Geom.*, **81** (2), 391–436, 2009. ← 249

[34] M. Mulase and B. Safnuk. Mirzakhani's recursion relations, Virasoro constraints and the KdV hierarchy. *Indian J. Math.*, **50** (1), 189–218, 2008. ← 243

[35] T. Nakanishi and M. Näätänen. Areas of two-dimensional moduli spaces. *Proc. Amer. Math. Soc.*, **129**(11), 3241–3252 (electronic), 2001. ← 230

[36] A. Okounkov and R. Pandharipande. Gromov-Witten theory, Hurwitz numbers, and matrix models, in: *Algebraic geometry—Seattle 2005. Part 1*, 325–414. *Proc. Sympos. Pure Math.*, **80**, Amer. Math. Soc., Providence, RI, 2009. ← 253

[37] A. Papadopoulos. (Ed.) Handbook of Teichmüller theory. Vol. I. *IRMA Lectures in Mathematics and Theoretical Physics*, **11**, European Mathematical Society (EMS), Zürich, 2007. ← 225

[38] A. Papadopoulos. (Ed.) Handbook of Teichmüller theory. Vol. II. *IRMA Lectures in Mathematics and Theoretical Physics*, **13**, European Mathematical Society (EMS), Zürich, 2009. ← 225

[39] R. C. Penner. Weil-Petersson volumes. *J. Differential Geom.*, **35** (3), 559–608, 1992. ← 229, 247

[40] G. Schumacher and S. Trapani. Estimates of Weil-Petersson volumes via effective divisors. *Comm. Math. Phys.*, **222** (1), 1–7, 2001. ← 247

[41] A. Cannas da Silva. Lectures on symplectic geometry. *Lecture Notes in Mathematics*, **1764**, Springer-Verlag, Berlin, 2001. ← 231

[42] S. P. Tan, Y. L. Wong, and Y. Zhang. Generalizations of McShane's identity to hyperbolic cone-surfaces. *J. Differential Geom.*, **72** (1), 73–112, 2006. ← 238, 244

[43] S. P. Tan, Y. L. Wong, and Y. Zhang. Necessary and sufficient conditions for McShane's identity and variations. *Geom. Dedicata*, **119**, 199–217, 2006. ← 238

[44] S. P. Tan, Y. L. Wong, and Y. Zhang. McShane's identity for classical Schottky groups. *Pacific J. Math.*, **237**(1), 183–200, 2008. ← 238

[45] W. P. Thurston. Three-dimensional geometry and topology, Vol. 1. *Princeton Mathematical Series*, **35**, Princeton University Press, Princeton, NJ, 1997. ← 224

[46] R. Vakil. The moduli space of curves and its tautological ring. *Notices Amer. Math. Soc.*, **50**(6), 647–658, 2003. ← 250

[47] R. Vakil. The moduli space of curves and Gromov-Witten theory, in: Enumerative invariants in algebraic geometry and string theory, 143–198. *Lecture Notes in Math.*, **1947**, Springer, Berlin, 2008. ← 250

[48] E. Witten. Two-dimensional gravity and intersection theory on moduli space, in: Surveys in differential geometry (Cambridge, MA, 1990), 243–310. Lehigh Univ., Bethlehem, PA, 1991. ← 252, 253

[49] S. Wolpert. On the homology of the moduli space of stable curves. *Ann. of Math. (2)*, **118** (3), 491–523, 1983. ← 229

[50] S. Wolpert. On the Kähler form of the moduli space of once punctured tori. *Comment. Math. Helv.*, **58** (2), 246–256, 1983. ← 229

[51] S. Wolpert. On the Weil-Petersson geometry of the moduli space of curves. *Amer. J. Math.*, **107** (4), 969–997, 1985. ← 226, 229, 250

[52] P. Zograf. The Weil-Petersson volume of the moduli space of punctured spheres, in: Mapping class groups and moduli spaces of Riemann surfaces (Göttingen, 1991/Seattle, WA, 1991), 367–372. *Contemp. Math.*, **150**, Amer. Math. Soc., Providence, RI, 1993. ← 219, 230

[53] P. Zograf. On the large genus asymptotics of Weil–Petersson volumes. `arXiv:0812.0544v1 [math.AG]`. ← 246

Department of Mathematics and Statistics, The University of Melbourne, Victoria 3010, Australia
Current address: School of Mathematical Sciences, Monash University, Victoria 3800, Australia
E-mail address: `normdo@gmail.com`

Equivariant geometry and the cohomology of the moduli space of curves

Dan Edidin

Abstract. In this chapter we give a categorical definition of the integral cohomology ring of a stack. For quotient stacks $[X/G]$ the categorical cohomology ring may be identified with the equivariant cohomology $H^*_G(X)$. Identifying the stack cohomology ring with equivariant cohomology allows us to prove that the cohomology ring of a quotient Deligne-Mumford stack is rationally isomorphic to the cohomology ring of its coarse moduli space. The theory is presented with a focus on the stacks $\mathcal{M}_g$ and $\overline{\mathcal{M}}_g$ of smooth and stable curves respectively.

Contents

2000 *Mathematics Subject Classification.* Primary 14D23, 14H10; Secondary 14C15, 55N91.
Key words and phrases. stacks, moduli of curves, equivariant cohomology.
The author was supported by NSA grant H98230-08-1-0059 while preparing this article.

1. Introduction

The study of the cohomology of the moduli space of curves has been a very rich research area for the last 30 years. Contributions have been made to the field from a remarkably broad range of researchers: algebraic geometers, topologists, mathematical physicists, hyperbolic geometers, etc. The goal of this article is to give an introduction to some of the foundational issues related to studying the cohomology of the moduli space from the algebro-geometric point of view.

In algebraic geometry the main difficulty in studying the cohomology of the moduli space is that the moduli space is not really a space but a *stack*. As a result, care is required in determining what the cohomology ring of the moduli space should mean. In the literature this difficulty is often dealt with by arguing that the stack of curves is an orbifold. Associated to an orbifold is an underlying rational homology manifold. The rational cohomology ring of the stack is then defined to be the cohomology ring of this underlying homology manifold.

There are two difficulties with this perspective. First, it can somewhat confusing to sort out the technicalities of intersection theory on orbifolds, and second, one abandons hope of obtaining a good theory with integer coefficients. In this article we will circumvent these difficulties by utilizing the categorical nature of stacks.

We propose a very natural functorial definition for the cohomology of a stack and explain how for quotient stacks $[X/G]$, (like $\mathcal{M}_g$) our functorial cohomology can be identified with equivariant cohomology of X. Because our techniques are algebraic we are also able to define the integral Chow ring of a stack. On $\mathcal{M}_g$ and $\overline{\mathcal{M}}_g$ the tautological classes all naturally live in the functorial cohomology ring. In addition, if $\mathcal{X}$ is a smooth quotient stack then functorial group $A^1(\mathcal{X})$ coincides with the Picard group of the "moduli problem" defined earlier by Mumford in [26].

Using techniques from equivariant cohomology we show that for smooth quotient stacks such as $\overline{\mathcal{M}}_g$, the functorial cohomology groups are rationally isomorphic to the cohomology groups of the underlying coarse moduli space. A similar results holds for Chow groups. This isomorphism defines an intersection product on the rational Chow groups of the projective, but singular, moduli scheme of curves $\overline{M}_g$.

To give a description of the cohomology of the stack of curves we obviously must consider a more basic question. *What is a stack?* In order to keep this chapter self-contained but still of a reasonable length we will give a very brief introduction to theory of Deligne-Mumford stacks via a series of examples. Most of our discussion will focus on quotient stacks, because the geometry of quotient stacks is, in essence, equivariant geometry of ordinary schemes. Most stacks that naturally arise in algebraic geometry, such as $\mathcal{M}_g$ and $\overline{\mathcal{M}}_g$ are in fact quotient stacks. For a further introduction to Deligne-Mumford stacks the reader is encouraged to look at Section 4 of Deligne and Mumford's paper [3] as well as Section 7 of Vistoli's paper [34]. The author's paper [4] gives an introduction to Deligne-Mumford stacks from the perspective of the moduli space of curves. The book by Laumon and Moret-Bailly [23] is the most comprehensive (and most technical) treatise on the theory of algebraic stacks.

The Chapter is organized as follows. In Section 2 we define and give examples of *categories fibred in groupoids* (CFGs). Our main focus is on *quotient CFGs* - that is CFGs arising from actions of linear algebraic groups on schemes. In Section 3 we define the cohomology and Chow rings of a CFG and prove that, for quotient CFGs, these rings can be identified with equivariant cohomology and Chow rings respectively. In Sections 4.1 and 4.2 we define Deligne-Mumford stacks and their coarse moduli spaces. Finally in Section 4.3 we explain the relationship between the cohomology ring of a Deligne-Mumford stack and its coarse moduli space.

Acknowledgments: The author is grateful to Damiano Fulghesu, Angelo Vistoli and the referee for a number of helpful comments on earlier versions of this Chapter. The author also thanks Andrew Kresch for sending him a copy of his preprint [22].

2. Categories fibred in groupoids (CFGs)

The purpose of this section is to give an introduction to the categorical underpinnings of the theory of stacks. We do not define stacks until Section 4. However, the categorical foundation is sufficient to define the cohomology and Chow rings of stacks, which we do in Section 3.

To begin, fix a base scheme S. For example, the analytically minded might take $S = \operatorname{Spec}\mathbb{C}$ and the arithmetically minded could consider $S = \operatorname{Spec}\mathbb{Z}$. Let $\mathcal{S}$ be the category of S-schemes.

Definition 2.1. [3, Definition 4.1] A *category fibred in groupoids* (CFG) over $\mathcal{S}$ is a category $\mathcal{X}$ together with a functor $\rho\colon \mathcal{X} \to \mathcal{S}$ satisfying the following conditions.

i) Given an object t of $\mathcal{X}$ let $T = \rho(t)$. If $f\colon T' \to T$ in $\mathcal{S}$ there exists a pullback object f^*t and a morphism $f^*t \to t$ in $\mathcal{X}$ whose image under the functor ρ is the morphism $T' \to T$. Moreover, f^*t is unique up to canonical isomorphism.

ii) If $\alpha\colon t_1 \to t_2$ is a map in $\mathcal{X}$ such that $\rho(\alpha) = T \stackrel{\mathrm{id}}{\to} T$ for some S-scheme T then α is an isomorphism.

Remark 2.2. The first condition is just stating that the category $\mathcal{X}$ has fibred products relative to S. If $\mathcal{X}$ satisfies i) then $\mathcal{X}$ is called a *fibred category*. The second conditions implies that if T is a fixed scheme then the subcategory $\mathcal{X}(T)$ consisting of objects mapping to T and morphisms mapping to the identity is a *groupoid*; that is all morphisms in $\mathcal{X}(T)$ are isomorphisms. Note that if t is an object of $\mathcal{X}$ and $\rho(t) = T$ for some S-scheme T, then t is an object of $\mathcal{X}(T)$. Hence every object of $\mathcal{X}$ is in $\mathcal{X}(T)$ for some scheme T.

The second condition may seem a little strange, but it is a natural one for moduli problems. The concept of category fibred groupoids over $\mathcal{S}$ is a generalization of the concept of contravariant functor from $\mathcal{S} \to$ Sets. A stack is a CFG that satisfies certain algebro-geometric conditions analogous to the conditions satisfied by a representable functor.

To give a feel for the theory of CFGs we will focus on three examples: CFGs of smooth and stable curves, representable CFGs, and quotient CFGs. We will show that the CFGs of smooth and stable curves are quotient CFGs. This implies that they are algebraic stacks, as quotient CFGs are always algebraic stacks (although we will not prove this here).

2.1. CFGs of curves

Throughout the rest of this paper a *curve* will denote a complete (hence projective) scheme of dimension 1.

Definition 2.3. (Smooth curves) For any $g \geqslant 2$, let $\mathcal{M}_g$ be the category whose objects are $\pi\colon X \to T$ where T is an S-scheme and π is a proper smooth morphism whose fibers are curves of genus g. A morphism from $X' \to T'$ to $X \to T$ is simply a cartesian diagram.

$$\begin{array}{ccc} X' & \to & X \\ \downarrow & & \downarrow \\ T' & \to & T \end{array}$$

Remark 2.4. Clearly $\mathcal{M}_g$ is a fibred category over $\mathcal{S}$ by construction. To see that it is fibered in groupoids observe that if

$$\begin{array}{ccc} X' & \to & X \\ \downarrow & & \downarrow \\ S & \stackrel{id}{\to} & S \end{array}$$

is cartesian then the map $X' \to X$ is an isomorphism.

Note that the fact that $\mathcal{M}_g$ is a category fibred in groupoids has essentially nothing to do with the fact that we are attempting to parametrize curves of given genus. The only thing we are using is that morphisms in $\mathcal{M}_g$ are cartesian diagrams. This shows that, like the notion of functor, the concept of category fibred in groupoids is very general.

Next we define the CFG of stable curves.

Definition 2.5. (Stable curves) A curve C of arithmetic genus $g \geqslant 2$ is *stable* if it is connected, has at worst nodes as singularities and if every rational component intersects the other components in at least 3 points.

Definition 2.6. If $g \geqslant 2$ then let $\overline{\mathcal{M}}_g$ be the category whose objects are $\pi\colon X \to T$ where T is an S-scheme and π is a finitely presented proper flat morphism whose fibers are stable curves of genus g. Morphisms are again cartesian diagrams.

Again $\overline{\mathcal{M}}_g$ is a CFG containing $\mathcal{M}_g$ as a full subcategory. We can relax the condition on rational components and obtain CFGs of *prestable* curves. However, if we do so then we will not obtain a Deligne-Mumford stack.

Definition 2.7. (Pointed curves) For any $g \geqslant 0$ we let $\mathcal{M}_{g,n}$ be the category whose objects are $(X \stackrel{\pi}{\to} T, \sigma_1, \ldots, \sigma_n)$ where $\pi \colon X \to T$ is a family of smooth curves and $\sigma_1, \ldots, \sigma_n$ are disjoint sections of π. A morphism

$$(X' \stackrel{\pi'}{\to} T', \sigma'_1, \ldots, \sigma'_n) \to (X' \stackrel{\pi}{\to} T, \sigma_1, \ldots, \sigma_n)$$

is a cartesian diagram

$$\begin{array}{ccc} X' & \stackrel{f}{\to} & X \\ \pi' \downarrow & & \pi \downarrow \\ T' & \stackrel{g}{\to} & T \end{array}$$

such that $f \circ \sigma'_i = \sigma_i \circ g$ for all i.

If $2g - 2 + n > 0$ then we define $\overline{\mathcal{M}}_{g,n}$ to be the CFG whose objects are $(X \stackrel{\pi}{\to} T, \sigma_1, \ldots, \sigma_n)$ where π is proper and flat and the fibers of π are connected curves with at worst nodes as singularities and the σ_i are again disjoint sections of π. We impose the additional condition that every rational component of a fiber must have a total of at least 3 points marked by the sections plus intersections with the other components. A morphism is once again cartesian diagram compatible with the sections.

The CFG $\mathcal{M}_{g,n}$ is a full subcategory of the CFG $\overline{\mathcal{M}}_{g,n}$ when $2g - 2 + n > 0$. We may also relax the stability condition to obtain the CFG of *prestable pointed curves* which also contains $\mathcal{M}_{g,n}$ as a full subcategory.

Remark 2.8. All CFGs of curves are subcategories of the the "universal" CFG $\mathrm{Mor}(\mathcal{S})$. Objects of $\mathrm{Mor}(\mathcal{S})$ are morphisms of S-schemes and morphisms in $\mathrm{Mor}(\mathcal{S})$ are cartesian diagrams of S-schemes. The CFGs $\mathcal{M}_g$ and $\overline{\mathcal{M}}_g$ are full subcategories of $\mathrm{Mor}(\mathcal{S})$ but $\mathcal{M}_{g,n}$ and $\overline{\mathcal{M}}_{g,n}$ are not.

2.2. Representable CFGs

Let $F \colon \mathcal{S} \to \mathrm{Sets}$ be a contravariant functor. There is an associated category fibred in groupoids $\underline{F}$. Given a scheme T in $\mathcal{S}$ an object of $\underline{F}(T)$ is simply an element of $F(T)$. Given $t' \in F(T')$ there is a morphism $t' \to t$ if and only if t' is the image of t under the map of sets $F(T') \to F(T)$. With this construction if T is an S-scheme then the groupoid $\underline{F}(T)$ is the category whose objects are the elements of $F(T)$ and all morphisms are identities.

In particular if X is an S-scheme then we can associate to its functor of points a CFG $\underline{X}$. This is simply the category of X-schemes viewed as a fibred category over the category of S-schemes.

Definition 2.9. A CFG $\mathcal{X}$ is *representable* if $\mathcal{X}$ is equivalent to a CFG $\underline{X}$ for some S-scheme X.

Remark 2.10. Yoneda's lemma implies that if X and Y are schemes then there is an isomorphism of schemes $X \to Y$ if and only if there is an equivalence of categories $\underline{X} \to \underline{Y}$.

Example 2.11. The existence of curves of every genus with non-trivial automorphism implies that the CFG $\mathcal{M}_g$ is not representable (i.e., not the CFG associated to a scheme) because if k is a field then the category $\mathcal{M}_g(\operatorname{Spec} k)$ is not equivalent to one where all morphisms are identities.

Proposition 2.12. *If $\mathcal{X}$ is a CFG and T is an S-scheme then to give an object of $\mathcal{X}(T)$ is equivalent to giving a functor $\underline{T} \to \mathcal{X}$ compatible with the projection functor to $\mathcal{S}$.*

Proof. Given an object t in $\mathcal{X}(T)$ define a functor $\underline{T} \to \mathcal{X}$ by mapping a T-scheme T' to a pullback of the object t via the morphism $f\colon T' \to T$ in $\mathcal{S}$. (Note that the definition of this functor requires a choice for each pullback. However, different choices give rise to equivalent functors). Conversely, given a functor $\underline{T} \to \mathcal{X}$ we set t to be the image of T in $\mathcal{X}$. □

Notation 2.13. If T is a scheme and $\mathcal{X}$ is a CFG we will streamline the notation by writing $T \to \mathcal{X}$ in lieu of $\underline{T} \to \mathcal{X}$.

Example 2.14. By Proposition 2.12 giving a family of smooth curves $X \to T$ of genus g is equivalent to giving a map $T \to \mathcal{M}_g$. In this way $\mathcal{M}_g$ is the "classifying space" for smooth curves. Similarly $\overline{\mathcal{M}}_g$ classifies stable curves.

2.3. Curves and quotient CFGs

Let G be a linear algebraic group; i.e., a closed subgroup of GL_n for some n. For simplicity we assume that G is smooth over S. When $S = \operatorname{Spec} \mathbb{C}$ this is automatic, but the assumption is necessary in positive or mixed characteristic.

Definition 2.15. Let T be a scheme. A G-torsor over T is a smooth morphism $p\colon E \to T$ where G acts freely on E, p is G-invariant and there is an isomorphism of G-spaces $E \times_T E \to G \times E$.

Definition 2.16. Let BG be the CFG whose objects are G-torsors $E \to T$ and whose morphism are cartesian diagrams

$$\begin{array}{ccc} E' & \to & E \\ \downarrow & & \downarrow \\ T' & \to & T \end{array}$$

with the added condition that the map $E' \to E$ is G-invariant.

More generally, if X is a scheme and G is an algebraic group acting on X then we define a CFG $[X/G]$ to be the category whose objects are pairs $(E \to T, E \xrightarrow{f} X)$ where $E \to T$ is a G-torsor and $f\colon E \to X$ is a G-equivariant map. A morphism

$(E' \to T', E' \xrightarrow{f'} X) \to (E \to T, E \xrightarrow{f} T)$ in $[X/G]$ is a cartesian diagram of torsors

$$\begin{array}{ccc} E' & \xrightarrow{h} & E \\ \downarrow & & \downarrow \\ T' & \to & T \end{array}$$

such that $f' = f \circ h$.

Definition 2.17. A CFG is $\mathcal{X}$ is a *quotient CFG* if $\mathcal{X}$ is equivalent to a CFG $[X/G]$ for some scheme X.

Example 2.18. If X is a scheme then $\underline{X}$ is equivalent to the quotient CFG $[(G \times X)/G]$ where G acts on $G \times X$ by the rule $g(g', x) = (gg', x)$.

Remark 2.19. Although we have not yet discussed the geometry of CFGs the geometry of a quotient CFG $[X/G]$ is the G-equivariant geometry of X. This point of view will be emphasized in our discussion of cohomology rings.

2.3.1. $\mathcal{M}_g$ is a quotient CFG Although the definition of $\mathcal{M}_g$ as a CFG is purely categorical, the fact that if $g \geqslant 2$ and $X \xrightarrow{\pi} T$ is a family of smooth curves then π is a *projective* morphism allows us to prove that $\mathcal{M}_g$ is a quotient CFG.

To state the result we introduce some notation.

Definition 2.20. Fix an integer $g \geqslant 2$ and let H be the Hilbert scheme of one dimensional subschemes of $\mathbb{P}^{5g-6}$ with Hilbert polynomial $(6t-1)(g-1)$. The action of PGL_{5g-5} on $\mathbb{P}^{5g-6}$ induces a corresponding action on the Hilbert scheme. If $g \geqslant 2$ then the canonical divisor on any curve C is ample and $3K_C$ is very ample. Let H_g be the locally closed subscheme of H corresponding to smooth curves $C \subset \mathbb{P}^{5g-6}$ with $\mathcal{O}_C(1) \simeq \omega_C^{\otimes 3}$.

Proposition 2.21. *There is an equivalence of categories $\mathcal{M}_g \to [H_g/PGL_{5g-5}]$. Hence $\mathcal{M}_g$ is a quotient CFG.*

Proof. We define a functor $p\colon \mathcal{M}_g \to [H_g/PGL_{5g-5}]$ as follows:

Given a family $X \xrightarrow{\pi} T$ of smooth curves consider the rank $(5g-6)$ projective space bundle $\mathbb{P}(\pi_*(\omega_{X/T}^{\otimes 3}))$ whose fiber at a point $p \in T$ is the complete linear series $|3K_{X_{\pi^{-1}(t)}}|$. Let $E \to T$ be the associated PGL_{5g-5} torsor. The pullback of $\mathbb{P}(\pi_*(\omega_{X/T}^{\otimes 3}))$ to E is trivial and defines an embedding of $X \times_T E \hookrightarrow \mathbb{P}_E^{5g-6}$. Hence we obtain a morphism $E \to H_g$. The construction is natural so the map $E \to H_g$ commutes with the natural PGL_{5g-5} action on E and H_g. Given a morphism

$$\begin{array}{ccc} X' & \to & X \\ \downarrow & & \downarrow \\ T' & \to & T \end{array}$$

our construction gives a morphism of PGL_{5g-5}-torsors

$$\begin{array}{ccc} E' & \to & E \\ \downarrow & & \downarrow \\ T' & \to & T \end{array}$$

compatible with the maps $E' \to H_g$ and $E \to H_g$.

We now check that our functor is an equivalence by defining a functor $q\colon [H_g/PGL_{5g-5}] \to \mathcal{M}_g$ as follows:

Given a torsor $E \to T$ and a map $E \to H_g$ we obtain a family of projective curves $Z \to E$. There is an action of PGL_{5g-5} on Z such that the morphism $Z \to E$ is PGL_{5g-5}-equivariant. By definition of E as the total space of a PGL_{5g-5}-torsor the action G on E is free. Since there is a G-equivariant morphism $Z \to E$ it follows that the action of G on Z is also free. We would like to let X be the quotient Z by the free PGL_{5g-5} action. Unfortunately, there is no *a priori* reason why the quotient of a scheme (even a projective or quasi-projective scheme) by the free action of an algebraic group exists in the category of schemes[1]. However, we are in a relatively special situation in that we already know that a quotient E/PGL_{5g-5} exists as a scheme (since it's equal to T) and the morphism $Z \to E$ is a *projective* morphism. Descent theory for projective morphisms (see [29, Proposition 7.1]) implies that there is a quotient $X = Z/G$ which is projective over $T = Z/G$ and such that the diagram

$$\begin{array}{ccc} Z & \to & E \\ \downarrow & & \downarrow \\ X & \to & T \end{array}$$

is cartesian. Hence $X \to T$ is a family of smooth curves over T. A similar analysis defines the image of a morphism in $[H_g/PGL_{5g-5}]$, and one can check that there are natural transformations $q \circ p \to Id_{\mathcal{M}_g}$ and $p \circ q \to Id_{[H_g/PGL_{5g-5}]}$. □

2.3.2. $\overline{\mathcal{M}}_g$ is a quotient CFG The key fact about stable curves, and the reason that the theory of stable curve is so elegant, is the following result of Deligne and Mumford.

Theorem 2.22. [3, Thm 2.1] *If $X \stackrel{\pi}{\to} T$ is a family of stable curves then the dualizing sheaf ω_π is locally free and relatively ample. Moreover, the sheaf $\omega_\pi^{\otimes 3}$ is relatively very ample; i.e., if p is a point of T and $X_p = \pi^{-1}(p)$ then the line bundle $\omega_{X_p}^{\otimes 3}$ is very ample.*

The same argument used in the proof of Proposition 2.21 can now be used to prove that the CFG $\overline{\mathcal{M}}_g$ is a quotient CFG.

Proposition 2.23. *Let $\overline{H}_g$ be the locally closed subscheme of H parametrizing embedded stable curves C where $\mathcal{O}_C(1) \simeq \omega_C^{\otimes 3}$. Then there is an equivalence of categories $\overline{\mathcal{M}}_g \to [\overline{H}_g/PGL(5g-5)]$.*

[1]One can prove, using a non-trivial theorem of Deligne and Mumford, that such a quotient automatically exists as an algebraic space.

2.3.3. Curves of very low genus In the previous section we only considered curves of genus $g \geqslant 2$. The purpose of this section is to briefly discuss the CFGs of curves of genus 0 and 1.

Example 2.24 (Curves of genus 0). Let $\mathcal{M}_0$ be the CFG of nodal curves of arithmetic genus 0. A nodal curve of genus 0 is necessarily a tree of $\mathbb{P}^1$s. However, there is no bound on the number of components. The CFG $\mathcal{M}_0$ can be stratified by the number of nodes (or equivalently irreducible components). Following [12] we denote by $\mathcal{M}_0^{\leqslant k}$ the CFG whose objects are families of rational nodal curves with at most k nodes. The CFG $\mathcal{M}_0^0$ of smooth rational curves is equivalent to $B\,PGL_2$. In [5, Proposition 6] it was shown that $\mathcal{M}_0^{\leqslant 1}$ is equivalent to the CFG $[X/GL_3]$ where X is the set of quadratic forms in 3 variables with rank at least 2. On the other hand there are families of rational curves $X \stackrel{\pi}{\to} T$ where π is not a projective morphism. In fact, Fulghesu constructed examples where T is a scheme and X is only an algebraic space. Despite the pathological behavior, Fulghesu [12] proved that $\mathcal{M}_0^{\leqslant k}$ is an algebraic stack for any k. Recently Kresch [22, Proposition 5.2] proved that for $k \geqslant 2$, $\mathcal{M}_0^{\leqslant k}$ is not equivalent to a quotient of the form $[Z/G]$ with Z an algebraic space; ie $\mathcal{M}_0^{\leqslant k}$ is not a quotient stack.

On the other hand, if $n \geqslant 3$ then the CFGs $\mathcal{M}_{0,n}$ and $\overline{\mathcal{M}}_{0,n}$ are well understood. They are represented by non-singular projective schemes [20].

Example 2.25 (Curves of genus 1). The CFG $\mathcal{M}_1$, of curves of genus 1, is rather strange. Its behavior highlights the distinction between curves of genus 1 and elliptic curves. An elliptic curve is a curve of genus 1 together with a point chosen to be the origin for the group law. In particular an elliptic curve is a projective algebraic group, while a curve of genus 1 is a torsor for this group. Given a curve C of genus 1 together with a choice of a point $O \in C$ the group scheme (C, O) acts on C. In particular, the automorphism group of a curve is not a linear algebraic group. Thus the CFG $\mathcal{M}_1$ cannot be equivalent to a CFG of the form $[X/G]$ with G a linear algebraic group.

However, the CFG $\mathcal{M}_{1,1}$ of elliptic curves behaves a lot like the CFGs $\mathcal{M}_g$ for $g \geqslant 2$. Specifically, if $(X \stackrel{\pi}{\to} T, \sigma)$ is a family of smooth curves of genus 1 with section σ the direct image $\pi_*(\mathcal{O}_X(\sigma)^{\otimes 3})$ is a locally free sheaf of rank 3. An argument similar to the Proof of Proposition 2.21 can be used to show that $\mathcal{M}_{1,1} = [U/PGL_3]$ where $U \subset \mathbb{P}^9$ is the open set of smooth plane cubics. Similarly $\overline{\mathcal{M}}_{1,1} = [W/PGL_3]$ where W is the open set of plane cubics with at worst nodes as singularities.

2.4. Fiber products of CFGs and universal curves

So far we have not discussed how universal curves fit into the CFG picture. In order to do so we need to introduce another categorical notion – the fiber product of two CFGs over a third CFG. Once this is done, we can explain why $\mathcal{M}_{g,1}$ is the universal curve over $\mathcal{M}_g$. Applying this fact inductively we can show that the CFGs $\mathcal{M}_{g,n}$ are $\overline{\mathcal{M}}_{g,n}$ are all quotient CFGs.

2.4.1. Fiber products of CFGs Since CFGs are categories the natural home for the class of all CFGs over $\mathcal{S}$ is called a *2-category*. A 2-category has objects (in our case the CFGs), morphisms (functors between the CFGs) and morphisms between morphisms (natural transformations of functors). For the most part this added level of complexity can be ignored, but it does show up in one of the most important categorical constructions, the fiber product of two CFGs.

Definition 2.26. Let $f\colon \mathcal{X} \to \mathcal{Z}$ and $g\colon \mathcal{Y} \to \mathcal{Z}$ be CFGs over our fixed category $\mathcal{S}$. We define the fiber product $\mathcal{X} \times_{\mathcal{Z}} \mathcal{Y}$ to be the CFG whose objects are triples (x, y, ϕ) where x is an object of $\mathcal{X}$, y is an object of $\mathcal{Y}$ and ϕ is an isomorphism in $\mathcal{Z}$ between $f(x)$ and $f(y)$.

A morphism in $\mathcal{X} \times_{\mathcal{Z}} \mathcal{Y}$ between (x', y', ϕ') and (x, y, ϕ) is given by a pair of morphisms $\alpha\colon x' \to x$, $\beta\colon y' \to y$ such that $\phi \circ f(\alpha) = g(\beta) \circ \phi'$.

Remark 2.27. A straightforward check shows that $\mathcal{X} \times_{\mathcal{Z}} \mathcal{Y}$ is indeed a CFG. There are also obvious functors $p_X\colon \mathcal{X} \times_{\mathcal{Z}} \mathcal{Y} \to \mathcal{X}$ and $p_Y\colon \mathcal{X} \times_{\mathcal{Z}} \mathcal{Y} \to \mathcal{Y}$ but the compositions $f \circ p_X$ and $g \circ p_Y$ are not equal as functors $\mathcal{X} \times_{\mathcal{Z}} \mathcal{Y} \to \mathcal{Z}$ since $f \circ p_X(x, y, \theta) = f(x)$ and $g \circ p_Y(x, y, \theta) = g(y)$ are isomorphic but not not necessarily the same objects of $\mathcal{Z}$. However, there is a natural transformation of functors $F\colon f \circ p_X \to g \circ p_Y$. For this reason we say that the diagram

$$\begin{array}{ccc} \mathcal{X} \times_{\mathcal{Z}} \mathcal{Y} & \stackrel{p_Y}{\to} & \mathcal{Y} \\ p_X \downarrow & & g \downarrow \\ \mathcal{X} & \stackrel{f}{\to} & \mathcal{Z} \end{array}$$

is *2-cartesian*. In general a diagram of CFGs

$$\begin{array}{ccc} \mathcal{W} & \stackrel{g}{\to} & \mathcal{X} \\ p \downarrow & & q \downarrow \\ \mathcal{Y} & \stackrel{f}{\to} & \mathcal{Z} \end{array}$$

is *2-commutative* if there is a natural transformation between the functors $q \circ g$ and $f \circ p$.

So far our discussion of CFGs has been very categorical. The next definition is the first step towards connecting the theory of CFGs with algebraic geometry.

Definition 2.28. A morphism of CFGs $f\colon \mathcal{Y} \to \mathcal{X}$ is *representable* if for every scheme T and morphism $T \to \mathcal{X}$ the fiber product of CFGs $T \times_{\mathcal{X}} \mathcal{Y}$ is represented by a scheme.

Remark 2.29. This definition is saying that although $\mathcal{Y}$ and $\mathcal{X}$ are not represented by schemes the fibers of the morphism are schemes. It also allows us to define algebro-geometric properties of representable morphisms of CFGs. If **P** is a property of morphisms of schemes which is preserved by base change[2] then a representable

[2]This covers many of the most common types of morphisms encountered in algebraic geometry. For example the properties of being separated, finite, proper, flat and smooth can are all preserved by base change.

morphism $\mathcal{Y} \to \mathcal{X}$ has property **P** if for every map of scheme $T \to \mathcal{X}$ the map of schemes $T \times_{\mathcal{X}} \mathcal{Y} \to T$ has property **P**.

Example 2.30. If G is an algebraic group acting on a scheme X then there is a morphism $X \to [X/G]$ defined as follows. Given a map of schemes $f \colon T \to X$ (an object of the category $\underline{X}$) we can consider the trivial torsor $G \times T \to T$ together with the G-equivariant map $(g, t) \mapsto gf(t)$ to define an object of $[X/G]$. From this definition it is clear that a morphism of X-schemes $T' \to T$ gives rise to a morphism in the CFG $[X/G]$.

We claim that the map $X \to [X/G]$ is representable. In fact if $T \to [X/G]$ is a morphism corresponding to a G-torsor $E \to T$ with equivariant map $f \colon E \to X$ then $T \times_{[X/G]} X$ is represented by the scheme E. Let us see this explicitly. An object of $(T \times_{[X/G]} X)(T')$ is a triple (f_1, f_2, ϕ) where $T' \xrightarrow{f_1} T$ makes T' a T-scheme, $T' \xrightarrow{f_2} X$ makes T an X-scheme and ϕ is an isomorphism in $[X/G]$ between the two images of T' in $[X/G]$. The image of $T' \xrightarrow{f_1} T$ is the torsor $E \times_T T' \to T'$ with G-equivariant map the composition $E \times_T T' \to E \to X$. The image of $T' \xrightarrow{f_2} X$ is the trivial torsor $G \times T' \to T'$ with equivariant map $(g, t') \mapsto gf_2(t')$. The isomorphism ϕ in $[X/G]$ gives an isomorphism between the trivial torsor $G \times T' \to T'$ and the pullback torsor $E \times_T T' \to T'$. In other words ϕ determines a trivialization of the pullback torsor $E \times_T T' \to T'$. The trivialization gives a section $T' \to E \times_T T'$. Composing the section with the projection $E \times_T T' \to E$ gives a map $T' \to E$; i.e., an object of $\underline{E}$. A similar, if more involved, analysis shows that morphisms in the category $T \times_{[X/G]} X$ determine morphisms of X-schemes. In this way we obtain a functor $T \times_{[X/G]} X \to \underline{E}$. This functor is in equivalence of categories. To see this note that if $T' \to X$ is an X-scheme then the map $T' \to X$ determines a trivialization of the pullback torsor $(G \times X)_X T' \to T'$ and thus an object of $(T \times_{[X/G]} X)(T')$.

Example 2.31. The fully faithful inclusion functor $\mathcal{M}_g \to \overline{\mathcal{M}}_g$ is represented by open immersions. Thus we say that $\mathcal{M}_g$ is an open subCFG of $\overline{\mathcal{M}}_g$.

2.4.2. Smooth pointed curves and the universal curve There is a an obvious morphism of CFGs $\mathcal{M}_{g,1} \to \mathcal{M}_g$ that forgets the section.

Proposition 2.32. *If $T \to \mathcal{M}_g$ is a morphism corresponding to a family of smooth curves of genus g, $X \to T$ then the fiber product $T \times_{\mathcal{M}_g} \mathcal{M}_{g,1}$ is represented by the scheme X. Hence the map $\mathcal{M}_{g,1} \to \mathcal{M}_g$ is representable and smooth and we can view $\mathcal{M}_{g,1}$ as the universal curve over $\mathcal{M}_g$.*

Sketch of proof. An object of the fiber product $(T \times_{\mathcal{M}_g} \mathcal{M}_{g,1})(T')$ is given by the following data: a morphism $T' \to T$, a family of pointed curves $X' \to T'$ with section σ', and isomorphism in $\mathcal{M}_g$ of the pullback family $X \times_T T' \to T'$ with the family $X' \to T'$. Since the map $T' \to X'$ has a section, the projection $X \times_T T' \to T'$ also has a section. Composing this section with the projection $X \times_T T' \to X$ gives a map $T' \to X$, i.e., an object $\underline{X}(T')$. Again, a straightforward (if tedious) check shows that this procedure maps morphisms in the category $T \times_{\mathcal{M}_g} \mathcal{M}_{g,1}$ to morphisms in the category $\underline{X}$. This defines a functor which is an equivalence of categories. □

2.4.3. Stable pointed curves and the universal curve A similar argument can be used to show that $\overline{\mathcal{M}}_{g,1}$ is the universal stable curve over $\overline{\mathcal{M}}_g$. However, this argument requires more care because the morphism of CFGs $\overline{\mathcal{M}}_{g,1} \to \overline{\mathcal{M}}_g$ is not as tautological as the corresponding morphism for smooth curves. The reason is that if $(X \to T, \sigma)$ is a family of pointed stable curves, the corresponding family of curves $X \to T$ need not be stable as it may have rational components which intersect the other components in only 2 points. To define a functor $\overline{\mathcal{M}}_{g,1} \to \overline{\mathcal{M}}_g$ we must show that given such family of curves $X \to T$, we may contract the offending rational components in the fibers of $X \to T$. This analysis was carried out by F. Knudsen in his paper [20].

Theorem 2.33 (Knudsen [20]). *If $2g - 2 + n > 0$ there is a representable contraction morphism $\overline{\mathcal{M}}_{g,n+1} \to \overline{\mathcal{M}}_{g,n}$ that makes $\overline{\mathcal{M}}_{g,n+1}$ into the universal curve over $\overline{\mathcal{M}}_{g,n}$.*

Remark 2.34. It is relatively easy to show that $\mathcal{M}_{0,3} = \overline{\mathcal{M}}_{0,3} = \operatorname{Spec} S$. It follows from Knudsen's results that for $n \geqslant 3$ the CFGs $\overline{\mathcal{M}}_{0,n}$ are all represented by schemes.

Combining Theorem 2.33 with the following proposition shows that CFGs of pointed stable curves of genus $g \geqslant 1$ are quotient CFGs.

Proposition 2.35. *If $\mathcal{X} = [X/G]$ is a quotient CFG and $\mathcal{Y} \to \mathcal{X}$ is a representable morphism then $\mathcal{Y}$ is equivalent to the CFG $[Y/G]$ where $Y = X \times_{\mathcal{X}} \mathcal{Y}$.*

Proof. (Sketch) Since $\mathcal{X} = [X/G]$ the fiber product $X \times_{\mathcal{X}} X$ is represented $G \times X$. Under this identification the two projection maps $X \times_{\mathcal{X}} X \to X$ correspond to the projection $p\colon G \times X \to X$ and the action morphism $\sigma\colon G \times X \to X$, $(g, x) \mapsto gx$. If Y represents $X \times_{\mathcal{X}} \mathcal{Y}$ then the fiber product $Y \times_{\mathcal{Y}} Y$ is represented by $G \times Y$. Under this identification, one of the projections $Y \times_{\mathcal{Y}} Y \to Y$ corresponds to the usual projection $G \times Y \to Y$ and the other gives an action map $G \times Y \to Y$ such that the map of schemes $Y \to X$ is G-equivariant.

Given a morphism $T \to \mathcal{Y}$ the composition with the morphism of CFGs $\mathcal{Y} \to \mathcal{X}$ gives a map $T \to \mathcal{X}$. Thus we obtain a G-torsor $E \to T$ together with an equivariant map $E \to X$. The total space E represents the fiber product $T \times_{\mathcal{X}} X$. Since the morphism $T \to \mathcal{X}$ factors through the morphism $\mathcal{Y} \to \mathcal{X}$, the fiber product $T \times_{\mathcal{X}} X$ is equivalent to the fiber product $T \times_{\mathcal{Y}} Y$. Thus we obtain a morphism $E \to Y$ which can be checked to be G-equivariant. Hence an object of $\mathcal{Y}(T)$ produces an object of $[Y/G](T)$. Further analysis shows that we can extend this construction to define a functor $\mathcal{Y} \to [Y/G]$.

The construction of a functor $[Y/G] \to \mathcal{Y}$ is more subtle. Given a G-torsor $E \to T$ and a G-equivariant map to $E \to Y$ the composition $E \to Y \to X$ is a G-equivariant map $E \to X$. Thus we obtain a morphism $T \to \mathcal{X} = [X/G]$. We would like to lift this to a morphism $T \to \mathcal{Y}$. Since $\mathcal{Y} \to \mathcal{X}$ is representable the fiber product $T \times_{\mathcal{X}} \mathcal{Y}$

is represented by a scheme T'. By construction there are projections $T' \to T$ and $T' \to \mathcal{Y}$, so to define a morphism $T \to \mathcal{Y}$ it suffices to construct a section of the projection map $T' \to T$. Let $E' = E \times_T T'$. Then we have a cartesian diagram where the horizontal maps are G-torsors

$$\begin{array}{ccc} E' & \to & T' \\ \downarrow & & \downarrow \\ E & \to & T \end{array}$$

Since $T' = T \times_{\mathcal{X}} \mathcal{Y}$ the fiber product $E' = E \times_T T'$ can be identified with the fiber product of schemes $E \times_X Y$. Moreover, the map $E \to X$ factors through the G-equivariant map $Y \to X$ so the projection $E' \to E$ has a G-equivariant section $E \to E'$. Hence there is an induced map of quotients $T \to T'$ which is a section to the map $T' \to T$. The composition $T \to T' \to \mathcal{Y}$ is our desired morphism. □

3. Cohomology of CFGs and equivariant cohomology

We now come to heart of this article - the definition of the integral cohomology ring of a CFG. Although this definition is very formal, the cohomology ring of the CFG $\overline{\mathcal{M}}_{g,n}$ naturally contains all tautological classes. Moreover, if $\mathcal{X} = [X/G]$ is a quotient CFG we will show that $H^*(\mathcal{X}) = H^*_G(X)$ where $H^*_G(X)$ is the G-equivariant cohomology ring of X.

Throughout this section we assume that the ground scheme S is the spectrum of a field. When we work with cohomology we assume $S = \operatorname{Spec} \mathbb{C}$ and the dimension of a variety is its complex dimension.

3.1. Motivation and definition

Since a CFG is not a space, we need an indirect method to define cohomology classes on a CFG. To do this we start with a simple observation. Since cohomology is a contravariant functor, a cohomology class $c \in H^*(X)$ determines a pullback cohomology class $c(t) \in H^*(T)$ for every map $T \xrightarrow{t} X$. (Here we can simply let T and X be topological spaces.) Moreover, functoriality also ensures that the classes $c(t)$ satisfy an obvious compatibility condition. Given morphisms $T' \xrightarrow{f} T \xrightarrow{t} X$ then $c(ft) = f^*c(t)$. With this motivation we make the following definition.

Definition 3.1. Let $\mathcal{X}$ be a CFG defined over $\mathbb{C}$. A cohomology class c on $\mathcal{X}$ is the data of a cohomology class $c(t) \in H^*(T)$ for every scheme T and every object t of $\mathcal{X}(T)$. The classes $c(t)$ should satisfy the following compatibility condition: Given schemes T' and T and objects t' in $\mathcal{X}(T')$, t in $\mathcal{X}(T)$ and a morphism $t' \to t$ whose image in $\mathcal{S}$ is a morphism $f \colon T' \to T$ then $f^*c(t) = c(t') \in H^*(T')$.

The cup product on cohomology of spaces guarantees that the collection of all cohomology classes on $\mathcal{X}$ forms a graded skew-commutative ring. We denote this ring by $H^*(\mathcal{X})$.

3.1.1. Chow cohomology of CFGs Since many naturally occurring classes in the cohomology of $\overline{\mathcal{M}}_g$ are algebraic we also define the Chow cohomology ring of a CFG defined over an arbitrary field.

Definition 3.2. [14, Definition 17.3] Let X be a scheme. A Chow cohomology class c of codimension i is an assignment, for every map $T \xrightarrow{t} X$, a map on Chow groups $c(t)\colon A_*(T) \to A_{*-i}(T)$ such that the $c(t)$ are compatible with the usual operations in intersection theory (flat pullback and lci pullback, proper pushforward, etc.). The group of Chow cohomology classes of codimension i is denoted $A^i(X)$.

If $\alpha \in A_k(T)$ and $c \in A^i(X)$ then we use the notation $c \cap \alpha$ to denote the image of α under the map $c(t)\colon A_k(T) \to A_{k-i}(T)$.

Composition of maps makes the collection of operations of all codimension into a graded ring which we denote $A^*(X) := \oplus A^i(X)$.

Remark 3.3. If X admits a resolution of singularities then $A^*(X)$ is commutative [14, Example 17.4.4].

The relationship between Chow cohomology and the usual Chow groups is given by the following proposition.

Proposition 3.4. *(Poincaré Duality* [14, Corollary 17.4]*) If X is a smooth variety of dimension n then the map* $A^i(X) \to A_{n-i}(X)$, $c \mapsto c(id) \cap [X]$ *is an isomorphism. Moreover, the ring structure on* $A^*(X)$ *given by composition of operations is compatible with the ring structure on* $A_*(X)$ *given by intersection product.*

Given a map $f\colon Y \to X$ there is a natural pullback $f^*\colon A^*(X) \to A^*(Y)$ given by $f^*c(T \xrightarrow{t} Y) = c(ft)$. This functoriality allows us to define the Chow cohomology ring of a CFG:

Definition 3.5. If $\mathcal{X}$ is a CFG, then a Chow cohomology class is an assignment for every scheme T and every object t of $\mathcal{X}(T)$ a Chow cohomology $c(t) \in A^*(T)$ satisfying the same compatibility conditions as in Definition 3.1.

Remark 3.6. If $p\colon \mathcal{Y} \to \mathcal{X}$ is a map of CFGs over $\mathcal{S}$ then there is a pullback homomorphism $p^*\colon H^*(\mathcal{X}) \to H^*(\mathcal{Y})$. If $c \in H^*(\mathcal{X})$ and t is an object of $\mathcal{Y}(T)$ then $p^*c(t) = c(p \circ t))$. In fancier language, H^* is a functor from the 2-category of CFGs to the category of skew-commutative rings. Similar functoriality holds for the Chow cohomology groups.

To connect our theory with the usual cohomology theory schemes over $\mathbb{C}$ we have:

Proposition 3.7. *If* $\mathcal{X} = \underline{X}$ *with X a scheme then* $H^*(\mathcal{X}) = H^*(X)$. *Likewise* $A^*(\mathcal{X}) = A^*(X)$.

Proof. Given a class $c \in H^*(X)$ we define a corresponding class in $H^*(\mathcal{X})$ by $c(t) = t^*c$ for every morphism $T \xrightarrow{t} X$. Conversely, given a class $c \in H^*(\mathcal{X})$ then $c(id)$ defines a class in $H^*(X)$. □

Example 3.8. If $\mathcal{X} = [X/G]$ is a quotient CFG and V is a G-equivariant vector bundle on X then V defines Chern classes $c_i(V) \in H^{2i}(\mathcal{X})$ (and $A^i(\mathcal{X})$) as follows: Suppose $T \xrightarrow{t} \mathcal{X}$ corresponds to a torsor $E \xrightarrow{\pi} T$ and equivariant map $E \xrightarrow{f} X$. Since $\pi: E \to T$ is a G-torsor the pullback π^* induces an equivalence of categories between vector bundles on T and G-equivariant vector bundles on E. Define $c_i(V)(t)$ to be the i-th Chern class of the vector bundle on T corresponding to the G-equivariant vector bundle f^*V on E.

3.1.2. Tautological classes and boundary classes The most familiar classes on the moduli space of curves are naturally defined as elements of the cohomology or Chow cohomology ring of the CFGs $\mathcal{M}_g$, $\overline{\mathcal{M}}_g$, etc. Precisely we have:

Construction 3.9. (λ and κ classes) The assignment which assigns to any family of stable curves $\pi: X \to T$ corresponding to an object t in $\overline{\mathcal{M}}_g(T)$ the class $\lambda_i(t) = c_i(\pi_*(\omega_{X/T}))$ and $\kappa_i(t) = \pi_*(c_1(\omega_{X/T})^{i+1})$ define classes ω_i and κ_i in $A^i(\overline{\mathcal{M}}_g)$ and $H^{2i}(\overline{\mathcal{M}}_g)$.

Functoriality of H^* implies that these classes pullback to classes in $H^*(\mathcal{M}_{g,n})$ and $\overline{\mathcal{M}}_{g,n}$.

Similarly on $\overline{\mathcal{M}}_{g,n}$ we have ψ-classes:

Construction 3.10. (ψ classes) The assignment to any family of stable curves $\pi: X \to T$ with sections $\sigma_1, \ldots, \sigma_n$ corresponding to an object t in $\overline{\mathcal{M}}_{g,n}(T)$ the classes $\psi_i(t) = c_1(\sigma_i^*\omega_{X/T})$ define classes ψ_i in $H^2(\overline{\mathcal{M}}_{g,n})$ and $A^1(\overline{\mathcal{M}}_{g,n})$.

Construction 3.11. (boundary classes) The deformation theory of curves implies that the locus in $\overline{H}_g$ parametrizing tri-canonically embedded curves with nodes is a normal crossing divisor [3, Corollary 1.9]. Since $\overline{H}_g$ is non-singular this is a Cartier divisor. Let $\Delta_0(\overline{H}_g)$ be the divisor parametrizing irreducible nodal curves, and for $1 \leqslant i \leqslant [g/2]$ let $\Delta_i(\overline{H}_g)$ the divisor parametrizing curves that are the union of a component of genus i and one of genus $g - i$. Each of these divisors is G-invariant and thus defines a G-equivariant line bundle $L(\Delta_i)$ on $\overline{H}_g$. Applying the construction of Example 3.8 yields boundary classes $\delta_i = c_1(L(\Delta_i))$ in $H^2(\overline{\mathcal{M}}_g)$ and $A^1(\overline{\mathcal{M}}_g)$.

Example 3.12. The λ_i classes can be also be defined via a vector bundle on $\overline{H}_g$ as was the case for the δ_i. Let $p_H: \overline{Z}_g \to \overline{H}_g$ be the universal family of tri-canonically embedded stable curves. Then $(p_H)_*(\omega_{\overline{Z}_g/\overline{H}_g})$ is a PGL_{5g-5}-equivariant vector bundle $\overline{H}_g$ and λ_i is the i-th Chern class of this bundle as defined in Example 3.8.

We will see in the next section that every class in $H^*(\overline{\mathcal{M}}_g)$ (resp. $A^*(\overline{\mathcal{M}}_g)$) may be defined in terms of equivariant cohomology (resp Chow) classes on $\overline{H}_g$.

3.2. Quotient CFGs and equivariant cohomology

The goal of this section is to show the cohomology ring of a CFG $\mathcal{X} = [X/G]$ is equal to the equivariant cohomology (resp. Chow) ring of X. We begin by recalling the definition of equivariant cohomology.

3.2.1. Equivariant cohomology Equivariant cohomology was classically defined using the *Borel construction*.

Definition 3.13. Let G be a topological group and let EG be a contractible space on which G acts freely. If X is a G-space then we define the equivariant cohomology ring $H^*_G(X)$ to be the cohomology of the quotient space $X \times_G EG$ where G acts on $X \times EG$ by the rule $g(x, v) = (gx, gv)$.

Remark 3.14. The definition is independent of the choice of the contractible space EG. If G acts freely on X with quotient X/G then there is an isomorphism $X \times_G EG \to X/G \times EG$. Since EG is contractible it follows that $H^*_G(X) = H^*(X/G)$. At the other extreme, if G acts trivially on X then $X \times_G EG = X \times BG$ where $BG = EG/G$ and $H^*_G(X) = H^*(X) \otimes H^*(BG)$. Note that because EG and BG are infinite dimensional, $H^k_G(X)$ can be non-zero for k arbitrarily large.

Example 3.15. (The localization theorem) If $G = \mathbb{C}^*$ we may take EG to be the limit as n goes to infinity of the spaces $\mathbb{C}^n \setminus \{0\}$ with usual free action of $\mathbb{C}^*$. The quotient is the topological space $\mathbb{CP}^\infty$ so $H^*_G(pt) = H^*_G(BG) = \mathbb{Z}[t]$ where t corresponds to $c_1(\mathcal{O}(1))$. Pullback along the projection to a point implies that for any $\mathbb{C}^*$ space X the equivariant cohomology $H^*_{\mathbb{C}^*}(X)$ is a $\mathbb{Z}[t]$-algebra. Let X^T be the fixed locus for the T-action. The inclusion of the fixed locus $X^T \to X$ induces a map in equivariant cohomology $H^*_{\mathbb{C}^*}(X) \to H^*_{\mathbb{C}^*}(X^T) = H^*(X^T) \otimes \mathbb{Z}[t]$. The *localization theorem* states that this map is an isomorphism after inverting the multiplicative set of homogenous elements in $\mathbb{Z}[t]$ of positive degree. The localization theorem is very powerful because it reduces certain calculations on X to calculations on the fixed locus X^T. In many situations the fixed locus of a space is quite simple - for example a finite number of points. In [21] Kontsevich developed the theory of stable maps and used the localization theorem (and its corollary, the Bott residue formula) on the moduli space of stable maps to give recursive formulas for the number of rational curves of a given degree on a general quintic hypersurface in $\mathbb{P}^4$. Subsequently, Graber and Pandharipande [16] proved a virtual localization formula for equivariant Chow classes. Their formula has been one of the primary tools in algebraic Gromov-Witten theory. For an introduction to this subject in algebraic geometry see [1, 8].

The relationship between equivariant cohomology and the cohomology of a quotient CFG is given by the following theorem.

Theorem 3.16. *If $\mathcal{X} = [X/G]$ is a quotient CFG where X is a complex variety and G an algebraic group then $H^*(\mathcal{X}) = H^*_G(X)$.*

3.2.2. Proof of theorem Theorem 3.16 Unfortunately even if G is an algebraic group, the topological spaces EG used in the construction of equivariant cohomology are not algebraic varieties. As a result the G-torsor $X \times EG \to X \times_G EG$ is not an object in the category $\mathcal{X}$, so it is difficult to directly compare $H^*_G(X) := H^*(X \times_G EG)$ and $H^*(\mathcal{X})$.

However Totaro [33] observed that EG and BG can be approximated by algebraic varieties. Precisely, if V is a complex representation of G containing an open set U on which G acts freely and $\operatorname{codim} V \setminus U > i$ then U/G approximates BG up to real codimension $2i$; i.e., $H^k(U/G) = H^k(BG)$ for $k \leqslant 2i$. Note that for every $i \geqslant 0$ there exists a complex representation V of G such that $\operatorname{codim} V \setminus U > i$. The reason is that G may be embedded in GL_n for some n and such representations are easily constructed for GL_n.

Proposition 3.17. *If* G, U, V *are as in the preceding paragraph then* $H^k(X \times_G U) = H^k_G(X)$ *for* $k \leqslant 2i$.

Proof. First observe that if $k \leqslant 2i + 1$ then $\pi_k(U) = 0$ since U is the complement of a subspace of real codimension at least $2i + 2$ in the contractible space V. Hence $H^k(U) = 0$ for $k \leqslant 2i$. In particular, $H^k(U) = H^k(EG) = 0$ for $k \leqslant 2i$. We claim that as a consequence the quotients $X \times_G U$ and $X \times_G EG$ have the same cohomology up to degree $2i$. To see this consider the quotient $Z = X \times_G (U \times EG)$. There is a projection $p_1 \colon Z \to X \times_G U$ which is an EG-fibration and a projection $p_2 \colon Z \to X \times_G EG$ which is a U-fibration. Since $H^k(EG) = 0$ for all k, p_1^* is an isomorphism and since $H^k(U) = 0$ for $k \leqslant 2i$, p_2^* is an isomorphism in degree $k \leqslant 2i$. Hence $H^k(X \times_G U)$ is isomorphic to $H^k(X \times_G EG)$ for $k \leqslant 2i$ as claimed. □

When X is a scheme then, with mild assumptions on X or G, we may always find U such that the quotient $X \times_G U$ is also a scheme[3].

In these cases our theorem now follows from the following proposition.

Proposition 3.18. *With* G, U, V *as above if* $\mathcal{X} = [X/G]$ *then the pullback map* $H^k(\mathcal{X}) \to H^k(X \times_G U)$ *is an isomorphism for* $k \leqslant 2i$.

Proof. Let $u \colon X \times_G U \to \mathcal{X}$ be the map associated to the torsor $X \times U \to X \times_G U$ and equivariant projection map $X \times U \to X$. If $c \in H^k(\mathcal{X})$ then the image of c under the pullback map is the cohomology class $c(u) \in H^k(X \times_G U)$. Assume that $c(u) = 0$. We wish to show that $c(t) = 0$ for every object t corresponding to a torsor $E \to T$ and equivariant map $E \to X$. Given such a torsor consider the cartesian diagram of torsors.

$$\begin{array}{ccc} E \times U & \to & E \\ \downarrow & & \downarrow \\ E \times_G U & \to & T \end{array}$$

Because the fiber of the map $E \times_G U \to E$ is U the map on quotients $E \times_G U \to T$ is also a U fibration. Since $H^k(U) = 0$ for $k \leqslant 2i$, the pullback map $H^k(T) \to H^k(E \times_G U)$ is an isomorphism when $k \leqslant 2i$. By definition of c as an element of $H^k(\mathcal{X})$, $c(t)$ and $c(u)$ have equal pullbacks in $H^k(E \times_G U)$. Since $c(u) = 0$ it follows that $c(t) = 0$ as well. Since t was arbitrary we see that $c = 0$, so the map is injective.

[3]In particular if X is quasi-projective or G is connected [7, Proposition 23].

The fact that the map $H^k(T) \to H^k(E \times_G U)$ is an isomorphism also implies that our map is surjective. Given a class $c(u) \in H^k(X \times_G U)$ we let $c(t)$ be the inverse image in $H^k(T)$ of the pullback of $c(u)$ to $H^k(E \times_G U)$. □

3.2.3. Equivariant homology Totaro's approximation of EG with open sets in representations allows one to define a corresponding equivariant homology theory which is dual to equivariant cohomology when X is smooth. Let G be a g-dimensional group and let V be an l-dimensional representation of G containing an open set U on which G acts freely and whose complement has codimension greater than $\dim X - i$.

Definition 3.19 (Equivariant homology). Let X be a G-space defined over $\mathbb{C}$ and define $H_i^G(X) = H_{i+l-g}^{BM}(X \times_G U)$, where H_*^{BM} indicates Borel-Moore homology (see [14, Section 19.1] for the definition of Borel-Moore homology).

Remark 3.20. Again this definition is independent of the choice of U and V as long as the codimension of $V \setminus U$ is sufficiently high. The reason we use Borel-Moore homology theory rather than singular homology is that on a non-compact manifold it is the theory naturally dual to singular cohomology. In equivariant theory, even if X is compact, the quotients $X \times_G U$ will in general not be compact. On compact spaces Borel-Moore homology coincides with the usual singular homology. Note that our grading convention means that $H_k^G(X)$ can be non-zero for arbitrarily negative k, but $H_k^G(X) = 0$ for $k \geqslant 2 \dim X$.

Theorem 3.21 (Equivariant Poincaré duality). *Let X be smooth n-dimensional G-space defined over $\mathbb{C}$. The map $H_G^k(X) \to H_{2n-k}^G(X)$, $c \mapsto c \cap [X]_G$ is an isomorphism.*

3.3. Equivariant Chow groups

When we work over an arbitrary field Totaro's approximation of EG lets us define equivariant Chow groups. The theory of equivariant Chow groups was developed in the paper [7].

Definition 3.22. [7] Let X be G-scheme defined over an arbitrary field. Then with the notation as in Definition 3.19 define $A_i^G(X) = A_{i+l-g}(X \times_G U)$.

Remark 3.23. Again the definition is independent of the choice of U, V.

Because equivariant Chow groups are defined as Chow groups of schemes they enjoy the same formal properties as ordinary Chow groups. In particular, if X is smooth then there is an intersection product on the equivariant Chow groups. Note, however, that cycles may have negative dimension; i.e. $A_k^G(X)$ can be non-zero for $k < 0$. As a consequence one cannot conclude that a particular equivariant intersection product is 0 for dimensional reasons.

For schemes over $\mathbb{C}$ the cycle maps $A_k(X \times_G U) \to H_{2k}^{BM}(X \times_G U)$ of [14, Chapter 19] define cycle maps on equivariant homology $cl\colon A_k^G(X) \to H_{2k}^G(X)$. When X is smooth of complex dimension n the latter group can be identified with $H_G^{2n-k}(X)$.

If X is a G-space then any G-invariant subvariety V of dimension k defines an equivariant fundamental class $[V]_G$ in $H^G_{2k}(X)$ and $A^G_k(X)$. However, unless G acts with finite stabilizers then $A^G_*(X)$ is not even rationally generated by fundamental classes of invariant subvarieties.

3.3.1. Equivariant Chow cohomology The definition of equivariant Chow cohomology is analogous to the definition of ordinary Chow cohomology.

Definition 3.24. An equivariant Chow cohomology class $c \in A^i_G(X)$ is an assignment of an operation $c(t): A^G_*(T) \to A^G_{*-i}(T)$ for every equivariant map $T \to X$ such that the $c(t)$ are compatible with the usual operations in equivariant intersection theory (flat pullback and lci pullback, proper pushforward, etc.). Composition makes the set of equivariant Chow cohomology classes into a graded ring which we denote $A^*_G(X) := \oplus_{i=0}^{\infty} A^i_G(X)$.

Remark 3.25. Because equivariant Chow groups can be non-zero in arbitrary negative degree $A^k_G(X)$ may be non-zero for arbitrarily large k.

Proposition 3.26. [7, Proposition 19] *If $\mathcal{X} = [X/G]$ is a quotient CFG then $A^*(\mathcal{X}) = A^*_G(X)$.*

Given a representation V and an open set U on which G acts freely such that $\operatorname{codim}(V \setminus U) > i$ then we can compare the Chow cohomology of $X \times_G U$ with $A^*_G(X)$.

Proposition 3.27. [7, Corollary 2] *If X has an equivariant resolution of singularities then $A^k_G(X) = A^k(X \times_G U)$ for $k < i$.*

Applying Proposition 3.27 and the Poincaré duality isomorphism between Chow groups and Chow cohomology we obtain.

Proposition 3.28. *If X is smooth of dimension n then there is an isomorphism $A^k(\mathcal{X}) \to A^G_{n-k}(X)$ where $\mathcal{X} = [X/G]$. Moreover, the ring structure on $A^*(\mathcal{X})$ given by composition of operations is compatible with the ring structure on $A^G_*(X)$ given by equivariant intersection product.*

Remark 3.29. Putting the various propositions together implies that if X is smooth, then a G-invariant subvariety $V \subset X$ defines a Chow cohomology class $c_V \in A^*([X/G])$. If G acts with finite stabilizers then the classes c_V generate $A^*([X/G])$ rationally [7, Proposition 13].

In addition when X is smooth over $\operatorname{Spec}\mathbb{C}$ there is a degree-doubling cycle class map $cl\colon A^*([X/G]) \to H^*([X/G])$ having the same formal properties as the cycle class map on smooth complex varieties.

3.3.2. Picard groups of moduli problems In [26, Section 5] Mumford defined the integral Picard group of what he called the *moduli problem* $\mathcal{M}_g$. As Mumford noted, the definition can be expressed in the language of fibered categories and makes sense for an arbitrary CFG over $\mathcal{S}$.

Definition 3.30. [26, p.64] Let $\mathcal{X}$ be a CFG. A line bundle on $\mathcal{X}$ is the assignment to every scheme T and every object t in $\mathcal{X}(T)$ a line bundle $L(t)$ on T. The line bundles $L(t)$ should satisfy the natural compatibility conditions with respect to pullbacks in the category $\mathcal{X}$. Tensor product makes the collection of line bundles on $\mathcal{X}$ into an abelian group $\mathrm{Pic}(\mathcal{X})$.

If L is a line on $\mathcal{X}$ then L has a first Chern class $c_1(L) \in A^1(\mathcal{X})$.

Proposition 3.31. [7, Proposition 18] *Let $\mathcal{X} = [X/G]$ be a quotient CFG with X smooth. Then the map* $\mathrm{Pic}(\mathcal{X}) \to A^1(\mathcal{X})$, $L \mapsto c_1(L)$ *is an isomorphism.*

Example 3.32. Since $\overline{\mathcal{M}}_g = [\overline{H}_g/PGL_{5g-5}]$ and $\overline{H}_g$ is smooth, the group $A^1(\overline{\mathcal{M}}_g)$ defined here may identified with the group $\mathrm{Pic}_{fun}(\overline{\mathcal{M}}_g)$ of [18, Definition 3.87].

3.4. Equivariant cohomology and the CFG of curves

The results of Sections 3.2 and 3.3 imply that the cohomology (resp. Chow) rings of the CFGs $\mathcal{M}_g$ and $\overline{\mathcal{M}}_g$ are identified with the PGL_{5g-6}-equivariant cohomology (resp. Chow) rings of the smooth varieties H_g and $\overline{H}_g$. In particular any invariant subvariety of H_g or $\overline{H}_g$ defines an integral class in $H^*(\mathcal{M}_g)$ or $H^*(\overline{\mathcal{M}}_g)$.

The description of the cohomology of $\mathcal{M}_g$ as the equivariant cohomology of H_g gives, in principal, a method for computing the integral cohomology ring $H^*(\mathcal{M}_g)$. Unfortunately, for $g > 2$ there are no effective methods for carrying out computations. Note that the rings $A^*(\mathcal{M}_g)$ and $H^*(\mathcal{M}_g)$ will have non-zero torsion in arbitrarily high degree. However, $H^k(\mathcal{M}_g) \otimes \mathbb{Q} = 0$ if $k > 6g - 6$ and $A^k(\mathcal{M}_g) \otimes \mathbb{Q} = 0$ if $k > 3g - 3$ (Proposition 4.39).

Example 3.33 (The stable cohomology ring of $\mathcal{M}_g$). The best results on the cohomology of $\mathcal{M}_g$ use the fact that its rational cohomology is the same as the rational cohomology of the mapping class group. In [17] Harer used topological methods to prove that $H^k(\mathcal{M}_g) \otimes \mathbb{Q}$ stabilizes for $g \geqslant 3k$. Thus we can define the stable cohomology ring of $\mathcal{M}_g$ as the limit of the $H^k(\mathcal{M}_g) \otimes \mathbb{Q}$ as $g \to \infty$. In [25] Madsen and Weiss proved Mumford's conjecture: the stable cohomology ring of $\mathcal{M}_g$ as $g \to \infty$ is isomorphic to the ring $\mathbb{Q}[\kappa_1, \kappa_2, \ldots]$.

Since the stable cohomology ring of $\mathcal{M}_g$ as $g \to \infty$ is algebraic, a natural question is the following:

Question 3.34. Is there a version of Harer's stability theorem for the Chow groups of $\mathcal{M}_g$?

To prove the stability theorem Harer used the fact that there are maps $H^k(\mathcal{M}_g) \to H^k(\mathcal{M}_{g+1})$ induced by the topological operation of cutting out two disks from a Riemann surface of genus g and gluing in a "pair of pants" to obtain a Riemann surface of genus $g + 1$. Since there is no algebraic counterpart to this operation, Question 3.34 seems to be out of reach of current techniques.

Example 3.35 (Integral Chow rings of curves of very low genus). In very low genus and for hyperelliptic curves the methods of equivariant algebraic geometry have been successfully used to compute integral Chow rings.

In genus 0 the Chow ring of $\mathcal{M}_0^0 = \mathrm{B\,PGL}_2$ was computed by Pandharipande [30]. The Chow ring of $\mathcal{M}_0^{\leqslant 1}$ was computed by Fulghesu and the author [5]. The rational Chow ring of $\mathcal{M}_0^{\leqslant 3}$ was computed by Fulghesu in [11].

In genus 1, the Chow rings of $\mathcal{M}_{1,1}$ and $\overline{\mathcal{M}}_{1,1}$ were computed in [7]. In [35] Vistoli computed the Chow ring of $\mathcal{M}_2$. Fulghesu, Viviani and the author [6, 13] extended Vistoli's techniques to compute, for all g, the Chow ring of the CFG $\mathcal{H}_g$ parametrizing hyperelliptic curves.

In each case the presentations of the Chow rings are relatively simple. The reader may refer to the cited papers for the precise statements.

In a sense, the greatest value of the equivariant point of view is psychological. It turns intersection theory on the CFGs $\mathcal{M}_g$ and $\overline{\mathcal{M}}_g$ into equivariant intersection theory on the smooth varieties H_g and $\overline{H}_g$. In this way we can avoid worrying about the intricate rational intersection theory on orbifolds and stacks developed by Mumford [28], Vistoli [34], and Gillet [15]. In addition, because the intersection theory has integer coefficients, we may obtain slightly stronger results. We illustrate with a simple example from [28].

Proposition 3.36. *The following relation holds in the integral Chow ring* $A^*(\mathcal{M}_g)$

$$(1 + \lambda_1 + \cdots + \lambda_g)(1 - \lambda_1 + \cdots + (-1)^g \lambda_g) = 1 \tag{3.37}$$

Proof. Apply the argument of [28] pp. 306 to the family of tri-canonically smooth curves $Z_g \stackrel{\pi}{\to} H_g$ to conclude that $c(\mathbb{E}_{H_g})c(\check{\mathbb{E}}_{H_g}) = 1$ where $\mathbb{E}_{H_g} = \pi_*(\omega_{Z_g/H_g})$ is the Hodge bundle on H_g. Under the identification $A^*(\mathcal{M}_g) = A^*_G(H_g)$ the tautological class λ_i identifies with $c_i(\mathbb{E}_{H_g})$. □

Mumford proved a number of other relations using the Grothendieck-Riemann-Roch theorem. Again these can be derived using the equivariant version of the Grothendieck-Riemann-Roch theorem for the map $\overline{Z}_g \to \overline{H}_g$. Since the formulas for the Chern character involves denominators, Mumford's formulas hold after inverting the primes dividing the denominators. A natural question is the following.

Question 3.38. Do the tautological relations obtained by Mumford in [28] hold in $A^*(\mathcal{M}_g)$ or $A^*(\overline{\mathcal{M}}_g)$ after minimally clearing denominators? Similarly, do Faber's relations [10] hold after minimally clearing denominators?

Remark 3.39. The author does not have any particular insight into this question. However, we point out that Mumford proved [27, Theorem 5.10] that the relation

$$12\lambda_1 = \kappa_1 + \delta \tag{3.40}$$

holds integrally in $\mathrm{Pic}(\overline{\mathcal{M}}_g) = A^1(\overline{\mathcal{M}}_g)$ where $\delta = \sum_{i=0}^{[g/2]} \delta_i$. Mumford was able to show this because he used transcendental methods to prove that $\mathrm{Pic}^G(\overline{H}_g) = \mathrm{Pic}(\overline{\mathcal{M}}_g) = A^1(\overline{\mathcal{M}}_g)$ is torsion free.

4. Stacks, moduli spaces and cohomology

The goal of this final section is to define Deligne-Mumford stacks and their moduli spaces. As mentioned in the introduction, stacks are CFGs with certain algebro-geometric properties. The CFGs $\mathcal{M}_g$, $\overline{\mathcal{M}}_g$, $\overline{\mathcal{M}}_{g,n}$ and $\overline{\mathcal{M}}_{g,n}$ are all Deligne-Mumford stacks whenever $2g-2+n>0$. Associated to a Deligne-Mumford stack is a *coarse moduli space*. In general the coarse moduli space of a Deligne-Mumford stack is an algebraic space, but the moduli spaces of the stacks of smooth and stable curves are in fact quasi-projective and projective varieties. In Section 4.3 we will construct an isomorphism between the rational Chow groups of a quotient Deligne-Mumford stack and the rational Chow groups of its moduli space. This isomorphism defines an intersection product on the Chow groups of the moduli space. A similar result also holds for homology.

4.1. Deligne-Mumford stacks

So far we have not imposed any algebro-geometric conditions on our CFGs. A Deligne-Mumford stack is a CFG which is, in an appropriate sense, "locally" a scheme.

Construction 4.1. Given a CFG $\mathcal{X}$ choose for each morphism of scheme $T' \xrightarrow{f} T$ and each object t in $\mathcal{X}(T)$ an object f^*t in $\mathcal{X}(T')$ such that there is a morphism $f^*t \to t$ whose image in $\mathcal{S}$ is the map $f\colon T' \to T$. The axioms of a CFG imply that the f^*t always exist and are unique up to canonical isomorphism.

Definition 4.2. Let $\mathcal{X}$ be a CFG. Let T be an S-scheme and let $\{T_i \xrightarrow{p_i} T\}$ be a collection of étale maps whose images cover T. Set $T_{ij} = T_i \times_T T_j$ and identify $T_{ij} = T_{ji}$ for all i, j. Similarly denote by T_{ijk} any of the products canonically isomorphic to $T_{ij} \times_{T_j} T_{jk}$. If t_i is an object $\mathcal{X}(T_i)$ let t_{ij} be the pullback of t_i to $\mathcal{X}(T_{ij})$ along the projection map $T_{ij} \to T_i$. A *descent datum* for the covering $\{T_i \to T\}$ is a collection of objects t_i in $\mathcal{X}(t_i)$ together with isomorphisms $\phi_{ij}\colon t_{ij} \to t_{ji}$ in $\mathcal{X}(T_{ij})$ which satisfy the cocycle condition when pulled back to $\mathcal{X}(T_{ijk})$; i.e., $\phi_{jk} \circ \phi_{ij} = \phi_{ik}$ after pullback to $\mathcal{X}(T_{ijk})$.

Definition 4.3. [3, Definition 4.6] A CFG $\mathcal{X}$ over $\mathcal{S}$ is a *Deligne-Mumford (DM) stack* if the following 3 conditions are satisfied.

i) The diagonal map $\mathcal{X} \to \mathcal{X} \times_S \mathcal{X}$ is representable, quasi-compact and and separated.[4]

[4]The representability of the diagonal map is equivalent to the condition that any morphism from a scheme $T \to \mathcal{X}$ is representable [34, Proposition 7.13].

ii) (Effectivity of descent) If T is an S-scheme and $\{T_i \xrightarrow{p_i} T\}$ is an étale cover of T then given descent datum relative to this covering there exists an object t in $\mathcal{X}$ and isomorphisms $\phi_i \colon p_i^* t \to t_i$ in $\mathcal{X}(T_i)$ such that after pullback to $\mathcal{X}(T_{ij})$ $\phi_{ij} = \phi_j \circ \phi_i^{-1}$.

iii) There exists an étale surjective morphism from a scheme $U \to \mathcal{X}$. (The representability of the diagonal implies that any $U \to \mathcal{X}$ is representable, so it makes sense to talk about this map being étale and surjective.)

Remark 4.4. In [3] the term algebraic stack was used for a CFG satisfying the conditions of Definition 4.3. However, recent works reserve the term *algebraic stack* for a more general class of stacks defined by Artin (also called Artin stacks). Note that we do not define a *stack* - only a *Deligne-Mumford stack.* A stack is a generalization to CFGs of the concept of an étale sheaf. The definition of a stack is given in [3, Definition 4.1]. A CFG satisfying conditions i) and ii) of Definition 4.3 is a stack in the sense of Definition 4.1 of [3] and all stacks satisfy condition (ii) of our definition. However, not all stacks satisfy condition (i).

Remark 4.5. If $\mathcal{X}$ is a CFG associated to a functor i) and ii) imply that $\mathcal{X}$ is a sheaf in the étale topology. A Deligne-Mumford stack which is a sheaf is called an *algebraic space.*

4.1.1. The diagonal and automorphisms To give a morphism $T \to \mathcal{X} \times \mathcal{X}$ is equivalent to giving a pair of objects t_1 in $\mathcal{X}(T)$ and t_2 in $\mathcal{X}(T)$. If $\mathcal{X}$ is a CFG with a representable diagonal then the fiber product $\mathcal{X} \times_{\mathcal{X} \times \mathcal{X}} T$ is the scheme which represents the functor Hom_{t_1,t_2}. This functor assigns to any T-scheme $T' \xrightarrow{f} T$ the set of isomorphisms in $\mathcal{X}(T')$ between f^*t_1 and f^*t_2. (Note that for any CFG and any scheme T the fiber product $\mathcal{X} \times_{\mathcal{X} \times \mathcal{X}} T$ is automatically the CFG associated to the functor Hom_{t_1,t_2} [23, Lemma 2.4.1.4].) In particular, if $t_1 = t_2 = t$ then $\mathcal{X} \times_{\mathcal{X} \times \mathcal{X}} T$ is the scheme which represents the automorphism group-scheme $\mathrm{Aut}(t) \to T$ whose fiber over a point $p \colon \operatorname{Spec} k \to T$ is the automorphism group of p^*t.

The condition that there is an étale surjective morphism from a scheme $U \to \mathcal{X}$ implies that the these automorphism groups are finite.

Proposition 4.6. [34, Proposition 7.15] *If* $\mathcal{X}$ *is a DM stack then the diagonal* $\mathcal{X} \to \mathcal{X} \times \mathcal{X}$ *is unramified. Equivalently, for every algebraically closed field* K *and every object* x *in* $\mathcal{X}(\operatorname{Spec} K)$ *the automorphism group* $\mathrm{Aut}(x)$ *is finite and reduced over* K.

4.1.2. Quotient CFGs and Deligne-Mumford stacks Let X be a scheme of finite type over S. If G is an smooth affine group scheme over S then it is relatively straightforward to show that the $[X/G]$ satisfies the effective descent condition of Definition 4.3 and the diagonal of $\mathcal{X} = [X/G]$ is representable, quasi-compact and separated. The only question is whether there exists an étale surjective morphism $U \to \mathcal{X}$. If G is étale over S (for example if S is a point and G is finite) then the map $X \to \mathcal{X}$ is étale so $[X/G]$ is a DM stack. Unfortunately, CFGs such as $\mathcal{M}_g$ are quotients by non-finite groups. The next theorem gives a criterion for $[X/G]$ to be a DM stack.

Theorem 4.7. *Let G be a smooth group scheme over S. A quotient CFG $\mathcal{X} = [X/G]$ is a DM stack if and only if the stabilizer of every geometric point of X is finite and reduced.*

Remark 4.8. Over a field of characteristic 0 every group scheme is smooth, so the the stabilizers of geometric points are always reduced. In this case $[X/G]$ is a DM stack if and only if G acts on X with finite stabilizers.

Proof sketch. Let $\mathcal{X} = [X/G]$ be a quotient CFG and let x be an object of $\mathcal{X}(\operatorname{Spec} K)$ where K is an algebraically closed field. Since any torsor $E \to \operatorname{Spec} K$ is trivial the isomorphism class of x is determined by the G-equivariant map $E \to X$. Since $E \to X$ is G-equivariant and E consists of a single G-orbit the image of E in X is an orbit. Hence objects of $\mathcal{X}(\operatorname{Spec} K)$ correspond to G-orbits. The automorphism group of an object of $\mathcal{X}(\operatorname{Spec} K)$ is the just the stabilizer of that orbit. Hence, a necessary condition for $[X/G]$ to be a DM stack is that the stabilizer of every point is finite and reduced.

The converse follows from the following theorem, originally stated by Deligne and Mumford. □

Theorem 4.9. [3, Thm 4.21] *Let $\mathcal{X}$ be a CFG which satisfies conditions i) and ii) of Definition 4.3. Assume in addition that*

i′) The diagonal $\mathcal{X} \to \mathcal{X} \times \mathcal{X}$ is unramified.

ii′) There exists a smooth surjective map from a scheme $X \to \mathcal{X}$.

Then $\mathcal{X}$ is a DM stack.

Remark 4.10. In characteristic 0 there is a relatively simple proof of Theorem 4.9 when G is reductive, X is smooth (or even normal) and covered by G-invariant affine open sets. The proof goes as follows. If x is a point of X then x is an contained in an affine G-invariant open set U_x. Conditions i) and ii) imply that the stabilizers of the G-action are finite so the G-orbits of closed points are also closed. The étale slice theorem [24] then implies that for closed points there is a smooth affine variety W_x and G-equivariant strongly étale map[5] $G \times_{G_x} W_x \to U_x$ (Here G_x is the stabilizer of $x \in X$). The collection of $\{W_x\}_{x \in X}$ forms an étale cover of $[X/G]$.

4.1.3. CFGs of stable curves are DM stacks

Proposition 4.11. *The CFGs $\mathcal{M}_g$, $\overline{\mathcal{M}}_g$ are DM stacks if $g \geqslant 2$.*

Proof. Since $\overline{\mathcal{M}}_g = [\overline{H}_g/\mathrm{PGL}_{5g-5}]$ we know that it satisfies conditions i) and ii) of Definition 4.3 as well as condition ii′) of Theorem 4.9. The only thing to show is that the diagonal is unramified. This is equivalent to showing that Aut(C) is finite and unramified for every stable curve C defined over an algebraically closed field K. The last fact follows from [3, Lemma 1.4] which implies that C does not have any infinitesimal automorphisms. □

[5] A G-equivariant map of affine varieties $\operatorname{Spec} A \to \operatorname{Spec} B$ is *strongly étale* if the induced map $\operatorname{Spec} A^G \to \operatorname{Spec} B^G$ is étale and $A = B \otimes_{B^G} A^G$. These conditions imply that the map $\operatorname{Spec} A \to \operatorname{Spec} B$ is also étale.

A similar argument shows that the CFG $\overline{\mathcal{M}}_{1,1}$ is a DM stack as well.

Proposition 4.12. *If* $2g - 2 + n > 0$ *then the CFGs* $\mathcal{M}_{g,n}$ *and* $\overline{\mathcal{M}}_{g,n}$ *are DM stacks.*

Proof. Recall that $\overline{\mathcal{M}}_{g,n+1} \to \overline{\mathcal{M}}_{g,n}$ is representable by Knudsen's theorem. The proof follows by induction and the following simple lemma. □

Lemma 4.13. *Let* $\mathcal{Y} \to \mathcal{X}$ *be a representable morphism of CFGs. If* $\mathcal{X}$ *is a DM stack then* $\mathcal{Y}$ *is as well.*

Example 4.14. (Curves of very low genus) The CFGs $\mathcal{M}_0$ and $\mathcal{M}_1$ are not DM stacks since the automorphism group of $\mathbb{P}^1$ is PGL_2 and the automorphism group of a curve E of genus 1 contains the curve itself.

4.1.4. Separated and proper DM stacks We briefly discuss what it means for a DM stack to be separated or proper over the ground scheme S. There is also a notion of separation and properness for arbitrary morphisms of DM stacks but we do not discuss this here. Note that we have already implicitly defined what it means for a representable morphism to be separated or proper, since these properties are preserved by base change (Remark 2.29).

Definition 4.15. A DM stack $\mathcal{X}$ is *separated* over S if the diagonal $\Delta\colon \mathcal{X} \to \mathcal{X} \times_S \mathcal{X}$ is proper.

Remark 4.16. Since the diagonal of a DM stack is unramified, a DM stack is separated if and only if the diagonal is finite.

Not surprisingly for quotient DM stacks the separation condition can be characterized in terms of the group action.

Proposition 4.17. *A quotient stack* $\mathcal{X} = [X/G]$ *is separated if and only if* G *acts* properly *on* X*; i.e. the map* $G \times X \to X \times X$, $(g, x) \mapsto (x, gx)$ *is a proper morphism.*

The following result from GIT shows that the properness of a group action is a relatively natural condition.

Theorem 4.18. [29, Converse 1.13] *Let* G *act on a projective variety and assume that the generic stabilizer is finite. Let* $X^s(L)$ *be the set of stable points with respect to an ample line bundle* L*. Then* G *acts properly on* $X^s(L)$.

To define the notion of a stack being proper over S we invoke the following result.

Theorem 4.19. [9, Theorem 2.7] *If* $\mathcal{X}$ *is a DM stack then there exits a finite surjective morphism from a scheme* $Y \to \mathcal{X}$.

Remark 4.20. When $\mathcal{X} = [X/G]$ is a quotient stack then Theorem 4.19 implies that there exists a finite surjective map $X' \to X$ such that G acts freely on X' and $X' \to Y$ is a G-torsor. This consequence was originally proved by Seshadri [31, Theorem 6.1] and the proof of Theorem 4.19 is an adaptation of Seshadri's argument to stacks.

Definition 4.21. A separated DM stack $\mathcal{X}$ is *proper* over S if there exists a finite surjective morphism from a scheme $Y \to \mathcal{X}$ with Y proper over S.

Remark 4.22. As is the case for morphisms of schemes there are valuative criteria for separation and properness of morphisms [3, Theorem 4.18, 4.19]. We do not state them here, but Deligne and Mumford used the valuative criteria together with the stable reduction theorem [3, Corollary 2.7] to prove the following theorem.

Theorem 4.23. *The DM stack $\overline{\mathcal{M}}_g$ is proper over* $\operatorname{Spec}\mathbb{Z}$.

4.2. Coarse moduli spaces of Deligne-Mumford stacks

The coarse moduli space of a DM stack is a space whose geometric points correspond to isomorphism classes of objects over the corresponding algebraically closed field. Before we give a definition we feel obliged to issue a warning.

Warning 4.24. Given a stack $\mathcal{X}$ such as $\overline{\mathcal{M}}_g$ one can define the *coarse moduli functor*. This is the contravariant functor $F_{\mathcal{X}}\colon \mathcal{S} \to$ Sets which assigns to any scheme T the set of isomorphism classes in the category $\mathcal{X}(T)$. **The coarse moduli space of $\mathcal{X}$ does not represent the coarse moduli functor** $F_{\mathcal{X}}$. For one thing the functor $F_{\mathcal{X}}$ is not in general a sheaf in étale topology. The reason is that two objects of a stack may become isomorphic after base change to an étale cover. For example, a trivial and iso-trivial family of curves are not isomorphic but become so after étale base change. For $F_{\mathcal{X}}$ to be represented by a scheme or algebraic space it would have to be an étale sheaf. One may attempt to replace $F_{\mathcal{X}}$ by its associated sheaf in the étale topology. However, the associated sheaf need not have an étale cover by a scheme.

Definition 4.25. Let $\mathcal{X}$ be a separated DM stack. A scheme M is a coarse moduli scheme for $\mathcal{X}$ if there is a map $p\colon \mathcal{X} \to M$ such that

i) p is universal for maps from $\mathcal{X}$ to schemes; i.e. given a morphism $q\colon \mathcal{X} \to Z$ with Z a scheme, there is a unique morphism $f\colon M \to Z$ such that $f \circ p = q$.

ii) If K is an algebraically closed field then there is a bijection between the points of M(K) and the set of isomorphism classes of objects in the groupoid $\mathcal{X}(K)$.

Proposition 4.26. *Let* G *be an algebraic group acting on a scheme* X. *If* $X \to M$ *is a geometric quotient in the sense of* [29, Definition 0.6] *then* M *is a coarse moduli scheme for* $\mathcal{X} = [X/G]$.

Proof. By definition, geometric points of M correspond to G-orbits in X. Since G-orbits in X correspond to isomorphism classes of objects in $\mathcal{X}$ condition ii) is satisfied. To give a map $[X/G] \to Y$ is equivalent to giving a G-invariant map $X \to Y$. By [29, Proposition 0.6] a geometric quotient is a categorical quotient which implies that p satisfies condition i). □

In general a DM stack need not have a coarse moduli scheme, but the universal property guarantees that M is unique if it exists. Keel and Mori [19] proved that every Artin stack with finite stabilizer [6] has a coarse moduli space in the category of algebraic spaces. Their result implies the following theorem for Deligne-Mumford stacks.

Theorem 4.27. [19, 2] *Given a separated DM stack $\mathcal{X}$ there exists a separated algebraic space* M *and a map* $p \colon \mathcal{X} \to M$ *such that:*

i) for every algebraically closed field K, p *induces a bijection between the set* $M(K)$ *and the set of isomorphism classes in* $\mathcal{X}(K)$

ii) p *is universal for maps to algebraic spaces.*

Moreover, if $\mathcal{X}$ is proper over S *then so is* M.

Definition 4.28. The algebraic space M associated to the separated DM stack $\mathcal{X}$ is called a coarse moduli space. Again, the universal property implies that M is unique up to isomorphism.

Since the universal property for maps to algebraic spaces is a stronger than the universal property for maps to schemes, it is a priori possible for a DM stack to have an algebraic space as a coarse moduli space and a different scheme as a coarse moduli scheme. Fortunately, this does not occur for separated DM stacks.

Proposition 4.29. *(cf.* [19, Proposition 9.1]*) If* M *is a coarse moduli scheme for a separated DM stack* $\mathcal{X}$ *then* M *is a coarse moduli space.*

Example 4.30 (Pathologies of non-separated stacks). If we relax the separation hypothesis then a DM stack may have a coarse moduli space which is not isomorphic to its coarse moduli scheme. This is essentially the phenomenon of Example 0.4 of [29]. In that example Mumford shows that $\mathbb{A}^1$ is the geometric quotient by SL_2 of an open set $X \subset \mathbb{A}^5$. The action of SL_2 is defined in such a way that it is free but not proper. In this case $\mathbb{A}^1$ is the coarse moduli scheme of the non-separated stack $[X/SL_2]$. Since the action is free, the stack $[X/SL_2]$ is in fact a non-separated algebraic space and so is its own coarse moduli space. By restricting our focus to *separated* stacks we avoid this difficulty.

4.2.1. Singularities of coarse moduli spaces It is well known that coarse moduli spaces of smooth stacks, like $\mathcal{M}_g$ and $\overline{\mathcal{M}}_g$, have quotient singularities. Using the language of stacks we can make this precise.

Theorem 4.31. *Let $\mathcal{X}$ be a DM stack with coarse moduli space* M. *Then for every point* m *of* M *there is an affine scheme* U, *a finite group* H, *a representable étale morphism*

---

[6] An algebraic stack $\mathcal{X}$ has *finite stabilizer* if the representable morphism $I_{\mathcal{X}} \to \mathcal{X}$ is finite, where $I_{\mathcal{X}} = \mathcal{X} \times_{\mathcal{X} \times \mathcal{X}} \mathcal{X}$ is the inertia stack. For quotient stacks this condition is equivalent to the requirement that the map $I_G(X) \to X$ is finite where $I_G(X) = \{(g, x) | gx = x\}$. Any separated stack has finite stabilizer but not every stack with finite stabilizer is separated.

$[U/H] \to \mathcal{X}$ *and a cartesian diagram*

$$\begin{array}{ccc} [U/H] & \to & \mathcal{X} \\ \downarrow & & \downarrow \\ U/H & \to & M \end{array}$$

such that the image of U/H *in* M *contains* m.

Remark 4.32. The result stated here is a special case of [19, Proposition 4.2]. In characteristic 0 if $\mathcal{X} = [X/G]$ with X smooth, G reductive and X covered by G-invariant affine open sets then Theorem 4.31 follows from the étale slice theorem (cf. Remark 4.10).

A corollary of Theorem 4.31 is the converse to Proposition 4.26

Corollary 4.33. *If* $\mathcal{X} = [X/G]$ *is a quotient DM and* $\mathcal{X} \to M$ *is a coarse moduli scheme then* M *is a geometric quotient of* X *by the action of* G.

4.2.2. Moduli spaces of curves The following is well known result was originally proved by Mumford and Knudsen.

Theorem 4.34. *If* $g \geqslant 2$ *then the DM stack* $\overline{\mathcal{M}}_g$ *has a coarse moduli scheme* $\overline{M}_g$ *which is a projective variety.*

Since Knudsen proved that the contraction morphisms $\overline{\mathcal{M}}_{g,n+1} \to \overline{\mathcal{M}}_{g,n}$ are representable and projective we can use induction to obtain the following Corollary of Theorem 4.34

Corollary 4.35. *If* $2g - 2 + n > 0$ *the DM stack* $\overline{\mathcal{M}}_{g,n}$ *stack has a projective coarse moduli scheme* $\overline{M}_{g,n}$.

The proof of theorem 4.34 is based on showing that a geometric quotient of the quasi-projective variety $\overline{H}_g$ by PGL_{5g-5} exists as a projective variety. The main technique is Mumford's geometric invariant theory. However, the proof is indirect and also uses techniques developed by Gieseker on asymptotic stability. Chapter 4 of the book [18] gives an excellent exposition of the proof of the projectivity of $\overline{M}_g$.

4.3. Cohomology of Deligne-Mumford stacks and their moduli spaces

In this final section we compare the cohomology/Chow ring of a quotient DM stack with the cohomology/Chow ring of its moduli space.

Definition 4.36. If $\mathcal{X}$ is a DM stack then we define $\dim \mathcal{X}$ to be $\dim U$ where $U \to \mathcal{X}$ is any étale surjective map from a scheme.

An immediate consequence of Theorem 4.31 is the following proposition.

Proposition 4.37. *If* $\mathcal{X}$ *is a separated DM stack and* $\mathcal{X} \to M$ *is a coarse moduli scheme, then* $\dim X = \dim M$.

For quotient stacks we can compute the dimension equivariantly.

Proposition 4.38. *If* $\mathcal{X} = [X/G]$ *is a quotient DM stack then* $\dim \mathcal{X} = \dim X - \dim G$.

The cohomology and Chow cohomology groups of CFG may be non-zero in arbitrarily high degree. The next results shows that the rational cohomology groups of a DM stack vanish in degree more than the real dimension of the stack.

Proposition 4.39. *Let* $\mathcal{X}$ *be a DM stack.*

i) If $\mathcal{X}$ *is defined over* $\mathbb{C}$, $H^k(\mathcal{X}) \otimes \mathbb{Q} = 0$ *for* $k > 2\dim_{\mathbb{C}} \mathcal{X}$.

ii) If $\mathcal{X}$ *is defined over an arbitrary field,* $A^k(\mathcal{X}) \otimes \mathbb{Q} = 0$ *for* $k > \dim \mathcal{X}$.

Proof. Suppose that $c \in H^k(\mathcal{X}) \otimes \mathbb{Q}$ with $k > 2\dim \mathcal{X}$. We will show that for any scheme T and map $T \to \mathcal{X}$ corresponding to an object t of $\mathcal{X}(T)$ the cohomology class c_T is 0. Let $Z \xrightarrow{z} \mathcal{X}$ be a finite surjective map from a scheme and let $Z_T = Z \times_{\mathcal{X}} T$ so we have a 2-cartesian diagram

$$\begin{array}{ccc} Z_T & \xrightarrow{z'} & T \\ t' \downarrow & & t \downarrow \\ Z & \xrightarrow{z} & \mathcal{X} \end{array}$$

Since $\dim Z = \dim \mathcal{X}$ we know that $c(z) = 0$. On the other hand, functoriality implies that $z'^*c(t) = t'^*c(z) = 0$. Now $Z_T \xrightarrow{z'} T$ is a finite surjective morphism of schemes. Hence it is a ramified covering of the underlying topological spaces. By [32] there is a transfer map $z'_*\colon H^*(Z_T, \mathbb{Q}) \to H^*(T, \mathbb{Q})$ such that $z'_*z'^*$ is multiplication by the degree of z'. Hence $c(t)$ is 0.

The proof for Chow rings is similar. □

For quotient DM stacks with a coarse moduli scheme we obtain a sharper result.

Theorem 4.40. *Let* $\mathcal{X} = [X/G]$ *be a DM quotient stack and let* $p\colon \mathcal{X} \to M$ *be a coarse moduli scheme.*

i) If $\mathcal{X}$ *is defined over* $\mathbb{C}$ *then there are isomorphisms* $H^*(\mathcal{X}) \otimes \mathbb{Q} \to H^*(M, \mathbb{Q})$ *and* $H_*(\mathcal{X}) \otimes \mathbb{Q} \to H_*(M, \mathbb{Q})$.

ii) In the algebraic category there analogous isomorphisms $A^*(\mathcal{X}) \otimes \mathbb{Q} \to A^*(M) \otimes \mathbb{Q}$ *and* $A_*(\mathcal{X}) \to A_*(M) \otimes \mathbb{Q}$.

Proof. The proof in cohomology is very simple. The coarse moduli space M is topologically the space of G-orbits X/G. Let $q\colon X \to X/G$ be the quotient map. There is a map of quotients $X \times_G EG \to X/G$ whose fiber at a point $m \in X/G$ is the quotient EG/G_x where G_x is the stabilizer of any point in the orbit $q^{-1}(m)$. Because G_x is finite and EG is acyclic the fiber EG/G_x is $\mathbb{Q}$-acyclic. Hence the pullback $H^*(X/G, \mathbb{Q}) \to H^*(X \times_G EG, \mathbb{Q})$ is an isomorphism.

The proof in intersection theory is more difficult. It makes use of the fact that there exists a finite surjective morphism form a scheme $Z \to \mathcal{X}$. The proof is given in Section 4 of [7] and the results is valid even if M is an algebraic space. □

As a corollary of Theorem 4.40 and equivariant Poincaré duality (Theorem 3.21) we conclude that there is an intersection product on the rational homology groups of the moduli space of a smooth DM quotient stack. Similarly, Proposition 3.28 implies that the rational Chow groups of the moduli space have an intersection product.

4.3.1. Algebraic cycles on DM stacks and their moduli Let $\mathcal{X} = [X/G]$ be a separated quotient DM stack with coarse moduli scheme M. The isomorphism $A_*(\mathcal{X}) \otimes \mathbb{Q} \to A_*(M) \otimes \mathbb{Q}$ of Theorem 4.40 can be explicitly described using equivariant cycles.

Lemma 4.41. [7, Proposition 13a] *Let* $\mathcal{X} = [X/G]$ *be a quotient DM stack. Every element of* $A_k(\mathcal{X}) \otimes \mathbb{Q} = A_*^G(X) \otimes \mathbb{Q}$ *can written as a* $\mathbb{Q}$*-linear combination* $\sum_i \alpha_i [V_i]_G$ *where the* $[V_i]_G$ *are the fundamental classes of* G*-invariant subvarieties of dimension* $k + \dim G$.

The next result follows from the proof of [7, Theorem 3a].

Proposition 4.42. *Let* $\mathcal{X} = [X/G]$ *be a quotient DM stack defined over a field of characteristic 0. Suppose that* $\mathcal{X}$ *has a coarse moduli scheme corresponding to a geometric quotient* $p: X \to M$. *Let* W *be a* k*-dimensional subvariety of* M *and let* $V = p^{-1}(W)$. *Let* e_V *be the order of the stabilizer at a general point of* V. *Then the isomorphism* $A_k(M) \otimes \mathbb{Q} \to A_k(\mathcal{X}) \otimes \mathbb{Q}$ *of Theorem 4.40 maps* $[W]$ *to* $e_V[V]_G$.

Example 4.43. When $\mathcal{X} = \overline{\mathcal{M}}_g$ then the map of Proposition 4.42 identifies with the map the map $\mathrm{Pic}(\overline{\mathcal{M}}_g) \otimes \mathbb{Q} \to \mathrm{Pic}_{fun}(\overline{\mathcal{M}}_g) \otimes \mathbb{Q}$ defined by Proposition 3.88 of [18][7].

Remark 4.44. The factor e_V in the statement of Proposition 4.42 can be understood as follows. Let V be a variety on which G acts properly and let $\pi: V \to W$ be a geometric quotient. By Theorem 4.19 there is finite surjective map $f: V' \to V$ and a torsor $\pi': V' \to W'$. It is easy to show that there is an induced map on quotients $W' \stackrel{h}{\to} W$ such that the following diagram is commutative (but not cartesian).

$$\begin{array}{ccc} V' & \stackrel{f}{\to} & V \\ \pi' \downarrow & & \pi \downarrow \\ W' & \stackrel{h}{\to} & W \end{array} \tag{4.45}$$

If $w \in W$ and $x \in \pi^{-1}(w)$ then $\pi^{-1}(w)$ can be identified with the orbit of x which is isomorphic to the quotient G/G_x. On the other hand, since G acts freely on V' the fibers of π' are all isomorphic to G. It follows from this observation that $\deg f = e_V \deg h$. By mapping $[W]$ to $e_V[V]_G$ we ensure that our map $A_*(M) \otimes \mathbb{Q} \to A_*^G(X) \otimes \mathbb{Q}$ commutes with proper pushforwards in diagrams such as (4.45).

[7]Note that in [18] $\overline{\mathcal{M}}_g$ refers to the coarse moduli scheme while here $\overline{\mathcal{M}}_g$ refers to the stack of stable curves.

Diagram (4.45) can also be reinterpreted in the language of stacks as saying the we have a sequence of finite surjective maps $W' \xrightarrow{p} [V/G] \xrightarrow{q} W$ such that $q \circ p = h$. Since W is a geometric quotient it is the coarse moduli space for the stack $[V/G]$. The map $p\colon W' \to [V/G]$ is finite and representable and its degree is the degree of the map $f\colon V' \to V$. Thus we may view q as a finite map, but if we require that $(\deg p)(\deg q) = \deg(q \circ p)$ then we come to the surprising conclusion that $\deg q = \frac{1}{e_V}$.

Example 4.46. It is possible for the coarse moduli scheme M of a smooth DM stack $\mathcal{X}$ to be smooth without $\mathcal{X}$ being representable. For example, if $\dim \mathcal{X} = 1$ then M is smooth, since Theorem 4.31 implies that is normal. However, it is not the case that there is an isomorphism of *integral* Chow rings $A^*(\mathcal{X})$ and $A^*(M)$. For example, if $\mathcal{X} = \overline{\mathcal{M}}_{1,1}$ then its coarse moduli scheme is $\overline{M}_{1,1} = \mathbb{P}^1$, but $A^*(\overline{\mathcal{M}}_{1,1}) = \mathbb{Z}[t]/24t^2$ while $A^*(\mathbb{P}^1) = \mathbb{Z}[h]/h^2$. Moreover, in the ring structure on $\mathbb{P}^1$ induced by the ring structure on $A^*(\overline{\mathcal{M}}_{1,1})$ the identity corresponds to $\frac{1}{2}[\mathbb{P}^1]$ because general elliptic curve has an automorphism group of order 2. If the automorphism group of a general point of $\mathcal{X}$ is trivial then the identity will be $[M]$.

We conclude with a discussion of the degree of a cycle on a complete DM quotient stack.

Definition 4.47. If $\mathcal{X}$ is complete n-dimensional DM stack with coarse moduli space M and $\alpha \in A_0(\mathcal{X})$ then we define $\deg \alpha$ to be the degree of its image under the isomorphism $A_0(\mathcal{X}) \otimes \mathbb{Q} \to A_0(M) \otimes \mathbb{Q}$. If $c \in A^n(\mathcal{X})$ we write $\int_{\mathcal{X}} c$ for the degree of the image $c \cap [\mathcal{X}]$ in $A_0(M) \otimes \mathbb{Q}$.

Note that the degree of an integral cycle on a DM stack need not be an integer. Our final example is an equivariant take on a well known calculation.

Example 4.48. We will show that $\int_{\overline{\mathcal{M}}_{1,1}} \lambda_1 = 1/24$ using equivariant methods. Consider the action of $\mathbb{C}^*$ on $\mathbb{A}^2$ with weights (4,6); i.e. $\lambda(x,y) = (\lambda^4 x, \lambda^6 y)$ and let $X = \mathbb{A}^2 \setminus \{0\}$. There is an equivalence of categories $\overline{\mathcal{M}}_{1,1} \to [X/\mathbb{C}^*]$ which associates to a family of elliptic curves $(C \xrightarrow{\pi} T, \sigma)$ the $\mathbb{C}^*$-bundle $E \to T$ where E is the complement of the zero section in the Hodge bundle $\mathbb{E} = \pi_*(\omega_{C/T})$. Under this identification the Hodge bundle $\mathbb{E}$ corresponds to the line bundle L_1 on X, where the total space of L_1 is $X \times \mathbb{C}$ and $\mathbb{C}^*$ acts on $\mathbb{C}$ with weight 1. Now $A^*([X/\mathbb{C}^*]) = \mathbb{Z}[t]/24t^2$ where $t = c_1(L_1)$. Thus, our problem is to compute $\int_{[X/\mathbb{C}^*]} t$. To do this we must represent t (or more precisely $t \cap [X]_{\mathbb{C}^*}$) in terms of $\mathbb{C}^*$-equivariant 0-cycles on X. There are two natural choices, the line[8] $x = 0$ or the line $y = 0$. The line $x = 0$ is cut out by an equation whose weight is 4 so $[x = 0] = 4t$ or $t = \frac{1}{4}[x = 0]$. A point on this line has coordinates $(0, a)$ for some $a \in \mathbb{C}^*$, and the stabilizer of such a point is μ_6. Hence the degree of $[x = 0]$ is $1/6$ so $\int_{[X/\mathbb{C}^*]} t = 1/4 \times 1/6 = 1/24$. Note that if

[8] A line in X determines an element of $A_0^{\mathbb{C}^*}(X)$ because we shift the degree by the dimension of the group when we define equivariant Chow groups.

we had chosen the line $y = 0$ then we would see that $t = \frac{1}{6}[y = 0]$ but the stabilizer at a point of the invariant subvariety $y = 0$ has order 4 - yielding 1/24 as well.

References

[1] Michel Brion. Equivariant cohomology and equivariant intersection theory, in: Representation theories and algebraic geometry (Montreal, PQ, 1997), 1–37, Notes by Alvaro Rittatore. *NATO Adv. Sci. Inst. Ser. C Math. Phys. Sci.,* bf 514, Kluwer Acad. Publ., Dordrecht, 1998. ← 274

[2] Brian Conrad. *Keel Mori theorem via stacks,* 2004. http://math.stanford.edu/~conrad/papers/coarsespace.pdf ← 285

[3] P. Deligne and D. Mumford. The irreducibility of the space of curves of given genus. *Inst. Hautes Études Sci. Publ. Math.,* (36), 75–109, 1969. ← 260, 261, 266, 273, 280, 281, 282, 284

[4] Dan Edidin. Notes on the construction of the moduli space of curves, in: Recent progress in intersection theory (Bologna, 1997), 85–113. *Trends Math.,* Birkhäuser Boston, Boston, MA, 2000. ← 260

[5] Dan Edidin and Damiano Fulghesu. The integral Chow ring of the stack of at most 1-nodal rational curves. *Comm. Algebra,* **36**(2), 581–594, 2008. ← 267, 279

[6] Dan Edidin and Damiano Fulghesu. The integral Chow ring of the stack of hyperelliptic curves of even genus. *Math. Res. Lett.,* **16**(1), 27–40, 2009. ← 279

[7] Dan Edidin and William Graham. Equivariant intersection theory. *Invent. Math.,* **131**(3), 595–634, 1998. ← 275, 276, 277, 278, 279, 287, 288

[8] Dan Edidin and William Graham. Localization in equivariant intersection theory and the Bott residue formula. *Amer. J. Math.,* **120**(3), 619–636, 1998. ← 274

[9] Dan Edidin, Brendan Hassett, Andrew Kresch, and Angelo Vistoli. Brauer groups and quotient stacks. *Amer. J. Math.,* **123**(4), 761–777, 2001. ← 283

[10] Carel Faber. A conjectural description of the tautological ring of the moduli space of curves, In: Moduli of curves and abelian varieties. *Aspects Math.,* **E33**, 109–129, Vieweg, Braunschweig, 1999. ← 279

[11] Damiano Fulghesu. The Chow ring of the stack of rational curves with at most 3 nodes. *Comm. Algebra,* **38** (3), 3125–3136, 2010. ← 279

[12] Damiano Fulghesu. The stack of rational curves. *Comm. Algebra,* **38**(7), 2405–2417, 2010. ← 267

[13] Damiano Fulghesu and F. Viviani. The Chow ring of the stack of cyclic covers of the projective line. *Ann. Inst. Fourier,* arXiv:1003:4091 (to appear) ← 279

[14] William Fulton. *Intersection theory,* Springer-Verlag, Berlin, 1984. ← 272, 276

[15] Henri Gillet. Intersection theory on algebraic stacks and Q-varieties. *Proceedings of the Luminy conference on algebraic* K-*theory (Luminy, 1983),* **34**, 193–240, 1984. ← 279

[16] T. Graber and R. Pandharipande. Localization of virtual classes. *Invent. Math.*, **135**(2), 487–518, 1999. ← 274

[17] John L. Harer. Stability of the homology of the mapping class groups of orientable surfaces. *Ann. of Math. (2)*, **121**(2), 215–249, 1985. ← 278

[18] Joe Harris and Ian Morrison. Moduli of curves. *Graduate Texts in Mathematics*, **187**, Springer-Verlag, New York, 1998. ← 278, 286, 288

[19] Seán Keel and Shigefumi Mori. Quotients by groupoids. *Ann. of Math. (2)*, **145**(1), 193–213, 1997. ← 284, 285, 286

[20] Finn F. Knudsen. The projectivity of the moduli space of stable curves. II. The stacks $M_{g,n}$. *Math. Scand.*, **52**(2), 161–199, 1983. ← 267, 270

[21] Maxim Kontsevich. Enumeration of rational curves via torus actions, in: The moduli space of curves (Texel Island, 1994), 335–368. *Progr. Math.*, **129**, Birkhäuser Boston, Boston, MA, 1995. ← 274

[22] Andrew Kresch. *Flattening stratifications and the stack of partial stabilizations of prestable curves*, preprint, 2011. ← 261, 267

[23] Gérard Laumon and Laurent Moret-Bailly. Champs algébriques, Ergebnisse der Mathematik und ihrer Grenzgebiete. 3. Folge. *A Series of Modern Surveys in Mathematics [Results in Mathematics and Related Areas. 3rd Series. A Series of Modern Surveys in Mathematics]*, **39**, Springer-Verlag, Berlin, 2000. ← 260, 281

[24] Domingo Luna. Slices étales, Sur les groupes algébriques. *Soc. Math. France*, 81–105, Paris, 1973. *Bull. Soc. Math. France*, Paris, Mémoire 33. ← 282

[25] Ib Madsen and Michael Weiss. The stable moduli space of Riemann surfaces: Mumford's conjecture. *Ann. of Math. (2)*, **165**(3), 843–941, 2007. ← 278

[26] David Mumford. Picard groups of moduli problems. *Arithmetical Algebraic Geometry (Proc. Conf. Purdue Univ., 1963)*, 33–81, Harper & Row, New York, 1965. ← 260, 277, 278

[27] David Mumford. Stability of projective varieties. *Enseignement Math. (2)*, **23**(1–2), 39–110, 1977. ← 279

[28] David Mumford. Towards an enumerative geometry of the moduli space of curves, In: Arithmetic and geometry, Vol. II. *Progr. Math.*, **36**, 271–328, Birkhäuser Boston, Boston, MA, 1983. ← 279

[29] D. Mumford, J. Fogarty, and F. Kirwan. *Geometric invariant theory*, third ed., Springer-Verlag, Berlin, 1994. ← 266, 283, 284, 285

[30] Rahul Pandharipande. Equivariant Chow rings of $O(k)$, $SO(2k+1)$, and $SO(4)$. *J. Reine Angew. Math.*, **496**, 131–148, 1998. ← 279

[31] C. S. Seshadri. Quotient spaces modulo reductive algebraic groups. *Ann. of Math. (2)*, **95**, 511–556, 1972; errata, *ibid. (2)*, **96**, 599, 1972. ← 283

[32] Larry Smith. Transfer and ramified coverings. *Math. Proc. Cambridge Philos. Soc.*, **93**(3), 485–493, 1983. ← 287

[33] Burt Totaro. The Chow ring of a classifying space, in: Algebraic K-theory (Seattle, WA, 1997). *Proc. Sympos. Pure Math.*, **67**, 249–281, Amer. Math. Soc., Providence, RI, 1999. ← 275

[34] Angelo Vistoli. Intersection theory on algebraic stacks and on their moduli spaces. *Invent. Math.*, **97**(3), 613–670, 1989. ← 260, 279, 280, 281

[35] Angelo Vistoli. The Chow ring of $\mathcal{M}_2$. Appendix to Equivariant intersection theory. *Invent. Math.*, **131**(3), 635–644, 1998. ← 279

University of Missouri
E-mail address: edidind@missouri.edu

Tautological and non-tautological cohomology of the moduli space of curves

C. Faber and R. Pandharipande

Abstract. After a short exposition of the basic properties of the tautological ring of $\overline{M}_{g,n}$, we explain three methods of detecting non-tautological classes in cohomology. The first is via curve counting over finite fields. The second is by obtaining length bounds on the action of the symmetric group Σ_n on tautological classes. The third is via classical boundary geometry. Several new non-tautological classes are found.

Contents

0. Introduction

0.1. Overview

Let $\overline{M}_{g,n}$ be the moduli space of Deligne-Mumford stable curves of genus g with n marked points. The cohomology of $\overline{M}_{g,n}$ has a distinguished subring of tautological classes

$$RH^*(\overline{M}_{g,n}) \subset H^*(\overline{M}_{g,n}, \mathbb{Q})$$

studied extensively since Mumford's seminal article [51]. While effective methods for exploring the tautological ring have been developed over the years, the structure of the non-tautological classes remains mysterious. Our goal here, after reviewing the basic definitions and properties of $RH^*(\overline{M}_{g,n})$ in Section 1, is to present three approaches for detecting and studying non-tautological classes.

2000 *Mathematics Subject Classification.* Primary 14H10; Secondary 14D10.
Key words and phrases. Moduli, tautological classes, cohomology.

We view $\overline{M}_{g,n}$ as a Deligne-Mumford stack. Not much is lost by considering the coarse moduli space (if the automorphism groups are not forgotten). We encourage the reader to take the coarse moduli point of view if stacks are unfamiliar.

0.2. Point counting and modular forms

Since the moduli space of Deligne-Mumford stable curves is defined over $\mathbb{Z}$, reduction to finite fields $\mathbb{F}_q$ is well-defined. Let $\overline{M}_{g,n}(\mathbb{F}_q)$ denote the set of $\mathbb{F}_q$-points. For various ranges of g, n, and q, counting the number of points of $\overline{M}_{g,n}(\mathbb{F}_q)$ is feasible. A wealth of information about $H^*(\overline{M}_{g,n}, \mathbb{Q})$ can be then obtained from the Lefschetz fixed point formula applied to Frobenius.

The first examples where point counting reveals non-tautological cohomology occur in genus 1. The relationship between point counting and elliptic modular forms is discussed in Section 2. By interpreting the counting results in genus 2, a conjectural description of the *entire* cohomology of $\overline{M}_{2,n}$ has been found by Faber and van der Geer [19] in terms of Siegel modular forms. The formula is consistent with point counting data for $n \leqslant 25$. In fact, large parts of the formula have been proven. The genus 2 results are presented in Section 3.

In genus 3, a more complicated investigation involving Teichmüller modular forms is required. The situation is briefly summarized in Section 3.7. As the genus increases, the connection between point counting and modular forms becomes more difficult to understand.

0.3. Representation theory

The symmetric group Σ_n acts on $\overline{M}_{g,n}$ by permuting the markings. As a result, an Σ_n-representation on $H^*(\overline{M}_{g,n}, \mathbb{Q})$ is obtained.

Studying the Σ_n-action on $H^*(\overline{M}_{g,n}, \mathbb{Q})$ provides a second approach to the non-tautological cohomology. In Section 4, we establish an upper bound for the length[1] of the irreducible Σ_n-representations occurring in the tautological ring $R^*(\overline{M}_{g,n})$. In many cases, the bound is sharp. Assuming the conjectural formulas for $H^*(\overline{M}_{2,n}, \mathbb{Q})$ obtained by point counting, we find several classes of Hodge type which are presumably algebraic (by the Hodge conjecture) but cannot possibly be tautological, because the length of the corresponding Σ_n-representations is too large. The first occurs in $\overline{M}_{2,21}$.

The proofs of the length bounds for the Σ_n-action on $R^*(\overline{M}_{g,n})$ are obtained by studying representations induced from the boundary strata. The strong vanishing of Proposition 2 of [23] with tautological boundary terms plays a crucial role in the argument.

[1] The number of parts in the corresponding partition.

0.4. Boundary geometry

The existence of non-tautological cohomology classes of Hodge type was earlier established by Graber and Pandharipande [36]. In particular, an explicit such class in the boundary of $\overline{M}_{2,22}$ was found. In Section 5, we revisit the non-tautological boundary constructions. The old examples are obtained in simpler ways and new examples in $\overline{M}_{2,21}$ are found. The method is by straightforward intersection theory on the moduli space of curves. Finally, we connect the boundary constructions in $\overline{M}_{2,21}$ to the representation investigation of Section 4.

0.5. Acknowledgements

We thank J. Bergström, C. Consani, G. van der Geer, T. Graber, and A. Pixton for many discussions related to the cohomology of the moduli space of curves.

C.F. was supported by the Göran Gustafsson foundation for research in natural sciences and medicine and grant 622-2003-1123 from the Swedish Research Council. R.P. was partially supported by NSF grants DMS-0500187 and DMS-1001154. The paper was completed while R.P. was visiting the Instituto Superior Técnico in Lisbon with support from a Marie Curie fellowship and a grant from the Gulbenkian foundation.

1. Tautological classes

1.1. Definitions

Let $\overline{M}_{g,n}$ be the moduli space of stable curves of genus g with n marked points defined over $\mathbb{C}$. Let $A^*(\overline{M}_{g,n}, \mathbb{Q})$ denote the Chow ring. The system of tautological rings is defined[2] to be the set of smallest $\mathbb{Q}$-subalgebras of the Chow rings,

$$R^*(\overline{M}_{g,n}) \subset A^*(\overline{M}_{g,n}, \mathbb{Q}),$$

satisfying the following two properties:

(i) The system is closed under push-forward via all maps forgetting markings:

$$\pi_* : R^*(\overline{M}_{g,n}) \to R^*(\overline{M}_{g,n-1}).$$

(ii) The system is closed under push-forward via all gluing maps:

$$\iota_* : R^*(\overline{M}_{g_1,n_1\cup\{\star\}}) \otimes_{\mathbb{Q}} R^*(\overline{M}_{g_2,n_2\cup\{\bullet\}}) \to R^*(\overline{M}_{g_1+g_2,n_1+n_2}),$$

$$\iota_* : R^*(\overline{M}_{g,n\cup\{\star,\bullet\}}) \to R^*(\overline{M}_{g+1,n}),$$

with attachments along the markings $\star$ and $\bullet$.

While the definition appears restrictive, natural algebraic constructions typically yield Chow classes lying in the tautological ring.

[2] We follow here the definition of tautological classes in Chow given in [23]. Tautological classes in cohomology are discussed in Section 1.7.

1.2. Basic examples

Consider first the cotangent line classes. For each marking i, let

$$\mathbb{L}_i \to \overline{M}_{g,n}$$

denote the associated cotangent line bundle. By definition

$$\psi_i = c_1(\mathbb{L}_i) \in A^1(\overline{M}_{g,n}).$$

Let π denote the map forgetting the last marking,

$$\pi : \overline{M}_{g,n+1} \to \overline{M}_{g,n}, \tag{1.1}$$

and let ι denote the gluing map,

$$\iota : \overline{M}_{g,\{1,2,\ldots,i-1,\star,i+1,\ldots,n\}} \times \overline{M}_{0,\{\bullet,i,n+1\}} \longrightarrow \overline{M}_{g,n+1}.$$

The $\mathbb{Q}$-multiples of the fundamental classes $[\overline{M}_{g,n}]$ are contained in the tautological rings (as $\mathbb{Q}$-multiples of the units in the subalgebras). A direct calculation shows:

$$-\pi_*\Big(\big(\iota_*([\overline{M}_{g,n}] \times [\overline{M}_{0,3}])\big)^2\Big) = \psi_i.$$

Hence, the cotangent line classes lie in the tautological rings,

$$\psi_i \in R^1(\overline{M}_{g,n}) .$$

Consider next the κ classes defined via push-forward by the forgetful map (1.1),

$$\kappa_j = \pi_*(\psi_{n+1}^{j+1}) \in A^j(\overline{M}_{g,n}) .$$

Since ψ_{n+1} has already been shown to be tautological, the κ classes are tautological by property (i),

$$\kappa_j \in R^j(\overline{M}_{g,n}) .$$

The ψ and κ classes are very closely related.

For a nodal curve C, let ω_C denote the dualizing sheaf. The Hodge bundle $\mathbb{E}$ over $\overline{M}_{g,n}$ is the rank g vector bundle with fiber $H^0(C, \omega_C)$ over the moduli point

$$[C, p_1, \ldots, p_n] \in \overline{M}_{g,n} .$$

The λ classes are defined by

$$\lambda_k = c_k(\mathbb{E}) \in A^k(\overline{M}_{g,n}) .$$

The Chern characters of $\mathbb{E}$ lie in the tautological ring by Mumford's Grothendieck-Riemann-Roch computation [51]. Since the λ classes are polynomials in the Chern characters,

$$\lambda_k \in R^k(\overline{M}_{g,n}) .$$

The ψ and λ classes are basic elements of the tautological ring $R^*(\overline{M}_{g,n})$ arising very often in geometric constructions and calculations. The ψ integrals,

$$\int_{\overline{M}_{g,n}} \psi_1^{a_1} \cdots \psi_n^{a_n}, \tag{1.2}$$

can be evaluated by KdV constraints due to Witten and Kontsevich [45, 65]. Other geometric approaches to the ψ integrals can be found in [49, 52]. In genus 0, the simple formula

$$\int_{\overline{M}_{0,n}} \psi_1^{a_1} \cdots \psi_n^{a_n} = \binom{n-3}{a_1, \ldots, a_n}$$

holds. Non-integer values occur for the first time in genus 1,

$$\int_{\overline{M}_{1,1}} \psi_1 = \frac{1}{24} .$$

Hodge integrals, when the λ classes are included,

$$\int_{\overline{M}_{g,n}} \psi_1^{a_1} \cdots \psi_n^{a_n} \lambda_1^{b_1} \ldots \lambda_g^{b_g},$$

can be reduced to ψ integrals (1.2) by Mumford's Grothendieck-Riemann-Roch calculation. An example of a Hodge integral evaluation[3] is

$$\int_{\overline{M}_g} \lambda_{g-1}^3 = \frac{|B_{2g}|}{2g} \frac{|B_{2g-2}|}{2g-2} \frac{1}{(2g-2)!} \tag{1.3}$$

The proof of (1.3) and several other exact formulas for Hodge integrals can be found in [20].

1.3. Strata

The boundary strata of the moduli spaces of curves correspond to *stable graphs*

$$A = (V, H, L, g : V \to \mathbb{Z}_{\geqslant 0}, a : H \to V, i : H \to H)$$

satisfying the following properties:

(1) V is a vertex set with a genus function g,
(2) H is a half-edge set equipped with a vertex assignment a and fixed point free involution i,
(3) E, the edge set, is defined by the orbits of i in H,
(4) (V, E) define a *connected* graph,
(5) L is a set of numbered legs attached to the vertices,
(6) For each vertex v, the stability condition holds:

$$2g(v) - 2 + n(v) > 0,$$

where $n(v)$ is the valence of A at v including both half-edges and legs.

The data of the topological type of a generic curve in a boundary stratum of the moduli space is recorded in the stable graph.

Let A be a stable graph. The genus of A is defined by

$$g = \sum_{v \in V} g(v) + h^1(A).$$

[3]Here, B_{2g} denotes the Bernoulli number.

Define the moduli space $\overline{M}_A$ by the product

$$\overline{M}_A = \prod_{v \in V(A)} \overline{M}_{g(v),n(v)}.$$

There is a canonical morphism[4]

$$\xi_A : \overline{M}_A \to \overline{M}_{g,n}$$

with image equal to the boundary stratum associated to the graph A. By repeated use of property (ii),

$$\xi_{A*}[\overline{M}_A] \in R^*(\overline{M}_{g,n}) .$$

We can now describe a set of additive generators for $R^*(\overline{M}_{g,n})$. Let A be a stable graph of genus g with n legs. For each vertex v of A, let

$$\theta_v \in R^*(\overline{M}_{g(v),n(v)})$$

be an arbitrary monomial in the ψ and κ classes of the vertex moduli space. The following result is proven in [36].

Theorem 1.4. $R^*(\overline{M}_{g,n})$ *is generated additively by classes of the form*

$$\xi_{A*}\Big(\prod_{v \in V(A)} \theta_v \Big).$$

By the dimension grading, the list of generators provided by Theorem 1.4 is finite. Hence, we obtain the following result.

Corollary 1.5. *We have* $\dim_{\mathbb{Q}} R^*(\overline{M}_{g,n}) < \infty$.

1.4. Further properties

The following two formal properties of the full system of tautological rings

$$R^*(\overline{M}_{g,n}) \subset A^*(\overline{M}_{g,n}, \mathbb{Q}),$$

are a consequence of properties (i) and (ii):

(iii) The system is closed under pull-back via the forgetting and gluing maps.

(iv) $R^*(\overline{M}_{g,n})$ is an $\mathbb{S}_n$-module via the permutation action on the markings.

Property (iii) follows from the well-known boundary geometry of the moduli space of curves. A careful treatment using the additive basis of Theorem 1.4 can be found in [36]. The meaning of property (iii) for the reducible gluing is that the Künneth components of the pull-backs of tautological classes are tautological. Since the defining properties (i) and (ii) are symmetric with respect to the marked points, property (iv) holds.

[4] To construct ξ_A, a family of stable pointed curves over $\overline{M}_A$ is required. Such a family is easily defined by attaching the pull-backs of the universal families over each of the $\overline{M}_{g(v),n(v)}$ along the sections corresponding to half-edges.

1.5. Pairing

Intersection theory on the moduli space $\overline{M}_{g,n}$ yields a canonical pairing

$$\mu : R^k(\overline{M}_{g,n}) \times R^{3g-3+n-k}(\overline{M}_{g,n}) \to \mathbb{Q}$$

defined by

$$\mu(\alpha, \beta) = \int_{\overline{M}_{g,n}} \alpha \cup \beta \, .$$

While the pairing μ has been speculated to be perfect in [21, 38, 53], very few results are known.

The pairing μ can be effectively computed on the generators of Theorem 1.4 by the following method. The pull-back property (iii) may be used repeatedly to reduce the calculation of μ on the generators to integrals of the form

$$\int_{\overline{M}_{h,m}} \psi_1^{a_1} \cdots \psi_m^{a_m} \cdot \kappa_1^{b_1} \cdots \kappa_r^{b_r} \, .$$

The κ classes can be removed to yield a sum of purely ψ integrals

$$\int_{\overline{M}_{h,m+r}} \psi_1^{a_1} \cdots \psi_{m+r}^{a_{m+r}},$$

see [3]. As discussed in Section 1.2, the ψ integrals can be evaluated by KdV constraints.

1.6. Further examples

We present here two geometric constructions which also yield tautological classes. The first is via stable maps and the second via moduli spaces of Hurwitz covers.

Let X be a nonsingular projective variety, and let $\overline{M}_{g,n}(X, \beta)$ be the moduli space of stable maps[5] representing $\beta \in H_2(X, \mathbb{Z})$. Let ρ denote the map to the moduli of curves,

$$\rho : \overline{M}_{g,n}(X, \beta) \to \overline{M}_{g,n}.$$

The moduli space $\overline{M}_{g,n}(X, \beta)$ carries a virtual class

$$[\overline{M}_{g,n}(X, \beta)]^{vir} \in A_*(\overline{M}_{g,n}(X, \beta))$$

obtained from the canonical obstruction theory of maps.

Theorem 1.6. *Let X be a nonsingular projective toric variety. Then,*

$$\rho_*[\overline{M}_{g,n}(X, \beta)]^{vir} \in R^*(\overline{M}_{g,n}).$$

---

[5]We refer the reader to [27] for an introduction to the subject. A discussion of obstruction theories and virtual classes can be found in [4, 5, 47].

The proof follows directly from the virtual localization formula of [35]. If $[C] \in M_g$ is a general moduli point, then

$$\rho_*[\overline{M}_g(C,1)]^{vir} = [C] \in A^{3g-3}(\overline{M}_g).$$

Since not all of $A^{3g-3}(\overline{M}_g)$ is expected to be tautological, Theorem 1.6 is unlikely to hold for C. However, the result perhaps holds for a much more general class of varieties X. In fact, we do not know[6] any nonsingular projective variety defined over $\bar{\mathbb{Q}}$ for which Theorem 1.6 is expected to be false.

The moduli spaces of Hurwitz covers of $\mathbb{P}^1$ also define natural classes on the moduli space of curves. Let $\mu^1, \ldots, \mu^m$ be m partitions of equal size d satisfying

$$2g - 2 + 2d = \sum_{i=1}^{m} \left(d - \ell(\mu^i)\right),$$

where $\ell(\mu^i)$ denotes the length of the partition μ^i. The moduli space of Hurwitz covers,

$$H_g(\mu^1, \ldots, \mu^m)$$

parameterizes morphisms,

$$f : C \to \mathbb{P}^1,$$

where C is a complete, connected, nonsingular curve with marked profiles $\mu^1, \ldots, \mu^m$ over m ordered points of the target (and no ramifications elsewhere). The moduli space of Hurwitz covers is a dense open set of the compact moduli space of admissible covers [39],

$$H_g(\mu^1, \ldots, \mu^m) \subset \overline{H}_g(\mu^1, \ldots, \mu^m).$$

Let ρ denote the map to the moduli of curves,

$$\rho : \overline{H}_g(\mu^1, \ldots, \mu^m) \to \overline{M}_{g, \sum_{i=1}^{m} \ell(\mu^i)}.$$

The following is a central result of [23].

Theorem 1.7. *The push-forwards of the fundamental classes lie in the tautological ring,*

$$\rho_*[\overline{H}_g(\mu^1, \ldots, \mu^m)] \in R^*(\overline{M}_{g, \sum_{i=1}^{m} \ell(\mu^i)}) .$$

The admissible covers in Theorem 1.7 are of $\mathbb{P}^1$. The moduli spaces of admissible covers of higher genus targets can lead to non-tautological classes [36].

---

[6] We do not know any examples at all of nonsingular projective X where

$$\rho_*[\overline{M}_{g,n}(X, \beta)]^{vir} \notin RH^*(\overline{M}_{g,n}).$$

See [46] for a discussion.

1.7. Tautological rings

The tautological subrings

(1.8) $$RH^*(\overline{M}_{g,n}) \subset H^*(\overline{M}_{g,n}, \mathbb{Q})$$

are defined to be the images of $R^*(\overline{M}_{g,n})$ under the cycle class map

$$A^*(\overline{M}_{g,n}, \mathbb{Q}) \to H^*(\overline{M}_{g,n}, \mathbb{Q})\ .$$

The tautological rings $RH^*(\overline{M}_{g,n})$ could alternatively be defined as the smallest system of subalgebras (1.8) closed under push-forward by all forgetting and gluing maps (properties (i) and (ii)). Properties (iii) and (iv) also hold for $RH^*(\overline{M}_{g,n})$.

Tautological rings in Chow and cohomology may be defined for all stages of the standard moduli filtration:

(1.9) $$\overline{M}_{g,n} \supset M^c_{g,n} \supset M^{rt}_{g,n}.$$

Here, $M^c_{g,n}$ is the moduli space of stable curves of compact type (curves with tree dual graphs, or equivalently, with compact Jacobians), and $M^{rt}_{g,n}$ is the moduli space of stable curves with rational tails (the preimage of the moduli space of nonsingular curves $M_g \subset \overline{M}_g$). The tautological rings

$$R^*(M^c_{g,n}) \subset A^*(M^c_{g,n}, \mathbb{Q}), \quad RH^*(M^c_{g,n}) \subset H^*(M^c_{g,n}, \mathbb{Q}),$$

$$R^*(M^{rt}_{g,n}) \subset A^*(M^{rt}_{g,n}, \mathbb{Q}), \quad RH^*(M^{rt}_{g,n}) \subset H^*(M^{rt}_{g,n}, \mathbb{Q})$$

are all defined as the images of $R^*(\overline{M}_{g,n})$ under the natural restriction and cycle class maps.

Of all the tautological rings, the most studied case by far is $R^*(M_g)$. A complete (conjectural) structure of $R^*(M_g)$ is proposed in [18] with important advances made in [42, 48, 50]. Conjectures for the compact type cases $R^*(M^c_{g,n})$ can be found in [21]. See [22, 54] for positive results for compact type. Tavakol [59, 60] has recently proved that $R^*(M^c_{1,n})$ and $R^*(M^{rt}_{2,n})$ are Gorenstein, with socles in degrees $n-1$ and n, respectively.

1.8. Hyperelliptic curves

As an example, we calculate the class of the hyperelliptic locus

$$H_g \subset M_g$$

following [51].

Over the moduli point $[C, p] \in M_{g,1}$ there is a canonical map

$$\phi : H^0(C, \omega_C) \to H^0(C, \omega_C/\omega_C(-2p))$$

defined by evaluating sections of ω_C at the marking $p \in C$. If we define $\mathbb{J}$ to be the rank 2 bundle over $M_{g,1}$ with fiber $H^0(C, \omega_C/\omega_C(-2p))$ over $[C, p]$, then we obtain a morphism

$$\phi : \mathbb{E} \to \mathbb{J}$$

over $M_{g,1}$. By classical curve theory, the map ϕ fails to be surjective precisely when C is hyperelliptic and $p \in C$ is a Weierstrass point. Let

$$\Delta \subset M_{g,1}$$

be the degeneracy locus of ϕ of pure codimension $g-1$. By the Thom-Porteous formula [26],

$$[\Delta] = \left(\frac{c(\mathbb{E}^*)}{c(\mathbb{J}^*)} \right)_{g-1} .$$

By using the jet bundle sequence

$$0 \to \mathbb{L}_1^2 \to \mathbb{J} \to \mathbb{L}_1 \to 0,$$

we conclude

$$[\Delta] = \left(\frac{1 - \lambda_1 + \lambda_2 - \lambda_3 + \ldots + (-1)^g \lambda_g}{(1-\psi_1)(1-2\psi_1)} \right)_{g-1} \in R^{g-1}(M_{g,1}) .$$

To find a formula for the hyperelliptic locus, we view

$$\pi : M_{g,1} \to M_g$$

as the universal curve over moduli space. Since each hyperelliptic curve has $2g+2$ Weierstrass points,

$$[H_g] = \frac{1}{2g+2} \pi_*([\Delta]) \in R^{g-2}(M_g) .$$

By Theorem 1.7, the class of the closure of the hyperelliptic locus is also tautological,

$$[\overline{H}_g] \in R^{g-2}(\overline{M}_g),$$

but no simple formula is known to us.

2. Point counting and elliptic modular forms

2.1. Elliptic modular forms

We present here an introduction to the close relationship between modular forms and the cohomology of moduli spaces. The connection is most direct between elliptic modular forms and moduli spaces of elliptic curves.

Classically, a modular form is a holomorphic function f on the upper half plane

$$\mathbb{H} = \{z \in \mathbb{C} : \operatorname{Im} z > 0\}$$

with an amazing amount of symmetry. Precisely, a modular form of *weight* $k \in \mathbb{Z}$ satisfies the functional equation

$$f\left(\frac{az+b}{cz+d} \right) = (cz+d)^k f(z)$$

for all $z \in \mathbb{H}$ and all $\left(\begin{smallmatrix} a & b \\ c & d \end{smallmatrix}\right) \in \mathsf{SL}(2, \mathbb{Z})$. By the symmetry for $\left(\begin{smallmatrix} -1 & 0 \\ 0 & -1 \end{smallmatrix}\right)$,

$$f(z) = (-1)^k f(z),$$

so k must be even. Using $\left(\begin{smallmatrix}1&1\\0&1\end{smallmatrix}\right)$, we see

$$f(z+1) = f(z),$$

so f can be written as a function of $q = \exp(2\pi i z)$. We further require f to be holomorphic at $q = 0$. The Fourier expansion is

$$f(q) = \sum_{n=0}^{\infty} a_n q^n.$$

If a_0 vanishes, f is called a *cusp* form (of weight k for $SL(2,\mathbb{Z})$).

Well-known examples of modular forms are the Eisenstein series[7]

$$E_k(q) = 1 - \frac{2k}{B_k}\sum_{n=1}^{\infty} \sigma_{k-1}(n) q^n$$

of weight $k \geqslant 4$ and the discriminant cusp form

$$\Delta(q) = \sum_{n=1}^{\infty} \tau(n) q^n = q \prod_{n=1}^{\infty} (1-q^n)^{24} \tag{2.1}$$

of weight 12 with the Jacobi product expansion (defining the Ramanujan τ-function). In both of the above formulas, the right sides are the Fourier expansions.

Let M_k be the vector space of holomorphic modular forms of weight k for $SL(2,\mathbb{Z})$. The ring of modular forms is freely generated [44, Exc. III.2.5] by E_4 and E_6,

$$\bigoplus_{k=0}^{\infty} M_k = \mathbb{C}[E_4, E_6]\,,$$

and the ideal $\bigoplus_k S_k$ of cusp forms is generated by Δ. Let $m_k = \dim M_k$ and $s_k = m_k - 1$.

For every positive integer n, there exists a naturally defined *Hecke operator*

$$T_n : M_k \to M_k\,.$$

The T_n commute with each other and preserve the subspace S_k. Moreover, S_k has a basis of simultaneous eigenforms, which are orthogonal with respect to the *Petersson inner product* and can be normalized to have $a_1 = 1$. The T_n-eigenvalue of such an eigenform equals its n-th Fourier coefficient [44, Prop. III.40].

2.2. Point counting in genus 1

While the moduli space $M_{1,1}$ of elliptic curves can be realized over $\mathbb{C}$ as the analytic space $\mathbb{H}/SL(2,\mathbb{Z})$, we will view $M_{1,1}$ here as an algebraic variety over a field k or a stack over $\mathbb{Z}$.

Over $\overline{k}$, the j-invariant classifies elliptic curves, so

$$M_{1,1} \cong_{\overline{k}} \mathbb{A}^1$$

---

[7] Recall the Bernoulli numbers B_k are given by $\frac{x}{e^x - 1} = \sum B_k \frac{x^k}{k!}$ and $\sigma_m(n)$ is the sum of the m-th powers of the positive divisors of n.

as a coarse moduli space. But what happens over a finite field $\mathbb{F}_p$, where p is a prime number? Can we calculate $\#M_{1,1}(\mathbb{F}_p)$? If we count elliptic curves E over $\mathbb{F}_p$ (up to $\mathbb{F}_p$-isomorphism) with weight factor

(2.2) $$\frac{1}{\#\mathrm{Aut}_{\mathbb{F}_p}(E)},$$

we obtain the expected result.

Proposition 2.3. *We have* $\#M_{1,1}(\mathbb{F}_p) = p$.

Proof. The counting is very simple for $p \neq 2$. Given an elliptic curve with a point $z \in E$ defined over $\mathbb{F}_p$, we obtain a degree 2 morphism

$$\phi : E \to \mathbb{P}^1$$

from the linear series associated to $\mathcal{O}_E(2z)$. If we view the image of z as $\infty \in \mathbb{P}^1$, the branched covering ϕ expresses E in Weierstrass form,

$$y^2 = x^3 + ax^2 + bx + c \quad \text{with } a, b, c \in \mathbb{F}_p,$$

where the cubic on the right has distinct roots in $\overline{\mathbb{F}}_p$. The number of monic cubics in x is p^3, the number with a single pair of double roots is $p^2 - p$, and the number with a triple root is p. Hence, the number of Weierstrass forms is $p^3 - p^2$.

Since only $\infty \in \mathbb{P}^1$ is distinguished, we must further divide by the group of affine transformations of

$$\mathbb{A}^1 = \mathbb{P}^1 \setminus \{\infty\}$$

over $\mathbb{F}_p$. Since the order of the affine transformation group is $p^2 - p$, we conclude

$$\#M_{1,1}(\mathbb{F}_p) = \frac{p^3 - p^2}{p^2 - p} = p\,.$$

We leave to the reader to check the above counting weights curves by the factor (2.2). We also leave the $p = 2$ case to the reader. □

Motivated by Proposition 2.3, we will study the number of points of $M_{1,n}(\mathbb{F}_p)$ weighted by the order of the $\mathbb{F}_p$-automorphism group of the marked curve. For example,

$$\#M_{1,6}(\mathbb{F}_2) = 0,$$

since an elliptic curve E over $\mathbb{F}_2$ contains at most 5 distinct points by the Weil bound [40, Exc. V.1.10],

$$|\#E(\mathbb{F}_2) - (2+1)| \leqslant 2\sqrt{2}\,.$$

To investigate the behavior of $M_{1,n}(\mathbb{F}_p)$, we can effectively enumerate n-pointed elliptic curves over $\mathbb{F}_p$ by computer. Interpreting the data, we find for $n \leqslant 10$,

(2.4) $$\#M_{1,n}(\mathbb{F}_p) = f_n(p),$$

where f_n is a monic polynomial of degree n with integral coefficients. The moduli spaces $M_{1,n}$ are rational varieties [6] for $n \leqslant 10$. The polynomiality (2.4) has been proved directly by Bergström [8, §15] using geometric constructions (as in the proof of Proposition 2.3). However for $n = 11$, the computer counts exclude the possibility of a degree 11 polynomial for $\#M_{1,11}(\mathbb{F}_p)$. The data suggest

$$\#M_{1,11}(\mathbb{F}_p) = f_{11}(p) - \tau(p). \tag{2.5}$$

For higher n, the p-th Fourier coefficients of the other Hecke cusp eigenforms eventually appear in the counting as well (multiplied with polynomials in p).

2.3. Cohomology of local systems

To understand what is going on, we recall the number of points of $M_{1,n}$ over $\mathbb{F}_p$ equals the trace of Frobenius on the Euler characteristic of the compactly supported ℓ-adic cohomology of $M_{1,n}$. The Euler characteristic has good additivity properties and the multiplicities in applying the Lefschetz fixed point formula to Frob_p are all equal to 1. The cohomology of projective space yields traces which are polynomial in p,

$$\#\mathbb{P}^n(\mathbb{F}_p) = p^n + p^{n-1} + \cdots + p + 1,$$

since the trace of Frobenius on the Galois representation $\mathbb{Q}_\ell(-i)$ equals p^i.

As is well-known, the fundamental class of an algebraic subvariety is of Tate type (fixed by the Galois group). The Tate conjecture essentially asserts the converse. If equation (2.5) holds for all p, then $\#M_{1,11}(\mathbb{F}_p)$ will fail to fit a degree 11 polynomial (even for all but finitely many p). We can then conclude $M_{1,11}$ possesses cohomology not represented by algebraic classes defined over $\mathbb{Q}$. Such cohomology, in particular, can not be tautological.

In fact, the moduli space $M_{1,n}(\mathbb{C})$ has cohomology related to cusp forms. The simplest connection is the construction of a holomorphic differential form on $M_{1,11}$ from the discriminant (2.1). We view $M_{1,11}$ as an open subset[8]

$$M_{1,11} \subset \frac{\mathbb{H} \times \mathbb{C}^{10}}{\mathsf{SL}(2,\mathbb{Z}) \ltimes (\mathbb{Z}^2)^{10}}.$$

The multiplication in the semidirect product $\mathsf{SL}(2,\mathbb{Z}) \ltimes (\mathbb{Z}^2)^{10}$ is defined by the following formula

$$\begin{aligned}\Big(\left(\begin{smallmatrix} a' & b' \\ c' & d' \end{smallmatrix}\right), (x_1', y_1'), \ldots, (x_{10}', y_{10}')\Big) \cdot \Big(\left(\begin{smallmatrix} a & b \\ c & d \end{smallmatrix}\right), (x_1, y_1), \ldots, (x_{10}, y_{10})\Big) = \\ \Big(\left(\begin{smallmatrix} a' & b' \\ c' & d' \end{smallmatrix}\right) \cdot \left(\begin{smallmatrix} a & b \\ c & d \end{smallmatrix}\right), (x_1', y_1') + \rho\left(\begin{smallmatrix} a' & b' \\ c' & d' \end{smallmatrix}\right) \cdot (x_1, y_1), \ldots, \\ (x_{10}', y_{10}') + \rho\left(\begin{smallmatrix} a' & b' \\ c' & d' \end{smallmatrix}\right)(x_{10}, y_{10})\Big)\end{aligned}$$

where ρ is conjugate to the standard representation

$$\rho\left(\begin{smallmatrix} a & b \\ c & d \end{smallmatrix}\right) = \left(\begin{smallmatrix} 1 & 0 \\ 0 & -1 \end{smallmatrix}\right) \cdot \left(\begin{smallmatrix} a & b \\ c & d \end{smallmatrix}\right) \cdot \left(\begin{smallmatrix} 1 & 0 \\ 0 & -1 \end{smallmatrix}\right) = \left(\begin{smallmatrix} a & -b \\ -c & d \end{smallmatrix}\right). \tag{2.6}$$

---

[8] In fact, $M_{1,11}$ is the open set obtained by removing lattice translates of zero sections and diagonal loci from the right side.

As usual, $SL(2,\mathbb{Z})$ acts on $\mathbb{H}$ via linear fractional transformations. At the point $(z,\zeta_1,\ldots,\zeta_{10}) \in \mathbb{H}\times\mathbb{C}^{10}$, the action is

$$\left(\begin{pmatrix} a & b \\ c & d\end{pmatrix}, (x_1,y_1),\ldots,(x_{10},y_{10})\right)\cdot(z,\zeta_1,\ldots,\zeta_{10}) =$$

$$\left(\left(\frac{az+b}{cz+d}\right), \frac{\zeta_1}{cz+d}+x_1+y_1\left(\frac{az+b}{cz+d}\right),\ldots,\frac{\zeta_{10}}{cz+d}+x_{10}+y_{10}\left(\frac{az+b}{cz+d}\right)\right).$$

A direct verification using the weight 12 functional equation for the discriminant shows the holomorphic 11-form

$$\Delta(e^{2\pi i z})\, dz\wedge d\zeta_1\wedge\ldots\wedge d\zeta_{10} \tag{2.7}$$

on $\mathbb{H}\times\mathbb{C}^{10}$ is invariant under the action and descends to $M_{1,11}$. Since the discriminant is a cusp form, the form (2.7) extends to a nontrivial element of $H^{11,0}(\overline{M}_{1,11},\mathbb{C})$. In fact,

$$H^{11,0}(\overline{M}_{1,11},\mathbb{C})\cong\mathbb{C}$$

by calculations of [32] (the Σ_{11}-representation on $H^{11,0}(\overline{M}_{1,11},\mathbb{C})$ is alternating). As a consequence, $\overline{M}_{1,11}$ is irrational.

Returning to the open locus $M_{1,n}$, we can consider the fibration

$$\pi: M_{1,n}\to M_{1,1}$$

with fibers given by open subsets of E^{n-1} (up to automorphisms). The only interesting cohomology of E is $H^1(E,\mathbb{Q})$. The motivic Euler characteristic of the cohomology of the fibers of π can be expressed in terms of the symmetric powers of $H^1(E,\mathbb{Q})$. Letting E vary, the a-th symmetric power gives rise to a *local system* $\mathbb{V}_a$ on $M_{1,1}$. The cohomology of these local systems was studied in detail by Shimura. A basic result [58] (see also [63, Thm. 4.2.6]) is the *Shimura isomorphism*,

$$H^1_!(M_{1,1},\mathbb{V}_a)\otimes\mathbb{C} = S_{a+2}\oplus\overline{S}_{a+2}\,.$$

Here, $H^i_!$ is the *inner cohomology*, the image of the cohomology H^i_c with compact support in the usual cohomology H^i. The Shimura isomorphism gives a connection between $M_{1,11}$ and the space of cusp forms S_{12}. The inner cohomology group $H^1_!(M_{1,1},\mathbb{V}_a)$ has a pure Hodge structure of weight $a+1$ and S_{a+2} has Hodge type $(a+1,0)$. We have found non-algebraic cohomology, but we still need to understand the contribution to the trace of Frobenius.

As shown by Deligne [16, Prop. 3.19], Hecke operators can be defined on the inner cohomology group $H^1_!(M_{1,1},\mathbb{V}_a)$ compatibly with the Shimura isomorphism and the earlier operators on the spaces of cusp forms. The Eichler-Shimura congruence relation [16, Thm. 4.9], [17, 58] establishes a connection between the Hecke operator T_p on $H^1_!(M_{1,1},\mathbb{V}_a)\otimes\mathbb{Q}_\ell$ and the Frobenius at p. More precisely, T_p equals the sum of Frobenius and Verschiebung at p (the adjoint of Frobenius with respect to the natural scalar product). As explained in [16, (5.7)], the results finally show the trace of Frobenius at p on $H^1_!(M_{1,1}, \mathbb{V}_a) \otimes \mathbb{Q}_\ell$ to be equal to the trace of the Hecke operator T_p on S_{a+2}, which is the sum of the p-th Fourier coefficients of the normalized Hecke cusp eigenforms.

We are now in a position to derive the mysterious counting formula

$$\#M_{1,11}(\mathbb{F}_p) = f_{11}(p) - \tau(p).$$

The kernel of the map

$$H^i_c(M_{1,1}, \mathbb{V}_a) \to H^i(M_{1,1}, \mathbb{V}_a)$$

is the *Eisenstein cohomology*. Like the inner cohomology, the Eisenstein cohomology for $a > 0$ is concentrated in degree 1. It consists of Hodge-Tate classes of weight 0 which over $\mathbb{C}$ are represented by suitably normalized Eisenstein series. Writing $S[a+2]$ for $H^1_!(M_{1,1}, \mathbb{V}_a)$, we obtain

$$e_c(M_{1,1}, \mathbb{V}_a) = -S[a+2] - 1 \tag{2.8}$$

for $a > 0$ even, while the Euler characteristic vanishes for a odd. A motivic construction of $S[a+2]$ can be found in [15, 57], see also [56]. By the discussion in the previous paragraph, the trace of Frobenius picks up $-\tau(p)$ from (2.8). As another consequence of (2.8), we find the following formula for the integer valued Euler characteristic,

$$E_c(M_{1,1}, \mathbb{V}_a) = -2s_{a+2} - 1.$$

In fact, formula (2.8) has been used by Getzler [32] to calculate the full Hodge decomposition of the cohomology of $\overline{M}_{1,n}$.

3. Point counting and Siegel modular forms

3.1. Summary

Siegel modular forms are related to moduli spaces of abelian varieties. As a consequence, a strong relationship between such modular forms and the cohomology of moduli spaces of curves of genus 2 is obtained, leading to a complete (conjectural) description of the cohomology of $\overline{M}_{2,n}$. Perhaps surprisingly, the resulting description in genus 3 is not complete. Elliptic and Siegel modular forms do not suffice to describe the cohomology of $M_{3,n}$ for n large enough — the Teichmüller modular forms introduced by Ichikawa [41] are needed. In genus 2 and 3, data from point counting over finite fields is used to formulate the conjectures and to explore the cohomology.

3.2. Siegel modular forms

In higher genus $g > 1$, the analogue of elliptic modular forms are *Siegel modular forms*. The latter are defined on $\mathbb{H}_g$, the space of $g \times g$ complex symmetric matrices with positive-definite imaginary part and satisfy the functional equation

$$f((az+b)(cz+d)^{-1}) = \rho(cz+d)f(z)$$

for all $z \in \mathbb{H}_g$ and all $\left(\begin{smallmatrix} a & b \\ c & d \end{smallmatrix}\right) \in \mathsf{Sp}(2g, \mathbb{Z})$. Here, ρ is a representation of $\mathsf{GL}(g, \mathbb{C})$. Classically,

$$\rho(M) = \mathrm{Det}^k(M),$$

but other irreducible representations of $\mathsf{GL}(g, \mathbb{C})$ are just as relevant. The form f is vector valued. Since z has positive definite imaginary part, $cz + d$ is indeed invertible.

Hecke operators can be defined on Siegel modular forms [1, 2]. The Fourier coefficients of an eigenform here contain much more information than just the Hecke eigenvalues:

(i) the coefficients are indexed by positive semi-definite $g \times g$ symmetric matrices with integers on the diagonal and half-integers elsewhere,
(ii) the coefficients themselves are vectors.

Cusp forms are Siegel modular forms in the kernel of the Siegel Φ-operator:

$$(\Phi f)(z') = \lim_{t \to \infty} f\left(\begin{smallmatrix} z' & 0 \\ 0 & it \end{smallmatrix}\right),$$

for $z' \in \mathbb{H}_{g-1}$. As before, the Hecke operators preserve the space of cusp forms.

3.3. Point counting in genus 2

Since genus 2 curves are double covers of $\mathbb{P}^1$, we can effectively count them over $\mathbb{F}_p$. As before, each isomorphism class over $\mathbb{F}_p$ is counted with the reciprocal of the number of $\mathbb{F}_p$-automorphisms. For small n, the enumeration of $M_{2,n}(\mathbb{F}_p)$ can be done by hand as in Proposition 2.3. For large n, computer counting is needed.

Bergström [9] has proven the polynomiality in p of $\#M_{2,n}(\mathbb{F}_p)$ for $n \leqslant 7$ by geometric stratification of the moduli space. Experimentally, we find $\#M_{2,n}(\mathbb{F}_p)$ is a polynomial of degree $n+3$ in p for $n \leqslant 9$. In the range $10 \leqslant n \leqslant 13$, the function $\tau(p)$ multiplied by a polynomial appears in $\#M_{2,n}(\mathbb{F}_p)$. More precisely, computer counting predicts

$$\#M_{2,10}(\mathbb{F}_p) = f_{13}(p) + (p-9)\tau(p)\,.$$

If the above formula holds, then the possibility of a degree 13 polynomial for $\#M_{2,10}(\mathbb{F}_p)$ is excluded, so we conclude $M_{2,10}$ has cohomology which is not algebraic over $\mathbb{Q}$ (and therefore non-tautological).

Higher weight modular forms appear in the counting for $n = 14$. Computer counts predict

$$\#M_{2,14}(\mathbb{F}_p) = f_{17}(p) + f_5(p)\tau(p) - 13c_{16}(p) - 429c_{18}(p),$$

where $c_{16}(p)$ and $c_{18}(p)$ are the Fourier coefficients of the normalized Hecke cusp eigenforms of weights 16 and 18. The appearance of $c_{16}(p)$ is perhaps not so surprising since

$$\tau = c_{12}$$

occurs in the function $\#M_{2,10}(\mathbb{F}_p)$. By counting Σ_{14}-equivariantly, we find the coefficient 13 arises as the dimension of the irreducible representation corresponding to the partition $[2\,1^{12}]$ (the tensor product of the standard representation and the alternating representation).

More surprising is the appearance of $c_{18}(p)$. Reasoning as before in Section 2.2, we would expect the motive S[18] to occur in the cohomology of $M_{2,14}$. Since the Hodge types are (17,0) and (0,17), the motive S[18] cannot possibly come from the boundary of the 17 dimensional moduli space, but must come from the interior. We would then expect a Siegel cusp form to be responsible for the occurrence, with Hecke eigenvalues closely related to those of the elliptic cusp form of weight 18. There indeed exists such a Siegel modular form: the classical Siegel cusp form χ_{10} of weight 10, the product of the squares of the 10 even theta characteristics, is a Saito-Kurokawa lift of $E_6\Delta$. In particular, the Hecke eigenvalues are related via

$$\lambda_{\chi_{10}}(p) = \lambda_{E_6\Delta}(p) + p^8 + p^9 = c_{18}(p) + p^8 + p^9.$$

The coefficient 429 is the dimension of the irreducible representation corresponding to the partition $[2^7]$, as a Σ_{14}-equivariant count shows.

The function $\#M_{2,15}(\mathbb{F}_p)$ is expressible in terms of τ, c_{16}, and c_{18} as expected. For $n = 16$, the Hecke eigenvalues of a vector valued Siegel cusp form appear for the first time: the unique form of type

$$\rho = \mathrm{Sym}^6 \otimes \mathrm{Det}^8$$

constructed by Ibukiyama (see [19]) arises. For $n \leqslant 25$, the computer counts of $\#M_{2,n}(\mathbb{F}_p)$ have to a large extent been successfully fit by Hecke eigenvalues of Siegel cusp forms [37].

3.4. Local systems in genus 2

To explain the experimental results for $\#M_{2,n}(\mathbb{F}_p)$ discussed above, we study the cohomology of local systems on the moduli space M_2. Via the Jacobian, there is an inclusion

$$M_2 \subset A_2$$

as an open subset in the moduli space of principally polarized abelian surfaces. In fact, the geometry of local systems on A_2 is a more natural object of study.

The irreducible representations $V_{a,b}$ of $\mathrm{Sp}(4,\mathbb{Z})$ are indexed by integers

$$a \geqslant b \geqslant 0.$$

More precisely, $V_{a,b}$ is the irreducible representation of highest weight occurring in $\mathrm{Sym}^{a-b}(R) \otimes \mathrm{Sym}^b(\wedge^2 R)$, where R is the dual of the standard representation. Since the moduli space of abelian surfaces arises as a quotient,

$$A_2 = \mathbb{H}_2/\mathrm{Sp}(4,\mathbb{Z}),$$

we obtain local systems $\mathbb{V}_{a,b}$ for $\mathrm{Sp}(4,\mathbb{Z})$ on A_2 associated to the representations $V_{a,b}$. The local system $\mathbb{V}_{a,b}$ is *regular* if $a > b > 0$.

Faltings and Chai [25, Thm. VI.5.5] relate the cohomology of $\mathbb{V}_{a,b}$ on A_2 to the space $S_{a-b,b+3}$ of Siegel cusp forms of type $\mathrm{Sym}^{a-b} \otimes \mathrm{Det}^{b+3}$. To start, $H^i_c(A_2, \mathbb{V}_{a,b})$ has a natural mixed Hodge structure with weights at most $a+b+i$, and $H^i(A_2, \mathbb{V}_{a,b})$ has a mixed Hodge structure with weights at least $a+b+i$. The inner cohomology $H^i_!(A_2, \mathbb{V}_{a,b})$ therefore has a pure Hodge structure of weight $a+b+i$. Faltings [24, Cor. of Thm. 9] had earlier shown that $H^i_!(A_2, \mathbb{V}_{a,b})$ is concentrated in degree 3 when the local system is regular. For the Hodge structures above, the degrees of the Hodge filtration are contained in

$$\{0, b+1, a+2, a+b+3\}.$$

Finally, there are natural isomorphisms

$$F^{a+b+3}H^3(A_2, \mathbb{V}_{a,b}) \cong M_{a-b,b+3},$$

$$F^{a+b+3}H^3_c(A_2, \mathbb{V}_{a,b}) = F^{a+b+3}H^3_!(A_2, \mathbb{V}_{a,b}) \cong S_{a-b,b+3}$$

with the spaces of Siegel modular (respectively cusp) forms of the type mentioned above. The quotients in the Hodge filtration are isomorphic to certain explicit coherent cohomology groups, see [31, Thm. 17].

One is inclined to expect the occurrence of motives $S[a-b, b+3]$ corresponding to Siegel cusp forms in $H^3_!(A_2, \mathbb{V}_{a,b})$, at least in the case of a regular weight. Each Hecke cusp eigenform should contribute a 4-dimensional piece with four 1-dimensional pieces in the Hodge decomposition of types

$$(a+b+3,0),\ (a+2, b+1),\ (b+1, a+2),\ (0, a+b+3).$$

By the theory of automorphic representations, the inner cohomology may contain other terms as well, the so-called endoscopic contributions. In the present case, the endoscopic contributions come apparently only[9] with Hodge types $(a+2, b+1)$ and $(b+1, a+2)$.

Based on the results from equivariant point counts, Faber and van der Geer [19] obtain the following explicit conjectural formula for the Euler characteristic of the compactly supported cohomology:

Conjecture. *For* $a > b > 0$ *and* $a+b$ *even,*

$$\begin{aligned} e_c(A_2, \mathbb{V}_{a,b}) = &-S[a-b, b+3] - s_{a+b+4}S[a-b+2]L^{b+1} \\ &+ s_{a-b+2} - s_{a+b+4}L^{b+1} - S[a+3] + S[b+2] + \tfrac{1}{2}(1+(-1)^a). \end{aligned}$$

The terms in the second line constitute the Euler characteristic of the Eisenstein cohomology. The second term in the first line is the endoscopic contribution. Just as in the genus 1 case, the hyperelliptic involution causes the vanishing of all cohomology when $a+b$ is odd. However, the above conjecture is not quite a precise

[9] However, our use of the term endoscopic is potentially non-standard.

statement. While L is the Lefschetz motive and $S[k]$ has been discussed in Section 2.3, the motives $S[a-b, b+3]$ have not been constructed yet. Nevertheless, several precise predictions of the conjecture will be discussed.

First, we can specialize the conjecture to yield a prediction for the integer valued Euler characteristic $E_c(A_2, \mathbb{V}_{a,b})$. The formula from the conjecture is

$$E_c(A_2, \mathbb{V}_{a,b}) = -4s_{a-b,b+3} - 2s_{a+b+4}s_{a-b+2} \\ + s_{a-b+2} - s_{a+b+4} - 2s_{a+3} + 2s_{b+2} + \tfrac{1}{2}(1 + (-1)^a).$$

for $a > b > 0$ and $a + b$ even. The lower case s denotes the dimension of the corresponding space of cusp forms. The dimension formula was proved by Grundh [37] for $b > 1$ using earlier work of Getzler [33] and a formula of Tsushima [61] for $s_{j,k}$, proved for $k > 4$ (and presumably true for $k = 4$). In fact, combining work of Weissauer [64] on the inner cohomology and of van der Geer [28] on the Eisenstein cohomology, one may deduce the implication of the conjecture obtained by taking the realizations as ℓ-adic Galois representations of all terms.

A second specialization of the conjecture yields a prediction for $\#M_{2,n}(\mathbb{F}_p)$ via the eigenvalues of Hecke cusp forms. The prediction has been checked for $n \leqslant 17$ and $p \leqslant 23$. The prediction for the trace of Frob_p on $e_c(M_2, \mathbb{V}_{a,b})$ has been checked in many more cases. Consider the local systems $\mathbb{V}_{a,b}$ with $a+b \leqslant 24$ and $b \geqslant 5$. There are 13 such local systems with $\dim S_{a-b,b+3} = 1$; the prediction has been checked for 11 of them, for $p \leqslant 23$. There are also 6 such local systems with $\dim S_{a-b,b+3} = 2$, the maximal dimension; the prediction has been checked for all 6 of them, for $p \leqslant 17$.

We believe the conjecture to be correct also when $a = b$ or $b = 0$ after a suitable re-interpretation of several terms. To start, we set $s_2 = -1$ and define

$$S[2] = -L - 1,$$

which of course is *not* equal to $H^1_!(M_{1,1}, \mathbb{V}_0)$ but does yield the correct answer for

$$e_c(M_{1,1}, \mathbb{V}_0) = L.$$

Let $S[k] = 0$ for odd k. Similarly, define

$$S[0,3] = -L^3 - L^2 - L - 1 .$$

By analogy, let $s_{0,3} = -1$, which equals the value produced by Tsushima's formula (the only negative value of $s_{j,k}$ for $k \geqslant 3$). We then obtain the correct answer for

$$e_c(A_2, \mathbb{V}_{0,0}) = L^3 + L^2.$$

Finally, $S[0,10]$ is defined as $L^8 + S[18] + L^9$, and more generally, $S[0, m+1]$ includes (for m odd) a contribution $SK[0, m+1]$ defined as $S[2m] + s_{2m}(L^{m-1} + L^m)$ — corresponding to the Saito-Kurokawa lifts, but not a part of $H^3_!(A_2, \mathbb{V}_{m-2,m-2})$.

3.5. Holomorphic differential forms

We can use the Siegel modular form χ_{10} to construct holomorphic differential forms on $M_{2,14}$. To start, we consider the seventh fiber product U^7 of the universal abelian surface

$$U \to A_2 .$$

We can construct U^7 as a quotient,

$$U^7 = \frac{\mathbb{H}_2 \times (\mathbb{C}^2)^7}{\mathsf{Sp}(4,\mathbb{Z}) \ltimes (\mathbb{Z}^4)^7} .$$

The semidirect product is defined by the conjugate of the standard representation following (2.6). We will take $z = \left(\begin{smallmatrix} z_{11} & z_{12} \\ z_{12} & z_{22} \end{smallmatrix}\right) \in \mathbb{H}_2$ and $\zeta_i = (\zeta_{i1}, \zeta_{i2}) \in \mathbb{C}^2$ to be coordinates. At the point

$$(z, \zeta_1, \dots, \zeta_7) \in \mathbb{H}_2 \times (\mathbb{C}^2)^7,$$

the action is

$$\begin{aligned} &\left(\left(\begin{smallmatrix} a & b \\ c & d \end{smallmatrix}\right), (x_1, y_1), \dots, (x_7, y_7)\right) \cdot (z, \zeta_1, \dots, \zeta_7) = \\ &\quad \Big((az+b)(cz+d)^{-1}, \zeta_1(cz+d)^{-1} + x_1 + y_1(az+b)(cz+d)^{-1}, \\ &\qquad \dots, \zeta_7(cz+d)^{-1} + x_7 + y_7(az+b)(cz+d)^{-1}\Big), \end{aligned}$$

where $x_i, y_i \in \mathbb{Z}^2$. A direct verification using the weight 10 functional equation for χ_{10} shows the holomorphic 17-form

$$\chi_{10}(z)\, dz_{11} \wedge dz_{12} \wedge dz_{22} \wedge d\zeta_{11} \wedge d\zeta_{12} \wedge \dots \wedge d\zeta_{71} \wedge d\zeta_{72} \tag{3.1}$$

on $\mathbb{H}_2 \times (\mathbb{C}^2)^7$ is invariant under the action and descends to U^7.

To obtain a 17-form on $M_{2,14}$, we consider the Abel-Jacobi map $M_{2,14} \to U^7$ defined by

$$(C, p_1, \dots, p_{14}) \mapsto (\mathrm{Jac}_0(C), \omega_C^*(p_1 + p_2), \dots, \omega_C^*(p_{13} + p_{14})) .$$

The pull-back of (3.1) yields a holomorphic 17-form on $M_{2,14}$. Since χ_{10} is a cusp form, the pull-back extends to a nontrivial element of $H^{17,0}(\overline{M}_{2,14}, \mathbb{C})$. Assuming the conjecture for local systems in genus 2,

$$H^{17,0}(\overline{M}_{2,14}, \mathbb{C}) \cong \mathbb{C}^{429}$$

with Σ_{14}-representation irreducible of type $[2^7]$. We leave as an exercise for the reader to show our construction of 17-forms naturally yields the representation $[2^7]$.

The existence of 17-forms implies the irrationality of $\overline{M}_{2,14}$. Both $M_{2,n\leqslant 12}$ and the quotient of $M_{2,13}$ obtained by unordering the last two points are rational by classical constructions [14]. Is $M_{2,13}$ rational?

3.6. The compactification $\overline{M}_{2,n}$

We now assume the conjectural formula for $e_c(A_2, \mathbb{V}_{a,b})$ holds for $a \geqslant b \geqslant 0$. As a consequence, we can compute the Hodge numbers of the compactifications $\overline{M}_{2,n}$ for all n.

We view, as before, $M_2 \subset A_2$ via the Torelli morphism. The complement is isomorphic to $\mathrm{Sym}^2 A_1$. The local systems $\mathbb{V}_{a,b}$ can be pulled back to M_2 and restricted to $\mathrm{Sym}^2 A_1$. The cohomology of the restricted local systems can be understood by combining the branching formula from Sp_4 to $\mathrm{SL}_2 \times \mathrm{SL}_2$ with an analysis of the effect of quotienting by the involution of $A_1 \times A_1$ which switches the factors, [13, 37, 55]. Hence, the Euler characteristic $e_c(M_2, \mathbb{V}_{a,b})$ may be determined. The motives $\wedge^2 S[k]$ and $\mathrm{Sym}^2 S[k]$ (which sum to $S[k]^2$) will occur. Note

$$\wedge^2 S[k] \cong L^{k-1}$$

when $s_k = 1$.

As observed by Getzler [31], the Euler characteristics $e_c(M_2, \mathbb{V}_{a,b})$ for $a+b \leqslant N$ determine and are determined by the Σ_n-equivariant Euler characteristics $e_c^{\Sigma_n}(M_{2,n})$ for $n \leqslant N$. Hence, the latter are (conjecturally) determined for all n. Next, by the work of Getzler and Kapranov [34], the Euler characteristics $e_c^{\Sigma_n}(\overline{M}_{2,n})$ are determined as well, since we know the Euler characteristics in genus at most 1. Conversely, the answers for $\overline{M}_{2,n}$ determine those for $M_{2,n}$. After implementing the Getzler-Kapranov formalism (optimized for genus 2), a calculation of $e_c^{\Sigma_n}(\overline{M}_{2,n})$ for all $n \leqslant 22$ has been obtained.[10] All answers satisfy Poincaré duality, which is a very non-trivial check.

The cohomology of $\overline{M}_{2,21}$ is of particular interest to us. The motives $\wedge^2 S[12]$ and $\mathrm{Sym}^2 S[12]$ appear for the first time for $n = 21$ (with a Tate twist). The coefficient of $L \wedge^2 S[12]$ in $e_c^{\Sigma_{21}}(\overline{M}_{2,21})$ equals

$$[3\,1^{18}] + [3\,2^2\,1^{14}] + [3\,2^4\,1^{10}] + [3\,2^6\,1^6] + [3\,2^8\,1^2] + [2\,1^{19}] + [2^2\,1^{17}]$$

$$+ [2^3\,1^{15}] + [2^4\,1^{13}] + [2^5\,1^{11}] + [2^6\,1^9] + [2^7\,1^7] + [2^8\,1^5] + [2^9\,1^3] + [2^{10}\,1^1]$$

as a Σ_{21}-representation. As mentioned,

$$\wedge^2 S[12] \cong L^{11}. \tag{3.2}$$

We find thus 1939938 independent Hodge classes L^{12}, which should be algebraic by the Hodge conjecture. However, 1058148 of these classes come with an irreducible representation of length at least 13. By Theorem 4.2 obtained in Section 4, the latter classes cannot possibly be tautological.

The 1058148 classes actually span a small fraction of the full cohomology $H^{12,12}(\overline{M}_{2,21})$. There are also

$$1243344485012727233336\,91$$

[10] Thanks to Stembridge's symmetric functions package SF.

classes L^{12} not arising via the isomorphism (3.2) — all the irreducible representations of length at most 12 arise in the coefficients.

3.7. Genus 3

In recent work [11], Bergström, Faber, and van der Geer have extended the point counts of moduli spaces of curves over finite fields to the genus 3 case. For all (n_1, n_2, n_3) and all prime powers $q \leqslant 17$, the frequencies

$$\sum_C \frac{1}{\# \mathrm{Aut}_{\mathbb{F}_q}(C)}$$

have been computed, where we sum over isomorphism classes of curves C over $\mathbb{F}_q$ with exactly n_i points over $\mathbb{F}_{q^i}$. The Σ_n-equivariant counts $\#M_{3,n}(\mathbb{F}_q)$ are then determined.

The isomorphism classes of Jacobians of nonsingular curves of genus 3 form an open subset of the moduli space A_3 of principally polarized abelian varieties of dimension 3.[11] The complement can be easily understood. Defining the $\mathsf{Sp}(6, \mathbb{Z})$ local systems $\mathbb{V}_{a,b,c}$ in expected manner, the traces of Frob_q on $e_c(M_3, \mathbb{V}_{a,b,c})$ and $e_c(A_3, \mathbb{V}_{a,b,c})$ can be computed for $q \leqslant 17$ and arbitrary $a \geqslant b \geqslant c \geqslant 0$.

Interpretation of the results has been quite successful for A_3 : an explicit conjectural formula for $e_c(A_3, \mathbb{V}_{a,b,c})$ has been found in [11], compatible with all known results (the results of Faltings [24] and Faltings-Chai [25] mentioned above, the dimension formula for the spaces of classical Siegel modular forms [62], and the numerical Euler characteristics [13, §8]). Two features of the formula are quite striking. First, it has a simple structure in which the endoscopic and Eisenstein contributions for genus 2 play an essential role. Second, it predicts the existence of many vector valued Siegel modular forms that are lifts, connected to local systems of *regular* weight[12], see [11].

Interpreting the results for M_3 is much more difficult. In fact, the stack M_3 cannot be correctly viewed as an open part of A_3. Rather, M_3 is a (stacky) double cover of the locus J_3 of Jacobians of nonsingular curves, branched over the locus of hyperelliptic Jacobians. The double covering occurs because

$$\mathrm{Aut}(\mathrm{Jac}_0(C)) = \mathrm{Aut}(C) \times \{\pm 1\}$$

for C a non-hyperelliptic curve, while equality of the automorphism groups holds for C hyperelliptic. As a result, the local systems $\mathbb{V}_{a,b,c}$ of odd weight $a + b + c$ will in general have non-vanishing cohomology on M_3, while their cohomology on A_3 vanishes. We therefore can not expect that the main part of the cohomology of $\mathbb{V}_{a,b,c}$ on M_3 can be explained in terms of Siegel cusp forms when $a + b + c$ is odd. On the other hand, the local systems of even weight provide no difficulties.

---

[11] For $g \geqslant 4$, Jacobians have positive codimension in A_g.

[12] An analogous phenomenon occurs for Siegel modular forms of genus 2 and level 2, see [10], §6.

Bergström [8] has proved that the Euler characteristics $e_c(M_3, \mathbb{V}_{a,b,c})$ are certain explicit polynomials in L, for $a + b + c \leqslant 7$.

In fact, calculations ([12], in preparation) show that motives not associated to Siegel cusp forms must show up in $e_c(M_3, \mathbb{V}_{11,3,3})$ and $e_c(M_3, \mathbb{V}_{7,7,3})$, and therefore in $e_c^{\Sigma_{17}}(\overline{M}_{3,17})$, with the irreducible representations of type $[3^3\, 1^8]$ and $[3^3\, 2^4]$. Teichmüller modular forms [41] should play an important role in accounting for the cohomology in genus 3 not explained by Siegel cusp forms.

4. Representation theory

4.1. Length bounds

Consider the standard moduli filtration

$$\overline{M}_{g,n} \supset M^c_{g,n} \supset M^{rt}_{g,n}$$

discussed in Section 1.7. We consider only pairs g and n which satisfy the stability condition

$$2g - 2 + n > 0 .$$

If $g = 0$, all three spaces are equal by definition

$$\overline{M}_{0,n} = M^c_{0,n} = M^{rt}_{0,n} .$$

If $g = 1$, the latter two are equal

$$M^c_{1,n} = M^{rt}_{1,n} .$$

For $g \geqslant 2$, all three are different.

The symmetric group Σ_n acts on $\overline{M}_{g,n}$ by permuting the markings. Hence, Σ_n-actions are induced on the tautological rings

$$R^*(\overline{M}_{g,n}),\ R^*(M^c_{g,n}),\ R^*(M^{rt}_{g,n}) . \tag{4.1}$$

By Corollary 1.5, all the rings (4.1) are finite dimensional representations of Σ_n.

We define the *length* of an irreducible representation of Σ_n to be the number of parts in the corresponding partition of n. The trivial representation has length 1, and the alternating representation has length n. We define the length $\ell(V)$ of a finite dimensional representation V of Σ_n to be the maximum of the lengths of the irreducible constituents.

Our main result bounds the lengths of the tautological rings in all cases (4.1).

Theorem 4.2. *For the tautological rings of the moduli spaces of curves, we have*

(i) $\ell(R^k(\overline{M}_{g,n})) \leqslant \min\left(k+1, 3g-2+n-k, \lfloor \frac{2g-1+n}{2} \rfloor\right)$,

(ii) $\ell(R^k(M^c_{g,n})) \leqslant \min\left(k+1, 2g-2+n-k\right)$,

(iii) $\ell(R^k(M^{rt}_{g,n})) \leqslant \min\left(k+1, g-1+n-k\right)$.

The length bounds in all cases are consistent with the conjectures of Poincaré duality for these tautological rings, see [18, 21, 38, 53]. For (i), the bound is invariant under

$$k \longleftrightarrow 3g-3+n-k\,.$$

For (ii), the bound is invariant under

$$k \longleftrightarrow 2g-3+n-k$$

which is consistent with a socle in degree $2g-3+n$. For (iii), the bound is invariant under

$$k \longleftrightarrow g-2+n-k$$

which is consistent with a socle in degree $g-2+n$ (for $g>0$).

In genus g, the bound $\lfloor \frac{2g-1+n}{2} \rfloor$ in case (i) improves upon the trivial length bound n only for $n \geqslant 2g$.

4.2. Induction

The proof of Theorem 4.2 relies heavily on a simple length property of induced representations of symmetric groups.

Let V_1 and V_2 be representations of Σ_{n_1} and Σ_{n_2} of lengths ℓ_1 and ℓ_2 respectively. For $n = n_1 + n_2$, we view

$$\Sigma_{n_1} \times \Sigma_{n_2} \subset \Sigma_n$$

in the natural way.

Proposition 4.3. *We have* $\ell\left(\mathrm{Ind}_{\Sigma_{n_1}\times\Sigma_{n_2}}^{\Sigma_n} V_1 \otimes V_2\right) = \ell_1 + \ell_2\,.$

Proof. To prove the result, we may certainly assume V_i is an irreducible representation V_{λ_i}. In fact, we will exchange V_{λ_i} for a representation more closely related to induction.

For $\alpha = (\alpha_1, \ldots, \alpha_k)$ a partition of m, let U_α be the representation of Σ_m induced from the trivial representation of the Young subgroup

$$\Sigma_\alpha = \Sigma_{\alpha_1} \times \cdots \times \Sigma_{\alpha_k}$$

of Σ_m. Young's rule expresses U_λ in terms of irreducible representations V_μ for which μ precedes λ in the lexicographic ordering,

$$U_\lambda \cong V_\lambda \oplus \bigoplus_{\mu > \lambda} K_{\mu\lambda} V_\mu\,. \tag{4.4}$$

The Kostka number $K_{\mu\lambda}$ does not vanish if and only if μ dominates λ, which implies $\ell(\mu) \leqslant \ell(\lambda)$. Therefore $V_\lambda - U_\lambda$ can be written in the representation ring as a $\mathbb{Z}$-linear combination of U_μ for which $\mu > \lambda$ and $\ell(\mu) \leqslant \ell(\lambda)$.

If we let $V_1 = U_{\lambda_1}$ and $V_2 = U_{\lambda_2}$, the induction of representations is easy to calculate,

$$\mathrm{Ind}_{\Sigma_{n_1}\times\Sigma_{n_2}}^{\Sigma_{n_1+n_2}} U_{\lambda_1} \otimes U_{\lambda_2} = \mathrm{Ind}_{\Sigma_{\lambda_1}\times\Sigma_{\lambda_2}}^{\Sigma_{n_1+n_2}} 1 \cong \mathrm{Ind}_{\Sigma_{\lambda_1+\lambda_2}}^{\Sigma_{n_1+n_2}} 1 = U_{\lambda_1+\lambda_2}\,,$$

where $\lambda_1 + \lambda_2$ is the partition of n consisting of the parts of λ_1 and λ_2, reordered. Hence, the Proposition is proven for U_{λ_1} and U_{λ_2}. By the decomposition (4.4), the Proposition follows for V_{λ_1} and V_{λ_2}. □

4.3. Proof of Theorem 4.2

4.3.1. Genus 0 The genus 0 result

$$\ell(R^k(\overline{M}_{0,n})) \leqslant \min(k+1, n-k-2)$$

plays an important role in the rest of the proof of Theorem 4.2 and will be proven first.

By Keel [43], we have isomorphisms

$$A^k(\overline{M}_{0,n}) = R^k(\overline{M}_{0,n}) = H^{2k}(\overline{M}_{0,n}).$$

Poincaré duality then implies

$$\ell(R^k(\overline{M}_{0,n})) = \ell(R^{n-3-k}(\overline{M}_{0,n})) \,. \tag{4.5}$$

The vector space $R^k(\overline{M}_{0,n})$ is generated by strata classes of curves with k nodes (and hence with $k+1$ components). The bound

$$\ell(R^k(\overline{M}_{0,n})) \leqslant k+1$$

follows by a repeated application of Proposition 4.3 or by observing

$$\ell(U_\lambda) = \ell(\lambda),$$

for any partition λ (here with at most $k+1$ parts). The bound

$$\ell(R^k(\overline{M}_{0,n})) \leqslant n-k-2$$

is obtained by using (4.5). □

4.3.2. Rational tails We prove next a length bound for the subrepresentations of $R^k(\overline{M}_{g,n})$ generated by the decorated strata classes of curves with rational tails, in other words, by the classes

$$\xi_{A*}\Big(\prod_{v\in V(A)} \theta_v\Big)$$

as in Theorem 1.4, where A is a stable graph of genus $g \geqslant 1$ with n legs and exactly one vertex of genus g (and all other vertices of genus 0).

Let α be such a class of codimension k and graph A. Let v be the vertex of genus g. We may assume the decoration is supported only at v, so

$$\alpha = \xi_{A*}\theta_v \,.$$

The graph A arises by attaching t rational trees T_i to v. Let T_i have (before attaching) k_i edges and m_i legs. Let $m \geqslant t$ denote the valence of v. After attaching the trees, v

has $m-t$ legs. Of the $m-t$ legs, let s_0 come without ψ class, s_1 come with ψ^1, s_2 come with ψ^2, and so forth, so

$$\sum_j s_j = m - t\,.$$

After discarding the s_j which vanish, we obtain a partition of $m-t$ with p positive parts. The $m-t$ legs contribute a class ζ_v of degree $\sum_j js_j$ to θ_v. Let η_v be the product of the κ classes at the vertex and the ψ classes at the t legs at which the trees are attached, so

$$\theta_v = \zeta_v \eta_v\,.$$

Let e be the degree of η_v. We clearly have

$$k = \sum_i k_i + t + e + \sum_j js_j \qquad \text{and} \qquad n = m + \sum_i m_i - 2t.$$

At the vertex v, we obtain a representation of Σ_{m-t} of length p. By the result for genus 0, each tree T_i generates a subrepresentation V_i of $R^{k_i}(\overline{M}_{0,m_i})$ with

$$\ell(V_i) \leqslant \min(k_i + 1, m_i - k_i - 2).$$

One of the m_i legs is used in attaching T_i to v. After restricting V_i to Σ_{m_i-1}, the length can only decrease. Applying Proposition 4.3, we see α generates a subrepresentation V of $R^k(\overline{M}_{g,n})$ with

$$\ell(V) \leqslant p + \sum_i \min(k_i + 1, m_i - k_i - 2).$$

Taking the first terms in the minimum expressions, we have

$$\ell(V) \leqslant p + \sum_i (k_i + 1) = p + k - e - \sum_j js_j\,. \tag{4.6}$$

Since

$$\sum_j js_j = \sum_{j\geqslant 1: s_j>0} js_j \geqslant \sum_{j\geqslant 1: s_j>0} 1 = p - 1 + \delta_{s_0,0}\,,$$

we see

$$p - \sum_j js_j \leqslant 1 - \delta_{s_0,0} \leqslant 1\,.$$

After substituting in (4.6) and using $e \geqslant 0$, we conclude

$$\ell(V) \leqslant k + 1. \tag{4.7}$$

Taking the second terms in the minimum expressions, we have

$$\begin{aligned}\ell(V) &\leqslant p + \sum_i (m_i - k_i - 2)\\ &= p + n - m + t + e + \sum_j js_j - k\\ &= n - k + (p - m + t) + e + \sum_j js_j\,.\end{aligned}$$

The bound

$$e + \sum_j j s_j \leqslant g - 1 \tag{4.8}$$

will be assumed here. Further, $p \leqslant m - t$, so

$$\ell(V) \leqslant g - 1 + n - k. \tag{4.9}$$

Combining the bounds (4.7) and (4.9) yields

$$\ell(V) \leqslant \min(k + 1, g - 1 + n - k)$$

for any subrepresentation V of $R^k(\overline{M}_{g,n})$ generated by a class supported on a stratum of curves with rational tails and respecting the bound (4.8).

If the bound (4.8) is violated, then θ_v can be expressed as a tautological boundary class at the vertex v by Proposition 2 of [23]. The boundary class will include terms which are supported on strata of curves with rational tails (to which the argument can be applied again) and terms which are not supported on rational tails strata (to which the argument can not be applied). Statement (iii) of Theorem 4.2 is an immediate consequence. □

4.3.3. Compact type We prove here a length bound for the subrepresentations of $R^k(\overline{M}_{g,n})$ generated by the decorated strata classes of curves of compact type with $g \geqslant 1$ and $n \geqslant 1$.

Let β be such a class of codimension k and graph B. Let B have b vertices v_a of positive genus g_a and valence $n_a \geqslant 1$. The graph B arises by attaching t rational trees T_i to the vertices v_a. Let T_i have (before attaching) k_i edges and m_i legs of which $l_i \geqslant 1$ legs are used in the attachment. We also allow the degenerate case in which T_i represents a single node, then

$$k_i = -1 \quad \text{and} \quad l_i = m_i = 2\,.$$

At the vertex v_a, the attachment of the trees uses $\widehat{n}_a$ of the n_a legs. Just as in the case of curves with rational tails, we keep track of the powers of ψ classes attached to the remaining $n_a - \widehat{n}_a$ legs. Let $s_{a,j}$ of those legs come with ψ^j. We obtain a partition of $n_a - \widehat{n}_a$ with length p_a. Finally, let e_a be the degree of the product of the κ classes at v_a and the ψ classes at the $\widehat{n}_a$ attachment legs. We have

$$g = \sum_a g_a\,, \qquad n = \sum_i m_i + \sum_a n_a - 2\sum_i l_i\,,$$

$$k = \sum_i k_i + \sum_i l_i + \sum_a e_a + \sum_{a,j} j s_{a,j}\,,$$

$$\sum_i l_i = \sum_a \widehat{n}_a = t + b - 1\,.$$

The last equality follows since B is a tree.

Applying Proposition 4.3, we see that β generates a subrepresentation V of $R^k(\overline{M}_{g,n})$ with

$$\ell(V) \leqslant \sum_a p_a + \sum_i \min(k_i + 1, m_i - k_i - 2).$$

Taking the first terms of the minimum expressions,

$$\begin{aligned}\ell(V) &\leqslant \sum_a p_a + \sum_i (k_i + 1) \\ &= \sum_a p_a + t + k - \sum_i l_i - \sum_a e_a - \sum_{a,j} js_{a,j} \\ &= k + 1 - b + \sum_a p_a - \sum_a e_a - \sum_{a,j} js_{a,j}\,.\end{aligned}$$

As in the rational tails case,

$$\sum_j js_{a,j} \geqslant p_a - 1, \qquad \text{so} \qquad \sum_{a,j} js_{a,j} \geqslant \sum_a p_a - b\,.$$

Therefore

$$\ell(V) \leqslant k + 1 - \sum_a e_a \leqslant k + 1\,.$$

Taking the second terms of the minimum expressions,

$$\begin{aligned}\ell(V) &\leqslant \sum_a p_a + \sum_i (m_i - k_i - 2) \\ &= \sum_a p_a + n - \sum_a n_a + 3\sum_i l_i + \sum_a e_a + \sum_{a,j} js_{a,j} - k - 2t \\ &= n - k + 2b - 2 + \sum_a (p_a - n_a + \widehat{n}_a) + \sum_a e_a + \sum_{a,j} js_{a,j} \\ &\leqslant n - k + 2b - 2 + g - b \\ &= n - k + g + b - 2 \\ &\leqslant n - k + 2g - 2.\end{aligned}$$

We have used the bound (4.8) at every vertex v_a. Therefore,

$$\ell(V) \leqslant \min(k + 1, 2g - 2 + n - k) \leqslant \left\lfloor \tfrac{2g-1+n}{2} \right\rfloor$$

for any subrepresentation V of $R^k(\overline{M}_{g,n})$ generated by a class supported on a stratum of curves of compact type satisfying the bound (4.8) everywhere. As before, Statement (ii) of Theorem 4.2 is an immediate consequence. □

4.3.4. Stable curves Finally, we prove Statement (i) by induction on the genus g. Statement (i) has been proven already for genus 0. We assume (i) is true for genus at most $g - 1$.

We have three bounds to establish to prove (i). The first bound to prove is

$$\ell(R^k(\overline{M}_{g,n})) \leqslant k + 1\,. \tag{4.10}$$

The bound holds for the subspace of $R^k(\overline{M}_{g,n})$ generated by decorated strata classes of compact type on $\overline{M}_{g,n}$ satisfying (4.8) at all vertices by the results of Section 4.3.3. If (4.8) is violated, we use Proposition 2 of [23] to express the vertex term via tautological boundary classes and repeat. If a class not of compact type arises, the class must occur as a push forward from $R^*(\overline{M}_{g-1,n+2})$. Here, we use the induction hypothesis, and conclude the bound (4.10) for all of $R^k(\overline{M}_{g,n})$. In fact, for the subspace in $R^k(\overline{M}_{g,n})$ generated by decorated strata not of compact type, the bound $\ell \leqslant k+1$ holds by induction.

The second bound in Statement (i) of Theorem 4.2 is

$$\ell(R^k(\overline{M}_{g,n})) \leqslant 3g-2+n-k\,.$$

Since $2g-2+n-k < 3g-2+n-k$, the bound holds for the subspace of $R^k(\overline{M}_{g,n})$ generated by decorated strata classes of compact type on $\overline{M}_{g,n}$ satisfying (4.8) at all vertices by the results of Section 4.3.3. We conclude as above. For a class not of compact type, hence pushed forward from $R^{k-1}(\overline{M}_{g-1,n+2})$, we have by induction

$$\ell \leqslant 3(g-1)-2+(n+2)-(k-1) = 3g-2+n-k$$

for the generated subspace.

The third and last bound to consider is

$$\ell(R^*(\overline{M}_{g,n})) \leqslant \left\lfloor \tfrac{2g-1+n}{2} \right\rfloor\,.$$

The result holds for the subspace of $R^*(\overline{M}_{g,n})$ generated by decorated strata classes of compact type on $\overline{M}_{g,n}$ satisfying (4.8) at all vertices. We conclude as above using the induction to control classes associated to decorated strata not of compact type. □

4.4. Sharpness

In low genera, the length bounds of Theorem 4.2 are often sharp. In fact, we have not yet seen a failure of sharpness in genus 0 or 1. In genus 2, the first failure occurs in $R^2(\overline{M}_{2,3})$. A discussion of the data is given here for $g \leqslant 2$.

In genus 0, we have $R^*(\overline{M}_{0,n}) = H^*(\overline{M}_{0,n},\mathbb{Q})$. Using the calculation of the Σ_n-representation on the latter space [29], the bound

$$\ell(R^k(\overline{M}_{0,n})) \leqslant \min(k+1, n-k-2)$$

has been verified to be sharp for $3 \leqslant n \leqslant 20$. However, the behavior is somewhat subtle. For $6 \leqslant n \leqslant 20$, the representation $[n-k-1, 2, 1^{k-1}]$ of length $k+1$ occurs for $1 \leqslant k \leqslant \lfloor \frac{n-3}{2} \rfloor$. For $10 \leqslant n \leqslant 20$, the representation $[n-k, 1^k]$ of length $k+1$ occurs for $0 \leqslant k \leqslant \lfloor \frac{n-3}{2} \rfloor$.

Consider next genus 1. Without too much difficulty, $R^*(\overline{M}_{1,n})$ can be shown to be Gorenstein (with socle in degree n) for $1 \leqslant n \leqslant 5$. Yang [66] has calculated the ranks of the intersection pairing on $R^*(\overline{M}_{1,n})$ for $n \leqslant 5$ and found them to coincide with the Betti numbers of $\overline{M}_{1,n}$ computed by Getzler [32]. Using Getzler's calculations of the cohomology groups as Σ_n-representations, the length bounds for $R^k(\overline{M}_{1,n})$ are seen to be sharp for $n \leqslant 5$.

Getzler has claimed $R^*(\overline{M}_{1,n})$ surjects onto the even cohomology on page 973 of [30] (though a proof has not yet been written). Assuming the surjection, we can use the Σ_n-equivariant calculation of the Betti numbers to check whether the length bounds for $R^k(\overline{M}_{1,n})$ are sharp for larger n. We have verified the sharpness for $n \leqslant 14$.

Tavakol [59] has proven $R^*(M^c_{1,n})$ is Gorenstein with socle in degree $n-1$. Yang [66] has calculated the ranks of the intersection pairing for $n \leqslant 6$. Using the results for $R^*(\overline{M}_{1,n})$, we have verified the length bounds for $R^k(M^c_{1,n})$ are sharp for $n \leqslant 5$. Probably, the Σ_n-action on $R^k(M^c_{1,n})$ can be analyzed more directly.

Finally, consider genus 2. The length bounds are easily checked to be sharp for $n = 2$. In fact,

$$R^*(M^{rt}_{2,2}),\ R^*(M^c_{2,2}),\ \text{and } R^*(\overline{M}_{2,2})$$

all are Gorenstein, with socles in degrees 2, 3, and 5 respectively.

The case $n = 3$ is more interesting. By Theorem 4.2, $R^2(\overline{M}_{2,3})$ and $R^4(\overline{M}_{2,3})$ have length at most 3. According to Getzler [31],

$$H^4(\overline{M}_{2,3}) \cong H^8(\overline{M}_{2,3})$$

has length 2. Yang [66] has shown that all the cohomology of $\overline{M}_{2,3}$ is tautological. The Gorenstein conjecture for $R^*(\overline{M}_{2,3})$ then implies both $R^2(\overline{M}_{2,3})$ and $R^4(\overline{M}_{2,3})$ have length 2. By analyzing separately the strata of compact type, the strata for which the dual graph has one loop, and the strata for which the dual graph has two loops, we can indeed prove the length 2 restriction. The length bound for $R^2(M^c_{2,3})$ is not sharp either.

The failure of sharpness signals unexpected symmetries among the tautological classes. Such symmetries can come from combinatorial symmetries of the strata or from unexpected relations. For the failure in $R^2(\overline{M}_{2,3})$, the origin is combinatorial symmetries in the strata. The nontrivial relation of [7] is not required. Also, relations can exist without the failure of sharpness: Getzler's relation [30] in $R^2(\overline{M}_{1,4})$ causes no problems.

By Tavakol [60], $R^*(M^{rt}_{2,n})$ is Gorenstein with socle in degree n. Using the Gorenstein property, we have verified the length bounds are sharp for $n \leqslant 4$ in the rational tails case.

5. Boundary geometry

5.1. Diagonal classes

We have seen the existence of non-tautological cohomology for $\overline{M}_{1,11}$,

$$H^{11,0}(\overline{M}_{1,11}, \mathbb{C}) = \mathbb{C}, \quad H^{0,11}(\overline{M}_{1,11}, \mathbb{C}) = \mathbb{C}\,.$$

As a result, the diagonal

$$\Delta_{11} \subset \overline{M}_{1,11} \times \overline{M}_{1,11}$$

has Künneth components which are not in $RH^*(\overline{M}_{1,11})$. Let

$$\iota : \overline{M}_{1,11} \times \overline{M}_{1,11} \to \overline{M}_{2,20}$$

be the gluing map. A natural question raised in [36] is whether

$$\iota_*[\Delta] \notin RH^*(\overline{M}_{2,20}) \ ? \tag{5.1}$$

Via a detailed analysis of the action of the ψ classes on $H^*(\overline{M}_{1,12}, \mathbb{C})$, the following result was proven in [36].

Theorem 5.2. *Let $\Delta_{12} \subset \overline{M}_{1,12} \times \overline{M}_{1,12}$ be the diagonal. After push-forward via the gluing map*

$$\iota : \overline{M}_{1,12} \times \overline{M}_{1,12} \to \overline{M}_{2,22} \ ,$$

we obtain a non-tautological class

$$\iota_*[\Delta] \notin RH^*(\overline{M}_{2,22}) \ .$$

While we are still unable to resolve (5.1), we give a simple new proof of Theorem 5.2 which has the advantage of producing new non-tautological classes in $H^*(\overline{M}_{2,21}, \mathbb{Q})$.

5.2. Left and right diagonals

We will study curves of genus 2 with 21 markings

$$[C, p_1, \ldots, p_{10}, q_1, \ldots, q_{10}, r] \in \overline{M}_{2,21} \ .$$

Consider the product $\overline{M}_{1,12} \times \overline{M}_{1,11}$ with the markings of the first factor given by $\{p_1, \ldots, p_{10}, r, \star\}$ and the markings of the second factor given by $\{q_1, \ldots, q_{10}, \bullet\}$. Define the *left diagonal*

$$\Delta_L \subset \overline{M}_{1,12} \times \overline{M}_{1,11}$$

to be the inverse image of the diagonal[13]

$$\Delta_{11} \subset \overline{M}_{1,11} \times \overline{M}_{1,11}$$

under the map forgetting the marking r,

$$\pi : \overline{M}_{1,12} \times \overline{M}_{1,11} \to \overline{M}_{1,11} \times \overline{M}_{1,11} \ .$$

The cycle Δ_L has dimension 12.

For the right diagonal, consider the product $\overline{M}_{1,11} \times \overline{M}_{1,12}$ with the markings of the first factor given by $\{p_1, \ldots, p_{10}, \star\}$ and the markings of the second factor given by $\{q_1, \ldots, q_{10}, r, \bullet\}$. Define

$$\Delta_R \subset \overline{M}_{1,11} \times \overline{M}_{1,12}$$

[13] The diagonal in $\overline{M}_{1,11} \times \overline{M}_{1,11}$ is defined by the bijection $p_i \leftrightarrow q_i$ and $\star \leftrightarrow \bullet$.

to be the inverse image of the diagonal Δ_{11} under the map forgetting the marking r,

$$\pi : \overline{M}_{1,11} \times \overline{M}_{1,12} \to \overline{M}_{1,11} \times \overline{M}_{1,11} ,$$

as before.

Our main result concerns the push-forwards $\iota_{L*}[\Delta_L]$ and $\iota_{R*}[\Delta_R]$ under the boundary gluing maps,

$$\iota_L : \overline{M}_{1,12} \times \overline{M}_{1,11} \to \overline{M}_{2,21} ,$$

$$\iota_R : \overline{M}_{1,11} \times \overline{M}_{1,12} \to \overline{M}_{2,21} ,$$

defined by connecting the markings $\{\star, \bullet\}$.

Theorem 5.3. *The push-forwards are non-tautological,*

$$\iota_{L*}[\Delta_L],\ \iota_{R*}[\Delta_R] \notin RH^*(\overline{M}_{2,21}) .$$

We view the markings of $\overline{M}_{2,22}$ as given by $\{p_1, \dots, p_{10}, r, q_1, \dots, q_{10}, s\}$ and the diagonal Δ_{12} as defined by the bijection

$$p_i \leftrightarrow q_i, \quad r \leftrightarrow s .$$

The cycle $\iota(\Delta_{12}) \subset \overline{M}_{2,22}$ maps birationally to $\iota_L(\Delta_L) \subset \overline{M}_{2,21}$ under the map

$$\pi : \overline{M}_{2,22} \to \overline{M}_{2,21}$$

forgetting s. Hence,

$$\pi_* \iota_* [\Delta_{12}] = \iota_{L*}[\Delta_L] .$$

In particular, Theorem 5.3 implies Theorem 5.2 since tautological classes are closed under π-push-forward.

5.3. Proof of Theorem 5.3

We first compute the class

$$\iota_L^* \iota_{L*}[\overline{M}_{1,12} \times \overline{M}_{1,11}] \in H^*(\overline{M}_{1,12} \times \overline{M}_{1,11}, \mathbb{Q}) .$$

The rules for such self-intersections are given in [36]. Since ι_L is an injection,

$$\iota_L^* \iota_{L*}[\overline{M}_{1,12} \times \overline{M}_{1,11}] = -\psi_\star - \psi_\bullet .$$

As a consequence, we find

$$\iota_L^* \iota_{L*}[\Delta_L] = (-\psi_\star - \psi_\bullet) \cdot [\Delta_L] . \tag{5.4}$$

If $\iota_{L*}[\Delta_L] \in RH^*(\overline{M}_{2,21})$, then the Künneth components of $\iota_L^* \iota_{L*}[\Delta_L]$ must be tautological cohomology by property (iii) of Section 1.4. Let

$$\pi : \overline{M}_{1,12} \times \overline{M}_{1,11} \to \overline{M}_{1,11} \times \overline{M}_{1,11} \tag{5.5}$$

be the map forgetting r in the first factor. If the Künneth components of $\iota_L^* \iota_{L*}[\Delta_L]$ are tautological in cohomology, then the Künneth components of

$$\pi_* \iota_L^* \iota_{L*}[\Delta_L] \in H^*(\overline{M}_{1,11} \times \overline{M}_{1,11}, \mathbb{Q})$$

must also be tautological in cohomology. We compute

$$\pi_*\Big((-\psi_\star - \psi_\bullet)\cdot[\Delta_L]\Big) = -\pi_*(\psi_\star\cdot[\Delta_L]) - \psi_\bullet\cdot(\pi_*[\Delta_L])$$
$$= -\pi_*(\psi_\star\cdot[\Delta_L])$$

since π restricted to Δ_L has fiber dimension 1.

The class $\psi_\star \in R^1(\overline{M}_{1,12})$ has a well-known boundary expression,

$$\psi_\star = \frac{1}{12}[\delta_{irr}] + \sum_{S\subset\{p_1,\ldots,p_{10},r\},\ S\neq\emptyset} [\delta_S]\ .$$

Here, δ_{irr} is the 'irreducible' boundary divisor (parameterizing nodal rational curves with 12 markings), and δ_S is the 'reducible' boundary divisor generically parameterizing 1-nodal curves

$$\mathbb{P}^1 \cup E$$

with marking $S\cup\{\star\}$ on $\mathbb{P}^1$ and the rest on E. The intersections

$$\delta_{irr}\times\overline{M}_{1,11}\cap\Delta_L,\quad \delta_S\times\overline{M}_{1,11}\cap\Delta_L$$

all have fiber dimension 1 with respect to (5.5) except when $S=\{r\}$. Hence,

$$-\pi_*(\psi_\star\cdot[\Delta_L]) = -\pi_*(\delta_{\{r\}}\times\overline{M}_{1,11}\cap\Delta_L)\ .$$

Since π maps $\delta_{\{r\}}\times\overline{M}_{1,11}\cap\Delta_L$ birationally onto $\Delta_{11}\subset\overline{M}_{1,11}\times\overline{M}_{1,11}$, we conclude

$$\pi_*\iota_L^*\iota_{L*}[\Delta_L] = -[\Delta_{11}]\ \in H^*(\overline{M}_{1,11}\times\overline{M}_{1,11},\mathbb{Q})\ .$$

Since $[\Delta_{11}]$ does not have a tautological Künneth decomposition, the argument is complete. The proof for Δ_R is identical. □

5.4. On $\overline{M}_{2,20}$

We finish with a remark about the relationship between question (5.1) and Theorem 5.3. Consider the map forgetting the marking labelled by r,

$$\pi:\overline{M}_{2,21}\to\overline{M}_{2,20}\ .$$

We easily see

$$\pi^*\left(\iota_*[\Delta_{11}]\right) = \iota_{L*}[\Delta_L] + \iota_{R*}[\Delta_R]\ .$$

The following push-forward relation holds by calculating the degree of the cotangent line $\mathbb{L}_r$ on the fibers of π,

$$\pi_*\Big(\psi_r\cdot(\iota_{L*}[\Delta_L]+\iota_{R*}[\Delta_R])\Big) = 22\cdot\iota_*[\Delta_{11}]\ .$$

As a consequence of the above two equations, we conclude the following result.

Proposition 5.6. *We have the equivalence:*

$$\iota_*[\Delta_{11}]\in RH^*(\overline{M}_{2,20}) \iff \iota_{L*}[\Delta_L]+\iota_{R*}[\Delta_R]\in RH^*(\overline{M}_{2,21})\ .$$

5.5. Connection with representation theory

Consider again the cycle $\Delta_L \subset \overline{M}_{1,12} \times \overline{M}_{1,11}$. Let

$$\Gamma_L \in H^{11,0}(\overline{M}_{1,12}) \otimes H^{0,11}(\overline{M}_{1,11}) \oplus H^{0,11}(\overline{M}_{1,12}) \otimes H^{11,0}(\overline{M}_{1,11})$$

be the Künneth component of Δ_L. We can consider the Σ_{21}-submodule

$$V \subset H^{12,12}(\overline{M}_{2,21})$$

generated by $\iota_{L*}(\Gamma_L)$. The class Γ_R can be defined in the same manner, and then $\iota_{R*}(\Gamma_R) \in V$.

The class Γ_L is alternating for the symmetric group Σ_{10} permuting the points $p_1, \ldots, p_{10}$. Similarly, Γ_L is alternating for the Σ_{10} permuting the points $q_1, \ldots, q_{10}$. Let

$$\Sigma_{10} \times \Sigma_{10} \subset \Sigma_{21}$$

be the associated subgroup. We can then consider the Σ_{21}-module defined by

$$\widetilde{V} = \operatorname{Ind}_{\Sigma_{10}\times\Sigma_{10}}^{\Sigma_{21}}(\alpha \otimes \alpha),$$

where α is the alternating representation. In fact, the representation $\widetilde{V}$ decomposes as

$$[1^{21}] + \sum_{i=0}^{9} [3\, 2^i\, 1^{18-2i}] + 2\sum_{j=1}^{10} [2^j\, 1^{21-2j}].$$

We can write the coefficient of $L \wedge^2 S[12]$ in $H^{12,12}(\overline{M}_{2,21})$ discussed in Section 3.6 as

$$\sum_{i=0,\ \text{even}}^{9} [3\, 2^i\, 1^{18-2i}] + \sum_{j=1}^{10} [2^j\, 1^{21-2j}].$$

We conjecture the canonical map

$$\widetilde{V} \to V$$

is simply a projection onto the above subspace of $H^{12,12}(\overline{M}_{2,21})$.

References

[1] A. N. Andrianov. Quadratic forms and Hecke operators. *Grundlehren der Mathematischen Wissenschaften*, **286**, Springer-Verlag, Berlin, 1987. ← 308

[2] T. Arakawa. Vector-valued Siegel's modular forms of degree two and the associated Andrianov L-functions. *Manuscripta Math.*, **44**(1–3), 155–185, 1983. ← 308

[3] E. Arbarello and M. Cornalba. Combinatorial and algebro-geometric cohomology classes on the moduli spaces of curves. *J. Algebraic Geom.*, **5**, 705–749, 1996. ← 299

[4] K. Behrend. Gromov-Witten invariants in algebraic geometry. *Invent. Math.*, **127**, 601–617, 1997. ← 299

[5] K. Behrend and B. Fantechi. The intrinsic normal cone. *Invent. Math.*, **128**, 45–88, 1997. ← 299

[6] P. Belorousski. Chow rings of moduli spaces of pointed elliptic curves. Ph.D. thesis. The University of Chicago, 1998. ← 304

[7] P. Belorousski and R. Pandharipande. A descendent relation in genus 2. *Ann. Scuola Norm. Sup. Pisa Cl. Sci.*, **29**, 171–191, 2000. ← 322

[8] J. Bergström. Cohomology of moduli spaces of curves of genus three via point counts. *J. Reine Angew. Math.*, **622**, 155–187, 2008. ← 305, 315

[9] J. Bergström. Equivariant counts of points of the moduli spaces of pointed hyperelliptic curves. *Doc. Math.*, **14**, 259–296, 2009. ← 308

[10] J. Bergström, C. Faber, and G. van der Geer. Siegel modular forms of genus 2 and level 2: cohomological computations and conjectures. *Int. Math. Res. Not. IMRN 2008*, Art. ID rnn 100, 20 pp. ← 314

[11] J. Bergström, C. Faber, and G. van der Geer. *Siegel modular forms of degree three and the cohomology of local systems*. `arXiv:1108.3731` ← 314

[12] J. Bergström, C. Faber and G. van der Geer. *Teichmüller modular forms and the cohomology of local systems on* M_3, in preparation. ← 315

[13] J. Bergström and G. van der Geer. The Euler characteristic of local systems on the moduli of curves and abelian varieties of genus three. *J. Topol.*, **1**(3), 651–662, 2008. ← 313, 314

[14] G. Casnati and C. Fontanari. On the rationality of moduli spaces of pointed curves. *J. Lond. Math. Soc. (2)*, **75**(3), 582–596, 2007. ← 312

[15] C. Consani and C. Faber. On the cusp form motives in genus 1 and level 1, in: Moduli spaces and arithmetic geometry, 297–314, *Adv. Stud. Pure Math.*, **45**, Math. Soc. Japan, Tokyo, 2006. ← 307

[16] P. Deligne. Formes modulaires et représentations ℓ-adiques. Séminaire Bourbaki 1968/69, no. 355, *Lecture Notes in Math.*, **179**, 139–172, Springer-Verlag, Berlin, 1971. ← 306

[17] M. Eichler. Eine Verallgemeinerung der Abelschen Integrale. *Math. Z.*, **67**, 267–298, 1957. ← 306

[18] C. Faber. A conjectural description of the tautological ring of the moduli space of curves, in: Moduli of Curves and Abelian Varieties (The Dutch Intercity Seminar on Moduli), C. Faber and E. Looijenga, eds., 109–129, *Aspects of Mathematics E 33*, Vieweg, Wiesbaden, 1999. ← 301, 316

[19] C. Faber and G. van der Geer. Sur la cohomologie des systèmes locaux sur les espaces des modules des courbes de genre 2 et des surfaces abéliennes, I, II. *C.R. Acad. Sci. Paris, Sér. I*, **338**, 381–384, 467–470, 2004. ← 294, 309, 310

[20] C. Faber and R. Pandharipande. Hodge integrals and Gromov-Witten theory. Invent. Math., **139**, 173–199, 2000. ← 297

[21] C. Faber and R. Pandharipande (with an appendix by D. Zagier). Logarithmic series and Hodge integrals in the tautological ring. *Michigan Math. J.*, **48**, 215–252, 2000. ← 299, 301, 316

[22] C. Faber and R. Pandharipande. Hodge integrals, partition matrices, and the λ_g conjecture. *Ann. of Math.*, **157**, 97–124, 2003. ← 301

[23] C. Faber and R. Pandharipande. Relative maps and tautological classes. *JEMS*, **7**, 13–49, 2005. ← 294, 295, 300, 319, 321

[24] G. Faltings. On the cohomology of locally symmetric Hermitian spaces. in: Paul Dubreil and Marie-Paule Malliavin algebra seminar, 35th year (Paris, 1982), 55–98, *Lecture Notes in Math.*, **1029**, Springer, Berlin, 1983. ← 310, 314

[25] G. Faltings and C.-L. Chai. Degeneration of abelian varieties. With an appendix by David Mumford. *Ergebnisse der Mathematik und ihrer Grenzgebiete (3)*, **22**, Springer-Verlag, Berlin, 1990. ← 310, 314

[26] W. Fulton. *Intersection theory*, Springer-Verlag, Berlin, 1984. ← 302

[27] W. Fulton and R. Pandharipande. Notes on stable maps and quantum cohomology, in: Proceedings of Symposia in Pure Mathematics: Algebraic Geometry Santa Cruz 1995. J. Kollár, R. Lazarsfeld, and D. Morrison, eds., **62**, Part 2, 45–96. ← 299

[28] G. van der Geer. Rank one Eisenstein cohomology of local systems on the moduli space of abelian varieties. *Sci. China Math.*, **54**(10), 1621–1634, 2011. ← 311

[29] E. Getzler. Operads and moduli spaces of genus 0 Riemann surfaces, in: *The moduli space of curves*, R. Dijkgraaf, C. Faber, and G. van der Geer, eds., 199–230, Birkhäuser: Basel, 1995. ← 321

[30] E. Getzler. Intersection theory on $\overline{M}_{1,4}$ and elliptic Gromov-Witten invariants. *J. Amer. Math. Soc.*, **10**(4), 973–998, 1997. ← 322

[31] E. Getzler. Topological recursion relations in genus 2. *Integrable systems and algebraic geometry (Kobe/Kyoto, 1997)*, 73–106, World Sci. Publ., River Edge, NJ, 1998. ← 310, 313, 322

[32] E. Getzler. The semi-classical approximation for modular operads. *Comm. Math. Phys.*, **194**(2), 481–492, 1998. ← 306, 307, 321

[33] E. Getzler. Euler characteristics of local systems on M_2. *Compositio Math*, **132**(2), 121–135, 2002. ← 311

[34] E. Getzler and M. M. Kapranov. Modular operads. *Compositio Math.*, **110**(1), 65–126, 1998. ← 313

[35] T. Graber and R. Pandharipande. Localization of virtual classes. *Invent. Math.*, **135**, 487–518, 1999. ← 300

[36] T. Graber and R. Pandharipande. Constructions of nontautological classes on moduli spaces of curves. *Michigan Math. J.*, **51**, 93–109, 2003. ← 295, 298, 300, 323, 324

[37] C. Grundh. Computations of vector valued Siegel modular forms. Ph.D. thesis, KTH, Stockholm, in preparation. ← 309, 311, 313

[38] R. Hain and E. Looijenga. Mapping class groups and moduli spaces of curves, in: Proceedings of Symposia in Pure Mathematics: Algebraic Geometry Santa Cruz 1995. J. Kollár, R. Lazarsfeld, and D. Morrison, eds., Volume 62, Part 2, 97–142. ← 299, 316

[39] J. Harris and D. Mumford. On the Kodaira dimension of the moduli space of curves. With an appendix by William Fulton. *Invent. Math.*, **67**(1), 23–88, 1982. ← 300

[40] R. Hartshorne. Algebraic geometry. *Graduate Texts in Mathematics*, **52**, Springer-Verlag, New York-Heidelberg, 1977. ← 304

[41] T. Ichikawa. Teichmüller modular forms of degree 3. *Amer. J. Math.*, **117**(4), 1057–1061, 1995. ← 307, 315

[42] E. Ionel. Relations in the tautological ring of M_g. *Duke Math J.*, **129**, 157–186, 2005. ← 301

[43] S. Keel. Intersection theory of moduli space of stable n-pointed curves of genus zero. *Trans. Amer. Math. Soc.*, **330**(2), 545–574, 1992. ← 317

[44] N. Koblitz. Introduction to elliptic curves and modular forms. *Graduate Texts in Mathematics*, **97**, Springer-Verlag, New York, 1984. ← 303

[45] M. Kontsevich. Intersection theory on the moduli space of curves and the matrix Airy function. *Comm. Math. Phys.*, **147**, 1–23, 1992. ← 297

[46] M. Levine and R. Pandharipande. Algebraic cobordism revisited. *Invent. Math.*, **176**, 63–130, 2009. ← 300

[47] J. Li and G. Tian. Virtual moduli cycles and Gromov-Witten invariants of algebraic varieties. *J. Amer. Math. Soc.*, **11**, 119–174, 1998. ← 299

[48] E. Looijenga. On the tautological ring of M_g. *Invent. Math.*, **121**, 411–419, 1995. ← 301

[49] M. Mirzakhani. Weil-Petersson volumes and intersection theory on the moduli space of curves. *J. Amer. Math. Soc.*, **20**, 1–23, 2007. ← 297

[50] S. Morita. Generators for the tautological algebra of the moduli space of curves. *Topology*, **42**, 787–819, 2003. ← 301

[51] D. Mumford. Towards an enumerative geometry of the moduli space of curves, in: *Arithmetic and Geometry*, M. Artin and J. Tate, eds., Part II, Birkhäuser, 1983, 271–328. ← 293, 296, 301

[52] A. Okounkov and R. Pandharipande. Gromov-Witten theory, Hurwitz numbers, and matrix models, in: Proceedings of Symposia in Pure Mathematics: Algebraic Geometry Seattle 2005. D. Abramovich, A. Bertram, L. Katzarkov, R. Pandharipande and M. Thaddeus. eds., Volume 80, Part 1, 325–414. ← 297

[53] R. Pandharipande. Three questions in Gromov-Witten theory, in: *Proceedings of the ICM (Beijing 2002)*, vol. II, 503–512, Higher Education Press, Beijing. ← 299, 316

[54] R. Pandharipande. The κ ring of the moduli of curves of compact type. *Acta Math.*, **208**(2), 335–388, 2012. ← 301

[55] D. Petersen. Euler characteristics of local systems on the loci of d-elliptic abelian surfaces. arXiv:1004.5462 ← 313

[56] D. Petersen. Cusp form motives and admissible G-covers. *Algebra Number Theory*, **6**(6), 1199–1221, 2012. ← 307

[57] A. J. Scholl. Motives for modular forms. *Invent. Math.*, **100**(2), 419–430, 1990. ← 307

[58] G. Shimura. Sur les intégrales attachées aux formes automorphes. *J. Math. Soc. Japan*, **11**, 291–311, 1959. ← 306

[59] M. Tavakol. The tautological ring of $M^{ct}_{1,n}$. arXiv:1007.3091 ← 301, 322

[60] M. Tavakol. The tautological ring of the moduli space $M^{rt}_{2,n}$. arXiv:1101.5242 ← 301, 322

[61] R. Tsushima. An explicit dimension formula for the spaces of generalized automorphic forms with respect to Sp(2, Z). *Proc. Japan Acad. Ser. A Math. Sci.*, **59**(4), 139–142, 1983. ← 311

[62] S. Tsuyumine. On Siegel modular forms of degree three. *Amer. J. Math.*, **108**(4), 755–862, 1001–1003, 1986. ← 314

[63] J.-L. Verdier. Sur les intégrales attachées aux formes automorphes (d'après Goro Shimura). *Séminaire Bourbaki*, Vol. **6**, Exp. No. 216, 149–175, Soc. Math. France, Paris, 1995. ← 306

[64] R. Weissauer. The trace of Hecke operators on the space of classical holomorphic Siegel modular forms of genus two. arXiv:0909.1744 ← 311

[65] E. Witten. Two dimensional gravity and intersection theory on moduli space. *Surveys in Diff. Geom.*, **1**, 243–310, 1991. ← 297

[66] S. Yang. Calculating intersection numbers on moduli spaces of pointed curves. arXiv:0808.1974 ← 321, 322

Department of Mathematics, KTH Royal Institute of Technology, Lindstedtsvägen 25, 100 44 Stockholm, Sweden
E-mail address: faber@math.kth.se

Department of Mathematics, ETH Zürich, Rämistrasse 101, 8092 Zürich, Switzerland
E-mail address: rahul@math.ethz.ch

Alternate compactifications of moduli spaces of curves

Maksym Fedorchuk and David Ishii Smyth

Abstract. We give an informal survey, emphasizing examples and open problems, of two interconnected research programs in moduli of curves: the systematic classification of modular compactifications of $M_{g,n}$, and the study of Mori chamber decompositions of $\overline{M}_{g,n}$.

Contents

1. Introduction

The purpose of this article is to survey recent developments in two interconnected research programs within the study of moduli of curves: the systematic construction and classification of modular compactifications of $M_{g,n}$, and the study

2000 *Mathematics Subject Classification.* Primary 14H10; Secondary 14E30.
Key words and phrases. moduli of curves, singularities, minimal model program.

of the Mori theory of $\overline{M}_{g,n}$. In this section we will informally describe the overall goal of each of these programs as well as outline the contents of this article.

One of the most fundamental and influential theorems in modern algebraic geometry is the construction of a modular compactification $M_g \subset \overline{M}_g$ for the moduli space of smooth complete curves of genus g [27]. In this context, the word *modular* indicates that the compactification $\overline{M}_g$ is itself a parameter space for a suitable class of geometric objects, namely stable curves of genus g.

Definition 1.1 (Stable curve). *A complete curve* C *is* stable *if*

(1) C *has only nodes* $(y^2 = x^2)$ *as singularities.*
(2) ω_C *is ample.*

The class of stable curves satisfies two essential properties, namely deformation openness (Definition 2.2) and the unique limit property (Definition 2.3). As we shall see in Theorem 2.4, any class of curves satisfying these two properties gives rise to a modular compactification of M_g.

While the class of stable curves gives a natural modular compactification of the space of smooth curves, it is not unique in this respect. The simplest example of an alternate class of curves satisfying these two properties is the class of pseudostable curves, introduced by Schubert [101].

Definition 1.2 (Pseudostable curve). *A curve* C *is* pseudostable *if*

(1) C *has only nodes* $(y^2 = x^2)$ *and cusps* $(y^2 = x^3)$ *as singularities.*
(2) ω_C *ample.*
(3) *If* $E \subset C$ *is any connected subcurve of arithmetic genus one, then* $|E \cap \overline{C \setminus E}| \geqslant 2$.

Notice that the definition of pseudostability involves a trade-off: cusps are exchanged for elliptic tails. It is easy to see how this trade-off comes about: As one ranges over all one-parameter smoothings of a cuspidal curve C, the associated limits in $\overline{M}_g$ are precisely curves of the form $\tilde{C} \cup E$, where $\tilde{C}$ is the normalization of C at the cusp and E is an elliptic curve (of arbitrary j-invariant) attached to $\tilde{C}$ at the point lying above the cusp (see Example 2.16). Thus, any separated moduli problem must exclude either elliptic tails or cusps. Schubert's construction naturally raises the question:

Problem 1.3. *Classify all possible stability conditions for curves, i.e. classes of possibly singular curves which are deformation open and satisfy the property that any one-parameter family of smooth curves contains a unique limit contained in that class.*

Of course, we can ask the same question with respect to classes of marked curves, i.e. modular compactifications of $M_{g,n}$, and the first goal of this article is to survey the various methods used to construct alternate modular compactifications of $M_{g,n}$. In Section 2, we describe constructions using Mori theory, the combinatorics of curves on surfaces, and geometric invariant theory, and, in Section 2.5, we assemble a (nearly) comprehensive list of the alternate modular compactifications that have so far appeared in the literature.

Since each stability condition for n-pointed curves of genus g gives rise to a proper birational model of $\overline{M}_{g,n}$, it is natural to wonder whether these models can be exploited to study the divisor theory of $\overline{M}_{g,n}$. This brings us to the second subject of this survey, namely the study of the birational geometry of $\overline{M}_{g,n}$ from the perspective of Mori theory. In order to explain the overall goal of this program, let us recall some definitions from birational geometry. For any normal, $\mathbb{Q}$-factorial, projective variety X, let $\mathrm{Amp}(X)$, $\mathrm{Nef}(X)$, and $\mathrm{Eff}(X)$ denote the cones of ample, nef, and effective divisors respectively (these are defined in Section 3.1). Recall that a *birational contraction* $\phi\colon X \dashrightarrow Y$ is simply a birational map to a normal, projective, $\mathbb{Q}$-factorial variety such that $\mathrm{codim}(\mathrm{Exc}(\phi^{-1})) \geqslant 2$. If ϕ is any birational contraction with exceptional divisors $E_1, \ldots, E_k$, then the *Mori chamber of* ϕ is defined as:

$$\mathrm{Mor}(\phi) := \phi^* \mathrm{Amp}(Y) + \sum_{a_i \geqslant 0} a_i E_i \subset \mathrm{Eff}(X).$$

The birational contraction ϕ can be recovered from any divisor D in $\mathrm{Mor}(\phi)$ as the contraction associated to the section ring of D, i.e. $X \dashrightarrow \mathrm{Proj} \bigoplus_{m \geqslant 0} H^0(X, mD)$. In good cases, e.g. when X is a toric or Fano variety, these Mori chambers are polytopes and there exist finitely many contractions ϕ_i which partition the entire effective cone, i.e. one has

$$\overline{\mathrm{Eff}}(X) = \overline{\mathrm{Mor}(\phi_1)} \cup \ldots \cup \overline{\mathrm{Mor}(\phi_k)}.$$

Understanding the Mori chamber decomposition of the effective cone of a variety X is therefore tantamount to understanding the classification of birational contractions of X. This collection of ideas, which will be amplified in Section 3.1, suggests the following problems concerning $\overline{M}_{g,n}$.

Problem 1.4.

(1) *Describe the nef and effective cone of* $\overline{M}_{g,n}$.
(2) *Describe the Mori chamber decomposition of* $\overline{\mathrm{Eff}}(\overline{M}_{g,n})$.

These questions merit study for two reasons: First, the attempt to prove results concerning the geometry of $\overline{M}_{g,n}$ has often enriched our understanding of algebraic curves in fundamental ways. Second, even partial results obtained along these lines may provide an interesting collection of examples of phenomena in higher-dimensional geometry. In Section 3.2, we will briefly survey the present state of knowledge concerning the nef and effective cones of $\overline{M}_{g,n}$, and many examples of partial Mori chamber decompositions will be given in Section 4. We should emphasize that Problem 1.4 (1) has generated a large body of work over the past 30 years. While we hope that Section 3.2 can serve as a reasonable guide to the literature, we will not cover these results in the detail they merit. The interested reader who wishes to push further in this direction is strongly encouraged to read the recent surveys of Farkas [36, 35] and Morrison [91].

A major focus of the present survey is the program initiated by Hassett and Keel to study certain log canonical models of $\overline{M}_g$ and $\overline{M}_{g,n}$. For any rational number α such that $K_{\overline{\mathcal{M}}_{g,n}} + \alpha\delta$ is effective, they define

$$\overline{M}_{g,n}(\alpha) = \operatorname{Proj} \bigoplus_{m \geqslant 0} H^0(\overline{\mathcal{M}}_{g,n}, m(K_{\overline{\mathcal{M}}_{g,n}} + \alpha\delta)),$$

where the sum is taken over m sufficiently divisible, and ask whether the spaces $\overline{M}_{g,n}(\alpha)$ admit a modular description. When g and n are sufficiently large so that $\overline{M}_{g,n}$ is of general type, $\overline{M}_{g,n}(1)$ is simply $\overline{M}_{g,n}$, while $\overline{M}_{g,n}(0)$ is the canonical model of $\overline{M}_{g,n}$. Thus, the series of birational transformations accomplished as α decreases from 1 to 0 constitute a complete minimal model program for $\overline{M}_{g,n}$. In terms of Problem 1.4 above, this program can be understood as asking:

(1) Describe the Mori chamber decomposition within the two dimensional subspace of $\mathrm{Eff}(\overline{M}_{g,n})$ spanned by $K_{\overline{\mathcal{M}}_{g,n}}$ and δ (where it is known to exist by theorems of higher-dimensional geometry [11]).
(2) Find modular interpretations for the resulting birational contractions.

Though a complete description of the log minimal model program for $\overline{M}_g$ $(g \gg 0)$ is still out of reach, the first two steps have been carried out by Hassett and Hyeon [61, 60]. Furthermore, the program has been completely carried out for $\overline{M}_2$, $\overline{M}_3$, $\overline{M}_{0,n}$, and $\overline{M}_{1,n}$ [59, 65, 46, 2, 108, 109], and we will describe the resulting chamber decompositions in Section 4. Finally, in Section 4.4 we will present some informal heuristics for making predictions about future stages of the program.

In closing, let us note that there are many fascinating topics in the study of birational geometry of $\overline{M}_{g,n}$ and related spaces which receive little or no attention whatsoever in the present survey. These include rationality, unirationality and rational connectedness of $\overline{M}_{g,n}$ [104, 105, 71, 73, 89, 113, 14], complete subvarieties of $M_{g,n}$ [28, 33, 52], birational geometry of Prym varieties [30, 17, 112, 114, 72, 29, 68, 39], Kodaira dimension of $\overline{M}_{g,n}$ for $n \geqslant 1$ [88, 37] and of the universal Jacobian over $\overline{M}_g$ [41], moduli spaces birational to complex ball quotients [82, 83], etc.

A truly comprehensive survey of the birational geometry of $\overline{M}_{g,n}$ should certainly contain a discussion of these studies, and these omissions reflect nothing more than (alas!) the inevitable limitations of space, time, and our own knowledge.

1.1. Notation

Unless otherwise noted, we work over a fixed algebraically closed field k of characteristic zero.

A *curve* is a complete, reduced, one-dimensional scheme of finite-type over an algebraically closed field. An *n-pointed curve* $(C, \{p_i\}_{i=1}^n)$ is a curve C, together with n distinguished smooth points $p_1, \dots, p_n \in C$. A *family of n-pointed curves*

$(f\colon \mathcal{C} \to T, \{\sigma_i\}_{i=1}^n)$ is a flat, proper, finitely-presented morphism $f\colon \mathcal{C} \to T$, together with n sections $\sigma_1, \dots, \sigma_n$, such that the geometric fibers are n-pointed curves, and with the total space $\mathcal{C}$ an algebraic space over k. A *class of curves* $\mathcal{S}$ is simply a set of isomorphism classes of n-pointed curves of arithmetic genus g (for some fixed $g, n \geqslant 0$), which is invariant under base extensions, i.e. if $k \hookrightarrow l$ is an inclusion of algebraically closed fields, then a curve C over k is in a class $\mathcal{S}$ iff $C \times_k l$ is. If $\mathcal{S}$ is a class of curves, a *family of $\mathcal{S}$-curves* is a family of n-pointed curves $(f\colon \mathcal{C} \to T, \{\sigma_i\}_{i=1}^n)$ whose geometric fibers $(C_{\bar{t}}, \{\sigma_i(\bar{t})\}_{i=1}^n)$ are contained in $\mathcal{S}$.

Δ will always denote the spectrum of a discrete valuation ring R with algebraically closed residue field l and field of fractions L. When we speak of a finite base change $\Delta' \to \Delta$, we mean that Δ' is the spectrum of a discrete valuation ring $R' \supset R$ with field of fractions L', where $L' \supset L$ is a finite separable extension. We use the notation

$$0 := \operatorname{Spec} l \to \Delta,$$
$$\Delta^* := \operatorname{Spec} L \to \Delta,$$

for the closed point and generic point respectively. The complex-analytically-minded reader will lose nothing by thinking of Δ as the unit disc in $\mathbb{C}$, and Δ^* as the punctured unit disc.

While a nuanced understanding of stacks should not be necessary for reading these notes, we will assume the reader is comfortable with the idea that moduli stacks represent functors. If the functor of families of $\mathcal{S}$-curves is representable by a stack, this stack will be denoted in script, e.g. $\overline{\mathcal{M}}_{g,n}[\mathcal{S}]$. If this stack has a coarse (or good) moduli space, it will be noted in regular font, e.g. $\overline{M}_{g,n}[\mathcal{S}]$. We call $\overline{\mathcal{M}}_{g,n}[\mathcal{S}]$ the *moduli stack of $\mathcal{S}$-curves* and $\overline{M}_{g,n}[\mathcal{S}]$ the *moduli space of $\mathcal{S}$-curves*.

Starting in Section 3, we will assume the reader is familiar with intersection on $\overline{\mathcal{M}}_{g,n}$ and $\overline{M}_{g,n}$, as described in [54] (the original reference for the theory is [55, Section 6]). The key facts which we use repeatedly are these: For any g, n satisfying $2g - 2 + n > 0$, there are canonical identifications

$$N^1(\overline{M}_{g,n}) \otimes \mathbb{Q} = \operatorname{Pic}(\overline{M}_{g,n}) \otimes \mathbb{Q} = \operatorname{Pic}(\overline{\mathcal{M}}_{g,n}) \otimes \mathbb{Q},$$

so we may consider any line bundle on $\overline{\mathcal{M}}_{g,n}$ as a numerical divisor class on $\overline{M}_{g,n}$ without ambiguity. Furthermore, by [5] and [6], $N^1(\overline{M}_{g,n}) \otimes \mathbb{Q}$ is generated by the classes of $\lambda, \psi_1, \dots, \psi_n$, and the boundary divisors δ_0 and $\delta_{h,S}$, which are defined as follows. If $\pi\colon \overline{\mathcal{M}}_{g,n+1} \to \overline{\mathcal{M}}_{g,n}$ is the universal curve with sections $\{\sigma_i\}_{i=1}^n$, then $\lambda = c_1(\pi_*\omega_\pi)$ and $\psi_i = \sigma_i^*(\omega_\pi)$. Let $\Delta = \overline{\mathcal{M}}_{g,n} \smallsetminus \mathcal{M}_{g,n}$ be the boundary with class δ. Nodal irreducible curves form a divisor Δ_0 whose class we denote δ_0. For any $0 \leqslant h \leqslant g$ and $S \subset \{1, \dots, n\}$, the boundary divisor $\Delta_{h,S} \subset \overline{\mathcal{M}}_{g,n}$ is the locus of reducible curves with a node separating a connected component of arithmetic genus h marked by the points in S from the rest of the curve, and we use $\delta_{h,S}$ to denote its class. Clearly, $\Delta_{h,S} = \Delta_{g-h,\{1,\dots,n\}\smallsetminus S}$ and $\Delta_{h,S} = \varnothing$ if $h = 0$ and $|S| \leqslant 1$. When

$g \geqslant 3$, these are the only relations and $\mathrm{Pic}\,(\overline{\mathcal{M}}_{g,n})$ is a free abelian group generated by $\lambda, \psi_1, \ldots, \psi_n$ and the boundary divisors [5, Theorem 2]. We refer the reader to [6, Theorem 2.2] for the list of relations in $g \leqslant 2$.

Finally, the canonical divisor of $\overline{\mathcal{M}}_{g,n}$ is given by the formula

$$K_{\overline{\mathcal{M}}_{g,n}} = 13\lambda - 2\delta + \psi. \tag{1.5}$$

This is a consequence of the Grothendieck-Riemann-Roch formula applied to the cotangent bundle of $\overline{\mathcal{M}}_{g,n}$

$$\Omega_{\overline{\mathcal{M}}_{g,n}} = \pi_*\big(\Omega_\pi(-\sum_{i=1}^{n} \sigma_i) \otimes \omega_\pi\big).$$

The computation in the unpointed case is in [55, Section 2], and the computation in the pointed case follows similarly. Alternately, the formula for $\overline{\mathcal{M}}_{g,n}$ with $n \geqslant 1$ can be deduced directly from that of $\overline{\mathcal{M}}_g$ as in [88, Theorem 2.6]. Note that we will often use $K_{\overline{\mathcal{M}}_{g,n}}$ to denote a divisor class on $\overline{M}_{g,n}$ using the canonical identification indicated above.

2. Compactifying the moduli space of curves

2.1. Modular and weakly modular birational models of $M_{g,n}$

What does it mean for a proper birational model X of $M_{g,n}$ to be modular? Roughly speaking, X should coarsely represent a geometrically defined functor of n-pointed curves. To make this precise, we introduce $\mathcal{U}_{g,n}$, the moduli stack of all n-pointed curves (hereafter, we always assume $2g - 2 + n > 0$). More precisely, if we define a functor from schemes over k to groupoids by

$$\mathcal{U}_{g,n}(T) := \left\{ \begin{array}{l} \text{Flat, proper, finitely-presented morphisms } \mathcal{C} \to T, \text{ with } n \text{ sections} \\ \{\sigma_i\}_{i=1}^{n}, \text{ and connected, reduced, one-dimensional geometric fibers} \end{array} \right\}$$

then $\mathcal{U}_{g,n}$ is an algebraic stack, locally of finite type over $\mathrm{Spec}\,k$ [107, Theorem B.1]; note that we allow $\mathcal{C}$ to be an algebraic space over k. Now we may make the following definition.

Definition 2.1. *Let X be a proper birational model of $M_{g,n}$. We say that X is* modular *if there exists a diagram*

$$\begin{array}{ccc} \mathcal{U}_{g,n} & & \\ \uparrow{\scriptstyle i} & & \\ \mathcal{X} & \xrightarrow[\pi]{} & X \end{array}$$

satisfying:

(1) *i is an open immersion,*
(2) *π is a coarse moduli map.*

Recall that if $\mathcal{X}$ is any algebraic stack, we say that $\pi\colon \mathcal{X} \to X$ is a *coarse moduli map* if π satisfies the following properties:

(1) π is categorical for maps to algebraic spaces, i.e. if $\phi\colon \mathcal{X} \to Y$ is any map from $\mathcal{X}$ to an algebraic space, then ϕ factors through π.

(2) π is bijective on k-points.

For a more complete discussion of the properties enjoyed by the coarse moduli space of an algebraic stack, we refer the reader to [76].

The key point of Definition 2.1 is that modular birational models of $M_{g,n}$ may be constructed purely by considering geometric properties of curves. More precisely, if $\mathcal{S}$ is any class of curves which is *deformation open* and satisfies the *unique limit property*, then there is an associated modular birational model $\overline{M}_{g,n}[\mathcal{S}]$. The precise definition of these two conditions is given below:

Definition 2.2 (Deformation open). *A class of curves $\mathcal{S}$ is* deformation open *if the following condition holds: Given any family of curves $(\mathcal{C} \to T, \{\sigma_i\}_{i=1}^n)$, the set $\{t \in T : (C_{\overline{t}}, \{\sigma_i(\overline{t})\}_{i=1}^n) \in \mathcal{S}\}$ is open in T.*

Definition 2.3 (Unique limit property). *A class of curves $\mathcal{S}$ satisfies the* unique limit property *if the following two conditions hold:*

(1) *(Existence of $\mathcal{S}$-limits) Let Δ be the spectrum of a discrete valuation ring. If $(\mathcal{C} \to \Delta^*, \{\sigma_i\}_{i=1}^n)$ is a family of $\mathcal{S}$-curves, there exists a finite base-change $\Delta' \to \Delta$, and a family of $\mathcal{S}$-curves $(\mathcal{C}' \to \Delta', \{\sigma_i'\}_{i=1}^n)$, such that*
$$(\mathcal{C}', \{\sigma_i'\}_{i=1}^n)|_{(\Delta')^*} \simeq (\mathcal{C}, \{\sigma_i\}_{i=1}^n) \times_\Delta (\Delta')^*.$$

(2) *(Uniqueness of $\mathcal{S}$-limits) Let Δ be the spectrum of a discrete valuation ring. Suppose that $(\mathcal{C} \to \Delta, \{\sigma_i\}_{i=1}^n)$ and $(\mathcal{C}' \to \Delta, \{\sigma_i'\}_{i=1}^n)$ are two families of $\mathcal{S}$-curves. Then any isomorphism over the generic fiber*
$$(\mathcal{C}, \{\sigma_i\}_{i=1}^n)|_{\Delta^*} \simeq (\mathcal{C}', \{\sigma_i'\}_{i=1}^n)|_{\Delta^*}$$
extends to an isomorphism over Δ:
$$(\mathcal{C}, \{\sigma_i\}_{i=1}^n) \simeq (\mathcal{C}', \{\sigma_i'\}_{i=1}^n).$$

Theorem 2.4. *Let $\mathcal{S}$ be a class of curves which is deformation open and satisfies the unique limit property. Then there exists a proper Deligne-Mumford stack of $\mathcal{S}$-curves $\overline{\mathcal{M}}_{g,n}[\mathcal{S}]$ and an associated coarse moduli space $\overline{M}_{g,n}[\mathcal{S}]$, which is a modular birational model of $M_{g,n}$.*

Proof. Set
$$\overline{\mathcal{M}}_{g,n}[\mathcal{S}] := \{t \in \mathcal{U}_{g,n} \mid (C_{\overline{t}}, \{\sigma_i(\overline{t})\}_{i=1}^n) \in \mathcal{S}\} \subset \mathcal{U}_{g,n}.$$
The fact that $\mathcal{S}$ is deformation open implies that $\overline{\mathcal{M}}_{g,n}[\mathcal{S}] \subset \mathcal{U}_{g,n}$ is open, so $\overline{\mathcal{M}}_{g,n}[\mathcal{S}]$ automatically inherits the structure of an algebraic stack over k.

Uniqueness of $\mathcal{S}$-limits implies that the automorphism group of each curve $[C, \{p_i\}_{i=1}^n] \in \overline{\mathcal{M}}_{g,n}[\mathcal{S}]$ is proper. If $\widetilde{C} \to C$ is the normalization and $\{q_i\}_{i=1}^m$ are points lying over the singularities of C, then we have an injection $\mathrm{Aut}(C, \{p_i\}_{i=1}^n) \hookrightarrow$

$\mathrm{Aut}(\widetilde{C}, \{q_i\}_{i=1}^m, \{p_i\}_{i=1}^n)$. Since none of the components of $\widetilde{C}$ is an unpointed genus 1 curve, the latter group is affine. We conclude that $\mathrm{Aut}(C, \{p_i\}_{i=1}^n)$ is finite, hence unramified (since we are working in characteristic 0). It follows by [85, Théorème 8.1] that $\overline{\mathcal{M}}_{g,n}[\mathcal{S}]$ is a Deligne-Mumford stack. The unique limit property implies that $\overline{\mathcal{M}}_{g,n}[\mathcal{S}]$ is proper, via the valuative criterion [85, Proposition 7.12]. Finally, $\overline{\mathcal{M}}_{g,n}[\mathcal{S}]$ has a coarse moduli space $\overline{M}_{g,n}[\mathcal{S}]$ by [76, Corollary 1.3], and it follows immediately that $\overline{M}_{g,n}[\mathcal{S}]$ is a modular birational model of $M_{g,n}$. □

While most of the birational models of $M_{g,n}$ that we construct will be modular in the sense of Definition 2.1, geometric invariant theory constructions (Section 2.4) give rise to models which are modular in a slightly weaker sense. Roughly speaking, these models are obtained from mildly non-separated functors of curves by imposing sufficiently many identifications to force a separated moduli space.

To formalize this, recall that if $\mathcal{X}$ is an algebraic stack over an algebraically closed field k, it is possible for one k-point of $\mathcal{X}$ to be in the closure of another. In the context of moduli stacks of curves, one can visualize this via the valuative criterion of specialization, i.e. if $[C, p]$ and $[C', p']$ are two points of $\mathcal{U}_{g,n}$, then $[C', p'] \in \overline{[C, p]}$ if and only if there exists an isotrivial specialization of (C, p) to (C', p'), i.e. a family $\mathcal{C} \to \Delta$ whose fibers over the punctured disc are isomorphic to (C, p) and whose special fiber is isomorphic to (C', p').

Exercise 2.5. Let $\mathcal{U}_{1,1}$ be the stack of all 1-pointed curves of arithmetic genus one. Let $[C, p] \in \mathcal{U}_{1,1}$ be the unique isomorphism class of a rational cuspidal curve with a smooth marked point. Show that if $(E, p) \in \mathcal{U}_{1,1}$ is a smooth curve, then we have $[C, p] \in \overline{[E, p]}$. (Hint: Consider a trivial family $E \times \Delta \to \Delta$, blow-up the point p in $E \times 0$, and contract the strict transform of $E \times 0$. See also Lemma 2.17 and Exercise 2.19 below.)

Now if $\mathcal{X}$ is an algebraic stack with a k-point in the closure of another, it is clear (for topological reasons) that $\mathcal{X}$ cannot have a coarse moduli map. However, if we let $\sim$ denote the equivalence relation generated by the relations $x \in \overline{y}$, then we might hope for a map $\mathcal{X} \to X$ such that the points of X correspond to equivalence classes of points of $\mathcal{X}$. We formalize this as follows: If $\mathcal{X}$ is any algebraic stack, we say that $\pi\colon \mathcal{X} \to X$ is a *good moduli map* if π satisfies the following properties:

(1) π is categorical for maps to algebraic spaces.
(2) π is bijective on closed k-points, i.e. if $x \in X$ is any k point, then $\phi^{-1}(x) \subset \mathcal{X}$ contains a unique closed k-point.

Here, a closed k-point is, of course, simply a k-point $x \in \mathcal{X}$ satisfying $\overline{x} = x$. The notion of a good moduli space was introduced and studied by Alper [3]. We should note that, for ease of exposition, we have adopted a slightly weaker definition than that appearing in [3]. Now we may define

Definition 2.6. *Let* X *be a proper birational model of* $M_{g,n}$*. We say that* X *is* weakly modular *if there exists a diagram*

$$\begin{array}{ccc} \mathcal{U}_{g,n} & & \\ {\scriptstyle i}\big\uparrow & & \\ \mathcal{X} & \xrightarrow[\pi]{} & X \end{array}$$

satisfying:

(1) i *is an open immersion,*
(2) π *is a good moduli map.*

Exercise 2.7 (4 points on $\mathbb{P}^1$). Consider the moduli space $M_{0,4}$ of 4 distinct ordered points on $\mathbb{P}^1$, up to isomorphism. Of course, $M_{0,4} \simeq \mathbb{P}^1 \smallsetminus \{0, 1, \infty\}$ where the isomorphism is given by the classical cross-ratio

$$(\mathbb{P}^1, \{p_i\}_{i=1}^4) \mapsto (p_1 - p_3)(p_2 - p_4)/(p_1 - p_2)(p_3 - p_4).$$

In this exercise, we will see that $\mathbb{P}^1$ can be considered as a weakly modular birational model of $M_{0,4}$.

The most naïve way to enrich the moduli functor of 4-pointed $\mathbb{P}^1$'s is to allow up to two points to collide, i.e. set

$$\mathcal{S} := \{(\mathbb{P}^1, \{p_i\}_{i=1}^4) \mid \text{no three of } \{p_i\}_{i=1}^4 \text{ coincident }\}.$$

The class of $\mathcal{S}$-curves is clearly deformation open so there is an associated moduli stack $\overline{M}_{0,4}[\mathcal{S}]$. The reader may check that the following assertions hold:

(1) There exists a diagram

$$\begin{array}{ccc} M_{0,4} & & \\ \big\downarrow & \searrow & \\ \overline{M}_{0,4}[\mathcal{S}] & \xrightarrow{\phi} & \mathbb{P}^1 \end{array}$$

extending the usual cross-ratio map.

(2) The fibers $\phi^{-1}(0)$, $\phi^{-1}(1)$, and $\phi^{-1}(\infty)$ comprise the set of all curves where two points coincide. Furthermore, each of these fibers contains a unique closed point, e.g. the unique closed point of $\phi^{-1}(0)$ is given by $p_1 = p_3$ and $p_2 = p_4$.

This shows that ϕ is a good moduli map, i.e. that $\mathbb{P}^1$ is a weakly proper birational model of $M_{0,4}$.

It is natural to wonder whether there is an analogue of Theorem 2.4 for weakly modular compactifications, i.e. whether one can formulate suitable conditions on a class of curves to guarantee that the associated stack has a good moduli space and that this moduli space is proper. For the purpose of this exposition, however, the only general method for constructing weakly modular birational models will be geometric invariant theory, as described in Section 2.4.

2.2. Modular birational models via MMP

How can we construct a class of curves which is deformation open and satisfies the unique limit property? From the point of view of general theory, the most satisfactory construction proceeds using ideas from higher dimensional geometry. Indeed, one of the foundational theorems of higher dimensional geometry is that the canonical ring of a smooth projective variety of general type is finitely generated ([11], [106], [87]). A major impetus for work on this theorem was the fact that finite generation of canonical rings in dimension $n+1$ gives a canonical limiting process for one-parameter families of canonically polarized varieties of dimension n. In what follows, we explain how this canonical limiting process works and why it leads (in dimension 1) to the standard definition of a stable curve.

Let us state the precise version of finite generation that we need:

Proposition 2.8. *Let $\pi\colon \mathcal{C} \to \Delta$ be a flat, projective family of varieties satisfying:*

(1) *$\mathcal{C}$ is smooth,*
(2) *The special fiber $C \subset \mathcal{C}$ is a normal crossing divisor,*
(3) *$\omega_{\mathcal{C}/\Delta}$ is ample on the generic fiber.*

Then

(1) *The relative canonical ring $\bigoplus_{m\geqslant 0} \pi_*(\omega^m_{\mathcal{C}/\Delta})$ is a birational invariant.*
(2) *The relative canonical ring $\bigoplus_{m\geqslant 0} \pi_*(\omega^m_{\mathcal{C}/\Delta})$ is a finitely generated $\mathcal{O}_\Delta$-algebra.*

Proof. This theorem is proved in [11]. We will sketch a proof in the case where the relative dimension of π is 1. To show that $\bigoplus_{m\geqslant 0} \pi_*\omega^m_{\mathcal{C}/\Delta}$ is a birational invariant, we must show that if $\pi\colon \mathcal{C} \to \Delta$ and $\pi'\colon \mathcal{C}' \to \Delta$ are two families satisfying hypotheses (1)-(3), and they are isomorphic over the generic fiber, then $\pi_*\omega^m_{\mathcal{C}/\Delta} = \pi'_*\omega^m_{\mathcal{C}'/\Delta}$ for each integer $m \geqslant 0$. Since $\mathcal{C}$ and $\mathcal{C}'$ are smooth surfaces, any birational map $\mathcal{C} \dashrightarrow \mathcal{C}'$ can be resolved into a sequence of blow-ups and blow-downs, and it suffices to show that the relative canonical ring does not change under a simple blow-up $\phi\colon \mathcal{C}' \to \mathcal{C}$. In this case, one has $\omega^m_{\mathcal{C}'/\Delta} = \phi^*\omega^m_{\mathcal{C}/\Delta}(mE)$ and so by the projection formula $\phi_*\omega^m_{\mathcal{C}'/\Delta} = \omega^m_{\mathcal{C}/\Delta}$.

To show that the ring is $\bigoplus_{m\geqslant 0} \pi_*\omega^m_{\mathcal{C}/\Delta}$ is finitely generated, we will show that it is the homogeneous coordinate ring of a certain birational model of $\mathcal{C}$, the so-called *canonical model* $\mathcal{C}^{can}$. To construct $\mathcal{C}^{can}$, we begin by contracting curves on which $\omega_{\mathcal{C}/\Delta}$ has negative degree. Observe that if $E \subset C$ is any irreducible curve in the special fiber on which $\omega_{\mathcal{C}/\Delta}$ has negative degree, then in fact $E \simeq \mathbb{P}^1$ and $\omega_{\mathcal{C}/\Delta} \cdot E = E^2 = -1$ (this is a simple consequence of adjunction). By Castelnuovo's contractibility criterion, there exists a birational morphism $\phi\colon \mathcal{C} \to \mathcal{C}'$, with $\mathrm{Exc}(\phi) = E$ and $\mathcal{C}'$

a smooth surface. After finitely many repetitions of this procedure, we obtain a birational morphism $\mathcal{C} \to \mathcal{C}^{min}$, such that the special fiber C^{min} has no smooth rational curves E satisfying $E^2 = -1$; we call $\mathcal{C}^{min}$ the *minimal model* of $\mathcal{C}$.

Now $\omega_{\mathcal{C}^{min}/\Delta}$ is big and relatively nef, so the Kawamata basepoint freeness theorem [81, Theorem 3.3] implies that $\omega_{\mathcal{C}^{min}/\Delta}$ is relatively base point free, i.e. there exists a projective, birational morphism over Δ

$$\mathcal{C}^{min} \to \mathcal{C}^{can} := \operatorname{Proj} \bigoplus_{m \geqslant 0} \pi_* \omega^m_{\mathcal{C}^{min}/\Delta}$$

contracting precisely those curves on which $\omega_{\mathcal{C}^{min}/\Delta}$ has degree zero. In particular, $\bigoplus_{m \geqslant 0} \pi_* \omega^m_{\mathcal{C}^{min}/\Delta}$ is finitely generated. □

Given Proposition 2.8, it is easy to describe the canonical limiting procedure. Let $\mathcal{C}^* \to \Delta^*$ be a family of smooth projective varieties over the unit disc whose relative canonical bundle $\omega_{\mathcal{C}^*/\Delta^*}$ is ample (in the case of curves, this condition simply means $g \geqslant 2$). We obtain a canonical limit for this family as follows:

(1) Let $\mathcal{C} \to \Delta$ be the flat limit with respect to an arbitrary projective embedding $\mathcal{C}^* \hookrightarrow \mathbb{P}^n_{\Delta^*} \subset \mathbb{P}^n_{\Delta}$.

(2) By the semistable reduction theorem [81, Theorem 7.17], there exists a finite base-change $\Delta' \to \Delta$ and a family $\mathcal{C}' \to \Delta'$ such that
   - $\mathcal{C}'|_{(\Delta')^*} \simeq \mathcal{C} \times_\Delta (\Delta')^*$
   - The special fiber $C \subset \mathcal{C}'$ is a reduced normal crossing divisor. In the case of curves, this simply means that C is a nodal curve.

(3) By Proposition 2.8, the $\mathcal{O}_{\Delta'}$-algebra $\bigoplus_{m \geqslant 0} \pi_* \omega^m_{\mathcal{C}'/\Delta'}$ is finitely generated. Thus, we may consider the natural rational map

$$\mathcal{C}' \dashrightarrow (\mathcal{C}')^{can} := \operatorname{Proj} \bigoplus_{m \geqslant 0} \pi_* \omega^m_{\mathcal{C}'/\Delta'},$$

which is an isomorphism over Δ'.

(4) The special fiber $C^{can} \subset (\mathcal{C}')^{can}$ is the desired canonical limit.

The fact that the section ring $\bigoplus_{m \geqslant 0} \pi_* \omega^m_{\mathcal{C}'/\Delta'}$ is a birational invariant implies that C^{can} does not depend on the chosen resolution of singularities. Furthermore, the reader may easily check that this construction is independent of the base change in the sense that if $(\mathcal{C}')^{can} \to \Delta'$ and $(\mathcal{C}'')^{can} \to \Delta''$ are families produced by this process using two different base-changes, then

$$(\mathcal{C}')^{can} \times_\Delta \Delta'' \simeq (\mathcal{C}'')^{can} \times_\Delta \Delta'.$$

Of course, this construction raises the question: which varieties actually show up as limits of smooth varieties in this procedure? When the relative dimension of $\mathcal{C}^* \to \Delta^*$ is 2 or greater, this is an extremely difficult question which has only been answered in certain special cases. In the case of curves, our explicit construction of the map $\mathcal{C} \dashrightarrow \mathcal{C}^{can}$ in the proof of Proposition 2.8 makes it easy to show that:

Proposition 2.9. *The set of canonical limits of smooth curves is precisely the class of stable curves (Definition 1.1).*

Proof. After steps (1) and (2) of the canonical limiting procedure, we have a family of curves $\mathcal{C} \to \Delta$ with smooth generic fiber, nodal special fiber, and smooth total space. In the proof of Proposition 2.8, we showed that the map $\mathcal{C} \dashrightarrow \mathcal{C}^{can}$ is actually regular, and is obtained in two steps: the map $\mathcal{C} \to \mathcal{C}^{min}$ was obtained by successively contracting smooth rational curves satisfying $E^2 = -1$ and the map $\mathcal{C}^{min} \to \mathcal{C}^{can}$ was obtained by contracting all smooth rational curves E satisfying $E^2 = -2$. It follows immediately that the special fiber $C^{can} \subset \mathcal{C}^{can}$ is a nodal curve which has no rational components meeting the rest of the curve in one or two points, i.e. C^{can} is stable.

Conversely, any stable curve arises as the canonical limit of a family of smooth curves. Indeed, given a stable curve C, let $\mathcal{C} \to \Delta$ be a smoothing of C with a smooth total space. Since $\mathcal{C}$ has no smooth rational curves satisfying $E^2 = -1$ or $E^2 = -2$, we see that $\omega_{\mathcal{C}/\Delta}$ has positive degree on every curve contained in a fiber, and is therefore relatively ample. Thus, $\mathcal{C} \simeq \operatorname{Proj} \bigoplus_{m \geqslant 0} \pi_* \omega_{\mathcal{C}/\Delta}^m$ is its own canonical model, and C is the canonical limit of the family $\mathcal{C}^* \to \Delta^*$. □

Corollary 2.10. *There exists a proper moduli stack $\overline{\mathcal{M}}_g$ of stable curves, with an associated coarse moduli space $\overline{M}_g$.*

Proof. The two conditions defining stable curves are evidently deformation open. The class of stable curves has the unique limit property by Proposition 2.9 so Theorem 2.4 gives a proper moduli stack of stable curves $\overline{\mathcal{M}}_g$, with the corresponding moduli space $\overline{M}_g$. □

Remark 2.11. *We only checked the existence and uniqueness of limits for one-parameter families with a smooth generic fiber. In general, to show that an algebraic stack $\mathcal{X}$ is proper, it is sufficient to check the valuative criterion for maps $\Delta \to \mathcal{X}$ which send the generic point of Δ into a fixed open dense subset $\mathcal{U} \subset \mathcal{X}$* [27, Theorem 4.19]. *Thus, when verifying that a stack of curves has the unique limit property, it is sufficient to consider generically smooth families.*

Next, we explain how the canonical limiting process can be adapted to the case of pointed curves. The basic idea is that, instead of using $\operatorname{Proj} \bigoplus_{m \geqslant 0} \pi_* \omega_{\mathcal{C}/\Delta}^m$ as the canonical completion of a family, we use $\operatorname{Proj} \bigoplus_{m \geqslant 0} \pi_*(\omega_{\mathcal{C}/\Delta}(\Sigma_{i=1}^n \sigma_i)^m)$. Again, the key point is that this is a finitely generated $\mathcal{O}_\Delta$-algebra and a birational invariant. In fact, there is a further variation, due to Hassett [58], worth considering. Namely, if we replace $\omega_{\mathcal{C}/\Delta}(\Sigma\sigma_i)$ by the $\mathbb{Q}$-line-bundle $\omega_{\mathcal{C}/\Delta}(\Sigma a_i \sigma_i)$ for any $a_i \in \mathbb{Q} \cap [0,1]$ satisfying $2g - 2 + \sum a_i \geqslant 0$, the same statements hold, i.e. $\bigoplus_{m \geqslant 0} \pi_*(\omega_{\mathcal{C}/\Delta}(\Sigma a_i \sigma_i)^m)$ is a finitely generated algebra and it is a birational invariant. This generalization may appear odd to those who are unversed in the yoga of higher dimensional geometry; the point is that it is precisely divisors of this form for which the general machinery of Kodaira vanishing, Kawamata basepoint freeness, etc. goes through to guarantee finite generation. For the sake of completeness, we will sketch the proof.

Proposition 2.12. *Let* $(\mathcal{C} \to \Delta, \{\sigma_i\}_{i=1}^n)$ *be a family of* n*-pointed curves satisfying:*

(a) $\mathcal{C}$ *is smooth,*
(b) $C \subset \mathcal{C}$ *is a nodal curve,*
(c) *The sections* $\{\sigma_i\}_{i=1}^n$ *are disjoint.*

Then for any $a_i \in \mathbb{Q} \cap [0,1]$

(1) *The* $\mathcal{O}_\Delta$*-algebra* $\bigoplus_{m \geqslant 0} \omega_{\mathcal{C}/\Delta}(\Sigma_i a_i \sigma_i)^m$ *is a birational invariant,*
(2) *The* $\mathcal{O}_\Delta$*-algebra* $\bigoplus_{m \geqslant 0} \omega_{\mathcal{C}/\Delta}(\Sigma_i a_i \sigma_i)^m$ *is finitely generated.*

Proof. We follow the steps in the proof of Proposition 2.8. To prove that the canonical algebra is invariant under birational transformations, we reduce to the case of a simple blow-up $\phi\colon \mathcal{C}' \to \mathcal{C}$. If the center of the blow-up does not lie on any section, then the proof proceeds as in Proposition 2.8. If the center of the blow-up lies on σ_i, then we have $\omega_{\mathcal{C}'/\Delta}(\sum a_i \sigma_i') = \phi^* \omega_{\mathcal{C}/\Delta}(\sum a_i \sigma_i) + (1 - a_i)E$. By the projection formula, $\phi_*(\omega_{\mathcal{C}'/\Delta}(\sum a_i \sigma_i')) = \omega_{\mathcal{C}/\Delta}(\sum a_i \sigma_i)$ if and only if $1 - a_i \geqslant 0$. This shows that the condition $a_i \leqslant 1$ is necessary and sufficient for (1).

For (2), we construct the canonical model $\mathcal{C}^{can}$ precisely as in the proof of Proposition 2.8. We have only to observe that if $E \subset C$ is any curve such that $\omega_{\mathcal{C}/\Delta}(\sum a_i \sigma_i) \cdot E < 0$, then the assumption that $a_i \geqslant 0$ implies that E is a smooth rational curve with $E \cdot E = -1$ and so can be blown down. Thus, we can construct a minimal model $\mathcal{C}^{min}$ on which $\omega_{\mathcal{C}/\Delta}(\sum a_i \sigma_i)$ is big and nef. On $\mathcal{C}^{min}$, we may apply the Kawamata basepoint freeness theorem to arrive at a family with $\omega_{\mathcal{C}/\Delta}(\sum a_i \sigma_i)$ relatively ample and (2) follows. □

For any fixed weight vector $\mathcal{A} = (a_1, \ldots, a_n) \in \mathbb{Q}^n \cap [0,1]$, we may now describe a canonical limiting process for a family $(\mathcal{C} \to \Delta^*, \{\sigma_i\}_{i=1}^n)$ of pointed curves over a punctured disk as follows:

(1) Let $(\mathcal{C} \to \Delta, \{\sigma_i\}_{i=1}^n)$ be the flat limit with respect to an arbitrary projective embedding $\mathcal{C}^* \hookrightarrow \mathbb{P}^n_{\Delta^*} \subset \mathbb{P}^n_\Delta$.
(2) By the semistable reduction theorem, there exists a finite base-change $\Delta' \to \Delta$ and a family $(\mathcal{C}' \to \Delta', \{\sigma_i\}_{i=1}^n)$ such that
   - $\mathcal{C}'|_{(\Delta')^*} \simeq \mathcal{C} \times_\Delta (\Delta')^*$
   - The special fiber $C' \subset \mathcal{C}'$ is a nodal curve and the sections meet the special fiber C' at distinct smooth points of C'.

(3) Proposition 2.12 implies that $\bigoplus_{m\geqslant 0} \pi_*(\omega_{\mathcal{C}'/\Delta'}(\Sigma_i a_i\sigma_i)^m)$ is a finitely generated $\mathcal{O}_{\Delta'}$-algebra. Thus, we may consider the natural rational map

$$\mathcal{C}' \dashrightarrow (\mathcal{C}')^{can} := \operatorname{Proj} \bigoplus_{m\geqslant 0} \pi_*(\omega_{\mathcal{C}'/\Delta'}(\Sigma a_i\sigma_i)^m).$$

(4) The special fiber $C^{can} \subset (\mathcal{C}')^{can}$ is the desired canonical limit.

Using the proof of Proposition 2.12, it is easy to see which curves actually arise as limits of smooth curves under this process, for a fixed weight vector $\mathcal{A}$. The key point is that the map

$$\mathcal{C}' \to \operatorname{Proj} \bigoplus_{m\geqslant 0} \pi_*(\omega_{\mathcal{C}'/\Delta'}(\Sigma a_i\sigma_i)^m)$$

successively contracts curves on which $\omega_{\mathcal{C}'/\Delta'}(\Sigma_i a_i\sigma_i)$ has non-positive degree. By adjunction, any such curve $E \subset C$ satisfies:

(1) E is smooth rational, $E \cdot E = -1$, and $\sum_{\{i:p_i\in E\}} a_i \leqslant 1$.
(2) E is smooth rational, $E \cdot E = -2$, and there are no marked points supported on E.

Contracting curves of the first type creates smooth points where the marked points $\{p_i \in E\}$ coincide, while contracting curves of the second type creates nodes. Thus, the special fibers $C^{can} \subset \mathcal{C}$ are precisely the *$\mathcal{A}$-stable curves*, defined below.

Definition 2.13 ($\mathcal{A}$-stable curves). *An n-pointed curve $(C, \{p_i\}_{i=1}^n)$ is $\mathcal{A}$-stable if it satisfies*

(1) *The only singularities of C are nodes,*
(2) *If $p_{i_1}, \dots, p_{i_k}$ coincide in C, then $\sum_{j=1}^k a_{i_j} \leqslant 1$,*
(3) *$\omega_C(\Sigma_{i=1}^n a_i p_i)$ is ample.*

Corollary 2.14 ([58]). *There exist proper moduli stacks $\overline{\mathcal{M}}_{g,\mathcal{A}}$ for $\mathcal{A}$-stable curves, with corresponding moduli spaces $\overline{M}_{g,\mathcal{A}}$.*

Proof. The conditions defining $\mathcal{A}$-stability are obviously deformation open, and we have just seen that they satisfy the unique limit property. The corollary follows from Theorem 2.4. □

Remark 2.15. *An stable n-pointed curve is simply an $\mathcal{A}$-stable curve with weights $\mathcal{A} = (1, \dots, 1)$, and the corresponding moduli space is typically denoted $\overline{M}_{g,n}$.*

Before moving on to the next section, let us take a break from all this high theory and do some concrete examples. We have just seen that it is always possible to fill in a one-parameter family of smooth curves with a stable limit. As we will see in subsequent sections, a question that arises constantly is how to tell *which* stable curve arises when applying this procedure to a family of smooth curves degenerating to a given singular curve. In practice, the difficult step in the procedure outlined above is finding a resolution of singularities of the total space of the family. Those cases in which it is possible to do stable reduction by hand typically involve a trick that allows one to bypass an explicit resolution of singularities. For example, here is a trick that allows one to perform stable reduction on any family of curves acquiring an A_{2k}-singularity ($y^2 = x^{2k+1}$).

Example 2.16 (Stable reduction for an A_{2k}-singularity). Begin with a generically smooth family $\mathcal{C} \to \Delta$ of pointed curves whose special fiber C_0 has an isolated singularity of type A_{2k} at a point $s \in C_0$, i.e. $\widehat{\mathcal{O}}_{C_0,s} \simeq k[[x,y]]/(y^2 - x^{2k+1})$. By the deformation theory of complete intersection singularities, the local equation of $\mathcal{C}$ near $s = (0,0)$ is

$$y^2 = x^{2k+1} + a_{2k-1}(t)x^{2k-1} + \cdots + a_1(t)x + a_0(t),$$

where $a_{2k-1}(0) = \cdots = a_0(0) = 0$. Therefore, we can regard $\mathcal{C}$ as a double cover of $\mathcal{Y} := \mathbb{A}^1_x \times \Delta$. After a finite flat base change we can assume that

$$x^{2k+1} + a_{2k-1}(t)x^{2k-1} + \cdots + a_1(t)x + a_0(t) = (x - \sigma_1(t)) \cdots (x - \sigma_{2k+1}(t)).$$

First, we perform stable reduction for the family $(\mathbb{A}^1_x \times \{t\}, \sigma_1(t), \dots, \sigma_{2k+1}(t))$ of pointed rational curves. Assume for simplicity that the sections $\sigma_i(t)$ intersect pairwise transversely at the point $(0,0) \in \mathbb{A}^1_x \times \Delta$. Let $\mathrm{Bl}_{(0,0)} \mathcal{Y} \to \mathcal{Y}$ be the ordinary blow-up with the center $(0,0)$ and the exceptional divisor E. Then the strict transforms $\overline{\sigma_i}$ of sections $\{\sigma_i\}_{i=1}^{2k+1}$ meet E in $2k+1$ distinct points. Unfortunately, the divisor $\sum_{i=1}^{2k+1} \overline{\sigma_i}$ is not divisible by 2 in $\mathrm{Pic}(\mathrm{Bl}_{(0,0)} \mathcal{Y})$ and so we cannot simply construct a double cover of $\mathrm{Bl}_{(0,0)} \mathcal{Y}$ branched over it. To circumvent this difficulty, we make another base change $\Delta' \to \Delta$ of degree 2 ramified over $0 \in \Delta$. The surface $\mathcal{Y}' := \mathrm{Bl}_{(0,0)} \mathcal{Y} \times_\Delta \Delta'$ has an A_1-singularity over the node in the central fiber of $\mathrm{Bl}_{(0,0)} \mathcal{Y} \to \Delta$. Make an ordinary blow-up $\mathcal{Y}'' \to \mathcal{Y}'$ to resolve this singularity. Denote by F the exceptional (-2)-curve, by Σ the preimage of $\sum_{i=1}^{2k+1} \overline{\sigma_i}$, and continue to denote by E the preimage of E. The divisor $\Sigma + F$ is easily seen to be divisible by 2 in $\mathrm{Pic}(\mathcal{Y}'')$, so we can construct the cyclic cover of degree 2 branched over Σ and F. The normalization of the cyclic cover is a smooth surface. Its central fiber over Δ' decomposes as $\widetilde{C}_0 \cup F' \cup T$, where $\widetilde{C}_0$ is the normalization of C_0 at s, F' is a smooth rational curve mapping $2:1$ onto F, and T is a smooth hyperelliptic curve of genus k mapping $2:1$ onto E and ramified over $E \cap (\Sigma \cup F)$. Blowing down F', we obtain the requisite stable limit $\widetilde{C}_0 \cup T$.

Note that *which* hyperelliptic curve actually arises as the tail of the stable limit depends on the initial family, i.e. on the slopes of the sections $\sigma_i(t)$. Varying the slopes of these sections, we can evidently arrange for any smooth hyperelliptic curve, attached to $\widetilde{C}_0$ along a Weierstrass point, to appear as the stable tail.

2.3. Modular birational models via combinatorics

While the moduli spaces $\overline{M}_{g,n}$ and $\overline{M}_{g,\mathcal{A}}$ are the most natural modular compactifications of $M_{g,n}$ from the standpoint of general Mori theory, they are far from unique in this respect. In this section, we will explain how to bootstrap from the stable reduction theorem to obtain many alternate deformation open classes of curves satisfying the unique limit property. The starting point for all these construction is the obvious, yet useful, observation that one can do stable reduction in reverse. For example, we have already seen that applying stable reduction to a family of curves acquiring a cusp has the effect of replacing the cusp by an elliptic tail (Example 2.16). Conversely, starting with a family of curves whose special fiber has an elliptic tail, one can contract the elliptic tail to get a cusp. More generally, we have

Lemma 2.17 (Contraction Lemma). *Let $\mathcal{C} \to \Delta$ be a family of curves with smooth generic fiber and nodal special fiber. Let $Z \subset C$ be a connected proper subcurve of the special fiber C. Then there exists a diagram*

$$\begin{array}{ccc} \mathcal{C} & \xrightarrow{\ \phi\ } & \mathcal{C}' \\ & {\scriptstyle\pi}\searrow \quad \swarrow{\scriptstyle\pi'} & \\ & \Delta & \end{array}$$

satisfying

(1) *ϕ is proper and birational morphism of algebraic spaces, with* $\mathrm{Exc}(\phi) = Z$.
(2) *$\pi'\colon \mathcal{C}' \to \Delta$ is a flat proper family of curves, with connected reduced special fiber C'.*
(3) *$\phi|_{\overline{C \setminus Z}}\colon \overline{C \setminus Z} \to C'$ is the normalization of C' at $p := \phi(Z)$.*
(4) *The singularity $p \in C'$ has the following numerical invariants:*

$$m(p) = |\overline{C \setminus Z} \cap Z|,$$

$$\delta(p) = p_a(Z) + m(p) - 1,$$

where $m(p)$ is the number of branches of p and $\delta(p)$ is the δ-invariant of p.

(For definition of $\delta(p)$ see [56, Ex. IV.1.8].)

Proof. If Z is any proper subcurve of the special fiber C, then any effective $\mathbb{Q}$-divisor supported on Z has negative self-intersection [107, Proposition 2.6]. Artin's contractibility criterion then implies the existence of a proper birational morphism $\phi\colon \mathcal{C} \to \mathcal{C}'$ with $\mathrm{Exc}(\phi) = Z$, to a normal algebraic space $\mathcal{C}'$ [9, Corollary 6.12]. Evidently, π factors through ϕ so we may regard $\pi'\colon \mathcal{C}' \to \Delta$ as a family of curves. (Flatness of π' is automatic since the generic point of $\mathcal{C}'$ lies over the generic point of Δ.)

To see that the special fiber C' is reduced, first observe that C' is a Cartier divisor in $\mathcal{C}'$, hence has no embedded points. On the other hand, no component of C' can be generically non-reduced because it is a birational image of some component of $\overline{C \setminus Z}$. This completes the proof of (1) and (2).

Conclusion (3) is immediate from the observation that $\overline{C \setminus Z}$ is smooth along the points $Z \cap \overline{C \setminus Z}$ and maps isomorphically to C' elsewhere. Since the number of branches of the singular point $p \in C'$ is, by definition, the number of points lying above p in the normalization, we have

$$m(p) = |\overline{C \setminus Z} \cap Z|.$$

Finally, the formula for $\delta(p)$ is a consequence of the fact that C and C' have the same arithmetic genus (since they occur in flat families with the same generic fiber). □

We would like to emphasize that even when applying the Contraction Lemma to a family $\mathcal{C} \to \Delta$ whose total space $\mathcal{C}$ is a scheme, we can expect $\mathcal{C}'$ to be only an algebraic space in general. (This explains our definition of a family in Section 1.1). Clearly, a necessary condition for $\mathcal{C}'$ to be a scheme is an existence of a line bundle in a Zariski open neighborhood of Z on $\mathcal{C}$ that restricts to a trivial line bundle on Z. Indeed, if $\mathcal{C}'$ is a scheme, the pullback of the trivial line bundle from an affine neighborhood of $p \in \mathcal{C}'$ restricts to $\mathcal{O}_Z$ on Z.

Exercise 2.18. Let C be a smooth curve of genus $g \geqslant 2$ and let $x \in C$ be a general point. Consider the family $\mathcal{C} \to \Delta$ obtained by blowing up the point x in the central fiber of the constant family $C \times \Delta$. Prove that the algebraic space obtained by contracting the strict transform of $C \times 0$ in $\mathcal{C}$ is not a scheme.

Lemma 2.17 raises the question: given a family $\mathcal{C} \to \Delta$ and a subcurve $Z \subset C$, how do you figure out which singularity $p \in C'$ arises from contracting Z? When contracting low genus curves, the easiest way to answer this question is simply to classify all singularities with the requisite numerical invariants. For example, Lemma 2.17 (3) implies that contracting an elliptic tail must produce a singularity with one branch and δ-invariant 1. However, the reader should have no difficulty verifying that the cusp is the unique singularity with these invariants (Exercise 2.19). Thus, Lemma 2.17 (3) implies that contracting an elliptic tail must produce a cusp!

Exercise 2.19. Prove that the unique curve singularity $p \in C$ with one branch and δ-invariant 1 is the standard cusp $y^2 = x^3$ (up to analytic isomorphism). (Hint: the hypotheses imply that $\widehat{\mathcal{O}}_{C,p} \subset k[[t]]$ is a codimension one k-subspace.)

For a second example, consider a family $\mathcal{C} \to \Delta$ with smooth generic fiber and nodal special fiber, and suppose $Z \subset C$ is an arithmetic genus zero subcurve meeting the complement $\overline{C \setminus Z}$ in m points. What singularity arises when we contract Z? Well, according to Lemma 2.17 (3), we must classify singularities with m branches and δ-invariant $m - 1$.

Exercise 2.20. Prove that the unique curve singularity $p \in C$ with m branches and δ-invariant $m - 1$ is simply the union of the m coordinates axes in m-space. More precisely,

$$\widehat{\mathcal{O}}_{C,p} \simeq k[[x_1, \dots, x_m]]/I_m,$$
$$I_m := (x_i x_j : 1 \leqslant i < j \leqslant m).$$

We call this singularity the *rational m-fold point*. Conclude that, in the situation described above, contracting $Z \subset C$ produces a rational m-fold point.

A special feature of these two examples is that the singularity produced by Lemma 2.17 depends only on the curve Z and not on the family $\mathcal{C} \to \Delta$. In general, this will not be the case. We can already see this in the case of elliptic bridges, i.e. smooth elliptic curves meeting the complement in precisely two points. According to Lemma 2.17, contracting an elliptic bridge must produce a singularity with two branches and δ-invariant 2. It is not difficult to show that there are exactly two singularities with these numerical invariants: the tacnode and the spatial singularity obtained by passing a smooth branch transversely through a planar cusp.

Definition 2.21. *We say that* $p \in C$ *is a* punctured cusp *if*

$$\hat{\mathcal{O}}_{C,p} \simeq k[[x,y,z]]/(zx, zy, y^2 - x^3).$$

We say that $p \in C$ *is a* tacnode *if*

$$\hat{\mathcal{O}}_{C,p} \simeq k[[x,y]]/(y^2 - x^4).$$

It turns out that contracting an elliptic bridge can produce *either* a punctured cusp or a tacnode, depending on the choice of smoothing. The most convenient way to express the relevant data in the choice of smoothing is the indices of the singularities of the total space at the attaching nodes of the elliptic bridge. More precisely, suppose $C = E \cup F$ is a curve with two components, where E is a smooth elliptic curve meeting F at two nodes, say p_1 and p_2, and let $\mathcal{C} \to \Delta$ be a smoothing of C. Let m_1, m_2 be the integers uniquely defined by the property that $\hat{\mathcal{O}}_{\mathcal{C},p_i} \simeq k[[x,y,t]]/(xy - t^{m_i})$. Then we have

Proposition 2.22. *With notation as above, apply Lemma 2.17 to produce a contraction* $\phi\colon \mathcal{C} \to \mathcal{C}'$ *with* $\mathrm{Exc}(\phi) = E$. *Then*

(1) *If* $m_1 = m_2$, $p := \phi(E) \in C'$ *is a tacnode.*
(2) *If* $m_1 \neq m_2$, $p := \phi(E) \in C'$ *is a punctured cusp.*

Proof. We will sketch an ad-hoc argument exploiting the fact that the tacnode is Gorenstein, while the punctured cusp is not [108, Appendix A]. This means that if C has a tacnode then the dualizing sheaf ω_C is invertible, while if C has a punctured cusp, it is not. First, let us show that if $m_1 \neq m_2$, $p := \phi(E) \in C'$ is a punctured cusp. Suppose, on the contrary that the contraction $\phi\colon \mathcal{C} \to \mathcal{C}'$ produces a tacnode. Then the relative dualizing sheaf $\omega_{\mathcal{C}'/\Delta}$ is invertible, and the pull back $\phi^*\omega_{\mathcal{C}'/\Delta}$ is an invertible sheaf on $\mathcal{C}$, canonically isomorphic to $\omega_{\mathcal{C}/\Delta}$ away from E. It follows that

$$\phi^*\omega_{\mathcal{C}'/\Delta} = \omega_{\mathcal{C}/\Delta}(D),$$

where D is a Cartier divisor supported on E and $\omega_{\mathcal{C}/\Delta}(D)|_E \simeq \mathcal{O}_E$. It is an easy to exercise to check that it is impossible to satisfy both these conditions if $m_1 \neq m_2$.

Conversely if $m_1 = m_2 = m$, then $\omega_{\mathcal{C}/\Delta}(mE)$ is a line bundle on $\mathcal{C}$ and $\omega_{\mathcal{C}/\Delta}(mE)|_E \simeq \mathcal{O}_E$. One can show that $\omega_{\mathcal{C}/\Delta}(mE)$ is semiample, so that the contraction ϕ is actually induced by a high power of $\omega_{\mathcal{C}/\Delta}(mE)$ [108, Lemma 2.12]. It follows that there is a line bundle $\mathcal{L}$ on $\mathcal{C}'$ which pulls back to $\omega_{\mathcal{C}/\Delta}(mE)$. We claim that $\mathcal{L} \simeq \omega_{\mathcal{C}'/\Delta}$, which implies that $\omega_{C'}$ is invertible, hence that $p \in C'$ is Gorenstein (hence a tacnode). To see this, simply note that $\mathcal{L} \simeq \omega_{\mathcal{C}'/\Delta}$ away from p and then use the fact that $\omega_{\mathcal{C}'/\Delta}$ is an S_2-sheaf [81, Corollary 5.69]. □

Using the contraction lemma, it is easy to use the original stable reduction theorem to construct new classes of curves which are deformation open and satisfy the unique limit property. For example, let us consider the class of pseudostable curves which was defined in the introduction (Definition 1.2).

Proposition 2.23. *For $g \geqslant 3$, the class of pseudostable curves is deformation open and satisfies the unique limit property.*

Proof. The conditions defining pseudostability are clearly deformation open, so it suffices to check the unique limit property. To prove existence of limits, let $\mathcal{C}^* \to \Delta^*$ be any family of smooth curves over the punctured disc, and let $\mathcal{C} \to \Delta$ be the stable limit of the family. Now, assuming $g \geqslant 3$, the elliptic tails in C are all disjoint. Let $Z \subset C$ be the union of the elliptic tails, and apply Lemma 2.17 to obtain a contraction $\phi\colon \mathcal{C} \to \mathcal{C}'$ replacing elliptic tails by cusps. The special fiber C' is evidently pseudostable, so $\mathcal{C}' \to \Delta$ is the desired family. The uniqueness of pseudostable limits follows immediately from the uniqueness of the stable limits (If C_1 and C_2 are two pseudostable limits to the same family $\mathcal{C}^* \to \Delta^*$, then their associated stable limits C_1^s and C_2^s are obtained by replacing the cusps by elliptic tails. Obviously, $C_1^s \simeq C_2^s$ implies $C_1 \simeq C_2$). □

Corollary 2.24 ([101, 61]). *For $g \geqslant 3$, there exists a proper Deligne-Mumford moduli stack $\overline{\mathcal{M}}_g^{ps}$ of pseudostable curves, with corresponding moduli space $\overline{M}_g^{ps}$.*

Proof. Immediate from Theorem 2.4 and Proposition 2.23. □

Remark 2.25 (Reduction morphism $\overline{M}_g \to \overline{M}_g^{ps}$). *For $g \geqslant 3$, there is a natural transformation of functors $\eta\colon \overline{\mathcal{M}}_g \to \overline{\mathcal{M}}_g^{ps}$ which is an isomorphism away from Δ_1. It induces a corresponding morphism $\overline{M}_g \to \overline{M}_g^{ps}$ on the coarse moduli spaces. The map is defined by sending a curve $C \cup E_1 \cup \ldots \cup E_m$, where E_i are elliptic tails attached to C at points $q_1, \ldots, q_m$ to the unique curve C' with cusps $q_1', \ldots, q_m' \in C'$ and pointed normalization $(C, q_1, \ldots, q_m)$. This replacement procedure defines η on points, and one can check that it extends to families.*

The class of pseudostable curves was first defined in the context of GIT, and we shall return to GIT construction of $\overline{M}_g^{ps}$ in Section 2.4. However, we can also use these ideas to construct examples which have no GIT construction (at least none currently known). One such example is the following, in which we use Exercise

2.20 to define an alternate modular compactification of $M_{0,n}$ using rational m-fold points.

Definition 2.26 (ψ-stable curves; cf. [12]). *Let $(C, \{p_i\}_{i=1}^n)$ be an n-pointed curve of arithmetic genus zero. We say that $(C, \{p_i\}_{i=1}^n)$ is ψ-stable if it satisfies:*

(1) *C has only nodes and rational m-fold points as singularities.*
(2) *Every component of C has at least three distinguished points (i.e. singular or marked points).*
(3) *Every component of C has at least one marked point.*

Our motivation for calling this condition ψ-stability will become clear in Section 4.2.

Proposition 2.27. *The class of ψ-stable curves is deformation open and satisfies the unique limit property.*

Proof. The conditions defining ψ-stability are evidently deformation open, so we need only check the unique limit property. Let $\mathcal{C}^* \to \Delta^*$ be any family of smooth curves over a punctured disk, and let $\mathcal{C} \to \Delta$ be the stable limit of the family. Now let $Z \subset C$ be the union of all the unmarked components of C, and apply Lemma 2.17 to obtain a contraction $\phi\colon \mathcal{C} \to \mathcal{C}'$ replacing these components by rational m-fold points. The special fiber C' is evidently ψ-stable, so $\mathcal{C}' \to \Delta$ is the desired limit family. The uniqueness of ψ-stable limits follows immediately from uniqueness of stable limits as in Proposition 2.23, and we leave the details to the reader. □

Corollary 2.28. *There exists a proper Deligne-Mumford moduli stack $\overline{\mathcal{M}}_{0,n}[\psi]$ of ψ-stable curves, and an associated moduli space $\overline{M}_{0,n}[\psi]$.*

Proof. Immediate from Theorem 2.4 and Proposition 2.27. □

Remark 2.29. *Since ψ-stable curves have no automorphisms, one actually has $\overline{\mathcal{M}}_{0,n}[\psi] \simeq \overline{M}_{0,n}[\psi]$, and so $\overline{M}_{0,n}[\psi]$ is a fine moduli space.*

Remark 2.30. *As in Remark 2.25, there is a natural transformation of functors $\psi\colon \overline{M}_{0,n} \to \overline{M}_{0,n}[\psi]$. If $[C] \in \overline{M}_{0,n}$, then $\psi([C])$ is obtained from C by contracting every maximal unmarked connected component of C meeting the rest of the curve in m points to the unique m-fold rational singularity. This replacement procedure defines ψ on points, and one can check that it extends to families. Clearly, positive dimensional fibers of ψ are families of curves with an unmarked moving component.*

At this point, the reader may be wondering how far one can take this method of simply selecting and contracting various subcurves of stable curves. This idea is investigated in [107], and we may summarize the conclusions of the study as follows:

(1) As g and n become large, this procedure gives rise to an enormous number of modular birational models of $\overline{M}_{g,n}$, involving all manner of exotic singularities.

(2) Nevertheless, this approach is not adequate for constructing moduli spaces of curves with certain natural classes of singularities, e.g. this procedure never gives rise to a class of curves containing only nodes ($y^2 = x^2$), cusps ($y^2 = x^3$), and tacnodes ($y^2 = x^4$).

Let us examine problem (2) a little more closely. What goes wrong if we simply consider the class of curves with nodes, cusps, and tacnodes, and disallow elliptic tails and elliptic bridges. Why doesn't this class of curves have the unique limit property? There are two issues: First, one cannot always replace elliptic bridges by tacnodes in a one-parameter family. As we saw in Proposition 2.22, contracting elliptic bridges sometimes gives rise to a punctured cusp. Second, whereas the elliptic tails of a stable curve are disjoint, so that one can contract them all simultaneously (at least when $g \geqslant 3$), elliptic bridges may interact with each other in such a way as to make this impossible. Consider, for example, a one-parameter family of smooth genus three curves specializing to a pair of elliptic bridges (Figure 1). How can one modify the special fiber to obtain a tacnodal limit for this family? Assuming the total space of the family is smooth, one can contract either E_1 or E_2 to obtain two non-isomorphic tacnodal special fibers, but there is no canonical way to distinguish between these two limits.

Neither of these problems is necessarily insoluble. Regarding the first problem, suppose that $\mathcal{C} \to \Delta$ is a one-parameter family of smooth curves acquiring an elliptic bridge in the special fiber. As in the set-up of Proposition 2.22, suppose that the elliptic bridge of the special fiber is attached by two nodes, and that one of these nodes is a smooth point of $\mathcal{C}$ while the other is a singularity of the form $xy - t^2$. Proposition 2.22 implies that contracting E outright would create a punctured cusp, but we can circumvent this problem as follows: First, desingularize the total space with a single blow-up, so that the strict transform of the elliptic bridge is connected at two nodes with smooth total space. Now we may contract the elliptic bridge to obtain a tacnodal special fiber, at the expense of inserting a $\mathbb{P}^1$ along one of the branches.

Regarding the second problem, one can try a similar trick: Given the family pictured in Figure 1, one may blow-up the two points of intersection $E_1 \cap E_2$, make a base-change to reduce the multiplicities of the exceptional divisors, and then contract *both* elliptic bridges to obtain a bi-tacnodal limit whose normalization comprises a pair of smooth rational curves. This limit curve certainly appears canonical, but it has an infinite automorphism group and contains the other two pictured limits as deformations. Of course, it is impossible to have a curve with an infinite automorphism group in a class of curves with the unique limit property. These examples suggest is that if we want a compact moduli space for curves with nodes, cusps, and tacnodes, we must make do with a weakly modular compactification, i.e. consider a slightly non-separated moduli functor. In Section 2.4, we will see that geometric invariant theory identifies just such a functor.

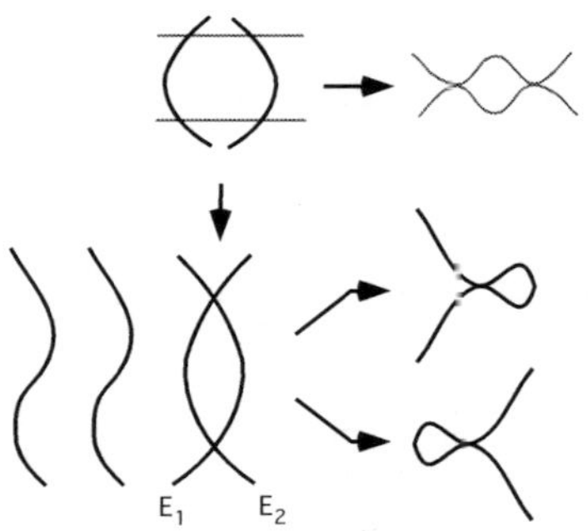

Figure 1. Three candidates for the tacnodal limit of a one-parameter family of genus three curves specializing to a pair of elliptic bridges.

It is worth pointing out that there is one family of examples, where the problem of interacting elliptic components does not arise and the naïve definition of a tacnodal functor actually gives rise to a class of curves with the unique limit property. This is the case of n-pointed curves of genus one. In fact, it turns out that there is a beautiful sequence of modular compactifications of $M_{1,n}$ which can be constructed by purely combinatorial methods, although the required techniques are somewhat more subtle than what we have described thus far. For the sake of completeness, and also because these spaces arise in the log MMP for $M_{1,n}$ (Section 4.3), we define this sequence of stability conditions below.

The singularities introduced in this sequence of stability conditions are the so-called elliptic m-fold points, whose name derives from the fact that they are the unique Gorenstein singularities which can appear on a curve of arithmetic genus one [108, Appendix A].

Definition 2.31 (The elliptic m-fold point). *We say that* $p \in C$ *is an* elliptic m-fold point *if*

$$\hat{\mathcal{O}}_{C,p} \simeq \begin{cases} k[[x,y]]/(y^2 - x^3) & m = 1 \ \textit{(ordinary cusp)} \\ k[[x,y]]/(y^2 - yx^2) & m = 2 \ \textit{(ordinary tacnode)} \\ k[[x,y]]/(x^2y - xy^2) & m = 3 \ \textit{(planar triple point)} \\ k[[x_1,\dots,x_{m-1}]]/J_m & m \geqslant 4, \textit{(} m \textit{ general lines through } 0 \in \mathbb{A}^{m-1}\textit{)}, \end{cases}$$

$$J_m := (x_h x_i - x_h x_j \ : \ i,j,h \in \{1,\dots,m-1\} \textit{ distinct}).$$

We now define the following stability conditions.

Definition 2.32 (m-stability). *Fix positive integers* $m < n$. *Let* $(C, \{p_i\}_{i=1}^n)$ *be an* n*-pointed curve of arithmetic genus one. We say that* $(C, \{p_i\}_{i=1}^n)$ *is* m-stable *if*

(1) C *has only nodes and elliptic* l*-fold points,* $l \leqslant m$*, as singularities.*

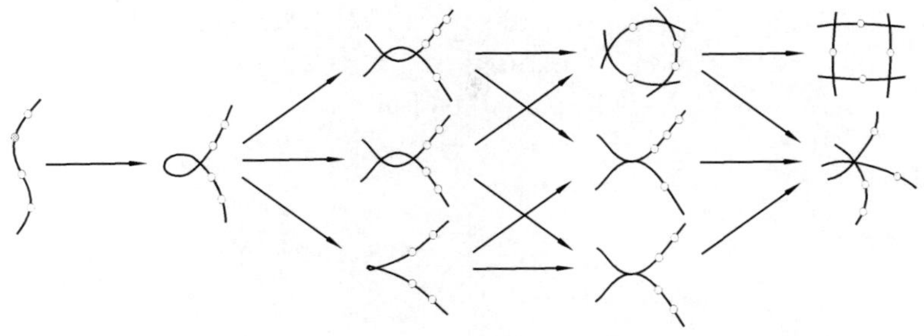

Figure 2. Topological types of curves in $\overline{M}_{1,4}(3)$. Every pictured component is rational, except the smooth elliptic curve pictured on the far left.

(2) *If* $E \subset C$ *is any connected subcurve of arithmetic genus one, then*

$$|E \cap \overline{C \setminus E}| + |\{p_i \,|\, p_i \in E\}| > m.$$

(3) $H^0(C, \Omega_C^\vee(-\Sigma_{i=1}^n p_i)) = 0$, *i.e.* $(C, \{p_i\}_{i=1}^n)$ *has no infinitesimal automorphisms.*

In order to provide some intuition for the definition of m-stability, Figure 2 displays all topological types of curves in $\overline{M}_{1,4}(3)$, as well as the specialization relations between them.

Proposition 2.33 **([108]).** *There exists a moduli stack* $\overline{\mathcal{M}}_{1,n}[m]$ *of* m*-stable curves, with corresponding moduli space* $\overline{M}_{1,n}[m]$.

While the proof of this theorem would take us too far afield, we may at least explain how to obtain the m-stable limit of a generically smooth family of n-pointed elliptic curves. The essential feature which makes this process more subtle than the constructions of $\overline{M}_g^{ps}$ and $\overline{M}_{0,n}[\psi]$ is that one must alternate between blowing-up and contracting, rather than simply contracting subcurves of the stable limit.

Given a family of smooth n-pointed curves of genus one over a punctured disc $(\mathcal{C}^* \to \Delta^*, \{\sigma_i\}_{i=1}^n)$, we obtain the m-stable limit as follows. (The process is pictured in Figure 3.)

(1) First, complete $(\mathcal{C}^* \to \Delta^*, \{\sigma_i\}_{i=1}^n)$ to a semistable family $(\mathcal{C} \to \Delta, \{\sigma_i\}_{i=1}^n)$ with smooth total space, i.e. take the stable limit and desingularize the total space.

(2) Isolate the minimal elliptic subcurve $Z \subset C$, i.e. the unique elliptic subcurve which contains no proper elliptic subcurves, blow-up the marked points on Z, then contract the strict transform Z.

(3) Repeat step (2) until the minimal elliptic subcurve satisfies

$$|Z \cap \overline{C \setminus Z}| + |\{p_i \,|\, p_i \in Z\}| > m.$$

(4) Stabilize, i.e. blow-down all smooth $\mathbb{P}^1$'s which meet the rest of the fiber in two nodes and have no marked points, or meet the rest of the fiber in a single node and have one marked point.

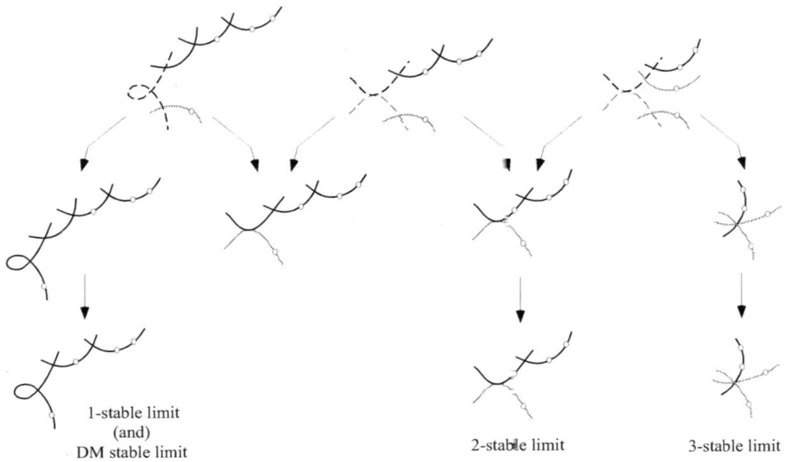

Figure 3. The process of blow-up/contraction/stabilization in order to extract the m-stable limit for each $m = 1,2,3$. Every irreducible component pictured above is rational. The left-diagonal maps are simple blow-ups along the marked points of the minimal elliptic subcurve, and exceptional divisors of these blow-ups are colored grey. The right-diagonal maps contract the minimal elliptic subcurve of the special fiber, and exceptional components of these contractions are dotted. The vertical maps are stabilization morphisms, blowing down all semistable components of the special fiber.

Finally, we should also note that the m-stability is compatible with the definition of $\mathcal{A}$-stability described above, thus yielding an even more general collection of stability conditions.

Definition 2.34 (($m, \mathcal{A}$)-stability). *Fix a pair of positive integers $m < n$, and let $\mathcal{A} = (a_1, \ldots, a_n) \in (0,1]^n$ be an n-tuple of rational weights. Let C be a connected reduced complete curve of arithmetic genus one, and let $p_1, \ldots, p_n \in C$ be smooth (not necessarily distinct) points of C. We say that $(C, p_1, \ldots, p_n)$ is $(m, \mathcal{A})$-stable if*

(1) *C has only nodes and elliptic l-fold points, $l \leqslant m$ as singularities.*
(2) *If $E \subset C$ is any connected subcurve of arithmetic genus one, then*

$$|E \cap \overline{C \setminus E}| + |\{p_i \,|\, p_i \in E\}| > n.$$

(3) $H^0(C, \Omega_C^\vee(-\Sigma_i p_i)) = 0.$

(4) *If* $p_{i_1} = \ldots = p_{i_k} \in C$ *coincide, then* $\sum_{j=1}^{k} a_{i_j} \leqslant 1$.
(5) $\omega_C(\sum_i a_i p_i)$ *is an ample* $\mathbb{Q}$*-divisor.*

In [108], it is shown that the class of $(m, \mathcal{A})$-stable curves is deformation open and satisfies the unique limit property. Thus, there exist corresponding moduli stacks $\overline{\mathcal{M}}_{1,\mathcal{A}}(m)$ and spaces $\overline{M}_{1,\mathcal{A}}(m)$.

2.4. Modular birational models via GIT

In this section, we discuss the use of geometric invariant theory (GIT) to construct modular and weakly modular birational models of $M_{g,n}$. Following Hassett and Hyeon [60], we will explain certain heuristics for interpreting GIT quotients as log canonical models of $\overline{M}_g$, and describe how to use these heuristics, in conjunction with intersection theory on $\overline{M}_g$, to predict the GIT-stability of certain Hilbert points.

GIT was invented by Mumford [96] to solve the following problem: Suppose G is a linearly reductive group acting on a normal projective variety X. We wish to form a projective quotient $X \to X/\!\!/G$, whose fibers are precisely the G-orbits of X. Unfortunately, there are obvious topological obstructions to the existence of such a quotient. For example, if one orbit is in the closure of another, both are necessarily mapped to the same point of $X/\!\!/G$. GIT gives a systematic method for constructing projective varieties which can be thought of as best- possible approximations to the desired quotient space. The GIT construction is not unique and depends on a choice of *linearization* of the action [96, p. 30]. A linearization of the G-action simply consists of an ample line bundle $\mathcal{L}$ on X, and a G-action on the total space of $\mathcal{L}$ which is compatible with the given G-action on X. More precisely, if the action on X is given by a morphism $G \times X \to X$, and the action on the line bundle by a morphism $G \times \underline{\mathrm{Spec}}_{\mathcal{O}_X} \mathrm{Sym}^* \mathcal{L} \to \underline{\mathrm{Spec}}_{\mathcal{O}_X} \mathrm{Sym}^* \mathcal{L}$, then we require the following diagram to commute:

$$\begin{array}{ccc} G \times \underline{\mathrm{Spec}}_{\mathcal{O}_X} \mathrm{Sym}^* \mathcal{L} & \longrightarrow & \underline{\mathrm{Spec}}_{\mathcal{O}_X} \mathrm{Sym}^* \mathcal{L} \\ \downarrow & & \downarrow \\ G \times X & \longrightarrow & X \end{array}$$

Given a linearization of the G-action on X, the GIT quotient is constructed as follows. First, note that since $\mathcal{L}$ is ample, we have

$$X = \mathrm{Proj} \bigoplus_{m \geqslant 0} H^0(X, \mathcal{L}^m).$$

The linearization gives a G-action on $\bigoplus_{m \geqslant 0} H^0(X, \mathcal{L}^m)$, so we may consider the ring of invariants

$$\bigoplus_{m \geqslant 0} H^0(X, \mathcal{L}^m)^G \subset \bigoplus_{m \geqslant 0} H^0(X, \mathcal{L}^m).$$

The GIT quotient is simply defined to be the associated rational map

$$q\colon X = \operatorname{Proj} \bigoplus_{m \geqslant 0} H^0(X, \mathcal{L}^m) \dashrightarrow \operatorname{Proj} \bigoplus_{m \geqslant q} H^0(X, \mathcal{L}^m)^G.$$

In order to describe the properties of this rational map, it is useful to make the following definitions.

Definition 2.35. *The* semistable locus *and* stable locus *of the linearized* G*-action on* X *are defined by*

$$X^{ss} := \{x \in X \mid \exists f \in H^0(X, \mathcal{L}^m)^G \text{ such that } f(x) \neq 0\},$$
$$X^{s} := \{x \in X^{ss} \mid G \cdot x \text{ is closed in } X^{ss} \text{ and } \dim G \cdot x = \dim G\}.$$

It is elementary to check that $X^s \subset X^{ss} \subset X$ is a sequence of G-invariant open immersions, and that X^{ss} is the locus where q is regular. For this reason, it is customary to denote the GIT quotient by $X^{ss}/\!/G$. Now we have a diagram:

(2.36)
$$\begin{array}{ccccc} X^s & \longrightarrow & X^{ss} & \longrightarrow & X \\ \big\downarrow q & & \big\downarrow q & \swarrow & \\ X^s/\!/G & \longrightarrow & X^{ss}/\!/G & & \end{array}$$

Here, q is a geometric quotient on the stable locus X^s and a categorical quotient on X^{ss} [96, pp. 4, 38]. This is made precise in the following proposition.

Proposition 2.37. *With notation as above, we have*

(1) *For any* $z \in X^s/\!/G$, *the fiber* $\phi^{-1}(z)$ *is a single closed* G*-orbit.*
(2) *For any* $z \in X^{ss}/\!/G$, *the fiber* $\phi^{-1}(z)$ *contains a unique closed* G*-orbit.*

Proof. Since G is linearly reductive, every G-representation is completely reducible. Hence, taking invariants is an exact functor on the category of G-representations. In particular, if Z_1, Z_2 are disjoint G-invariant closed subschemes of a distinguished open affine X_f, for some $f \in H^0(X, \mathcal{L}^k)^G$, then the surjection of G-representations

$$H^0(X_f, \mathcal{L}^m) \to H^0(Z_1, \mathcal{L}^m|_{Z_1}) \oplus H^0(Z_2, \mathcal{L}^m|_{Z_2}) \to 0$$

gives rise to

$$H^0(X_f, \mathcal{L}^m)^G \to H^0(Z_1, \mathcal{L}^m|_{Z_1})^G \oplus H^0(Z_2, \mathcal{L}^m|_{Z_2})^G \to 0.$$

By considering the preimages of (1,0) and (0,1), and scaling by a power of f, we deduce the existence of G-invariant sections of $\mathcal{L}^n$, $n \gg 0$, separating Z_1 and Z_2.

Note that if $x \in X^s$, the orbit $G \cdot x$ is closed of maximal dimension inside some X_f, hence for any other point $y \notin G \cdot x$, we have $G \cdot x \cap \overline{G \cdot y} = \varnothing$. Both (1) and (2) now follow from the just established fact that G-invariant sections of $\mathcal{L}^n$, $n \gg 0$, separate any two closed disjoint G-invariant subschemes of X^{ss}. □

While Definition 2.35 in theory determines the semistable locus X^{ss}, it requires the knowledge of *all* invariant sections of $\mathcal{L}^m$ that we rarely possess in practice. In order to use GIT, we clearly need a more algorithmic method to determine the stability of points in X. It was for this purpose that Mumford developed the so-called *numerical criterion for stability* [96, Chapter 2.1]. The first step is to observe that $x \in X$ is semistable if and only if for every nonzero lift $\tilde{x}$ of x to the affine cone $\operatorname{Spec} \bigoplus_{m \geqslant 0} H^0(X, \mathcal{L}^m)$, the closure of the G-orbit of $\tilde{x}$ does not contain the origin. Mumford's key insight was that *this* property can be tested by looking at one-parameter subgroups of G, i.e. all subgroups $k^* \subset G$. More precisely, if we are given any one-parameter subgroup $\rho\colon k^* \to G$, we may diagonalize the action of ρ on $H^0(X, \mathcal{L})$ and hence on the $\mathcal{L}$-coordinates $(x_0, \dots, x_n)$ of $\overline{x}$. We then define the *Hilbert-Mumford index* of x with respect to ρ by

$$\mu_\rho(x) := \min_{x_i \neq 0}\{w \ : \ \rho(t) \cdot x_i = t^w x_i\},$$

and we say that $x \in X$ is *stable* (resp. *semistable, nonsemistable*) *with respect to* ρ if $\mu_\rho(x) < 0$ (resp. $\mu_\rho(x) \leqslant 0$, $\mu_\rho(x) > 0$). With this terminology, Mumford's numerical criterion is easy to state.

Proposition 2.38 (Hilbert-Mumford numerical criterion). *A point* $x \in X$ *is* stable *(resp.* semistable*) if and only if it is stable (resp. semistable) with respect to all one-parameter subgroups of* G. *A point* $x \in X$ *is* nonsemistable *if it is* ρ*-nonsemistable for some one-parameter subgroup* ρ.

In words, $x \in X$ is stable if *every* one-parameter subgroup of G acts on the nonzero $\mathcal{L}$-coordinates of x with both positive and negative weights, and $x \in X$ is nonsemistable if there exists a one-parameter subgroup of G acting on the nonzero $\mathcal{L}$-coordinates of x with either all positive or all negative weights. We will see an example of how to use the numerical criterion in Example 2.49. First, however, let us step back and explain how the entire geometric invariant theory is applied to construct compactifications of the moduli space of curves.

First, fix an integer $n \geqslant 2$. If C is any smooth curve of genus $g \geqslant 2$, a choice of basis for the vector space $H^0(C, \omega_C^n)$ determines an embedding

$$|\omega_C^n| : C \hookrightarrow \mathbb{P}^N,$$

where $N = (2n-1)(g-1) - 1$. The subscheme $C \subset \mathbb{P}^N$ determines a point $[C] \in \mathrm{Hilb}_{P(x)}(\mathbb{P}^N)$, the Hilbert scheme parametrizing subschemes of $\mathbb{P}^N$ with Hilbert polynomial $P(x) := n(2g-2)x + (1-g)$. Now let

$$\mathrm{Hilb}_{g,n} \subset \mathrm{Hilb}_{P(x)}(\mathbb{P}^N)$$

denote the locally closed subscheme of all such n-canonically embedded smooth curves, and let $\overline{\mathrm{Hilb}}_{g,n}$ denote the closure of $\mathrm{Hilb}_{g,n}$ in $\mathrm{Hilb}_{P(x)}(\mathbb{P}^N)$.

The natural action of $SL(N+1)$ on $\mathbb{P}^N$ induces an action on $\mathrm{Hilb}_{P(x)}(\mathbb{P}^N)$, hence also on $\overline{\mathrm{Hilb}}_{g,n}$ and $\mathrm{Hilb}_{g,n}$.

Exercise 2.39. Verify that the action of $SL(N+1)$ on $Hilb_{g,n}$ is proper and the stack quotient $[Hilb_{g,n} / SL(N+1)]$ is canonically isomorphic to $\mathcal{M}_g$.

Our plan therefore is to apply GIT to the action of $SL(N+1)$ on the projective variety $\overline{Hilb}_{g,n}$ and *hope* that $Hilb_{g,n}$ is contained in the corresponding stable locus $\overline{Hilb}^{s}_{g,n}$. If this can be verified then the GIT quotient $\overline{Hilb}^{ss}_{g,n} /\!\!/ SL(N+1)$ will give a compactification of M_g.

In order to apply GIT, we must linearize the $SL(N+1)$-action on $\overline{Hilb}_{g,n}$. This is accomplished as follows. First, by Castelnuovo-Mumford regularity [94, Lecture 14], there exists an integer $m \geqslant 0$ such that $H^1(\mathbb{P}^N, I_C(m)) = 0$ for *any* curve $C \subset \mathbb{P}^N$ with Hilbert polynomial $P(x)$. Now set

$$W_m = \bigwedge^{P(m)} H^0\left(\mathbb{P}^N, \mathcal{O}_{\mathbb{P}^N}(m)\right).$$

We obtain an embedding

$$Hilb_{P(x)}(\mathbb{P}^N) \hookrightarrow \mathbb{P}W_m$$

as follows: Corresponding to a point $[C] \subset Hilb_{P(x)}(\mathbb{P}^N)$, we have the surjection

$$H^0(\mathbb{P}^N, \mathcal{O}_{\mathbb{P}^N}(m)) \to H^0(C, \mathcal{O}_C(m)) \to 0, \tag{2.40}$$

which is called the mth *Hilbert point* of $C \hookrightarrow \mathbb{P}^N$. Taking the $P(m)^{th}$ exterior power of this sequence gives a one-dimensional quotient of W_m, hence a point of $\mathbb{P}(W_m)$. Now $\mathcal{O}_{\mathbb{P}(W_m)}(1)$ restricts to an ample line bundle on $Hilb_{P(x)}(\mathbb{P}^N)$, and the $SL(N+1)$-action extends to an action on the total space of $\mathcal{O}_{\mathbb{P}(W_m)}(1)$ because W_m is an $SL(N+1)$-representation. Corresponding to this linearization, we obtain a GIT quotient $\overline{Hilb}^{ss,m}_{g,n} /\!\!/ SL(N+1)$.

This is an undeniably slick construction, but one glaring question remains: Can we actually determine the stable and semistable locus $\overline{Hilb}^{s,m}_{g,n} \subset \overline{Hilb}^{ss,m}_{g,n} \subset \overline{Hilb}_{g,n}$? Before addressing this issue, let us take a step back and review the following two aspects of the construction that we've just presented

(1) What choices went into the construction of $\overline{Hilb}^{ss,m}_{g,n} /\!\!/ SL(N+1)$?
(2) To what extent is $\overline{Hilb}^{ss,m}_{g,n} /\!\!/ SL(N+1)$ modular?

Regarding (1), the construction depended on two numerical parameters: the integer n which determined the Hilbert scheme, and the integer m which determined the linearization. An interesting problem, which will be addressed presently, is to understand how the quotient $\overline{Hilb}^{ss,m}_{g,n} /\!\!/ SL(N+1)$ changes as one varies the parameters n and m.

Regarding (2), it is not a priori obvious that one should be able to identify $\left[\overline{Hilb}^{ss,m}_{g,n} / SL(N+1)\right]$ with an open substack of $\mathcal{U}_g$, the stack of genus g curves. It turns out, however, that under relatively mild hypotheses on $\overline{Hilb}^{ss,m}_{g,n}$, this will be the case. More precisely, we have

Proposition 2.41. *Assume that*

(1) $\overline{\mathrm{Hilb}}^{ss,m}_{g,n} \neq \emptyset$,

(2) *Every point* $[C] \in \overline{\mathrm{Hilb}}^{ss,m}_{g,n}$ *corresponds to a Gorenstein curve.*

Then we have

(1) $\overline{\mathrm{Hilb}}^{ss,m}_{g,n} /\!/ \mathrm{SL}(N+1)$ *is a weakly modular birational model of* $M_{g,n}$,

(2) *If* $\overline{\mathrm{Hilb}}^{ss,m}_{g,n} = \overline{\mathrm{Hilb}}^{s,m}_{g,n}$, *then* $\overline{\mathrm{Hilb}}^{ss,m}_{g,n} /\!/ \mathrm{SL}(N+1)$ *is a modular birational model of* $M_{g,n}$.

Proof. Consider the universal curve

$$\begin{array}{ccc} \mathcal{C} & \longrightarrow & \mathbb{P}^N \times \overline{\mathrm{Hilb}}^{ss,m}_{g,n} \\ \big\downarrow{\scriptstyle \pi} & \swarrow & \\ \overline{\mathrm{Hilb}}^{ss,m}_{g,n} & & \end{array}$$

Since the fibers of π are Gorenstein, the relative dualizing sheaf $\omega_{\mathcal{C}/\overline{\mathrm{Hilb}}^{ss,m}_{g,n}}$ is a line bundle, and we claim that $\omega^n_{\mathcal{C}/\overline{\mathrm{Hilb}}^{ss,m}_{g,n}} \simeq \mathcal{O}_{\mathbb{P}^N}(1) \otimes \pi^*\mathcal{L}$ for some line bundle $\mathcal{L}$ on $\overline{\mathrm{Hilb}}^{ss,m}_{g,n}$. This is immediate from the fact that $\omega^n_{\mathcal{C}/\overline{\mathrm{Hilb}}^{ss,m}_{g,n}} \simeq \mathcal{O}_{\mathbb{P}^N}(1)$ on the fibers of $\mathcal{C}$ over the nonempty open set $\mathrm{Hilb}_{g,n} \cap \overline{\mathrm{Hilb}}^{ss,m}_{g,n}$.

Now consider the natural map from $\overline{\mathrm{Hilb}}^{ss,m}_{g,n}$ to the stack $\mathcal{U}_g$ (of all reduced connected genus g curves) induced by the universal family. Since any two n-canonically embedded curves $C \subset \mathbb{P}^N$ and $C' \subset \mathbb{P}^N$ are isomorphic iff there exists a projective linear transformation taking C to C', this map factors through the $\mathrm{SL}(N+1)$-quotient to give $\left[\overline{\mathrm{Hilb}}^{ss,m}_{g,n} / \mathrm{SL}(N+1)\right] \to \mathcal{U}_g$. We claim that this map is an open immersion. Equivalently, that $\overline{\mathrm{Hilb}}^{ss,m}_{g,n}$ parameterizes a deformation open class of curves. For this, we must see that if $[C] \in \overline{\mathrm{Hilb}}^{ss,m}_{g,n}$ then every abstract deformation of C can be realized as an embedded deformation of $C \subset \mathbb{P}^N$. By [107, Corollary B.8], this follows from the fact that $H^1(C, \mathcal{O}_C(1)) = H^1(C, \omega_C^n) = 0$ for $n \geqslant 2$.

Thus, we may identify $\left[\overline{\mathrm{Hilb}}^{ss,m}_{g,n} / \mathrm{SL}(N+1)\right]$ with an open substack of the stack of curves, and we have a diagram

$$\begin{array}{ccc} \mathcal{U}_g & & \\ \big\uparrow{\scriptstyle i} & & \\ \left[\overline{\mathrm{Hilb}}^{ss,m}_{g,n} / \mathrm{SL}(N+1)\right] & \xrightarrow{\ \phi\ } & \overline{\mathrm{Hilb}}^{ss,m}_{g,n} /\!/ \mathrm{SL}(N+1) \end{array}$$

It only remains to check that ϕ is a good moduli map. The first property of good moduli maps, that they are categorical with respect to algebraic spaces, is a general feature of GIT [96], [3]. The second property, namely that $\phi^{-1}(x)$ contains a unique closed point for each $x \in \overline{\mathrm{Hilb}}^{ss,m}_{g,n} /\!/ \mathrm{SL}(N+1)$ is an immediate consequence of Proposition 2.37. □

Finally, it remains to consider the problem of how one actually computes the semistable locus

$$\overline{\mathrm{Hilb}}_{g,n}^{ss,m} \subset \mathrm{Hilb}_{P(x)}(\mathbb{P}^N).$$

Historically, the important breakthrough was Gieseker's asymptotic stability result: Gieseker showed that for any $n \geqslant 10$, a generic smooth n-canonically embedded curve $C \subset \mathbb{P}^N$ is asymptotically Hilbert stable [51] and Mumford extended this result[1] to 5-canonically embedded curves [95]:

Theorem 2.42. *For* $n \geqslant 5$ *and* $m \gg 0$,

$$\overline{\mathrm{Hilb}}_{g,n}^{ss,m} = \overline{\mathrm{Hilb}}_{g,n}^{s,m} = \{[C] \in \mathrm{Hilb}_{P(x)}(\mathbb{P}^N) \,|\, C \subset \mathbb{P}^N \text{ is stable and } \mathcal{O}_C(1) \simeq \omega_C^n\}.$$

Equivalently, $\overline{\mathrm{Hilb}}_{g,n}^{ss,m} /\!/ \mathrm{SL}(N+1) \simeq \overline{M}_g$.

Rather than describe the proof of this theorem, which is well covered in other surveys (e.g. [54] and [92]), we shall consider here the natural remaining question: what happens for smaller values of n and m? Recently, Hassett, Hyeon, Lee, Morrison, and Swinarski have taken the first steps toward understanding this problem, and a beautiful picture is emerging in which the GIT quotients corresponding to various values of n and m admit natural interpretations as log canonical models of $\overline{M}_g$ [61, 60, 65, 62, 93]. Before describing their results in detail, however, let us present an important heuristic for predicting which curves $C \subset \mathbb{P}^N$ *ought* to be semistable for given values of n and m. It is worth nothing that, in practice, having an educated guess for what the semistable locus $\overline{\mathrm{Hilb}}_{g,n}^{ss,m}$ ought to be is the most important step in actually describing it.

To begin, let us assume that $\overline{\mathrm{Hilb}}_{g,n}^{ss,m}$ satisfies the hypotheses

(1) $\overline{\mathrm{Hilb}}_{g,n}^{ss,m} \neq \varnothing$,
(2) Every point $[C] \in \overline{\mathrm{Hilb}}_{g,n}^{ss,m}$ corresponds to a Gorenstein curve,
(3) The locus of moduli non-stable curves in $\overline{\mathrm{Hilb}}_{g,n}^{ss,m} /\!/ \mathrm{SL}(N+1)$ has codimension at least two.

Under assumptions (1) and (2), Proposition 2.41 says that we may identify the quotient stack $\left[\overline{\mathrm{Hilb}}_{g,n}^{ss,m} / \mathrm{SL}(N+1)\right]$ with a weakly modular birational model of $\overline{M}_g$. Assumption (3) implies that we have a birational contraction (cf. Definition 3.1)

$$\phi\colon \overline{M}_g \dashrightarrow \overline{\mathrm{Hilb}}_{g,n}^{ss,m} /\!/ \mathrm{SL}(N+1).$$

The key idea is to interpret this contraction as the rational map associated to a certain divisor on $\overline{M}_g$. To set this up, first note that we may define *Weil divisor* classes λ and δ on $\overline{\mathrm{Hilb}}_{g,n}^{ss,m} /\!/ \mathrm{SL}(N+1)$ simply by pushing forward the corresponding classes from $\overline{M}_g$. (The fact that ϕ is a birational contraction implies that push-forward of Weil divisors is well-defined; since we have *a priori* no control over the singularities of $\overline{\mathrm{Hilb}}_{g,n}^{ss,m} /\!/ \mathrm{SL}(N+1)$, λ and δ may very well fail to be Cartier.) Next, we observe that

[1] Strictly speaking, Mumford works with the Chow variety.

the divisor class of the polarization of $\overline{\mathrm{Hilb}}_{g,n}^{ss,m} /\!/ \mathrm{SL}(N+1)$ coming from the GIT construction can be computed as an explicit linear combination of λ and δ.

Proposition 2.43 ([60, Section 5]). *Suppose that assumptions (1)–(3) above hold. Then there is an ample line bundle $\Lambda_{m,n}$ on $\overline{\mathrm{Hilb}}_{g,n}^{ss,m} /\!/ \mathrm{SL}(N+1)$ with divisor class given by:*

(2.44)

$$\Lambda_{m,n} = \begin{cases} \lambda + (m-1)\left[(4g+2)m - g + 1)\lambda - \frac{gm}{2}\delta\right], & \text{if } n = 1, \\ (m-1)(g-1)\left[(6mn^2 - 2mn - 2n + 1)\lambda - \frac{mn^2}{2}\delta\right], & \text{if } n > 1. \end{cases}$$

Proof. Consider the universal curve

$$\begin{array}{ccc} \mathcal{C} & \longrightarrow & \mathbb{P}^N \times \overline{\mathrm{Hilb}}_{g,n}^{ss,m} \\ \big\downarrow \pi & \swarrow & \\ \overline{\mathrm{Hilb}}_{g,n}^{ss,m} & & \end{array}$$

As we have described above, the linearization on the Hilbert scheme comes from its embedding into $\mathbb{P}(W_m)$, where $W_m = \bigwedge^{P(m)} H^0\left(\mathbb{P}^N, \mathcal{O}_{\mathbb{P}^N}(m)\right))$. It follows from the construction of this embedding that the tautological quotient line bundle $\mathcal{O}_{\mathbb{P}(W_m)}(1)$ pulls back to the line bundle $\bigwedge^{P(m)} \pi_*\mathcal{O}_{\mathcal{C}}(m)$ on $\overline{\mathrm{Hilb}}_{g,n}$. Hence, $\bigwedge^{P(m)} \pi_*\mathcal{O}_{\mathcal{C}}(m)$ is an $\mathrm{SL}(N+1)$-equivariant line bundle on $\overline{\mathrm{Hilb}}_{g,n}^{ss,m}$ which descends to an ample line bundle, denoted $\Lambda_{m,n}$, on the quotient. To compute $c_1\left(\bigwedge^{P(m)} \pi_*\mathcal{O}_{\mathcal{C}}(m)\right)$, and hence the class of $\Lambda_{m,n}$, one observes that

(a) $\mathcal{O}_{\mathcal{C}}(1) = \omega_\pi^n \otimes \pi^*\mathcal{L}$ for some line bundle $\mathcal{L}$ on $\overline{\mathrm{Hilb}}_{g,n}^{ss,m}$ (this is simply the definition of being a family of n-canonically embedded curves),

(b) $\pi_*\mathcal{O}_{\mathcal{C}}(1)$ is a trivial $\mathbb{P}^N$-bundle (this says that all curves in $\overline{\mathrm{Hilb}}_{g,n}^{ss,m}$ are embedded by complete linear systems).

As we've remarked in the proof of Proposition 2.41, assumptions (1)-(2) imply that $R^1\pi_*\mathcal{O}_{\mathcal{C}}(1) = 0$. The class of $\Lambda_{m,n}$ now can be computed using the Grothendieck-Riemann-Roch formula. □

Corollary 2.45. *If $\Lambda_{m,n}$ denotes the divisor class on $\overline{M}_g$ as above, then $\phi_*\Lambda_{m,n}$ is Cartier and ample on $\overline{\mathrm{Hilb}}_{g,n}^{ss,m} /\!/ \mathrm{SL}(N+1)$.*

Proof. This is just a restatement of the proposition. □

According to the corollary, $\overline{M}_g \dashrightarrow \overline{\mathrm{Hilb}}_{g,n}^{ss,m} /\!/ \mathrm{SL}(N+1)$ will be precisely the contraction associated to $\Lambda_{m,n}$ once we verify that

(2.46) $$\Lambda_{m,n} - \phi^*\phi_*\Lambda_{m,n} \geqslant 0 \in N^1(\overline{M}_g).$$

We may now state our heuristic informally as:

Principle 2.47 (GIT Heuristic). *The singularities of curves appearing in the GIT quotient of n-canonically embedded curves using the mth Hilbert point polarization should be precisely those whose variety of stable limits in $\overline{M}_g$ is covered by curves on which $\Lambda_{m,n}$ is non-positive.*

For a more precise description of the *variety of stable limits* associated to a singularity, as well as a more detailed justification for this heuristics, we refer the reader to Section 4.4. Here, let us just give a simple example. Using the formulae for $\Lambda_{m,n}$, the reader may easily check that

$$\lim_{m\to\infty} \Lambda_{m,3} \sim \frac{32}{3}\lambda - \delta.$$

Now it is easy to see that this line bundle is negative on the family of varying elliptic tails (cf. Example 3.10), which is precisely the variety of stable limits of the cusp (cf. Example 2.16). Our heuristic thus suggests that for $m \gg 0$, cuspidal curves should be contained in the semistable locus $\overline{\mathrm{Hilb}}_{g,3}^{ss,m}$. In Section 4.4.2, we describe the variety of stable limits of any ADE singularity (among others) and describe the threshold slopes at which these varieties are covered by curves on which $s\lambda - \delta$ is negative. Using these computations, it is easy to compute the threshold values of m and n at which we may expect curves with given singularities to become Hilbert (semi) stable.

The intersection theory heuristic, however useful, can be used only as a guide and will not by itself establish stability of any given curve. Methods used by Gieseker to prove stability of 5-canonically embedded stable curves are asymptotic in nature and are of no help in the case of finite Hilbert stability. Having said this, we note that there is also a precise if slightly unwieldy method for determining stability of Hilbert points of an embedded curve C. By tracing through the construction of the mth Hilbert point in Equation (2.40) and the definition of the resulting linearization on $\overline{\mathrm{Hilb}}_{g,n}^{m}$, and applying Proposition 2.38 we deduce the following result (see [54, Proposition 4.23]).

Proposition 2.48 (Stability of Hilbert points). *The mth Hilbert point of an embedded curve $C \subset \mathbb{P}^N$ is stable (resp. semistable) if and only if for every one-parameter subgroup ρ of $SL(N+1)$ and a basis $(x_0, \dots, x_N)$ of $H^0(C, \mathcal{O}_C(1))$ diagonalizing the action of ρ, there is a* monomial basis *of $H^0(C, \mathcal{O}_C(m))$ consisting of degree m monomials in $x_0, \dots, x_N$ such that the sum of the ρ-weights of these monomials is negative (resp. non-positive).*

While the criterion of Proposition 2.48 gives in theory a way to prove stability, in practice it can be efficiently used only to prove nonsemistability or to prove stability with respect to a fixed torus. Occasionally, in the presence of nontrivial automorphisms, the theory of worst one-parameter subgroups developed by Kempf [77] allows to reduce verification of stability to a fixed torus, where the numerical criterion can be applied. This technique is used in [93] to prove finite Hilbert stability of certain low genus bicanonical curves. We will settle here for an illustration of how the numerical criterion of Proposition 2.48 can be used to *destabilize* a curve. We do so on the example of a generically non-reduced curve of arithmetic genus 4.

Example 2.49 (Genus 4 ribbon). A ribbon is a non-reduced curve locally isomorphic to $\mathbb{A}_1 \times \operatorname{Spec} k[\epsilon]/(\epsilon^2)$. We refer to [10] for the general theory of such curves and focus here on the ribbon C of genus 4 defined by the ideal

$$I_C = (xz - y^2, z^3 + xw^2 - 2yzw)$$

in $\mathbb{P}^4$. This curve is interesting because it is a flat limit of one-parameter families of canonically embedded curves whose stable limit is a hyperelliptic curve in $\overline{\mathcal{M}}_4$. Since hyperelliptic curves are not canonically embedded, C is a natural candidate for a semistable point that replaces hyperelliptic curves in the GIT quotient $\overline{\mathrm{Hilb}}_{4,1}^{m,ss}$. However, as the following application of the numerical criterion shows, C has nonsemistable m^{th} Hilbert point for all m. (This was first observed by Fong [47, p.298]; see also [93, Example 8.10].)

The generators of I are homogeneous with respect to the one-parameter subgroup $\rho\colon \operatorname{Spec} k[\lambda, \lambda^{-1}] \to SL(4)$ acting by

$$\lambda \cdot (x, y, z, w) = (\lambda^{-3}x, \lambda^{-1}y, \lambda z, \lambda^3 w).$$

Since ρ is a subgroup of $\operatorname{Aut}(C) \subset SL(4)$, every monomial basis of $H^0(C, \mathcal{O}_C(m))$ has the same ρ-weight, which we now compute. We pick the basis of $H^0(C, \mathcal{O}_C(m))$ consisting of monomials not divisible by y^2 and xw^2:

$$\{x^i z^{m-i}\}_{i=0}^{m}, \{x^i z^{m-i-1} w\}_{i=0}^{m-1}, \{z^{m-i} w^i\}_{i=2}^{m},$$
$$\{x^i y z^{m-i-1}\}_{i=0}^{m-1}, \{x^i y z^{m-i-2} w\}_{i=0}^{m-2}, \{y z^{m-i-1} w^i\}_{i=2}^{m-1}.$$

The ρ-weights of the monomials in this basis sum up to $3m - 4 > 0$. We conclude by Proposition 2.48 that the mth Hilbert point of X is *nonsemistable for all* m.

Having discussed heuristics for verifying finite Hilbert stability and having seen one example of the numerical criterion in action, we now summarize what is known and what is yet to be done. We display the known results on the GIT quotients of n-canonically embedded curves and the predictions based on Principle 2.47 in Table 1. In order to understand the results displayed there, we need only to make three additional remarks: First, one can allow fractional values of the polarization; these are discussed in [93, Remark 3.2]. Second, rather than using the Hilbert scheme, one can also use the Chow variety of n-canonically embedded curves. Following a procedure similar to that which we have sketched above, one obtains the following ample divisor on $\overline{\mathrm{Chow}}_n^{ss} /\!/ SL(N+1)$ (see. [95, Theorem 5.15] or [60, Proposition 5.2]):

$$\begin{cases} (4g+2)\lambda - \frac{g}{2}\delta, & \text{if } n = 1, \\ (g-1)n\left[(6n-2)\lambda - \frac{n}{2}\delta\right], & \text{if } n > 1. \end{cases} \tag{2.50}$$

Third, we must describe the meaning of the notation $A_k^\dagger$ and $D_k^\dagger$ that appears in Table 1. Note that, in general, the semistability of an n-canonically embedded curve is not a local analytic question: It depends not only on the singularities of the curve but also on their global arrangement. For example, the classic stability analysis of plane quartics shows that a quartic with a node (A_1) and an oscnode (A_5) is semistable if it is a union of two conics, and not semistable if it is a union of a cubic and its flex line [96, Chapter 3.2]. In fact, Principle 2.47 can be refined to understand the thresholds at which different global arrangements of singularities appear, provided one understands the different varieties of stable limits associated to these different arrangements. For the purposes of Table 1, the one essential global distinction is the following:

Definition 2.51 (Dangling A_{2k-1} and D_{2k+2}). *We say that a proper curve C of genus g has a* dangling A_{2k-1}-singularity (or simply, an $A_{2k-1}^\dagger$-singularity) *if $C = R \cup S$ where $R \simeq \mathbb{P}^1$, $p := R \cap S$ in a smooth point of S and the intersection multiplicity of R and S is k ($k \geqslant 3$). Similarly, we say that a proper curve C of genus g has a* dangling D_{2k+2}-singularity (or simply, a $D_{2k+2}^\dagger$-singularity) *if $C = R \cup S$ where $R \simeq \mathbb{P}^1$, $p := R \cap S$ is a node of S and the intersection multiplicity of R with one of the branches of S at p is k ($k \geqslant 2$).*

Exercise 2.52. Show that a general genus $g \geqslant k$ curve C with an $A_{2k-1}^\dagger$ or $D_{2k+2}^\dagger$-singularity has ω_C ample and $\mathrm{Aut}(C)$ finite.

It is not difficult to see that the locus of stable limits of a curve with an $A_{2k-1}^\dagger$ or $D_{2k+2}^\dagger$-singularity is covered by curves with different numerical properties than the locus of stable limits of an irreducible curve with an A_{2k-1} or D_{2k+2}-singularity. This is why these two curves must be considered separately and arise at different threshold values in Table 1.

2.4.1. Established results on GIT As noted in Table 1, the semistable locus has $\overline{\mathrm{Hilb}}_{g,n}^{ss,m}$ has actually been completely determined in a number of cases by Hassett, Hyeon, Lee, Morrison and Schubert. To finish out our discussion of GIT, we present their descriptions of these loci and the connections between the resulting moduli spaces.

Definition 2.53. *Let C be any curve. We say that*

(1) C has an elliptic tail *if there exists a morphism $i\colon (E, p) \hookrightarrow C$ satisfying*
- *(E, p) is a 1-pointed curve of arithmetic genus one,*
- *$i|_{E-p}$ is an open immersion.*

(2) C has an elliptic bridge *if there exists a morphism $i\colon (E, p, q) \hookrightarrow C$ satisfying*
- *(E, p, q) is a 2-pointed curve of arithmetic genus one,*
- *$i|_{E-\{p,q\}}$ is an open immersion,*
- *$i(p)$, $i(q)$ are nodes of C.*

Table 1. GIT quotients of n-canonically embedded curves

GIT quotient	Status	Moduli space	Singularities
$\overline{\mathrm{Hilb}}_{g,n}^{ss,m} /\!\!/ SL(N+1)$, $n \geqslant 5$, $m \gg 0$	classical	$\overline{M}_g$	A_1
$\overline{\mathrm{Chow}}_{g,4}^{ss} /\!\!/ SL(7g-7)$	established [66]	$\overline{M}_g(\frac{9}{11}) = \overline{M}_g[A_2^*]$	$A_1 - A_2$
$\overline{\mathrm{Hilb}}_{g,4}^{ss,m} /\!\!/ SL(7g-7)$, $m \gg 0$	established [66]	$\overline{M}_g(\frac{9}{11} - \epsilon) = \overline{M}_g[A_2]$	$A_1 - A_2$
$\overline{\mathrm{Chow}}_{g,3}^{ss} /\!\!/ SL(5g-5)$	established [101]	$\overline{M}_g(\frac{25}{32}) = \overline{M}_g[A_2]$	$A_1 - A_2$
$\overline{\mathrm{Hilb}}_{g,3}^{ss,m} /\!\!/ SL(5g-5)$, $m \gg 0$	conjectural	$\overline{M}_g(\frac{25}{32} - \epsilon) = \overline{M}_g[A_2]$	$A_1 - A_2$
$\overline{\mathrm{Chow}}_{g,2}^{ss} /\!\!/ SL(3g-3)$	established [60]	$\overline{M}_g(\frac{7}{10}) = \overline{M}_g[A_3^*]$	$A_1 - A_3$
$\overline{\mathrm{Hilb}}_{g,2}^{ss,m} /\!\!/ SL(3g-3)$, $m \gg 0$	established [60]	$\overline{M}_g(\frac{7}{10} - \epsilon) = \overline{M}_g[A_3]$	$A_1 - A_3$
$\overline{\mathrm{Hilb}}_{g,2}^{ss,6} /\!\!/ SL(3g-3)$	conjectural	$\overline{M}_g(\frac{2}{3})$	$A_1 - A_4$
$\overline{\mathrm{Hilb}}_{g,2}^{ss,\frac{9}{2}} /\!\!/ SL(3g-3)$	conjectural	$\overline{M}_g(\frac{19}{29})$	$A_1 - A_4, A_5^\dagger$
$\overline{\mathrm{Hilb}}_{g,2}^{ss,\frac{9}{4}} /\!\!/ SL(3g-3)$	conjectural	$\overline{M}_g(\frac{17}{28})$	$A_1 - A_5$
$\overline{\mathrm{Hilb}}_{g,2}^{ss,\frac{27}{14}} /\!\!/ SL(3g-3)$	conjectural	$\overline{M}_g(\frac{49}{83})$	$A_1 - A_6$
$\overline{\mathrm{Hilb}}_{g,2}^{ss,\frac{3}{2}} /\!\!/ SL(3g-3)$	conjectural	$\overline{M}_g(\frac{5}{9})$	$A_1 - A_6, D_4, D_5^\dagger, D_6^\dagger$
	...	...	...
$\overline{\mathrm{Chow}}_{g,1}^{ss} /\!\!/ SL(g)$	conjectural	$\overline{M}_g(\frac{3g+8}{8g+4})$	ADE, Ribbons, X_9, J_{10}, E_{12}, Others?
$\overline{\mathrm{Hilb}}_{g,1}^{ss,m} /\!\!/ SL(g)$, $m \gg 0$	conjectural	$\overline{M}_g(\frac{3g+8}{8g+4} - \epsilon)$	ADE, Ribbons, X_9, J_{10}, E_{12}, Others?

(3) C has an elliptic chain *if there exists a morphism* $i\colon \bigcup_{i=1}^{k}(E_i, p_i, q_i) \hookrightarrow C$ *satisfying*

- (E_i, p_i, q_i) *is a 2-pointed curve of arithmetic genus one,*
- $i|_{E_i - \{p_i, q_i\}}$ *is an open immersion.*
- $i(q_i) = i(p_{i+1}) \in C$ *is a tacnode of* C*, for* $i = 1, \dots, k-1$.
- $i(p_1)$ *and* $i(q_k) \in C$ *are nodes of* C.

Definition 2.54 (Stability conditions). *If* C *is any curve with* ω_C *ample, we say that*

(1) C *is* A_2^**-stable if*
  (a) C *has only nodes and cusps as singularities.*
(2) C *is* A_2*-stable if*
  (a) C *has only nodes and cusps as singularities,*
  (b) C *has no elliptic tails.*
(3) C *is* A_3^**-stable if*
  (a) C *has only nodes, cusps, and tacnodes as singularities,*

   (b) *C has no elliptic tails.*
(4) *C is A_3-stable if*
   (a) *C has only nodes, cusps, and tacnodes as singularities,*
   (b) *C has no elliptic tails and no elliptic bridges or chains.*

Remark 2.55. *These stability conditions appear with different names in the literature: pseudostability for A_2-stability, c-semistability for A_3^*-stability, and h-semistability for A_3-stability. Pseudostability was introduced in [101], as we discussed in Section 2.3, while the latter two stability conditions were introduced in [60]. Our change of notation is motivated by the desire to have the name of the stability condition indicate the singularities allowed.*

The main results of GIT so far, as indicated in Table 1 is the following theorem, complementing the results of Gieseker and Mumford.

Theorem 2.56. *For any $g \geqslant 3$:*
(1) *For $n = 4$:*

$$\overline{\mathrm{Chow}}^{ss}_{g,n} = \{[C] \in \mathrm{Chow}(\mathbb{P}^N) \mid C \subset \mathbb{P}^N \text{ is } A_2^*\text{-stable and } \mathcal{O}_C(1) \simeq \omega_C^n\}.$$

Equivalently, $\overline{\mathrm{Chow}}^{ss}_{g,n} /\!/ \mathrm{SL}(N+1) \simeq \overline{M}_g[A_2^*]$.
(2) *For $n = 3$:*

$$\overline{\mathrm{Chow}}^{ss}_{g,n} = \{[C] \in \mathrm{Chow}(\mathbb{P}^N) \mid C \subset \mathbb{P}^N \text{ is } A_2\text{-stable and } \mathcal{O}_C(1) \simeq \omega_C^n\}.$$

Equivalently, $\overline{\mathrm{Chow}}^{ss}_{g,n} /\!/ \mathrm{SL}(N+1) \simeq \overline{M}_g[A_2]$ *for $m \gg 0$.*
(3) *For $n = 2$:*

$$\overline{\mathrm{Chow}}^{ss}_{g,n} = \{[C] \in \mathrm{Chow}(\mathbb{P}^N) \mid C \subset \mathbb{P}^N \text{ is } A_3^*\text{-stable and } \mathcal{O}_C(1) \simeq \omega_C^n\}.$$

Equivalently, $\overline{\mathrm{Chow}}^{ss}_{g,n} /\!/ \mathrm{SL}(N+1) \simeq \overline{M}_g[A_3^*]$.
(4) *For $n = 2$ and $m \gg 0$:*

$$\overline{\mathrm{Hilb}}^{ss,m}_{g,2} = \{[C] \in \mathrm{Hilb}_{P(x)}(\mathbb{P}^N) \mid C \subset \mathbb{P}^N \text{ is } A_3\text{-stable and } \mathcal{O}_C(1) \simeq \omega_C^n\}.$$

Equivalently, $\overline{\mathrm{Hilb}}^{ss,m}_{g,n} /\!/ \mathrm{SL}(N+1) \simeq \overline{M}_g[A_3]$ *for $m \gg 0$.*

Proof. The case of $n = 4$ is in [66]; the case of $n = 3$ is in [101]; the case of $n = 2$ is in [60]. □

For the purposes of the log MMP for $\overline{M}_g$, the important point is how these spaces relate to each other. The essential points are summarized in the following proposition:

Proposition 2.57.
(1) *There exists a birational contraction $\eta\colon \overline{M}_g \to \overline{M}_g[A_2]$ with $\mathrm{Exc}(\eta) = \Delta_1$. Furthermore, the restriction of η to Δ_1 contracts the fiber of the projection*

$$\Delta_1 \simeq \overline{M}_{g-1,1} \times \overline{M}_{1,1} \to \overline{M}_{g-1,1}.$$

(2) *There exists a small birational contraction* $\phi^-\colon \overline{M}_g[A_2] \to \overline{M}_g[A_3^*]$ *for which* $\mathrm{Exc}(\phi^{-1})$ *is the locus of elliptic bridges, defined to be the union of the images of the following maps:*

$$\overline{M}_{1,2} \times \overline{M}_{g-2,2} \to \overline{M}_g \to \overline{M}_g[A_2]$$
$$\overline{M}_{1,2} \times \overline{M}_{i,1} \times \overline{M}_{j,1} \to \overline{M}_g \to \overline{M}_g[A_2], \quad i+j=g-1.$$

Furthermore, the restriction of ϕ^- *to each of these strata contracts the (images under* η *of) fibers of projections*

$$\overline{M}_{1,2} \times \overline{M}_{g-2,2} \to \overline{M}_{g-2,2}$$
$$M_{1,2} \times \overline{M}_{i,1} \times \overline{M}_{j,1} \to \overline{M}_{i,1} \times \overline{M}_{j,1}, \quad i+j=g-1.$$

(3) *There exists a small birational contraction* $\phi^+\colon \overline{M}_g[A_3] \to \overline{M}_g[A_3^*]$ *with* $\mathrm{Exc}(\phi^+)$ *precisely the locus of tacnodal curves.*

Proof. For the full proof of this fact, we must refer the reader to [60]. Nevertheless, we will sketch the argument. For (1), the key observation is that there exists a natural transformation at the level of stacks $\overline{\mathcal{M}}_g \to \overline{\mathcal{M}}_g[A_2]$, replacing elliptic tails by cusps, which induces the map η. The fact that η is an isomorphism outside Δ_1 is then clear, while the fact that $\eta|_{\Delta_1}$ induces the given contraction is a consequence of the fact that the data of the j-invariant of the elliptic tail is not retained in the cuspidal curve which replaces it.

The analysis of ϕ^- and ϕ^+ is more subtle. At the level of stacks, there are obvious open immersions

$$\overline{\mathcal{M}}_g[A_2] \hookrightarrow \overline{\mathcal{M}}_g[A_3^*] \hookleftarrow \overline{\mathcal{M}}_g[A_3].$$

Yet at the level of coarse moduli spaces, these maps induce birational contractions. How can this be? The key point is that although $\overline{\mathcal{M}}_g[A_3^*]$ contains more curves than $\overline{\mathcal{M}}_g[A_2]$ or $\overline{\mathcal{M}}_g[A_3]$, it is only *weakly* modular and many curves which represent separate points of $\overline{M}_g[A_2]$ or $\overline{M}_g[A_3]$ are identified in $\overline{M}_g[A_3^*]$.

Indeed, all curves of the form $C \cup E$, where (C, p_1, p_2) is a fixed smooth 2-pointed curve and (E, q_1, q_2) ranges over all smooth 2-pointed elliptic curves glued along $p_i \sim q_i$, become identified in $\overline{M}_g[A_3^*]$ because they all specialize isotrivially to the curve $C \cup E_0$, where E_0 is the union of two $\mathbb{P}^1$s meeting tacnodally (See Figure 2.4.1). This is the essential reason that ϕ^- induces the stated contractions.

To show that ϕ^- and ϕ^+ are isomorphisms outside the locus of elliptic bridges and tacnodes respectively, one shows that if C is any A_2-stable curve (resp. A_3-stable curve) without elliptic bridges (resp. without tacnodes) then C does not specialize isotrivially to any A_3^*-stable curve, i.e. there is no curve through C that is contracted in the above manner. This requires a careful analysis of the isotrivial degenerations of A_3^*-stable curves, which is carried out in [60]. Finally, the fact that ϕ^- and ϕ^+ are small, i.e. isomorphisms in codimension one, follows immediately from a simple

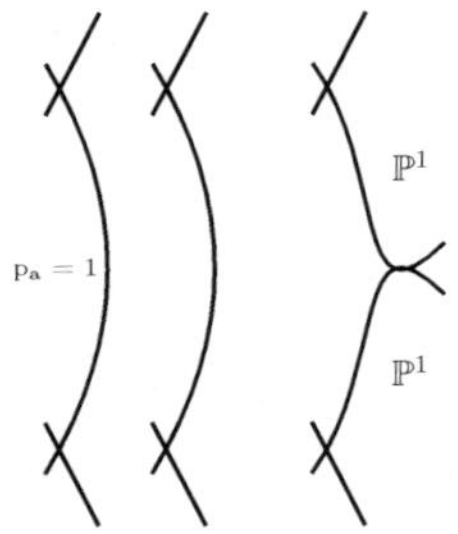

Figure 4. Isotrivial specialization of an elliptic bridge to a maximally degenerate tacnodal curve

dimension count: the locus of elliptic bridges (resp. tacnodal curves) evidently has codimension two (resp. three) in moduli. □

2.5. Looking forward

A comprehensive list of all the modular and weakly modular compactifications we have encountered thus far is displayed in Table 2. Staring at this table, the curious

Table 2. Modular and Weakly Modular Compactifications

Modular Compactifications	Weakly Modular Compactifications
$\overline{M}_{g,n}$, $\overline{M}_{g,\mathcal{A}}$, $\overline{M}_g[A_2]$ $(g \geqslant 3)$,	$\overline{M}_2[A_2]$, $\overline{M}_g[A_2^*]$, $\overline{M}_g[A_3^*]$ $(g \geqslant 3)$,
$\overline{M}_{0,n}[\psi]$, $\overline{M}_{1,n}[m]$, $\overline{M}_4[A_3]$	$\overline{M}_g[A_3]$ $(g = 3, g \geqslant 5)$

reader is bound to wonder: How comprehensive is it? Is it possible to obtain a complete classification of modular and weakly modular birational models of $\overline{M}_{g,n}$ for each g and n? On the one hand, there is certainly no reason to expect this list to be comprehensive. There are many natural classes of singularities (e.g. deformation open classes of ADE singularities, deformation open classes of toric singularities) generalizing the class {nodes, cusps, tacnodes} and we see no obvious obstruction to constructing stability conditions for every such class. On the other hand, judging by the phenomena encountered in the tacnodal case, it seems clear that, in formulating the appropriate stability conditions for these classes of singularities, we will almost certainly have to settle for weakly modular compactifications.

Since the only general method for constructing weakly modular compactifications is GIT, the most straightforward approach for producing new stability conditions is to push the GIT machinery further. On the other hand, there are compactifications of $M_{g,n}$ for which no GIT construction is known ($\overline{M}_{1,n}[m]$ being one example). For this reason, it would be interesting to have a more systematic method of constructing weakly modular birational models of $\overline{M}_{g,n}$, along the lines of what we have sketched in Section 2.3. Indeed, using the ideas of Section 2.3, the second author has come very close to giving a complete classification of modular compactifications of $M_{g,n}$ [107]. Given systematic methods for constructing mildly non-separated functors combinatorially, we might approach such a classification for weakly modular compactifications. The natural starting point for such a program would be the construction of pointed versions of the stability conditions discussed above, i.e. the construction of moduli spaces $\overline{M}_{g,n}[A_2^*]$, $\overline{M}_{g,n}[A_2]$, $\overline{M}_{g,n}[A_3^*]$, $\overline{M}_{g,n}[A_3]$.

Finally, we should discuss the projectivity of these alternate birational models. All the models in the displayed table are projective, though the methods of proof vary. The models $\overline{M}_g[A_2^*]$, $\overline{M}_g[A_2]$, $\overline{M}_g[A_3^*]$, and $\overline{M}_g[A_3]$ are constructed via GIT and therefore automatically come with a polarization. For the models $\overline{M}_{g,\mathcal{A}}$, Kollár's semipositivity provides a direct proof as we will see in Section 3.2.5. On the other hand, for the models $\overline{M}_{1,n}[m]$ the only proofs of projectivity rely on direct intersection theory techniques. In light of these differences, it seems natural to wonder whether Kollár's techniques can be generalized to curves with worse-than-nodal singularities. For example, could they be used to give a direct proof of the projectivity of $\overline{M}_g[A_2]$ or $\overline{M}_{1,n}[m]$? Even more ambitiously, we can ask whether they could be applied in the case of weakly modular compactifications, e.g. to give a direct proof of the projectivity of $\overline{M}_g[A_3]$. Such techniques would be interesting inasmuch as they might be applicable to moduli functors where no GIT construction exists.

3. Birational geometry of moduli spaces of curves

3.1. Birational geometry in a nutshell

The purpose of this section is to recall those definitions from high-dimensional geometry which will be used repeatedly in the sequel. Let X be a normal, projective, $\mathbb{Q}$-factorial variety. We would like to classify the set of all projective birational models of X. Of course, no such classification is feasible if we allow birational models dominating X, so it is reasonable to consider models which are, in some sense, smaller than X. This gives rise to the following definition.

Definition 3.1. *A* birational contraction *is a birational map* $\phi\colon X \dashrightarrow Y$ *such that* Y *is a normal projective variety and* $\mathrm{Exc}(\phi^{-1}) \subset Y$ *has codimension at least* 2. *A birational contraction is* $\mathbb{Q}$-factorial *if* Y *is.*

Two birational contractions $\phi_1\colon X \dashrightarrow Y_1$ and $\phi_2\colon X \dashrightarrow Y_2$ are equivalent if there is an isomorphism $Y_1 \simeq Y_2$ making the obvious diagram commute. Remarkably, the set of all birational contractions up to equivalence, while not necessarily finite, can in some sense be classified purely in terms of the numerical divisor theory of X. To explain this, we recall a few standard definitions from higher dimensional geometry.

If C_1, C_2 are proper curves in the Chow group $A_1(X)$, we say that C_1 and C_2 are numerically equivalent (and write $C_1 \equiv C_2$) if $\mathcal{L}.C_1 = \mathcal{L}.C_2$ for all line bundles $\mathcal{L} \in \mathrm{Pic}(X)$. Dually, we say that two line bundles $\mathcal{L}_1, \mathcal{L}_2 \in \mathrm{Pic}(X)$ are numerically equivalent (and write $\mathcal{L}_1 \equiv \mathcal{L}_2$) if $\mathcal{L}_1.C = \mathcal{L}_2.C$ for all curves $C \subset X$. We set

$$N_1(X) := A_1(X) \otimes \mathbb{Q}/ \equiv$$
$$N^1(X) := \mathrm{Pic}(X) \otimes \mathbb{Q}/ \equiv$$

It follows immediately from the definition, that there is a perfect pairing

$$N^1(X) \times N_1(X) \to \mathbb{Q},$$

induced by the intersection pairing $(\mathcal{L}, C) \to \mathcal{L}.C$, so $N^1(X)$ and $N_1(X)$ are dual. The theorem of the base asserts that $N^1(X)$ and $N_1(X)$ are finite-dimensional vector spaces [86, Proposition 1.1.16].

For understanding the birational geometry of X, two closed convex cones in $N^1(X)$ play a central role, namely the *nef cone* $\mathrm{Nef}(X)$ and the *pseudoeffective cone* $\overline{\mathrm{Eff}}(X)$. The nef cone is simply the closed convex cone in $N^1(X)$ generated by all nef divisor classes[2], while the pseudoeffective cone is the closed convex cone in $N^1(X)$ generated by all effective divisor classes. Note that the cone generated by all effective divisor classes (typically denoted $\mathrm{Eff}(X)$) is, in general, neither open nor closed so the pseudoeffective cone is the closure of $\mathrm{Eff}(X)$. A standard result characterizes the interior of the nef cone as the open cone $\mathrm{Amp}(X)$ generated by classes of ample divisors, and the interior of the pseudoeffective cone as the open cone $\mathrm{Big}(X)$ generated by classes of big divisors [3], i.e. we have $\overline{\mathrm{Amp}(X)} = \mathrm{Nef}(X)$ and $\overline{\mathrm{Big}}(X) = \overline{\mathrm{Eff}}(X)$ [86, Theorems 1.4.23 and 2.2.26]. The reader who is not familiar with these definitions can find several examples of these cones in [86, Sections 1.5 and 2.3].

Now if $\phi\colon X \dashrightarrow Y$ is any $\mathbb{Q}$-factorial birational contraction, there exist open sets $U \subset X$, $V \subset Y$ satisfying

- ϕ restricts to a regular morphism $\phi|_U\colon U \to V$
- $\mathrm{codim}(X - U) \geqslant 2$ and $\mathrm{codim}(Y - V) \geqslant 2$.

Then there are natural homomorphisms $\phi_*\colon N^1(X) \to N^1(Y)$ induced by restricting a codimension one cycle to U, pushing forward to V and then taking the closure in Y. Similarly, there is a pull-back $\phi^*\colon N^1(Y) \to N^1(X)$ induced by restricting a $\mathbb{Q}$-line-bundle on Y to V, pulling back along $\phi|_U$ and then extending to a $\mathbb{Q}$-line-bundle on X. It is an exercise to check that these operations respect numerical equivalence [100, 1.2.12 and 1.2.18].

[2]Recall that a $\mathbb{Q}$-divisor D is *nef* if $D.C \geqslant 0$ for all curves $C \subset X$.

[3]Recall that a $\mathbb{Q}$-divisor D is *big* if $\dim H^0(X, mD) = cm^n + O(m^{n-1})$ for some constant c and all $m \gg 0$ sufficiently divisible.

Definition 3.2. *Let $\phi\colon X \dashrightarrow Y$ be a birational contraction with exceptional divisors $E_1,\dots,E_m$. The* Mori chamber of ϕ *is defined as:*

$$\mathrm{Mor}(\phi) := \{\phi^*D + \sum_{i=1}^{m} a_i E_i \mid a_i \geqslant 0,\ D \in \mathrm{Amp}(Y)\} \subset N^1(X).$$

Lemma 3.3.

(1) $\overline{\mathrm{Mor}(\phi)}$ *is a convex set of dimension* $\dim N^1(Y)$.
(2) $\phi_1 \sim \phi_2$ *iff* $\mathrm{Mor}(\phi_1) \cap \mathrm{Mor}(\phi_2) \neq \varnothing$ *iff* $\mathrm{Mor}(\phi_1) = \mathrm{Mor}(\phi_2)$.
(3) $D \in \mathrm{Mor}(\phi)$ *iff* $R(X,D)$ *is finitely generated and* $Y = \mathrm{Proj}\, R(X,D)$.

Proof. This is standard, see e.g. [64]. □

This lemma suggests an approach to the classification of birational contractions of X. If one can find a set of birational contractions $\phi_j\colon X \dashrightarrow Y_j$ such that the associated Mori chambers partition $\overline{\mathrm{Eff}}(X)$:

$$\overline{\mathrm{Eff}}(X) = \overline{\mathrm{Mor}(\phi_1)} \cup \ldots \cup \overline{\mathrm{Mor}(\phi_k)},$$

then Lemma 3.3 guarantees that these must be all the birational contractions of X and this decomposition of the pseudoeffective cone $\overline{\mathrm{Eff}}(X)$ is called the *Mori chamber decomposition*. If a Mori chamber decomposition exists for the entire pseudoeffective cone $\overline{\mathrm{Eff}}(X)$, we call X a *Mori dream space* [64]. Such a chamber decomposition may fail to exist, however, for two reasons. First, X may have infinitely many birational contractions, in which case the chamber decomposition is not finite. Second, a divisor D may fail to have a finitely generated section ring, in which case it belongs to no Mori chamber [86, Example 2.1.30]. The following powerful result (which will be used repeatedly in the sequel) gives a sufficient condition for basepoint freeness.

Theorem 3.4 (Kawamata basepoint freeness theorem). *Let (X,Δ) be a proper klt pair with Δ effective. Let D be a nef Cartier divisor such that $aD - K_X - \Delta$ is nef and big for some rational $a > 0$. Then $|mD|$ is basepoint free for $m \gg 0$.*

We refer to [81, Section 3] for the definition of klt pairs, the proof of Theorem 3.4, and the history of this result.

While it is too much to hope, in general, for a chamber decomposition of the entire effective cone of a variety, the theory of higher dimensional geometry, culminating in the recent work of Siu [106] and Birkar, Cascini, Hacon, and McKernan [11] shows that there *is* a certain region within the effective cone where it is possible to give a Mori chamber decomposition. For our purposes, the most relevant result is:

Theorem 3.5. *Let $\Delta = \sum_{i\in S} \Delta_i \subset \overline{M}_{g,n}$ be the boundary, where S is the indexing set for irreducible boundary components. For a rational positive number $\alpha \in (0,1]$, let $\Delta(\alpha) \subset N^1(\overline{M}_{g,n})$ be the polytope*

$$\{K_{\overline{M}_{g,n}} + \sum_{i\in S} \alpha_i \Delta_i \mid \alpha_i \in [\alpha,1]\}$$

Then $\Delta(\alpha)$ admits a finite Mori chamber decomposition. More precisely, there exists a finite set of birational contractions $\phi_j : X \dashrightarrow Y_j$ such that

(1) $\overline{\mathrm{Mor}(\phi_i)} \cap [0,1]^{|S|}$ *is a piecewise linear polytope.*
(2) $[\alpha, 1]^{|S|} = \overline{\mathrm{Mor}(\phi_1)} \cup \ldots \cup \overline{\mathrm{Mor}(\phi_k)}$.

Proof. See [11, Corollary 1.2.1]. □

This theorem raises the question: can we compute this chamber decomposition explicitly? Of course, the Hassett-Keel log minimal model program is just the special case, in which one restricts attention to the ray $\alpha_0 = \alpha_1 = \ldots = \alpha_{\lfloor g/2 \rfloor}$. We will address what is known about this problem in Section 4.

3.2. Effective and nef cone of $\overline{M}_{g,n}$

Table 3. Effective and nef cones of $\overline{M}_{g,n}$

	$\overline{M}_g$	$\overline{M}_{0,n}$	$\overline{M}_{g,n}$
Rays of restricted effective cone	$\delta, s\lambda - \delta$, $\frac{60}{g+4} < s \leqslant 6 + \frac{12}{g+1}$	$-\frac{n-1}{\lfloor \frac{n}{2} \rfloor (n - \lfloor \frac{n}{2} \rfloor)} \psi + \delta$, $\frac{n-1}{2(n-2)} \psi - \delta$	unknown
Full effective cone	unknown	Castravet-Tevelev Conjecture	unknown
Rays of restricted nef cone	$\lambda, 11\lambda - \delta$	ψ $(n \geqslant 8)$ and $\frac{2}{3}\psi - \delta$ $(n \geqslant 5)$	$\lambda, \psi, 11\lambda + \frac{2}{3}\psi - \delta$ $(n \geqslant 3)$
Full nef cone	F-conjecture	F-conjecture	F-conjecture
Mori Dream Space?	'Yes' for $g = 2$, unknown for $g \geqslant 3$	'Yes' for $n \leqslant 6$, unknown for $n \geqslant 7$	'No' for $g \geqslant 3$, $n = 1$ and for $g \geqslant 2$, $n \geqslant 2$.

In this section, we describe most of what is known about the nef cone and the effective cone of $\overline{M}_{g,n}$. Since much of this material is well-covered in the existing surveys (e.g. [36], [91]), we will focus on statements of results, emphasizing mainly those details that are necessary for the results of Section 4 on Mori chamber decompositions of $\mathrm{Eff}(\overline{M}_{g,n})$. To begin with, we should emphasize that neither the nef cone nor the effective cone of $\overline{M}_{g,n}$ is known in full generality. When seeking partial results, it is common to focus on the *restricted effective cone* $\mathrm{Eff}_r(\overline{M}_{g,n})$ (respectively, the *restricted nef cone* $\mathrm{Nef}_r(\overline{M}_{g,n})$), which we define to be the intersection of the effective cone (respectively, the nef cone) with the subspace $\mathbb{Q}\{\lambda, \psi, \delta\} \subset N^1(\overline{M}_{g,n})$. Note that in the special cases $n = 0$ (resp. $g = 0$), we have $\psi = 0$ (resp. $\lambda = 0$), so these restricted cones are simply two-dimensional convex cones in a two-dimensional vector space. Our present state of knowledge concerning all these cones is displayed in Table 3.

Before elaborating on the individual entries in Table 3, let us briefly recount the historical progression of ideas which preceded and motivated this work. Long before the concepts and vocabulary of Mori theory became the standard approach to birational geometry, the problem of finding rational or unirational parametrizations for M_g vexed (and inspired) many mathematicians. Perhaps the first recognizable foray into the birational geometry of M_g was made in the context of invariant theory. For example, the fact that every genus two curve admits a unique double cover of $\mathbb{P}^1$ branched at 6 points implies that M_2 is naturally birational to the orbit space of binary sextic forms under $SL(2)$. The generators and relations for the corresponding ring of invariants were computed by Clebsch in 1872 [24], who thus gave explicit coordinates on M_2 (analogous to the Weierstrass j-function for M_1) and proved the rationality of M_2 as a by-product (see also [67]). Similar results for M_g, $g \leqslant 6$, have been obtained only recently, with Shepherd-Barron proving rationality of M_4 and M_6 [104, 105] and Katsylo proving rationality of M_3 and M_5 [73, 71].

While approaches to the rationality of M_g remain essentially algebraic, Severi realized that the weaker problem of proving unirationality (i.e. of exhibiting merely a dominant rational, rather than birational, map $\mathbb{P}^N \dashrightarrow M_g$) was accessible by geometric methods [103]. Severi used the variety $V_{d,g}$ parameterizing plane curves of degree d and geometric genus g (now called *Severi varieties*) to prove the unirationality of M_g for $g \leqslant 10$. Severi's idea was to represent a general curve of genus g as a plane curve of degree $d(g) = \lceil \frac{2g+6}{3} \rceil$ with exactly $\delta(g) = \binom{d-1}{2} - g$ nodes. For genus $g \leqslant 10$, a simple dimension count shows the existence of plane curves of degree $d(g)$ with $\delta(g)$ nodes in general position in $\mathbb{P}^2$. Thus, $V_{d,g}$ is birational to a projective bundle over $(\mathbb{P}^2)^{\delta(g)}$, hence rational. The natural dominant rational map $V_{d,g} \dashrightarrow M_g$ then proves the unirationality of M_g. As explained in [7], Severi's argument made implicit use of several unproven assumptions about $V_{d,g}$ – Arbarello and Sernesi proceed to fill the gaps in Severi's proof thus giving a rigorous proof of unirationality of $\overline{M}_g$ for $g \leqslant 10$.

The question of whether M_g remains unirational for all $g \geqslant 10$ was open until the pioneering work of Mumford and Harris [55], who showed that $\overline{M}_g$ is of general type for odd $g \geqslant 25$ and has Kodaira dimension at least zero for $g = 23$. This work was later generalized by Eisenbud and Harris to show that $\overline{M}_g$ is of general type for *all* $g \geqslant 24$ [32]. In particular, M_g is not unirational for any $g \geqslant 23$. From a historical perspective, this work is a marvelous reflection (and application) of the emerging understanding of the importance of Kodaira dimension in the study of birational classification problems. Note that since Kodaira dimension is meaningful only for projective varieties, this perspective could not be brought to bear on the unirationality problem without the existence of a suitable projective compactification $M_g \subset \overline{M}_g$. The moduli space of stable curves, constructed less than 15 years previously by Deligne and Mumford [27], provided the requisite compactification.

Now, let us briefly outline the argument of Harris and Mumford. The two key ingredients are:

(1) $\overline{M}_g$ has canonical singularities for all $g \geqslant 2$.
(2) For odd $g \geqslant 25$, the canonical divisor $K_{\overline{M}_g}$ is big.

(2) says that the canonical divisor has lots of sections, and (1) says that these sections extend to an arbitrary desingularization of $\overline{M}_g$. Together, they imply that $\overline{M}_g$ is of general type for odd $g \geqslant 25$.

The proof of (1) is a straight-forward (albeit arduous) application of the Reid-Tai criterion (cf. [99, 110]), which gives an explicit method for determining whether a finite quotient singularity is canonical. To explain the proof of (2), it is useful to make a preliminary definition: If $D \subset \overline{M}_g$ is any Weil divisor, we define the *slope of* D to be

$$s(D) := \inf \left\{ \frac{a}{b} \ : \ a\lambda - b\delta - D \geqslant 0 \right\}. \tag{3.6}$$

For any irreducible non-boundary effective divisor $D \subset \overline{M}_g$, we have

$$D \equiv a\lambda - \sum_{i=0}^{\lfloor g/2 \rfloor} b_i \delta_i \in N^1(\overline{M}_g),$$

where a and b_i are positive rational numbers. Consequently,

$$s(D) = \frac{a}{\min_{0 \leqslant i \leqslant \lfloor g/2 \rfloor} b_i}.$$

The proof of (2) is connected with the problem of understanding slopes of effective divisors by the following two observations: First, λ is a big divisor (indeed, it is semiample and defines the Torelli morphism $\tau \colon \overline{M}_g \to \overline{\mathcal{A}}_g$ to the Satake compactification of principally polarized abelian varieties [97]). Second, since $K_{\overline{\mathcal{M}}_g} = 13\lambda - 2\delta$ and $\overline{\mathcal{M}}_g \to \overline{M}_g$ is ramified along Δ_1 (see [55, Section 2]), the canonical class of the coarse moduli space is given by the formula

$$K_{\overline{M}_g} = 13\lambda - 2\delta_0 - 3\delta_1 - 2 \sum_{i=2}^{\lfloor g/2 \rfloor} \delta_i.$$

It follows that if $D \subset \overline{M}_g$ is any effective divisor of slope $s(D) < 13/2$, we have a numerical proportionality

$$K_{\overline{M}_g} \equiv D + a\lambda + \sum_i b_i \delta_i,$$

where a and b_i are positive rational constants. In particular, since λ is big and $D + \sum_i b_i \delta_i$ is effective, $K_{\overline{M}_g}$ is big. In other words, to show that $K_{\overline{M}_g}$ is big, it is sufficient to find an effective divisor of slope less than 13/2.

The effective divisor $D \subset \overline{M}_g$ originally considered by Mumford and Harris was the closure of the locus of d-gonal curves where $d = \frac{g+1}{2}$. Subsequently, Eisenbud and Harris considered the more general *Brill-Noether divisors* $D_{r,d} \subset \overline{M}_g$, defined as the closure of the locus of curves possessing linear series g^r_d satisfying $g - (r+1)(g + r - d) = -1$ (as well as the so-called Gieseker-Petri divisors whose definition we omit). In each case, the key technical difficulty is the computation of the numerical class of these divisors, i.e. the coefficients a and $\{b_i\}_{i=0}^{\lfloor g/2 \rfloor}$ such that $D_{r,d} \equiv a\lambda + \sum_{i=0}^{\lfloor g/2 \rfloor} b_i \delta_i$. One can extract these coefficients by the method of test curves, provided one knows which singular curves in $\overline{M}_g$ actually lie in D_d or $D_{r,d}$. To answer this question, Harris and Mumford developed the theory of admissible covers [55], and Eisenbud and Harris developed the theory of limit linear series [31]. The final result of this computation is a numerical proportionality

$$D_{r,d} \sim \left(6 + \frac{12}{g+1}\right)\lambda - \delta_0 - \sum_{i=1}^{\lfloor g/2 \rfloor} a_i\delta_i, \quad a_i \geqslant 0.$$

In particular, for $g \geqslant 24$, $s(D_{r,d}) < 13/2$ so $\overline{M}_g$ is of general type as desired.

The discovery that $\overline{M}_g$ is of general type for $g \geqslant 24$ spurred the interest to the intermediate cases of $11 \leqslant g \leqslant 23$. The $g = 12$ had actually been settled by Sernesi prior to the work of Harris and Mumford [102], ' but Chang and Ran used a more general method to establish the unirationality of $\overline{M}_g$ for $g = 11, 12, 13$ [18]. In addition, they showed that the slopes of $\overline{M}_{15}$ and $\overline{M}_{16}$ are greater than $6\frac{1}{2}$, thus proving that $\overline{M}_{15}$ and $\overline{M}_{16}$ have negative Kodaira dimension [19, 20]. It follows that these spaces are uniruled [13], and Bruno and Verra proved the stronger result that $\overline{M}_{15}$ is rationally connected [14]. The unirationality question has been recently revisited by Verra who proved that $\overline{M}_{14}$ is unirational [113]. As we discuss below, Farkas has shown that $\overline{M}_{22}$ is of general type [35]. As far as we know, the Kodaira dimension of $\overline{M}_g$ is completely unknown for $17 \leqslant g \leqslant 21$.

With this historical background in place, let us return to Table 3 and explain each of the rows in greater detail.

3.2.1. Effective cones The work of Eisenbud, Harris, and Mumford gave rise to an intensive study of the restricted effective of $\overline{M}_g$. Since the boundary δ obviously lies on the edge of the effective cone for $g \geqslant 3$, the key question is: what is the *slope*

$$s_g := \inf\{s(D) \ : \ D \subset \overline{M}_g \text{ irreducible, not contained in the boundary}\}$$

i.e. the slope of lower boundary of the effective cone $\mathrm{Eff}(\overline{M}_g)$. For a long time, the Brill-Noether divisors $D_{r,d}$ were the divisors of the smallest known slope. This led Harris and Morrison to formulate the *Slope Conjecture*, namely that *every effective divisor has slope at least* $6 + \frac{12}{g+1}$ [53]. The obvious way to test the conjecture is by computing the slopes of various effective divisors in $\overline{M}_g$, and ultimately Farkas and Popa produced a counterexample by doing just that [40]. Their counterexample is given by the divisor $K_{10} \subset \overline{M}_{10}$ of curves isomorphic to a hyperplane section of a K3 surface, first introduced by Cukierman and Ulmer [26]. Farkas and Popa computed the class of K_{10} and showed that $s(K_{10}) = 7 < 6 + \frac{12}{11}$. Subsequently, more counterexamples to the Slope Conjecture were produced by Farkas in [34, 37] and Khosla [78, 79]. In particular, in [37] it is shown that all known effective divisors of small slope (Brill-Noether, Gieseker-Petri, K_{10}, Khosla's divisors) can be obtained from a single syzygy-theoretic construction. These so-called *Koszul divisors* provide an infinite sequence of counterexamples to the Slope Conjecture. Furthermore, when $g = 22$, there is a Koszul divisor of slope $17121/2636 < 6\frac{1}{2}$ making $\overline{M}_{22}$ a variety of general type [35, Theorem 7.1]. Curiously, all Koszul divisors have slopes no smaller than $6 + \frac{10}{g}$ which leaves open the possibility of a positive genus-independent lower bound for the slopes s_g. As far as we know, no one has constructed an effective divisor of slope less than 6.

In a different direction, there have been several attempts to produce lower bounds for the slope via the method of moving curves. Here, the key idea is that if C is a numerical curve class whose deformations cover $\overline{M}_g$ (such curve classes are called *moving*), then evidently $D.C \geqslant 0$ for any effective divisor $D = a\lambda - \sum_{i=0}^{\lfloor g/2 \rfloor} b_i\delta_i$. In particular,

$$s(D) = \inf_{i=0}^{\lfloor g/2 \rfloor} a/b_i \geqslant \delta.C/\lambda.C.$$

Thus, if we define the *slope* of a curve C to be the ratio $\delta.C/\lambda.C$, then the slope of any moving curve class gives a lower bound for the slope s_g. Harris and Morrison exploited this idea by using admissible covers to compute the slopes of one-parameter families of k-gonal curves [53]. When $k \geqslant (g+1)/2$, such families define moving curves in $\overline{M}_g$ and using this the authors establish the Slope Conjecture for $g \leqslant 5$, and give an asymptotic lower bound $s_g \geqslant \frac{576}{5g}$ for $g \gg 0$. The idea of using families of admissible covers to obtain lower bounds on s_g was recently revisited by Dawei Chen who writes down infinite Zariski dense families of branched covers of an elliptic curve in [22, 23] and branched covers of $\mathbb{P}^1$ in [21]. Somewhat surprisingly, the resulting asymptotic lower bound on s_g in [23] turns out to be $\frac{576}{5g}$ as well, which led Chen to conjecture that $s_g \sim \frac{576}{5g}$ for $g \to \infty$.

There are many other approaches to producing moving curves on $\overline{M}_g$. For example, taking linear sections of the Severi variety $V_{d,g}$ gives moving curves on $\overline{M}_g$ when $d \geqslant \lfloor \frac{2g+8}{3} \rfloor$. The corresponding curve classes were computed in the first author's thesis [42], and yield an asymptotic bound of the form $o(1/g)$. In a different vein, if $\pi\colon \overline{M}_{g,1} \to \overline{M}_g$ is the universal curve, the fact that ψ_1 is the limit of ample divisors implies that $\pi_*(\psi_1^{3g-4})$ is the limit of moving curve classes on $\overline{M}_g$. Pandharipande [98] computed the slope of this curve class $\pi_*(\psi_1^{3g-4})$ to be $\frac{60}{g+4}$ using Hodge integrals, thus giving a lower bound on the slope for all g. Alas, there is a large gap between $6 + \frac{10}{g}$ and $\frac{60}{g+4}$, especially when g is large, and the precise boundary of the (restricted) effective cone remains unknown.

The study of the effective cone of $\overline{M}_{0,n}$ has a much different flavor; since $\lambda = 0$ on $\overline{M}_{0,n}$, the question simply becomes: what linear combinations of boundary divisors are effective? Since every boundary divisor of $\overline{M}_{0,n}$ clearly lies on the boundary of the effective cone, it is natural to hope that the effective cone is simply spanned by the boundary divisors. However, Keel and Vermeire independently produced divisors in $\overline{M}_{0,6}$ which are not effective sums of boundary divisors [111]. Hassett and Tschinkel proved that Keel-Vermeire divisors together with the boundary divisors generate $\mathrm{Eff}(\overline{M}_{0,6})$ [63]. Recently, Castravet and Tevelev [16] have constructed a plethora of new examples of effective divisors which are not effective sums of boundary divisors – their so-called *hypertree divisors*. They have conjectured that the hypertree divisors, together with the boundary divisors, should span the entire effective cone.

As for the restricted effective cone of $\overline{M}_{0,n}$, it is an easy consequence of a theorem of Keel and McKernan, who showed that every effective divisor on the symmetrized moduli space $\overline{M}_{0,n}/S_n$ *is* numerically equivalent to an effective sum of S_n-invariant boundary divisors [75, Theorem 1.3]. Since

$$\psi = \sum_{i=1}^{\lfloor n/2 \rfloor} \frac{i(n-i)}{n-1} \Delta_i$$

is an S_n-invariant sum of boundary divisors, it follows that any divisor of the form $a\psi + b\delta$ is effective iff the coefficients of Δ_i's in the sum

$$\sum_{i=1}^{\lfloor n/2 \rfloor} \left(a + b\frac{i(n-i)}{n-1} \right) \Delta_i$$

are positive. From this, one sees that the restricted effective cone of $\overline{M}_{0,n}$ is spanned by the two rays indicated in Table 3.

Finally, while Logan [88] and Farkas [37] have extended the work of Eisenbud, Harris, and Mumford to $\overline{M}_{g,n}$ (see [37, Theorem 1.10] for a comprehensive list of (g, n) where $\overline{M}_{g,n}$ is known to be of general type), no analogues of the slope conjecture have been offered or investigated for $\overline{M}_{g,n}$. Needless to say, there is no reason to expect the restricted effective cone of $\overline{M}_{g,n}$ to be easier to study than that of $\overline{M}_g$.

3.2.2. Nef cones Next, we consider the nef cone of $\overline{M}_g$. The Hodge class λ is well-known to be nef on $\overline{M}_g$; indeed it is the pullback of the canonical polarization on $\overline{\mathcal{A}}_g$ via the extended Torelli morphism $\tau\colon \overline{M}_g \to \overline{\mathcal{A}}_g$. The nefness of $11\lambda - \delta$ is the result of Cornalba and Harris [25], who exploit the fact that the canonical embedding of a non-hyperelliptic smooth curve of genus g is asymptotically Hilbert stable to deduce that the divisor $\left(8 + \frac{4}{g}\right)\lambda - \delta$ is positive on families with smooth non-hyperelliptic generic fibers. Combining this with a separate analysis of the hyperelliptic and nodal case gives the following.

Theorem 3.7 ([25, Theorem 1.3]). *The divisor $s\lambda - \delta$ is ample for $s > 11$. Moreover, $11\lambda - \delta$ has degree 0 precisely on curves whose only moving components are elliptic tails.*

In fact, Cornalba and Harris proved a stronger result regarding generically smooth families of stable curves. We do not state it here since it has been since sharpened by Moriwaki [90] as follows.

Proposition 3.8. *The divisor*

$$(8g+4)\lambda - g\delta_0 - \sum_{i=1}^{\lfloor g/2 \rfloor} 4i(g-i)\delta_i$$

has non-negative degree on every complete one-parameter family of generically smooth curves of genus $g \geqslant 2$.

The full nef cone of $\overline{M}_g$ is not known, but there is a fascinating conjectural description – the so-called F-conjecture – which would in principle determine $\mathrm{Nef}(\overline{M}_{g,n})$ for all g and n. Furthermore, the F-conjecture, which is described in greater detail in Section 3.2.4 below, would imply that the nef cone of $\overline{M}_{g,n}$ is finite polyhedral for all g and n, which is surprising from a purely Mori-theoretic point of view. Nevertheless, two recent results comprise substantial evidence for the F-conjecture: First, Gibney, Keel, and Morrison have shown that if the F-conjecture holds for $\overline{M}_{0,g+n}/S_g$, then it holds for $\overline{M}_{g,n}$ [50]. In particular, it is sufficient to prove the conjecture for $\overline{M}_{0,n}$ for all n, where it appears somewhat more plausible. Second, Gibney has developed efficient computational methods for checking the conjecture, and has proved that it holds for $\overline{M}_g$ up to $g = 24$ [49].

3.2.3. Mori dream spaces Finally, we briefly address the question of whether $\overline{M}_{g,n}$ is a Mori dream space. Keel [74, Corollary 3.1] showed that the divisor class ψ on $\overline{M}_{g,1}$ is big and nef but not semiample (in characteristic zero) for $g \geqslant 3$. The same argument shows that ψ is big and nef but not semiample for $(g, n) \geqslant (2,2)$. Hence $\overline{M}_{g,n}$ cannot be a Mori dream space if $g \geqslant 3$ and $n = 1$ or $g \geqslant 2$ and $n \geqslant 2$. This leaves the cases of $\overline{M}_g$, $\overline{M}_{0,n}$, $\overline{M}_{1,n}$, $\overline{M}_{2,2}$. It is easy to see by a direct computation that $\overline{M}_2$ and $\overline{M}_{0,5}$ are Mori dream spaces (see Sections 4.1.1 and 4.2). A direct proof that $\overline{M}_{0,6}$ is a Mori dream space was given by Castravet in [15]. That $\overline{M}_{0,n}$ is a Mori dream space for $n \leqslant 6$ also follows from the fact that it is log Fano and from the results of [11]. The remaining cases are, so far as we know, open.

Obviously, a full treatment of the results mentioned in the preceding paragraphs would require a survey in itself, so we confine ourselves here to just a few highlights: In Section 3.2.4, we shall describe the F-conjecture in greater detail and explain the partial results which are used in the log MMP for $\overline{M}_g$. In Section 3.2.5, we shall sketch a simplified version of the Kollár's method of producing nef divisors on $\overline{M}_{g,\mathcal{A}}$. These methods do not reproduce the sharp results of Cornalba-Harris (Theorem 3.7) and Moriwaki (Proposition 3.8), but they do give a feel for the basic ideas involved. Furthermore, the resulting nef divisors allow us to interpret $\overline{M}_{g,\mathcal{A}}$ as log canonical models of $\overline{M}_{g,n}$ and they also play a key role in the log MMP for $\overline{M}_{0,n}$.

3.2.4. F-conjecture Recall that $\overline{M}_{g,n}$ admits a natural stratification by topological type, in which the locus of curves with m nodes form the strata of codimension m. Thus, the closure of the locus of curves with $3g - 4 + n$ nodes comprises a collection of irreducible curves on $\overline{M}_{g,n}$, the so-called *F-curves*. The F-conjecture simply asserts

Conjecture 3.9 (F-conjecture). *A divisor* D *on* $\overline{M}_{g,n}$ *is nef if and only if* D *is F-nef, i.e.* D *intersects every F-curve non-negatively.*

There are finitely many F-curves, and it is straightforward to compute their intersection numbers with the divisor classes of $\overline{M}_{g,n}$. Let us give some explicit examples of F-curves that will be relevant in Section 4.

Example 3.10 (Elliptic tails and elliptic bridges). We define the F-curve of elliptic tails, denoted T_1, by taking a pencil of plane cubics with a marked base point section – a degree 12 cover of the universal family over $\overline{M}_{1,1}$ – and attaching a fixed maximally degenerate genus $g-1$ curve (i.e. a genus $g-1$ curve with $3g-7$ nodes) along the section. One easily computes the intersection numbers:

$$\begin{aligned} \lambda \cdot T_1 &= 1, & \delta_0 \cdot T_1 &= 12, \\ \delta_1 \cdot T_1 &= -1, & \delta_i \cdot T_1 &= 0,\ i \neq 0,1. \end{aligned} \tag{3.11}$$

We define the F-curve of separating elliptic bridges (resp. non-separating elliptic bridges), denoted EB_i^s (resp. EB^{ns}) as follows: Take a 4-pointed rational curve $(\mathbb{P}^1, p_1, p_2, p_3, p_4)$ and two 1-pointed maximally degenerate curves (E_1, q_1) of genus i and (E_2, q_2) of genus $g-i-1$ (resp. a 2-pointed maximally degenerate curve (C, q_1, q_2) of genus $g-2$). Identify $p_1 \sim q_1$, $p_4 \sim q_2$ and $p_2 \sim p_3$, and consider the family obtained by varying the 4 points on $\mathbb{P}^1$.

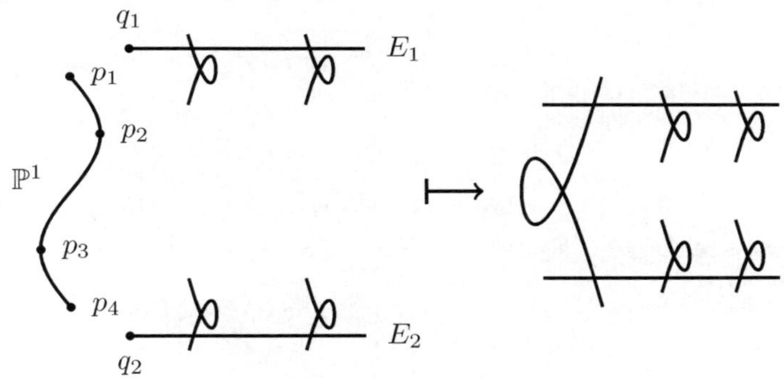

Figure 5. Separating elliptic bridge

The non-trivial intersection numbers of these F-curves are easily computed as

$$\lambda \cdot EB_i^s = 0, \quad \delta_1 \cdot EB_i^s = 1, \quad \delta_i \cdot EB_i^s = -1, \quad \delta_{g-i-1} \cdot EB_i^s = -1$$

in the separating case, and

$$\lambda \cdot EB^{ns} = 0, \quad \delta_1 \cdot EB^{ns} = 1, \quad \delta_0 \cdot EB^{ns} = -2,$$

in the non-separating case.

Finally, a general *family of elliptic bridges* is obtained by taking a one-parameter family of 2-pointed genus 1 curves in $\overline{\mathcal{M}}_{1,2}$ and attaching constant curves to obtain a family of genus g curves. Because of the relation $\lambda = \delta_0/12$ and $\psi = 2\lambda + 2\delta_{0,\{1,2\}}$ in $\operatorname{Pic}(\overline{\mathcal{M}}_{1,2})$, the curve class of every one-parameter family of elliptic bridges is an effective linear combination of classes T_1 and EB_i^s if the bridge is separating, and classes T_1 and EB^{ns} if the bridge is non-separating.

Continuing as in the example, one may assemble a list of the intersection numbers of F-curves. Thus, the F-conjecture may be formulated as saying that a divisor $D = a\lambda - \sum_{i=0}^{\lfloor g/2 \rfloor} b_i\delta_i$ is nef if a certain finite set of linear inequalities holds (see [50, Theorem 2.1] for a comprehensive list). In particular, the F-conjecture would imply the cone is finite polyhedral.

As we remarked above, perhaps the most striking evidence in favour of the F-conjecture is the theorem of Gibney, Keel and Morrison which says that if the F-conjecture holds for $\overline{M}_{0,g+n}/S_g$, then it holds for $\overline{M}_{g,n}$. Their argument proceeds by considering the map

$$i\colon \overline{M}_{0,g+n}/S_g \to \overline{M}_{g,n},$$

obtained by attaching fixed genus 1 curves onto the first g (unordered) marked points. They then prove [50, Theorem 0.3]:

Theorem 3.12. *A divisor* D *is nef on* $\overline{M}_{g,n}$ *if and only if* D *is F-nef and* i^*D *is nef on* $\overline{M}_{0,g+n}$.

From this, they obtain two corollaries.

Corollary 3.13. *If the F-conjecture holds for all* $\overline{M}_{0,n}$, *then it holds for all* $\overline{M}_{g,n}$.

Proof. If D is F-nef on $\overline{M}_{g,n}$, then i^*D is F-nef on $\overline{M}_{0,g+n}$. If the F-conjecture holds for $\overline{M}_{0,n}$, then i^*D is nef so the theorem implies D is nef. □

We pause here to remark that the F-conjecture for $\overline{M}_{0,n}$ is a special case of a conjecture of Fulton stating that any effective k-cycle on $\overline{M}_{0,n}$ is an *effective* combination of k-dimensional strata of curves with $n-3-k$ nodes. As we have discussed above, the divisors constructed by Keel and Vermeire show that the conjecture is false for $k = n-4$. However, there is a similar statement, dubbed *Fulton's conjecture,* which if true would imply Conjecture 3.9.

Conjecture 3.14 (Fulton's conjecture). *Every F-nef divisor on* $\overline{M}_{0,n}$ *is an effective combination of boundary divisors.*

This conjecture was verified for $n = 5, 6$ by Farkas and Gibney [38] and for $n = 7$ by Larsen [84]. In the special case of S_m-invariant ($m \geqslant n-3$) divisors it was proved by Fontanari [48].

We now return to Theorem 3.12 and deduce a result that enables first several steps of the log minimal model program of $\overline{M}_g$.

Corollary 3.15 ([50, Proposition 6.1]). *Suppose that a divisor* $D = a\lambda - \sum_{i=0}^{\lfloor g/2 \rfloor} b_i\delta_i$ *is F-nef on* $\overline{M}_g$, *and for each* $i \geqslant 1$, *either* $b_i = 0$ *or* $b_i \geqslant b_0$, *then* D *is nef.*

Proof. Consider the natural morphism $f\colon \overline{M}_{0,2g} \to \overline{M}_g$ defined by identifying g pairs of points. The image of f includes the locus of genus g curves obtained from $\overline{M}_{0,g}$ by gluing on a fixed *nodal* elliptic curve. Hence by Theorem 3.12 it suffices to show that f^*D is nef. Suppose that $b_i \geqslant b_0$, for all $i \geqslant 1$. Then the divisor f^*D is of the form $K_{\overline{M}_{0,2g}} + \sum_i a_i\Delta_i$ with $a_i \in [0,1]$. Applying [75, Theorem 1.2(2)], we conclude that f^*D is nef. We refer to [50] for details of the proof in the case $b_i = 0$ for some i. □

This result forms the main input into the following Proposition, which will be used in Section 4.

Proposition 3.16. *We have that in* $N^1(\overline{M}_g)$:

(1) *The divisor* $11\lambda - \delta$ *is nef and has degree* 0 *precisely on the families of elliptic tails, i.e. the F-curve* T_1 *of Example 3.10.*
(2) *The divisor* $10\lambda - \delta - \delta_1$ *is nef and has degree* 0 *precisely on the families of elliptic bridges. These are effective linear combinations of F-curves* T_1, EB^{ns}, *and* EB_i^s *by Example 3.10.*

Proof. These divisors satisfy the assumptions of Corollary 3.15, hence are nef. Note also that (1) follows from Theorem 3.7. We proceed to prove (2). One easily checks that the listed are the only F-curves on which $10\lambda - \delta - \delta_1$ is zero. It remains to show that for any curve $B \subset \overline{M}_g$, the only moving components of the stable family over B on which $10\lambda - \delta - \delta_1$ has degree zero are elliptic tails or elliptic bridges.

Let $\mathcal{X} \to B$ be a moving component with a smooth generic fiber of genus h and n sections. The divisor $10\lambda - \delta - \delta_1$ restricts to the divisor

$$D := 10\lambda - \delta + \psi - \delta_{1,\varnothing} + \psi'$$

on $\overline{\mathcal{M}}_{h,n}$, where ψ' corresponds to attaching sections of elliptic tails. If $h \geqslant 2$, we apply to Exercise 3.17 below to observe that the degree of D is greater than the degree of $10\lambda - \delta_0 - \delta_1$ on an unpointed family of stable curves of genus h. It then follows from Proposition 3.8 that $(10\lambda - \delta_0 - \delta_1) \cdot B \geqslant 0$, and we are done. If $h = 0$, then $\delta_{1,\varnothing} = 0$ and so D is positive on B because it is a sum of $\psi - \delta$ and ψ' which are, respectively, ample and nef on $\overline{M}_{0,n}$. Finally, suppose that $h = 1$. By (3.11), $10\lambda - \delta - \delta_1$ has degree 0 on a family of elliptic tails. Suppose now $n \geqslant 2$, then, as before, $10\lambda - \delta - \delta_1$ restricts to $D = 10\lambda - \delta + \psi - \delta_{1,\varnothing} + \psi'$, where $\psi' \cdot B \geqslant 0$. Using the following relations in $\mathrm{Pic}(\overline{\mathcal{M}}_{1,n})$ [6, Theorem 2.2]

$$\lambda = \delta_0/12, \qquad \psi = n\lambda + \sum_S |S|\delta_{0,S},$$

and noting that $\delta_{1,\varnothing} = \delta_{0,\{1,\dots,n\}}$, we rewrite

$$10\lambda - \delta + \psi - \delta_{1,\varnothing} = (n-2)\lambda + \sum_{S: 2 \leqslant |S| < n} (|S|-1)\delta_{0,S} + (n-2)\delta_{0,\{1,\dots,n\}}.$$

Evidently this can be zero only if $n = 2$, i.e. when $\mathcal{X} \to B$ is a family of elliptic bridges. □

Exercise 3.17. Suppose $B \subset \overline{\mathcal{M}}_{g,n}$ is a one-parameter family of generically smooth n-pointed ($n \geqslant 1$) curves of genus $g \geqslant 2$, and let B' be the family in $\overline{\mathcal{M}}_g$ obtained by forgetting marked points and stabilizing. Then

$$(10\lambda - \delta - \delta_{1,\varnothing} + \psi) \cdot B > (10\lambda - \delta - \delta_1) \cdot B'.$$

For later reference, we record the loci swept out by curves numerically equivalent to an effective combination of the F-curves T_1, EB^{ns}, and EB^{s}_{i}.

Corollary 3.18.

(1) *A curve $C \subset \overline{M}_g$ is numerically equivalent to T_1 iff every moving component of C is an elliptic tail.*

(2) *A curve $C \subset \overline{M}_g$ is an effective linear combination of T_1, EB^{ns}, EB^{s}_{i} iff every moving component of C is an elliptic tail or an elliptic bridge.*

Proof. This is a restatement of Proposition 3.16. □

3.2.5. Kollár's Semipositivity and nef divisors on $\overline{M}_{g,\mathcal{A}}$ Here, we explain a simple method for producing nef and ample divisors on spaces $\overline{M}_{g,\mathcal{A}}$ introduced in Definition 2.13, which will be essential for describing the Mori chambers of $\overline{M}_{g,n}$ corresponding to these models. The main idea, due to Kollár [80], is to exploit the positivity of the canonical polarization. More concretely, if $(f\colon \mathcal{C} \to B, \{\sigma_i\}_{i=1}^{n})$ is any one-parameter family of $\mathcal{A}$-stable curves, Kollár deduces that $\omega_{\mathcal{C}/B}(\sum a_i\sigma_i)$ is nef from the semipositivity of $f_*((\omega_{\mathcal{C}/B}(\sum a_i\sigma_i))^{\ell})$, where $\ell \geqslant 2$ is such that $\ell a_i \in \mathbb{Z}$; see [80, Corollary 4.6 and Proposition 4.7]. By intersecting with other curve classes on $\mathcal{C}$ and pushing forward, one gets a variety of nef divisor classes on $\overline{M}_{g,n}$, as we shall see in Corollary 3.20.

Proposition 3.19. *Let $\pi\colon (\mathcal{C}; \sigma_1, \dots, \sigma_n) \to \overline{\mathcal{M}}_{g,\mathcal{A}}$ be the universal family. Then the line bundle $\omega_\pi(\sum_{i=1}^{n} a_i\sigma_i)$ is nef on $\mathcal{C}$.*

Proof. See [45] for a direct, elementary proof in the spirit of [74, Theorem 0.4]. □

Using Proposition 3.19, we can describe several nef divisor classes on $\overline{M}_{g,\mathcal{A}}$. First, however, we need a bit of notation. For each i, j such that $a_i + a_j \leqslant 1$, there is an irreducible boundary divisor $\Delta_{i,j} \subset \overline{M}_{g,\mathcal{A}}$ parameterizing curves where the marked point p_i coincides with the marked point p_j. Indeed, if $\pi\colon \mathcal{C} \to \overline{M}_{g,\mathcal{A}}$ is the universal curve with universal sections $\{\sigma_i\}_{i=1}^{n}$, then $\Delta_{i,j} := \pi_*(\sigma_i \cap \sigma_j)$. The remaining boundary divisors of $\overline{\mathcal{M}}_{g,\mathcal{A}}$ parameterize nodal curves and for this reason we denote the total class of such divisors by δ_{nodal}. Note that Mumford's relation gives $\kappa := \omega_\pi^2 = 12\lambda - \delta_{nodal}$ [6].

Corollary 3.20. *The following divisors*

$$A = A(a_1, \dots, a_n) = 12\lambda - \delta_{nodal} + \psi + \sum_{i<j} (a_i + a_j)\Delta_{ij},$$

$$B = B(a_1, \dots, a_n) = 12\lambda - \delta_{nodal} + \sum (2a_i - a_i^2)\psi_i + \sum_{i<j} (2a_i a_j)\Delta_{ij},$$

$$C = C(a_1, \dots, a_n) = \sum (1 - a_i)\psi_i + \sum_{i<j} (a_i + a_j)\Delta_{ij},$$

are nef on $\overline{\mathcal{M}}_{g,\mathcal{A}}$. Moreover, the divisor A is ample.

Proof. Let $f\colon \mathcal{X} \to T$ be a family over a complete smooth curve. The divisor $L := \omega_{\mathcal{X}/T} + \sum a_i \sigma_i$ is nef by Proposition 3.19, hence pseudoeffective and has a non-negative self-intersection. We will show that the intersection numbers of T with A, B and C are non-negative by expressing each of them as an intersection of L with an effective curve class on $\mathcal{X}$. For A, we note that $\omega + \sum_{i=1}^n \sigma_i$ is an effective combination of L and σ_i, $1 \leqslant i \leqslant n$. Therefore,

$$0 \leqslant (\omega + \sum_{i=1}^{n} \sigma_i) \cdot L = \kappa + \psi + \sum_{i<j} (a_i + a_j)\Delta_{ij}.$$

For B, we have

$$0 \leqslant L^2 = (\omega + \sum_{i=1}^{n} a_i\sigma_i)^2 = \kappa + \sum_{i=1}^{n} (2a_i - a_i^2)\psi_i + \sum_{i<j} (2a_i a_j)\Delta_{ij}.$$

For C, we have

$$0 \leqslant L \cdot \sum_{i=1}^{n} \sigma_i = \sum (1 - a_i)\psi_i + \sum_{i<j} (a_j + a_j)\Delta_{ij}.$$

For the proof of ampleness in the case of A, we refer to [45], where it is established using Kleiman's criterion on $\overline{M}_{g,\mathcal{A}}$. □

Now we can describe all $\overline{\mathcal{M}}_{g,\mathcal{A}}$ as log canonical models of $\overline{\mathcal{M}}_{g,n}$. While these models do not (except in the case $g = 0$) appear in the Hassett-Keel minimal model program for $\overline{\mathcal{M}}_{g,n}$, they do correspond to fairly natural Mori chambers.

Corollary 3.21.

$$\overline{M}_{g,\mathcal{A}} = \operatorname{Proj} \bigoplus_{m \geqslant 0} H^0(\overline{\mathcal{M}}_{g,n}, \lfloor m(K_{\overline{\mathcal{M}}_{g,n}} + \sum_{i<j: a_i + a_j \leqslant 1} (a_i + a_j - 1)\Delta_{0,\{i,j\}} + \delta) \rfloor),$$

where, δ is the total boundary of $\overline{\mathcal{M}}_{g,n}$.

Proof. Let $D := K_{\overline{\mathcal{M}}_{g,n}} + \sum_{i<j:a_i+a_j\leqslant 1}(a_i+a_j-1)\Delta_{0,\{i,j\}} + \delta$. Consider the birational reduction morphism $\phi\colon \overline{\mathcal{M}}_{g,n} \to \overline{\mathcal{M}}_{g,\mathcal{A}}$; see [58, Section 4]. Then

$$\phi_* D := K_{\overline{\mathcal{M}}_{g,\mathcal{A}}} + \sum_{i<j:a_i+a_j\leqslant 1} (a_i + a_j)\Delta_{i,j} + \delta_{nodal}$$

may be expressed as a sum of tautological classes using the Grothendieck-Riemann-Roch formula (see [58, Section 3.1.1]) $K_{\overline{\mathcal{M}}_{g,\mathcal{A}}} = 13\lambda - 2\delta_{nodal} + \psi$.

We must show that

(1) $\phi_* D$ is ample on $\overline{\mathcal{M}}_{g,\mathcal{A}}$,
(2) $D - \phi^*\phi_* D$ is effective.

Indeed, (1) implies $\overline{\mathcal{M}}_{g,\mathcal{A}} = \operatorname{Proj} \bigoplus_{m\geqslant 0} H^0(\overline{\mathcal{M}}_{g,\mathcal{A}}, \lfloor m\phi_* D\rfloor)$, and (2) implies

$$\bigoplus_{m\geqslant 0} H^0(\overline{\mathcal{M}}_{g,n}, \lfloor mD\rfloor) = \bigoplus_{m\geqslant 0} H^0(\overline{\mathcal{M}}_{g,\mathcal{A}}, \lfloor m\phi_* D\rfloor),$$

so together they yield the desired statement.

For (1), simply observe that $D = A + \lambda$, where A is as in Corollary 3.20. It follows that D is a sum of an ample divisor A and a semiample divisor λ, so is ample.

For (2), the morphism ϕ sends a stable n-pointed curve to an $\mathcal{A}$-stable curve obtained by collapsing all rational components on which $\omega_\pi(\sum a_i\sigma_i)$ has non-positive degree. From this, one easily sees that the exceptional divisors of ϕ are given by $\operatorname{Exc}(\phi) = \bigcup \Delta_{0,S}$ for all $S \subset \{1,\ldots,n\}$ such that $|S| \geqslant 3$ and $\sum_{i\in S} a_i \leqslant 1$. Thus, we may write

$$D - \phi^*\phi_* D = \sum_{|S|\geqslant 3} a_S \delta_{0,S},$$

where $a_S \in \mathbb{Q}$ is the *discrepancy* of $\Delta_{0,S}$. A simple computation with test curves shows that $a_S = (|S|-1)\left(1 - \sum_{i\in S} a_i\right) \geqslant 0$, which completes the proof of (2). □

4. Log minimal model program for moduli spaces of curves

In this section, we will describe what is currently known regarding the Mori chamber decomposition of $\operatorname{Eff}(\overline{M}_{g,n})$, with special emphasis on those chambers corresponding to modular birational models of $\overline{M}_{g,n}$. As we have seen in Section 3.2, the full effective cone of $\overline{M}_{g,n}$ is completely unknown, so it is reasonable to focus our attention on the *restricted effective cone* $\operatorname{Eff}_r(\overline{M}_{g,n})$, i.e. the intersection of the effective cone with the subspace $\mathbb{Q}\{\lambda, \psi, \delta\} \subset N^1(\overline{M}_{g,n})$. Consequently, most of the results in this section will concern the restricted effective cone.

In Section 4.1, we focus on the Mori chamber decomposition of $\overline{M}_g$: In Sections 4.1.1 and 4.1.2, we give complete Mori chamber decompositions for the restricted effective cones of $\overline{M}_2$ and $\overline{M}_3$, following Hassett and Hyeon-Lee [59, 65]. In Section 4.1.3, we describe two chambers of the restricted effective cone of $\overline{M}_g$ (for all $g \geqslant 3$) corresponding to the first two steps of the log minimal model program for $\overline{M}_g$, as carried out by Hassett and Hyeon [61, 60].

In Section 4.2, we turn our attention to the Mori chamber decomposition of $\overline{M}_{0,n}$. We will give a complete Mori chamber decomposition for half the restricted effective cone (divisors of the form $s\psi - \delta$) while the other half (divisors of the form $s\psi + \delta$) remains largely mysterious. We will see that every chamber in the first half-space corresponds to a modular birational model of $\overline{M}_{0,n}$, namely $\overline{M}_{0,\mathcal{A}}$ for a suitable weight vector $\mathcal{A}$. Finally, in Section 4.3, we will tackle $\overline{M}_{1,n}$, the first example where the restricted effective cone is three-dimensional. As with $\overline{M}_{0,n}$, we will give a complete Mori-chamber decomposition for half the restricted effective cone. We will see that every chamber corresponds to a modular birational model of $\overline{M}_{1,n}$, with both $\overline{M}_{1,\mathcal{A}}$, the moduli spaces of weighted stable curves, and $\overline{M}_{1,n}(m)$, the moduli spaces of m-stable curves, making an appearance.

Before proceeding, let us indicate the general strategy of proof which is essentially the same in each of the cases considered. Given a divisor $D \in N^1(\overline{M}_{g,n})$ and a birational contraction $\phi\colon \overline{M}_{g,n} \dashrightarrow \overline{M}$, where $\overline{M}$ is some alternate modular compactification, to prove that $D \in \text{Mor}(\phi)$ requires two calculations. First, one must show that $D - \phi^*\phi_*D \geqslant 0$, so that $H^0(\overline{M}_{g,n}, mD) = H^0(\overline{M}, m\phi_*D)$ for all $m \geqslant 0$. Second, one must show that ϕ_*D is ample on $\overline{M}$. Together, these two facts immediately imply that

$$\overline{M} = \text{Proj}\, R(\overline{M}, \phi_*D) = \text{Proj}\, R(\overline{M}_{g,n}, D),$$

so $D \in \text{Mor}(\phi)$ as desired.

To carry out the first step, one typically uses the method of test curves: Write

$$D - \phi^*\phi_*D = \sum a_i D_i,$$

where D_i are generators for the Picard group of $\overline{M}_{g,n}$ and $a_i \in \mathbb{Q}$ are undetermined coefficients. If C is any curve which is both contained in the locus where ϕ is regular and is contracted by ϕ, then we necessarily have $\phi^*\phi_*D \cdot C = 0$. Thus, if the intersection numbers $D.C$ and $D_i.C$ can be determined, one may solve for the coefficients a_i.

In order to show that ϕ_*D is ample, there are essentially two strategies. If ϕ is regular, one may consider the pull-back $\phi^*\phi_*D$ to $\overline{M}_{g,n}$. If $\phi^*\phi_*D$ is nef, has degree zero only on ϕ-exceptional curves, and can be expressed as $K_{\overline{M}_{g,n}} + \sum D_i$ with $(\overline{M}_{g,n}, \sum D_i)$ a klt pair, then one may conclude ϕ_*D is ample using the Kawamata basepoint freeness theorem (Theorem 3.4). If ϕ is not regular, one may show ϕ_*D is ample using Kleiman's criterion, i.e. essentially by proving that ϕ_*D is positive on every curve in $\overline{M}$. The key point is that $\overline{M}$ is modular, so one can typically write ϕ_*D as a linear combination of tautological classes on $\overline{M}$ whose intersection with one-parameter families can be evaluated by geometric methods.

4.1. Log minimal model program for $\overline{M}_g$

The restricted effective cone of $\overline{M}_g$ is two-dimensional, and we have

$$\mathrm{Eff}_r(\overline{M}_2) = \mathbb{Q}\{\delta_1, \delta_0\}$$
$$\mathrm{Eff}_r(\overline{M}_g) = \mathbb{Q}\{\delta, s_{\mathrm{eff}}\lambda - \delta\}, \text{ for } g \geqslant 3,$$

where the slope of the effective cone s_{eff} satisfies $\frac{60}{g+4} < s_{\mathrm{eff}} \leqslant 6 + \frac{12}{g+1}$ (as we have discussed in Section 3.2, s_{eff} remains unknown for all but finitely many g).

Let us note that one chamber of $\mathrm{Eff}(\overline{M}_g)$ is easily accounted for. By [97], the Torelli morphism $\tau\colon M_g \hookrightarrow A_g$ extends to give a regular birational map

$$\tau\colon \overline{M}_g \to \tau(\overline{M}_g) \subset \overline{\mathcal{A}}_g,$$

where $\overline{\mathcal{A}}_g$ is the Satake compactification of the moduli space of polarized abelian varieties of dimension g. The map τ contracts the entire boundary of $\overline{M}_g$ (resp. δ_0) when $g \geqslant 3$ (resp. $g = 2$), and satisfies $\tau^*\mathcal{O}_{\overline{\mathcal{A}}_g}(1) = \lambda$, where $\mathcal{O}_{\overline{\mathcal{A}}_g}(1)$ is the canonical polarization on $\overline{\mathcal{A}}_g$ coming from its realization as the Proj of the graded algebra of Siegel modular forms [97]. It follows immediately that

$$\mathrm{Mor}(\tau) = \mathbb{Q}\{\lambda, \delta_0\}, \;\; g = 2,$$
$$\mathrm{Mor}(\tau) = \mathbb{Q}\{\lambda, \delta\}, \;\; g \geqslant 3.$$

Thus, the only interesting part of the restricted effective cone lies in the quadrant spanned by λ and $-\delta$. Any divisor in this quadrant is proportional to a uniquely defined divisor of the form $s\lambda - \delta$, where s is the *slope* of the divisor. On the other hand, since $K_{\overline{\mathcal{M}}_g} = 13\lambda - 2\delta$, we also have

$$s\lambda - \delta \sim K_{\overline{\mathcal{M}}_g} + \alpha\delta$$

for $\alpha = 2 - 13/s$, so we may use either the parameter α or s to express the boundary thresholds in the Mori chamber decomposition of $\mathrm{Eff}_r(\overline{M}_g)$. From the point of view of higher-dimensional geometry, α is more natural, while from the point of view of past developments in moduli of curves, the slope is more natural. In order to have all information available, we will label our figures with both.

4.1.1. Mori chamber decomposition for $\overline{M}_2$. The geometry of $\overline{M}_2$ is sufficiently simple that it is possible to give a complete Mori chamber decomposition of $\mathrm{Eff}(\overline{M}_2)$. We have encountered precisely two alternate birational models of $\overline{M}_2$ so far, $\overline{\mathcal{A}}_2$ and $\overline{M}_2[A_2] \simeq \overline{M}_2^{\,ps}$.

Remark 4.1. *While the construction of $\overline{M}_g^{\,ps} = \overline{M}_g[A_2]$ in Corollary 2.24 was restricted to $g \geqslant 3$, one may still consider the stack $\overline{\mathcal{M}}_2[A_2]$ of A_2-stable curves of genus two. This stack is not separated as one can see from the fact that one point of $\overline{\mathcal{M}}_2[A_2]$, namely the unique rational curve with cusps at 0 and ∞, has automorphism group $\mathbb{G}_m$. Nevertheless, $\overline{\mathcal{M}}_2[A_2]$ gives rise to a weakly modular compactification $\overline{M}_2[A_2]$. The good moduli map $\overline{\mathcal{M}}_2[A_2] \to \overline{M}_2[A_2]$ maps all cuspidal curves to a single point $p \in \overline{M}_2[A_2]$, and the* *aforementioned rational bicuspidal curve is the unique closed point in* $\phi^{-1}(p)$*. For a full discussion of these matters, we refer the reader to* [59]*. For our purposes, the essential fact we need is the analogue of Proposition 2.57, i.e. the existence of a birational contraction* $\eta\colon \overline{M}_2 \to \overline{M}_2[A_2]$ *with* $\mathrm{Exc}(\eta) = \Delta_1$ *and* $\eta(\Delta_1) = p$.

We proceed to describe the Mori chambers associated to the models $\overline{M}_2[A_2]$ and $\overline{\mathcal{A}}_2$. Recall that $N^1(\overline{M}_2)$ is two-dimensional, generated by δ_0 and δ_1, with the relation $\lambda = \frac{1}{10}\delta_0 + \frac{1}{5}\delta_1$ [6]. Thus, the Mori chamber associated to any $\mathbb{Q}$-factorial rational contraction will be a two-dimensional polytope, spanned by two extremal rays.

Lemma 4.2. $\mathrm{Mor}(\overline{\mathcal{A}}_2)$ *is spanned by* λ *and* δ_0.

Proof. Since $\tau\colon \overline{M}_2 \to \overline{\mathcal{A}}_2$ is a divisorial contraction, $N^1(\overline{\mathcal{A}}_2)$ is one-dimensional, generated by $\mathcal{O}_{\overline{\mathcal{A}}_2}(1)$. Since $\tau^*\mathcal{O}_{\overline{\mathcal{A}}_2}(1) = \lambda$ and δ_0 is τ-exceptional, $\mathrm{Mor}(\overline{\mathcal{A}}_2)$ is spanned by λ and δ_0. □

Lemma 4.3. $\mathrm{Mor}(\overline{M}_2[A_2])$ *is spanned by* $11\lambda - \delta$ *and* δ_1.

Proof. By Remark 4.1, there is a divisorial contraction $\eta\colon \overline{M}_2 \to \overline{M}_2[A_2]$ contracting δ_1, so it is sufficient to show that $11\lambda - \delta$ is semiample, pulled back from an ample divisor on $\overline{M}_2[A_2]$.

Since η is an extremal divisorial contraction, $\overline{M}_2[A_2]$ is $\mathbb{Q}$-factorial [81, Corollary 3.18]. Thus, $\lambda := \eta_*\lambda$ and $\delta := \eta_*\delta$ make sense as numerical divisor classes. We claim that

$$\begin{aligned} \eta^*\lambda &= \lambda + \delta_1, \\ \eta^*\delta &= \delta_0 + 12\delta_1, \\ \eta^*(11\lambda - \delta) &= 11\lambda - \delta. \end{aligned} \tag{4.4}$$

To see this, note that the morphism η contracts a family E of elliptic tails whose intersection numbers (see Example 3.10) are $E \cdot \lambda = 1$, $E \cdot \delta_0 = 12$, and $E \cdot \delta_1 = -1$. Writing $\eta^*(\lambda) = \lambda + a\delta_1$ and intersecting with E, we obtain $a = 1$. The second formula is proved in similar fashion. The third follows immediately from the first two.

Now the divisor $11\lambda - \delta$ is nef and has degree zero only on curves lying in Δ_1 by Proposition 3.16. Since

$$11\lambda - \delta \sim K_{\overline{\mathcal{M}}_2} + \frac{9}{11}\delta,$$

the Kawamata basepoint freeness theorem 3.4 implies that $11\lambda - \delta$ is semiample. In other words, $\eta^*(11\lambda - \delta)$ is semiample and contracts only η-exceptional curves. It follows that $11\lambda - \delta$ is ample on $\overline{M}_2[A_2]$ as desired. □

Corollary 4.5.

(1) $\mathrm{Eff}(\overline{M}_2)$ *is spanned by* δ_0 *and* $\delta_1 = \frac{10}{13}(K_{\overline{\mathcal{M}}_2} + \frac{7}{10}\delta)$.

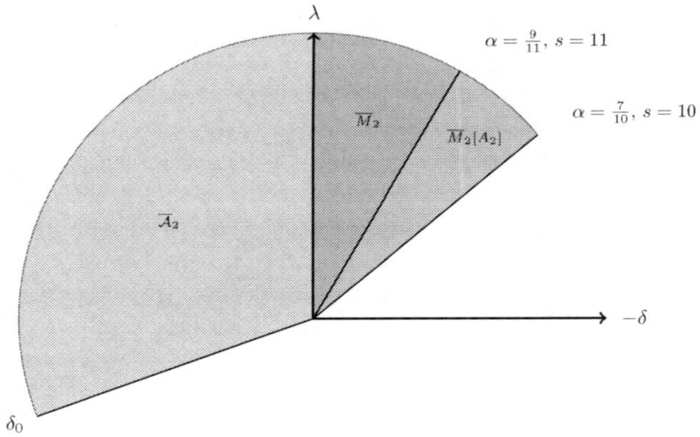

Figure 6. Restricted effective cone of $\overline{M}_2$.

(2) $\mathrm{Nef}(\overline{M}_2)$ *is spanned by* $11\lambda - \delta$ *and* λ.

Proof. (1) Since δ_0 and δ_1 are each contracted by a divisorial contraction, they are extremal rays of the effective cone. That $\delta_1 = \frac{10}{13}(K_{\overline{\mathcal{M}}_2} + \frac{7}{10}\delta)$ follows from relations $K_{\overline{\mathcal{M}}_2} = 13\lambda - 2\delta$ and $\lambda = \frac{1}{10}\delta_0 + \frac{1}{5}\delta_1$.

(2) We have seen that $11\lambda - \delta$ and λ are semiample, and since $\mathrm{Nef}(\overline{M}_2)$ is two-dimensional, they span the nef cone. □

Corollary 4.6. *The Mori chamber decomposition has precisely three chambers.*

Proof. Evidently, the Mori chambers of $\overline{\mathcal{A}}_2$, $\overline{M}_2[A_2]$, and $\overline{M}_2$ span the effective cone. □

The full Mori chamber decomposition of $\overline{M}_2$ is displayed in Figure 6. For the sake of future comparison, let us formulate this result in terms of the log minimal model program for $\overline{M}_2$. This is, of course, equivalent to listing the Mori chamber decomposition for the first quadrant, using the parameter α. We have

$$\overline{M}_2(\alpha) = \begin{cases} \overline{M}_2 & \text{iff } \alpha \in (9/11, 1], \\ \overline{M}_2[A_2] & \text{iff } \alpha \in (7/10, 9/11], \\ \text{point} & \text{iff } \alpha = 7/10. \end{cases} \tag{4.7}$$

4.1.2. Mori chamber decomposition for $\overline{M}_3$. The birational geometry of $\overline{M}_3$ is far more intricate than that of $\overline{M}_2$. For example, as we mentioned in Section 3.2, the rationality of $\overline{M}_3$ was not established until 1996 [73]. Even before the advent of the Hassett-Keel program, several alternate birational models of $\overline{M}_3$ had been constructed and studied, including the Satake compactification $\tau(\overline{M}_3)$, the Mumford's GIT quotient of the space of plane quartics, and the Schubert's space of pseudostable curves. A systematic study of the Mori chamber decomposition of $\overline{M}_3$ was undertaken by Rulla [100], who showed (among other things) the existence of a divisorial contraction $\overline{M}_3 \to X$ contracting Δ_1 that is *not* isomorphic to the contraction $\overline{M}_3 \to \overline{M}_3^{ps}$. Here, X can be realized as the image of $\overline{M}_3$ under the natural map to the Alexeev's space of semi-abelic pairs [1]. Around the same time, Kondo proved that M_3 is birational to a complex ball quotient [82]. This gives rise to a proper birational model of $\overline{M}_3$ by taking the Baily-Borel compactification of the ball quotient. Using the theory of K3 surfaces, Artebani constructed an alternative compactification of M_3 that admits a regular morphism to the Kondo's space [8]. These developments were rounded off by Hyeon and Lee who described all the log canonical models $\overline{M}_3(\alpha)$ and their relation with previously known birational models of $\overline{M}_3$. In particular, their work implies that the Artebani's compactification is isomorphic to $\overline{M}_3[A_3]$ and that the Kondo's compactification is isomorphic to $\overline{M}_3[A_3^*]$.

To describe where the Mori chambers of the known birational models of $\overline{M}_3$ fall inside the effective cone, we begin with the well-known fact that the effective cone inside $N^1(\overline{M}_3) = \mathbb{Q}\{\lambda, \delta_0, \delta_1\}$ is generated by $9\lambda - \delta_0 - 3\delta_1$, δ_0, and δ_1, and where $9\lambda - \delta_0 - 3\delta_1$ is the class of the hyperelliptic divisor [100]). In particular, the restricted effective cone is spanned by δ and $9\lambda - \delta$, or, equivalently, by δ and $K_{\overline{\mathcal{M}}_3} + \frac{5}{9}\delta$. The following proposition, due to Hyeon and Lee [65], shows that weakly modular birational models account for a complete Mori chamber decomposition of the restricted effective cone of $\overline{M}_3$ (see Figure 7).

Proposition 4.8.

$$\overline{M}_3(\alpha) = \begin{cases} \overline{M}_3 & \textit{iff } \alpha \in (9/11, 1], \\ \overline{M}_3[A_2] & \textit{iff } \alpha \in (7/10, 9/11], \\ \overline{M}_3[A_3^*] & \textit{iff } \alpha = 7/10, \\ \overline{M}_3[A_3] & \textit{iff } \alpha = (17/28, 7/10), \\ \overline{M}_3[Q] & \textit{iff } \alpha = (5/9, 17/28], \\ \textit{point} & \textit{iff } \alpha = 5/9. \end{cases}$$

These log canonical models and the morphisms between them fit into the following diagram

(4.9)
$$\begin{array}{ccccccc} \overline{M}_3 & & & & & & \\ \downarrow \eta & & & & & & \\ \overline{M}_3[A_2] & & & & \overline{M}_3[A_3] & & \\ & \searrow^{\phi^-} & & \swarrow^{\phi^+} & & \searrow^{\psi} & \\ & & \overline{M}_3[A_3^*] & & & & \overline{M}_3[Q] \end{array}$$

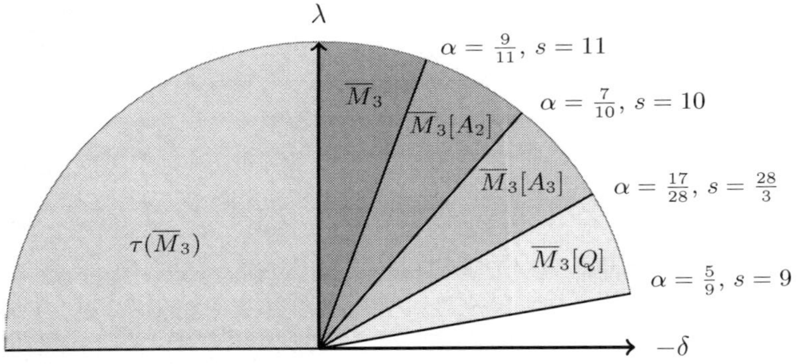

Figure 7. Restricted effective cone of $\overline{M}_3$.

Recall that the spaces $\overline{M}_3[A_2]$, $\overline{M}_3[A_3^*]$, $\overline{M}_3[A_3]$ were described using GIT in Section 2.4 and the maps η, ϕ, ϕ^+ have been described in Proposition 2.57. By contrast, the space $\overline{M}_3[Q]$, where Q stands for 'quartic,' has not yet been described. It is the GIT quotient of the space of degree 4 plane curves, i.e. $\mathrm{Hilb}^{ss}_{3,1} /\!/ \mathrm{SL}(3) = \mathbb{P}\mathrm{H}^0(\mathbb{P}^2, \mathcal{O}_{\mathbb{P}^2}(4)) /\!/ \mathrm{SL}(3)$. Whereas the GIT analysis of semistable points for the Hilbert scheme of canonically embedded curves has never been carried out for all g, the genus 3 case is a classic example, and the stable and semistable points $\mathrm{Hilb}^{ss}_{3,1} \subset \mathbb{P}\mathrm{H}^0(\mathbb{P}^2, \mathcal{O}_{\mathbb{P}^2}(4))$ are described in [96, Chapter 3.2]. The nonsemistable points are precisely either quartics with a point of multiplicity 3 or quartics which are a union of a plane cubic with its flex line. The stable points are plane quartics with at worst cusps. The rest are semistable: these are plane quartics with at best A_3 and worst A_7 singularities, as well as double smooth conics. Note that since all strictly semistable curves specialize isotrivially to a double conic, and any two double conics are projectively equivalent, all strictly semistable points correspond to a unique point $p \in \overline{M}_3[Q]$.

It is straightforward to see that the natural rational map $\psi\colon \overline{M}_3[A_3] \dashrightarrow \overline{M}_3[Q]$ is regular, contracts the hyperelliptic locus in $\overline{M}_3[A_3]$ to p, and is an isomorphism elsewhere. That it is an isomorphism away from the hyperelliptic locus is an immediate consequence of the fact that any A_3-stable curve which is not hyperelliptic is still Q-stable, i.e. contained in the semistable locus $\mathrm{Hilb}^{ss}_{3,1}$. To see that ψ maps the hyperelliptic divisor to p, it suffices to observe that the stable limit of any family of quartics degenerating to a *GIT stable* quartic does not lie in the hyperelliptic locus. Therefore, when a smooth non-hyperelliptic genus 3 curve degenerates to a hyperelliptic curve, the GIT semistable limit in $\overline{M}_3[Q]$ is p.

Proof of Proposition 4.8. Throughout the proof, we will abuse notation by using the symbols λ, δ_0, and δ_1 to denote the divisor classes obtained by pushing-forward to each of the birational models the divisor classes of the same name on $\overline{M}_3$.

We proceed chamber by chamber. First, observe that

$$K_{\overline{\mathcal{M}}_3} + \alpha\delta = 13\lambda - (2-\alpha)\delta \sim \frac{13}{2-\alpha}\lambda - \delta.$$

For $\alpha > 9/11$, this divisor has slope greater than 11, hence is ample by Theorem 3.7. This immediately implies that

$$\overline{M}_3(\alpha) \simeq \overline{M}_3 \text{ for } \alpha \in (9/11, 1].$$

At $\alpha = \frac{9}{11}$, the divisor is numerically proportional to $11\lambda - \delta$, which is nef and has degree zero precisely on T_1 – the curve class of elliptic tails (see Proposition 3.16). Applying the Kawamata basepoint freeness theorem 3.4 to $K_{\overline{\mathcal{M}}_3} + \frac{9}{11}\delta$, we obtain a birational contraction[4]

$$\overline{M}_3 \to \overline{M}_3(9/11),$$

contracting the curve class T_1. Since this is precisely the curve class contracted by the map $\eta\colon \overline{M}_3 \to \overline{M}_3[A_2]$ (Proposition 2.57), we may identify $\overline{M}_3(\frac{9}{11}) \simeq \overline{M}_3[A_2]$. In particular, $K_{\overline{\mathcal{M}}_3[A_2]} + \frac{9}{11}\delta$ is Cartier and ample on $\overline{M}_3[A_2]$. In order to determine the chamber of $\overline{M}_3[A_2]$, the key question is: As we scale α down from $\frac{9}{11}$, how long does $K_{\overline{\mathcal{M}}_3[A_2]} + \alpha\delta$ remain ample? The simplest way to answer this question is to compute the pull back

$$\eta^*(K_{\overline{\mathcal{M}}_3[A_2]} + \alpha\delta) = 13\lambda - (2-\alpha)\delta_0 - (11-12\alpha)\delta_1 = K_{\overline{\mathcal{M}}_3} + \alpha\delta + (11\alpha - 9)\delta_1 \tag{4.10}$$

using Equation (4.4) (which holds for $g \geqslant 2$), and observe that $13\lambda - (2-\alpha)\delta_0 - (11-12\alpha)\delta_1$ is a positive linear combination of $11\lambda - \delta$ and $10\lambda - \delta - \delta_1$ for $\alpha \in (\frac{7}{10}, \frac{9}{11})$. It follows from Proposition 3.16 that for $\alpha \in (\frac{7}{10}, \frac{9}{11})$, the divisor $\eta^*(K_{\overline{M}_3[A_2]} + \alpha\delta)$ is nef and has degree zero precisely on the curve class T_1, hence descends to an ample divisor on $\overline{M}_3[A_2]$. We conclude that

$$\overline{M}_3(\alpha) \simeq \overline{M}_3[A_2] \text{ for } \alpha \in (7/10, 9/11].$$

Furthermore, at $\alpha = \frac{7}{10}$,

$$\eta^*(K_{\overline{\mathcal{M}}_3[A_2]} + \alpha\delta) = 10\lambda - \delta - \delta_1.$$

This divisor has degree zero on the curve classes described in Proposition 3.16 (2), namely the class T_1 of the family of elliptic tails and the curve classes EB_i^s, EB^{ns} of families of elliptic bridges. Thus, $K_{\overline{\mathcal{M}}_3[A_2]} + \frac{7}{10}\delta$ is nef on $\overline{M}_3[A_2]$ and has degree zero precisely on the curve classes $\eta(EB_i^s)$ and $\eta(EB^{ns})$. Applying the Kawamata basepoint freeness theorem to $K_{\overline{\mathcal{M}}_3[A_2]} + \frac{7}{10}\delta$, we obtain a map $\overline{M}_3[A_2] \to \overline{M}_3(\frac{7}{10})$ which contracts these curve classes. By Corollary 3.18, these curve classes sweep out precisely the locus of elliptic bridges, i.e. the image of the natural gluing map

[4]Since $\overline{M}_3$ is mildly singular, it is not completely obvious that $(\overline{M}_3, \frac{9}{11}\Delta)$ is a klt pair; this is verified in [65]. In order to simplify exposition, we will omit standard discrepancy calculations needed to justify the use of the basepoint freeness theorem.

$\overline{M}_{1,2} \times \overline{M}_{1,2} \to \overline{M}_3 \to \overline{M}_3[A_2]$. Thus, $\overline{M}_3[A_2] \to \overline{M}_3(\frac{7}{10})$ contracts the locus of the elliptic bridges to a point, and is an isomorphism elsewhere.

Now consider the map $\phi^- : \overline{M}_3[A_2] \to \overline{M}_3[A_3^*]$. By Proposition 2.57, ϕ^- also contracts precisely the locus of elliptic bridges, so we obtain an identification

$$\overline{M}_3(7/10) \simeq \overline{M}_3[A_3^*].$$

Next, we must compute the Mori chamber of $\overline{M}_3[A_3]$, i.e. we must show

$$\overline{M}_3(\alpha) = \overline{M}_3[A_3] \text{ for } \alpha \in (17/28, 7/10).$$

Note that knowing the ample cone of $\overline{M}_3$ can no longer help us here, since the map $\overline{M}_3 \dashrightarrow \overline{M}_3[A_3]$ is only rational. Instead, we use a slightly ad-hoc argument, applicable only in genus three. We claim that it is sufficient to prove that $K_{\overline{\mathcal{M}}_3[A_3]} + \frac{7}{10}\delta$ is nef, with degree zero precisely on the locus of tacnodal curves, and that $K_{\overline{\mathcal{M}}_3[A_3]} + \frac{17}{28}\delta$ is nef, with degree zero precisely on the locus of hyperelliptic curves. Indeed, one can check that the intersection of the locus of tacnodal curves with the closure of the hyperelliptic curves on $\overline{M}_3[A_3]$ consists of a single point, so that a positive linear combination of $K_{\overline{\mathcal{M}}_3[A_3]} + \frac{7}{10}\delta$ and $K_{\overline{\mathcal{M}}_3[A_3]} + \frac{17}{28}\delta$ has positive degree on all curves, hence is ample.[5] Furthermore, for any $\alpha < \frac{9}{11}$, we have

$$R(\overline{\mathcal{M}}_3[A_3], K_{\overline{\mathcal{M}}_3[A_3]} + \alpha\delta) = R(\overline{\mathcal{M}}_3[A_2], K_{\overline{\mathcal{M}}_3[A_2]} + \alpha\delta) = R(\overline{\mathcal{M}}_3, K_{\overline{\mathcal{M}}_3} + \alpha\delta),$$

the first equality holding because $\overline{M}_3[A_2] \dashrightarrow \overline{M}_3[A_3]$ is an isomorphism in codimension one, and the second holding by the discrepancy computation in Equation (4.10). Together these observations immediately imply $\overline{M}_3(\alpha) \simeq \overline{M}_3[A_3]$ for $\alpha \in (17/28, 7/10)$.

Now the fact that $K_{\overline{\mathcal{M}}_3[A_3]} + \frac{7}{10}\delta$ is nef, with degree zero precisely on the locus of tacnodal curves, is immediate from the fact that, since ϕ^+ is small, $(\phi^+)^*(K_{\overline{\mathcal{M}}_3[A_3^*]} + \frac{7}{10}\delta) = K_{\overline{\mathcal{M}}_3[A_3]} + \frac{7}{10}\delta$. To obtain the nefness of $K_{\overline{\mathcal{M}}_3[A_3]} + \frac{17}{28}\delta$, we write

$$K_{\overline{\mathcal{M}}_3[A_3]} + \frac{17}{28}\delta = \frac{5}{42}\left(K_{\overline{\mathcal{M}}_3[A_3]} + \frac{7}{10}\delta\right) + \frac{13}{52}H,$$

where $H \equiv 9\lambda - \delta_0$ is the hyperelliptic divisor. It follows that if $K_{\overline{\mathcal{M}}_3[A_3]} + \frac{17}{28}\delta$ has negative degree on any curve, that curve must lie in the hyperelliptic locus. On the other hand, using the natural map $\overline{M}_{0,8} \to H \subset \overline{M}_3[A_3]$ one can verify by a direct calculation that $K_{\overline{\mathcal{M}}_3[A_3]} + \frac{17}{28}\delta$ restricts to a trivial divisor on H [65, Proposition 19].

Finally, since $K_{\overline{\mathcal{M}}_3[A_3]} + \frac{17}{28}\delta$ is trivial on the hyperelliptic divisor, applying the Kawamata basepoint freeness theorem 3.4 to $K_{\overline{\mathcal{M}}_3[A_3]} + \frac{17}{28}\delta$ yields a map $\overline{M}_3[A_3] \to \overline{M}_3(\frac{17}{28})$, contracting the hyperelliptic divisor to a point. Since this is precisely the locus contracted by $\overline{M}_3[A_3] \to \overline{M}_3[Q]$, we conclude that $\overline{M}_3[Q] \simeq \overline{M}_3(\frac{17}{28})$. To complete the proof of the proposition, we make the following two observations: Since $\overline{M}_3[Q]$ has Picard number one by its GIT construction, $N^1(\overline{M}_3[Q])$ is generated

[5] To check that a klt divisor is ample, it suffices, by the Kawamata basepoint freeness theorem 3.4, to prove positivity on curves.

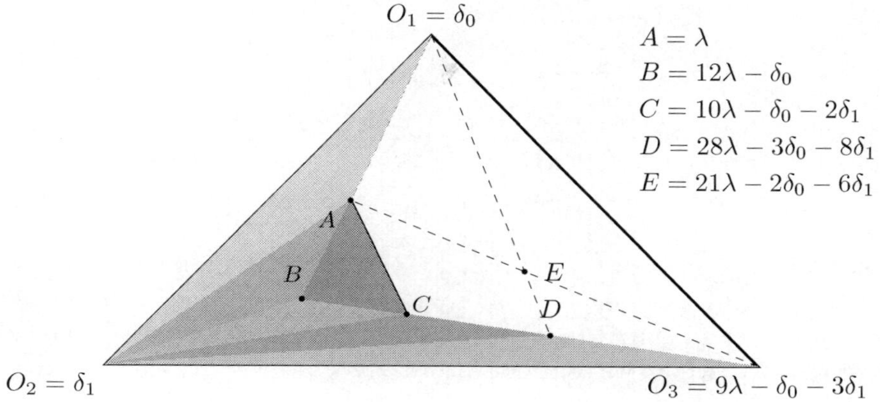

Figure 8. Effective cone of $\overline{M}_3$

by δ. Second, since ψ contracts the hyperelliptic divisor, we must have $9\lambda - \delta = 0 \in N^1(\overline{M}_3[Q])$. It follows that

$$K_{\overline{M}_3[Q]} + \alpha\delta \sim (\alpha - 5/9)\delta,$$

so that $K_{\overline{M}_3[Q]} + \alpha\delta$ is ample iff $\alpha > \frac{5}{9}$ and becomes trivial at $\alpha = \frac{5}{9}$. □

The preceding proposition shows that the modular birational models of $\overline{M}_3$ give a complete Mori chamber decomposition of the restricted effective cone. Since we actually know the full effective cone of $\overline{M}_3$, it is interesting to see what parts of the effective cone are not accounted for by these birational models. In Figure 4.1.2, we have drawn a cross section of the effective cone, spanned by δ_0, δ_1, and $9\lambda - \delta_0 - 3\delta_1$. By [100, Section 2.4], the nef cone is ABC and the moving cone is ABDE. The chambers O_1AO_2, O_2BC, O_2CD and O_2DO_3 correspond to $\tau(\overline{M}_3)$, $\overline{M}_3[A_2]$, $\overline{M}_3[A_3]$, and $\overline{M}_3[Q]$, respectively. However, we do not know much about the birational models accounting for the remaining portion of the effective cone: what are the small modifications of $\overline{M}_3$ coming from the divisors in ACDE and do they have a modular meaning? In particular, the question of whether $\overline{M}_3$ is a Mori dream space remains open.

4.1.3. Mori chamber decomposition for $\overline{M}_g$. Hassett and Hyeon have proved that the first two steps of the log MMP for $\overline{M}_g$ proceed exactly as in the case $g = 3$, i.e. we have

Theorem 4.11 ([61] and [60]).

$$\overline{M}_g(\alpha) = \begin{cases} \overline{M}_g & \textit{iff } \alpha \in (9/11, 1], \\ \overline{M}_g[A_2] & \textit{iff } \alpha \in (7/10, 9/11], \\ \overline{M}_g[A_3^*] & \textit{iff } \alpha = 7/10, \\ \overline{M}_g[A_3] & \textit{iff } \alpha = (7/10 - \epsilon, 7/10]. \end{cases}$$

Proof. The first steps of the proof proceed exactly as in the genus three case. By Proposition 3.16, $K_{\overline{\mathcal{M}}_g} + \frac{9}{11}\delta = 11\lambda - \delta$ is nef and has degree zero precisely on the curve class of elliptic tails. But this is precisely the curve class contracted by $\eta\colon \overline{M}_g \to \overline{M}_g[A_2]$, so we conclude that $\overline{M}_g(\frac{9}{11}) \simeq \overline{M}_g[A_2]$. Exactly as in the proof of Proposition 4.8, one shows that $K_{\overline{\mathcal{M}}_g} + \alpha\delta$ remains ample for $\alpha \in (\frac{7}{10}, \frac{9}{11})$, and that at $\alpha = \frac{7}{10}$, we have

$$\eta^*(K_{\overline{\mathcal{M}}_g[A_2]} + \frac{7}{10}\delta) = 10\lambda - \delta - \delta_1.$$

Applying the Kawamata basepoint freeness theorem 3.4 to $K_{\overline{\mathcal{M}}_g[A_2]} + \frac{7}{10}\delta$ gives a birational morphism $\overline{M}_g[A_2] \to \overline{M}_g(\frac{7}{10})$, and Proposition 3.16 implies that the curves contracted by this morphism are precisely the curves in which every moving component is an elliptic bridge.

If we knew that the map $\overline{M}_g[A_2] \to \overline{M}_g[A_3^*]$ contracted the same curves, we would immediately conclude that $\overline{M}_g(\frac{7}{10}) = \overline{M}_g[A_3^*]$. But here, the case $g \gg 0$ is more complicated than the case $g = 3$. While it is clear from Proposition 2.57 that $\mathrm{Exc}(\phi^-)$ is the locus of the elliptic bridges and that any curve in which every moving component is an elliptic bridge is contracted by ϕ^-, it is not clear *a priori* that every curve contracted by ϕ^- is simply a curve in which elliptic bridges are varying in moduli. The problem is that the geometry of the exceptional locus $\mathrm{Exc}(\phi^-)$ is extremely complicated and does not admit any simple description as in the genus three case. While it should, in principle, be possible to precisely characterize the contracted curves of ϕ^- by refining the analysis of Proposition 2.57, we can avoid this altogether by taking advantage of the natural polarization coming from the GIT construction of $\overline{M}_g[A_3^*]$.

Since $\overline{M}_g[A_3^*]$ was constructed as the Chow quotient of bicanonically embedded curves, the Chow analogue of Corollary 2.45 implies that $\overline{M}_g[A_3^*]$ possesses an ample line bundle (cf. Equation (2.50)) numerically equivalent to

$$\lim_{m\to\infty} \Lambda_{2,m} = 10\lambda - \delta \sim K_{\overline{\mathcal{M}}_g[A_3^*]} + \frac{7}{10}\delta.$$

Furthermore,

$$R(\overline{\mathcal{M}}_g[A_3^*], K_{\overline{\mathcal{M}}_g[A_3^*]} + \frac{7}{10}\delta) = R(\overline{\mathcal{M}}_g[A_2], K_{\overline{\mathcal{M}}_g[A_2]} + \frac{7}{10}\delta) = R(\overline{\mathcal{M}}_g, K_{\overline{\mathcal{M}}_g} + \frac{7}{10}\delta)$$

The first equality holds because $\overline{\mathcal{M}}_g[A_3^*]$ is isomorphic to $\overline{\mathcal{M}}_g[A_2]$ in codimension one (Proposition 2.57), and the second by a simple discrepancy calculation (see Equation (4.10)). It follows immediately $\overline{M}_g(\frac{7}{10}) \simeq \overline{M}_g[A_3^*]$.

Finally, to show that $\overline{M}_g(\alpha) \simeq \overline{M}_g[A_3]$ for $\alpha \in (\frac{7}{10} - \epsilon, \frac{7}{10})$, it suffices to prove that $K_{\overline{\mathcal{M}}_g[A_3]} + (\frac{7}{10} - \epsilon)\delta$ is ample on $\overline{M}_g[A_3]$ for some $\epsilon > 0$. Indeed, since ϕ^+ is an isomorphism in codimension one, we have $(\phi^+)^*(K_{\overline{\mathcal{M}}_g[A_3^*]} + \frac{7}{10}\delta) = K_{\overline{\mathcal{M}}_g[A_3]} + \frac{7}{10}\delta$ is nef. Furthermore, for any $\alpha \leqslant 7/10$,

$$R(\overline{\mathcal{M}}_g[A_3^*], K_{\overline{\mathcal{M}}_g[A_3^*]} + \alpha\delta) = R(\overline{\mathcal{M}}_g[A_2], K_{\overline{\mathcal{M}}_g[A_2]} + \alpha\delta) = R(\overline{\mathcal{M}}_g, K_{\overline{\mathcal{M}}_g} + \alpha\delta)$$

for precisely the same reasons as above. To obtain the requisite ample divisor, we simply use the construction of $\overline{M}_g[A_3]$ as the GIT quotient of the asymptotically linearized Hilbert scheme of bicanonically embedded curves. Corollary 2.45 implies that the line bundle $\Lambda_{2,m}$ is ample for a sufficiently large m, and Equation (2.44) shows that for a large m, $\Lambda_{2,m} = (10-\epsilon)\lambda - \delta \sim K_{\overline{\mathcal{M}}_g[A_3]} + (\frac{7}{10} - \epsilon)\delta$, as desired. □

4.2. Log minimal model program for $\overline{M}_{0,n}$

In this section, we turn our attention to $\overline{M}_{0,n}$, which has a much different flavor that $\overline{M}_g$. As noted in Section 3.2, the restricted effective cone is spanned by $-\frac{n-1}{\lfloor n/2\rfloor(n-\lfloor n/2\rfloor)}\psi + \Delta$ and $\frac{n-1}{2(n-2)}\psi - \Delta$. As in the case of $\overline{M}_g$, the log canonical divisors

$$K_{\overline{M}_{0,n}} + \alpha\Delta = \psi - (2-\alpha)\Delta$$

fill out the first quadrant of the restricted effective cone (see Figure 9). But whereas in the case $\overline{M}_g$, the Mori chamber decomposition of the second quadrant was easily disposed of using the Torelli morphism while the Mori chamber decomposition of the first quadrant is still largely unknown, the case of $\overline{M}_{0,n}$ is exactly the opposite. While we have no understanding of the chamber decomposition of the second quadrant, we can give a complete description of the log minimal model program for $\overline{M}_{0,n}$. As usual, we set

$$\overline{M}_{0,n}(\alpha) = \operatorname{Proj} \bigoplus_{m\geqslant 0} H^0(\overline{M}_{0,n}, m(K_{\overline{M}_{0,n}} + \alpha\Delta)).$$

The following result, pictured in Figure 9 and proved in [2, 46], gives a complete description of the models $\overline{M}_{0,n}(\alpha)$.

Theorem 4.12.

(1) *If* $\alpha \in \mathbb{Q} \cap (\frac{2}{k+2}, \frac{2}{k+1}]$ *for some* $k = 1, \dots, \lfloor\frac{n-1}{2}\rfloor$, *then* $\overline{M}_{0,n}(\alpha) \simeq \overline{M}_{0,\mathcal{A}}$, *the moduli space of* $\mathcal{A}$*-stable curves, with* $\mathcal{A} = \underbrace{(1/k, \dots, 1/k)}_{n}$.

(2) *If* $\alpha \in \mathbb{Q} \cap (\frac{2}{n-1}, \frac{2}{\lfloor n/2\rfloor+1}]$, *then* $\overline{M}_{0,n}(\alpha) = (\mathbb{P}^1)^n /\!/ SL(2)$.

Here, $(\mathbb{P}^1)^n /\!/ SL(2)$ is the classical GIT quotient of n points on $\mathbb{P}^1$, taken with the symmetric linearization [96, Chapter 3]; in the case of odd n, it is still of the form $\overline{M}_{0,\mathcal{A}}$.

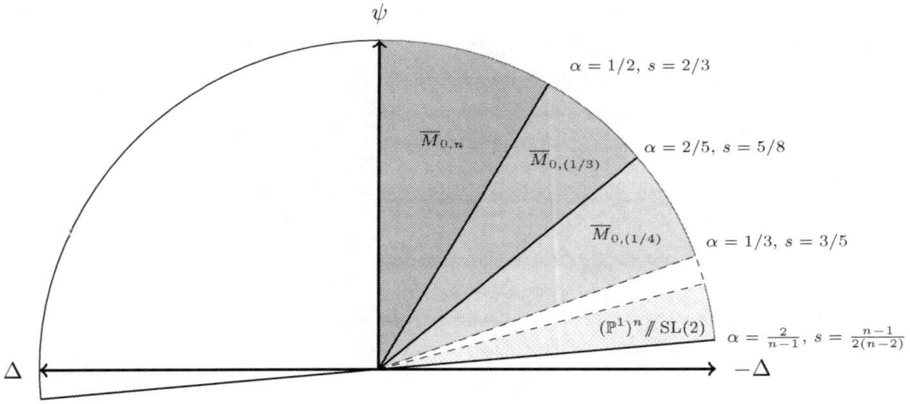

Figure 9. Mori chamber decomposition for $\overline{M}_{0,n}$.

Proof of Theorem 4.12: The proof of (2) is elementary, and given the results of Section 3.2.5, the proof of (1) is fairly quick as well. Fixing $\mathcal{A} = (1/k, \dots, 1/k)$ and $\alpha \in \mathbb{Q} \cap (\frac{2}{k+2}, \frac{2}{k+1}]$, we must show that:

i $(K_{\overline{M}_{0,n}} + \alpha\Delta) - \phi^*\phi_*(K_{\overline{M}_{0,n}} + \alpha\Delta) \geqslant 0$,
ii $\phi_*(K_{\overline{M}_{0,n}} + \alpha\Delta)$ is ample on $\overline{M}_{0,\mathcal{A}}$,

where $\phi\colon \overline{M}_{0,n} \to \overline{M}_{0,\mathcal{A}}$ is the natural reduction morphism. For (i), since the exceptional divisors of ϕ are the boundary divisors Δ_m with $3 \leqslant m \leqslant k$ (see the discussion preceding Corollary 3.20), we have

$$(K_{\overline{M}_{0,n}} + \alpha\Delta) - \phi^*\phi_*(K_{\overline{M}_{0,n}} + \alpha\Delta) = \sum_{3\leqslant m\leqslant k} a_i\Delta_i,$$

for some undetermined coefficients a_i. In Exercise 4.13 below, we construct a curve $T_m \subset \overline{M}_{0,n}$ which is contracted by ϕ, has intersection numbers $T_m \cdot \psi = m(m-2)$, $T_m \cdot \Delta_2 = \binom{m}{2}$, $T_m \cdot \Delta_m = -1$, and avoids all other boundary divisors. Intersecting T_m with the above equation gives

$$a_m = -(K_{\overline{M}_{0,n}} + \alpha\delta) \cdot T_m = (m-2)(1 - \frac{m+1}{2}\alpha) \geqslant 0,$$

since $m \leqslant k$ and $\alpha \in (\frac{2}{k+2}, \frac{2}{k+1}]$. It remains to show that $\phi_*(K_{\overline{M}_{0,n}} + \alpha\Delta)$ is ample on $\overline{M}_{0,\mathcal{A}}$.

Here, the key point is that the divisor $\phi_*(K_{\overline{M}_{0,n}} + \alpha\Delta)$ lies in the convex hull of the nef divisors constructed in Corollary 3.20. Defining $\Delta_{i,j}$ and Δ_{nodal} as in the discussion preceding Corollary 3.20, we may rewrite

$$\phi_*(K_{\overline{M}_{0,n}} + \alpha\Delta) \sim c\psi + (2c-1)\sum_{i,j} \Delta_{i,j} - \Delta_{nodal},$$

where $c = 1/(2-\alpha) \in \left(\frac{k+2}{2(k+1)}, \frac{k+1}{2k}\right]$. Setting $D(c) := c\psi + (2c-1)\sum_{i,j}\Delta_{i,j} - \Delta_{nodal}$, one easily checks that, in the notation of Corollary 3.20, we have

$$D\left(\frac{k+1}{2k}\right) = B\left(\frac{1}{k}, \dots, \frac{1}{k}\right) + \frac{k-2}{2k}C\left(\frac{1}{k}, \dots, \frac{1}{k}\right),$$
$$D\left(\frac{k+2}{2(k+1)}\right) = B\left(\frac{1}{k+1}, \dots, \frac{1}{k+1}\right) + \frac{k-1}{2(k+1)}C\left(\frac{1}{k+1}, \dots, \frac{1}{k+1}\right).$$

Corollary 3.20 now implies that $D(c) \sim K_{\overline{M}_{0,\mathcal{A}}} + \alpha\Delta$ is nef and has positive degree on every complete curve. To see that $K_{\overline{M}_{0,\mathcal{A}}} + \alpha\Delta$ is actually ample (not merely nef), simply observe that $(\overline{M}_{0,\mathcal{A}}, \alpha\Delta)$ is a big and nef klt pair, so the Kawamata basepoint freeness theorem 3.4 says that $K_{\overline{M}_{0,\mathcal{A}}} + \alpha\Delta$ is semiample. Since $K_{\overline{M}_{0,\mathcal{A}}} + \alpha\Delta$ has positive degree on every curve, it is also ample. □

Exercise 4.13. Let $m \geqslant 3$. Construct a curve $T_m \subset \overline{M}_{0,n}$ contracted by the morphism $\overline{M}_{0,n} \to \overline{M}_{0,(1/k,\dots,1/k)}$ for all $k \geqslant m$ as follows: Consider $\mathbb{P}^2$ blown up at a point and marked by m lines of self-intersection 1. A curve $T_m \subset \overline{M}_{0,n}$ is obtained by attaching a constant family of $(n-m)$-pointed curves along the (-1)-curve. Compute that

$$\psi \cdot T_m = m(m-2), \qquad \Delta_2 \cdot T_m = \binom{m}{2}, \qquad \Delta_m \cdot T_m = -1. \tag{4.14}$$

4.2.1. The contraction associated to ψ In this subsection, we make a small digression to discuss the geometry of the contraction associated to ψ. As seen in Figure 9, ψ lies on the left-most side of the restricted nef cone of $\overline{M}_{0,n}$. Here, we will show that ψ is semiample and defines a morphism $\overline{M}_{0,n} \to \overline{M}_{0,n}[\psi]$, where $\overline{M}_{0,n}[\psi]$ is the moduli space of ψ-stable curves constructed in Corollary 2.28, obtaining the projectivity of $\overline{M}_{0,n}[\psi]$ as a byproduct.

By Remark 2.30, there is a morphism $\psi\colon \overline{M}_{0,n} \to \overline{M}_{0,n}[\psi]$ contracting the locus of curves possessing an unmarked component with at least 4 nodes. In the following lemma, we will show that this morphism is actually projective.

Lemma 4.15. *ψ is semiample and descends to an ample line bundle on $\overline{M}_{0,n}[\psi]$.*

Proof. First, note that the morphism $\psi\colon \overline{M}_{0,n} \to \overline{M}_{0,n}[\psi]$ contracts precisely those families of curves on which the line bundle ψ has degree 0, namely those families in which every moving component is unmarked. Moreover, $\psi = \psi^*\mathcal{L}$, where $\mathcal{L}$ is the analogue of the ψ class for $\overline{M}_{0,n}[\psi]$, i.e. the line bundle $\otimes_{i=1}^n \sigma_i^*\mathcal{O}_{\mathcal{C}}(-\sigma_i)$, where $\{\sigma_i\}_{i=1}^n$ are the sections of the universal curve $\mathcal{C} \to \overline{M}_{0,n}[\psi]$. Finally, ψ is semiample on $\overline{M}_{0,n}$ by Exercise 4.16 below. Since the semiample line bundle ψ has degree zero precisely on the curves contracted by the morphism $\psi\colon \overline{M}_{0,n} \to \overline{M}_{0,n}[\psi]$ defined in Remark 2.30, we conclude that $\overline{M}_{0,n}[\psi]$ is projective with normalization equal to $\operatorname{Proj} R(\overline{M}_{0,n}, \psi)$. □

Exercise 4.16. Let $\mathcal{A} = (1, 1/(n-2), \dots, 1/(n-2))$. Show that

(1) $\overline{M}_{0,\mathcal{A}} \simeq \mathbb{P}^{n-3}$ and $\psi_1 \simeq \mathcal{O}_{\mathbb{P}^{n-3}}(1)$.

(2) Show that if $\xi\colon \overline{M}_{0,n} \to \overline{M}_{0,\mathcal{A}}$ is the natural morphism, then $\psi_1 = \xi^*(\psi_1)$. Conclude that ψ_i is semiample for each $i = 1, \dots, n$, hence that ψ is semiample.

Remark 4.17. *The morphism $\xi\colon \overline{M}_{0,n} \to \overline{M}_{0,\mathcal{A}} \simeq \mathbb{P}^{n-3}$ of Exercise 4.16 factors into a sequence of blow-ups with smooth centers and constitutes the Kapranov's construction of $\overline{M}_{0,n}$; see* [69, 70] *and* [58, Section 6.1].

Note that since $\overline{M}_{0,n}[\psi]$ is not $\mathbb{Q}$-factorial, the chamber corresponding to the contraction defined by ψ is not full dimensional; within the restricted effective cone of $\overline{M}_{0,n}$, it is simply the extremal ray spanned by ψ. However, if $\overline{M}_{0,n}$ is a Mori dream space, it should be possible to flip this contraction to obtain a $\mathbb{Q}$-factorial birational model whose chamber lies on the opposite side of this ray. This motivates the question whether ψ can be flipped.

Question 4.18. *What are the normal, projective, $\mathbb{Q}$-factorial varieties X admitting a morphism $\psi^+\colon X \to \overline{M}_{0,n}[\psi]$ such that $\overline{M}_{0,n} \dashrightarrow X$ is an isomorphism in codimension one and the line bundle ψ is nef on X?*

4.3. Log minimal model program for $\overline{M}_{1,n}$

In this section, we present a complete Mori chamber decomposition for a restricted effective cone of $\overline{M}_{1,n}$. In contrast to the case of $\overline{M}_g$ and $\overline{M}_{0,n}$, $\{K_{\overline{\mathcal{M}}_{1,n}} + \alpha\delta : \alpha \geqslant 0\}$ is not the most natural ray of the restricted effective cone to consider, so we will adopt slightly different notation for the log canonical models of $\overline{M}_{1,n}$. Set

$$D(s,t) := s\lambda + t\psi - \delta,$$
$$R(s,t) := \oplus_{m\geqslant 0} H^0(\overline{M}_{1,n}, mD(s,t)), \tag{4.19}$$
$$\overline{M}_{1,n}^{s,t} := \operatorname{Proj} R(s,t).$$

Evidently, any divisor in the half-space $\{s\lambda + t\psi + u\delta : u \leqslant 0\}$ is numerically proportional to a divisor of the form $D(s,t)$. Since $K_{\overline{\mathcal{M}}_{1,n}} + \alpha\delta = 13\lambda + \psi - (2-\alpha)\delta$, we have $\overline{M}_{1,n}(\alpha) = \overline{M}_{1,n}\left(\frac{13}{2-\alpha}, \frac{1}{2-\alpha}\right)$, and the birational models of the Hassett-Keel log minimal model program are a special class of the models that we study. As we shall see, however, s and t are the most convenient parameters for describing the thresholds in the chamber decomposition.

The half-space $\operatorname{Eff}_r(\overline{M}_{1,n}) \cap \{s\lambda + t\psi + u\delta : u \leqslant 0\}$ is determined in Part (1) of the proposition, while the Mori chamber decomposition of this half-space is determined in Part (2).

Proposition 4.20.

(1) $D(s,t)$ *is effective iff* $s + nt - 12 \geqslant 0$ *and* $t \geqslant 1/2$.

(2) $\overline{M}_{1,n}^{s,t} = \overline{M}_{1,\mathcal{A}_n^k}(m)$ *for all* $(s,t) \in B_{k,m}$, *where the polytope* $B_{k,m} \subset \mathbb{Q}\{s,t\}$ *is defined by:*

	$k=1$	$k=1,\dots,n-1$	$k=n$
$m=0$	$(11,\infty)\times(\frac{3}{4},\infty)$	$(11,\infty)\times(\frac{k+2}{2k+2},\frac{k+1}{2k}]$	$(11,\infty)\times(\frac{1}{2},\frac{n+1}{2n})$
$m=1$	$(10,11]\times(\frac{3}{4},\infty)$	$(10,11]\times(\frac{k+2}{2k+2},\frac{k+1}{2k}]$	$(10,11]\times(\frac{1}{2},\frac{n+1}{2n})$
$m\geqslant 2$	$(11-m,12-m)\times(\frac{3}{4},\infty)$	$(11-m,12-m)\times(\frac{k+2}{2k+2},\frac{k+1}{2k}]$	$(11-m,12-m)\times(\frac{1}{2},\frac{n+1}{2n})$

We will not discuss the proof of this result, except to note that the strategy is, as usual, to prove that $s\lambda+t\psi-\delta$ becomes ample on the appropriate model $\overline{M}_{1,\mathcal{A}}[m]$ by showing that it has positive intersection on every curve. Methods for proving the positivity of such divisor classes on $\overline{M}_{1,n}[m]$ are developed in [109] – in which this Proposition is proved in the special case $t=1$. These can be combined with the methods of Section 3.2.5 for proving positivity of certain divisor classes on $\overline{M}_{1,\mathcal{A}}$ to obtain the result.

Our main purpose in stating this result is to indicate how the results of the log MMP for $\overline{M}_g$ (in which successive singularities are introduced into the moduli functors $\overline{\mathcal{M}}_g(\alpha)$) and the log MMP for $\overline{M}_{0,n}$ (in which successively more marked points are allowed to collide) may be expected to blend when considering the Mori chamber decomposition for $\overline{M}_{g,n}$. In the case of $\overline{M}_{1,n}$, we see that scaling λ relative to $\psi-\delta$ has the effect of introducing singularities (at the same slope thresholds as in the log MMP for $\overline{M}_g$), while scaling ψ relative to $\psi-\delta$ has the effect of allowing marked points to collide (at the same slope thresholds as in the log MMP for $\overline{M}_{0,n}$.) More generally, we expect to the same pattern to hold for the Mori chamber decomposition of $\overline{M}_{g,n}$.

4.4. Heuristics and predictions

In this section, we will discuss a heuristic method for predicting which singularities should arise in future stages of the log MMP for $\overline{M}_g$, as well as the critical α-values at which these various singularities appear. As we shall see, these predictions are connected with a number of fascinating problems around the geometry and deformation theory of curves, many of which are worth of exploration in their own right.

The heuristic operates as follows: Let X be an irreducible proper curve of arithmetic genus $g\geqslant 2$ with a single isolated singularity $p\in X$. Since X has a finite automorphism group (cf. proof of Theorem 2.4), there is a universal deformation space $\mathrm{Def}(X)$. Under very mild assumptions on the singularity, $\mathrm{Def}(X)$ is irreducible of dimension $3g-3$ and the generic point of the versal family corresponds to a smooth curve of genus g. Thus, we obtain a rational map $\mathrm{Def}(X)\dashrightarrow\overline{\mathcal{M}}_g$. Now we define $\mathcal{T}_{X,p}$, the *space of stable limits of* X, as follows: Consider a birational resolution

$$\begin{array}{ccc} & W & \\ {}^{p}\swarrow & & \searrow^{q} \\ \mathrm{Def}(X) & \dashrightarrow & \overline{\mathcal{M}}_g \end{array}$$

and let $\mathcal{T}_{X,p} := q(p^{-1}(0))$. Intuitively, $\mathcal{T}_{X,p}$ is simply the locus of stable curves appearing as stable limits of smoothings of X. Note that if $m(p)$ is the number of branches of $p \in X$ and $\delta(p)$ is the δ-invariant, then the curves lying in $\mathcal{T}_{X,p}$ have form $\widetilde{X} \cup T$, where $(\widetilde{X}, q_1, \ldots, q_{m(p)})$ is the pointed normalization of X and $(T, p_1, \ldots, p_{m(p)})$ is an $m(p)$-pointed curve attached to $\widetilde{X}$ by identifying p_i with q_i. Clearly, $p_a(T) = \delta(p) - m(p) + 1$. (We call $(T, p_1, \ldots, p_{m(p)})$ the tail of the stable limit.) We set

$$\alpha(X) := \sup_{\alpha \in \mathbb{Q}} \{\mathcal{T}_{X,p} \text{ is covered by } (K_{\overline{\mathcal{M}}_g} + \alpha\delta)\text{-negative curves}\}.$$

Remark 4.21. *Even though the set of tails of stable limits of a given curve* (X, p) *depends only on* $\widehat{\mathcal{O}}_{X,p}$, *the variety* $\mathcal{T}_{X,p}$ *may depend on the global geometry of* X. *See Remark 4.31 below.*

We can now state our main heuristic principle:

Principle 4.22. *If* $\overline{M}_g(\alpha)$ *is weakly modular, then the curve* X *appears in the corresponding moduli functor* $\overline{\mathcal{M}}_g(\alpha)$ *at* $\alpha = \alpha(X)$.

The intuition behind this heuristic is that once X appears in the moduli functor, the unique limit property dictates that all stable limits associated to X should *not* appear. Hence $\mathcal{T}_{X,p}$ should be in the exceptional locus of the map

$$|m(K_{\overline{\mathcal{M}}_g} + \alpha\delta)|\colon \overline{M}_g \dashrightarrow \overline{M}_g(\alpha), \quad m \gg 0.$$

Under mild assumptions, this implies that $\mathcal{T}_{X,p}$ is covered by curves which $K_{\overline{\mathcal{M}}_g} + \alpha\delta$ intersects non-positively.

Assuming this heuristic is reliable, we may predict when a given singularity will appear, provided that we know:

(1) The variety $\mathcal{T}_{X,p}$ of stable limits.
(2) The extremal curve classes on $\mathcal{T}_{X,p}$, i.e. covering families of maximal slope, where the slope of a curve class is defined by $s(C) = \delta \cdot C / \lambda \cdot C$.

Note the usual conversion between slope and α. If a family of curves covers $\mathcal{T}_{X,p}$ with slope s, then $\mathcal{T}_{X,p}$ must lie in the stable base locus of $K_{\overline{\mathcal{M}}_g} + \alpha\delta$ for $\alpha \leqslant 2 - 13/s$.

In what follows, we will apply Principle 4.22 to make three kinds of predictions. First, we will describe which rational m-fold points (Exercise 2.20) and elliptic m-fold points (Definition 2.31) should appear in the log MMP. This calculation raises several interesting questions regarding the qualitative nature of the stacks $\overline{\mathcal{M}}_g(\alpha)$ which should appear in the course of this program. Second, we will give a formula for the expected α-value of an arbitrary quasitoric singularity $x^p = y^q$. In particular, we will see that we expect only two infinite families of singularities, namely A and D singularities, to appear before the hyperelliptic threshold $\alpha = \frac{3g+8}{8g+4}$, as well as a finite number of others, including E_6, E_7 and E_8. Finally, we will explain why certain non-reduced schemes are expected to arise in $\overline{\mathcal{M}}_g(\frac{3g+8}{8g+4})$.

4.4.1. Rational m-fold points and elliptic m-fold points As a warm-up application of our heuristic, let us show that, for $m \geqslant 3$, the rational m-fold points do not appear in the moduli problems $\overline{\mathcal{M}}_g(\alpha)$ for any α.

Lemma 4.23. *Let X be an irreducible curve with a single rational m-fold point $p \in X$. Then the variety of stable limits of X is:*

$$i\colon \mathcal{T}_{X,p} \simeq \overline{M}_{0,m} \hookrightarrow \overline{M}_g,$$

where the map i is defined by mapping a curve $(C, \{p_i\}_{i=1}^m)$ to the point $\widetilde{X} \cup (C, \{p_i\}_{i=1}^m)$, where the points $\{p_i\}_{i=1}^m$ are attached to $\widetilde{X}$ at the m points lying above $p \in X$.

Proof. It is clear, by genus considerations, that any stable limit of X must be of the form $\widetilde{X} \cup (C, \{p_i\}_{i=1}^m)$ with $(C, \{p_i\}_{i=1}^m) \in \overline{M}_{0,m}$. The fact that *any* $(C, \{p_i\}_{i=1}^m) \in \overline{M}_{0,m}$ does arise as a stable limit is an easy consequence of Exercise 2.20. Indeed, given any curve of the form $\widetilde{X} \cup (C, \{p_i\}_{i=1}^m)$, we may consider a smoothing $\mathcal{C} \to \Delta$, and the Contraction Lemma 2.17 in conjunction with the exercise shows that contracting C produces a special fiber isomorphic to X. Thus, $\widetilde{X} \cup (C, \{p_i\}_{i=1}^m) \in \mathcal{T}_{X,p}$. □

To analyze the associated invariant $\alpha(X)$, we observe that

$$i^*\lambda = 0,$$
$$i^*(-\delta) = \psi - \delta,$$

and that $\psi - \delta$ is ample on $\overline{M}_{0,m}$. It follows that $K_{\overline{\mathcal{M}}_g} + \alpha\delta \sim \frac{13}{2-\alpha}\lambda - \delta$ has positive degree on every curve in $\mathcal{T}_{X,p}$ for all values of α! Hence, X should never appear in the moduli functors $\overline{\mathcal{M}}_g(\alpha)$.

For our next example, let us compute the α-invariants associated to elliptic m-fold points. To begin, we have the following analogue to Lemma 4.23.

Lemma 4.24. *Let X be an irreducible curve with a single elliptic m-fold point $p \in X$. Then the variety of stable limits of X is:*

$$i\colon \mathcal{T}_{X,p} \simeq \overline{M}_{1,m} \hookrightarrow \overline{M}_g,$$

where the map i is defined by mapping a curve $(C, \{p_i\}_{i=1}^m)$ to the point $\widetilde{X} \cup (C, \{p_i\}_{i=1}^m)$, where the points $\{p_i\}_{i=1}^m$ are attached to $\widetilde{X}$ at the m points lying above $p \in X$.

Proof. It is clear, by genus considerations, that any stable limit of X must be of the form $\widetilde{X} \cup (C, \{p_i\}_{i=1}^m)$ with $(C, \{p_i\}_{i=1}^m) \in \overline{M}_{1,m}$, and we only need to check that every $(C, \{p_i\}_{i=1}^m)$ arises as a tail. For tacnodes, we can check this using Proposition 2.22. Indeed, Proposition 2.22 shows that if $\mathcal{C} \to \Delta$ is any smoothing of $\widetilde{X} \cup (C, \{p_i\}_{i=1}^m)$

with smooth total space, then contracting C produces a special fiber with a tacnode. Essentially the same argument works for the general elliptic m-fold point (see [108, Lemma 2.12]). □

Now let us analyze the associated invariant $\alpha(X)$. We have

$$i^*\lambda = \lambda,$$
$$i^*(-\delta) = \psi - \delta,$$
$$i^*(s\lambda - \delta) = s\lambda + \psi - \delta.$$

Using Proposition 4.20, we see that $s\lambda + \psi - \delta$ is big until $s = 12 - m$, so that any covering family of $\overline{M}_{1,m}$ must have slope less than $12 - m$. On the other hand, it is easy to see that $\overline{M}_{1,m}$ is covered by curves of slope $12 - m$. (For $m \leqslant 9$, one can construct these families explicitly using pencils of cubics.) Since slope $12 - m$ corresponds to $\alpha = \frac{11-2m}{12-m}$, we conclude that $\overline{M}_{1,m}$ is covered by $(K_{\overline{\mathcal{M}}_g} + \alpha\delta)$-non-positive curves for the first time at $\alpha = \frac{11-2m}{12-m}$, i.e. *elliptic m-fold points should arise in the moduli functor $\overline{\mathcal{M}}_g(\alpha)$ at $\alpha = \frac{11-2m}{12-m}$*. Note that this is consistent with the result for cusps and tacnodes. Furthermore, we see that $\alpha > 0$ iff $m \leqslant 5$, i.e. we only expect elliptic m-fold points to appear for $m \leqslant 5$.

It is natural to wonder whether there is some intrinsic property of these singularities which determines whether or not they appear in the moduli functors $\overline{\mathcal{M}}_g(\alpha)$. An interesting observation in this regard is that rational m-fold points are not Gorenstein (for $m \geqslant 3$), while elliptic m-fold points are. Furthermore, an elliptic m-fold point has unobstructed deformations iff $m \leqslant 5$. This raises the following questions:

Question 4.25.

(1) *Is there any reason the moduli stacks $\overline{\mathcal{M}}_g(\alpha)$ should involve only Gorenstein singularities?*

(2) *Is there any reason that the moduli stacks $\overline{\mathcal{M}}_g(\alpha)$ should be smooth for $\alpha > 0$.*

4.4.2. A_k and D_k singularities In this section, we apply our heuristic to compute the threshold α-values at which A_k $(y^2 = x^{k+1})$ and D_k $(xy^2 = x^{k-1})$ singularities should appear. The results are displayed in Table 4. Note, in particular, that all A_k and D_k singularities are expected to appear before the hyperelliptic threshold $\alpha = \frac{3g+8}{8g+4}$.

The variety of stable limits associated to these singularities is described in [57] and we recall this description below.

Proposition 4.26 (Varieties $\mathcal{T}_{A_k}$ and $\mathcal{T}_{D_k}$).

(1) *Let X be an irreducible curve of genus g with a single A_k singularity. Then the stable limits of X are as follows:*

(a) *If k is even, a curve $\widetilde{X} \cup (T, p)$ with T hyperelliptic of arithmetic genus $\lfloor k/2 \rfloor$, p a Weierstrass point.*

(b) *If* k *is odd, a curve* $\widetilde{X} \cup (T, p, q)$ *with* T *hyperelliptic of arithmetic genus* $\lfloor k/2 \rfloor$, p *and* q *conjugate points.*

(2) *Let* X *be an irreducible curve of genus* g *with a single* D_k *singularity. Then the stable limits of* X *are as follows:*

(a) *If* k *is odd, a curve* $\widetilde{X} \cup (T, p, q)$ *with* T *hyperelliptic of arithmetic genus* $\lfloor (k-1)/2 \rfloor$, p *a Weierstrass point,* r *a free point.*

(b) *If* k *is even, a curve* $\widetilde{X} \cup (T, p, q, r)$ *with* T *hyperelliptic of arithmetic genus* $\lfloor (k-1)/2 \rfloor$, p *and* q *conjugate points,* r *a free point.*

In order to understand when the varieties $\mathcal{T}_{A_k}$ and $\mathcal{T}_{D_k}$ fall into the base locus of $K_{\overline{\mathcal{M}}_g} + \alpha\delta$, we must construct covering families for these varieties of maximal slope. The key step is the following construction of hyperelliptic families.

Proposition 4.27 (Families of hyperelliptic curves). *Let* $k \geqslant 1$ *be an integer.*

(1) *There exists a complete one-parameter family* T_k *of* 1*-pointed curves of genus* k *such that the generic fiber is a smooth hyperelliptic curve with a marked Weierstrass point. Furthermore,*

$$\begin{aligned} \lambda \cdot T_k &= k^2, \\ \delta_0 \cdot T_k &= 8k^2 + 4k, \\ \psi \cdot T_k &= 1, \\ \delta_1 \cdot T_k &= \cdots = \delta_{\lfloor k/2 \rfloor} \cdot T_k = 0. \end{aligned} \tag{4.28}$$

(2) *There exists a complete one-parameter family* B_k *of* 2*-pointed curves of genus* k *such that the generic fiber is a smooth hyperelliptic curve with marked points conjugate under the hyperelliptic involution. Furthermore,*

$$\begin{aligned} \lambda \cdot B_k &= (k^2 + k)/2, \\ \delta_0 \cdot B_k &= 4k^2 + 6k + 2, \\ \psi_1 \cdot B_k &= \psi_2 \cdot B_k = 1, \\ \delta_1 \cdot B_k &= \cdots = \delta_{\lfloor k/2 \rfloor} \cdot B_k = 0. \end{aligned} \tag{4.29}$$

Proof. There is an elementary way to write down a family of hyperelliptic curves marked with a Weierstrass section. One begins with the Hirzebruch surface $\mathbb{F}_2$ realized as a $\mathbb{P}^1$-bundle $\mathbb{F}_2 \to B$ over $B \simeq \mathbb{P}^1$. Denote by E the unique (-2)-section. Next, choose $2k+1$ general divisors $S_1, \dots, S_{2k+1}$ in the linear system $|E + 2F|$ (these are sections of $\mathbb{F}_2 \to B$ of self-intersection 2). The divisor $E + \sum_{i=1}^{2k+1} S_i$ is divisible by 2 in $\operatorname{Pic}(\mathbb{F}_2)$ and so there is a cyclic degree 2 branched cover $X \to \mathbb{F}_2$ branched over $E + \sum_{i=1}^{2k+1} S_i$. Denote by Σ the preimage of E. Then $(X, \Sigma) \to B$ is a family of at worst nodal hyperelliptic curves of genus k with a marked Weierstrass point. It is easy to verify that the constructed family has the required intersection numbers.

The proof of the second part proceeds in an analogous manner. The only modification being is that one needs to consider the double cover of $\mathbb{F}_1$ branched over $2k+2$ sections of self-intersection 1. □

If $g \geqslant k+1$, we abuse notation by using T_k to denote the family of stable curves obtained by gluing the family constructed in Part (1) of Proposition 4.27 to a constant family of 1-pointed genus $g-k$ curves. We call the resulting family of stable genus g curves the *hyperelliptic Weierstrass tails* of genus k. Using the intersection numbers in Proposition 4.27, one easily computes the slope of T_k to be $\frac{\delta \cdot T_k}{\lambda \cdot T_k} = \frac{8k^2+4k-1}{k^2}$.

If $g \geqslant k+2$, we abuse notation by using B_k to denote the family of stable curves obtained by gluing the family of 2-pointed hyperelliptic genus k curves of Proposition 4.27 (2) to a constant 2-pointed curve of genus $g-k-1$. We call the resulting family of stable genus g curves the *hyperelliptic conjugate bridges* of genus k. Using Proposition 4.27, one easily computes $\frac{\delta \cdot B_k}{\lambda \cdot B_k} = \frac{8k+12}{k+1}$.

Remark 4.30. *By exploiting the existence of a regular birational morphism from $\mathcal{T}_{A_k}$ to a variety of Picard number one, constructed in* [43, Main Theorem 1], *it is easy to deduce that the family T_k (resp. B_k) of Proposition 4.27 is a covering family of $\mathcal{T}_{A_{2k}}$ (resp. $\mathcal{T}_{A_{2k+1}}$) of the maximal slope.*

This elementary construction of hyperelliptic tails and bridges can be modified to produce the following families (see [4, Section 6.1]):

(1) The *hyperelliptic bridges* attached at a Weierstrass and a non-Weierstrass point. These arise from stable reduction of a D_{2k+1}-singularity. We denote this family by BW_k.

(2) The *hyperelliptic triboroughs* attached at a free point and two conjugate points. These arise from stable reduction of a D_{2k+2}-singularity. We denote this family by Tri_k.

(3) The *hyperelliptic tails* attached at generically non-Weierstrass points. These arise from stable reduction of a curve with a dangling A_{2k+1}-singularity (also dubbed $A^{\dagger}_{2k+1}$; see Definition 2.51). We denote this family by H_k.

Remark 4.31. *The distinction between a curve having an A_{2k+1} or an $A^{\dagger}_{2k+1}$ singularity is seen only at the level of a complete curve. The reason that $A^{\dagger}_{2k+1}$ singularities must be treated separately is that they give rise to a different variety of stable limits. The key point is that the normalization of a curve with $A^{\dagger}_{2k+1}$ singularity has a 1-pointed $\mathbb{P}^1$ which disappears after stabilization. Consequently, the stable limit is simply a nodal union of hyperelliptic genus k curve and a genus $g-k$ curve, in contrast with Proposition 4.26 (1.b).*

The intersection numbers of these curves can also be computed, and are displayed in Table 4.

4.4.3. Toric singularities In this section, we observe that similar methods can be used to compute expected α-values for toric singularities. Let $1 < q < p$ be the coprime integers. Consider a genus g curve X with an isolated toric singularity $x^{pb} - y^{qb} = 0$. The space of stable limits of X has been described by Hassett [57].

Table 4. Families of hyperelliptic tails

Curve in $N_1(\overline{\mathcal{M}}_g)$	λ	δ	Singularity	α
T_k	k^2	$8k^2+4k-1$	A_{2k}	$\frac{3k^2+8k-2}{8k^2+4k-1}$
B_k	$(k+1)k/2$	$k(4k+6)$	A_{2k+1}	$\frac{3k+11}{8k+12}$
H_k $(k \geqslant 2)$	$(k+1)k/2$	$2(k+1)(2k+1)-1$	$A^{\dagger}_{2k+1}$	$\frac{3k^2+11k+4}{8k^2+12k+2}$
BW_k	k^2	$8k^2+2k$	D_{2k+1}	$\frac{3k+4}{8k+2}$
Tri_k	$(k+1)k/2$	$k(4k+5)$	D_{2k+2}	$\frac{3k+7}{8k+10}$

Proposition 4.32. *The variety of stable limits of* X *are of the form* $\widetilde{X} \cup T$, *where the tail* $(T, p_1, \dots, p_b)$ *is a* b*-pointed curve of genus* $g = (pqb^2 - pb - qb - b + 2)/2$. *Moreover,*

(1) $K_T = (pqb - p - q - 1)(p_1 + \cdots + p_b)$.
(2) T *is* qb*-gonal with* g^1_{qb} *given by* $|q(p_1 + \cdots + p_b)|$.

The construction of covering families for this locus is much more delicate than in the case of hyperelliptic tails and hyperelliptic bridges and appears in [4, Proposition 6.6]. The final result is

Proposition 4.33. *There is a family of tails* $(T, p_1, \dots, p_b)$ *inside* $\mathcal{T}_X \subset \overline{\mathcal{M}}_{g,b}$ *whose intersection numbers are*

$$\lambda = \frac{b}{12}\left((pqb - p - q)^2 + pq(pqb^2 - pb - qb + 1) - 1\right),$$
$$\delta_0 = pqb(pqb^2 - pb - qb + 1),$$
$$\psi = b.$$

Corollary 4.34. *Suppose* p *and* q *are coprime. Then there is a family of tails of stable limits of* $x^p = y^q$ *of slope*

$$s = 12\frac{pq(p-1)(q-1) - 1}{(p-1)(q-1)(2pq - p - q - 1)}.$$

We expect these families to provide maximal slope covering families, in which case the expected α-values at which the singularity $x^p = y^q$ should arise is:

$$\alpha = \frac{(p-1)(q-1)(-2pq + 13p + 13q + 13) - 24}{12(pq(p-1)(q-1) - 1)}. \tag{4.35}$$

This formula suggests that there are only finitely many infinite families of toric singularities that should make an appearance in the intermediate log canonical models $\overline{M}_g(\alpha)$ for $\alpha \in [0,1]$. This raises the question: is there an intrinsic characterization of those singularities for which the above α-invariant is non-negative?

4.4.4. The hyperelliptic locus and ribbons A fascinating question is what happens to $\overline{\mathcal{M}}_g(\alpha)$ at $\alpha = \frac{3g+8}{8g+4}$, at which point the closure of the hyperelliptic locus falls into the base locus of $K_{\overline{\mathcal{M}}_g} + \alpha\delta$. (This is an immediate consequence of the fact that the maximal slope of a covering family of the hyperelliptic locus is $8 + \frac{4}{g}$, as proved in [25].) If smooth hyperelliptic curves must be removed from the moduli functor $\overline{\mathcal{M}}_g(\alpha)$, what should replace them? One likely candidate, suggested to us by Joe Harris, is *ribbons*, i.e. non-reduced curves of multiplicity two whose underlying reduced curve is the rational normal curve in $\mathbb{P}^{g-1}$ [10]. This prediction is certainly consonant with the final step of the log MMP in genus three, where we saw hyperelliptic curves replaced by a double plane conic. More generally, ribbons arise as the flat limit of canonically embedded smooth curves specializing (abstractly) to smooth hyperelliptic curves [47]. Using this it is easy to see that the variety of stable limits of a ribbon in $\mathbb{P}^{g-1}$ is precisely the locus of hyperelliptic curves in $\overline{M}_g$, which makes the suggestion plausible.

However, Example 2.49 shows that in genus four ribbons alone cannot be the whole story: the canonical genus 4 ribbon has nonsemistable m^{th} Hilbert point for every $m \geqslant 3$. We can get insight into other candidates by considering the GIT quotient $|\mathcal{O}_{\mathbb{P}^1\times\mathbb{P}^1}(3,3)|^{ss}/\!\!/(SL(2) \times SL(2)) \rtimes \mathbb{Z}_2$, which is in fact equal to $\overline{\mathcal{M}}_4(29/60)$ [44]. A simple application of the Hilbert-Mumford numerical criterion (Proposition 2.38) shows that the union of a $(2,1)$-curve and a $(1,2)$-curve on $\mathbb{P}^1 \times \mathbb{P}^1$ meeting at an A_9-singularity ($y^2 = x^{10}$) and an irreducible $(3,3)$-curve with a unique A_8-singularity ($y^2 = x^9$) are *stable*. These examples suggest that for higher genera the singular curves which replace hyperelliptic curves in $\overline{\mathcal{M}}_g(\alpha)$ with $\alpha < \frac{3g+8}{8g+4}$ should be, in the order of generality: rational curves with an A_{2g} singularity, two rational normal curves meeting in an A_{2g+1} singularity, and certain semistable ribbons. All these curves arise as flat limits of canonically embedded smooth curves specializing abstractly to smooth hyperelliptic curves.

We should also remark that there is no reason to preclude the possibility of non-reduced curves showing up prior to $\alpha = \frac{3g+8}{8g+4}$. Needless to say, the prospect of constructing and classifying modular birational models involving non-reduced curves raises a host of questions regarding degenerations and deformations of non-reduced schemes, pushing far beyond the methods of this survey.

References

[1] Valery Alexeev. Complete moduli in the presence of semiabelian group action. *Ann. of Math. (2)*, **155**(3), 611–708, 2002. ← 389

[2] Valery Alexeev and David Swinarski. Nef divisors on $\overline{M}_{0,n}$ from GIT, in: Geometry and Arithmetic, 1–21. *EMS Series of Congress Reports*, European Mathematical Society (EMS), Zürich, 2012. ← 334, 395

[3] Jarod Alper. Good moduli spaces for Artin stacks, to appear in *Ann. Inst. Fourier (Grenoble)*, 2008. `arXiv:0810.1677 [math.AG]` ← 338, 359

[4] Jarod Alper, Maksym Fedorchuk, and David Smyth. Singularities with $\mathbb{G}_m$-action and the log minimal model program for $\overline{M}_g$, 2010. arXiv: 1010.3751v2 [math.AG] ← 404, 405

[5] Enrico Arbarello and Maurizio Cornalba. The Picard groups of the moduli spaces of curves. *Topology*, **26**(2), 153–171, 1987. ← 335, 336

[6] Enrico Arbarello and Maurizio Cornalba. Calculating cohomology groups of moduli spaces of curves via algebraic geometry. *Inst. Hautes Études Sci. Publ. Math.*, (88), 97–127 (1999), 1998. ← 335, 336, 381, 383, 387

[7] Enrico Arbarello and Edoardo Sernesi. The equation of a plane curve. *Duke Math. J.*, **46**(2), 469–485, 1979. ← 373

[8] Michela Artebani. A compactification of $\mathcal{M}_3$ via K3 surfaces. *Nagoya Math. J.*, **196**, 1–26, 2009. ← 389

[9] M. Artin. Algebraization of formal moduli. II. Existence of modifications. *Ann. of Math. (2)*, **91**, 88–135, 1970. ← 346

[10] Dave Bayer and David Eisenbud. Ribbons and their canonical embeddings. *Trans. Amer. Math. Soc.*, **347**(3), 719–756, 1995. ← 363, 406

[11] Caucher Birkar, Paolo Cascini, Christopher D. Hacon, and James M^c^Kernan. Existence of minimal models for varieties of log general type. *J. Amer. Math. Soc.*, **23**(2), 405–468, 2010. ← 334, 340, 371, 372, 378

[12] Marco Boggi. Compactifications of configurations of points on $\mathbb{P}^1$ and quadratic transformations of projective space. *Indag. Math. (N.S.)*, **10**(2), 191–202, 1999. ← 350

[13] S. Boucksom, J.-P. Demailly, M. Paun, and T. Peternell. The pseudo-effective cone of a compact Kähler manifold and varieties of negative Kodaira dimension, 2004. To appear in *J. Algebraic Geom.*, doi: 10.1090/S1056-3911-2012-00574-8 ← 375

[14] Andrea Bruno and Alessandro Verra. $\mathcal{M}_{15}$ is rationally connected, in: *Projective varieties with unexpected properties*, 51–65. Walter de Gruyter GmbH & Co. KG, Berlin, 2005. ← 334, 375

[15] Ana-Maria Castravet. The Cox ring of $\overline{M}_{0,6}$. *Trans. Amer. Math. Soc.*, **361**(7), 3851–3878, 2009. ← 378

[16] Ana-Maria Castravet and Jenia Tevelev. Hypertrees, projections, and moduli of stable rational curves. To appear in *J. Reine Angew. Math.*, 2010, arXiv:1004.2553 [math.AG] ← 376

[17] F. Catanese. On the rationality of certain moduli spaces related to curves of genus 4, in: *Algebraic geometry (Ann Arbor, Mich., 1981)*, 30–50. *Lecture Notes in Math.*, **1008**, Springer, Berlin, 1983. ← 334

[18] Mei-Chu Chang and Ziv Ran. Unirationality of the moduli spaces of curves of genus 11, 13 (and 12). *Invent. Math.*, **76**(1), 41–54, 1984. ← 375

[19] Mei-Chu Chang and Ziv Ran. The Kodaira dimension of the moduli space of curves of genus 15. *J. Differential Geom.*, **24**(2), 205–220, 1986. ← 375

[20] Mei-Chu Chang and Ziv Ran. On the slope and Kodaira dimension of $\overline{\mathcal{M}}_g$ for small g. *J. Differential Geom.*, **34**(1), 267–274, 1991. ← 375

[21] Dawei Chen. Covers of the projective line and the moduli space of quadratic differentials. *Geom. Dedicata*, 2012. doi: 10.1007/s10711-012-9737-x ← 376

[22] Dawei Chen. Covers of elliptic curves and the moduli space of stable curves. *J. Reine Angew. Math.*, **649**, 167–205, 2010. ← 376

[23] Dawei Chen. Square-tiled surfaces and rigid curves on moduli spaces. *Adv. Math.*, **228**(2), 1135–1162, 2011. ← 376

[24] A. Clebsch. *Theorie der binären algebraischen Formen*. Verlag von B.G. Teubner, Leipzig, 1872. ← 373

[25] Maurizio Cornalba and Joe Harris. Divisor classes associated to families of stable varieties, with applications to the moduli space of curves. *Ann. Sci. École Norm. Sup. (4)*, **21**(3), 455–475, 1988. ← 377, 406

[26] Fernando Cukierman and Douglas Ulmer. Curves of genus ten on K3 surfaces. *Compositio Math.*, **89**(1), 81–90, 1993. ← 375

[27] P. Deligne and D. Mumford. The irreducibility of the space of curves of given genus. *Inst. Hautes Études Sci. Publ. Math.*, (36), 75–109, 1969. ← 332, 342, 373

[28] Steven Diaz. A bound on the dimensions of complete subvarieties of $\mathcal{M}_g$. *Duke Math. J.*, **51**(2), 405–408, 1984. ← 334

[29] Igor V. Dolgachev. Rationality of $\mathcal{R}_2$ and $\mathcal{R}_3$. *Pure Appl. Math. Q.*, **4**(2, part 1), 501–508, 2008. ← 334

[30] Ron Donagi. The unirationality of $\mathcal{A}_5$. *Ann. of Math. (2)*, **119**(2), 269–307, 1984. ← 334

[31] David Eisenbud and Joe Harris. Limit linear series, the irrationality of M_g, and other applications. *Bull. Amer. Math. Soc. (N.S.)*, **10**(2), 277–280, 1984. ← 375

[32] David Eisenbud and Joe Harris. The Kodaira dimension of the moduli space of curves of genus $\geqslant 23$. *Invent. Math.*, **90**(2), 359–387, 1987. ← 373

[33] Carel Faber and Gerard van der Geer. Complete subvarieties of moduli spaces and the Prym map. *J. Reine Angew. Math.*, **573**, 117–137, 2004. ← 334

[34] Gavril Farkas. Syzygies of curves and the effective cone of $\overline{\mathcal{M}}_g$. *Duke Math. J.*, **135**(1), 53–98, 2006. ← 375

[35] Gavril Farkas. Birational aspects of the geometry of $\overline{\mathcal{M}}_g$. In: *Surveys in differential geometry. Vol. XIV. Geometry of Riemann surfaces and their moduli spaces. Surv. Differ. Geom.*, pages 57–110. Int. Press, Somerville, MA, 2009. ← 334, 375, 376

[36] Gavril Farkas. The global geometry of the moduli space of curves, in: *Algebraic geometry—Seattle 2005. Part 1*, 125–147. *Proc. Sympos. Pure Math.*, **80**, Amer. Math. Soc., Providence, RI, 2009. ← 334, 372

[37] Gavril Farkas. Koszul divisors on moduli spaces of curves. *Amer. J. Math.*, **131**(3), 819–867, 2009. ← 334, 375, 377

[38] Gavril Farkas and Angela Gibney. The Mori cones of moduli spaces of pointed curves of small genus. *Trans. Amer. Math. Soc.*, **355**(3), 1183–1199 (electronic), 2003. ← 380

[39] Gavril Farkas and Katharina Ludwig. The Kodaira dimension of the moduli space of Prym varieties. *J. Eur. Math. Soc. (JEMS)*, **12**(3), 755–795, 2010. ← 334

[40] Gavril Farkas and Mihnea Popa. Effective divisors on $\overline{\mathcal{M}}_g$, curves on K3 surfaces, and the slope conjecture. *J. Algebraic Geom.*, **14**(2), 241–267, 2005. ← 375

[41] G. Farkas and A. Verra. The classification of universal Jacobians over the moduli space of curves. To appear in *Commentarii Mathematici Helvetici*, 2011. `arXiv:1005.5354 [math.AG]` ← 334

[42] Maksym Fedorchuk. Severi varieties and the moduli space of curves, 2008. Ph.D. thesis, Harvard University. ← 376

[43] Maksym Fedorchuk. Moduli spaces of hyperelliptic curves with A and D singularities, 2010. `arXiv:1007.4828 [math.AG]` ← 404

[44] Maksym Fedorchuk. The final log canonical model of the moduli space of stable curves of genus 4. *Int. Math. Res. Not. IMRN*, 2012. ← 406

[45] Maksym Fedorchuk. Moduli of weighted pointed stable curves and log canonical models of $\overline{\mathcal{M}}_{g,n}$. *Math. Res. Lett.*, **18**(4), 663–675, 2011. ← 382, 383

[46] Maksym Fedorchuk and David Ishii Smyth. Ample divisors on moduli spaces of pointed rational curves. *J. Algebraic Geom.*, **20**(4), 599–629, 2011. ← 334, 395

[47] Lung-Ying Fong. Rational ribbons and deformation of hyperelliptic curves. *J. Algebraic Geom.*, **2**(2), 295–307, 1993. ← 363, 406

[48] Claudio Fontanari. Positive divisors on quotients of $\overline{M}_{0,n}$ and the Mori cone of $\overline{M}_{g,n}$. *J. Pure Appl. Algebra*, **213**(4), 454–457, 2009. ← 380

[49] Angela Gibney. Numerical criteria for divisors on $\overline{M}_g$ to be ample. *Compos. Math.*, **145**(5), 1227–1248, 2009. ← 378

[50] Angela Gibney, Sean Keel, and Ian Morrison. Towards the ample cone of $\overline{M}_{g,n}$. *J. Amer. Math. Soc.*, **15**(2), 273–294 (electronic), 2002. ← 378, 380, 381

[51] D. Gieseker. *Lectures on moduli of curves*, volume 69 of *Tata Institute of Fundamental Research Lectures on Mathematics and Physics*. Published for the Tata Institute of Fundamental Research, Bombay, 1982. ← 360

[52] Samuel Grushevsky and Igor Krichever. The universal Whitham hierarchy and the geometry of the moduli space of pointed Riemann surfaces, in: Surveys in differential geometry. Vol. XIV. Geometry of Riemann surfaces and their moduli spaces, 111–129. *Surv. Differ. Geom.*,**14** , Int. Press, Somerville, MA, 2009. ← 334

[53] J. Harris and I. Morrison. Slopes of effective divisors on the moduli space of stable curves. *Invent. Math.*, **99**(2), 321–355, 1990. ← 375, 376

[54] Joe Harris and Ian Morrison. Moduli of curves. *Graduate Texts in Mathematics*, **187**, Springer-Verlag, New York, 1998. ← 335, 360, 362

[55] Joe Harris and David Mumford. On the Kodaira dimension of the moduli space of curves. *Invent. Math.*, **67**(1), 23–88, 1982. With an appendix by William Fulton. ← 335, 336, 373, 374, 375

[56] Robin Hartshorne. Algebraic geometry. *Graduate Texts in Mathematics*, **52**, Springer-Verlag, New York, 1977. ← 346

[57] Brendan Hassett. Local stable reduction of plane curve singularities. *J. Reine Angew. Math.*, **520**, 169–194, 2000. ← 402, 404

[58] Brendan Hassett. Moduli spaces of weighted pointed stable curves. *Adv. Math.*, **173**(2), 316–352, 2003. ← 342, 344, 384, 398

[59] Brendan Hassett. Classical and minimal models of the moduli space of curves of genus two, in: *Geometric methods in algebra and number theory*, 169–192. *Progr. Math.*, **235**, Birkhäuser Boston, Boston, MA, 2005. ← 334, 384, 387

[60] Brendan Hassett and Donghoon Hyeon. Log minimal model program for the moduli space of curves: the first flip, To appear in *Ann. of Math.*, 2008. `arXiv:0806.3444 [math.AG]` ← 334, 355, 360, 361, 363, 365, 366, 367, 384, 394

[61] Brendan Hassett and Donghoon Hyeon. Log canonical models for the moduli space of curves: the first divisorial contraction. *Trans. Amer. Math. Soc.*, **361**(8), 4471–4489, 2009. ← 334, 349, 360, 384, 394

[62] Brendan Hassett, Donghoon Hyeon, and Yongnam Lee. Stability computation via Gröbner basis. *J. Korean Math. Soc.*, **47**(1), 41–62, 2010. ← 360

[63] Brendan Hassett and Yuri Tschinkel. On the effective cone of the moduli space of pointed rational curves, in: Topology and geometry: commemorating SISTAG, 83–96. *Contemp. Math.*, **314**, Amer. Math. Soc., Providence, RI, 2002. ← 376

[64] Yi Hu and Seán Keel. Mori dream spaces and GIT. *Michigan Math. J.*, **48**, 331–348, 2000. Dedicated to William Fulton on the occasion of his 60th birthday. ← 371

[65] Donghoon Hyeon and Yongnman Lee. Log minimal model program for the moduli space of stable curves of genus three. *Math. Res. Lett.*, **17**(4), 625–636, 2010. ← 334, 360, 384, 389, 391, 392

[66] Donghoon Hyeon and Ian Morrison. Stability of tails and 4-canonical models. *Math.Res.Lett.*, **17**(04), 721–729, 2010. ← 365, 366

[67] Jun-Ichi Igusa. Arithmetic variety of moduli for genus two. *Ann. of Math. (2)*, **72**, 612–649, 1960. ← 373

[68] Elham Izadi, Marco Lo Giudice, and Gregory Kumar Sankaran. The moduli space of étale double covers of genus 5 curves is unirational. *Pacific J. Math.*, **239**(1), 39–52, 2009. ← 334

[69] M. M. Kapranov. Chow quotients of Grassmannians. I, in: *I. M. Gelfand Seminar*, 29–110. *Adv. Soviet Math.*, **16**, Amer. Math. Soc., Providence, RI, 1993. ← 398

[70] M. M. Kapranov. Veronese curves and Grothendieck-Knudsen moduli space $\overline{M}_{0,n}$. *J. Algebraic Geom.*, **2**(2), 239–262, 1993. ← 398

[71] P. I. Katsylo. Rationality of the variety of moduli of curves of genus 5. *Mat. Sb.*, **182**(3), 457–464, 1991. ← 334, 373

[72] P. I. Katsylo. On the unramified 2-covers of the curves of genus 3, in: *Algebraic geometry and its applications (Yaroslavl, 1992)*, 61–65. *Aspects Math.*, **E25**, Vieweg, Braunschweig, 1994. ← 334

[73] P. I. Katsylo. Rationality of the moduli variety of curves of genus 3. *Comment. Math. Helv.*, **71**(4), 507–524, 1996. ← 334, 373, 388

[74] Seán Keel. Basepoint freeness for nef and big line bundles in positive characteristic. *Ann. of Math. (2)*, **149**(1), 253–286, 1999. ← 378, 382

[75] Seán Keel and James McKernan. Contractible extremal rays on $\overline{M}_{0,n}$, Handbook of moduli, Vol. II (Gavril Farkas and Ian Morrison eds.), 113–128. *Advanced Lectures in Mathematics*, **25**, Higher Education Press & International Press, Beijing-Boston, 2012. ← 377, 381

[76] Seán Keel and Shigefumi Mori. Quotients by groupoids. *Ann. of Math. (2)*, **145**(1), 193–213, 1997. ← 337, 338

[77] George R. Kempf. Instability in invariant theory. *Ann. of Math. (2)*, **108**(2), 299–316, 1978. ← 362

[78] Deepak Khosla. Moduli spaces of curves with linear series and the slope conjecture. Ph.D. thesis, Harvard University, Cambridge, MA, 2005. ← 375

[79] Deepak Khosla. Tautological classes on moduli spaces of curves with linear series and a push-forward formula when $\rho = 0$, 2007. `arXiv:0704.1340 [math.AG]` ← 375

[80] János Kollár. Projectivity of complete moduli. *J. Differential Geom.*, **32**(1), 235–268, 1990. ← 382

[81] János Kollár and Shigefumi Mori. Birational geometry of algebraic varieties. *Cambridge Tracts in Mathematics*, **134**, Cambridge University Press, Cambridge, 1998. With the collaboration of C. H. Clemens and A. Corti, Translated from the 1998 Japanese original. ← 341, 349, 371, 387

[82] Shigeyuki Kondō. A complex hyperbolic structure for the moduli space of curves of genus three. *J. Reine Angew. Math.*, **525**, 219–232, 2000. ← 334, 389

[83] Shigeyuki Kondō. The moduli space of curves of genus 4 and Deligne-Mostow's complex reflection groups, in: *Algebraic geometry 2000, Azumino (Hotaka)*, 383–400. *Adv. Stud. Pure Math.*, **36**, Math. Soc. Japan, Tokyo, 2002. ← 334

[84] Paul L. Larsen. Fulton's conjecture for $\overline{M}_{0,7}$. *Journal of the London Mathematical Society*, **85**(1), 1–21, 2012. ← 380

[85] Gérard Laumon and Laurent Moret-Bailly. Champs algébriques, in: Ergebnisse der Mathematik und ihrer Grenzgebiete. 3. Folge.*A Series of Modern Surveys in Mathematics,* **39**, Springer-Verlag, Berlin, 2000. ← 338

[86] Robert Lazarsfeld. Positivity in algebraic geometry. I. *Ergebnisse der Mathematik und ihrer Grenzgebiete. 3. Folge. A Series of Modern Surveys in Mathematics,* **48**, Springer-Verlag, Berlin, 2004. Classical setting: line bundles and linear series. ← 370, 371

[87] V. Lazic. Adjoint rings are finitely generated, 2009, `arXiv: 0905.2707 [math.AG]` ← 340

[88] Adam Logan. The Kodaira dimension of moduli spaces of curves with marked points. *Amer. J. Math.*, **125**(1), 105–138, 2003. ← 334, 336, 377

[89] Shigefumi Mori and Shigeru Mukai. The uniruledness of the moduli space of curves of genus 11, in: *Algebraic geometry (Tokyo/Kyoto, 1982)*, 334–353. *Lecture Notes in Math.*, **1016**, Springer, Berlin, 1983. ← 334

[90] Atsushi Moriwaki. Relative Bogomolov's inequality and the cone of positive divisors on the moduli space of stable curves. *J. Amer. Math. Soc.*, **11**(3), 569–600, 1998. ← 377

[91] Ian Morrison. Mori theory of moduli spaces of stable curves. Projective Press, New York, 2009. ← 334, 372

[92] Ian Morrison. GIT constructions of moduli spaces of stable curves and maps, in: *Surveys in differential geometry. Vol. XIV. Geometry of Riemann surfaces and their moduli spaces*, 315–369. *Surv. Differ. Geom.*, **14**, Int. Press, Somerville, MA, 2009. ← 360

[93] Ian Morrison and David Swinarski. Gröbner techniques for low-degree Hilbert stability. *Exp. Math.*, **20**(1), 34–56, 2011. ← 360, 362, 363

[94] David Mumford. Lectures on curves on an algebraic surface, With a section by G. M. Bergman. *Annals of Mathematics Studies,***59**, Princeton University Press, Princeton, N.J., 1966. ← 358

[95] David Mumford. Stability of projective varieties. *Enseignement Math. (2)*, **23**(1-2), 39–110, 1977. ← 360, 363

[96] D. Mumford, J. Fogarty, and F. Kirwan. Geometric invariant theory. *Ergebnisse der Mathematik und ihrer Grenzgebiete (2)*, **34**, Springer-Verlag, Berlin, third edition, 1994. ← 355, 356, 357, 359, 364, 390, 395

[97] Yukihiko Namikawa. On the canonical holomorphic map from the moduli space of stable curves to the Igusa monoidal transform. *Nagoya Math. J.*, **52**, 197–259, 1973. ← 374, 386

[98] Rahul Pandharipande. Descendent bounds for effective divisors on $\overline{M}_g$. *J. Algebraic Geom.*, **21**(2), 299–303, 2012. ← 376

[99] Miles Reid. Canonical 3-folds, in: *Journées de Géometrie Algébrique d'Angers, Juillet 1979/Algebraic Geometry, Angers, 1979*, 273–310. Sijthoff & Noordhoff, Alphen aan den Rijn, 1980. ← 374

[100] William Rulla. The birational geometry of $\overline{M}_{2,1}$ and $\overline{M}_3$. Ph.D. thesis, University of Texas at Austin, 2001. ← 370, 389, 393

[101] David Schubert. A new compactification of the moduli space of curves. *Compositio Math.*, **78**(3), 297–313, 1991. ← 332, 349, 365, 366

[102] Edoardo Sernesi. Unirationality of the variety of moduli of curves of genus twelve. *Ann. Scuola Norm. Sup. Pisa Cl. Sci. (4)*, **8**(3), 405–439, 1981. ← 375

[103] Francesco Severi. *Vorlesungen über Algebraische Geometrie*. Teubner, Leipzig, 1921. ← 373

[104] N. I. Shepherd-Barron. The rationality of certain spaces associated to trigonal curves, in: *Algebraic geometry, Bowdoin, 1985 (Brunswick, Maine, 1985)*, 165–171. *Proc. Sympos. Pure Math.*, **46**, Amer. Math. Soc., Providence, RI, 1987. ← 334, 373

[105] N. I. Shepherd-Barron. Invariant theory for S_5 and the rationality of M_6. *Compositio Math.*, **70**(1), 13–25, 1989. ← 334, 373

[106] Yum-Tong Siu. Finite generation of canonical ring by analytic method. *Sci. China Ser. A*, **51**(4), 481–502, 2008. ← 340, 371

[107] David I. Smyth. Towards a classification of modular compactifications of the moduli space of curves. *Invent. Math.*, 2012. With an appendix by Jack Hall, doi: 10.1007/s00222-012-0416-1 ← 336, 346, 350, 359, 369

[108] David I. Smyth. Modular compactifications of the space of pointed elliptic curves I. *Compositio Math.*, **147**(3), 877–913, 2011. ← 334, 348, 349, 352, 353, 355, 402

[109] David I. Smyth. Modular compactifications of the space of pointed elliptic curves II. *Compositio Mathematica*, **147**(06), 1843–1884, 2011. ← 334, 399

[110] Yung-Sheng Tai. On the Kodaira dimension of the moduli space of abelian varieties. *Invent. Math.*, **68**(3), 425–439, 1982. ← 374

[111] Peter Vermeire. A counterexample to Fulton's conjecture on $\overline{M}_{0,n}$. *J. Algebra*, **248**(2), 780–784, 2002. ← 376

[112] Alessandro Verra. A short proof of the unirationality of $\mathcal{A}_5$. *Nederl. Akad. Wetensch. Indag. Math.*, **46**(3), 339–355, 1984. ← 334

[113] Alessandro Verra. The unirationality of the moduli spaces of curves of genus 14 or lower. *Compos. Math.*, **141**(6), 1425–1444, 2005. ← 334, 375

[114] Alessandro Verra. On the universal abelian variety of dimension 4. *Contemp. Math.*, **465**, 253–274, 2008. ← 334

Department of Mathematics, Columbia University, 2990 Broadway, New York, NY 10027

Current address: Department of Mathematics, Boston College, 140 Commonwealth Avenue, Chestnut Hill, MA 02467

E-mail address: maksym.fedorchuk@bc.edu

Department of Mathematics, Harvard University, One Oxford Street, Cambridge, MA 01238

E-mail address: dsmyth@math.harvard.edu

The cohomology of the moduli space of Abelian varieties

Gerard van der Geer

Contents

Introduction

That the moduli spaces of abelian varieties are a rich source of arithmetic and geometric information slowly emerged from the work of Kronecker, Klein, Fricke and many others at the end of the 19th century. Since the 20th century we know that the first place to dig for these hidden treasures is the cohomology of these moduli spaces.

In this survey we are concerned with the cohomology of the moduli space of abelian varieties. Since this is an extensive and widely ramified topic that connects to many branches of algebraic geometry and number theory we will have to limit ourselves. I have chosen to stick to the moduli spaces of principally polarized abelian

2000 *Mathematics Subject Classification.* Primary 14; Secondary: 11G, 14F, 14K, 14J10, 10D.
Key words and phrases. Abelian Varieties, Moduli, Modular Forms.

varieties, leaving aside the moduli spaces of abelian varieties with non-principal polarizations and the other variations like the moduli spaces with extra structure (like conditions on their endomorphism rings) since often the principles are the same, but the variations are clad in heavy notation.

The emphasis in this survey is on the tautological ring of the moduli space of principally polarized abelian varieties. We discuss the cycle classes of the Ekedahl-Oort stratification, that can be expressed in tautological classes, and discuss differential forms on the moduli space. We also discuss complete subvarieties of $\mathcal{A}_g$. Finally, we discuss Siegel modular forms and its relations to the cohomology of these moduli spaces. We sketch the approach developed jointly with Faber and Bergström to calculate the traces of the Hecke operators by counting curves of genus $\leqslant 3$ over finite fields, an approach that opens a new window on Siegel modular forms.

1. The moduli space of principally polarized Abelian varieties

We shall assume the existence of the moduli space of principally polarized abelian varieties as given. So throughout this survey $\mathcal{A}_g$ will denote the Deligne-Mumford stack of principally polarized abelian varieties of dimension g. It is a smooth Deligne-Mumford stack over $\mathrm{Spec}(\mathbb{Z})$ of relative dimension $g(g+1)/2$, see [25].

Over the complex numbers this moduli space can be described as the arithmetic quotient (orbifold) $\mathrm{Sp}(2g,\mathbb{Z})\backslash\mathfrak{H}_g$ of the Siegel upper half space by the symplectic group. This generalizes the well-known description of the moduli of complex elliptic curves as $\mathrm{SL}(2,\mathbb{Z})\backslash\mathfrak{H}$ with $\mathfrak{H}=\mathfrak{H}_1$ the usual upper half plane of the complex plane. We refer to Milne's account in this Handbook for the general theory of Shimura varieties.

The stack $\mathcal{A}_g$ comes with a universal family of principally polarized abelian varieties $\pi:\mathcal{X}_g\to\mathcal{A}_g$. Since abelian varieties can degenerate the stack $\mathcal{A}_g$ is not proper or complete.

The moduli space $\mathcal{A}_g$ admits several compactifications. The first one is the Satake compactification or Baily-Borel compactification. It is defined by considering the vector space of Siegel modular forms of sufficiently high weight, by using these to map $\mathcal{A}_g$ to projective space and then by taking the closure of the image of $\mathcal{A}_g$ in the receiving projective space. This construction was first done by Satake and by Baily-Borel over the field of complex numbers, cf. [5]. The Satake compactification $\mathcal{A}_g^*$ is very singular for $g\geqslant 2$. It has a stratification

$$\mathcal{A}_g^*=\mathcal{A}_g\sqcup\mathcal{A}_{g-1}^*=\mathcal{A}_g\sqcup\mathcal{A}_{g-1}\sqcup\cdots\sqcup\mathcal{A}_1\sqcup\mathcal{A}_0.$$

In an attempt to construct non-singular compactifications of arithmetic quotients, such as $\mathrm{Sp}(2g,\mathbb{Z})\backslash\mathfrak{H}_g$, Mumford with a team of co-workers created a theory of so-called toroidal compactifications of $\mathcal{A}_g$ in [3], cf. also the new edition [4]. These compactifications are not unique but depend on a choice of combinatorial data, an admissible cone decomposition of the cone of positive (semi)-definite symmetric bilinear forms in g variables. Each such toroidal compactification admits a morphism $q : \tilde{\mathcal{A}}_g \to \mathcal{A}_g^*$. Faltings and Chai showed how Mumford's theory of compactifying quotients of bounded symmetric domains could be used to extend this to a compactification of $\mathcal{A}_g$ over the integers, see [24, 25, 3]. This also led to the Satake compactification over the integers.

There are a number of special choices for the cone decompositions, such as the second Voronoi decomposition, the central cone decomposition or the perfect cone decomposition. Each of these choices has its advantages and disadvantages. Moreover, Alexeev has constructed a functorial compactification of $\mathcal{A}_g$, see [1, 2]. It has as a disadvantage that for $g \geqslant 4$ it is not irreducible but possesses extra components. The main component of this corresponds to $\mathcal{A}_g^{\mathrm{Vor}}$, the toroidal compactification defined by the second Voronoi compactification. Olsson has adapted Alexeev's construction using log structures and obtained a functorial compactification. The normalization of this compactification is the compactification $\mathcal{A}_g^{\mathrm{Vor}}$, cf [59].

The toroidal compactification $\mathcal{A}_g^{\mathrm{perf}}$ defined by the perfect cone decomposition is a canonical model of $\mathcal{A}_g$ for $g \geqslant 12$ as was shown by Shepherd-Barron, see [67] and his contribution to this Handbook.

The partial desingularization of the Satake compactification obtained by Igusa in [44] coincides with toroidal compactification corresponding to the central cone decomposition. We refer to a survey by Grushevsky ([37]) on the geometry of the moduli space of abelian varieties.

These compactifications (2nd Voronoi, central and perfect cone) agree for $g \leqslant 3$, but are different for higher g. For $g = 1$ one has $\tilde{\mathcal{A}}_1 = \mathcal{A}_1^* = \overline{\mathcal{M}}_{1,1}$, the Deligne-Mumford moduli space of 1-pointed stable curves of genus 1. For $g = 2$ the Torelli morphism gives an identification $\tilde{\mathcal{A}}_2 = \overline{\mathcal{M}}_2$, the Deligne-Mumford moduli space of stable curves of genus 2. The open part $\mathcal{A}_2$ corresponds to stable curves of genus 2 of compact type. But please note that the Torelli morphism of stacks $\mathcal{M}_g \to \mathcal{A}_g$ is a morphism of degree 2 for $g \geqslant 3$, since every principally polarized abelian variety possesses an automorphism of order 2, but the generic curve of genus $g \geqslant 3$ does not.

We shall use the term Faltings-Chai compactifications for the compactifications (over rings of integers) defined by admissible cone decompositions.

2. The compact dual

The moduli space $\mathcal{A}_g(\mathbb{C})$ has the analytic description as $\mathrm{Sp}(2g, \mathbb{Z})\backslash\mathfrak{H}_g$. The Siegel upper half space $\mathfrak{H}_g$ can be realized in various ways, one of which is as an open subset of the so-called compact dual and the arithmetic quotient inherits various properties from this compact quotient. For this reason we first treat the compact dual at some length.

To construct $\mathfrak{H}_g$, we start with an non-degenerate symplectic form on a complex vector space, necessarily of even dimension, say $2g$. To be explicit, consider the vector space $\mathbb{Q}^{2g}$ with basis $e_1, \ldots, e_{2g}$ and symplectic form $\langle , \rangle$ given by $J(x,y) = x^t J y$ with

$$J = \begin{pmatrix} 0 & 1_g \\ -1_g & 0 \end{pmatrix}.$$

We let $G = \mathrm{GSp}(2g, \mathbb{Q})$ be the algebraic group over $\mathbb{Q}$ of symplectic similitudes of this symplectic vector space

$$G = \{g \in \mathrm{GL}(2g, \mathbb{Q}) : J(gx, gy) = \eta(g) J(x,y)\},$$

where $\eta : G \to \mathbb{G}_m$ is the so-called multiplier. Then G is the group of matrices $\gamma = (A\,B; C\,D)$ with A, B, C, D integral $g \times g$-matrices with

$$A^t \cdot C = C^t \cdot A,\ B^t \cdot D = D^t \cdot B \quad \text{and} \quad A^t \cdot D - C^t \cdot B = \eta(g) 1_{2n}.$$

We denote the kernel of η by G^0.

Let Y_g be the Lagrangian Grassmannian

$$Y_g = \{L \subset \mathbb{C}^{2g} : \dim(L) = g,\ J(x,y) = 0 \text{ for all } x, y \in L\}$$

that parametrizes all totally isotropic subspaces of our complex symplectic vector space. This is a homogeneous manifold of complex dimension $g(g+1)/2$ for the action of $G(\mathbb{C}) = \mathrm{GSp}(2g, \mathbb{C})$; in fact this group acts transitively and the quotient of $\mathrm{GSp}(2g, \mathbb{C})$ by the central $\mathbb{G}_m(\mathbb{C})$ acts effectively. We can write Y_g as a quotient $Y_g = G(\mathbb{C})/Q$, where Q is a parabolic subgroup of $G(\mathbb{C})$. More precisely, if we fix a point $y_0 = e_1 \wedge \ldots \wedge e_g \in Y_g$ then we can write Q as the group of matrices $(A\,B; C\,D)$ in $\mathrm{GSp}(2g, \mathbb{C})\}$ with $C = 0$. This parabolic group Q has a Levi decomposition as $Q = M \ltimes U$ with M the subgroup of G that respects the decomposition of the symplectic space as $\mathbb{Q}^g \oplus \mathbb{Q}^g$; the matrices of M are of the form $(A\,0; 0\,D)$ and is isomorphic to $\mathrm{GL}(g) \times \mathbb{G}_m$, while those of U are of the form $(1_g\,B; 0\,1_g)$ with B symmetric.

There is an embedding of $\mathfrak{H}_g$ into Y_g as follows. We consider the group $G^0(\mathbb{R})$ and the maximal compact subgroup K of elements that fix $\sqrt{-1}\,1_g$. Its elements are described as $(A\ -B; B\,A)$ and assigning to it the element $A + \sqrt{-1}B$ gives an isomorphism of K with the unitary group $U(g)$. One way to describe the Siegel upper half space is as the orbit $X_g = G^0(\mathbb{R})/K$ under the action of G^0; this can be embedded in $Y_g = (G/Q)(\mathbb{C})$ as the set of all maximal isotropic subspaces V such that $-\sqrt{-1}\langle v, \bar{v}\rangle$ is positive definite on V. Each such subspace has a basis consisting of the columns of the transpose of the matrix $(-1_g\,\tau)$ for a unique $\tau \in \mathfrak{H}_g$. The subgroup $G^+(\mathbb{R})$ leaves this subset invariant and this establishes the embedding of the domain $\mathfrak{H}_g$ in its Y_g and this space Y_g is called the *compact dual* of $\mathfrak{H}_g$. It contains $X_g \sim \mathfrak{H}_g$ as an open subset. The standard example (for $g = 1$) is that of the upper half plane contained in $\mathbb{P}^1$.

For later use we extend this a bit by looking not only at maximal isotropic subspaces, but also at symplectic filtrations on our standard symplectic space.

Consider for $i = 1, \dots, g$ the (partial) flag variety $U_g^{(i)}$ of symplectic flags $E_i \subset \dots \subset E_{g-1} \subset E_g$, of linear subspaces E_j of $\mathbb{C}^{2g}$ with $\dim(E_j) = j$ and E_g totally isotropic. We have $U_g^{(g)} = Y_g$. There are natural maps $\pi_i : U_g^{(i)} \to U_g^{(i+1)}$ and the fibre of π_i is a Grassmann variety of dimension i. We can represent $U_g^{(1)}$ as a quotient G/B, where B is a Borel subgroup of G. These spaces $U_g^{(i)}$ come equipped with universal flags $E_i \subset E_{i+1} \subset \dots \subset E_g$.

The manifold Y_g possesses, as all Grassmannians do, a cell decomposition. To define it, choose a fixed isotropic flag $\{0\} = Z_0 \subsetneq Z_1 \subsetneq Z_2 \subsetneq \dots \subsetneq Z_g$; that is, $\dim Z_i = i$ and Z_g is an isotropic subspace of our symplectic space. We extend the filtration by setting $Z_{g+i} = (Z_{g-i})^{\perp}$ for $i = 1, \dots, g$. For general $V \in Y_g$ we expect that $V \cap Z_j = \{0\}$ for $j \leqslant g$. Therefore, for $\mu = (\mu_1, \dots, \mu_r)$ with μ_i non-negative integers satisfying

$$0 \leqslant \mu_i \leqslant g, \quad \mu_i - 1 \leqslant \mu_{i+1} \leqslant \mu_i \tag{1}$$

we put

$$W_\mu = \{V \in Y_g : \dim(V \cap Z_{g+1-\mu_i}) = i\}.$$

This gives a cell decomposition with cells only in even real dimension. The cell W_μ has (complex) codimension $\sum \mu_i$. Denote the set of n-tuples $\mu = (\mu_1, \dots, \mu_g)$ satisfying (1) by M_g. Then $\# M_g = 2^g$. Moreover, we have

$$W_\mu \subseteq \overline{W}_\nu \iff \mu_i \geqslant \nu_i \quad \text{for} \quad 1 \leqslant i \leqslant g.$$

From the cell decomposition we find the homology of Y_g.

Proposition 2.1. *The integral homology of Y_g is generated by the cycle classes of $[\overline{W}_\mu]$ of the closed cells with $\mu \in M_g$. The Poincaré-polynomial of Y_g is given by the summation $\sum b_{2i}\, t^i = (1+t)(1+t^2)\cdots(1+t^g)$.*

Note that Y_g is a rational variety and that the Chow ring R_g of Y_g is the isomorphic image of the cohomology ring of Y_g under the usual cycle class map. On Y_g we have a sequence of tautological vector bundles $0 \to E \to H \to Q \to 0$, where H is the trivial bundle of rank $2g$ defined by our fixed symplectic space and where the fibre of E_y of E over y is the isotropic subspace of dimension g corresponding to y. The bundle E corresponds to the standard representation of $K = U(g)$, see also Section 3. The tangent space to a point $e = [E]$ of Y_g is $\mathrm{Hom}^{\mathrm{sym}}(E, Q)$, the space of symmetric homomorphisms from E to Q; indeed, usually the tangent space of a Grassmannian is described as $\mathrm{Hom}(E, Q)$ ("move E a bit and it moves infinitesimally out of the kernel of $H \to Q$"), but we have to preserve the symplectic form that identifies E with the dual of Q; therefore we have to take the 'symmetric' homomorphisms. We can identify this with $\mathrm{Sym}^2(E^\vee)$.

We consider the Chern classes $u_i = c_i(E) \in R_g = CH^*(Y_g)$ for $i = 1, \dots, g$. We call them *tautological classes*. The symplectic form J on H can be used to identify E

with the dual of the quotient bundle Q. The triviality of the bundle H implies the following relation for the Chern classes of E in R_g:

$$(1+u_1+u_2+\ldots+u_g)(1-u_1+u_2+\ldots+(-1)^g u_g)=1\,.$$

These relations may be succinctly stated as

$$\mathrm{ch}_{2k}(E)=0 \qquad (k\geqslant 1), \tag{2}$$

where ch_{2k} is the part of degree $2k$ of the Chern character.

Proposition 2.2. *The Chow ring R_g of Y_g is the ring generated by the u_i with as only relations $\mathrm{ch}_{2k}(E)=0$ for $k\geqslant 1$.*

One can check algebraically that the ring which is the quotient of $\mathbb{Z}[u_1,\ldots,u_g]$ by the relation (2) has Betti numbers as given in Prop. 2.1. This description implies that this ring after tensoring with $\mathbb{Q}$ is in fact generated by the u_j with j odd. Furthermore, by using induction one obtains the relations

$$u_g u_{g-1}\cdots u_{k+1}u_k^2=0 \qquad \text{for } k=g,\ldots,1. \tag{3}$$

It follows that this ring has the following set of 2^g basis elements

$$\prod_{i\in I} u_i\,, \qquad I\subseteq\{1,\ldots,g\}.$$

The ring R_g is a so-called Gorenstein ring with socle $u_g u_{g-1}\cdots u_1$.

Using $R_g/(u_g)\cong R_{g-1}$ one finds the following properties.

Lemma 2.3. *In $R_g/(u_g)$ we have $u_1^{g(g-1)/2}\neq 0$ and $u_1^{g(g-1)/2+1}=0$.*

Define now classes in the Chow ring of the flag spaces $U_g^{(i)}$ as follows:

$$v_j=c_1(E_j/E_{j-1}) \quad \in CH^*(U_g^{(i)}) \quad j=i,\ldots,g.$$

Moreover, we set

$$u_j^{(i)}=c_j(E_i) \quad \in CH^*(U_g^{(i)}) \quad j=1,\ldots,i.$$

We can view $u_j^{(i)}$ as the jth symmetric function in $v_1,\ldots,v_i$. Then we have the relations under the forgetful maps $\pi_i: U_g^{(i)}\to U_g^{(i+1)}$

$$\pi_i^*(u_j^{(i+1)})=u_j^{(i)}+v_{i+1}u_{j-1}^{(i)} \qquad j=1,\ldots,i+1, \tag{4}$$

and

$$v_j^j-u_1^{(j)}v_j^{j-1}+\ldots+(-1)^j u_j^{(j)}=0 \qquad j=1,\ldots,g. \tag{5}$$

The Chow ring of $U_g^{(1)}$ has generators

$$v_1^{\eta_1}v_2^{\eta_2}\cdots v_g^{\eta_g}u_g^{\epsilon_g}u_{g-1}^{\epsilon_{g-1}}\cdots u_1^{\epsilon_1}$$

with $0\leqslant \eta_i<i$ and $\epsilon_i=0$ or 1.

Lemma 2.4. *The following Gysin formulas hold:*

(1) $(\pi_i)_*(c)=0$ for all $c\in CH^*(U_g^{(i+1)})$;

(2) $(\pi_i)_*(v^i_{i+1}) = 1$;
(3) $(\pi_i)_*(u^{(i)}_i) = (-1)^i$.

These formulas together with (4) and (5) completely determine the image of the Gysin map $(\pi)_* = (\pi_{g-1} \cdots \pi_1)_*$.

For a sequence of $r \leqslant g$ positive integers μ_i with $\mu_i \geqslant \mu_{i+1}$ we define an element of R_g:

$$\Delta_\mu(u) = \det\left((u_{\mu_i - i + j})_{1 \leqslant i \leqslant r; 1 \leqslant j \leqslant r}\right).$$

We also define so-called Q-polynomials by the formula:

$$Q_{i,j} = u_i u_j - 2u_{i+1}u_{j-1} + \ldots + (-1)^j 2u_{i+j} \text{ for } i < j.$$

We have $Q_{i0} = u_i$ for $i = 1, \ldots, g$. Let μ be a strict partition. For r even we define an anti-symmetric matrix $[x_\mu] = [x_{ij}]$ as follows. Let

$$x_{i,j} = Q_{\mu_i \mu_j}(u) \quad \text{for } 1 \leqslant i < j \leqslant r.$$

We then set for even r

$$\Xi_\mu = \mathrm{Pf}([x_\mu]),$$

while for r odd we define $\Xi_\mu = \Xi_{\mu_1 \ldots \mu_r 0}$. These expressions may look a bit artificial, but their purpose is clearly demonstrated by the following Theorem, due to Pragacz [64, 31].

Theorem 2.5. *(Pragacz's formula) The class of the cycle $[\overline{W}_\mu]$ in the Chow ring is given by (a multiple of) $\Xi_\mu(u)$.*

From (2.4) we have the property: if we are given partitions μ and ν with $\sum \mu_i + \sum \nu_i = g(g+1)/2$, then we have

$$\Xi_\mu \Xi_\nu = \begin{cases} 1 & \text{if } \nu = \rho - \mu, \\ 0 & \text{if } \nu \neq \rho - \mu, \end{cases}$$

where $\rho = \{g, g-1, \ldots, 1\}$. We have the following relations in $CH^*(Y_g)$:

$$\text{i)} \quad u_g u_{g-1} \ldots u_1 = 1,$$

$$\text{ii)} \quad u_1^N = N! \prod_{k=1}^{g} \frac{1}{(2k-1)!!},$$

where 1 represents the class of a point and $N = g(g+1)/2$.

Proof. For i) We have $T_{Y_g} \cong \mathrm{Sym}^2(E^\vee)$, thus the Chern classes of the tangent bundle T_{Y_g} are expressed in the u_i. By [29], Ex. 14.5.1, p. 265) for the top Chern class of Sym^2 of a vector bundle we have

$$e(Y_g) = 2^g = 2^g \Delta_{(g,g-1,\ldots,1)}(u) = 2^g \det \begin{pmatrix} u_g & 0 & 0 & \ldots & 0 \\ u_{g-2} & u_{g-1} & u_g & 0 \ldots & \\ & & \ldots & & \\ & & & & u_1 \end{pmatrix}.$$

Developing the determinant gives

$$1 = u_g u_{g-1} \dots u_1 + u_g^2 A(u) + u_g u_{g-1}^2 B(u) + \dots ,$$

from which the desired identity results by using (3). Alternatively, one may use Pragacz's formula above for the degeneracy locus given by $Z_g = V$ for some subspace V. □

3. The Hodge bundle

Since the space $\mathcal{A}_g$ comes with a universal principally polarized abelian variety $\pi : \mathcal{X}_g \to \mathcal{A}_g$ we have a rank g vector bundle or locally free sheaf

$$\mathbb{E} = \mathbb{E}_g = \pi_*(\Omega^1_{\mathcal{X}_g/\mathcal{A}_g}),$$

called the *Hodge Bundle.* It has an alternative definition as

$$\mathbb{E} = s^*(\Omega^1_{\mathcal{X}_g/\mathcal{A}_g}),$$

i.e., as the cotangent bundle to the zero section $s : \mathcal{A}_g \to \mathcal{X}_g$. If $\mathcal{X}_g^t$ is the dual abelian variety (isomorphic to $\mathcal{X}_g$ because we stick to a principal polarization) then the Hodge bundle $\mathbb{E}^t$ of $\mathcal{X}_g^t$ satisfies

$$(\mathbb{E}^t)^\vee = \mathrm{Lie}(\mathcal{X}_g^t) \cong R^1\pi_*\mathcal{O}_{\mathcal{X}_g}.$$

The Hodge bundle $\mathbb{E}$ can be extended to any toroidal compactification $\tilde{\mathcal{A}}_g$ of Faltings-Chai type. In fact, over $\tilde{\mathcal{A}}_g$ we have a universal family of semi-abelian varieties and one takes the dual of $\mathrm{Lie}(\tilde{\mathcal{X}})$, cf. [25]. We shall denote it again by $\mathbb{E}$.

If we now go to a fine moduli space, say $\mathcal{A}_g[n]$ with $n \geqslant 3$, the moduli space of principally polarized abelian varieties with a level n structure, and take a smooth toroidal compactification then we can describe the sheaf of holomorphic 1-forms in terms of the Hodge bundle. With $D = \tilde{\mathcal{A}}_g[n] - \mathcal{A}_g[n]$, the divisor at infinity, we have the important result:

Proposition 3.1. *The Hodge bundle $\mathbb{E}$ on $\tilde{\mathcal{A}}_g[n]$ for $n \geqslant 3$ satisfies the identity*

$$\mathrm{Sym}^2(\mathbb{E}) \cong \Omega^1_{\tilde{\mathcal{A}}_g[n]}(\log D).$$

This result extends the description of the tangent space to $\mathfrak{H}_g$ in the compact dual. It is proven in the general setting in [25], p. 86.

Recall the description of Y_g as the symplectic Grassmannian $G(\mathbb{C})/Q(\mathbb{C})$ with $G = \mathrm{GSp}(2g, \mathbb{Z})$ and Q the parabolic subgroup with Levi decomposition $M \ltimes U$ with $M = \mathrm{GL}(g) \times \mathbb{G}_m$. The Siegel upper half space $\mathfrak{H}_g$ can be viewed as an open subset of Y_g. Put $G_0 = \mathrm{Sp}(2g, \mathbb{Z})$, $Q_0 = Q \cap G_0$ and $M_0 = M \cap G_0$. Then if $\rho : Q_0 \to \mathrm{GL}(V)$ is a finite-dimensional complex representation, we define an equivariant vector bundle $\mathcal{V}_\rho$ on Y_g by

$$\mathcal{V}_\rho = G_0(\mathbb{C}) \times^{Q_0(\mathbb{C})} V,$$

where the contracted product is defined by the usual equivalence relation $(g, v) \sim (gq, \rho(q)^{-1}v)$ for all $g \in G_0(\mathbb{C})$ and $q \in Q_0(\mathbb{C})$. Then our group $\mathrm{Sp}(2g, \mathbb{Z})$

acts on the bundle $\mathcal{V}_\rho$ and the quotient is a vector bundle V_ρ in the orbifold sense on $\mathcal{A}_g(\mathbb{C}) = \mathrm{Sp}(2g, \mathbb{Z})\backslash\mathfrak{H}_g$.

A representation of GL(g) can be lifted to a representation of Q_0 by letting it act trivially on the unipotent radical U of M_0. Carrying out this construction with the standard (tautological) representation of GL(g) produces the Hodge bundle.

The Hodge bundle can be extended to a bundle over our toroidal compactification $\tilde{\mathcal{A}}_g$. Since any bundle V_ρ is obtained by applying Schur functors to powers of the Hodge bundle (see [30]) we can extend the bundle V_ρ by applying the Schur functor to the extended power of the Hodge bundle. In this way we obtain a canonical extension to $\tilde{\mathcal{A}}_g$ for all equivariant holomorphic bundles V_ρ.

4. The tautological ring of $\mathcal{A}_g$

The moduli space $\mathcal{A}_g$ is a Deligne-Mumford stack or orbifold and as such it has a Chow ring. To be precise, consider the moduli space $\mathcal{A}_g \otimes k$ over an algebraically closed field k. For simplicity we shall write $\mathcal{A}_g$ instead.

We are interested in the Chow rings $\mathrm{CH}(\mathcal{A}_g)$ and $\mathrm{CH}_\mathbb{Q}(\mathcal{A}_g)$ of this moduli space. In general these rings seems unattainable. However, they contain subrings that we can describe well and that play an important role.

We denote the Chern classes of the Hodge bundle by

$$\lambda_i := c_i(\mathbb{E}), \qquad i = 1, \ldots, g.$$

We define the *tautological subring* of the Chow ring $\mathrm{CH}_\mathbb{Q}(\mathcal{A}_g)$ as the subring generated by the Chern classes of the Hodge bundle $\mathbb{E}$.

The main result is a description of the tautological ring in terms of the Chow ring R_{g-1} of the compact dual Y_{g-1} (see [32]).

Theorem 4.1. *The tautological subring T_g of the Chow ring $\mathrm{CH}_\mathbb{Q}(\mathcal{A}_g)$ generated by the λ_i is isomorphic to the ring R_{g-1}.*

This implies that a basis of the codimension i part is given by the monomials

$$\lambda_1^{e_1}\lambda_2^{e_2}\cdots\lambda_{g-1}^{e_{g-1}} \qquad e_j \in \{0,1\} \text{ and } \sum_{j=1}^{g-1} j\, e_j = i.$$

This theorem follows from the following four results, each interesting in its own right.

Theorem 4.2. *The Chern classes λ_i of the Hodge bundle $\mathbb{E}$ satisfy the relation*

$$(1+\lambda_1+\cdots+\lambda_g)(1-\lambda_1+\cdots+(-1)^g\lambda_g) = 1.$$

Theorem 4.3. *The top Chern class λ_g of the Hodge bundle $\mathbb{E}$ vanishes in $\mathrm{CH}_\mathbb{Q}(\mathcal{A}_g)$.*

Theorem 4.4. *The first Chern class λ_1 of $\mathbb{E}$ is ample on $\mathcal{A}_g$.*

Theorem 4.5. *In characteristic $p > 0$ the moduli space $\mathcal{A}_g \otimes \mathbb{F}_p$ contains a complete subvariety of codimension g.*

To deduce Theorem 4.1 from Thms 4.2 - 4.5 we argue as follows. By Theorem 4.2 the tautological subring T_g is a quotient ring of R_g via $u_i \mapsto \lambda_i$, and then by Theorem 4.3 also of $R_{g-1} \cong R_g/(u_g)$. The ring R_{g-1} is a Gorenstein ring with socle $u_1 u_2 \cdots u_{g-1}$ and this is a non-zero multiple of $u_1^{g(g-1)/2}$. By Theorem 4.4 the ample class λ_1 satisfies $\lambda_1^j \cdot [V] \neq 0$ on any complete subvariety V of codimension j in $\mathcal{A}_g$. The existence of a complete subvariety of codimension g in positive characteristic implies that $\lambda_1^{g(g-1)/2}$ does not vanish in $CH_{\mathbb{Q}}(\mathcal{A}_g \otimes \mathbb{F}_p)$, hence the class $\lambda_1^{g(g-1)/2}$ does not vanish in $CH_{\mathbb{Q}}(\mathcal{A}_g)$. If the map $T_g \to R_{g-1}$ defined by $\lambda_i \mapsto u_i$ would have a non-trivial kernel then $\lambda_1^{g(g-1)/2}$ would have to vanish.

Theorem 4.2 was proved in [32]. The proof is obtained by applying the Grothendieck-Riemann-Roch formula to the theta divisor on the universal abelian variety $\mathcal{X}_g$ over $\mathcal{A}_g$. In fact, take a line bundle L on $\mathcal{X}_g$ that provides on each fibre X a theta divisor and normalize L such that it vanishes on the zero section of $\mathcal{X}_g$ over $\mathcal{A}_g$. Then the Grothendieck-Riemann-Roch Theorem tells us that

$$\begin{aligned} ch(\pi_! L) &= \pi_*(ch(L) \cdot Td^{\vee}(\Omega^1_{\mathcal{X}_g/\mathcal{A}_g})) \\ &= \pi_*(ch(L) \cdot Td^{\vee}(\pi^*(\mathbb{E}_g))) \qquad (1) \\ &= \pi_*(ch(L)) \cdot Td^{\vee}(\mathbb{E}), \end{aligned}$$

where we used $\Omega^1_{\mathcal{X}_g/\mathcal{A}_g} \cong \pi^*(\mathbb{E})$ and the projection formula. Here $Td^{\vee}$ is defined for a line bundle with first Chern class γ by $\gamma/(e^{\gamma} - 1)$. But $R^i\pi_* L = 0$ for our relatively ample L for $i > 0$ and so $\pi_! L$ is represented by a vector bundle, and because L defines a principal polarization, by a line bundle. We write $\theta = c_1(\pi_! L)$. This gives the identity

$$\sum_{k=0}^{\infty} \frac{\theta^k}{k!} = \pi_*(\sum_{k=0}^{\infty} \frac{\Theta^{g+k}}{(g+k)!}) Td^{\vee}(\mathbb{E}). \qquad (2)$$

Recall that $Td^{\vee}(\mathbb{E}) = 1 - \lambda_1/2 + (\lambda_1^2 + \lambda_2)/12 + \ldots$. We can compare terms of equal codimension; the codimension 1 term gives

$$\theta = -\lambda_1/2 + \pi_*(\Theta^{g+1}/(g+1)!). \qquad (3)$$

If we now look how both sides of (2) behave when we replace L by L^n, we see immediately that the term Θ^{g+k} of the right hand side changes by a factor n^{g+k}. As to the left hand side, for a principally polarized abelian variety X the space of sections $H^0(X, L_X^{\otimes n})$ is a representation of the Heisenberg group; it is the irreducible representation of degree n^g. This implies that $ch(\pi_! L) = n^g ch(\pi_! L)$. We see

$$n^g \sum_{k=0}^{\infty} \frac{\theta^k}{k!} = \pi_* \left(\sum_{k=0}^{\infty} \frac{n^{g+k}\Theta^{g+k}}{(g+k)!} \right) \cdot Td^{\vee}(\mathbb{E}).$$

Comparing the coefficients leads immediately to the following result.

Corollary 4.6. *In* $CH_{\mathbb{Q}}(\mathcal{A}_g)$ *we have* $\pi_*(ch(L)) = 1$ *and* $ch(\pi_* L) = Td^\vee(\mathbb{E})$.

In particular $\pi_*(\Theta^{g+1}) = 0$. Substituting this in (3) gives the following result.

Corollary 4.7. *(Key Formula) We have* $2\theta = -\lambda_1$.

Corollary 4.8. *We have* $ch_{2k}(\mathbb{E}) = 0$ *for* $k \geqslant 1$.

Proof. The relation (1) reduces now to

$$e^{-\lambda_1/2} = Td^\vee(\mathbb{E}).$$

If, say $\rho_1, \ldots, \rho_g$ are the Chern roots of $\mathbb{E}$, so that our λ_i is the ith elementary symmetric function in the ρ_j, then this relation is

$$\prod_{i=1}^{g} \frac{e^{\rho_i} - e^{-\rho_i}}{\rho_i} = 1.$$

This is equivalent with $ch_{2k}(\mathbb{E}) = 0$ for $k > 0$ and also with $Td(\mathbb{E} \oplus \mathbb{E}^\vee) = 0$. □

The proof of Theorem 4.3 is also an application of Grothendieck-Riemann-Roch, this time applied to the structure sheaf (i.e., $n = 0$ in the preceding case). We find

$$ch(\pi_! \mathcal{O}_{\mathcal{X}_g}) = \pi_*(ch(\mathcal{O}_{\mathcal{X}_g}) \cdot Td^\vee \pi^*(\mathbb{E}_g)) = \pi_*(1) Td^\vee(\mathbb{E}_g) = 0.$$

For an abelian variety X the cohomology group $H^i(X, \mathcal{O}_X)$ is the ith exterior power of $H^1(X, \mathcal{O}_X)$ and we thus see that the left hand side equals (global Serre duality)

$$ch(1 - \mathbb{E}^\vee + \wedge^2 \mathbb{E}^\vee - \cdots + (-1)^g \wedge^g \mathbb{E}^\vee).$$

But by some general yoga (see [10]) we have for a vector bundle B of rank r the relation $\sum_{j=0}^{r} (-1)^j ch(\wedge^j B^\vee) = c_r(B) Td(B)^{-1}$. So the left hand side is $\lambda_g \, Td(\mathbb{E})^{-1}$ and since $Td(\mathbb{E})$ starts with $1 + \ldots$ it follows that λ_g is zero.

Theorem 4.4 is a classical result in characteric zero. If we present our moduli space $\mathcal{A}_g$ over $\mathbb{C}$ as the quotient space $Sp(2g, \mathbb{Z}) \backslash \mathfrak{H}_g$ then the determinant of the Hodge bundle is a (n orbifold) line bundle that corresponds to the factor of automorphy

$$\det(c\tau + d) \qquad \text{for } \tau \in \mathfrak{H}_g \text{ and } \begin{pmatrix} a & b \\ c & d \end{pmatrix} \in Sp(2g, \mathbb{Z}).$$

In other words, a section of the kth power of $\det(\mathbb{E})$ gives by pull back to $\mathfrak{H}_g$ a holomorphic function $f : \mathfrak{H}_g \to \mathbb{C}$ with the property that

$$f((a\tau + b)(c\tau + d)^{-1}) = \det(c\tau + d)^k f(\tau).$$

A very well-known theorem by Baily and Borel (see [5]) says that modular forms of sufficiently high weight k define an embedding of $Sp(2g, \mathbb{Z}) \backslash \mathfrak{H}_g$ into projective space. The idea is that one can construct sufficiently many modular forms to separate points and tangent vectors on $\mathcal{A}_g(\mathbb{C})$. So λ_1 is an ample class. Clearly this holds then also in characteristic p if p is sufficiently large. So if we have a complete subvariety of of codimension j in $\mathcal{A}_g \otimes \mathbb{F}_p$ for every p then λ_j cannot vanish when restricted to this subvariety.

Theorem 4.4 was extended to all characteristics by Moret-Bailly, see [54].

5. The tautological ring of $\tilde{\mathcal{A}}_g$

In this section, just as in the preceding one, we will write $\mathcal{A}_g$ for $\mathcal{A}_g \otimes k$ with k an algebraically closed field.

The class λ_g vanishes in the rational Chow ring of $\mathcal{A}_g$. However, in a suitable compactification of $\mathcal{A}_g$ this is no longer true and then, instead of the quotient $R_{g-1} = R_g/(u_g)$, we find a copy of R_g in the Chow ring of such a compactification.

We consider a smooth toroidal compactification $\tilde{\mathcal{A}}_g$ of $\mathcal{A}_g$ of the type constructed by Faltings and Chai. Over $\tilde{\mathcal{A}}_g$ we have a 'universal' semi-abelian variety $\mathcal{G}$ with a zero-section s, see [25]. Then $s^*\mathrm{Lie}(\mathcal{G})$ defines in a canonical way an extension of the Hodge bundle $\mathbb{E}$ on $\mathcal{A}_g$ to a vector bundle $\tilde{\mathcal{A}}_g$. We will denote this extension again by $\mathbb{E}$.

The relation (1) of the preceding section can now be extended to $\tilde{\mathcal{A}}_g$. A proof of this extension was given by Esnault and Viehweg, see [21]. They show that in characteristic 0 for the Deligne extension of the cohomology sheaf $\mathcal{H}^1$ with its Gauss-Manin connection the Chern character vanishes in degree $\neq 0$ by applying Grothendieck-Riemann-Roch to a log extension and using the action of -1 to separate weights. By specializing one finds the following result.

Theorem 5.1. *The Chern classes λ_i of the Hodge bundle $\mathbb{E}$ on $\tilde{\mathcal{A}}_g$ satisfy the relation*

$$(1+\lambda_1+\cdots+\lambda_g)(1-\lambda_1+\cdots+(-1)^g\lambda_g)=1.$$

This implies the following result:

Theorem 5.2. *The tautological subring $\tilde{T}_g$ of $\mathrm{CH}_{\mathbb{Q}}(\tilde{\mathcal{A}}_g)$ is isomorphic to R_g*

Proof. Clearly, by the relation of 5.1 the tautological ring is a quotient of R_g; since λ_1 is ample on an open dense part ($\mathcal{A}_g$) the socle

$$\lambda_g\lambda_{g-1}\cdots\lambda_1 = \frac{1}{(g(g+1)/2)!}\left(\prod_{k=1}^{g}(2k-1)!!\right)\lambda_1^{g(g+1)/2}$$

does not vanish, and the tautological ring must be isomorphic to R_g. □

The Satake compactification $\mathcal{A}_g^*$ of $\mathcal{A}_g$, defined in general as the proj of the ring of Siegel modular forms, possesses a stratification

$$\mathcal{A}_g^* = \mathcal{A}_g \sqcup \mathcal{A}_{g-1} \sqcup \cdots \sqcup \mathcal{A}_1 \sqcup \mathcal{A}_0.$$

By the natural map $q : \tilde{\mathcal{A}}_g \to \mathcal{A}_g^*$ this induces a stratification of $\tilde{\mathcal{A}}_g$: we let $\mathcal{A}_g^{(t)}$ be the inverse image $q^{-1}(\mathcal{A}_g \sqcup \mathcal{A}_{g-1} \sqcup \cdots \mathcal{A}_{g-t})$, the moduli space of abelian varieties of torus rank $\leqslant t$. We have the following extension of Theorems 4.3 and 4.5.

Theorem 5.3. *For* $t < g$ *we have the relation* $\lambda_g \lambda_{g-1} \cdots \lambda_{g-t} = 0$ *in the Chow group* $CH^m_{\mathbb{Q}}(\mathcal{A}_g^{(t)})$ *with* $m = \sum_{i=0}^{t} g - i$.

Theorem 5.4. *The moduli space* $\mathcal{A}_g^{(t)} \otimes \mathbb{F}_p$ *in characteristic* $p > 0$ *contains a complete subvariety of codimension* $g - t$*, namely the locus of abelian varieties of* p*-rank* $\leqslant t$.

Corollary 5.5. *We have* $\lambda_1^{g(g-1)/2+t} \neq 0$ *on* $\mathcal{A}_g^{(t)}$.

Proposition 5.6. *For* $0 \leqslant k \leqslant g$ *and* $r > 0$ *we have on the boundary* $\mathcal{A}_k^* \subset \mathcal{A}_g^*$ *of the Satake compactification*

$$\lambda_1^{k(k+1)/2} \neq 0 \qquad \textit{and} \qquad \lambda_1^{k(k+1)/2+r} = 0.$$

This follows from the fact that λ_1 is an ample class on $\mathcal{A}_g^*$, that $\lambda_1|\mathcal{A}_k^*$ is again λ_1 (now of course defined on $\mathcal{A}_k^*$) and the fact that $\dim \mathcal{A}_k^* = k(k+1)/2$.

6. The proportionality principle

The result on the tautological ring has the following immediate corollary, a special case of the so-called Proportionality Principle of Hirzebruch-Mumford:

Theorem 6.1. *The characteristic numbers of the Hodge bundle are proportional to those of the tautological bundle on the 'compact dual'* Y_g:

$$\lambda_1^{n_1} \cdots \lambda_g^{n_g}([\tilde{\mathcal{A}}_g]) = (-1)^{g(g+1)/2} \frac{1}{2^g} \left(\prod_{k=1}^{g} \zeta(1-2k) \right) \cdot u_1^{n_1} \cdots u_g^{n_g}([Y_g])$$

for all $(n_1, \ldots, n_g)$ *with* $\sum n_i = g(g+1)/2$.

Indeed, any top-dimensional class $\lambda_1^{n_1} \cdots \lambda_g^{n_g}$ is a multiple $m(n_1, \ldots, n_g)$ times the top-dimensional class $\lambda_1 \cdots \lambda_g$ where the coefficient depends only on the structure of R_g. So the proportionality principle is clear, and to make it explicit one must find the value of one top-dimensional class, say $\lambda_1 \cdots \lambda_g$ on $\tilde{\mathcal{A}}_g$. The top Chern class of the bundle $\mathrm{Sym}^2(\mathbb{E})$ on $\mathcal{A}_g$ equals

$$2^g \lambda_1 \cdots \lambda_g \,.$$

We can interpret $2^g \deg(\lambda_1 \cdots \lambda_g)$ up to a sign $(-1)^{g(g+1)/2}$ as the log version of the Euler number of $\mathcal{A}_g$. It is known via the work of Siegel and Harder:

Theorem 6.2. *(Siegel-Harder) The Euler number of* $\mathcal{A}_g$ *is equal to*

$$\zeta(-1)\zeta(-3) \cdots \zeta(1-2g).$$

Example 6.3. For $g = 1$ the Euler number of $\mathcal{A}_1(\mathbb{C})$ is $-1/12$. It is obtained by integrating

$$(-1)\frac{1}{2\pi i} \frac{dx \wedge dy}{y^2}$$

over the usual fundamental domain ($\{\tau = x + iy \in \mathfrak{H} : |x| \leqslant 1/2, x^2 + y^2 \geqslant 1\}$) for the action of $SL(2, \mathbb{Z})$ and (which gives $2\zeta(-1) = -1/6$) and then dividing by 2 since the group $SL(2, \mathbb{Z})$ does not act effectively (-1 acts trivially).

The Proportionality Principle immediately extends to our equivariant bundles V_ρ since these are obtained by applying Schur functors to powers of the Hodge bundle.

Corollary 6.4. *(The Proportionality Principle of Hirzebruch-Mumford) The characteristic numbers of the equivariant vector bundle V_ρ on $\tilde{\mathcal{A}}_g$ are proportional to those of the corresponding bundle on the compact dual Y_g*

The proportionality factor here is

$$p(g) = (-1)^{g(g+1)/2} \prod_{j=1}^{g} \frac{\zeta(1-2j)}{2}$$

We give a few values.

g	1	2	3	4	5
1/p(g)	24	5760	2903040	1393459200	367873228800

A remark about the history of this 'principle'. Hirzebruch, inspired by a paper of Igusa, found in 1958 (see [42]) that for a discrete torsion-free group Γ of automorphisms of a bounded symmetric domain D such that the quotient $X_\Gamma = \Gamma \backslash D$ is compact, the Chern numbers of the quotient are proportional to the Chern numbers of the compact dual $\hat{D}$ of D. In the case of a subgroup $\Gamma \subset Sp(2g, \mathbb{R})$ that acts freely on $\mathfrak{H}_g$ this means that there is a proportionality factor e_Γ such that for all $n = (n_1, \ldots, n_{g(g+1)/2})$ with $n_i \in \mathbb{Z}_{\geqslant 0}$ and $\sum n_i = g(g+1)/2$ we have

$$c^n(X_\Gamma) = e_\Gamma \, c^n(Y_g),$$

where c^n stands for the top Chern class $\prod_i c_i^{n_i}$ and Y_g is the compact dual of $\mathfrak{H}_g$. His argument was that the top Chern classes of X are represented by $G(\mathbb{R})$-invariant differential forms of top degree and these are proportional to each other on D and on $\hat{D}$. The principle extends to all equivariant holomorphic vector bundles on $\mathfrak{H}_g$. Mumford has obtained (see [56]) an extension of Hirzebruch's Proportionality Principle that applies to his toroidal compactifications of $\mathcal{A}_g$.

7. The Chow rings of $\tilde{\mathcal{A}}_g$ for $g = 1,2,3$

For low values of g the Chow rings of $\mathcal{A}_g$ and their toroidal compactifications can be explicitly described.

Theorem 7.1. *The Chow ring $CH_{\mathbb{Q}}(\tilde{\mathcal{A}}_1)$ is isomorphic to $\mathbb{Q}[\lambda_1]/(\lambda_1^2)$.*

Mumford determined in [57] the Chow ring of $\tilde{\mathcal{A}}_2$.

Theorem 7.2. *The Chow ring* $CH_{\mathbb{Q}}(\tilde{\mathcal{A}}_2)$ *is generated by* λ_1, λ_2 *and* σ_1 *and isomorphic to* $\mathbb{Q}[\lambda_1, \lambda_2, \sigma_1]/(I_2)$, *with* I_2 *the ideal generated by*

$$(1+\lambda_1+\lambda_2)(1-\lambda_1+\lambda_2)-1, \quad \lambda_2\sigma_1, \quad \sigma_1^2 - 22\lambda_1\sigma_1 + 120\lambda_1^2.$$

The ranks of the Chow groups are 1,2,2,1.

The intersection numbers in this ring are given by

2\1	λ_1	σ_1
λ_1^2	1/2880	0
$\lambda_1\sigma_1$	0	−1/24

while the degrees of the top classes in the sigmas are

$$\deg(\sigma_3) = 1/4, \quad \deg(\sigma_2\sigma_1) = -1/4, \quad \deg(\sigma_1^3) = -11/12.$$

For genus 3 the result is (cf. [34]):

Theorem 7.3. *The tautological ring of* $\mathcal{A}_3$ *is generated by the Chern classes* λ_i *and isomorphic to* $\mathbb{Q}[\lambda_1]/(\lambda_1^4)$.

Theorem 7.4. *The Chow ring* $CH_{\mathbb{Q}}(\tilde{\mathcal{A}}_3)$ *of* $\tilde{\mathcal{A}}_3$ *is generated by the* λ_i, $i = 1,2,3$ *and the* σ_i, $i = 1,2$ *and is isomorphic to the graded ring (subscript is degree)*

$$\mathbb{Q}[\lambda_1, \lambda_2, \lambda_3, \sigma_1, \sigma_2]/I_3,$$

with I_3 *the ideal generated by the relations*

$$\begin{aligned}
&(1+\lambda_1+\lambda_2+\lambda_3)(1-\lambda_1+\lambda_2-\lambda_3) = 1,\\
&\lambda_3\sigma_1 = \lambda_3\sigma_2 = \lambda_1^2\sigma_2 = 0,\\
&\sigma_1^3 = 2016\,\lambda_3 - 4\,\lambda_1^2\sigma_1 - 24\,\lambda_1\sigma_2 + \frac{11}{3}\,\sigma_2\sigma_1,\\
&\sigma_2^2 = 360\,\lambda_1^3\sigma_1 - 45\,\lambda_1^2\sigma_1^2 + 15\,\lambda_1\sigma_2\sigma_1,\\
&\sigma_1^2\sigma_2 = 1080\,\lambda_1^3\sigma_1 - 165\,\lambda_1^2\sigma_1^2 + 47\,\lambda_1\sigma_2\sigma_1.
\end{aligned}$$

The ranks of the Chow groups are: 1, 2, 4, 6, 4, 2, 1.

For the degrees of the top dimensional elements we refer to [34].

8. Some intersection numbers

As stated in section 1 the various compactifications employed for $\mathcal{A}_g$ each have their own merits. For example, the toroidal compactification associated to the perfect cone decomposition has the advantage that its boundary is an irreducible divisor $D = D_g$.

By a result of Borel [9] it is known that in degrees $\leqslant g-2$ the cohomology of $Sp(2g,\mathbb{Q})$ (and in fact for any finite index subgroup Γ) is generated by elements in degree $2k+4$ for $k=0,1,\dots,(g-6)/2$:

$$H^*(\Gamma,\mathbb{Q}) = \mathbb{Q}[x_2, x_6, x_{10}, \dots].$$

In particular in degree 2 there is one generator. We deduce (at least for $g \geqslant 6$ and with special arguments also for $2 \leqslant g \leqslant 5$):

Proposition 8.1. *The Picard group of $\mathcal{A}_g$ for $g \geqslant 2$ is generated by the determinant λ_1 of the Hodge bundle.*

By observing that on $\mathcal{A}_g^{(1)}$ and $\mathcal{A}_g^{\mathrm{Perf}}$ the boundary divisor is irreducible we get:

Corollary 8.2. *The Picard group of $\mathcal{A}_g^{(1)}$ and of $\mathcal{A}_g^{\mathrm{Perf}}$ is generated by λ_1 and the class of* D.

It would be nice to know the top intersection numbers $\lambda_1^n D^{G-n}$ with $G = g(g+1)/2 = \dim \mathcal{A}_g$. It seems that these numbers are zero when n is not of the form $k(k+1)/2$. In fact, Erdenberger, Grushevsky and Hulek formulated this as a conjecture, see [20].

Conjecture 8.3. *The intersection number* $\deg \lambda_1^n D^{G-n}$ *on* $\mathcal{A}_g^{\mathrm{Perf}}$ *vanishes unless n is of the form $n = k(k+1)/2$ for $k \leqslant g$.*

From our results above it follows that

$$\deg \lambda_1^G = (-1)^G \frac{G!}{2^g} \prod_{k=1}^{g} \frac{\zeta(1-2k)}{(2k-1)!!}.$$

Erdenberger, Grushevsky and Hulek calculated the next two cases. We quote only the first

$$\deg \lambda_1^{g(g-1)/2} D^g = (-1)^{G-1} \frac{(g-1)!(G-g)!}{2} \prod_{k=1}^{g-1} \frac{\zeta(1-2k)}{(2k-1)!!}$$

and we refer for the intersection number $\deg \lambda^{(g-2)(g-1)/2} D^{2g-1}$ to loc. cit.

9. The top class of the Hodge bundle

As before, we shall write here $\mathcal{A}_g$ for $\mathcal{A}_g \otimes k$ with k an algebraically closed field.

The cycle class λ_g vanishes in $CH^g_{\mathbb{Q}}(\mathcal{A}_g)$. However, it does not vanish on $CH^g_{\mathbb{Q}}(\tilde{\mathcal{A}}_g)$, for example because $\lambda_1\lambda_2\cdots\lambda_g$ is a non-zero multiple of $\lambda_1^{g(g+1)/2}$ that has positive degree since λ_1 is ample on $\mathcal{A}_g$. This raises two questions. First, what is the order of λ_g in $CH^g(\mathcal{A}_g)$ as a torsion class? Second, since up to torsion λ_g comes from the boundary $\tilde{\mathcal{A}}_g - \mathcal{A}_g$, one can ask for a natural supporting cycle for this class in the boundary. Since we work on stacks one has to use intersection theory on stacks; the theory is still in its infancy, but we use Kresch's approach [49]. Mumford answered the first question for $g = 1$ in [55] .

In joint work with Ekedahl ([17, 18]) we considered these two questions. Let us begin with the torsion order of λ_g in $CH^g(\mathcal{A}_g)$. A well-known formula of Borel and Serre (used above in Section 3) says that the Chern character of the alternating sum of the exterior powers of the Hodge bundle satisfies

$$\mathrm{ch}(\wedge^* \mathbb{E}) = (-1)^g \lambda_g \mathrm{Td}(\mathbb{E})^{-1}$$

and this implies that its terms of degree $1, \ldots, g-1$ vanish and that, in degree g, it equals $-(g-1)!\lambda_g$.

Lemma 9.1. *Let* p *be a prime and* $\pi : A \to S$ *be a family of abelian varieties of relative dimension* g *and* L *a line bundle on* A *that is of order* p *on all fibres of* π. *If* $p > \min(2g, \dim S + g)$ *then* $p\,(g-1)!\lambda_g = 0$.

Indeed, we may assume after twisting by a pull back from the base S that L^p is trivial. Let $[L]$ be its class in $K_0(A)$. Then

$$0 = [L]^p - 1 = ([L]-1)^p + p\,([L]-1)(1 + \frac{p-1}{2}([L]-1) + \ldots).$$

Now $[L]-1$ is nilpotent because it has support of codimension $\geqslant 1$ and so it follows that $p\,([L]-1)$ lies in the ideal generated by $([L]-1)^p$ which has codimension $\geqslant p$. Now if $p > 2g$ or $p > \dim S + g$ the codimension is $> 2g$, hence the image under π has codimension $> g$ or is zero. We thus can safely remove it and may assume that $p\,[L] = p$ in $K_0(A)$. Consider now the Poincaré bundle P on $A \times \hat{A}$; we know that $H^i(X \times \hat{X}, P)$ for an abelian variety X is zero for $i \neq g$ and 1-dimensional for $i = g$. So the derived pullback of $R\pi_* P$ along the zero section of $A \times \hat{A}$ over A is $R\pi_* \mathcal{O}_A$. We know that $p\,[P] = p[L \otimes P]$ and $p\,[R\pi_* P] = p\,[R\pi_*(L \otimes P)]$ and $R\pi_*(L \otimes P)$ has support along the inverse of the section of $\hat{A}$ corresponding to L. This section is everywhere disjoint from the zero section, so the pull back of $R\pi_*(L \otimes P)$ along the zero section is trivial: $p\,[R\pi_* \mathcal{O}_a] = 0$ and since $R\pi_* \mathcal{O}_A = \wedge^* R^1 \pi_* \mathcal{O}_A = \wedge^* \mathbb{E}$ we find $-p(g-1)!\lambda_g = 0$.

We put

$$n_g := \gcd\{p^{2g} - 1 : \text{ p running through the primes} > 2g+1\}.$$

Note that for a prime ℓ we have $\ell^k | n_g$ if and only if the exponent of $(\mathbb{Z}/\ell^k\mathbb{Z})^*$ divides $2g$. By Dirichlet's prime number theorem we have for $p > 2$

$$\mathrm{ord}_p(n_g) = \begin{cases} 0 & (p-1) \not| 2g \\ \max\{k : p^{k-1}(p-1)|2g\} & (p-1)|2g, \end{cases}$$

while

$$\mathrm{ord}_2(n_g) = \max\{k : 2^{k-2}|2g\}.$$

For example, we have

$$n_1 = 24, \quad n_2 = 240, \quad n_3 = 504, \quad n_4 = 480.$$

Theorem 9.2. *Let* $\pi : \mathcal{X}_g \to \mathcal{A}_g$ *be the universal family. Then* $(g-1)!n_g\lambda_g = 0$.

For the proof we consider the commutative diagram

$$\begin{array}{ccc} \mathcal{X}'_g & \longrightarrow & \mathcal{X}_g \\ \downarrow & & \downarrow \\ \mathcal{A}'_g & \longrightarrow & \mathcal{A}_g \end{array}$$

where $\mathcal{A}'_g \to \mathcal{A}_g$ is the degree $p^{2g} - 1$ cover obtained by adding a line bundle of order p. It follows that $(g-1)!p(p^{2g}-1)\lambda_g$ vanishes. Then the rest follows from the definition of n_g.

In [55] Mumford proved that the order of λ_1 is 12. So our result is off by a factor 2.

In the paper [17] we also gave the vanishing orders for the Chern classes of the de Rham bundle $\mathcal{H}^1_{dR}$ for the universal family $\mathcal{X}_g \to \mathcal{A}_g$. This bundle is provided with an integrable connection and so its Chern classes are torsion classes in integral cohomology. If l is a prime different from the characteristic these Chern classes are torsion too by the comparison theorems and using specialization. We denote these classes by $r_i \in H^{2i}(\mathcal{A}_g, \mathbb{Z}_l)$. We determined their orders up to a factor 2. Note that r_i vanishes for i odd as $\mathcal{H}^1_{dR}$ is a symplectic bundle.

Theorem 9.3. *For all i the class r_{2i+1} vanishes. For $1 \leqslant i \leqslant g$ the order of r_{2i} in integral (resp. l-adic) cohomology equals (resp. equals the l-part of) $n_i/2$ or n_i.*

We now turn to the question of a representative cycle in the boundary for λ_g in $CH_{\mathbb{Q}}(\tilde{\mathcal{A}}_g)$.

It is a very well-known fact that there is a cusp form Δ of weight 12 on $SL(2, \mathbb{Z})$ and that it has a only one zero, a simple zero at the cusp. This leads to the relation $12\,\lambda_1 = \delta$, where δ is the cycle representing the class of the cusp. This formula has an analogue for higher g.

Theorem 9.4. *In the Chow group $CH_{\mathbb{Q}}(\mathcal{A}_g^{(1)})$ of codimension g cycles on the moduli stack $\mathcal{A}_g^{(1)}$ of rank $\leqslant 1$ degenerations the top Chern class λ_g satisfies the formula*

$$\lambda_g = (-1)^g \zeta(1-2g)\,\delta_g,$$

with δ_g the $\mathbb{Q}$-class of the locus Δ_g of semi-abelian varieties which are trivial extensions of an abelian variety of dimension $g-1$ by the multiplicative group $\mathbb{G}_m$.

The number $\zeta(1-2g)$ is a rational number $-b_{2g}/2g$ with b_{2g} the $2g$th Bernoulli number. For example, we have

$$12\,\lambda_1 = \delta_1, \quad 120\,\lambda_2 = \delta_2, \quad 252\,\lambda_3 = \delta_3, \quad 240\,\lambda_4 = \delta_4, \quad 132\,\lambda_5 = \delta_5$$

One might also wish to work on the Satake compactification $\mathcal{A}_g^*$, singular though as it is. Every toroidal compactification of Faltings-Chai type $\tilde{\mathcal{A}}_g$ has a canonical morphism $q : \tilde{\mathcal{A}}_g \to \mathcal{A}_g^*$. Then we can define a class ℓ_α where α is a subset of $\{1,2,\ldots,g\}$ by

$$\ell_\alpha = q_*(\lambda_\alpha) \quad \text{with } \lambda_\alpha \text{ equal to } \prod_{i\in\alpha} \lambda_i \in CH_d^{\mathbb{Q}}(\tilde{\mathcal{A}}_g)$$

with $d = d(\alpha) = \sum_{i\in\alpha} i$ the degree of α and $CH_d^{\mathbb{Q}}(\mathcal{A}_g^*)$ the Chow homology group. Note that this push forward does not depend on the choice of toroidal compactification as such toroidal compactifications always allow a common refinement and the λ_i are compatible with pull back. One can ask for example whether the classes of the boundary components $\mathcal{A}_j^*$ lie in the $\mathbb{Q}$-vector space generated by the ℓ_α with $d(\alpha) = \text{codim}_{\mathcal{A}_g^*}(\mathcal{A}_j^*)$. In [18] we made the following conjecture.

Conjecture 9.5. *In the group* $CH_d^{\mathbb{Q}}(\mathcal{A}_g^*)$ *with* $d = g(g+1)/2 - k(k+1)/2$ *we have*

$$[\mathcal{A}_{g-k}^*] = \frac{(-1)^k}{\prod_{i=1}^k \zeta(2k-1-2g)} \ell_{g,g-1,\ldots,g+1-k}$$

The evidence that Ekedahl and I gave is:

Theorem 9.6. *Conjecture 9.5 is true for* $k = 1$ *and* $k = 2$ *and if* $\text{char}(k) = p > 0$ *then for all* k.

We shall see later that a multiple of the class λ_g has a beautiful representative cycle in $\mathcal{A}_g \otimes \mathbb{F}_p$, namely the locus of abelian varieties of p-rank 0.

10. Cohomology

By a result of Borel [9] the stable cohomology of the symplectic group is known; this implies that in degrees $\leqslant g-2$ the cohomology of $Sp(2g,\mathbb{Q})$ (and in fact for any finite index subgroup Γ) is generated by elements in degree $2k+4$ for $k = 0,1,\ldots,(g-6)/2$:

$$H^*(\Gamma,\mathbb{Q}) = \mathbb{Q}[x_2, x_6, x_{10}, \ldots].$$

The Chern classes λ_{2k+1} of the Hodge bundle provide these classes.

There are also some results on the stable homology of the Satake compactification, see [12]; besides the λ_{2k+1} there are other classes α_{2k+1} for $k \geqslant 1$.

Van der Kallen and Looijenga proved in [46] that the rational homology of the pair $(\mathcal{A}_g, \mathcal{A}_{g,\text{dec}})$ with $\mathcal{A}_{g,\text{dec}}$ the locus of decomposable principally polarized abelian varieties, vanishes in degree $\leqslant g-2$.

For low values of g the cohomology of $\mathcal{A}_g$ is known. For $g = 1$ one has $H^0(\mathcal{A}_1,\mathbb{Q}) = \mathbb{Q}$ and $H^1 = H^2 = (0)$. For $g = 2$ one can show that $H^0(\mathcal{A}_2,\mathbb{Q}) = \mathbb{Q}$, $H^2(\mathcal{A}_2,\mathbb{Q}) = \mathbb{Q}(-1)$.

Hain determined the cohomology of $\mathcal{A}_3$. His result ([38]) is:

Theorem 10.1. *The cohomology* $H^*(\mathcal{A}_3, \mathbb{Q})$ *is given by*

$$H^j(\mathcal{A}_3, \mathbb{Q}) \cong \begin{cases} \mathbb{Q} & j = 0 \\ \mathbb{Q}(-1) & j = 2 \\ \mathbb{Q}(-2) & j = 4 \\ E & j = 6 \\ 0 & \textit{else} \end{cases}$$

where E *is an extension*

$$0 \to \mathbb{Q}(-3) \to E \to \mathbb{Q}(-6) \to 0$$

We find for the compactly supported cohomology a similar result

$$H^j_c(\mathcal{A}_3, \mathbb{Q}) \cong \begin{cases} \mathbb{Q}(-6) & j = 12 \\ \mathbb{Q}(-5) & j = 10 \\ \mathbb{Q}(-4) & j = 8 \\ E' & j = 6 \\ 0 & \text{else} \end{cases}$$

where E' is an extension $0 \to \mathbb{Q} \to E' \to \mathbb{Q}(-3) \to 0$.

The natural map $H^*_c \to H^*$ is the zero map. Indeed, the classes in H^j_c for $j = 12$, 10 and 8 are $\lambda_3\lambda_1^3$, $\lambda_3\lambda_1^2$ and $\lambda_3\lambda_1$, while λ_3 gives a non-zero class in H^6_c. On the other hand, 1, λ_1, λ_1 and λ_1^3 give non-zero classes in H^0, H^2, H^4 and H^6. Looking at the weight shows that the map is the zero map.

By calculating the cohomology of $\tilde{A}_3$ Hulek and Tommasi proved in [43] that the cohomology of the Voronoi compactification for $g \leqslant 3$ coincides with the Chow ring (known by the Theorems of Section 7):

Theorem 10.2. *The cycle class map gives an isomorphism*

$$CH^*_{\mathbb{Q}}(\mathcal{A}_g^{\mathrm{Vor}}) \cong H^*(\mathcal{A}_g^{\mathrm{Vor}}) \quad \textit{for } g = 1,2,3\,.$$

11. Siegel modular forms

The cohomology of $\mathcal{A}_g$ itself and of the universal family $\mathcal{X}_g$ and its powers $\mathcal{X}_g^n$ is closely linked to modular forms. We therefore pause to give a short description of these modular forms on $\mathrm{Sp}(2g, \mathbb{Z})$. For a general reference we refer to the book of Freitag [27] and to [33] and the references there.

Siegel modular forms generalize the notion of usual (elliptic) modular forms on $\mathrm{SL}(2, \mathbb{Z})$ and its (finite index) subgroups. We first need to generalize the notion of the weight of a modular form. To define it we need a finite-dimensional complex representation $\rho : \mathrm{Gl}(g, \mathbb{C}) \to \mathrm{GL}(V)$ with V a finite-dimensional complex vector space.

Definition 11.1. A holomorphic map $f : \mathfrak{H}_g \to V$ is called a Siegel modular form of weight ρ if

$$f(\gamma(\tau)) = \rho(c\tau + d)f(\tau) \quad \text{for all } \gamma = \begin{pmatrix} a & b \\ c & d \end{pmatrix} \in \mathrm{Sp}(2g, \mathbb{Z}) \text{ and all } \tau \in \mathfrak{H}_g,$$

plus for $g = 1$ the extra requirement that f is holomorphic at the cusp. (The latter means that f has a Fourier expansion $f = \sum_{n \geqslant 0} a(n)q^n$ with $q = e^{2\pi i \tau}$.)

Modular forms of weight ρ form a $\mathbb{C}$-vector space $M_\rho(\mathrm{Sp}(2g, \mathbb{Z}))$; as it turns out this space is finite-dimensional. This vector space can be identified with the space of holomorphic sections of the vector bundle V_ρ defined before (see Section 3). Since we are working on an orbifold, one has to be careful; we could replace $\mathrm{Sp}(2g, \mathbb{Z})$ by a normal congruence subgroup Γ of finite index that acts freely on $\mathfrak{H}_g$, take the space $M_\rho(\Gamma)$ of modular forms on Γ and consider the invariant subspace under the action of $\mathrm{Sp}(2g, \mathbb{Z})/\Gamma$.

Since we can decompose the representation ρ into irreducible representations it is no restriction of generality to limit ourselves to the case where ρ is irreducible. The irreducible representations ρ of $\mathrm{GL}(g, \mathbb{C})$ are given by g-tuples integers $(a_1, a_2, \ldots, a_g)$ with $a_i \geqslant a_{i+1}$, the highest weight of the representation.

A special case is where ρ is given by $a_i = k$, in other words

$$\rho(c\tau + d) = \det(c\tau + d)^k .$$

In this case the Siegel modular forms are called *classical Siegel modular forms* of weight k.

A modular form f has a Fourier expansion

$$f(\tau) = \sum_{n \text{ half-integral}} a(n) e^{2\pi i \mathrm{Tr}(n\tau)},$$

where n runs over the symmetric $g \times g$ matrices that are half-integral (i.e. $2n$ is an integral matrix with even entries along the diagonal) and

$$\mathrm{Tr}(n\tau) = \sum_{i=1}^{g} n_{ii}\tau_{ii} + 2 \sum_{1 \leqslant i < j \leqslant g} n_{ij}\tau_{ij}.$$

A classical result of Koecher (cf. [27]) asserts that $a(n) = 0$ for n that are not semi-positive. This is a sort of Hartogs extension theorem.

We shall use the suggestive notation q^n for $e^{2\pi i \mathrm{Tr}(n\tau)}$ and then can write the Fourier expansion as

$$f(\tau) = \sum_{n \geqslant 0, \text{ half-integral}} a(n)\, q^n.$$

We observe the invariance property

$$a(u^t n u) = \rho(u^t) a(n) \quad \text{for all } u \in \mathrm{GL}(g, \mathbb{Z}) \tag{1}$$

This follows by the short calculation

$$a(u^t n u) = \int_{x \bmod 1} f(\tau) e^{-2\pi i \mathrm{Tr}(u^t n u \tau)} dx$$
$$= \rho(u^t) \int_{x \bmod 1} f(u\tau u^t) e^{-2\pi i \mathrm{Tr}(n\, u\tau u^t)} dx = \rho(u^t) a(n),$$

where we wrote $\tau = x + iy$.

There is a way to extend Siegel modular forms on $\mathcal{A}_g$ to modular forms on $\mathcal{A}_g \sqcup \mathcal{A}_{g-1}$ and inductively to $\mathcal{A}_g^*$ by means of the so-called Φ-operator of Siegel. For $f \in M_\rho$ one defines

$$\Phi(f)(\tau') = \lim_{t \to \infty} f(\begin{smallmatrix} \tau' & 0 \\ 0 & it \end{smallmatrix}) \qquad \tau' \in \mathfrak{H}_{g-1}.$$

The limit is well-defined and gives a function that satisfies

$$(\Phi f)(\gamma'(\tau')) = \rho(\begin{smallmatrix} c\tau + d & 0 \\ 0 & 1 \end{smallmatrix})(\Phi f)(\tau'),$$

where $\gamma' = (a\, b; c\, d) \in \mathrm{Sp}(2g-2, \mathbb{Z})$ is embedded in $\mathrm{Sp}(2g, \mathbb{Z})$ as the automorphism group of the symplectic subspace $\langle e_i, f_i : i = 1, \ldots, g-1 \rangle$. That is, we get a linear map

$$M_\rho \longrightarrow M_{\rho'}, \quad f \mapsto \Phi(f),$$

for some representation ρ'; in fact, with the representation $\rho' = (a_1, a_2, \ldots, a_{g-1})$ for irreducible $\rho = (a_1, a_2, \ldots, a_g)$, cf. the proof of 11.4.

Definition 11.2. A Siegel modular form $f \in M_\rho$ is called a *cusp form* if $\Phi(f) = 0$.

We can extend this definition by

Definition 11.3. A non-zero modular form $f \in M_\rho$ has co-rank k if $\Phi^k(f) \neq 0$ and $\Phi^{k+1}(f) = 0$.

That is, f has co-rank k if it survives (non-zero) till $\mathcal{A}_{g-k}$ and no further. So cusp forms have co-rank 0.

For an irreducible representation ρ of GL_g with highest weight $(a_1, \ldots, a_g)$ we define the co-rank as

$$\mathrm{co-rank}(\rho) = \#\{i : 1 \leqslant i \leqslant g, a_i = a_g\}$$

Weissauer proved in [70] the following result.

Theorem 11.4. *Let ρ be irreducible. For a non-zero Siegel modular form $f \in M_\rho$ one has* $\mathrm{corank}(f) \leqslant \mathrm{corank}(\rho)$.

Proof. We give Weissauer's proof. Suppose that $f : \mathfrak{H}_g \to V$ is a form of co-rank k with Fourier series $\sum a(n)q^n$; then there exists a semi-definite half-integral matrix n such that

$$n = \begin{pmatrix} n' & 0 \\ 0 & 0 \end{pmatrix} \quad \text{with } n' \text{ a } (g-k) \times (g-k) \text{ matrix}$$

such that $a(n) \neq 0$. The identity (1) implies that $a(n) = \rho(u)a(n)$ for all u in the group

$$G_{g,k} = \{ \begin{pmatrix} 1_{g-k} & b \\ 0 & d \end{pmatrix} : d \in GL_k \}$$

This group is a semi-direct product $GL_k \ltimes N$ with N the unipotent radical. The important remark now is that the Zariski closure of $GL(k, \mathbb{Z})$ in $G_{g,k}$ contains $SL_k(\mathbb{C}) \ltimes N$ and we have $a(n) = \rho(u)a(n)$ for all $u \in SL_k(\mathbb{C}) \ltimes N$. So the Fourier coefficient $a(n)$ lies in the subspace V^N of N-invariants. The parabolic group $P = \{(a\, b; 0\, d); a \in GL_{n-k}, d \in GL_k\}$, which is also a semi-direct product $(GL_{n-k} \times GL_k) \ltimes N$, acts on V^N via its quotient $GL_{n-k} \times GL_k$. Since this is a reductive group V^N decomposes into irreducible representations, each of which is a tensor product of an irreducible representation of GL_{n-k} times an irreducible representation of GL_k. If U_k denotes the subgroup of upper triangular unipotent matrices in $GL(k, \mathbb{C})$, then the highest weight space is $(V^N)^{U_{g-k} \times U_k}$ and this equals V^{U_g}, and this is 1-dimensional. Thus V^N is an irreducible representation of $GL_{g-k} \times GL_k$ and one checks that it is given by $(a_1, \dots, a_{g-k}) \otimes (a_{g-k+1}, \dots, a_g)$. The space of invariants $V^{G_{g,k}} \subset V^N$ under $G_{g,k}$ can only contain $SL_k(\mathbb{C})$-invariant elements if the GL_k-representation is 1-dimensional. Therefore $V^{G_{g,k}}$ is zero unless $(a_{g-k+1}, \dots, a_g)$ is a 1-dimensional representation, hence $a_{g-k+1} = \dots = a_g$. In that case the representation for GL_{g-k} is given by $(a_1, \dots, a_{g-k})$. Hence the Fourier coefficients of f all have to vanish if the corank of f is greater than the corank of ρ. This proves the result. □

12. Differential forms on $\mathcal{A}_g$

Here we look first at the moduli space over the field of complex numbers $\mathcal{A}_g(\mathbb{C}) = Sp(2g, \mathbb{Z})\backslash\mathfrak{H}_g$. We are interested in differential forms living on $\tilde{\mathcal{A}}_g$. If η is a holomorphic p form on $\tilde{\mathcal{A}}_g$ then we can pull the form back to $\mathfrak{H}_g$. It will there be a section of some exterior power of $\Omega^1_{\mathfrak{H}_g}$, hence of some exterior power of the second symmetric power of the Hodge bundle. Such forms can be analytically described by vector-valued Siegel modular forms.

As we saw in 3.1, the bundle $\Omega^1(\log D)$ is associated to the second symmetric power $\mathrm{Sym}^2(\mathbb{E})$ of the Hodge bundle (at least in the stacky interpretation) and we are led to ask which irreducible representations occur in the exterior powers of the second symmetric power of the standard representation of $GL(g)$. The answer is that these are exactly those irreducible representations ρ that are of the form $w\eta - \eta$, where $\eta = (g, g-1, \dots, 1)$ is half the sum of the positive roots and w runs through the 2^g Kostant representatives of $W(GL_g)\backslash W(Sp_{2g})$, where W is the Weyl group. They have the property that they send dominant weights for Sp_{2g} to dominant weights for GL_g.

To describe this explicitly, recall that the Weyl group W of $Sp(2g, \mathbb{Z})$ is the semi-direct product $S_g \ltimes (\mathbb{Z}/2\mathbb{Z})^g$ of signed permutations. The Weyl group of GL_g is equal to the symmetric group S_g. Every left coset $S_g \backslash W$ contains exactly one Kostant element w. Such an element w corresponds one to one to an element of $(\mathbb{Z}/2\mathbb{Z})^g$ that we view a g-tuple $(\epsilon_1, \ldots, \epsilon_g)$ with $\epsilon_i \in \{\pm 1\}$ and the action of this element w on the root lattice is then via

$$(a_1, \ldots, a_g) \xrightarrow{w} (\epsilon_{\sigma(1)} a_{\sigma(1)}, \ldots, \epsilon_{\sigma(g)} a_{\sigma(g)})$$

for all $a_1 \geqslant a_2 \geqslant \cdots \geqslant a_g$ and with σ the unique permutation such that

$$\epsilon_{\sigma(1)} a_{\sigma(1)} \geqslant \cdots \geqslant \epsilon_{\sigma(g)} a_{\sigma(g)}.$$

If f is a classical Siegel modular form of weight $k = g + 1$ on the group Γ_g then $f(\tau) \prod_{i \leqslant j} d\tau_{ij}$ is a top differential on the smooth part of quotient space $\Gamma_g \backslash \mathcal{H}_g = \mathcal{A}_g(\mathbb{C})$. It can be extended over the smooth part of the rank-1 compactification $\mathcal{A}_g^{(1)}$ if and only if f is a cusp form. It is not difficult to see that this form can be extended as a holomorphic form to the whole smooth compactification $\tilde{\mathcal{A}}_g$.

Proposition 12.1. *The map that associates to a classical Siegel modular cusp form* $f \in S_{g+1}(\Gamma_g)$ *of weight* $g + 1$ *the top differential* $\omega = f(\tau) \prod_{i \leqslant j} d\tau_{ij}$ *defines an isomorphism between* $S_{g+1}(\Gamma_g)$ *and the space of holomorphic top differentials* $H^0(\tilde{\mathcal{A}}_g, \Omega^{g(g+1)/2})$ *on any smooth compactification* $\tilde{\mathcal{A}}_g$.

Freitag and Pommerening proved in [28] the following extension result.

Theorem 12.2. *Let* $p < g(g + 1)/2$ *and* ω *a holomorphic* p*-form on the smooth locus of* $Sp(2g, \mathbb{Z}) \backslash \mathfrak{H}_g$. *Then it extends uniquely to a holomorphic* p*-form on any smooth toroidal compactification* $\tilde{\mathcal{A}}_g$.

But for $g > 1$ the singular locus has codimension $\geqslant 2$. This implies:

Corollary 12.3. *For* $p < g(g + 1)/2$ *holomorphic* p*-forms on* $\tilde{\mathcal{A}}_g$ *correspond one-one to* $Sp(2g, \mathbb{Z})$*-invariant holomorphic* p*-forms on* $\mathfrak{H}_g$:

$$\Gamma(\tilde{\mathcal{A}}_g, \Omega^p) \cong (\Omega^p(\mathfrak{H}_g))^{Sp(2g, \mathbb{Z})}.$$

When can holomorphic p-forms exist on $\tilde{\mathcal{A}}_g$? Weissauer proved in [70] a vanishing theorem of the following form.

Theorem 12.4. *Let* $\tilde{\mathcal{A}}_g$ *be a smooth compactification of* $\mathcal{A}_g$. *If* p *is an integer* $0 \leqslant p < g(g + 1)/2$ *then the space of holomorphic* p*-forms on* $\tilde{\mathcal{A}}_g$ *vanishes unless* p *is of the form* $g(g + 1)/2 - r(r + 1)/2$ *with* $1 \leqslant r \leqslant g$ *and then* $H^0(\tilde{\mathcal{A}}_g, \Omega^p_{\tilde{\mathcal{A}}_g}) \cong M_\rho(\Gamma_g)$ *with* $\rho = (g + 1, \ldots, g + 1, g - r, \ldots, g - r)$ *with* $g - r$ *occurring* r *times.*

Weissauer deduced this from the following Vanishing Theorem for Siegel modular forms (cf. loc. cit.).

Theorem 12.5. *Let* $\rho = (a_1, \ldots, a_g)$ *be irreducible with* $\text{co} - \text{rank}(\rho) < g - a_g$. *If we have*

$$\#\{1 \leqslant i \leqslant g : a_i = a_g + 1\} < 2(g - a_g - \text{co-rank}(\rho))$$

then $M_\rho = (0)$.

13. The Kodaira dimension of $\mathcal{A}_g$

The Kodaira dimension of the moduli space $\mathcal{A}_g$ over $\mathbb{C}$, the least integer $\kappa = \kappa(\mathcal{A}_g)$ such that the sequence P_m/m^κ with $P_m(\mathcal{A}_g) = \dim H^0(\mathcal{A}_g, K^{\otimes m})$, is bounded, is an important invariant. In terms of Siegel modular forms this comes down to the growth of the dimension of the space of classical Siegel modular forms f of weight $k(g+1)$ that extend to holomorphic tensors $f(\tau)\eta^{\otimes k}$ with $\eta = \wedge_{1 \leqslant i \leqslant j \leqslant g} d\tau_{ij}$ on $\tilde{\mathcal{A}}_g$. The first condition is that f vanishes with multiplicity $\geqslant k$ along the divisor at infinity. Then one has to deal with the extension over the quotient singularities. Reid and Tai independently found a criterion for the extension of pluri-canonical forms over quotient singularities.

Proposition 13.1. *(Reid-Tai Criterion) A pluri-canonical form* η *on* $\mathbb{C}^n$ *that is invariant under a finite group* G *acting linearly on* $\mathbb{C}^n$ *extends to a resolution of singularities if for every non-trivial element* $\gamma \in G$ *and every fixed point* x *of* γ *we have*

$$\sum_{j=1}^{n} \alpha_j \geqslant 1,$$

where the action of γ *on the tangent space of* x *has eigenvalues* $e^{2\pi i \alpha_j}$. *If* G *does not possess pseudo-reflections then it extends if and only if* $\sum_{j=1}^{n} \alpha_j \geqslant 1$.

Tai checked (cf. [68]) that if $g \geqslant 5$ then every fixed point of a non-trivially acting element of $Sp(2g, \mathbb{Z})$ satisfies the requirement of the Reid-Tai criterion and thus these forms extend over the quotient singularities of $\mathcal{A}_g$. He also checked that the extension over the singularities in the boundary did not present difficulties. Thus he showed by calculating the dimension of the space of sections of $K^{\otimes k}$ on a level cover of $\mathcal{A}_g$ and calculating the space over invariants under the action of $Sp(2g, \mathbb{Z}/\ell\mathbb{Z})$ that for $g \geqslant 9$ the space $\mathcal{A}_g$ was of general type. He thus improved earlier work of Freitag ([26]).

Mumford extended this result in [58] and proved:

Theorem 13.2. *The moduli space* $\mathcal{A}_g$ *is of general type if* $g \geqslant 7$.

His approach was similar to the method used in his joint paper with Harris on the Kodaira dimension of $\mathcal{M}_g$. First he works on the moduli space $\mathcal{A}_g^{(1)}$ of rank 1-degenerations and observes that the Picard group $\mathrm{Pic}(\mathcal{A}_g^{(1)}) \otimes \mathbb{Q}$ is generated by two divisors: λ_1 and the class δ of the boundary. By Proposition 3.1 for the coarse moduli space we know the canonical class.

Proposition 13.3. *Let U be the open subset of* $\mathcal{A}_g^{(1)}$ *of semi-abelian varieties of torus rank* $\leqslant 1$ *with automorphism group* $\{\pm\}$. *Then the canonical class of U is given by* $(g+1)\lambda_1 - \delta$.

Now there is an explicit effective divisor N_0 in $\mathcal{A}_g^{(1)}$, namely the divisor of principally polarized abelian varieties (X, Θ) that have a singular theta divisor. By a tricky calculation in a 1-dimensional family Mumford is able to calculate the divisor class of N_0.

Theorem 13.4. *The divisor class of* $\bar{N}_0$ *is given by*

$$[\bar{N}_0] = \left(\frac{(g+1)!}{2} + g!\right)\lambda_1 - \frac{(g+1)!}{12}\delta .$$

The divisor N_0 is in general not irreducible and splits off the divisor θ_0 of principally polarized abelian varieties whose symmetric theta divisor has a singularity at a point of order 2. This divisor is given by the vanishing of an even theta constant (Nullwert) and so this divisor class can be easily computed. The divisor class of θ_0 equals

$$[\theta_0] = 2^{g-2}(2^g+1)\,\lambda_1 - 2^{2g-5}\,\delta .$$

Proof. (of Theorem 13.2) We have $[\bar{N}_0] = \theta_0 + 2\,[R]$ with R an effective divisor given as

$$[R] = \left(\frac{(g+1)!}{4} + \frac{g!}{2} - 2^{g-3}(2^g+1)\right)\lambda_1 - \left(\frac{(g+1)!}{24} - 2^{2g-6}\right)\delta .$$

For an expression $a\lambda_1 - b\delta$ we call the ratio a/b its slope. If R is effective with a slope $\leqslant$ slope of the expression for the canonical class we know that the canonical class is ample. From the formulas one sees that the slope of $\bar{N}_0$ is $6 + 12/(g+1)$ and deduces that the inequality holds for $g \geqslant 7$. □

From the relatively easy equation for the class $[\theta_0]$ one deduces that the even theta constants provide a divisor with slope $8 + 2^{3-g}$, and this gives that $\mathcal{A}_g$ is of general type for $g \geqslant 8$; cf. [27].

For $g \leqslant 5$ we know that $\mathcal{A}_g$ is rational or unirational. For $g = 1$ and 2 rationality was known in the 19th century; Katsylo proved rationality for $g = 3$, Clemens proved unirationality for $g = 4$ and unirationality for $g = 5$ was proved by several people (Mori-Mukai, Donagi, Verra). But the case $g = 6$ is still open.

For some results on the nef cone we refer to the survey [37] of Grushevsky.

14. Stratifications

In positive characteristic the moduli space $\mathcal{A}_g \otimes \mathbb{F}_p$ possesses stratifications that can tell us quite a lot about the moduli space; it is not clear what the characteristic zero analogues of these stratifications are. This makes this moduli space in some sense more accessible in characteristic p than in characteristic zero, a fact that may sound counterintuitive to many.

The stratifications we are referring to are the Ekedahl-Oort stratification and the Newton polygon stratification. Since the cycle classes are not known for the latter one we shall stick to the Ekedahl-Oort stratification, E-O for short. It was originally defined by Ekedahl and Oort in terms of the group scheme $X[p]$, the kernel of multiplication by p for an abelian variety in characteristic p. It has been studied intensively by Oort and many others, see e.g. [62] and [63].

The alternative definition using degeneracy loci of vector bundles was given in my paper [32] and worked out fully in joint work with Ekedahl, see [19]. In this section we shall write $\mathcal{A}_g$ for $\mathcal{A}_g \otimes \mathbb{F}_p$. We start with the universal principally polarized abelian variety $\pi : \mathcal{X}_g \to \mathcal{A}_g$. For an abelian variety X over a field k the de Rham cohomology $H^1_{\mathrm{dR}}(X)$ is a vector space of rank $2g$. We get by doing this in families a cohomology sheaf $\mathcal{H}^1_{\mathrm{dR}}(\mathcal{X}_g/\mathcal{A}_g)$, the hyperdirect image $R^1\pi_*(\mathcal{O}_{\mathcal{X}_g} \to \Omega^1_{\mathcal{X}_g/\mathcal{A}_g})$. Because of the polarization it comes with a symplectic pairing $\langle\, , \,\rangle : \mathcal{H}^1_{\mathrm{dR}} \times \mathcal{H}^1_{\mathrm{dR}} \to \mathcal{O}_{\mathcal{A}_g}$. We have the Hodge filtration of $\mathcal{H}^1_{\mathrm{dR}}$:

$$0 \to \pi_*(\Omega^1_{\mathcal{X}_g/\mathcal{A}_g}) \to \mathcal{H}^1_{\mathrm{dR}} \to R^1\pi_*\mathcal{O}_{\mathcal{X}_g} \to 0,$$

where the first non-zero term is the Hodge bundle $\mathbb{E}_g$. In characteristic $p > 0$ we have additionally two maps

$$F : \mathcal{X}_g \to \mathcal{X}_g^{(p)}, \quad V : \mathcal{X}_g^{(p)} \to \mathcal{X}_g,$$

relative Frobenius and the Verschiebung satisfying $F \circ V = p \cdot \mathrm{id}_{\mathcal{X}_g^{(p)}}$ and $V \circ F = p \cdot \mathrm{id}_{\mathcal{X}_g}$.

Look at the simplest case $g = 1$. Since Frobenius is inseparable the kernel $X[p]$ of multiplication by p is not reduced and has either 1 or p physical points. An elliptic curve in characteristic $p > 0$ is called *supersingular* if and only if $X[p]_{\mathrm{red}} = (0)$. Equivalently this means that V is also inseparable, hence the kernel of F and V (for $\mathcal{X}_g$ instead of $\mathcal{X}_g^{(p)}$) coincide. There are finitely many points on the j-line $\mathcal{A}_1$ that correspond to the supersingular elliptic curves.

So in general we will compare the relative position of the kernel of F and of V inside $\mathcal{H}^1_{\mathrm{dR}}$. As it turns out, it is better to work with the flag space $\mathcal{F}_g$ of symplectic flags on $\mathcal{H}^1_{\mathrm{dR}}$; by this we mean that we consider the space of flags $(E_i)_{i=1}^{2g}$ on $\mathcal{H}^1_{\mathrm{dR}}$ such that $\mathrm{rank}(E_i) = i$, $E_g = \mathbb{E}$ and $E_{g-i} = E_j^{\perp}$. We can then introduce a second flag on $\mathcal{H}^1_{\mathrm{dR}}$, say $(D_i)_{i=1}^{2g}$ defined by setting

$$D_g = \ker(V) = V^{-1}(0), \quad D_{g+i} = V^{-1}(E_i^{(p)},$$

and complementing by

$$D_{g-i} = D_{g+i}^{\perp} \qquad \text{for } i = 1, \ldots, g.$$

This is called the conjugate flag. For an abelian variety X we define the *canonical flag* of X as the coarsest flag that is stable under F: if G is a member, then also $F(G^{(p)})$; usually this will not be a full flag. The conjugate flag D is a refinement of the canonical flag.

So starting with one filtration we end up with two filtrations. We then compare these two filtrations. In order to do this we need the Weyl group W_g of the symplectic group $Sp(2g, \mathbb{Z})$. This group is isomorphic to the semi-direct product

$$W_g \cong S_g \ltimes (\mathbb{Z}/2\mathbb{Z})^g = \{\sigma \in S_{2g} : \sigma(i) + \sigma(2g+1-i) = 2g+1,\ i = 1, \dots, g\}.$$

As is well-known each element $w \in W_g$ has a *length* $\ell(w)$. Let

$$W'_g = \{\sigma \in W_g : \sigma\{1,2,\dots,g\} = \{1,2,\dots,g\}\} \cong S_g .$$

Now for every coset aW'_g there is a unique element w of minimal length in aW'_g such that every $w' \in aW'_g$ can be written as $w' = wx$ with $\ell(w') = \ell(w) + \ell(x)$.

There is a partial order on W_g, the Bruhat-Chevalley order:

$$w_1 \geqslant w_2 \quad \text{if and only if} \quad r_{w_1}(i,j) \leqslant r_{w_2}(i,j) \quad \text{for all } i, j,$$

where the function $r_w(i,j)$ is defined as

$$r_w(i,j) = \#\{n \leqslant i : w(n) \leqslant j\}.$$

There are 2^g so-called *final elements* in W_g: these are the minimal elements of the cosets. In another context (see Section 12) these are called Kostant representatives. An element $\sigma \in W_g$ is final if and only if $\sigma(i) < \sigma(j)$ for $i < j \leqslant g$. These final elements correspond one-to-one to so-called final types: these are increasing surjective maps

$$\nu : \{0,1,\dots,2g\} \to \{0,1,\dots,2g\}$$

satisfying $\nu(2g-i) = \nu(i) + (g-i)$ for $0 \leqslant i \leqslant g$. The bijection is gotten by associating to $w \in W_g$ the final type ν_w with $\nu_w(i) - i - r_w(g,i)$.

Now the stratification on our flag space $\mathcal{F}_g$ is defined by

$$(E_., D_.) \in U_w \iff \dim(E_i \cap D_j) \geqslant r_w(i,j) .$$

There is also a scheme-theoretic definition: the two filtrations mean that we have two sections s, t of a G/B-bundle T (with structure group G) over the base with $G = Sp_{2g}$ and B a Borel subgroup; locally (in the étale topology) we may assume that t is a trivializing section; then we can view s as a map of the base to G/B and we simply take U_w (resp. $\overline{U}_w$) to be the inverse image of the B-orbit BwB (resp. its closure). This definition is independent of the chosen trivialization. We thus find global subschemes U_w and $\bar{U}_w$ for the $2^g(g!)$ elements of W_g.

It will turn out that for *final* elements the projection map $\mathcal{F}_g \to \mathcal{A}_g$ restricted to U_w defines a finite morphism to its image in $\mathcal{A}_g$, and these will be the E-O strata. But the strata on $\mathcal{F}_g$ behave in a much better way. That is why we first study these on $\mathcal{F}_g$. We can extend the strata to strata over a compactification, see [19]: the Hodge bundle extends and so does the de Rham sheaf, namely as the logarithmic de Rham sheaf $R^1\pi_*(\Omega^{\cdot}_{\tilde{\mathcal{X}}_g/\tilde{\mathcal{A}}_g})$ and then we play the same game as above.

There is a smallest stratum: $\bar{U}_1$ associated to the identity element of W_g.

Proposition 14.1. *Any irreducible component of any $\bar{U}_w$ in $\mathcal{F}_g$ contains a point of $\bar{U}_1$.*

The advantage of working in the flag space comes out clearly if we consider the local structure of our strata. The idea is that locally our moduli space (flag space) looks like the space FL_g of complete symplectic flags in a $2g$-dimensional symplective space.

We need the notion of *height* 1 *maps* in charactistic $p > 0$. A closed immersion $S \to S'$ is called a height 1 map if $I^{(p)} = (0)$ with I the ideal sheaf defining S and $I^{(p)}$ the ideal generated by p-th powers of elements. If $S = \mathrm{Spec}(R)$ and $x : \mathrm{Spec}(k) \to S$ is a closed point then the height 1 neighborhood is $\mathrm{Spec}(R/m^{(p)})$. We can then introduce the notions of *height* 1 *isomorphic* and *height* 1 *smooth* in an obvious way.

Our basic result of [19] about our strata is now:

Theorem 14.2. *Let k be a perfect field. For every k-point x of $\mathcal{F}_g$ there exists a k-point of FL_g such that their height* 1 *neighborhoods are isomorphic as stratified spaces.*

Here the stratification on the flag space FL_g is the usual one given by the Schubert cells. The idea of the proof is to trivialize the de Rham bundle over a height 1 neighborhood and then use infinitesimal Torelli for abelian varieties. Our theorem has some immediate consequences.

Corollary 14.3. (1) *Each stratum U_w is smooth of dimension $\ell(w)$.*

(2) *Each stratum $\bar{U}_w$ is Cohen-Macaulay, reduced and normal; moreover $\bar{U}_w$ is the closure of U_w.*

(3) *For w a final element the projection $\mathcal{F}_g \to \mathcal{A}_g$ induces a finite étale map $U_w \to V_w$, with V_w the image ofU_w.*

We now descend from $\mathcal{F}_g$ to $\mathcal{A}_g$ to define the Ekedahl-Oort stratification.

Definition 14.4. For a final element $w \in W_g$ the E-O stratum V_w is defined to be the image of U_w under the projection $\mathcal{F}_g \to \mathcal{A}_g$.

Over a E-O stratum $V_w \subset \mathcal{A}_g$ the type of the canonical filtration (i.e., the dimensions of the filtration steps that occur) are constant.

Proposition 14.5. *The image of any U_w under the projection $\mathcal{F}_g \to \mathcal{A}_g$ is a union of strata V_v; the image of any $\bar{U}_w$ is a union of strata $\bar{V}_v$.*

We also have an irreducibility result (cf. [19]). Each final element $w \in W_g$ corresponds to a partition and a Young diagram.

Theorem 14.6. *If $w \in W_g$ is a final element whose Young diagram Y does not contain all rows of length i with $\lceil (g+1)/2 \rceil \leqslant i \leqslant g$ then $\bar{V}_w$ is irreducible and $U_w \to V_w$ is a connected étale cover.*

Harashita proved in [39] that the other ones are in general reducible (with exceptions maybe for small characteristics).

A stratification on a space is not worth much unless one knows the cycle classes of the strata. In our case one can calculate these.

On the flag space the Chern classes λ_i of the Hodge bundle decompose in their roots:

$$c_1(E_i) = l_1 + \ldots + l_i .$$

We then have

$$c_1(D_{g+i}) - c_1(D_{g+1-i}) = p\, l_i .$$

Theorem 14.7. *The cycle classes of $\bar{U}_w$ are polynomials in the classes l_i with coefficients that are polynomials in p.*

We refer to [19] for an explicit formula. Using the analogues of the maps π_i of Section 2 and Lemma 2.4 we can calculate the cycle classes of the E-O strata on $\mathcal{A}_g$.

Instead of giving a general formula we restrict to giving the formulas for some important strata. For example, there are the p-rank strata

$$V_f := \{[X] \in \mathcal{A}_g : \#X[p](\bar{k}) \leqslant p^f\}$$

for $f = g, g-1, \ldots, 0$. Besides these there are the a-number strata

$$T_a := \{[X] \in \mathcal{A}_g : \dim_k \mathrm{Hom}(\alpha_p, X) \geqslant a\} .$$

Recall that the p-rank $f(X)$ of an abelian variety is f if and only if $\#X[p](\bar{k}) = p^f$ and $0 \leqslant f \leqslant g$ with $f = g$ being the generic case. Similarly, the a-number of $a(X)$ of X is $\dim_k(\alpha_p, X)$ and this equals the rank of $\ker(V) \cap \ker(F)$; so $0 \leqslant a(X) \leqslant g$ with $a(X) = 0$ being the generic case. The stratum V_f has codimension $g - f$ while the stratum T_a has codimension $a(a+1)/2$. These codimensions were originally calculated by Oort and follow here easily from 14.3.

Theorem 14.8. *The cycle class of the p-rank stratum V_f is given by*

$$[V_f] = (p-1)(p^2-1)\cdots(p^{g-f}-1)\lambda_{g-f} .$$

For example, for $g = 1$ the stratum V_0 is the stratum of supersingular elliptic curves. We have $[V_0] = (p-1)\lambda_1$. By the cycle relation $12\lambda_1 = \delta$ with δ the cycle of the boundary we find

$$[V_0] = \frac{p-1}{12}\delta .$$

Since the degenerate elliptic curve (rational nodal curve) has two automorphisms we find (using the stacky interpretation of our formula) the Deuring Mass Formula

$$\sum_{E/\bar{k} \text{ supersingular}} \frac{1}{\#\mathrm{Aut}_k(E)} = \frac{p-1}{24}$$

for the number of supersingular elliptic curves in characteristic p. One may view all the formulas for the cycle classes as a generalization of the Deuring Mass Formula.

The formulas for the a-number strata are given in [32] and [19]. We have

Theorem 14.9. *The cycle class of the locus* T_a *of abelian varieties with* a*-number* $\geqslant a$ *is given by*

$$\sum Q_\beta(\mathbb{E}^{(p)} \cdot Q_{\rho(a)-\beta}(\mathbb{E}^\vee),$$

where Q_μ *is defined as in Section 2 and the sum is over all partitions* β *contained in* $\rho(a) = \{a, a-1, \ldots, 1\}$.

For example, the formula for the stratum T_1 is $p\lambda_1 - \lambda_1$, which fits since a-number $\geqslant 1$ means exactly that the p-rank is $\leqslant g-1$. For $a = 2$ the formula is $[T_2] = (p-1)(p^2+1)(\lambda_1\lambda_2) - (p^3-1)2\lambda_3$. For $a = g$ the formula reads

$$[T_g] = (p-1)(p^2+1)\cdots(p^g+(-1)^g)\lambda_1\lambda_2\cdots\lambda_g.$$

This stratum is of maximal codimension; this formula is visibly a generalization of the Deuring Mass Formula and counts the number of superspecial abelian varieties; this number was first calculated by Ekedahl in [16].

We formulate a corollary of our formulas for the p-rank strata.

Corollary 14.10. *The Chern classes* λ_i *of the Hodge bundle are represented by effective* $\mathbb{Q}$*-classes.*

Another nice aspect of our formulas is that when we specialize $p = 0$ in our formulas, that are polynomials in the l_i and λ_i with coefficients that are polynomials in p, we get back the formulas for the cycle classes of the Schubert cells both on the Grassmannian and the flag space, see [19].

15. Complete subvarieties of A_g

The existence of complete subvarieties of relative small codimension can give us interesting information about a non-complete variety. In the case of the moduli space $\mathcal{A}_g \otimes k$ (for some field k) we know that λ_1 is an ample class and that $\lambda_1^{g(g-1)/2+1}$ vanishes. Since for a complete subvariety X of dimension d we must have $\lambda_1^d|X \neq 0$, this implies immediately a lower bound on the codimension.

Theorem 15.1. *The minimum possible codimension of a complete subvariety of* $\mathcal{A}_g \otimes k$ *is* g.

This lower bound can be realized in positive characteristic as was noted by Koblitz and Oort, see [48], [60]. The idea is simple: a semi-abelian variety with a positive torus rank has points of order p. So by requiring that our abelian varieties have p-rank 0 we stay inside $\mathcal{A}_g$ and this defines the required complete variety.

Theorem 15.2. *The moduli stack* $\mathcal{A}_g \otimes k$ *with* $\mathrm{char}(k) = p > 0$ *contains a complete substack of codimension* g*: the locus* V_0 *of abelian varieties with* p*-rank zero.*

A generalization is:

Theorem 15.3. *The partial Satake compactification $\mathcal{A}_g^* - \mathcal{A}_{g-(t+1)}^*$ in characteristic $p > 0$ of degenerations of torus rank t contains a complete subvariety of codimension $g - t$, namely the locus V_t of p-rank $\leqslant t$.*

In characteristic 0 there is no such obvious complete subvariety of codimension g, at least if $g \geqslant 3$. Oort conjectured in [61] that it should not exist for $g \geqslant 3$. This was proved by Keel and Sadun in [47].

Theorem 15.4. *If $X \subset \mathcal{A}_g(\mathbb{C})$ is a complete subvariety with the property that $\lambda_i|X$ is trivial in cohomology for some $1 \leqslant i \leqslant g$ then $\dim X \leqslant i(i-1)/2$ with strict inequality if $i \geqslant 3$.*

This theorem implies the following corollary.

Corollary 15.5. *For $g \geqslant 3$ the moduli space $A_g \otimes \mathbb{C}$ does not possess a complete subvariety of codimension g.*

So the question arises what the maximum dimension of a complete subvariety of $\mathcal{A}_g(\mathbb{C})$ is.

One might conclude that the analogue of $\mathcal{A}_g(\mathbb{C})$ in positive characteristic in some sense is rather the locus of principally polarized abelian varieties with maximal p-rank ($= g$) than $\mathcal{A}_g \otimes \mathbb{F}_p$.

16. Cohomology of local systems and relations to modular forms

There is a close connection between the cohomology of moduli spaces of abelian varieties and modular forms. This connection was discovered in the 19th century and developed further in the work of Eichler, Shimura, Kuga, Matsushima and many others. It has developed into a central theme involving the theory of automorphic representations and the Langlands philosophy. We shall restrict here to just one aspect of this. This is work in progress that is being developed in joint work with Jonas Bergström and Carel Faber.

Let us start with $g = 1$. The space of cusp forms S_{2k} of weight $2k$ on $SL(2, \mathbb{Z})$ has a cohomological interpretation. To describe it we consider the universal elliptic curve $\pi : \mathcal{X}_1 \to \mathcal{A}_1$ and let $V := R^1\pi_*\mathbb{Q}$ be the local system of rank 2 with as fibre over $[X]$ the cohomology $H^1(X, \mathbb{Q})$ of the elliptic curve X

From V we can construct other local systems: define for $a \geqslant 1$

$$V_a := \mathrm{Sym}^a(V).$$

This is a local system of rank $a + 1$. There is also the l-adic analogue $V^{(l)} = R^1\pi_*\mathbb{Q}_l$ for l-adic étale cohomology and its variants $V_a^{(l)}$. The basic result of Eichler and Shimura says that for $a > 0$ and a even there is an isomorphism

$$H_c^1(\mathcal{A}_1 \otimes \mathbb{C}, V_a \otimes \mathbb{C}) = S_{a+2} \oplus \bar{S}_{a+2} \oplus \mathbb{C}.$$

So we might say that as a mixed Hodge module the compactly supported cohomology of the local system $V_a \otimes \mathbb{C}$ equals $S_{a+2} \oplus \bar{S}_{a+2} \oplus \mathbb{C}$.

But this identity can be stretched further. The left hand side may be replaced by other flavors of cohomology, for example by l-adic étale cohomology $H^1(\mathcal{A}_1 \otimes \bar{\mathbb{Q}}, V_a^{(l)})$ that comes with a natural Galois action of $\mathrm{Gal}(\bar{\mathbb{Q}}/\mathbb{Q})$. In view of this, we replace the left hand side by an Euler characteristic

$$e_c(\mathcal{A}_1, V_a) := \sum_{i=0}^{2} [H_c^i(\mathcal{A}_1, V_a)],$$

where now the cohomomology groups of compactly supported cohomology are to be interpreted in an appropriate Grothendieck group, e.g. of mixed Hodge structures when we consider complex cohomology $H_c^*(\mathcal{A}_1 \otimes \mathbb{C}, V_a \otimes \mathbb{C})$ with its mixed Hodge structure, or in the Grothendieck group of Galois representations when we consider compactly supported étale l-adic cohomology $H_{c,et}^*(\mathcal{A}_1 \otimes \bar{\mathbb{Q}}, V_a^{(l)})$. On the other hand, for the right hand side Scholl defined in [65] a Chow motive $S[2k]$ associated to the space of cusp forms S_{2k} for $k > 1$. Then a sophisticated form of the Eichler-Shimura isomorphism asserts that we have an isomorphism

$$e_c(\mathcal{A}_1, V_a) = -S[a+2] - 1 \qquad a \geqslant 2 \text{ even}.$$

We have a natural algebra of operators, the Hecke operators, acting on the spaces of cusp forms, but also on the cohomology since the Hecke operators are defined by correspondences. Then the isomorphism above is compatible with the action of the Hecke operators.

But our moduli space $\mathcal{A}_1$ is defined over $\mathbb{Z}$. We thus can study the cohomology over $\mathbb{Q}$ by looking at the fibres $\mathcal{A}_1 \otimes \mathbb{F}_p$ and the corresponding local systems $V_a^{(l)} \otimes \mathbb{F}_p$ (for $l \neq p$) by using comparison theorems. Now in characteristic p the Hecke operator is defined by the correspondence $X_0(p)$ of (cyclic) p-isogenies $\phi : X \to X'$ between elliptic curves (i.e. we require $\deg \phi = p$); it allows maps $q_i : X_0(p) \to \mathcal{A}_1$ ($i = 1,2$) by sending ϕ to its source X and target X'. In characteristic p the correspondence decomposes into two components

$$X_0(p) \otimes \mathbb{F}_p = F_p + F_p^t, \qquad \text{(the congruence relation)}$$

where F_p is the correspondence $X \mapsto X^{(p)}$ and F_p^t its transpose. This follows since for such a p-isogeny we have that $X' \cong X^{(p)}$ or $X \cong (X')^{(p)}$. This implies a relation between the Hecke operator $T(p)$ and the action of Frobenius on $H^1(\mathcal{A}_1 \otimes \bar{\mathbb{F}}_p, V_a^{(\ell)})$; The result is then a relation between the action of Frobenius on étale cohomology and the action of a Hecke operator on the space of cusp forms ([14], Prop. 4.8)

$$\mathrm{Tr}(F_p, H_c^1(\mathcal{A}_1 \otimes \overline{\mathbb{F}}_p, V_a)) = \mathrm{Tr}(T(p), S_{a+2}) + 1\,.$$

We can calculate the traces of the Frobenius by counting points over finite fields. In fact, if we make a list of all elliptic curves over $\mathbb{F}_p$ up to isomorphism over $\mathbb{F}_p$ and we calculate the eigenvalues $\alpha_X, \bar{\alpha}_X$ of F_p acting on the étale cohomology $H^1(X, \mathbb{Q}_l)$ then we can calculate the trace of F_p on the cohomology $H_c^1(\mathcal{A}_1 \otimes \bar{\mathbb{F}}_p, V_a^{(\ell)})$ by summing the expression

$$\frac{\alpha_X^a + \alpha_X^{a-1}\bar{\alpha}_X + \ldots + \bar{\alpha}_X^a}{\#\mathrm{Aut}_{\mathbb{F}_p}(X)}$$

over all X in our list. So by counting elliptic curves over a finite field $\mathbb{F}_p$ we can calculate the trace of the Hecke operator $T(p)$ on S_{a+2}; in fact, once we have our list of elliptic curves over $\mathbb{F}_p$ together with the eigenvalues $\alpha_X, \bar{\alpha}_X$ and the order of their automorphism group, we can calculate the trace of $T(p)$ on the space of cusp forms S_{a+2} for *all* a.

The term -1 in the Eichler-Shimura identity can be interpreted as coming from the kernel

$$-1 = \sum (-1)^i [\ker H_c^i(\mathcal{A}_1, V_a) \to H^i(\mathcal{A}_1, V_a)].$$

So to avoid this little nuisance we might replace the compactly supported cohomology by the image of compactly supported cohomology in the usual cohomology, i.e. define the *interior* cohomology $H_!^i$ as the image of compactly supported cohomology in the usual cohomology. Then the result reads

$$e_!(\mathcal{A}_1, V_a) = -S[a+2] \qquad \text{for } a > 2 \text{ even.}$$

Some words about the history of the Eichler-Shimura result may be in order here. Around 1954 Eichler showed (see [15]) that for some congruence subgroup Γ of $SL(2,\mathbb{Z})$ the p-part of the zeta function of the corresponding modular curve X_Γ in characteristic p is given by the Hecke polynomial for the Hecke operator $T(p)$ acting on the space of cusp forms of weight 2 for Γ. This was generalized to some other groups by Shimura. M. Sato observed in 1962 that by combining the Eichler-Selberg trace formula for modular forms with the congruence relation (expressing the Hecke correspondence in terms of the Frobenius correspondence and its transpose) one could extend Eichler's results by expressing the Hecke polynomials in terms of the zeta functions of $\mathcal{M}_{1,n} \otimes \mathbb{F}_p$, except for problems due to the non-completeness of the moduli spaces. A little later Kuga and Shimura showed that Sato's idea worked for compact quotients of the upper half plane (parametrizing abelian surfaces). Ihara then made Sato's idea reality in 1967 by combining the Eichler-Selberg trace formula with results of Deuring and Hasse on zeta functions of elliptic curves, cf. [45], where one also finds references to the history of this problem. A year later Deligne solved in [14] the problems posed by the non-completeness of the moduli spaces. Finally Scholl proved the existence of the motive $S[k]$ for even k in 1990. A different construction of this motive was given by Consani and Faber in [13].

That the approach sketched above for calculating traces of Hecke operators by counting over finite fields is not used commonly, is due to the fact that we have an explicit formula for the traces of the Hecke operators, the Eichler-Selberg trace formula, cf. [72]. But for higher genus g, i.e. for modular forms on the symplectic group $Sp(2g,\mathbb{Z})$ with $g \geqslant 2$, no analogue of the trace formula for Siegel modular forms is known. Moreover, a closer inspection reveals that vector-valued Siegel modular forms are the good analogue for higher g of the modular forms on $SL(2,\mathbb{Z})$ (rather than classical Siegel modular forms only). This suggests to try the analogue of the point counting method for genus 2 and higher. That is what Carel Faber and I did for $Sp(4,\mathbb{Z})$ and in joint work with Bergström extended to genus 2 and level 2 and also to genus 3, see [22, 6, 7].

There are alternative methods that we should point out. In general one tries to compare a trace formula of Selberg type (Arthur trace formula) with the Grothendieck-Lefschetz fixed point formula. There is a topological trace formula for the trace of Hecke operators acting on the compactly supported cohomology, see for example [40], esp. the letter to Goresky and MacPherson there. Laumon gives a spectral decomposition of the cohomology of local systems for $g = 2$ (for the trivial one in [50] and in general in [51]); cf. also the work of Kottwitz in general. We also refer to Sophie Morel's book [53]. But though these methods in principle could lead to explicit results on Siegel modular forms, as far as I know it has not yet done that.

So start with the universal principally polarized abelian variety $\pi : \mathcal{X}_g \to \mathcal{A}_g$ and form the local system $V := R^1\pi_*\mathbb{Q}$ and its l-adic counterpart $R^1\pi_*\mathbb{Q}_l$ for étale cohomology. By abuse of notation we will write simply V. We consider π as a morphism of stacks. For any irreducible representation λ of Sp_{2g} we can construct a corresponding local system V_λ; it is obtained by applying a Schur functor, cf. e.g. [30]. So if we denote λ by its highest weight $\lambda_1 \geqslant \lambda_2 \geqslant \cdots \geqslant \lambda_g$ then $V = V_{1,0,\ldots,0}$ and $\mathrm{Sym}^a(V) = V_{a,0,\ldots,0}$.

We now consider

$$e_c(\mathcal{A}_g, V_\lambda) := \sum_i (-1)^i [H^i_c(\mathcal{A}_g, V_\lambda)],$$

where as before the brackets indicate that we consider this in the Grothendieck group of the appropriate category (mixed Hodge modules, Galois representations). It is a basic result of Faltings [25, 23] that $H^i(\mathcal{A}_g \otimes \mathbb{C}, V_\lambda \otimes \mathbb{C})$ and $H^i_c(\mathcal{A}_g \otimes \mathbb{C}, V_\lambda \otimes \mathbb{C})$ carry a mixed Hodge structure. Moreover, the interior cohomology $H^i_!$ carries a pure Hodge structure with the weights equal to the 2^g sums of any of the subsets of $\{\lambda_1 + g, \lambda_2 + g - 1, \ldots, \lambda_g + 1\}$. He also shows that for regular λ, that is, when $\lambda_1 > \lambda_2 \cdots > \lambda_g > 0$, the cohomology $H^i_!$ vanishes when $i \neq g(g+1)/2$.

These results of Faltings are the analogue for the non-compact case of results of Matsushima and Murakami in [52]; they could use Hodge theory to deduce the vanishing of cohomology groups and the decompositions indexed by pairs of elements in the Weyl group.

For $g = 2$ a local system V_λ is specified by giving $\lambda = (a, b)$ with $a \geqslant b \geqslant 0$. Then the Hodge filtration is

$$F^{a+b+3} \subseteq F^{a+2} \subseteq F^{b+1} \subseteq F^0 = H^3_!(\mathcal{A}_2 \otimes \mathbb{C}, V_\lambda \otimes \mathbb{C}).$$

Moreover, there is an identification of the first step in the Hodge filtration

$$F^{a+b+3} \cong S_{a-b,b+3},$$

with the space $S_{a-b,b+3}$ of Siegel modular cusp forms whose weight ρ is the representation $\mathrm{Sym}^{a-b}\mathrm{St} \otimes \det(\mathrm{St})^{b+3}$ with St the standard representation of $\mathrm{GL}(2,\mathbb{C})$.

Again, as for $g = 1$, we have an algebra of Hecke operators induced by geometric correspondences and it acts both on the cohomology and the modular forms compatible with the isomorphism.

Given our general ignorance of vector-valued Siegel modular forms for $g > 1$ the obvious question at this point is whether we can mimic the approach sketched above for calculating the traces of the Hecke operators by counting over finite fields.

Note that by Torelli we have morphism $\mathcal{M}_2 \to \mathcal{A}_2 = \mathcal{M}_2^{ct}$, where $\mathcal{M}_2^{ct}$ is the moduli space of curves of genus 2 of compact type. This means that we have to include besides the smooth curves of genus 2 the stable curves of genus 2 that are a union of two elliptic curves.

Can we use this to calculate the traces of the Hecke operator $T(p)$ on the spaces of vector-valued Siegel modular forms?

Two problems arise. The first is the so-called Eisenstein cohomology. This is

$$e_{\mathrm{Eis}}(\mathcal{A}_2, V_{a,b}) := \sum (-1)^i (\ker H_c^i \to H^i)(\mathcal{A}_2, V_{a,b}).$$

In the case of genus 1 this expression was equal to the innocent -1, but for higher g this is a more complicated term. The second problem is the *endoscopy*: the terms in the Hodge filtration that do not see the first and last step of the Hodge filtration. For torsion-free groups there is work by Schwermer (see [66]) on the Eisenstein cohomology; cf. also the work of Harder [40]. And there is an extensive literature on endoscopy. But explicit formulas were not available.

On the basis of numerical calculations Carel Faber and I guessed in [22] a formula for the Eisenstein cohomology. I was able to prove this formula in [35] for regular λ. We also made a guess for the endoscopic term. Putting this together we get the following conjectural formula for $(a, b) \neq (0,0)$.

Conjecture 16.1. *The trace of the Hecke operator* $T(p)$ *on the space* $S_{a-b,b+3}$ *of cusp forms on* $\mathrm{Sp}(4,\mathbb{Z})$ *equals*

$$-\mathrm{Tr}(F_p, e_c(\mathcal{A}_2 \otimes \mathbb{F}_p, V_{a,b})) + \mathrm{Tr}(F_p, e_{2,\mathrm{extra}}(a, b)),$$

where the term $e_{2,\mathrm{extra}}(a, b)$ *is defined as*

$$s_{a-b+2} - s_{a+b+4}(S[a-b+2]+1)\mathbb{L}^{b+1} + \begin{cases} S[b+2]+1 & a \textit{ even}, \\ -S[a+3] & a \textit{ odd}, \end{cases}$$

and $s_n = \dim S_n(\mathrm{SL}(2,\mathbb{Z}))$ *is the dimension of the space of cusp forms on* $\mathrm{SL}(2,\mathbb{Z})$ *and* $\mathbb{L} = h^2(\mathbb{P}^1)$ *is the Lefschetz motive.*

If $a > b = 0$ or $a = b > 0$ then one should put $s_2 = -1$ and $S[2] = -1 - \mathbb{L}$ in the formula.

In the case of regular local systems (i.e. $a > b > 0$) our conjecture can be deduced from results of Weissauer as he shows in his preprint [71]. In the case of local system of non-regular highest weight the conjecture is still open.

We have counted the curves of genus 2 of compact type for all primes $p \leqslant 37$. This implies that we can calculate the traces of the Hecke operator $T(p)$ for $p \leqslant 37$ on the space of cusp forms $S_{a-b,b+3}$ *for all* $a > b > 0$, and assuming the conjecture also for $a > b = 0$ and $a = b > 0$.

Our results are in accordance with results on the numerical Euler characteristic $\sum(-1)^i \dim H^i_c(\mathcal{A}_2, V_{a,b})$ due to Getzler ([36]) and with the dimension formula for the space of cusp form $S_{a-b,b+3}$ due to Tsushima ([69]).

We give an example. For $(a, b) = (11,5)$ we have

$$e_c(\mathcal{A}_2, V_{11,5}) = -\mathbb{L}^6 - S[6,8]$$

with $S[6,8]$ the hypothetical motive associated to the (1-dimensional) space of cusp forms $S_{6,8}$. We list a few eigenvalues $\lambda(p)$ and $\lambda(p^2)$ of the Hecke operators. This allows us to give the characteristic polynomial of Frobenius (and the Euler factor of the spinor L-function of the Siegel modular form)

$$1 - \lambda(p)X + (\lambda(p)^2 - \lambda(p^2) - p^{a+b+2})X^2 - \lambda(p)p^{a+b+3}X^3 + p^{2(a+b+3)}X^4$$

and its slopes.

p	$\lambda(p)$	$\lambda(p^2)$	slopes
2	0	−57344	13/2,25/2
3	−27000	143765361	3,7,12,16
5	2843100	−7734928874375	2,7,12,17
7	−107822000	4057621173384801	0,6,13,19

Another indication that the computer counts of curves of genus 2 are correct comes from a conjecture of Harder about congruences between Hecke eigenvalues of cusp forms for genus 1 and genus 2. Harder had the idea that there should be such congruences many years ago (cf. [40]), but our results motivated him to make his conjecture precise and explicit. He conjectured that if a (not too small, or better an ordinary) prime ℓ divides a critical value s of the L-function of an eigenform on $SL(2, \mathbb{Z})$ of weight r then there should be a vector-valued Siegel modular form of prescribed weight depending on s and r and a congruence modulo ℓ between the eigenvalues under the Hecke operator $T(p)$ for f and F. We refer to [11], the papers by Harder [41] and van der Geer [33] there, for an account of this fascinating story. These congruences generalize the famous congruence

$$\tau(n) \equiv p^{11} + 1 \pmod{691}$$

for the Hecke eigenvalues $\tau(p)$ (p a prime) of the modular form $\Delta = \sum_{n>0} \tau(n) q^n$ of weight 12 on $SL(2, \mathbb{Z})$. One example of such a congruence is the congruence

$$\lambda(p) \equiv p^8 + a(p) + p^{13} \pmod{41}$$

where $f = \sum a(n) q^n$ is the normalized ($a(1) = 1$) cusp form of weight 22 on $SL(2, \mathbb{Z})$ and the $\lambda(p)$ are the Hecke eigenvalues of the genus 2 Siegel cusp form $F \in S_{4,10}$. We checked this congruence for all primes p with $p \leqslant 37$. For example, $a(37) = 22191429912035222$ and $\lambda(37) = 11555498201265580$.

In joint work with Bergström and Faber we extended this to level 2. One considers the moduli space $\mathcal{A}_2[2]$ of principally polarized abelian surfaces of level 2. This moduli space contains as a dense open subset the moduli space $\mathcal{M}_2[w^6]$ of curves of genus 2 together with six Weierstrass points. It comes with an action of $S_6 \cong Sp(4, \mathbb{Z}/2\mathbb{Z})$. We formulated an analogue of conjecture 16.1 for level 2. Assuming this conjecture we can calculate the traces of the Hecke operators $T(p)$ for $p \leqslant 37$ for the spaces of cusp forms of all level 2. Using these numerical data we could observe liftings from genus 1 to genus 2 and could make precise conjectures about such liftings; and again we could predict and verify numerically congruences between genus 1 and genus 2 eigenforms.

We give an example. For $(a, b) = (4,2)$ we find

$$e_c(\mathcal{A}_2[2], V_{4,2}) = -45\mathbb{L}^3 + 45 - S[\Gamma_2[2], (2,5)]$$

and assuming our conjecture we can calculate the traces of the Hecke operators on the space of cusp forms of weight $(2,5)$ on the level 2 congruence subgroup $\Gamma_2[2]$ of $Sp(4, \mathbb{Z})$; this space is a representation of type $[2^2, 1^2]$ for the symmetric group S_6 and is generated by one Siegel modular form; for this Siegel modular form we have the eigenvalue $\lambda(23) = -323440$ for $T(23)$.

It is natural to ask how the story continues for genus 3. The first remark is that the Torelli map $\mathcal{M}_3 \to \mathcal{A}_3$ is a morphism of degree 2 in the sense of stacks. This is due to the fact that every principally polarized abelian variety has a non-trivial automorphism $-\mathrm{id}$, while the general curve of genus 3 has a trivial automorphism group. This has as a consequence that for local systems $V_{a,b,c}$ with $a + b + c$ odd the cohomology on $\mathcal{A}_3$ vanishes, but on $\mathcal{M}_3$ it need not; and in fact in general it does not.

In joint work with Bergström and Faber we managed to formulate an analogue of 16.1 for genus 3, see [7]. Assuming the conjecture we are able to calculate the traces of Hecke operators $T(p)$ on the space of cusp forms $S_{a-b,b-c,c+4}$ for all primes p for which we did the counting (at least $p \leqslant 19$). Again the numerically data fit with calculations of dimensions of spaces of cusp forms and numerical Euler characteristics. (These numerical Euler characteristics were calculated in [8].) And again we could observe congruences between eigenvalues for $T(p)$ for $g = 3$

eigenforms and those of genus 1 and 2. We give two examples:

$$e_c(\mathcal{A}_3, V_{10,4,0}) = -\mathbb{L}^7 + \mathbb{L} + S[6,8],$$

the same S[6,8] for genus 2 we met above. And for example

$$e_c(\mathcal{A}_3, V_{8,4,4}) = -S[12]\,\mathbb{L}^6 + S[12] + S[4,0,8],$$

where genuine Siegel modular forms (of weight (4,0,8)) of genus 3 do occur. We refer to [7] for the details and to the Chapter by Faber and Pandharipande for the cohomology of local systems and modular forms on the moduli spaces of curves.

Acknowledgement The author thanks Jonas Bergström and Carel Faber for the many enlightening discussions we had on topics dealt with in this survey. Thanks are due to T. Katsura for inviting me to Japan where I found the time to finish this survey. I am also greatly indebted to Torsten Ekedahl from whom I learned so much. I was a great shock to hear that he passed away; his gentle and generous personality and sharp intellect will be deeply missed.

References

[1] V. Alexeev. Complete moduli in the presence of semiabelian group action. *Ann. of Math. (2)*, **155**, 611–708, 2002. ← 417

[2] V. Alexeev and I. Nakamura. On Mumford's construction of degenerating abelian varieties. *Tohoku Math. J. (2)*, **51**, 399–420, 1999. ← 417

[3] A. Ash, D. Mumford, M. Rapoport, and Y. Tai. Smooth compactification of locally symmetric varieties. *Lie Groups: History, Frontiers and Applications* , Vol. **IV**. Math. Sci. Press, Brookline, Mass., 1975. iv+335 pp. ← 416, 417

[4] A. Ash, D. Mumford, M. Rapoport, and Y. Tai. Smooth compactification of locally symmetric varieties. Second edition. With the collaboration of Peter Scholze. *Cambridge Mathematical Library*. Cambridge University Press, Cambridge, 2010. ← 416

[5] W. Baily and A. Borel. Compactification of arithmetic quotients of bounded symmetric domains. *Ann. of Math. (2)*, **84**, 442–528, 1966. ← 416, 425

[6] J. Bergström, C. Faber, and G. van der Geer. Siegel modular forms of genus 2 and level 2: cohomological computations and conjectures. IMRN doi 10.1093/imrn/rnn100, 2008. `arXiv:0803.0917` ← 449

[7] J. Bergström, C. Faber, and G. van der Geer. Siegel modular forms of degree three and the cohomology of local systems. `arXiv:1108:3731` ← 449, 452, 453

[8] J. Bergström and G. van der Geer. The Euler characteristic of local systems on the moduli of curves and abelian varieties of genus three. *J. Topol.*, **1**(3), 651–662, 2008. ← 452

[9] A. Borel. Stable real cohomology of arithmetic groups II, in: *Manifolds and Lie groups*, Birkhäuser-Boston, 1981. ← 430, 433

[10] A. Borel and J.-P. Serre. Le théorème de Riemann-Roch. *Bull. Soc. Math. France*, **86**, 97–136, 1958. ← 425

[11] J. Bruinier, G. van der Geer, G. Harder and D. Zagier. *The* $1-2-3$ *of Modular Forms*. Springer Verlag, 2008. ← 451

[12] R. Charney and R. Lee. Cohomology of the Satake compactification. *Topology*, **22**, 389–423, 1983. ← 433

[13] C. Consani and C. Faber. On the cusp form motives in genus 1 and level 1, in: Moduli spaces and arithmetic geometry, 297–314. *Adv. Stud. Pure Math.*, **45**, Math. Soc. Japan, Tokyo, 2006. ← 448

[14] P. Deligne. *Formes modulaires et représentations* ℓ*-adiques*. Séminaire Bourbaki 1968-1969, exp. 355. ← 447, 448

[15] M. Eichler. Quaternäre quadratische Formen und die Riemannsche Vermutung für die Kongruenzzetafunktion. *Arch. Math.*, **5**, 1954. ← 448

[16] T. Ekedahl. On supersingular curves and abelian varieties. *Math. Scand.*, **60**, 151–178, 1987. ← 445

[17] T. Ekedahl and G. van der Geer. The order of the top Chern class of the Hodge bundle on the moduli space of abelian varieties. *Acta Math.*, **192**, 95–109, 2004. ← 431, 432

[18] T. Ekedahl and G. van der Geer. Cycles representing the top Chern class of the Hodge bundle on the moduli space of abelian varieties. *Duke Math. J.*, **129**, 187–199, 2005. ← 431, 433

[19] T. Ekedahl and G. van der Geer. Cycle classes of the E-O stratification on the moduli of abelian varieties, in: Algebra, Arithmetic, and Geometry: in honor of Yu. I. Manin. Vol. I, 567–636. *Progr. Math.*, **269**, Birkhäuser Boston, Inc., Boston, MA, 2009. ← 441, 442, 443, 444, 445

[20] C. Erdenberger, S. Grushevsky, and K. Hulek. Some intersection numbers of divisors on toroidal compactifications of $\mathcal{A}_g$. *J. Algebraic Geom.*, **19**, 99–132, 2010. ← 430

[21] H. Esnault and E. Viehweg. Chern classes of Gauss-Manin bundles of weight 1 vanish. *K-theory*, **26**, 287–305, 2002. ← 426

[22] C. Faber and G. van der Geer. Sur la cohomologie des systèmes locaux sur les espaces des modules des courbes de genre 2 et des surfaces abéliennes, I, II. *C.R. Acad. Sci. Paris, Sér. I*, **338**, 381–384, 467–470, 2004. ← 449, 450

[23] G. Faltings. On the cohomology of locally symmetric Hermitian spaces, in: Paul Dubreil and Marie-Paule Malliavin algebra seminar, 35th year (Paris, 1982), 55–98. *Lecture Notes in Math.*, **1029**, Springer, Berlin, 1983. ← 449

[24] G. Faltings. Arithmetische Kompaktifizierung des Modulraums der abelschen Varietäten, in; Workshop Bonn 1984 (Bonn, 1984), 321–383. *Lecture Notes in Math.*, **1111**, Springer, Berlin, 1985. ← 417

[25] G. Faltings and C-L. Chai. Degeneration of abelian varieties. *Ergebnisse der Mathematik*, **22**, Springer Verlag, Berlin, 1990. ← 416, 417, 422, 426, 449

[26] E. Freitag. Die Kodairadimension von Körpern automorpher Funktionen. *J. reine und angewandte Math.*, **296**, 162-170, 1977. ← 439

[27] E. Freitag. Siegelsche Modulfunktionen. *Grundlehren der Mathematischen Wissenschaften*, **254**, Springer-Verlag, Berlin, 1983. ← 434, 435, 440

[28] E. Freitag and K. Pommerening. Reguläre Differentialformen des Körpers der Siegelschen Modulfunktionen. *J. Reine Angew. Math.*, **331**, 207–220, 1982. ← 438

[29] W. Fulton. Intersection Theory. Second edition. *Ergebnisse der Mathematik und ihrer Grenzgebiete. 3. Folge*, Springer-Verlag, Berlin, 1998. ← 421

[30] W. Fulton and J. Harris. Representation theory. A first course. *Graduate Texts in Mathematics*, **129**, Readings in Mathematics. Springer-Verlag, New York, 1991. ← 423, 449

[31] W. Fulton and P. Pragacz. Schubert varieties and degeneracy loci. *Lecture Notes in Math.*, **1689**, Springer Verlag, New York, 1998. ← 421

[32] G. van der Geer. Cycles on the moduli space of abelian varieties, in: Moduli of curves and abelian varieties. Editors: C. Faber and E. Looijenga. *Aspects Math.*, **E33**, 65–89, 1999. ← 423, 424, 441, 444

[33] G. van der Geer. Siegel modular forms and their applications, in: The 1–2–3 of Modular Forms, 181–245, *Universitext*, Springer, Berlin, 2008. ← 434, 451

[34] G. van der Geer. The Chow ring of the moduli space of abelian threefolds. *J. Algebraic Geom.*, **7**, 753–770, 1998; Corrigendum: *J. Algebraic Geom.*, **18**, 795–796, 2009. ← 429

[35] G. van der Geer. Rank one Eisenstein cohomology of local systems on the moduli space of abelian varieties. *Sci. China Math.*, **54**, 2011. doi 10.1007/s11425-010-4159-4 ← 450

[36] E. Getzler. Euler characteristics of local systems on $\mathcal{M}_2$. *Compositio Math.*, **132**, 121–135, 2002. ← 451

[37] S. Grushevsky. Geometry of $\mathcal{A}_g$ and its compactifications, in: Algebraic geometry, Seattle 2005. Part 1, 193–234; *Proc. Sympos. Pure Math.*, **80** , Part 1, Amer. Math. Soc., Providence, RI, 2009. ← 417, 440

[38] R. Hain. The rational cohomology ring of the moduli space of abelian 3-folds. *Math. Res. Lett.*, **9**, 473–491, 2002. ← 433

[39] S. Harashita. Ekedahl-Oort strata and the first Newton slope strata. *J. Algebraic Geom.*, **16**, 171–199, 2007. ← 443

[40] G. Harder. Eisensteinkohomologie und die Konstruktion gemischter Motive. *Lecture Notes in Mathematics*, **1562**, Springer-Verlag, Berlin, 1993. ← 449, 450, 451

[41] G. Harder. A congruence between a Siegel and an elliptic modular form, in: The 1-2-3 of Modular Forms, 247–262. *Universitext*, Springer, Berlin, 2008. ← 451

[42] F. Hirzebruch. Automorphe Formen und der Satz von Riemann-Roch, in: 1958 Symposium internacional de topología algebraica, 129–144. Universidad Nacional Autónoma de México, Mexico City. ← 428

[43] K. Hulek and O. Tommasi. Cohomology of the toroidal compactification of $\mathcal{A}_3$, in: Vector bundles and complex geometry, 89–103. *Contemp. Math.*, **522**, Amer. Math. Soc., Providence, RI, 2010. ← 434

[44] J. Igusa. A desingularization problem in the theory of Siegel modular functions. *Math. Ann.*, **168**, 228–260, 1967. ← 417

[45] Y. Ihara. Hecke polynomials as congruence ζ functions in elliptic modular case. *Annals of Math.*, **85**, 267–295, 1967. ← 448

[46] W. van der Kallen and E. Looijenga. Spherical complexes attached to symplectic lattices, to appear in *Geometriae Dedicata*. arXiv:1001.0883 ← 433

[47] S. Keel and L. Sadun. Oort's conjecture for $\mathcal{A}_g \otimes \mathbb{C}$. *JAMS*, **16**, 887–900, 2003. ← 446

[48] N. Koblitz. p-adic variation of the zeta function over families of varieties defined over finite fields. *Comp. Math.*, **31**, 119–218, 1975. ← 445

[49] A. Kresch. Cycle groups for Artin stacks. *Invent. Math.*, **138**, 495–536, 1999. ← 430

[50] G. Laumon. Sur la cohomologie à supports compacts des variétés de Shimura pour $\mathrm{GSp}(4)_{\mathbb{Q}}$. *Compositio Math.*, **105**, 267–359, 1997. ← 449

[51] G. Laumon. Fonctions zêtas des variétés de Siegel de dimension trois, in: Formes automorphes. II. Le cas du groupe GSp(4), 1–66 *Astérisque*, **302**, 2005. ← 449

[52] Y. Matsushima and S. Murakami. On vector bundle valued harmonic forms and automorphic forms on symmetric riemannian manifolds. *Ann. of Math.(2)*, **78**, 365–416, 1963. ← 449

[53] S. Morel. On the cohomology of certain noncompact Shimura varieties. With an appendix by Robert Kottwitz. *Annals of Mathematics Studies*, **173**, Princeton University Press, Princeton, NJ, 2010. ← 449

[54] L. Moret-Bailly. Pinceaux de variétés abéliennes. *Astérisque*, **129**, 266, 1995. ← 426

[55] D. Mumford. Picard groups of moduli problems, in: Arithmetical Algebraic Geometry (Proc. Conf. Purdue Univ., 1963), 33–81, Harper & Row, New York, 1965. ← 430, 432

[56] D. Mumford. Hirzebruch's Proportionality Theorem in the non-compact case, *Inv. Math.*, **42**, 239–272, 1977. ← 428

[57] D. Mumford. Towards an enumerative geometry of the moduli space of curves, in: Arithmetic and geometry, Vol. II, 271–328, *Progr. Math.*, **36**, Birkhäuser Boston, Boston, MA, 1983. ← 428

[58] D. Mumford. On the Kodaira dimension of the Siegel modular variety., in: Algebraic geometry: open problems (Ravello, 1982), 348–375. *Lecture Notes in Math.*, **997**, Springer, Berlin, 1983. ← 439

[59] M. Olsson. Compactifying moduli spaces for abelian varieties. *Lecture Notes in Mathematics*, **1958**, Springer-Verlag, Berlin, 2008. ← 417

[60] F. Oort. Subvarieties of moduli spaces. *Inv. Math.*, **24**, 95–119, 1974. ← 445

[61] F. Oort. Complete subvarieties of moduli spaces, in: Abelian Varieties (Egloffstein, 1993), 225-235, de Gruyter, Berlin, 1995. ← 446

[62] F. Oort. A stratification of a moduli space of abelian varieties, in: Moduli of Abelian varieties (Texel Island), Editors: C. Faber, G. van der Geer, F. Oort., 345–416. *Progress in Math.*, **195**, Birkhäuser 195, Basel, 2001. ← 441

[63] F. Oort. Foliations in moduli spaces of abelian varieties. *J. Amer. Math. Soc.*, **17**, 267–296, 2004. ← 441

[64] P. Pragacz. Enumerative geometry of degeneray loci. *Ann. Scient. Ec. Norm. Sup.*, **21**, 413–454, 1988. ← 421

[65] A. J. Scholl. Motives for modular forms. *Invent. Math.*, **100**, 419–430, 1990. ← 447

[66] J. Schwermer. Eisenstein series and cohomology of arithmetic groups: the generic case. *Invent. Math.*, **116**, 481–511, 1994. ← 450

[67] N. Shepherd-Barron. Perfect forms and the moduli space of abelian varieties. *Invent. Math.*, **163**, 25–45, 2006. ← 417

[68] Y-S. Tai. On the Kodaira dimension of the moduli space of abelian varieties. *Invent. Math.*, **68**, 425–439, 1982. ← 439

[69] R. Tsushima. An explicit dimension formula for the spaces of generalized automorphic forms with respect to $\mathrm{Sp}(2, \mathbb{Z})$. *Proc. Japan Acad. Ser. A, Math. Sci.*, **59**, 139–142, 1983. ← 451

[70] R. Weissauer. Vektorwertige Siegelsche Modulformen kleinen Gewichtes. *J. Reine Angew. Math.*, **343**, 184–202, 1983. ← 436, 438

[71] R. Weissauer. The trace of Hecke operators on the space of classical holomorphic Siegel modular forms of genus two. `arXiv:0909.1744` ← 451

[72] D. Zagier. Correction to: The Eichler-Selberg trace formula on $\mathrm{SL}_2(Z)$ (Introduction to modular forms, Appendix, 44–54, Springer, Berlin, 1976) by S. Lang, in: Modular functions of one variable, VI (Bonn, Bonn, 1976), 171–173. *Lecture Notes in Math.*, **627**, Springer, Berlin, 1977. ← 448

University of Amsterdam
E-mail address: geer@science.uva.nl

Moduli of K3 surfaces and irreducible symplectic manifolds

V. Gritsenko, K. Hulek, and G.K. Sankaran

Abstract. The name "K3 surfaces" was coined by A. Weil in 1957 when he formulated a research programme for these surfaces and their moduli. Since then, irreducible holomorphic symplectic manifolds have been introduced as a higher dimensional analogue of K3 surfaces. In this paper we present a review of this theory starting from the definition of K3 surfaces and going as far as the global Torelli theorem for irreducible holomorphic symplectic manifolds as recently proved by M. Verbitsky.

For many years the last open question of Weil's programme was that of the geometric type of the moduli spaces of polarised K3 surfaces. We explain how this problem has been solved. Our method uses algebraic geometry, modular forms and Borcherds automorphic products. We collect and discuss the relevant facts from the theory of modular forms with respect to the orthogonal group $O(2, n)$. We also give a detailed description of quasi pull-back of automorphic Borcherds products. This part contains previously unpublished results.

Contents

2000 *Mathematics Subject Classification.* Primary 14J10; Secondary 14D99.
Key words and phrases. moduli, K3 surfaces, hyperkähler, modular form.

1. Introduction

The history of K3 surfaces goes back to classical algebraic geometry. In modern complex algebraic geometry a *K3 surface* is a compact complex surface S whose canonical bundle is trivial, i.e. $K_S \cong \mathcal{O}_S$, and whose irregularity $q(S) = h^1(S, \mathcal{O}_S) = 0$. These two facts immediately imply that $H^0(S, \Omega_S^1) = H^0(S, T_S) = H^2(S, T_S) = 0$.

The easiest examples of algebraic K3 surfaces are smooth quartic surfaces in $\mathbb{P}^3$. Further examples are complete intersections of type (2,3) in $\mathbb{P}^4$ and of type (2,2,2) in $\mathbb{P}^5$. Another classical example is the *Kummer surfaces*, i.e. the (desingularised) quotient of a 2-dimensional torus A by the involution $\iota\colon x \mapsto -x$.

The modern name "K3 surface" was coined by A. Weil in his famous "Final report on research contract AF 18(603)-57" [131, Vol II, pp 390–395]. In his comments on this report [131, Vol II, p 546] Weil writes: "Dans la seconde partie de mon rapport, il s'agit des variétés kählériennes dites K3, ainsi nommées en l'honneur de Kummer, Kodaira, Kähler et de la belle montagne K2 au Cachemire."

In that report the following conjectures, due to Andreotti and Weil, were stated:

(i) K3 surfaces form one family;
(ii) all K3 surfaces are Kähler;
(iii) the period map is surjective;
(iv) a form of global Torelli theorem holds.

Weil also remarked that the structure of the moduli space of (polarised) K3 surfaces must be closely related to the theory of automorphic forms.

By now all of these questions have been answered positively and much progress has been made in understanding the moduli spaces of K3 surfaces. Conjecture (i) was proved by Kodaira [71, Part I, theorem 19]. Conjecture (ii) was first shown by Siu [122, Section 14]. Nowadays it can be derived from the more general theorem, due to Buchdahl and Lamari, that a compact complex surface is Kähler if and only if its first Betti number is even ([6, Theorem IV.3.1], [19],[78]). Surjectivity of the period map (conjecture (iii)) was proved for special K3 surfaces in various papers by Shah [117], [118], [119] and by Horikawa [61]. Kulikov [77] gave a proof for projective K3 surfaces, with some points clarified by Persson and Pinkham [105]. The general result for Kähler K3 surfaces was proved by Todorov [126] and Looijenga [80]. The strong Torelli theorem (i.e. conjecture (iv)) for algebraic K3 surfaces was first proved by Piatetskii-Shapiro and Shafarevich [108] with amendments by Rapoport and Shioda. It was proved for Kähler K3 surfaces (and hence for all) by Burns and Rapoport [20]. A detailed and supplemented account of the original proof was written by Looijenga and Peters [81]. Finally, Friedman [34] gave a proof using degenerations.

It should be noted, though, that Weil's definition of K3 surface was different from the standard definition used nowadays. For him a surface was K3 if its underlying differentiable structure was that of a quartic surface in $\mathbb{P}^3$. Using results from Seiberg-Witten theory one can indeed show that any compact complex surface diffeomorphic to a quartic is a K3 surface.

The crucial ingredient is the fact that the plurigenera of a compact complex surface are invariant under orientation preserving diffeomorphism: see [6, Theorem IX.9.4 and Theorem IX.9.5]. This was formulated as a question in [103, p. 244] and became known as the Van de Ven conjecture. A first (partial) proof of this was published by Brussee [18] (see also [35]), and an elegant argument using Seiberg-Witten invariants can be found in Dürr's thesis [29]. It follows that the only surfaces which admit an orientation preserving diffeomorphism to a quartic are K3 surfaces or possibly tori, but the latter are excluded because they are not simply connected. If a surface is diffeomorphic to a quartic surface, but with an orientation changing diffeomorphism, then its signature is -16 and its Euler number equals $c_2(S) = 24$. By the Thom-Hirzebruch index theorem $c_1^2(S) = 96$, but that contradicts the Miyaoka-Yau inequality, so such a surface cannot exist.

Moduli spaces of polarised K3 surfaces are a historically old subject, studied by the classical Italian geometers (starting with the spaces of double sextics and plane quartics in $\mathbb{P}^3$). Mukai extended the classical constructions and unirationality results for the moduli spaces $\mathcal{F}_{2d}$ parametrising polarised K3 surfaces of degree $2d$ to many more cases, going as high as $d = 19$. See [89], [90], [91], [92] and [93] for details. Kondo [74] proved that the moduli spaces $\mathcal{F}_{2p^2}$ are of general type for sufficiently large prime numbers p, but without giving an effective bound. Finally, it was shown in [46] that the moduli spaces $\mathcal{F}_{2d}$ are of general type for $d > 61$ and

$d = 46,50,54,57,58,60$, and later it was noticed by A. Peterson [106] that this proof also works for $d = 52$: see Theorem 6.1 below. For an account of these results see also Voisin's Bourbaki exposé [129].

In higher dimension K3 surfaces can be generalised in two ways, namely to Calabi-Yau varieties or to irreducible symplectic manifolds (or hyperkähler manifolds). In fact, together with tori, these three classes of manifolds are the building blocks of all compact Kähler manifolds with trivial first Chern class (over $\mathbb{R}$), also called Ricci flat manifolds: every such manifold admits a finite étale cover which is a product of tori, Calabi-Yau manifolds and irreducible symplectic manifolds (see [8], [9]).

The first examples of irreducible symplectic manifolds, by now regarded as classical, were studied by Beauville in [8], namely the Hilbert schemes of points on K3 surfaces and generalised Kummer varieties (and their deformations). Two further classes of examples were later found by O'Grady [99], [100]. The theory of irreducible symplectic manifolds, which started from work of Bogomolov, Beauville and Fujiki, was significantly advanced by Huybrechts [63] who, among other results, proved surjectivity of the period map. It was noticed early on by Debarre [24] and Namikawa [95] that the obvious generalisations of the Torelli theorem are false. Nevertheless, one can use the period map to exhibit moduli spaces of polarised irreducible symplectic manifolds as dominant finite to one covers of quotients of type IV domains by arithmetic groups. This was used in [50] and [51] to obtain general type results for many of these moduli spaces (Theorem 6.2 and Theorem 6.3 below).

Very recently Verbitsky [127] has announced a Torelli theorem for irreducible symplectic manifolds. The consequences of Verbitsky's result for the moduli problem of polarised irreducible symplectic manifolds were worked out in detail by Markman [86]. We also refer the reader to Huybrecht's Bourbaki talk [64].

The theory of K3 surfaces and irreducible symplectic manifolds is a fascinating and vast subject. We have started this introduction by giving the definition of K3 surfaces in complex geometry (which also allows for non-algebraic surfaces). The notion of K3 surface also makes perfect sense in algebraic geometry over arbitrary fields: a K3 surface is an irreducible smooth algebraic surface S defined over a field k with trivial canonical bundle and irregularity $q = h^1(S, \mathcal{O}_S) = 0$. In this article we shall, however, concentrate on the theory over the complex numbers, and especially on moduli problems. We are fully aware that we thus exclude a large area of interesting questions. Concerning K3 surfaces in positive characteristic we refer the reader to the papers by Artin [2], Ogus [102] and Nygaard [98] for a first introduction. Another aspect, which we will not touch upon, is the arithmetic of K3 surfaces. An introduction to this field can be found in Schütt's survey paper [116].

In this article we mainly survey known results. Theorem 8.18 is new, however (special cases occur in the literature) and so is Proposition 9.5.

2. Periods of K3 surfaces and the Torelli theorem

In this section we will discuss the period domain and the period map for K3 surfaces, and the local and global Torelli theorems. The case of K3 surfaces presents several special features and the existence of a straightforward global Torelli theorem is one of them.

2.1. Lattices

Let S be a K3 surface. By Noether's formula $b_2(S) = 22$, and since S is simply connected (being diffeomorphic to a quartic hypersurface), $H^2(S, \mathbb{Z})$ is a free $\mathbb{Z}$-module of rank 22. The intersection form defines a non-degenerate symmetric bilinear form on $H^2(S, \mathbb{Z})$, giving it the structure of a lattice, which has signature $(3,19)$ by the Hodge index theorem. As the intersection form is even and unimodular, this lattice is, independently of the surface S, the *K3 lattice*

$$L_{K3} := 3U \oplus 2E_8(-1) \tag{2.1}$$

where U is the hyperbolic plane (the unique even unimodular lattice of signature $(1,1)$) and E_8 is the unique even unimodular positive definite rank 8 lattice. For any lattice L and $m \in \mathbb{Z}$, the notation $L(m)$ indicates that the quadratic form is multiplied by m, so $E_8(-1)$ is negative definite.

It was proven by Siu [122] that every K3 surface is Kähler. Today, it is known that a compact complex surface is Kähler if and only if its first Betti number is even [6, Theorem IV.3.1], which immediately implies Siu's result. Hence $H^2(S, \mathbb{C})$ has a Hodge decomposition

$$H^2(S, \mathbb{C}) = H^{2,0}(S) \oplus H^{1,1}(S) \oplus H^{0,2}(S).$$

Since $K_S \cong \mathcal{O}_S$ it follows that $H^{2,0}(S) = H^0(S, K_S)$ is 1-dimensional and thus $H^{1,1}(S)$ has dimension 20.

Since the second cohomology has no torsion we can, via the universal coefficient theorem, consider $H^2(S, \mathbb{Z})$ as a lattice in $H^2(S, \mathbb{C})$. The *Néron-Severi group* is the intersection $NS(S) = H^2(S, \mathbb{Z}) \cap H^{1,1}(S)$, which, in this case, coincides with the Picard group $\mathrm{Pic}(S)$. The *transcendental lattice* is defined as the smallest lattice whose complexification contains a generator ω of $H^0(S, K_S)$. If the intersection pairing on $NS(S)$ is nondegenerate, e.g. if S is projective, then $T(S)$ is the orthogonal complement in $H^2(S, \mathbb{Z})$ of the Néron-Severi group.

Note that for a general K3 surface S we have no algebraic classes, i.e. $T(S) = H^2(S, \mathbb{Z})$. The *Picard number* $\rho(S)$ of S is the rank of the Néron-Severi group $NS(S)$.

For future use we also need the *Kähler cone* of a K3 surface, which is the cone of classes of Kähler $(1,1)$-forms. This lives in $H^{1,1}(S, \mathbb{R}) = H^{1,1}(S) \cap H^2(S, \mathbb{R})$. The restriction of the intersection product to $H^{1,1}(S, \mathbb{R})$ has signature $(1,19)$. Let

$$\mathcal{C}_S \subset \{x \in H^{1,1}(S, \mathbb{R}) \mid (x, x) > 0\}$$

be the connected component that contains one (and hence all) Kähler classes. This is called the *positive cone* of S.

A class in $H^2(S,\mathbb{Z})$ is called *effective* if it is represented by an effective divisor. By a *nodal* class we mean the class δ of an effective divisor D of self-intersection $D^2 = -2$. We denote by Δ the set of all nodal classes. Every nodal class $\delta \in \Delta$ defines a reflection

$$s_\delta \colon H^2(S,\mathbb{Z}) \to H^2(S,\mathbb{Z})$$

given by $s_\delta(x) = x + (x,\delta)\delta$ and called the *Picard-Lefschetz reflection* defined by δ. We shall denote the $\mathbb{R}$- and $\mathbb{C}$-linear extensions of s_δ by the same symbol. Clearly s_δ is the reflection in the hyperplane H_δ orthogonal to δ. The set of effective classes on S is the semi-group generated by the integral points in the closure of $\mathcal{C}_S$ and the nodal classes. The connected components of the set $\mathcal{C}_S \setminus \bigcup_{\delta\in\Delta} H_\delta$ are called the *chambers* of $\mathcal{C}_S$. The chamber

$$\mathcal{C}_S^+ = \{x \in \mathcal{C}_S \mid (x,\delta) > 0 \text{ for all effective } \delta \in \Delta\} \tag{2.2}$$

is equal to the Kähler cone of S according to [6, Corollary VIII.3.9].

2.2. Markings and the period map

A *marking* of a K3 surface is an isometry $\phi\colon H^2(S,\mathbb{Z}) \to L_{K3}$ and we refer to a pair (S,ϕ) as a *marked* K3 surface. An isomorphism between marked K3 surfaces is an isomorphism between the surfaces that commutes with the markings. If ω is a non-zero 2-form on S then $\mathbb{C}\phi(\omega) = \phi(H^{2,0}(S))$ is a line in the complex vector space $L_{K3} \otimes \mathbb{C}$.

For any indefinite lattice L we define

$$\Omega_L = \{[x] \in \mathbb{P}(L_{K3} \otimes \mathbb{C}) \mid (x,x) = 0,\ (x,\overline{x}) > 0\}. \tag{2.3}$$

In the case of the K3 lattice, $\Omega = \Omega_{L_{K3}}$ is a connected complex manifold of dimension 20, called the *period domain of K3 surfaces*. Since $(\omega,\omega) = 0$ and $(\omega,\overline{\omega}) > 0$ it follows that $[\phi(\omega)] \in \Omega$. This is the *period point* of the marked K3 surface (S,ϕ).

Let $p\colon \mathcal{S} \to U$ be a flat family of K3 surfaces over some sufficiently small contractible open set U. If $\phi_0\colon H^2(S_0,\mathbb{Z}) \to L_{K3}$ is a marking then this can be extended to a marking $\phi_U\colon R^2p_*\mathbb{Z}_U \to (L_{K3})_U$ where $(L_{K3})_U$ is the constant sheaf with fibre L_{K3} on U. This defines a holomorphic map $\pi_U\colon U \to \Omega$, called the *period map* defined by the family $p\colon \mathcal{S} \to U$.

2.3. The Torelli theorem

The Torelli problem asks how much information about an algebraic variety can be reconstructed from its Hodge structure. In the case of K3 surfaces this means whether one can recover a K3 surface S from a period point. This question can be made precise in different ways. As we shall see, one can prove a very strong form of the Torelli theorem in the case of K3 surfaces.

We start by discussing the local Torelli theorem. For this let $p\colon \mathcal{S} \to \mathfrak{U}$ be a representative of the Kuranishi family (or versal deformation) of S. Since $H^0(S, T_S) = H^2(S, T_S) = 0$ the base space of the Kuranishi family is smooth of dimension $h^1(S, T_S) = h^{1,1}(S) = 20$. Choosing any marking of the central fibre of the Kuranishi family defines a marking for the entire family (we shall choose $\mathfrak{U}$ sufficiently small and contractible) and hence a period map $\pi\colon \mathfrak{U} \to \Omega$.

Theorem 2.4. (Local Torelli) *The base space of the Kuranishi family is smooth of dimension 20. It is universal for all points in a sufficiently small neighbourhood of the origin. The differential of the period map is an isomorphism and thus the period map is a local isomorphism.*

Proof. See [6, Theorem VIII.7.3]. □

In order to discuss the global Torelli theorem we need the notion of Hodge isometry. Let S and S′ be K3 surfaces. An isomorphism of $\mathbb{Z}$-modules $\Phi\colon H^2(S, \mathbb{Z}) \to H^2(S', \mathbb{Z})$ is called a *Hodge isometry* if it is an isometry and if its $\mathbb{C}$-linear extension $\Phi_{\mathbb{C}}\colon H^2(S, \mathbb{C}) \to H^2(S', \mathbb{C})$ preserves the Hodge decomposition. It is moreover called *effective* if it maps $\mathcal{C}_S$ to $\mathcal{C}'_S$ and maps effective classes to effective classes.

Proposition 2.5. *Let* S *and* S′ *be* K3 *surfaces. Then the following are equivalent for a Hodge isometry* $\Phi\colon H^2(S, \mathbb{Z}) \to H^2(S', \mathbb{Z})$:

(i) Φ *is effective,*
(ii) $\Phi(\mathcal{C}_S^+) \subset \mathcal{C}_{S'}^+$, *i.e.* Φ *maps the Kähler cone of* S *into that of* S′.
(iii) Φ *maps one element of the Kähler cone of* S *into the Kähler cone of the surface* S′.

Proof. See [6, Proposition VIII.3.11]. □

The crucial result for the theory of moduli of K3 surfaces is the following.

Theorem 2.6. (Strong Torelli) *Let* S *and* S′ *be two* K3 *surfaces and suppose that* $\Phi\colon H^2(S', \mathbb{Z}) \to H^2(S, \mathbb{Z})$ *is an effective Hodge isometry. Then there is a unique isomorphism* $f\colon S \to S'$ *that induces* Φ, *i.e. such that* $\Phi = f^*$.

A proof of this theorem can be found in [6, Sections VIII.7–VIII.11]. Very roughly speaking the idea is to prove the Torelli theorem for projective Kummer surfaces first. The second step then consists of showing that the period points of marked Kummer surfaces are dense (in the complex topology) in the period domain Ω. The final step is then to prove the Torelli theorem for all K3 surfaces by approximating them by Kummer surfaces and taking suitable limits in the Barlet topology.

The following weaker form of the Torelli theorem is still useful.

Theorem 2.7. (Weak Torelli theorem) *Two* K3 *surfaces* S *and* S′ *are isomorphic if and only if there is a Hodge isometry* $H^2(S', \mathbb{Z}) \to H^2(S, \mathbb{Z})$.

Proof. Assume that $\Phi\colon H^2(S',\mathbb{Z}) \to H^2(S,\mathbb{Z})$ is a Hodge isometry. Let W_S be the group of isometries of $H^2(S,\mathbb{Z})$ generated by the Picard-Lefschetz reflections. This group acts on the positive cone $\mathcal{C}_S$ properly discontinuously. The closure of the Kähler cone in the positive cone is a fundamental domain for this action (cf. [6, Proposition VIII.3.10]). Hence we can choose an element $w \in W_S$ such that for a suitable choice of sign $\pm w \circ \Phi$ is an effective Hodge isometry by Proposition 2.5 and thus is induced by an isomorphism $f\colon S \to S'$ by Theorem 2.6. □

2.4. The universal family of marked K3 surfaces

For each K3 surface S we choose a representative $p\colon \mathcal{S} \to U$ of the Kuranishi family with U contractible and sufficiently small such that the following hold:

(i) $p\colon \mathcal{S} \to U$ is the Kuranishi family for each point $s \in U$.

(ii) If $\phi\colon R^2p_*\mathbb{Z}_{\mathcal{S}} \to (L_{K3})_U$ is a marking, then the associated period map $\pi\colon U \to \Omega$ is injective.

We consider all marked Kuranishi families, i.e. pairs $(p\colon \mathcal{S} \to U, \phi)$, having the above properties. We can glue the various copies of U by identifying those points where the marked K3 surfaces are isomorphic. This defines a space M_1 all of whose points have neighbourhoods isomorphic to some U. Hence M_1 is a 20-dimensional analytic manifold, but possibly not Hausdorff. It is also possible to show (cf. [6, Theorem VIII.10.6]) that one can glue the Kuranishi families to obtain a global family of K3 surfaces over M_1.

It turns out that the space M_1 is indeed not Hausdorff. This follows from an example due to Atiyah, also referred to as Atiyah's flop and nowadays crucial in higher-dimensional birational geometry. Consider, for example, the following 1-parameter family S_t of quartic (and hence K3) surfaces in $\mathbb{P}^3$, which is given in affine coordinates by

$$x^2(x^2-2)+y^2(y^2-2)+z^2(z^2-2)=2t^2.$$

Let the parameter t vary in a small neighbourhood B of the origin. For $t \neq 0$ these surfaces are smooth, whereas the surface S_0 has an ordinary double point at (0,0,0). This is also the only singularity, again an ordinary double point, of the total space $\mathcal{S}$. Blowing up this node one obtains a smooth 3-fold $\widetilde{\mathcal{S}}$ which contains a quadric $\mathbb{P}^1 \times \mathbb{P}^1$ as exceptional divisor E. The proper transform $\hat{S}_0$ of S_0 is a smooth K3 surface intersecting the exceptional divisor E in a rational curve of bidegree (1,1). This is a nodal curve on $\hat{S}_0$. The two rulings on E can each be contracted giving rise to smooth 3-dimensional spaces $p_1\colon \mathcal{S}_1 \to B$ and $p_2\colon \mathcal{S}_2 \to B$. These families are by construction identical over $B \setminus 0$. They are, however, not identical over all of B, since the identity on $\mathcal{S} \setminus S_0$ would, otherwise, have to extend to an automorphism of the total space acting non-trivially on the tangent cone of the double point, which is clearly impossible. Now choose a marking for p_1. This also defines a marking for p_2. Since the families coincide outside the origin, the period maps π_1 and π_2 also coincide away from 0. However, the markings differ for the central fibre (namely by the Picard-Lefschetz reflection defined by the nodal curve on $\hat{S}_0$). This shows that M_1 cannot be Hausdorff. In fact all non-Hausdorff behaviour comes from the existence of different resolutions of double points in families: see [20, Theorem 1′ and Section 7].

There is another formulation of the Torelli theorem, which we will now describe. For this we consider

$$K\Omega = \{(\kappa, [\omega]) \in (L_{K3} \otimes \mathbb{R}) \times \Omega \mid (\kappa, \mathrm{Re}(\omega)) = (\kappa, \mathrm{Im}(\omega)) = 0,\ (\kappa, \kappa) > 0\}$$

and define $E(\kappa, \omega)$ as the oriented 3-dimensional space spanned by the ordered basis $\{\kappa, \mathrm{Re}(\omega), \mathrm{Im}(\omega)\}$. By mapping each point $(\kappa, [\omega])$ to the space $E(\kappa, \omega)$ we obtain a fibration

$$\Pi\colon K\Omega \to \mathrm{Gr}^+(3, L_{K3} \otimes \mathbb{R})$$

over the Grassmannian $\mathrm{Gr}^+(3, L_{K3} \otimes \mathbb{R})$ of oriented 3-planes in $L_{K3} \otimes \mathbb{R}$ on which the form $(\ ,\ )$ is positive definite. This is an $SO(3, \mathbb{R})$-fibre bundle and the projection Π is equivariant with respect to the action of the orthogonal group $\mathrm{Aut}(L_{K3} \otimes \mathbb{R}) \cong O(3,19)$. We define

$$(K\Omega)^0 := \{(\kappa, [\omega]) \in K\Omega \mid (\kappa, d) \neq 0 \text{ for } d \in L_{K3}, (d, d) = -2, (\omega, d) = 0\}.$$

This is an open subset of $K\Omega$.

For every point $\omega \in \Omega$ we consider the cone

$$C_\omega = \{x \in L_{K3} \otimes \mathbb{R} \mid (x, \omega) = 0,\ (x, x) > 0\}.$$

This has two connected components and varies differentiably with ω. Since Ω is connected and simply connected, we can globally choose one of these components, say C^+_ω. Let (S, ϕ) be a marked K3 surface and let $\kappa \in H^{1,1}(S, \mathbb{R})$ be a Kähler class. Then we say that (S, κ), or more precisely $((S, \phi), \kappa)$, is a *marked pair* if $\phi_{\mathbb{C}}(\kappa) \in C^+_\omega$, where ω is the period point defined by (S, ϕ).

Let M'_2 be the real analytic vector bundle with fibre $H^{1,1}(S_t)$ over the base M_1 of the universal family of marked K3 surfaces and let $M_2 \subset M'_2$ be the subset of Kähler classes. This is open by [72, Theorem 15]. In particular, M_2 is real analytic of dimension $60 = 40 + 20$, where 40 is the real dimension of the base M_1 and 20 is the dimension of the fibre. We can now define a real-analytic map

$$\pi_2\colon M_2 \to (K\Omega)^0$$

by mapping $\kappa \in H^{1,1}(S_t), t \in M_1$ to $\pi_2(\kappa) = (\phi_{\mathbb{C}}(\kappa), \pi(t))$. This is called the *refined period map*. In this way we obtain the obvious commutative diagram

$$\begin{array}{ccc} M_2 & \xrightarrow{\pi_2} & (K\Omega)^0 \\ \downarrow & & \downarrow \\ M_1 & \xrightarrow{\pi_1} & \Omega. \end{array}$$

The Torelli theorem can now be reformulated as follows.

Theorem 2.8. *The map π_2 is injective (and thus M_2 is Hausdorff).*

Using this one can finally prove that the period map is surjective.

Theorem 2.9. (Surjectivity of the period map) *The refined period map π_2 is surjective. In particular, every point of Ω appears as the period point of some marked K3 surface.*

Proof. This is proven in [6, Section VIII.14]. □

The surjectivity of the period map was one of the major questions in the theory of K3 surfaces. A. Todorov [126] was the first to give a proof; the argument given in [6] is due to Looijenga.

In view of Theorem 2.7 and Theorem 2.9 we now have

Theorem 2.10. *The set $O(L_{K3})\backslash\Omega$ is in $1:1$ correspondence with the set of isomorphism classes of K3 surfaces.*

We can thus think of $O(L_{K3})\backslash\Omega$ as the "moduli space of K3 surfaces". It must be pointed out, however, that the action of the group $O(L_{K3})$ is not well behaved. In particular, it is not properly discontinuous. We shall now turn to the case of polarised K3 surfaces where we shall see that the situation is much better.

2.5. Moduli spaces of polarised K3 surfaces

A *polarisation* on a K3 surface is an ample line bundle $\mathcal{L}$. Since the irregularity of K3 surfaces is 0 and the Picard group has no torsion we can identify a line bundle $\mathcal{L}$ with its first Chern class $h = c_1(\mathcal{L}) \in H^2(S, \mathbb{Z})$. An ample line bundle $\mathcal{L}$ is nef and big, and conversely a nef and big line bundle on a K3 surface is ample if there are no (-2)-curves C with $h.C = 0$. This can be seen from the description of the Kähler cone (2.2) and Nakai's criterion, or from Reider's theorem [6, Theorem IV.11.4] or from [113]. Throughout, we shall only consider primitive polarisations, i.e. we shall assume that the class h is non-divisible in the K3 lattice. The *degree* of a polarisation is $\deg(\mathcal{L}) = h^2 = 2d$. Note that the degree is always even.

We denote by $\langle -2d \rangle$ the rank 1 lattice whose generator has square $-2d$.

Lemma 2.11. *Suppose $h \in L_{K3}$ is a primitive vector with $h^2 = 2d > 0$. Then the orthogonal complement $L_h = h^{\perp}_{L_{K3}}$ of h is isometric to L_{2d}, where L_{2d}, the lattice L_{2d} associated with K3 surfaces with a polarisation of degree $2d$, is defined by*

$$L_{2d} = 2U \oplus 2E_8(-1) \oplus \langle -2d \rangle. \tag{2.12}$$

Proof. It follows from Eichler's criterion (see Lemma 7.8 and Example 7.9 below) that there is a unique $O(L_{K3})$-orbit of primitive vectors h of given degree in the K3 lattice $L_{K3} = 3U \oplus 2E_8(-1)$. Hence we can assume that h is in one of the copies of the hyperbolic plane and that $h = e + df$ where e, f are a basis of a copy of U with $e^2 = f^2 = 0$ and $(e, f) = 1$. The structure of L_h is clear in this case. □

The lattice L_{2d} is an even lattice of signature $(2,19)$. The period domain $\Omega_{2d} = \Omega_{L_{2d}}$ has two connected components, $\mathcal{D}_{2d}$ and $\mathcal{D}'_{2d}$, interchanged by complex conjugation. The domain $\mathcal{D}_{2d}$ is a 19-dimensional symmetric homogeneous domain of type IV: see [114, Appendix 6]. One can also describe Ω_{2d} as the intersection of the domain Ω with the hyperplane $h^{\perp}_{L_{K3}}$.

We shall fix $h \in L_{K3}$ once and for all. For each polarised K3 surface $(S, \mathcal{L})$ of degree $2d$ we can consider polarised markings, i.e. markings $\phi\colon H^2(S,\mathbb{Z}) \to L_{K3}$ with $\phi(c_1(\mathcal{L})) = h$. Any two such markings differ by an element in the group

$$O(L_{K3}, h) = \{g \in O(L_{K3}) \mid g(h) = h\}. \tag{2.13}$$

This group leaves the orthogonal complement L_{2d} of h invariant and hence is a subgroup of $O(L_{2d})$.

For any lattice L we denote by $L^{\vee} = \operatorname{Hom}(L,\mathbb{Z})$ its dual lattice. The *discriminant group* $D(L) = L^{\vee}/L$ is a finite abelian group of order $|\det L|$ and carries a discriminant quadratic form q_L (if L is even) and a discriminant bilinear form b_L, with values in $\mathbb{Q}/2\mathbb{Z}$ and $\mathbb{Q}/\mathbb{Z}$ respectively (see [97, Section 1.3]). The *stable orthogonal group* of an even lattice L is defined as the kernel

$$\widetilde{O}(L) = \ker(O(L) \to O(D(L))). \tag{2.14}$$

If L is indefinite, the subgroup $O^+(L)$ is defined to be the group of elements of real spinor norm 1 (see [49] for a definition of the real spinor norm). We define

$$\widetilde{O}^+(L) = \widetilde{O}(L) \cap O^+(L) \tag{2.15}$$

and generally, for any subgroup $G < O(L)$, we use G^+ to denote the kernel of the real spinor norm and $\widetilde{G}$ to denote the stable subgroup, the kernel of $G \to O(D(L))$.

For $h \in L_{K3}$ with $h^2 = 2d$, it follows from Nikulin's theory [97, Corollary 1.5.2] that

$$O(L_{K3}, h) = \widetilde{O}(L_{2d})$$

considered as subgroups of $O(L_{2d})$. The two connected components $\mathcal{D}_{2d}$ and $\mathcal{D}'_{2d}$ are interchanged by the group $O(L_{2d})$. The index 2 subgroup that fixes the components is $O^+(L_{2d})$. Finally we define

$$\mathcal{F}_{2d} = \widetilde{O}(L_{2d})\backslash\Omega_{2d} = \widetilde{O}^+(L_{2d})\backslash\mathcal{D}_{2d}. \tag{2.16}$$

It is important to note that the situation is much better here than in the non-polarised case. The reason lies in the change of signature which is now $(2,19)$ rather than $(3,19)$. We are thus dealing with locally symmetric hermitian domains and, as a result, the action of the group $O(L_{2d})$ on Ω_{2d} is properly discontinuous. Hence the quotient space of Ω_{2d} by any subgroup of $O(L_{2d})$ of finite index (i.e. an *arithmetic subgroup*) is a complex space with only finite quotient singularities (also sometimes called a V-manifold). By a famous result of Baily and Borel [5] these quotients are quasi-projective and thus carry a natural structure of an algebraic variety. We shall discuss various compactifications of these quasi-projective varieties below in Section 5.

In order to describe the moduli space of polarised K3 surfaces we need one more ingredient. For $h \in L_{K3}$ we define

$$\Delta_h = \{\delta \in L_{K3} \mid \delta^2 = -2,\ (\delta, h) = 0\}.$$

For each $\delta \in \Delta_h$ we define the hyperplane $H_\delta = \delta^\perp_{L_{K3}}$, i.e. the hyperplane fixed by the Picard-Lefschetz reflection defined by δ. We set

$$\Omega^0_{2d} = \Omega_{2d} \setminus \bigcup_{\delta \in \Delta_h} (H_\delta \cap \Omega_{2d}).$$

There are only finitely many $\widetilde{O}(L_{2d})$-orbits in Δ_h. This follows immediately from Lemma 7.8, below: in fact there are at most two orbits by [48, Proposition 2.4(ii)]. Since the group acts properly discontinuously on Ω the hyperplanes H_δ for $\delta \in \Delta_h$ form a locally finite collection. Clearly, the action of the group $\widetilde{O}(L_{2d})$ on Ω_{2d} restricts to an action on Ω^0_{2d}. We define

$$\mathcal{F}^0_{2d} = \widetilde{O}(L_{2d})\backslash\Omega^0_{2d}.$$

Note that this is again a quasi-projective variety (it arises from $\mathcal{F}_{2d}$ by removing finitely many divisors) with only finite quotient singularities.

Theorem 2.17. *The variety $\mathcal{F}^0_{2d}$ is the moduli space of polarised* K3 *surfaces of degree* $2d$*, i.e. its points are in* $1:1$ *correspondence with polarised* K3 *surfaces of degree* $2d$.

Proof. Let $(S, \mathcal{L})$ be a polarised K3 surface with $\deg(\mathcal{L}) = 2d$. We consider polarised markings $\phi\colon H^2(S, \mathbb{Z}) \to L_{K3}$ with $\phi(c_1(\mathcal{L})) = h$. Since $(\omega_S, c_1(\mathcal{L})) = 0$ we find that $\phi(\omega_S) \in \Omega_{2d}$. In fact, since $\mathcal{L}$ is ample it has positive degree on all (-2)-curves and hence $\phi(\omega_S)$ lies in Ω^0_{2d}. Any two polarised markings differ by an element in $O(L_{K3}, h) = \widetilde{O}(L_{2d})$ and hence we obtain a well-defined map which associates to an isomorphism class $(S, \mathcal{L})$ a point in $\mathcal{F}^0_{2d}$. This map is injective: assume that a point in $\mathcal{F}^0_{2d}$ arises from two polarised surfaces $(S, \mathcal{L})$ and $(S', \mathcal{L}')$. Then there exists a Hodge isometry $H^2(S', \mathbb{Z}) \to H^2(S, \mathbb{Z})$ mapping the ample class $c_1(\mathcal{L}')$ to $c_1(\mathcal{L})$. It then follows from the strong Torelli Theorem (Theorem 2.6) that this map is induced by an isomorphism $(S, \mathcal{L}) \cong (S', \mathcal{L}')$. Thus we get an injective map from the set of isomorphism classes of degree $2d$ polarised K3 surfaces into $\mathcal{F}^0_{2d}$. Finally, the surjectivity of this map follows from the surjectivity of the period map, Theorem 2.9. □

In the literature one often finds references to $\mathcal{F}_{2d}$ as the moduli space of polarised K3 surfaces. One can interpret the points in the complement of $\mathcal{F}^0_{2d}$ as weakly- or semi-polarised K3 surfaces, i.e. $\mathcal{L}$ has positive degree and is nef, but not ample, as it has degree 0 on some nodal class(es). Alternatively, one can consider ample line bundles on K3 surfaces with rational double points. There is still a version of the strong Torelli theorem for weakly polarised K3 surfaces due to D. Morrison [88], but the precise formulation is subtle. For purposes of the birational geometry of these spaces, it obviously does not matter whether one works with $\mathcal{F}_{2d}$ or its open part $\mathcal{F}^0_{2d}$.

The notion of polarised K3 surfaces was generalised by Nikulin [96] to that of *lattice-polarised* K3 surfaces: see also Dolgachev's paper [26] for a concise account, in particular in connection with mirror symmetry. To describe this, we fix a lattice M of signature $(1, t)$, which we assume can be embedded primitively into the K3-lattice L_{K3}. The cone $V_M = \{x \in M_{\mathbb{R}} \mid (x, x) > 0\}$ has two connected components: we fix one and denote it by $\mathcal{C}_M$. Let

$$\Delta_M = \{d \in M \mid (d, d) = -2\}$$

and choose a decomposition $\Delta_M = \Delta_M^+ \cup (-\Delta_M^+)$. Moreover let

$$\mathcal{C}_M^+ = \{h \in V(M)^+ \cap M \mid (h, d) > 0 \text{ for all } d \in \Delta_M^+\}.$$

An M-*polarised* K3 surface is then a pair (S, j) where S is a K3 surface and $j \colon M \hookrightarrow \operatorname{Pic}(S)$ is a primitive embedding. An isomorphism between M-polarised K3 surfaces is an isomorphism between the surfaces that commutes with the primitive embeddings. We call (S, j) ample (or pseudo-ample), if $j(\mathcal{C}_M^+)$ contains an ample (or pseudo-ample) class. The classical case of polarised K3 surfaces is the case where $t = 0$ and $M = \langle 2d \rangle$.

The theory of moduli of polarised K3 surfaces carries over naturally to lattice polarised K3 surfaces. For this, one has to consider the domain $\Omega_M = \{\omega \in \Omega \mid (w, M) = 0\}$ and the group $O(L_{K3}, M)$ of orthogonal transformations of the K3 lattice which fixes M. The role of the variety $\mathcal{F}_{2d}$ is then taken by the quotient

$$\mathcal{F}_M = O(L_{K3}, M) \backslash \Omega_M. \tag{2.18}$$

Lattice polarised K3 surfaces play a role in mirror symmetry. For this we consider *admissible* lattices, i.e. lattices M which admit an embedding $j \colon M \hookrightarrow L_{K3}$ such that

$$M^{\perp}_{L_{K3}} \cong U(m) \oplus \widetilde{M}.$$

The choice of such a splitting determines a primitive embedding $\tilde{j} \colon \widetilde{M} \hookrightarrow L_{K3}$. The lattice $\widetilde{M}$ is hyperbolic: more precisely, its signature is $(1, 18 - t)$. The variety $\mathcal{F}_{\widetilde{M}}$ is then a mirror family to $\mathcal{F}_M$. There are more ingredients to the concept of mirror symmetry which we will not describe here, such as the Yukawa coupling and the mirror map. For details we refer to [26]: see also [52], [56].

Finally, we want to comment on the relationship between the construction of moduli spaces of K3 surfaces as quotients of homogeneous domains and GIT-like constructions. By Viehweg's results [128] moduli spaces of polarised varieties with trivial canonical bundle exist and can be constructed as group quotients. For this we first fix a Hilbert polynomial $P(m)$ of a line bundle on a K3 surface. This is of the form $P(m) = m^2 d + 2$ where the degree of the line bundle is $2d$. Let $\mathcal{M}_{2d}$ be the GIT moduli space of degree $2d$ polarised K3 surfaces. We want to relate this to $\mathcal{F}_{2d}$.

We first note that for any ample line bundle $\mathcal{L}$ on a K3 surface S its third power $\mathcal{L}^{\otimes 3}$ is very ample. For this, see [113], but in general one can use Matsusaka's big theorem ([87], [79]) to show that there is a positive integer m_0 such that $\mathcal{L}^{\otimes m_0}$ is very ample for all polarised varieties $(X, \mathcal{L})$ with fixed Hilbert polynomial. Now let $m_0 \geqslant 3$ be sufficiently big. Then we have embeddings $f_{|\mathcal{L}^{\otimes m_0}|} \colon S \to \mathbb{P}^{N-1}$ where $N = h^0(S, \mathcal{L}^{\otimes m_0}) = P(m_0)$. Such an embedding depends on the choice of a basis of $H^0(S, \mathcal{L}^{\otimes m_0})$. Let H be an irreducible component of the Hilbert scheme $\mathrm{Hilb}_P(\mathbb{P}^{N-1})$ containing at least one point corresponding to a smooth K3 surface S. Let H_{sm} be the open part of H parametrising smooth surfaces. Then it is easy to prove that H_{sm} is smooth and that every point in H_{sm} parametrises a K3 surface. There exists a universal family $\mathcal{S}_{sm} \to H_{sm}$. The group $SL(N, \mathbb{C})$ acts on H_{sm} and every irreducible component of the GIT moduli space of degree 2d polarised K3 surfaces is of the form $SL(N, \mathbb{C})\backslash H_{sm}$. Let $\mathcal{M}'_{2d}$ be such a component. Choosing local polarised markings for the universal family one can construct a map $H_{sm} \to \mathcal{F}^0_{2d}$, which clearly factors through the action of $SL(N, \mathbb{C})$, i.e. gives rise to a map $\pi \colon \mathcal{M}'_{2d} \to \mathcal{F}^0_{2d}$. By construction this is a holomorphic map. On the other hand both $\mathcal{M}'_{2d}$ and $\mathcal{F}_{2d}$ are quasi-projective varieties. It then follows from a theorem of Borel [15] that π is a morphism of quasi-projective varieties.

We claim that π is an isomorphism and that $\mathcal{M}_{2d}$ has only one component. First of all we note that one can, as in the proof of [50, Theorem 1.5], take a finite étale cover $H'_{sm} \to H_{sm}$ such that the action of $SL(N+1, \mathbb{C})$ lifts to a free action on H'_{sm} as well as on the pullback $\mathcal{S}'_{sm} \to H'_{sm}$ of the universal family. This gives a quotient family over $Z_{sm} = SL(N+1, \mathbb{Z})\backslash H'_{sm}$ which is smooth and maps finite-to-one to $\mathcal{M}_{2d}$. By the local Torelli theorem the natural map $Z_{sm} \to \mathcal{F}^0_{2d}$ has discrete fibres and hence the same is true for π. But then π must be dominant. Now we can use Theorem 2.17 to conclude that $\mathcal{M}_{2d}$ is irreducible and that π is a bijection. Since $\mathcal{F}^0_{2d}$ is a normal variety, it also follows that π is an isomorphism. We can thus summarise our discussion as follows.

Theorem 2.19. *There is an isomorphism $\mathcal{M}_{2d} \cong \mathcal{F}^0_{2d}$, i.e. the GIT moduli space $\mathcal{M}_{2d}$ is isomorphic to the modular variety $\mathcal{F}^0_{2d}$.*

3. Irreducible symplectic manifolds

In this section we recall the main properties of irreducible symplectic manifolds, discuss the Torelli theorem and give basic facts about moduli spaces of polarised symplectic manifolds.

3.1. Basic theory of irreducible symplectic manifolds

The theory of irreducible symplectic manifolds is less developed than that of K3 surfaces. Nevertheless, several results have been proved over the last 30 years.

Definition 3.1. A complex manifold X is called an *irreducible symplectic manifold* or *hyperkähler manifold* if the following conditions are fulfilled:

(i) X is a compact Kähler manifold;
(ii) X is simply-connected;
(iii) $H^0(X, \Omega_X^2) \cong \mathbb{C}\omega$ where ω is an everywhere nondegenerate holomorphic 2-form.

According to the Bogomolov decomposition theorem [9], irreducible symplectic manifolds are one of the building blocks for compact Kähler manifolds with trivial canonical bundle: see also [8, Théorème 2]. The others are abelian varieties and Calabi-Yau varieties (here we mean Calabi-Yau in its strictest sense, i.e. a compact Kähler manifold X such that $\pi_1(X) = 1$ and $H^0(X, \Omega_X^i) = 0$ for $0 < i < \dim X$).

In dimension two the only irreducible symplectic manifolds are K3 surfaces. Although irreducible symplectic manifolds have now been studied for nearly 30 years, only four classes of such manifolds have so far been discovered and it is a wide open problem whether other types exist or not. The known examples are:

(i) The length n Hilbert scheme $S^{[n]} = \mathrm{Hilb}^n(S)$ for a K3 surface S, and its deformations. Note that the deformation space of such a variety has dimension 21 if $n \geqslant 2$ and that, since K3 surfaces only depend on 20 parameters, a general deformation will not itself be of the form $S^{[n]}$. We shall refer to these varieties as irreducible symplectic manifolds of *deformation* $K3^{[n]}$ *type* or *deformation* $K3^{[n]}$ *manifolds*.

(ii) Let A be a 2-dimensional complex torus and consider the length-$(n+1)$ Hilbert scheme $A^{[n+1]} = \mathrm{Hilb}^{n+1}(A)$ together with the morphism $p\colon A^{[n+1]} \to A$ given by addition. Then $X = p^{-1}(0)$ is an irreducible symplectic manifold, called a *generalised Kummer variety* (even though it is not necessarily algebraic). The deformation space of these manifolds has dimension 5 if $n \geqslant 2$, again one more than for 2-dimensional complex tori.

(iii) O'Grady's irreducible symplectic manifolds of dimension 6, described in [100]. These are deformations of (desingularised) moduli spaces of sheaves on an abelian surface and depend on 6 parameters.

(iv) O'Grady's irreducible symplectic manifolds of dimension 10, described in [99]. These are deformations of (desingularised) moduli spaces of sheaves on a K3 surface and depend on 22 parameters.

Other moduli spaces of sheaves, apart from those considered in [99] and [100], cannot be desingularised symplectically: see [67] and also [132].

In many ways irreducible symplectic manifolds behave like K3 surfaces, but there are also important differences, as we shall see later. We first notice that it follows immediately from the definition that X must have even dimension $2n$ over $\mathbb{C}$ and that its canonical bundle ω_X is trivial (an n-fold exterior power of a generator ω of $H^0(X, \Omega_X^2)$ will define a trivialisation of the canonical bundle). Clearly $h^{2,0}(X) = h^{0,2}(X) = 1$ and $h^{1,0}(X) = h^{0,1}(X) = 0$. By a result of Bogomolov [10], the deformation space of X is unobstructed. This result was generalised to Ricci-flat manifolds by Tian [125] and Todorov [126], and algebraic proofs were given by Kawamata [68] and Ran [109] (see also [36]). Since

$$T_{[0]}\operatorname{Def}(X) \cong H^1(X, T_X) \cong H^1(X, \Omega_X^1)$$

the dimension of the deformation space is $b_2(X) - 2$.

As in the K3 case we have a Hodge decomposition $H^2(X, \mathbb{C}) = H^{2,0} \oplus H^{1,1} \oplus H^{0,2}$ with $H^{2,0}$ and $H^{0,2}$ both 1-dimensional. Unlike the K3 case the intersection form does not immediately provide $H^2(S, \mathbb{Z})$ with the structure of a lattice. It was, however, shown by Beauville [8] that $H^2(X, \mathbb{Z})$ does carry a natural structure as a lattice. To define this, let $\omega \in H^{2,0}(X)$ be such that $\int_X (\omega\overline{\omega})^n = 1$ and define

$$q_X'(\alpha) = \frac{n}{2}\int_X \alpha^2(\omega\overline{\omega})^{n-1} + (1-n)\left(\int_X \alpha\omega^{n-1}\overline{\omega}^n\right)\left(\int_X \overline{\alpha}\omega^n\overline{\omega}^{n-1}\right).$$

After multiplication by a positive constant γ the quadratic form $q_X = \gamma q_X'$ defines an indivisible integral symmetric bilinear form $(\ ,\ )_X$ on $H^2(X, \mathbb{Z})$: this is the Beauville form. Clearly $(\omega, \omega)_X = 0$ and $(\omega, \overline{\omega})_X > 0$.

There is another way to introduce the Beauville form. For this let $\nu(\alpha) = \alpha^{2n}$ be given by the cup product. Then, by a result of Fujiki [36, Theorem 4.7], there is a positive rational number c, the *Fujiki invariant*, such that

$$\nu(\alpha) = c q_X(\alpha)^n$$

for all $\alpha \in H^2(X, \mathbb{Z})$. In this sense the Beauville form can be derived from the cup product of the cohomology.

Proposition 3.2. *The Beauville lattices and Fujiki invariants for the known examples of irreducible symplectic manifolds are as follows:*

(i) *The Beauville lattice of a deformation* $K3^{[n]}$ *manifold is* $L_{K3,2n-2} = 3U \oplus 2E_8(-1) \oplus \langle -2(n-1)\rangle$. *It has rank 23, one more than the K3 lattice, to which it is closely related. If* $X = S^{[n]}$ *for a K3 surface S, then* $3U \oplus 2E_8(-1)$ *comes from* $H^2(S, \mathbb{Z})$ *and the summand* $\langle -2(n-1)\rangle$ *is generated (over* $\mathbb{Q}$*) by the exceptional divisor E, which is the blow-up of the diagonal in the symmetric product* $S^{(n)}$. *As an element in the Picard group the divisor E is 2-divisible. The Beauville lattice remains constant under deformations. The Fujiki invariant is* $c = (2n)!/(n!2^n)$.

(ii) *The Beauville lattice of a generalised Kummer variety (or deformation thereof) is* $3U \oplus \langle -2(n+1)\rangle$ *and the Fujiki invariant is* $c = (n+1)(2n)!/(n!2^n)$.

(iii) *The Beauville lattice of the 6-dimensional example of O'Grady is* $3U \oplus \langle -2\rangle \oplus \langle -2\rangle$ *and the Fujiki invariant is* $c = 60$.

(iv) *The Beauville lattice of the 10-dimensional example of O'Grady is* $3U \oplus 2E_8(-1) \oplus A_2(-1)$ *and the Fujiki invariant is* $c = 945$.

For proofs, see [8], [111] and [110]. Note that all these lattices are even. It is, however, not known whether this is a general fact for irreducible symplectic manifolds.

Fix an abstract lattice L that is isomorphic to $(H^2(X,\mathbb{Z}), q_X)$ for some irreducible symplectic manifold X. A *marking* is an isomorphism of lattices $\phi\colon H^2(X,\mathbb{Z}) \xrightarrow{\sim} L$. Let $p\colon \mathcal{X} \to U$ be a representative of the Kuranishi family of deformations of X with sufficiently small and contractible base. Note that by unobstructedness the base space of the Kuranishi family is smooth and of dimension $b_2(X) - 2$. The marking ϕ for X defines a marking for $\mathcal{X}$ and we obtain a period map $\pi_U\colon U \to \Omega_L$ to the period domain defined by (2.3). As in the K3 case we have a local Torelli theorem.

Theorem 3.3. (Beauville) *The differential of the period map defined by the Kuranishi family is an isomorphism and thus the period map is a local isomorphism.*

Proof. See [9], [8]. □

As in the K3 case one can define a moduli space of marked irreducible symplectic manifolds (of a given type). Again, this will not be Hausdorff. Another result which carries over from the K3 case is surjectivity of the period map.

Theorem 3.4. (Huybrechts) *Let* L *be a lattice of an irreducible symplectic manifold and let* Ω_L *be the associated period domain. If* $\mathcal{M}'_L$ *is a non-empty component of the moduli space* $\mathcal{M}_L$ *of irreducible symplectic manifolds with Beauville lattice* L, *then the period map* $\pi\colon \mathcal{M}'_L \to \Omega_L$ *is surjective.*

Proof. A proof can be found in [63, Section 8]. □

3.2. Hodge theoretic Torelli theorem

So far, many results from K3 surfaces have carried over to other irreducible symplectic manifolds. The situation changes when it comes to the global Torelli theorem. The first counterexample to this is due to Debarre [24]. He showed the following: let S be a K3 surface containing only one curve, which is a smooth rational curve C, and consider the Hilbert scheme $X = S^{[n]}$ with $n \geqslant 2$. Then X contains $S^n(C) \cong \mathbb{P}^n$. One can perform an elementary transformation on X by first blowing up $S^n(C)$ and then contracting the exceptional divisor in another direction. This gives another compact complex manifold X' which is bimeromorphic but not isomorphic to X. In general X' need not be Kähler, but Debarre produced an example where X' does have a Kähler structure and thus is an irreducible symplectic manifold.

Since the natural bimeromorphic transformation $f\colon X' \dashrightarrow X$ defines a Hodge isometry $f^*\colon H^2(X,\mathbb{Z}) \to H^2(X',\mathbb{Z})$ this gives a counterexample to the global Torelli theorem. It should be noted, though, that neither the surface S nor the varieties X and X' are projective. Moreover, the existence of (-2)-curves on a K3 surface S is exactly the cause for the failure of the Hausdorff property for the base of the universal family.

Debarre's counterexample would still allow for a version of the Torelli theorem where the existence of a Hodge isometry only implies birational equivalence, not an isomorphism (for K3 surfaces the two notions coincide). However, this is also ruled out by the following counterexample which is due to Y. Namikawa [95]. For this one starts with a generic abelian surface A with a polarisation of type (1,3). Then the dual abelian surface $\widehat{A}$ also carries a (1,3)-polarisation. Let $X = \mathrm{Km}^{[2]} A$ and $\widehat{X} = \mathrm{Km}^{[2]}(\widehat{A})$ be the associated generalised Kummer varieties of dimension 4. Then X and $\widehat{X}$ are birationally equivalent if and only if A and $\widehat{A}$ are isomorphic abelian surfaces. The reason for this is the following: every birational isomomorphism must send the exceptional divisor E on X to the exceptional divisor $\widehat{E}$ on $\widehat{X}$. Since the Albanese of E and $\widehat{E}$ are A and $\widehat{A}$ respectively, this implies that A and $\widehat{A}$ are isomorphic. This is not the case for general A. This shows that Namikawa's example gives a counterexample even to the birational global Torelli theorem. Moreover, one can easily make this into a counterexample to the polarised Torelli theorem. This can be done by choosing polarisations of the form $mL - \delta$ and $m\widehat{L} - \widehat{\delta}$, where m is sufficiently large, L and $\widehat{L}$ are induced from the polarisation on A and $\widehat{A}$ respectively, $\delta = 2E$ and $\widehat{\delta} = 2\widehat{E}$ (the exceptional divisors are 2-divisible in the Picard group). The Hodge isomorphism respects these polarisations.

At first, these counterexamples seem to indicate that there is no chance of proving a version of the global Torelli theorem for irreducible symplectic manifolds. However, the above example is not as surprising as it seems at a first glance. It is well known, and also well understood in terms of period domains and arithmetic groups, that A and $\widehat{A}$ are not isomorphic as polarised abelian surfaces (their period points in the Siegel space are inequivalent under the paramodular group), but that the associated Kummer surfaces Km(A) and Km($\widehat{A}$) are isomorphic (and their period points in the corresponding type IV domain are equivalent under the orthogonal group). Details can be found in [43]. An analysis of this situation suggests that a version of the Torelli theorem could hold if one considers Hodge isometries with extra conditions.

Verbitsky [127] has announced a global Torelli theorem for irreducible symplectic manifolds (see also Huybrecht's Bourbaki talk [64]). His results were further elucidated by Markman [86]. The crucial idea here is to use monodromy operators and parallel transport operators. Markman first noticed the importance of these operators for the study of irreducible symplectic manifolds, developing his ideas in a series of papers [84], [83], [85]. To define them, let X_1, X_2 be irreducible symplectic manifolds. We say that $f\colon H^*(X_1, \mathbb{Z}) \to H^*(X_2, \mathbb{Z})$ is a *parallel transport operator* if there exists a smooth, proper flat family $\pi\colon \mathcal{X} \to B$ of irreducible symplectic manifolds together with points $b_1, b_2 \in B$ such that there are isomorphisms $\alpha_i\colon X_i \xrightarrow{\sim} \mathcal{X}_{b_i}$ and a continuous path $\gamma\colon [0,1] \to B$ with $\gamma(0) = b_1$ and $\gamma(1) = b_2$, such that the parallel transport in the local system $R\pi_*\mathbb{Z}$ along γ induces the isomorphism $(\alpha_2^{-1})^* \circ f \circ \alpha_1^*\colon H^*(\mathcal{X}_{b_1}, \mathbb{Z}) \to H^*(\mathcal{X}_{b_2}, \mathbb{Z})$.

For a single irreducible symplectic manifold X, we call an automorphism $f\colon H^*(X, \mathbb{Z}) \to H^*(X, \mathbb{Z})$ a *monodromy operator* if it is a parallel transport operator (with $X_1 = X_2 = X$). The *monodromy group* Mon(X) is defined as the subgroup of $GL(H^*(X,\mathbb{Z}))$ generated by monodromy operators. Restricting the group action to the second cohomology group we obtain a subgroup $Mon^2(X)$ of $GL(H^2(X,\mathbb{Z}))$. Since monodromy operators preserve the Beauville form we obtain a subgroup $Mon^2(X) \subset O(H^2(X,\mathbb{Z}))$.

Based on Verbitsky's results [127], Markman [86] has formulated the following Hodge theoretic global Torelli theorem.

Theorem 3.5. (Hodge theoretic Torelli) *Suppose that X and Y are irreducible symplectic manifolds.*

(i) *If* $f\colon H^2(Y,\mathbb{Z}) \to H^2(X,\mathbb{Z})$ *is an isomorphism of integral Hodge structures which is a parallel transport operator, then X and Y are bimeromorphic.*

(ii) *If, moreover, f maps a Kähler class of Y to a Kähler class of X, then X and Y are isomorphic.*

Proof. The first part of the theorem follows easily from Verbitsky's results. The second part uses in addition results on the Kähler cone of irreducible symplectic manifolds. For a more detailed discussion see [86, Theorem 1.3] and [86, Section 3.2]. □

3.3. Moduli spaces of polarised irreducible symplectic manifolds

We shall now turn to the case of polarised irreducible symplectic manifolds. In the K3 case we saw that the degree is the only invariant of a polarisation, or, equivalently, that there is only one $O(L_{K3})$-orbit of primitive vectors of given length. This is no longer true in general, as can already be seen in the case of $S^{[2]}$. Recall that the Beauville lattice in this case is isomorphic to $L_{K3,2} = 3U \oplus 2E_8(-1) \oplus \langle -2\rangle$. If h is a primitive vector the number $\mathrm{div}(h)$ (the divisor of h: see Equation (7.4) below) is the positive generator of the ideal $(h, L_{K3,2})$, which is the biggest positive integer by which one can divide h as a vector in the dual lattice $L_{K3,2}^{\vee}$. Since $L_{K3,2}$ is not unimodular, but has determinant 2, the divisor $\mathrm{div}(h)$ can be 1 or 2. Indeed, both of these happen when $d \equiv -1 \mod 4$ and accordingly we have one $O(L_{K3,2})$-orbit if $d \not\equiv -1 \mod 4$ and two if $d \equiv -1 \mod 4$. Details of this can be found in [50]. These two cases are referred to as the *split* case ($\mathrm{div}(h) = 1$) and the *non-split* case ($\mathrm{div}(h) = 2$). The reason for this terminology is the behaviour of the orthogonal lattice: if $h^2 = 2d$ the possibilities for $L_h = h^{\perp}_{L_{K3,2}}$ are (see Example 7.12 below)

$$L_h = 2U \oplus 2E_8(-1) \oplus \langle -2\rangle \oplus \langle -2d\rangle \text{ for } \mathrm{div}(h) = 1 \tag{3.6}$$

and

$$L_h = 2U \oplus 2E_8(-1) \oplus \begin{pmatrix} -2 & 1 \\ 1 & -\frac{d+1}{2} \end{pmatrix} \text{ for } \mathrm{div}(h) = 2. \tag{3.7}$$

For the higher dimensional case of $S^{[n]}$ the situation becomes more involved as the possibilities for the divisor of h increase in number. Whenever the primitive vectors of length d form more than one orbit, the moduli space of polarised irreducible symplectic manifolds of degree d will not be connected: see [50].

In order to discuss moduli spaces of polarised irreducible symplectic manifolds we first fix some discrete data: the dimension $2n$, the Beauville lattice (considered as an abstract lattice L), and the Fujiki invariant c. Together L and c define the *numerical type* of the irreducible symplectic manifold, denoted by $\mathbf{N}$. Next we choose a *polarisation type*, i.e. an $O(L)$-orbit of a primitive vector $h \in L$. Viehweg's theory gives us the existence of a moduli space $\mathcal{M}_{n,\mathbf{N},h}$ parametrising polarised irreducible symplectic manifolds $(X, \mathcal{L})$ of dimension $2n$ with the chosen Beauville lattice, Fujiki invariant and polarisation type. This is a quasi-projective variety and can be constructed as a GIT quotient as in the K3 case, the only difference being that we must here invoke Matsusaka's big theorem [87] and a result of Kollár and Matsusaka [73] to be guaranteed a uniform bound N_0 such that $\mathcal{L}^{\otimes N_0}$ is very ample for all pairs $(X, \mathcal{L})$.

Although there is not a Torelli theorem as in the K3 case, these moduli spaces are still related to quotients of homogeneous domains of type IV. Let Ω_L be the period domain defined by the lattice L and let $L_h = h_L^\perp$. This is a lattice of signature $(2, \mathrm{rk}(L) - 3)$ and defines a homogeneous domain Ω_{L_h} of type IV. Let $O(L, h)$ be the stabiliser of h in $O(L)$. This can be considered as a subgroup of $O(L_h)$. The domain Ω_{L_h} has two connected components, of which we choose one, which we denote by $\mathcal{D}_{L_h}$. Again $O^+(L, h)$, the subgroup of $O(L, h)$ of real spinor norm 1, is the subgroup fixing the components of Ω_{L_h}.

Theorem 3.8. *For every component $\mathcal{M}'_{n,\mathbf{N},h}$ of the moduli space $\mathcal{M}_{n,\mathbf{N},h}$ there exists a finite to one dominant morphism*

$$\psi\colon \mathcal{M}'_{n,\mathbf{N},h} \to O^+(L, h)\backslash\mathcal{D}_{L_h}.$$

The proof of this is analogous to the proof of Theorem 2.19. There are, however, differences compared to the K3 case. In general, ψ will not be injective (see the discussion below). There is also a difference concerning the image of ψ. In the K3 case we know that a big and nef line bundle is ample if and only if it has positive degree on the nodal curves. Hence it is necessary and sufficient to remove the hyperplanes orthogonal to the nodal classes. So far, no complete analogue is known for the higher dimensional case, but Hassett and Tschinkel have proved partial results for $n = 2$ in [58], and more precise results in a special case in [59].

Nevertheless, Theorem 3.8 is enough to prove results on the Kodaira dimension of moduli spaces of polarised irreducible symplectic manifolds. This was done in [50], [51]: see Theorem 6.2 and Theorem 6.3 below.

Very recent work of Verbitsky [127] and Markman [86] improves Theorem 3.8, by using a polarised analogue of the monodromy group $\mathrm{Mon}^2(X) \subset O(H^2(X), \mathbb{Z})$. Let H be an ample divisor on X. We call an element in $f \in \mathrm{Mon}(X)$ a *polarised parallel transport operator of the pair* (X, H) if it is a parallel transport operator for a family $\pi\colon \mathcal{X} \to B$ with base point $b_0 \in B$ and isomorphism $\alpha\colon X \to \mathcal{X}_{b_0}$ which fixes $c_1(H)$, such that there exists a flat section h of $R^2\pi_*\mathbb{Z}$ with $h(b_0) = \alpha_*(c_1(H))$ and

$h(b)$ an ample class in $H^2(\mathcal{X}_b, \mathbb{Z})$ for all $b \in B$. These operators define a subgroup $\mathrm{Mon}^2(X, H) \subset O(H^2(X), \mathbb{Z})$. It was shown by Markman [86, Proposition 1.9] that $\mathrm{Mon}^2(X, H)$ is the stabiliser of $c_1(H)$ in $\mathrm{Mon}^2(X)$. Given a marking $\phi\colon H^2(X, \mathbb{Z}) \to L$ this defines a subgroup

$$\Gamma = \phi(\mathrm{Mon}^2(X, H)) \subset O(L, h), \tag{3.9}$$

which can be shown to be independent of the marking ϕ: see [86, Section 7.1].

Let $\mathcal{M}'_{n,\mathbf{N},h}$ be a component of the moduli space of polarised irreducible symplectic manifolds with fixed numerical type and polarisation type. Given an element (X, H) in this component one thus obtains a group $\Gamma \subset O(L, h)$ as above, which is also independent of the chosen pair (X, H) by the results of [86, Section 7.1]. In fact $\Gamma \subset O^+(L, h)$, as monodromy operators are obviously orientation-preserving: see [86, Section 1.2]. Thus Γ acts on the homogeneous domain $\mathcal{D}_{L_h}$.

Theorem 3.10. *The map ψ from Theorem 3.8 lifts to an open immersion*

$$\widetilde{\psi}\colon \mathcal{M}'_{n,\mathbf{N},h} \to \Gamma\backslash\mathcal{D}_{L_h},$$

where Γ is as in Equation (3.9).

Proof. It is easy to see that the map $\psi\colon \mathcal{M}'_{n,\mathbf{N},h} \to O^+(L, h)\backslash\mathcal{D}_{L_h}$ lifts to a map $\widetilde{\psi}\colon \mathcal{M}'_{n,\mathbf{N},h} \to \Gamma\backslash\mathcal{D}_{L_h}$ (see the beginning of the proof of [50, Theorem 2.3]). The hard part is the injectivity of $\widetilde{\psi}$ and this is where the Torelli theorem for irreducible symplectic manifolds is used. For details we refer the reader to [86]. □

Remark 3.11. We note that in general the projective group $\Gamma/\pm 1$ is a proper subgroup of $O^+(L, h)/\pm 1$ and thus Theorem 3.10 is a substantial improvement of Theorem 3.8.

In the case of irreducible symplectic manifolds of $K3^{[n]}$ type this can be made explicit. For an even lattice L we define $\mathrm{Ref}(L)$ to be the subgroup generated by -2-reflections and the negative of $+2$-reflections. This is a subgroup of $O^+(L)$. If X is an irreducible symplectic manifold of $K3^{[n]}$ type, then we define $\mathrm{Ref}(X)$ accordingly.

Theorem 3.12. (Markman) *If X is a deformation $K3^{[n]}$ manifold then*

$$\mathrm{Mon}^2(X) = \mathrm{Ref}(X).$$

Proof. This is proved in [83, Theorem 1.2]. □

For a lattice L we let

$$\widehat{O}(L) = \{g \in O(L) \mid g|_{L^\vee/L} = \pm \mathrm{id}_{L^\vee/L}\}$$

and given an element $h \in L$ we set $\widehat{O}(L, h) = \{g \in \widehat{O}(L) \mid g(h) = h\}$. Recall the convention of Equation (2.15) and that by Proposition 3.2(i) the Beauville lattice is $L_{K3,2n-2}$.

It then follows from Theorem 3.12 in conjunction with Kneser's result [70, Satz 4] that

$$\operatorname{Ref}(L_{K3,2n-2}) = \widehat{O}^+(L_{K3,2n-2}). \tag{3.13}$$

Combining this with Theorem 3.10 we thus obtain

Theorem 3.14. *Let $\mathcal{M}'_h$ be an irreducible component of the moduli space of polarised deformation $K3^{[n]}$ manifolds. Then the map ψ of Theorem 3.8 factors through the finite cover $\widehat{O}^+(L_{K3,2n-2}, h)\backslash\mathcal{D}_{L_h} \to O^+(L_{K3,2n-2}, h)\backslash\mathcal{D}_{L_h}$ that is, there is a commutative diagram*

$$\begin{array}{ccc} \mathcal{M}'_h & \xrightarrow{\widetilde{\psi}} & \widehat{O}^+(L_{K3,2n-2}, h)\backslash\mathcal{D}_{L_h} \\ & \searrow^{\psi} & \downarrow \\ & & O^+(L_{K3,2n-2}, h)\backslash\mathcal{D}_{L_h}. \end{array}$$

Moreover the map $\widetilde{\psi}$ is an open immersion.

Remark 3.15. In [50, Proposition 2.3] we stated that the map ψ lifts to the quotient $\widetilde{O}^+(L_{K3,2n-2}, h)\backslash\mathcal{D}_{L_h}$. This is not correct since, contrary to what was said in the proof, the projective groups $\widetilde{O}^+(L_{K3,2n-2}, h)/\pm 1$ and $\widehat{O}^+(L_{K3,2n-2}, h)/\pm 1$ are not identical if $n > 2$. In that case, in fact, $\widetilde{O}^+(L_{K3,2n-2}, h)/\pm 1$ is an index 2 subgroup of $\widehat{O}^+(L_{K3,2n-2}, h)/\pm 1$. If, however, $n = 2$, then the two groups coincide since $\widetilde{O}^+(L_{K3,2}) = \widehat{O}^+(L_{K3,2})$ and thus the results of [50] are not affected by this error.

Remark 3.16. Theorem 3.14 gives an affirmative answer to [50, Question 2.6] (with the correct group).

Remark 3.17. The results discussed so far do not give an answer to the question whether moduli spaces of polarised irreducible symplectic manifolds of given deformation type and given type of polarisation are always connected. Apostolov [1] has obtained some results on this. For example in the $\mathrm{Hilb}^{[n]}$ case these moduli spaces are always connected for $n = 2$ (both in the split and the non-split case), but in general there can be more than one component.

4. Projective models

Besides the abstract theory of moduli spaces there is a vast literature which deals with concrete geometric descriptions of K3 surfaces, and to a much lesser degree, also of irreducible symplectic manifolds of higher dimension. The easiest example is degree 4 surfaces in $\mathbb{P}^3$. Any smooth quartic surface is a K3 surface and counting parameters one obtains a family of dimension $34 - 15 = 19$, because 34 is the number of quartics and 15 is the dimension of $\mathrm{PGL}(4, \mathbb{C})$. This argument shows that the moduli space $\mathcal{F}_4$ of polarised K3 surfaces of degree 4 is unirational. The same approach yields unirationality for degrees $2d = 2$, 6 and 8, as these correspond to double covers of the projective plane branched along a sextic curve, complete intersections of type (2,3) in $\mathbb{P}^4$, and complete intersections of type (2,2,2) in $\mathbb{P}^5$ respectively.

In general it can be very hard to decide whether a moduli space of polarised K3 surfaces of low degree is unirational or not. Mukai ([89], [90], [91], [92], [93]) has contributed most significantly to this problem. So far there are three approaches to proving unirationality.

(1) Describing the K3 surfaces as complete intersections in homogeneous spaces (this can be used for $1 \leqslant d \leqslant 9$, $d = 11, 12, 17, 19$).
(2) Using non-abelian Brill-Noether theory of vector bundles over algebraic curves (here one obtains results for $d = 6, 8, 10, 16$).
(3) Using specific geometric constructions for certain degrees ($d = 11, 12, 15, 19$). An example is Mukai's most recent work ([93]) for $d = 15$ where he describes K3 surfaces of degree 30 as complete intersections in a certain rank 10 vector bundle on the Ellingsrud-Piene-Strømme moduli space of twisted cubics.

There are only a few results about rationality for these cases. Shepherd-Barron proves rationality for the cases $d = 3$ in [121] and $d = 9$ in [120].

For a discussion of low degree cases we also refer the reader to Voisin's Bourbaki exposé [130]. One can summarise the results as follows.

Theorem 4.1. *The moduli spaces $\mathcal{F}_{2d}$ of polarised* K3 *surfaces of degree* $2d$ *are unirational for* $1 \leqslant d \leqslant 12$ *and* $d = 15, 16, 17, 19$.

Some other special moduli spaces related to K3 surfaces have also been studied. Perhaps the most notable of these is the moduli space of Enriques surfaces, which is a special case of the lattice polarised K3 moduli spaces given in (2.18). This approach was first seen in Horikawa's announcement [62] of a Torelli theorem for Enriques surfaces. It was shown by Kondo [75] that the moduli space of Enriques surfaces is rational.

It is also natural in this context to consider moduli of K3 surfaces with automorphisms. There is an extensive literature on such surfaces, much of it touching on moduli problems: for example [37] and the recent work of Ma and Yoshikawa [82].

Many of these moduli spaces turn out to be unirational or even rational, typically because the symmetry exhibits the K3 surfaces as covers of $\mathbb{P}^2$ and the family of possible ramification curves can be parametrised.

Some moduli of K3 surfaces with extra structure can be described as ball quotients: see for example [27]. Again there is an extensive literature on this subject. For a guide, we refer the reader to the survey [28] by Dolgachev and Kondo. In the introduction to [28] it is conjectured that all Deligne-Mostow arithmetic complex ball quotients are moduli spaces of K3 surfaces.

Much less is known in the case of irreducible symplectic manifolds, but some cases have been studied.

Example 4.2. A classical case is the Fano variety of lines contained in a cubic fourfold, which was studied in detail by Voisin [129]. These are varieties of $K3^{[2]}$ type. In our terminology this corresponds to the degree 6 non-split case and the lattice L_h orthogonal to the polarisation vector is isomorphic to $2U \oplus 2E_8(-1) \oplus A_2(-1)$.

Example 4.3. O'Grady studied double covers of Eisenbud-Popescu-Walter sextics in [101]. This is the case of split polarisation of minimal degree (degree 2) for the $K3^{[2]}$-type. The lattice L_h is $2U \oplus 2E_8(-1) \oplus \langle -2 \rangle \oplus \langle -2 \rangle$.

Example 4.4. Iliev and Ranestad ([65], [66]) have shown that the variety of sums of powers VSP(F, 10) of presentations of a general cubic form F in 6 variables as a sum of 10 cubes is an irreducible symplectic 4-fold. These are deformations of length 2 Hilbert schemes of K3 surfaces with a degree 14 polarisation. The precise nature of the polarisation of the irreducible symplectic manifold is unknown.

Example 4.5. Debarre and Voisin ([25]) have constructed examples in the Grassmannian Gr(6, V) where V is a 10-dimensional complex vector space.Starting with a sufficiently general form $\sigma\colon \bigwedge^3 V \to \mathbb{C}$ they show that the subspace of Gr(6, V) consisting of 6-planes L such that $\sigma|_{\bigwedge^3 L} = 0$ is an irreducible symplectic fourfold of $K3^{[2]}$-type. This defines a 20-dimensional family with polarisation of non-split Beauville degree $2d = 22$: the lattice L_h is $2U \oplus 2E_8(-1) \oplus \begin{pmatrix} -2 & 1 \\ 1 & -6 \end{pmatrix}$.

Many authors have asked about the construction of geometrically meaningful compactifications of moduli spaces of polarised K3 surfaces. There are few general results known about this. For small degree, some results can be found in [118], [119], [123] and [115]. A partial compactification is discussed in [34] and there is an approach via log geometry in [104].

5. Compactifications

The spaces $\mathcal{F}_{2d}$ (defined by (2.16) above) and the other quotients of period domains described in Section 4 are complex analytic spaces by construction. We observed above that they are in fact quasi-projective varieties by the results of Baily and Borel, and the GIT moduli spaces of polarised K3 surfaces and of irreducible symplectic manifolds are quasi-projective by the general results of Viehweg. Nevertheless, we require projective models and preferably smooth, or nearly smooth, models also.

In this section we describe the most commonly used compactifications and we give some results about the singularities that arise. We begin by describing the class of spaces we wish to compactify.

5.1. Modular varieties of orthogonal type

As usual we let L be an integral lattice of signature $(2, n)$, $n \geqslant 3$, and consider the symmetric space

$$\mathcal{D}_L = \{x \in \mathbb{P}(L \otimes \mathbb{C}) \mid (x, x) = 0,\ (x, \bar{x}) > 0\}^+ \tag{5.1}$$

where the superscript $^+$ denotes a choice of one of the two connected components of Ω_L. We let Γ be a subgroup of finite index in $O^+(L)$. Any such Γ acts properly discontinuously on $\mathcal{D}_L$, but in general there are elements of finite order in Γ and they have fixed points in $\mathcal{D}_L$.

The quotient

$$\mathcal{F}_L(\Gamma) = \Gamma \backslash \mathcal{D}(L) \tag{5.2}$$

is called a *modular variety of orthogonal type* or *orthogonal modular variety*. In particular it is a locally symmetric variety, i.e. a variety that is the quotient of a symmetric space by a discrete group of automorphisms. It is not compact and is by its construction a complex analytic space: if Γ is torsion-free it is a complex manifold. In fact it is a quasi-projective variety by [5].

Some particularly important examples of orthogonal modular varieties are:

(a) the moduli spaces of polarised K3 surfaces (the signature is (2,19));
(b) the moduli spaces of lattice-polarised K3 surfaces (signature $(2, n)$, with $n < 19$);
(c) the moduli spaces of polarised abelian or Kummer surfaces (signature (2,3));
(d) the moduli space of Enriques surfaces (signature (2,10));
(e) quotients of the period domains of polarised irreducible symplectic varieties (signature (2,4), (2,5), (2,20) and (2,21) in the known cases).

For nearly all the orthogonal modular varieties $\mathcal{F}_L(\Gamma)$ that occur in this article, Γ is not torsion-free. The fixed points can lead to singularities of $\mathcal{F}_L(\Gamma)$. Since the stabiliser of any point of $\mathcal{D}_L$ is finite, the singularities are finite quotient singularities: that is, locally analytically they are isomorphic to a quotient of $\mathbb{C}^n$ by a finite subgroup G of $GL(n, \mathbb{C})$. They are not arbitrary finite quotient singularities, though, and we give some details about them in Section 5.6.

The quotient $\mathbb{C}^n/G$ may in fact be smooth, however. This happens, by a result of Chevalley [22], if and only if G is generated by quasi-reflections (see Definition 5.19). More importantly for us, the ramification divisors of $\mathcal{D}_L \to \mathcal{F}_L(\Gamma)$ are precisely the fixed loci of elements of Γ acting as quasi-reflections on the tangent space. In fact the only quasi-reflections that occur for orthogonal modular varieties are reflections. (Let us emphasise that here we are discussing the action of an element of G, not of Γ: being a reflection on the tangent space to $\mathcal{D}_L$ at a fixed point is not the same as being a reflection as an element of $O(L)$.)

The existence of elements acting as reflections, and thus of ramification divisors, is a significant feature of orthogonal modular varieties. For Siegel modular varieties (the symplectic group rather than the orthogonal group) there are no ramification divisors, except in the case of Siegel modular 3-folds. Siegel modular 3-folds, however, may also be regarded as orthogonal modular varieties because of the isogeny between Sp_2 and SO(2,3).

Differential forms on $\mathcal{F}_L(\Gamma)$ may be interpreted as modular forms for Γ: see Section 6.1 for more details. Therefore arithmetic information (modular forms) may be used to obtain geometric information about $\mathcal{F}_L(\Gamma)$. In particular we can use modular forms to decide whether $\mathcal{F}_L(\Gamma)$ is of general type, or more generally to try to determine its Kodaira dimension. If Y is a connected smooth projective variety of dimension n, the *Kodaira dimension* $\kappa(Y)$ of Y is defined by

$$\kappa(Y) = \text{tr.deg}\Big(\bigoplus_{k\geqslant 0} H^0(Y, kK_Y)\Big) - 1,$$

or $-\infty$ if $H^0(Y, kK_Y) = 0$ for all $k > 0$. Thus $h^0(Y, kK_Y) \sim k^{\kappa(Y)}$ for k sufficiently divisible. The possible values of $\kappa(Y)$ are $-\infty, 0, 1, \dots, n = \dim Y$, and we say Y is of *general type* if $\kappa(Y) = n$. Kodaira dimension is a bimeromorphic invariant so it makes sense to extend the definition to arbitrary irreducible quasi-projective varieties X by putting $\kappa(X) = \kappa(\widetilde{X})$ for $\widetilde{X}$ a desingularisation of a compactification of X.

With this in mind, we now turn to describing some algebraic compactifications of $\mathcal{F}_L(\Gamma)$, and the singularities that can occur. Much further information about compactifications of locally symmetric varieties (not all algebraic) may be found in the book [16], especially in [16, Part III], and the references there. However, the emphasis there is on the geometry and topology of symmetric and locally symmetric spaces as real manifolds.

We are interested in two kinds of compactification: the Baily-Borel compactification $\mathcal{F}_L(\Gamma)^*$ and the toroidal compactifications $\overline{\mathcal{F}_L(\Gamma)}$. We shall describe the construction and some of the properties of each.

Remark 5.3. The constructions can also be made if Γ is an arithmetic subgroup of $O^+(L \otimes \mathbb{Q})$; that is, if $\Gamma < O(L \otimes \mathbb{Q})$ and $\Gamma \cap O^+(L)$ is of finite index in both Γ and $O^+(L)$.

In some important ways the generalisation to arithmetic subgroups does not change things much. If we are willing to change the lattice, by Proposition 5.4 we can always assume that Γ is contained in O(L), so that we do not need rational entries in the matrices. In particular the results of Sections 5.5–5.7 still hold.

Proposition 5.4. *If $\Gamma < O(L \otimes \mathbb{Q})$ is arithmetic then there exists a lattice $M \subset L \otimes \mathbb{Q}$ such that $\Gamma < O(M)$.*

Proof. Let us consider g(L) for all $g \in \Gamma$. The index $[\Gamma : \Gamma \cap O(L)]$ is finite, therefore the number of different copies g(L) of L is finite. Therefore the $\mathbb{Z}$-module generated by the union of all g(L) ($g \in \Gamma$) is finitely generated. Denote this lattice by M. Then Γ is a subgroup of O(M) by the definition of M, and the quadratic form on M is induced by the quadratic form on L. If the quadratic form is not even integral, then we can make it integral taking a renormalisation by an integral constant c, i.e. we set $(u,v)_M = c(u,v)_L$. Doing so does not change the orthogonal group. It follows that Γ can be considered as a subgroup of $O(M)$ for some even integral lattice M. □

Notice, though, that the renormalisation by c does change the stable orthogonal group.

5.2. The Baily-Borel compactification

The Baily-Borel compactification, which in this context is often also referred to as the Satake compactification, can be defined very quickly as $\operatorname{Proj} \bigoplus M_k(\Gamma, 1)$, where $M_k(\Gamma,1)$ denotes the space of weight k modular forms with trivial character: see Definition 6.4. A priori, however, it is not clear that the ring of modular forms is finitely generated, nor that the modular forms separate points of $\mathcal{F}_L(\Gamma)$. Nor does that description immediately give a picture of the boundary $\mathcal{F}_L(\Gamma)^* \setminus \mathcal{F}_L(\Gamma)$. Instead the approach of Baily and Borel is to synthesise $\mathcal{F}_L(\Gamma)$ by topological and analytic methods, adding boundary components, and to show that the resulting space is a projective variety. Full details are given in [5], and a more detailed sketch than we give here may also be found in [16].

By writing $\mathcal{D}_L$ in the form (5.1) we have exhibited it as a Hermitian domain of type IV. As a Riemannian domain, $\mathcal{D}_L \cong SO_0(2,n)/SO(2) \times SO(n)$, where $SO_0(2,n)$ is the identity component. The Baily-Borel compactification is defined for any Hermitian symmetric space $\mathcal{D} = G/K$ (instead of $\mathcal{D}_L$) and for any quotient $X = \Gamma\backslash\mathcal{D}$ of $\mathcal{D}$ by an arithmetic group Γ. We describe the construction in general but we keep in mind the case $\mathcal{D} = \mathcal{D}_L$ as above.

An irreducible symmetric space G/K is Hermitian if and only if the centre of the maximal compact subgroup K has positive dimension: this explains why we consider only lattices of signature $(2,n)$, because $SO(m) \times SO(n)$ has discrete centre unless $n = 2$ or $m = 2$. Any Hermitian symmetric space of noncompact type can be embedded as a bounded symmetric domain in the holomorphic tangent space $T_K\mathcal{D}$ (the Harish-Chandra embedding). For these facts see [16, Prop. I.5.9] or [60].

The Baily-Borel compactification $\mathcal{D}^{BB}$ of $\mathcal{D}$ is simply the closure of $\mathcal{D}$ in the Harish-Chandra embedding, which in this case is the closure $\overline{\mathcal{D}}_L$ of $\mathcal{D}_L$ in $\mathbb{P}(L \otimes \mathbb{C})$; this in turn is contained in the compact dual, which in this case is the quadric

$$\check{\mathcal{D}} = \{x \in \mathbb{P}(L \otimes \mathbb{C}) \mid (x,x) = 0\}.$$

The tangent space to G/K at K is identified, as a complex manifold, with an open subset of $\check{\mathcal{D}}$.

A subset of $\mathcal{D}^{BB}$ is called a *boundary component* if it is an analytic arc component: that is, an equivalence class under the relation $x \sim y$ if there exist finitely many holomorphic maps $f_i\colon \Delta = \{z \in \mathbb{C} \mid |z| < 1\} \to \mathcal{D}^{BB}$ such that $x \in f_1(\Delta)$, $y \in f_k(\Delta)$ and $f_i(\Delta) \cap f_{i+i}(\Delta) \neq \emptyset$ for $1 \leqslant i < k$. The Baily-Borel compactification $\mathcal{D}^{BB}$

decomposes as a disjoint union of boundary components F_P, which are themselves symmetric spaces associated with certain parabolic subgroups p of G:

(5.5) $$\mathcal{D}^{BB} = \mathcal{D} \amalg \coprod_P F_P.$$

Not all parabolic subgroups occur, but only those associated with certain collections of strongly orthogonal roots. See [16, Section I.5] for precise details.

By construction, G acts on $\mathcal{D}^{BB}$. The normaliser of F_P,

(5.6) $$\mathcal{N}(F_P) = \{g \in G \mid g(F_P) = F_P\}$$

is a maximal parabolic subgroup of G (the parabolic subgroup P is in general not maximal). We shall later also need to consider the centraliser

(5.7) $$\mathcal{Z}(F_P) = \{g \in G \mid g|_{F_P} = \mathrm{id}\}.$$

To construct the Baily-Borel compactification of X – in our case, of $\mathcal{F}_L(\Gamma) = \Gamma\backslash\mathcal{D}_L$ – one must first restrict to rational boundary components.

Definition 5.8. A boundary component F is called a *rational boundary component* if

(i) the normaliser $\mathcal{N}(F)$ of F in G is a parabolic subgroup defined over $\mathbb{Q}$, and
(ii) the centraliser $\mathcal{Z}(F)$ contains a cocompact subgroup, normal in $\mathcal{N}(F)$, which is an algebraic subgroup defined over $\mathbb{Q}$.

A boundary component satisfying (i) is called a weakly rational boundary component. It is shown in [5, Theorem 3.7] that for the Baily-Borel compactification (ii) follows from (i), so that weakly rational boundary components are automatically rational.

G acts on $\mathcal{D}^{BB}$ but of course does not preserve rational boundary components. However Γ, being an arithmetic subgroup, does take rational boundary components to rational boundary components, so it acts on $\mathcal{D}^* := \mathcal{D} \amalg \coprod_{F_P \text{ rational}} F_P$. The effect of condition (ii) is that $\Gamma_P = \Gamma \cap \mathcal{N}(F_P)$ is again a discrete group. Moreover, F_P is again a Hermitian symmetric space and Γ_P is an arithmetic group acting on F_P.

We obtain the Baily-Borel compactification $(\Gamma\backslash\mathcal{D})^*$ by taking the quotient of $\mathcal{D}^*$ by the action of Γ. Each boundary component $\Gamma_P\backslash F_P$ has the structure of a complex analytic space and it is shown in [5] that these structures can be glued together to give an analytic structure on $(\Gamma\backslash\mathcal{D})^*$ extending the analytic structure on $\Gamma\backslash\mathcal{D}$.

It is at this stage that modular forms enter the picture. Each boundary component is an analytic space by construction but to show that their union is also an analytic space one must exhibit local analytic functions, and to show that the resulting space is projective we need analytic functions that separate points. Baily and Borel do this by using the Siegel domain realisation of $\mathcal{D}$ over a boundary, which we describe below (Section 8.2) for the cases we need here. For a more general description, see [114]. In these coordinates one may write down suitable series (Poincaré-Eisenstein series) that define modular forms having the required properties.

Theorem 5.9. *The Baily-Borel compactification $(\Gamma\backslash\mathcal{D})^*$ is an irreducible normal projective variety over $\mathbb{C}$. It contains $\Gamma\backslash\mathcal{D}$ (which in our case is $\mathcal{F}_L(\Gamma)$) as a Zariski-open subset, and may be decomposed as*

$$(\Gamma\backslash\mathcal{D})^* = \Gamma\backslash\mathcal{D} \amalg \coprod_P \Gamma_P\backslash X_P, \tag{5.10}$$

where P runs through representatives of Γ-equivalence classes of parabolic subgroups determining rational boundary components.

Although we defined rational boundary components in terms of $\mathcal{D}^{BB}$, they determine, according to condition (i) above, rational maximal parabolic subgroups of G. If G is simple, which will always be the case for us, the boundary components of $\mathcal{D}^{BB}$ correspond precisely to the maximal real parabolic subgroups: see [3, III.3, Proposition 3.9].

At least for the classical groups, the rational maximal parabolic subgroups can be described in combinatorial terms. In the case of $O(L)$ with L of signature $(2, n)$ they are the stabilisers of isotropic subspaces of $L \otimes \mathbb{Q}$. Because of the signature, such spaces have dimension 2 or 1 (or 0, corresponding to the "boundary" component $\Gamma\backslash\mathcal{D}_L$).Therefore we obtain the following description of $\mathcal{F}_L(\Gamma)^*$.

Theorem 5.11. *$\mathcal{F}_L(\Gamma)^*$ decomposes into boundary components as*

$$\mathcal{F}_L(\Gamma)^* = \mathcal{F}_L(\Gamma) \amalg \coprod_\Pi X_\Pi \amalg \coprod_\ell Q_\ell, \tag{5.12}$$

where ℓ and Π run through representatives of the finitely many Γ-orbits of isotropic lines and isotropic planes in $L \otimes \mathbb{Q}$ respectively. Each X_Π is a modular curve, each Q_ℓ is a point, and Q_ℓ is contained in the closure of X_Π if and only if the representatives may be chosen so that $\ell \subset \Pi$.

X_Π and Q_ℓ are usually referred to as 1-dimensional and 0-dimensional boundary components, or corank 1 and corank-2 boundary components. The boundary components of the Baily-Borel compactification are also known as the *cusps*.

5.3. Toroidal compactifications

Toroidal compactifications in general are described in the book [3]. They are made by adding a divisor at each cusp. Locally in the analytic topology near a cusp, the toroidal compactification is a quotient of an open part of a toric variety over the cusp: this variety is determined by a choice of admissible fan in a suitable cone, and the choices must be made so as to be compatible with inclusions among the closures of the Baily-Borel boundary components. A summary may be found in [3, III.5].

The case we are concerned with, of $O(2, n)$, is simpler than the general case because only the 0-dimensional cusps need any attention. However, we shall begin by describing the general theory, starting with $\mathcal{D} = G/K$ and an action of an arithmetic group Γ.

Let F be a boundary component: we may as well assume immediately that it is a rational boundary component. In general one has a description of $\mathcal{D}$ as a Siegel domain, an analytic open subset inside

$$\mathcal{D}(F) := F \times V(F) \times U(F)_{\mathbb{C}}. \tag{5.13}$$

In this decomposition, $U(F)$ is the centre of the unipotent radical $W(F)$ of $\mathcal{N}(F)$, the normaliser of F in G, and $V(F) \cong W(F)/U(F)$ is an abelian Lie group. This is not a holomorphic decomposition ($V(F)$ does not have a natural complex structure) but $U(F)_{\mathbb{C}}$ acts holomorphically on $\mathcal{D}(F)$ and in the diagram

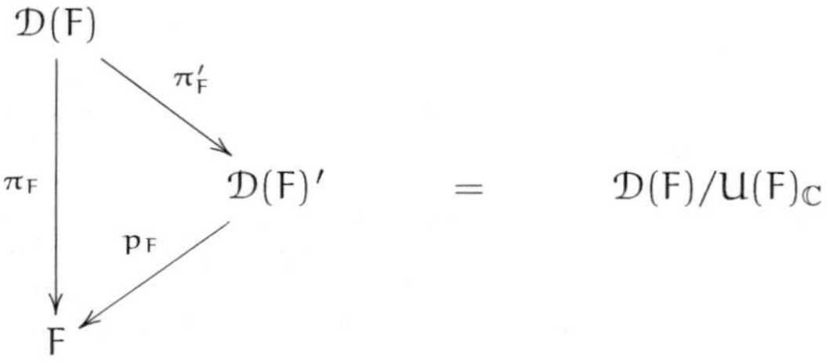

all the maps are holomorphic.

Now $\mathcal{D}$ is given by a tube domain condition: there is a cone $C(F) \subset U(F)$ such that

$$\mathcal{D} = \{x \in \mathcal{D}(F) \mid \mathrm{Im}(\mathrm{pr}_U(x)) \in C(F)\}. \tag{5.14}$$

where $\mathrm{pr}_U \colon \mathcal{D}(F) \to U(F)_{\mathbb{C}}$ is the projection map from the decomposition (5.13): this is a holomorphic map, even though (5.13) is not a holomorphic product decomposition.

In fact, there is a holomorphic product decomposition of $\mathcal{D}(F)$ which is (perhaps confusingly) similar:

$$\mathcal{D}(F) \cong U(F)_{\mathbb{C}} \times \mathbb{C}^k \times F, \tag{5.15}$$

where, of course, $k = \frac{1}{2}\dim_{\mathbb{R}} V(F)$ but $\mathbb{C}^k$ is not naturally identified with $V(F)$.

Denoting the map $x \in \mathcal{D}(F) \mapsto \mathrm{Im}(\mathrm{pr}_U(x)) \in U(F)$ by ϕ_F, as in [3], we have the diagram

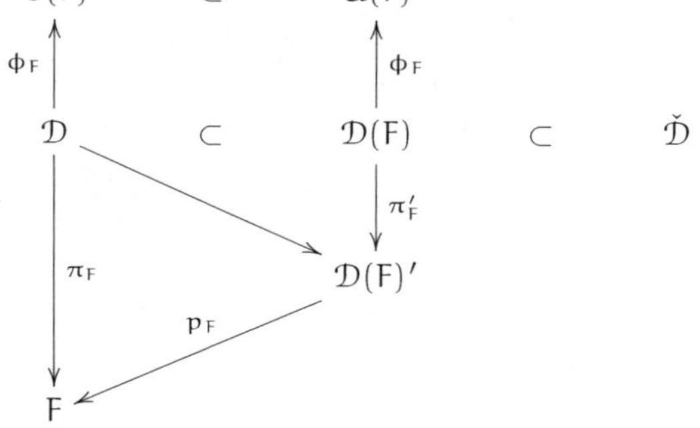

in which $\pi'_F\colon \mathcal{D}(F) \to \mathcal{D}(F)'$ and $p_F\colon \mathcal{D}(F)' \to F$ are principal homogeneous spaces for $U(F)_{\mathbb{C}}$ and $V(F)$ respectively.

When Γ acts, the group that acts on $\mathcal{D}(F)$ and on $\mathcal{D}$ is $\mathcal{N}(F)_{\mathbb{Z}} = \Gamma \cap \mathcal{N}(F)$, which is a discrete group because F is a rational boundary component. So, looking at the action of $U(F)_{\mathbb{Z}} = \Gamma \cap U(F)$, we get a principal fibre bundle

$$\mathcal{D}(F)/U(F)_{\mathbb{Z}} \longrightarrow \mathcal{D}(F)' \tag{5.16}$$

whose fibre is $T(F) = U(F)_{\mathbb{C}}/U(F)_{\mathbb{Z}}$, an algebraic torus over $\mathbb{C}$.

Toroidal compactification proceeds by replacing this torus with a toric variety $X_{\Sigma(F)}$ and taking the closure of $\mathcal{D}/U(F)_{\mathbb{Z}}$ in the $X_{\Sigma(F)}$-bundle over $\mathcal{D}(F)'$ that results. Doing this for each cusp F separately one can then take the quotients of each such $X_{\Sigma(F)}$ by $\mathcal{N}(F)_{\mathbb{Z}}$ and (under suitable conditions) glue the resulting pieces together by identifying the copies of $\mathcal{D}/\Gamma$ contained in each one.

In this process $\Sigma(F)$ is in general very far from unique. It is a fan in $C(F)$ (more precisely, of the convex hull of the rational points of the closure of $C(F)$), i.e. a decomposition of $C(F)$ into rational polyhedral cones (the integral structure is given by the lattice $U(F)_{\mathbb{Z}} \subset U(F)$), which is required to be $\mathcal{N}(F)_{\mathbb{Z}}$-equivariant and locally finite, but is not itself finite except in trivial cases: thus $X_{\Sigma(F)}$ is locally Noetherian, but not Noetherian.

In general, in order for the gluing procedure to work, the fans must satisfy a compatibility condition between fans for different cusps that arises when one cusp is in the closure of another, but in the case of $O(2, n)$ the condition is automatically satisfied. The reason is that the 1-dimensional cusps have $\dim_{\mathbb{R}} U(F) = 1$ and $C(F) = \mathbb{R}_+$, and the cone decomposition is therefore unique, and trivial. At the 0-dimensional cusps, in contrast, one has $\dim_{\mathbb{R}} U(F)$, so $\mathcal{D}$ is actually a tube domain in $\mathcal{D}(F)$: we describe this situation explicitly in Section 8.2.

At the 0-dimensional cusps, therefore, many different choices of compactification are possible. Below we shall choose one that suits our purpose.

In the end we need to take the quotient by $\mathcal{N}(F)_{\mathbb{Z}}$, not just $U(F)_{\mathbb{Z}}$. This has two consequences. First, this is why it is necessary to choose $\Sigma(F)$ and hence $X_{\Sigma(F)}$ in such a way that $N(F)_{\mathbb{Z}}$ still acts; secondly, even if $X_{\Sigma(F)}$ is chosen to be smooth, the action of $N(F)_{\mathbb{Z}}$ may reintroduce quotient singularities into the finished toroidal compactification. It is easy to choose $X_{\Sigma(F)}$ to be smooth, by the usual method of subdivision to resolve toric singularities.

Theorem 5.17. *A suitable choice of fans $\{\Sigma(F)\}$ for rational boundary components F determines a toroidal compactification $\overline{\mathcal{D}/\Gamma}$ of $\mathcal{D}/\Gamma$. This compactification may be chosen to be projective, and to have at worst finite quotient singularities.*

Proof. The only part not described above is the assertion that the compactification may be chosen to be projective. This is shown in [3, IV.2] with the extra assumption that Γ is neat, which is harmless because one may work with a neat normal subgroup

$\Gamma' < \Gamma$ of finite index and then use the Γ/Γ'-action. See also [31, V.5] for more details in the Siegel (symplectic group) case, in a more arithmetic framework. □

5.4. Canonical singularities

In this part, we give sufficient conditions for the moduli space or a suitable toroidal compactification of it to have canonical singularities. We outline the proof, from [46, Section 2], that orthogonal modular varieties of dimension $n \geqslant 9$ satisfy these conditions.

Definition 5.18. A normal complex variety X is said to have *canonical singularities* if it is $\mathbb{Q}$-Gorenstein and for some (hence any) resolution of singularities $f\colon \widetilde{X} \to X$ the discrepancy $\Delta = K_{\widetilde{X}} - f^*K_X$ is an effective Weil $\mathbb{Q}$-divisor.

Recall that X being $\mathbb{Q}$-Gorenstein means that for some $r \in \mathbb{N}$, if K_X is a canonical (Weil) divisor on X then rK_X is Cartier. Therefore f^*K_X makes sense: by definition it is the $\mathbb{Q}$-Cartier divisor $\frac{1}{r}f^*(rK_X)$.

$\Delta = \sum \alpha_i E_i$ is supported on the irreducible exceptional divisors E_i for f, so X has canonical singularities if and only if the rational numbers α_i are all non-negative. Equivalently, X has canonical singularities if and only if on any open set $U \subset X$, any pluricanonical form (i.e. section of rK_x for some r) on the smooth part of U extends holomorphically to the whole of $\widetilde{U}$. For more detail on canonical singularities, see [112].

A point $P \in X$ is said to be a canonical singularity if some neighbourhood X_0 of P has canonical singularities.

As we saw in Section 5.1, the singularities of $\mathcal{F}_L(\Gamma)$ are finite quotient singularities, arising at the images of points of $\mathcal{D}$ whose stabiliser in Γ is a non-trivial finite group. Any such action can be linearised locally [21]: we therefore consider the action of a finite subgroup $G < GL(n, \mathbb{C})$ on $\mathbb{C}^n$ and the singularities of the quotient $X = \mathbb{C}^n/G$. Any element $g \in G$ can be diagonalised since it is of finite order, and the eigenvalues of g are roots of unity.

Definition 5.19. An element g is a *quasi-reflection* if exactly one of the eigenvalues is different from 1. It is a *reflection* if that eigenvalue is -1.

For a cyclic subgroup $\langle g\rangle \subset GL(n, \mathbb{C})$ of finite order $m > 1$, we choose a primitive mth root of unity ζ (without loss of generality, $\zeta = e^{2\pi i/m}$) and we define the Reid-Tai sum

$$\Sigma(g) = \sum \left\{\frac{a_i}{m}\right\} \tag{5.20}$$

where the eigenvalues of g are ζ^{a_i} and $\{\}$ denotes the fractional part, $0 \leqslant \{q\} < 1$. For convenience we set $\Sigma(1) = 1$. The usual form of the Reid-Tai criterion is the following.

Proposition 5.21. *Suppose that* G *is a finite subgroup of* $GL(n, \mathbb{C})$ *containing no quasi-reflections. Then* $\mathbb{C}^n/G$ *has canonical singularities if and only if* $\Sigma(g) \geqslant 1$ *for all* $g \in G$.

This is sufficient if one wants to classify singularities, since any quotient singularity is isomorphic to a quotient singularity where there are no quasi-reflections. In our situation, the isotropy groups sometimes do contain quasi-reflections, so we want a version of the criterion that can be applied directly in that case. First, we state a lemma that will allow us to consider the elements of G one at a time.

Lemma 5.22. *Suppose* $G \subset GL(n, \mathbb{C})$ *is a finite group. If* $\mathbb{C}^n/\langle g\rangle$ *has canonical singularities for every* $g \in G$*, then* $\mathbb{C}^n/G$ *has canonical singularities.*

Proof. Let η be a form on $(\mathbb{C}^n/G)_{\mathrm{reg}}$ and let $\pi\colon \mathbb{C}^n \to \mathbb{C}^n/G$ be the quotient map. Then $\pi^*(\eta)$ is a G-invariant regular form on $\mathbb{C}^n \setminus \pi^{-1}(\mathbb{C}^n/G)_{\mathrm{sing}}$. Since $\pi^{-1}(\mathbb{C}^n/G)_{\mathrm{sing}}$ has codimension at least 2, the form $\pi^*(\eta)$ extends by Hartog's theorem to a G-invariant regular form on all of $\mathbb{C}^n$. Now the claim follows from [124, Proposition 3.1], which says that a G-invariant form on $\mathbb{C}^n$ extends to a desingularisation of $\mathbb{C}^n/G$ if and only if it extends to a desingularisation of $\mathbb{C}^n/\langle g\rangle$ for every $g \in G$. □

The converse of Lemma 5.22 is false.

Suppose that $g \in GL(n, \mathbb{C})$ is of order $m = sk$, where k is the smallest positive integer such that g^k is either a quasi-reflection or the identity. Order the eigenvectors so that the first $n-1$ eigenvalues of g^k are equal to 1, so that the last eigenvalue is a primitive sth root of unity. We define a modified Reid-Tai sum

$$\Sigma'(g) = \left\{\frac{sa_n}{m}\right\} + \sum \left\{\frac{a_i}{m}\right\} \tag{5.23}$$

where again the eigenvalues of g are ζ^{a_i}, and put $\Sigma'(1) = 1$. The idea (originating in an observation of Katharina Ludwig) is that this enables us to handle the quasi-reflections correctly.

Proposition 5.24. *Suppose that* G *is a finite subgroup of* $GL(n, \mathbb{C})$*. Then* $\mathbb{C}^n/G$ *has canonical singularities if* $\Sigma'(g) \geqslant 1$ *for all* $g \in G$.

Proof. If no power of g is a quasi-reflection then $s = 1$ and the usual Reid-Tai criterion (Proposition 5.21) shows that $\mathbb{C}^n/\langle g\rangle$ has canonical singularities.

Otherwise, consider g with $g^k = h$ a quasi-reflection as above. The eigenvalues of g are $\zeta^{a_1}, \ldots, \zeta^{a_n}$, where ζ is a primitive mth root of unity, $\mathrm{hcf}(s, a_n) = 1$ and $s | a_i$ for $i < n$. The group $\langle h\rangle$ is generated by quasi-reflections so $\mathbb{C}^n/\langle h\rangle \cong \mathbb{C}^n$, and we need to look at the action of the group $\langle g\rangle/\langle h\rangle$ on $\mathbb{C}^n/\langle h\rangle$. The eigenvalues of the differential of $g^l\langle h\rangle$ on $\mathbb{C}^n/\langle h\rangle$ are $\zeta^{la_1}, \ldots, \zeta^{la_{n-1}}, \zeta^{sla_n}$, so

$$\Sigma(g^l\langle h\rangle) = \Sigma'(g^l) \geqslant 1. \tag{5.25}$$

Thus $(\mathbb{C}^n/\langle h\rangle/\langle g\langle h\rangle\rangle \cong \mathbb{C}^n/\langle g\rangle$ has canonical singularities and the result follows by Lemma 5.22. □

5.5. Singularities of modular varieties

We are interested primarily in the singularities of $\mathcal{F}_{2d}$ and the other spaces mentioned in Section 4, but we may more generally consider the singularities of compactified locally symmetric varieties associated with the orthogonal group of a lattice of signature $(2, n)$. Unless n is small, it turns out that the compactification may be chosen to have canonical singularities.

Theorem 5.26. *Let L be a lattice of signature $(2, n)$ with $n \geqslant 9$, and let $\Gamma < O^+(L)$ be a subgroup of finite index. Then there exists a projective toroidal compactification $\overline{\mathcal{F}}_L(\Gamma)$ of $\mathcal{F}_L(\Gamma) = \Gamma\backslash\mathcal{D}_L$ such that $\overline{\mathcal{F}}_L(\Gamma)$ has canonical singularities and there are no branch divisors in the boundary. The branch divisors in $\mathcal{F}_L(\Gamma)$ arise from the fixed divisors of $\pm$reflections.*

There are three parts to the proof of this theorem. One must first consider the singularities of the open part, which means working out some details of the action of Γ on $\mathcal{D}_L$. Then there are the possible singularities over the 0-dimensional cusps: these lead to toric questions, and here one must choose the toroidal compactification appropriately. Finally, in order to deal with the 1-dimensional cusps we need a full description of the geometry there.

5.6. Singularities in the interior

Here we are interested in the singularities that arise at fixed points of the action of Γ on $\mathcal{D}_L$. Let $\boldsymbol{w} \in L_{\mathbb{C}}$ and let $G \subset \Gamma$ be the stabiliser of $[\boldsymbol{w}] \in \mathcal{D}_L$. For $[\boldsymbol{w}] \in \mathcal{D}_L$ we define $\mathbb{W} = \mathbb{C}\boldsymbol{w}$. Then G acts on $\mathbb{W}$, so $g(\boldsymbol{w}) = \alpha(g)\boldsymbol{w}$ for some character $\alpha\colon G \to \mathbb{C}^*$, and we put $G_0 = \ker\alpha$. We also put $S = (\mathbb{W} \oplus \overline{\mathbb{W}})^{\perp} \cap L$ (possibly $S = \{0\}$) and $T = S^{\perp} \subset L$. In the case of polarised K3 surfaces, S is the primitive part of the Picard lattice and T is the transcendental lattice of the surface corresponding to the period point $\boldsymbol{w}$.

It is easy to check that $S_{\mathbb{C}} \cap T_{\mathbb{C}} = \{0\}$ and that G acts on S and on T: moreover G_0 acts trivially on $T_{\mathbb{Q}}$.

Since $G/G_0 \subset \operatorname{Aut}\mathbb{W} \cong \mathbb{C}^*$ it is a cyclic group: we denote its order by $r_{\boldsymbol{w}}$. So by the above, $\mu_{r_w} \cong G/G_0$ acts on $T_{\mathbb{Q}}$. (By μ_r we mean the group of rth roots of unity in $\mathbb{C}$.)

For any $r \in \mathbb{N}$ there is a unique faithful irreducible representation of μ_r over $\mathbb{Q}$, which we call $\mathcal{V}_r$. The dimension of $\mathcal{V}_r$ is $\varphi(r)$, where φ is the Euler φ function and, by convention, $\varphi(1) = \varphi(2) = 1$. The eigenvalues of a generator of μ_r in this representation are precisely the primitive rth roots of unity: $\mathcal{V}_1$ is the 1-dimensional trivial representation. Note that $-\mathcal{V}_d = \mathcal{V}_d$ if d is even and $-\mathcal{V}_d = \mathcal{V}_{2d}$ if d is odd.

Using the fact that $S_{\mathbb{C}} \cap T_{\mathbb{C}} = 0$, we may check that $T_{\mathbb{Q}}$ splits as a direct sum of irreducible representations $\mathcal{V}_{r_w}$ (in particular, $\varphi(r_{\boldsymbol{w}}) | \dim T_{\mathbb{Q}}$) and that if $g \in G$ and $\alpha(g)$ is of order r (so $r | r_{\boldsymbol{w}}$), then $T_{\mathbb{Q}}$ splits as a g-module into a direct sum of irreducible representations $\mathcal{V}_r$ of dimension $\varphi(r)$.

We are interested in the action of G on the tangent space to $\mathcal{D}_L$. We have a natural isomorphism

$$T_{[w]}\mathcal{D}_L \cong \mathrm{Hom}(\mathbb{W}, \mathbb{W}^\perp/\mathbb{W}) =: V.$$

Suppose $g \in G$ is of order m and $\alpha(g)$ is of order r: as usual we take $\zeta = e^{2\pi i/m}$, and henceforth we think of g as an element of $GL(V)$, with eigenvalues $\zeta^{a_1}, \ldots, \zeta^{a_n}$.

If $\varphi(r)$ is not very small, the copy of $\mathcal{V}_r$ containing $\boldsymbol{w}$ already contributes at least 1 to $\Sigma(g)$. The cases $r = 1$ and $r = 2$ are also simple.

Proposition 5.27. *Assume that* $g \in G$ *does not act as a quasi-reflection on* V *and that* $\varphi(r) > 4$. *Then* $\Sigma(g) \geqslant 1$.

Proof. As ξ runs through the mth roots of unity, $\xi^{m/r}$ runs through the rth roots of unity. We denote by $k_1, \ldots, k_{\varphi(r)}$ the integers such that $0 < k_i < r$ and $(k_i, r) = 1$, in no preferred order. Without loss of generality, we assume $\alpha(g) = \zeta^{mk_2/r}$ and $\overline{\alpha(g)} = \alpha(g)^{-1} = \zeta^{mk_1/r}$, with $k_1 \equiv -k_2 \bmod r$.

One of the $\mathbb{Q}$-irreducible subrepresentations of g on $L_\mathbb{C}$ contains the eigenvector $\boldsymbol{w}$: we call this $\mathbb{V}_r^{\boldsymbol{w}}$ (it is the smallest g-invariant complex subspace of $L_\mathbb{C}$ that is defined over $\mathbb{Q}$ and contains $\boldsymbol{w}$). It is a copy of $\mathcal{V}_r \otimes \mathbb{C}$: to distinguish it from other irreducible subrepresentations of the same type we write $\mathbb{V}_r^{\boldsymbol{w}} = \mathcal{V}_r^{\boldsymbol{w}} \otimes \mathbb{C}$.

If $\boldsymbol{v}$ is an eigenvector for g with eigenvalue $\zeta^{mk_i/r}$, $i \neq 1$ (in particular $\boldsymbol{v} \notin \overline{\mathbb{W}}$), then $\boldsymbol{v} \in \mathbb{W}^\perp$ since $(\boldsymbol{v}, \boldsymbol{w}) = (g(\boldsymbol{v}), g(\boldsymbol{w})) = \zeta^{mk_i/r}\alpha(g)(\boldsymbol{v}, \boldsymbol{w})$. Therefore the eigenvalues of g on $\mathbb{V}_r^{\boldsymbol{w}} \cap \mathbb{W}^\perp/\mathbb{W}$ include $\zeta^{mk_i/r}$ for $i \geqslant 3$, so the eigenvalues on $\mathrm{Hom}(\mathbb{W}, \mathbb{V}_r^{\boldsymbol{w}} \cap \mathbb{W}^\perp/\mathbb{W}) \subset V$ include $\zeta^{mk_1/r}\zeta^{mk_i/r}$ for $i \geqslant 3$. So

$$\Sigma(g) \geqslant \sum_{i=3}^{\varphi(r)} \left\{ \frac{k_1}{r} + \frac{k_i}{r} \right\} \geqslant 1, \tag{5.28}$$

where the last inequality is an elementary verification. □

The cases $r = 1$ and $r = 2$ are simple to deal with because one always finds two conjugate eigenvalues, which between them contribute 1 to $\Sigma(g)$.

So far we have needed no hypothesis on the dimension, but the remaining cases ($r = 3, 4, 5, 6, 8, 10$ or 12) do require such a condition because the contributions to $\Sigma(g)$ from each $\mathcal{V}_r$ are small and we need to have enough of them. We refer to [46] for details. In the end we find

Theorem 5.29. *Assume that* $g \in G$ *does not act as a quasi-reflection on* V *and that* $n \geqslant 6$. *Then* $\Sigma(g) \geqslant 1$.

One must then carry out a similar analysis for quasi-reflections. One more dimension is needed to guarantee $\Sigma'(g) \geqslant 1$, because the a_n term does not help us. The analysis also has the corollary that the quasi-reflections in the tangent space V that arise are in this case always reflections, and moreover that the elements of Γ that they come from are themselves (up to sign) reflections, considered as elements of $O(L)$.

Corollary 5.30. *If* $n \geqslant 7$ *then* $\mathcal{F}_L(\Gamma)$ *has canonical singularities.*

5.7. Singularities at the cusps

We now consider the boundary $\overline{\mathcal{F}}_L(\Gamma)\setminus\mathcal{F}_L(\Gamma)$. Cusps, or boundary components in the Baily-Borel compactification, correspond to orbits of totally isotropic subspaces $E \subset L_{\mathbb{Q}}$. Since L has signature $(2, n)$, the dimension of E is 1 or 2, corresponding to dimension 0 and dimension 1 boundary components respectively.

A toroidal compactification over a cusp F coming from an isotropic subspace E corresponds to an admissible fan Σ in some cone $C(F) \subset U(F)$. We have, as in [3]

$$\mathcal{D}_L(F) := U(F)_{\mathbb{C}}\mathcal{D}_L \subset \check{\mathcal{D}}_L$$

where $\check{\mathcal{D}}_L$ is the compact dual of $\mathcal{D}_L$ (see [3, III.2]).

The case $\dim E = 1$, that is, isotropic vectors in L, is the case of 0-dimensional cusps in the Baily-Borel compactification and leads to a purely toric problem. In this case we have

$$\mathcal{D}_L(F) \cong F \times U(F)_{\mathbb{C}} = U(F)_{\mathbb{C}}.$$

Put $M(F) = U(F)_{\mathbb{Z}}$ and define the torus $T(F) = U(F)_{\mathbb{C}}/M(F)$. In general $(\mathcal{D}_L/M(F))_\Sigma$ is by definition the interior of the closure of $\mathcal{D}_L/M(F)$ in $\mathcal{D}_L(F)/M(F) \times_{T(F)} X_\Sigma(F)$, i.e. in $X_\Sigma(F)$ in this case, where $X_\Sigma(F)$ is the torus embedding corresponding to the torus $T(F)$ and the fan Σ. We may choose Σ so that $X_\Sigma(F)$ is smooth and $G(F) := N(F)_{\mathbb{Z}}/U(F)_{\mathbb{Z}}$ acts on $(\mathcal{D}_L/M(F))_\Sigma$: this is also implicit in [3] and explained in [31, p.173]. The toroidal compactification is locally isomorphic to $X_\Sigma(F)/G(F)$. Thus the problem of determining the singularities is reduced to a question about toric varieties, which is answered by Theorem 5.31, below.

We take a lattice M of dimension n and denote its dual lattice by N. A fan Σ in $N \otimes \mathbb{R}$ determines a toric variety X_Σ with torus $T = \mathrm{Hom}(M, \mathbb{C}^*) = N \otimes \mathbb{C}^*$.

Theorem 5.31. *Let* X_Σ *be a smooth toric variety and suppose that a finite group* $G < \mathrm{Aut}(T) = \mathrm{GL}(M)$ *of torus automorphisms acts on* X_Σ*. Then* X_Σ/G *has canonical singularities.*

Proof. This is [46, Theorem 2.17]. The proof also shows (with a little modification) that there are no branch divisors contained in the boundary over 0-dimensional cusps either. □

It remains to consider the dimension 1 cusps. We consider a rank 2 totally isotropic subspace $E_{\mathbb{Q}} \subset L_{\mathbb{Q}}$, corresponding to a dimension 1 boundary component F of $\mathcal{D}_L$. The idea is to choose standard bases for $L_{\mathbb{Q}}$ so as to be able to identify $U(F)$, $U(F)_{\mathbb{Z}}$ and $N(F)_{\mathbb{Z}}$ explicitly, as is done in [115] for maximal K3 lattices, where $n = 19$. Then, following Kondo [74] one can analyse the group action in coordinates, using the Siegel domain realisation of $\mathcal{D}$ associated with the given cusp. Both in [74] and in [115] there are special features that allow one to work over $\mathbb{Z}$, but in general one must work over $\mathbb{Q}$. For details we refer to [46].

6. Modular forms and Kodaira dimension

One of the main tools in the study of the geometry of the orthogonal modular varieties $\mathcal{F}_L(\Gamma)$ is the theory of modular forms with respect to an orthogonal group of type $O(2, n)$. One application is to prove that $\mathcal{F}_L(\Gamma)$ is often of general type. The methods described here were used in [46] to prove the following result.

Theorem 6.1. *The moduli space $\mathcal{F}_{2d}$ of K3 surfaces with a polarisation of degree $2d$ is of general type for any $d > 61$ and for $d = 46, 50, 52, 54, 57, 58$ and 60.*

If $d \geqslant 40$ and $d \neq 41, 44, 45$ or 47 then the Kodaira dimension of $\mathcal{F}_{2d}$ is non-negative.

Similar methods apply to irreducible symplectic manifolds and their polarisations, discussed in Section 3.3. For deformations of length 2 Hilbert schemes of K3 surfaces with polarisation of split type (see Equation (3.6)) there is the following result, from [50].

Theorem 6.2. *The variety $\mathcal{M}_{2d}^{[2],split}$ is of general type if $d \geqslant 12$. Moreover its Kodaira dimension is non-negative if $d = 9$ and $d = 11$.*

For the ten-dimensional O'Grady case [99], there are again split and non-split polarisations, and a fairly complete general type result in the split case was proved in [51].

Theorem 6.3. *Let d be a positive integer not equal to 2^n with $n \geqslant 0$. Then every component of the moduli space of ten-dimensional polarised O'Grady varieties with split polarisation h of Beauville degree $h^2 = 2d \neq 2^{n+1}$ is of general type.*

We do not attempt to prove Theorem 6.3 here but the theory we develop in the rest of this article will give proofs (though not with full details) of Theorem 6.1 and Theorem 6.2.

6.1. Modular forms of orthogonal type

In Definition 6.4 below we follow [11]. An "affine" definition similar to the one usually given for of SL(2) can be found in [40]. The *affine cone* over $\mathcal{D}_L$ is $\mathcal{D}_L^\bullet = \{y \in L \otimes \mathbb{C} \mid x = \mathbb{C}^* y \in \mathcal{D}_L\}$.

Definition 6.4. Suppose that L has signature $(2, n)$, with $n \geqslant 3$. Let $k \in \mathbb{Z}$ and let $\chi\colon \Gamma \to \mathbb{C}^*$ be a character of a subgroup $\Gamma < O^+(L)$ of finite index. A holomorphic function $F\colon \mathcal{D}_L^\bullet \to \mathbb{C}$ is called a *modular form* of *weight* k and character χ for the group Γ if

$$F(tZ) = t^{-k} F(Z) \quad \forall t \in \mathbb{C}^*,$$

$$F(gZ) = \chi(g) F(Z) \quad \forall g \in \Gamma.$$

A modular form is called a *cusp form* if it vanishes at every cusp.

The weight as defined here is what is sometimes called *arithmetic weight*. Some authors prefer to use the geometric weight, which is k/n, normally only in contexts where $n|k$. We shall always use the arithmetic weight. One may choose a complex volume form dZ on $\mathcal{D}_L$ such that if F is a modular form of weight mn and character $\det^m$ for Γ then $F(dZ)^m$ is a Γ-invariant section of $mK_{\mathcal{D}_L}$: see [7] for a precise account.

If $n < 3$ then one has to add to Definition 6.4 the condition that F is holomorphic at the boundary. According to Koecher's principle (see [4], [32], [107]) this condition is automatically fulfilled if the dimension of a maximal isotropic subspace of $L \otimes \mathbb{Q}$ is smaller than n. In particular, this is always true if $n \geqslant 6$.

We denote the linear spaces of modular and cusp forms of weight k and character χ by $M_k(\Gamma, \chi)$ and $S_k(\Gamma, \chi)$ respectively. If $M_k(\Gamma, \chi)$ is nonzero then one knows that $k \geqslant (n-2)/2$ (see [40]). The minimal weight $k = (n-2)/2$ is called *singular*. Modular forms of singular weight are very special. The first example of such forms for orthogonal groups was constructed in [38]. Cusp forms are possible only if $k > (n-2)/2$. The weight $k = \dim(\mathcal{F}_L(\Gamma))$ is called *canonical* because by a lemma of Freitag

$$S_n(\Gamma, \det) \cong H^0(\widetilde{\mathcal{F}}_L(\Gamma), K_{\widetilde{\mathcal{F}}_L(\Gamma)}),$$

where $\widetilde{\mathcal{F}}_L(\Gamma)$ is a smooth compact model of the modular variety $\mathcal{F}_L(\Gamma)$ and $K_{\widetilde{\mathcal{F}}_L(\Gamma)}$ is the sheaf of canonical differential forms (see [32, Hilfssatz 2.1, Kap. 3]). Therefore we have the following important formula for the geometric genus of the modular variety:

$$p_g(\widetilde{\mathcal{F}}_L(\Gamma)) = \dim S_n(\Gamma, \det). \tag{6.5}$$

Remark 6.6. Below (see Section 8.3) we describe the property of being holomorphic at the boundary (needed only if $n \leqslant 2$) in terms of the Fourier expansions.

Remark 6.7. In this article we usually assume that $n \geqslant 3$. In this case the order of any character χ in Definition 6.4 is finite according to Kazhdan's property (T) (see [69]).

Remark 6.8. If the lattice L contains two orthogonal copies of the hyperbolic plane $U \cong \left(\begin{smallmatrix} 0 & 1 \\ 1 & 0 \end{smallmatrix}\right)$ and if its reduction modulo 2 (respectively 3) is of rank at least 6 (respectively 5) then $\widetilde{O}^+(L)$ has only one non-trivial character, namely det (see [49]). In particular the modular group $\widetilde{O}^+(L_{2d})$ related to the polarised K3 surfaces has only one non-trivial character.

Remark 6.9. If $L_{2t}^{(5)} = 2U \oplus \langle -2t \rangle$, of signature $(2,3)$, then the modular forms with respect to $\widetilde{SO}^+(L_{2t}^{(5)})$ coincide with Siegel modular forms with respect to the paramodular group Γ_t (see [39], [43], [53], [55]). In particular, if $t = 1$ we obtain the Siegel modular forms with respect to $Sp_2(\mathbb{Z})$. In contrast to Remark 6.8 the group $\widetilde{SO}^+(L_{2t}^{(5)})$ has non-trivial characters. They were described in [44, Section 2]. One can construct important cusp forms of the minimal possible weight 1 with non-trivial character for the full modular group $\widetilde{\mathrm{SO}}^+(L_{2t}^{(5)}) \cong \Gamma_t/\{\pm 1\}$ for some t (see [55]).

6.2. Rational quadratic divisors

For any $v \in L \otimes \mathbb{Q}$ such that $v^2 = (v, v) < 0$ we define the *rational quadratic divisor*

$$\mathcal{D}_v = \mathcal{D}_v(L) = \{[Z] \in \mathcal{D}_L \mid (Z, v) = 0\} \cong \mathcal{D}_{v_L^\perp} \tag{6.10}$$

where $v_L^\perp$ is an even integral lattice of signature $(2, n-1)$. Therefore $\mathcal{D}_v$ is also a homogeneous domain of type IV. We note that $\mathcal{D}_v(L) = \mathcal{D}_{tv}(L)$ for any $t \neq 0$. The theory of automorphic Borcherds products (see [13]) gives a method of constructing automorphic forms with rational quadratic divisors. Special divisors of this type (the reflective divisors defined below) play an important role in the theory of moduli spaces.

The reflection with respect to the hyperplane defined by a non-isotropic vector r is given by

$$\sigma_r \colon l \longmapsto l - \frac{2(l, r)}{(r, r)} r. \tag{6.11}$$

If r is primitive in L and the reflection σ_r fixes L, i.e. $\sigma_r \in O(L)$, then we say that r is a *reflective vector*, also known as a *root*. If $(r, r) = d$ we say that r is a d-vector or (if it is a root) a d-root. A 2-vector or a -2-vector is always a root.

If $v \in L^\vee$ and $(v, v) < 0$, the divisor $\mathcal{D}_v(L)$ is called a *reflective divisor* if $\sigma_v \in O(L)$. It was proved in [46, Corollary 2.13] that for $n \geqslant 3$ the branch divisor of the modular projection

$$\pi_\Gamma \colon \mathcal{D}_L \to \Gamma \setminus \mathcal{D}_L$$

is the union of the reflective divisors with respect to Γ:

$$\mathrm{Bdiv}(\pi_\Gamma) = \bigcup_{\mathbb{Z}r \subset L,\ \sigma_r \in \Gamma \cup -\Gamma} \mathcal{D}_r(L). \tag{6.12}$$

Note that here we have to allow r such that $-\sigma_r \in \Gamma$ as well as those with $\sigma_r \in \Gamma$: compare Remark 6.16 below, concerning modular forms.

6.3. Low weight cusp form trick

The next theorem, proved in [46, Theorem 1.1], is called the *low weight cusp form trick*. It plays a crucial role in the application of modular forms to moduli problems. If F is a modular form (of any weight or character) then the *divisor* div F in $\mathcal{D}_L$ is given by the equation $F(Z) = 0$: this is well-defined in view of Definition 6.4.

Theorem 6.13. *Let L be an integral lattice of signature $(2, n)$, $n \geqslant 9$. The modular variety $\mathcal{F}_L(\Gamma)$ is of general type if there exists a non-zero cusp form $F_a \in S_a(\Gamma, \chi)$ of small weight $a < n$ vanishing with order at least 1 at infinity such that* $\operatorname{div} F_a \geqslant \mathrm{Bdiv}(\pi_\Gamma)$.

Proof. We let $\overline{\mathcal{F}}_L(\Gamma)$ be a projective toroidal compactification of $\mathcal{F}_L(\Gamma)$ with canonical singularities and no ramification divisors at infinity, which exists by Theorem 5.26. We take a smooth projective model Y of $\mathcal{F}_L(\Gamma)$ by taking a resolution of singularities of $\overline{\mathcal{F}}_L(\Gamma)$. We want to show the existence of many pluricanonical forms on Y.

Suppose that $F_{nk} \in M_{nk}(\Gamma, \det^k)$. By choosing a 0-dimensional cusp we may realise $\mathcal{D}_L$ as a tube domain (see Section 8.2 for details) and use this to select a holomorphic volume element dZ. Then the differential form $\Omega(F_{nk}) = F_{nk}\,(dZ)^k$ is Γ-invariant and therefore determines a section of the pluricanonical bundle $kK = kK_Y$ away from the branch locus of $\pi\colon \mathcal{D}_L \to \mathcal{F}_L(\Gamma)$ and the cusps: see [3, p. 190] (but note that weight 1 in the sense of [3] corresponds to weight n in our definition).

In general $\Omega(F_{nk})$ will not extend to a global section of kK. We distinguish three kinds of obstruction to its doing so. There are *elliptic obstructions,* arising because of singularities given by elliptic fixed points of the action of Γ; *reflective obstructions,* arising from the ramification divisors in $\mathcal{D}_L$; and *cusp obstructions,* arising from divisors at infinity.

In order to deal with these obstructions we consider a neat normal subgroup Γ' of Γ of finite index and set $G := \Gamma/\Gamma'$. Let $X := \mathcal{F}_L(\Gamma')$ and let $\overline{X} := \overline{\mathcal{F}}_L(\Gamma')$ be the toroidal compactification of $\mathcal{F}_L(\Gamma')$ given by the same choice of fan as for $\overline{\mathcal{F}}_L(\Gamma)$. Then $\overline{X}$ is a smooth projective manifold with $\overline{\mathcal{F}}_L(\Gamma) = \overline{X}/G$. Let $D := \overline{X} \setminus X$ be the boundary divisor of $\overline{X}$. For any element $g \in G$ we define its fixed locus $\overline{X}^g := \{x \in \overline{X} \mid g(x) = x\}$ and denote its divisorial part by $\overline{X}^g_{(1)}$. Then $R := \bigcup_{g \neq 1} \overline{X}^g_{(1)}$ is the ramification divisor of the map $\pi\colon \overline{X} \to \overline{X}/G$.

The results of Section 5.5 (see Theorem 5.26 and Theorem 5.31) can be summarised as follows:

(i) R does not contain a component of D;
(ii) the ramification index of $\pi\colon \overline{X} \to \overline{X}/G$ along R is 2;
(iii) $\overline{X}/G$ has canonical singularities.

We will now apply the low-weight cusp form trick, used for example in [32] (for Siegel modular forms), [40], [42] and [57]. The main point is to use special cusp forms. For this let the order of χ be N and assume that k is a multiple of $2N$. Then we consider forms $F^0_{nk} \in S_{nk}(\Gamma, \det^k) = S_{nk}(\Gamma, 1)$ of the form

$$F^0_{nk} = F^k_a F_{(n-a)k}$$

where $F_{(n-a)k} \in M_{(n-a)k}(\Gamma, 1)$ is a modular form of weight $(n-a)k \geqslant k$. We claim that the corresponding forms $\Omega(F^0_{nk})$ give rise to pluricanonical forms on Y. To see this, we deal with the three kinds of obstruction in turn.

Cusp obstructions. By definition, $\Omega(F^0_{nk})$ is a G-invariant holomorphic section of kK_X. Since F_a is a cusp form of weight $a < n$, the form F^0_{nk} has zeroes of order k along the boundary D and hence extends to a G-invariant holomorphic section of $kK_{\overline{X}}$ by [3, IV.1, Theorem 1.1].

Reflective obstructions. Since $R \subset \operatorname{div}(F_a)$ by assumption, $\Omega(F^0_{nk})$ has zeroes of order k on $R \setminus D$. By (i) above, $\Omega(F^0_{nk})$ actually has zeroes of order k along all of R. By (ii) the form $\Omega(F^0_{nk})$ descends to a holomorphic section of $kK_{(\overline{X}/G)_{\text{reg}}}$ where $(\overline{X}/G)_{\text{reg}}$ is the regular part of $\overline{X}/G$.
Elliptic obstructions. By (iii) the form $\Omega(F^0_{nk})$ extends to a holomorphic section of kK_Y.

Therefore $F^k_a M_{(n-a)k}(\Gamma,1)$ is a subspace of $H^0(Y, kK_Y)$. The theorem now follows because by Hirzebruch-Mumford proportionality (see [94]), $\dim M_{(n-a)k}(\Gamma,1)$ grows like k^n. □

Remark 6.14. There is another way to deal with the reflective obstructions, which works even if a cusp form with the right properties cannot be found. Among forms of very high weight there must be some that vanish along the reflective divisors, because $\dim M_k(\Gamma,1)$ grows faster with k than the space of obstructions, which are sections in some bundles on the reflective divisors. In [48] we estimate these dimensions using Hirzebruch-Mumford proportionality. This method can be used to produce general type results even in cases where special forms constructed by quasi pull-back are not available, but if the quasi pull-back method is available it normally produces much stronger results.

6.4. Reflective modular forms

For Theorem 6.13 we used cusp forms of low weight ($k < n$) with large divisor ($\operatorname{div} F \geqslant \operatorname{Bdiv}(\pi_\Gamma)$). We construct such modular forms for the moduli spaces of polarised K3 surfaces and other holomorphic symplectic varieties in Section 8. Modular forms of high weight ($k \geqslant n$) with small divisor ($\operatorname{div} F \leqslant \operatorname{Bdiv}(\pi_\Gamma)$) also have applications to the theory of moduli spaces such as Theorem 6.19 below.

Definition 6.15. A modular form $F \in M_k(\Gamma,\chi)$ is called *reflective* if

$$\operatorname{Supp}(\operatorname{div} F) \subset \bigcup_{\mathbb{Z}r \subset L,\ \sigma_r \in \Gamma \cup -\Gamma} \mathcal{D}_r(L) = \operatorname{Bdiv}(\pi_\Gamma).$$

We call F *strongly reflective* if the multiplicity of any irreducible component of $\operatorname{div} F$ is equal to one.

Remark 6.16. In the definition of reflective modular forms given in [54] only the condition $\sigma_r \in \Gamma$ was considered. The present definition, allowing $-\sigma_r \in \Gamma$, is explained by equation (6.12).

Example 6.17. The most famous example of a strongly reflective modular form is the *Borcherds modular form* $\Phi_{12} \in M_{12}(O^+(II_{2,26}), \det)$ (see [11]). This is the unique modular form of singular weight 12 with character det with respect to the orthogonal group $O^+(II_{2,26})$ of the even unimodular lattice $II_{2,26} \cong 2U \oplus 3E_8(-1)$ of signature $(2,26)$. The form Φ_{12} is the Kac-Weyl-Borcherds denominator function of the Fake

Monster Lie algebra. For any (-2)-vector $r \in II_{2,26}$ we have

$$\Phi_{12}(\sigma_r(Z)) = \det(\sigma_r)\Phi_{12}(Z) = -\Phi_{12}(Z).$$

Therefore Φ_{12} vanishes along $\mathcal{D}_r(II_{2,26})$. According to [11] the order of vanishing is 1 and the full divisor of this modular form is the union of the mirrors of such reflections:

$$\mathrm{div}_{\mathcal{D}(II_{2,26})}\,\Phi_{12} = \sum_{\substack{\pm r\in II_{2,26}\\ r^2=-2}} \mathcal{D}_r(II_{2,26}).$$

According to Eichler's criterion (see Lemma 7 8) all (-2)-vectors of $II_{2,26}$ constitute one $\widetilde{O}^+(II_{2,26})$-orbit. In other words, the ramification divisor of the 26-dimensional modular variety $\mathcal{F}_{II_{2,26}}(\widetilde{O}^+(II_{2,26}))$ is irreducible.

Remark 6.18. Modular forms of weight n (known as canonical weight) have special properties. Suppose L has signature $(2, n)$ and $F \in M_n(\Gamma, \det)$. If $\sigma_r \in \Gamma$, then $F(\sigma_r(Z)) = -F(Z)$. Hence F vanishes along $\mathcal{D}_r(L)$. If $-\sigma_r \in \Gamma$, then

$$(-1)^n F(\sigma_r(Z)) = F((-\sigma_r)(Z)) = \det(-\sigma_r)F(Z) = (-1)^{n+1}F(Z)$$

and F also vanishes along $\mathcal{D}_r(L)$. Therefore any Γ-modular form of canonical weight with character det vanishes along $\mathrm{Bdiv}(\pi_\Gamma)$.

If $S_n(\Gamma, \det) \neq 0$ then the Kodaira dimension of $\overline{\mathcal{F}}_L(\Gamma)$ is non-negative (with no restriction needed on the dimension n), because if $F_n \in S_n(\Gamma, \det)$ then by Freitag's lemma $F_n(Z)dZ$ defines an element of $H^0(\overline{\mathcal{F}}_L(\Gamma), K_{\overline{\mathcal{F}}_L(\Gamma)})$. Therefore $p_g(\overline{\mathcal{F}}_L(\Gamma)) \geqslant 1$ and the plurigenera do not all vanish.

The next theorem was proved in [41] and contrasts with Theorem 6.13.

Theorem 6.19. *Suppose that L has signature $(2, n)$, with $n \geqslant 3$. Let $F_k \in M_k(\Gamma, \chi)$ be a strongly reflective modular form of weight k and character χ for a subgroup $\Gamma < O^+(L)$ of finite index. Then*

$$\kappa(\mathcal{F}_L(\Gamma)) = -\infty \tag{6.20}$$

if $k > n$, or if k and F_k is not a cusp form. If k and F_n is a cusp form whose order of zero at infinity is at least 1 then

$$\kappa(\Gamma_\chi\backslash\mathcal{D}_L) = 0, \tag{6.21}$$

where $\Gamma_\chi = \ker(\chi \cdot \det)$ is a subgroup of Γ.

Proof. The requirement that the cusp form should have order of vanishing at least 1 is almost always satisfied: see [49].

To prove (6.20) we have to show that there are no pluricanonical differential forms on $\overline{\mathcal{F}}_L(\Gamma)$. Any such differential form can be obtained using a modular form (see the proof of Theorem 6.13). The differential form $F_{nm}(dZ)^m$ is Γ-invariant and it determines a section of the pluricanonical bundle mK over a smooth open part of the modular variety away from the branch locus of $\pi\colon \mathcal{D}_L \to \mathcal{F}_L(\Gamma)$ and the cusps. In the proof of Theorem 6.13 we indicated three kinds of obstruction to extending $F_{nm}(dZ)^m$ to a global section of mK. In the proof of this theorem we use the reflective obstruction, arising from the ramification divisor in $\mathcal{D}_L$ by $\pm$ reflections in Γ (see Equation (6.12)). Therefore if F_{nm} determines a global section then F_{nm} has zeroes of order at least m on $\mathrm{Bdiv}(\pi_\Gamma)$. The modular form $F_k \in M_k(\Gamma, \chi)$ is strongly reflective of weight $k \geqslant n$. Hence F_{nm}/F_k^m is a holomorphic modular form of weight $m(n-k) \leqslant 0$. According to Koecher's principle ($n \geqslant 3$) this function is constant. Therefore $F_{nm} \equiv 0$ if $k > n$ or $F_{nm} = C \cdot F_n^m$ if k. If the strongly reflective form F_n is non-cuspidal of weight n, then $F_n^m(dZ)^{\otimes m}$ cannot be extended to the compact model because of cusp obstructions (F_n^m should have zeroes of order at least m along the boundary). If F_n is a cusp form of weight k then we can consider F_n as a cusp form with respect to the subgroup Γ_χ.

Then $F_n(Z)\,dZ$ is Γ_χ-invariant and, according to Freitag's lemma, it can be extended to a global section of the canonical bundle $\Omega_{\overline{\mathcal{F}}_L(\Gamma_\chi)}$ for any smooth compact model $\overline{\mathcal{F}}_L(\Gamma_\chi)$ of $\mathcal{F}_L(\Gamma_\chi)$. Moreover Koecher's principle shows that any m-pluricanonical form is equal, up to a constant, to $F_n^m(dZ)^{\otimes m}$, proving (6.21). The strongly reflective cusp form of canonical weight determines essentially the unique m-pluricanonical differential form on $\mathcal{F}_L(\Gamma_\chi)$. □

We can apply Theorem 6.19 to find examples of moduli spaces of lattice-polarised K3 surfaces having $\kappa = -\infty$ and $\kappa = 0$: see [41]. Meanwhile a stronger statement concerning uniruledness of orthogonal varieties has been proven, see [45].

7. Orthogonal groups and reflections

The material in this and subsequent sections is not so easily found in the literature, so from here on we shall give slightly more detail.

For applications, the most important subgroups of $O(L)$ are the stable orthogonal groups $\widetilde{O}(L)$, $\widetilde{O}^+(L)$ and $\widetilde{SO}^+(L)$, as defined in Equations (2.14) and (2.15). The reason for using the word "stable" to describe $\widetilde{O}(L)$ is the following property.

Lemma 7.1. *For any sublattice S of a lattice L the group $\widetilde{O}(S)$ can be considered as a subgroup of $\widetilde{O}(L)$.*

Proof. Let $S^\perp$ be the orthogonal complement of S in L. We have

$$S \oplus S^\perp \subset L \subset L^\vee \subset S^\vee \oplus (S^\perp)^\vee \tag{7.2}$$

where $S \oplus S^\perp$ is a sublattice of finite index in L. We can extend $g \in \widetilde{O}(S)$ on $S \oplus S^\perp$ putting $g|_{S^\perp} \equiv \mathrm{id}$. It is clear that $g \in \widetilde{O}(S \oplus S^\perp)$. We consider $g \in \widetilde{O}(S \oplus S^\perp)$ as an element of $O(S^\vee \oplus (S^\perp)^\vee)$. For any $l^\vee \in L^\vee$ we have $g(l^\vee) \in l^\vee + (S \oplus S^\perp)$. In particular, $g(l) \in L$ for any $l \in L$ and $g \in \widetilde{O}(L)$. □

Let S be a primitive sublattice of L. We define the groups

$$O(L,S) = \{g \in O(L) \mid g|_S \in \widetilde{O}(S)\} \quad \text{and} \quad \widetilde{O}(L,S) = O(L,S) \cap \widetilde{O}(L).$$

Note that $O(L,\mathbb{Z}h) = O(L,h)$ if $h^2 \neq \pm 2$. The technique of discriminant forms developed by Nikulin in [97] is very useful here, and we describe the main ideas behind it below. For simplicity we assume that all the lattices we consider are even.

Let $S^\perp$ be the orthogonal complement of a primitive nondegenerate sublattice S in L. As in the proof of Lemma 7.1 we have the inclusions (7.2). The overlattice L is defined by the finite subgroup

$$H = L/(S^\perp \oplus S) < (S^\perp)^\vee/S^\perp \oplus S^\vee/S = D(S^\perp) \oplus D(S) \tag{7.3}$$

which is an isotropic subgroup of $D(S^\perp) \oplus D(S)$. Moreover $L/(S \oplus S^\perp) \cong L^\vee/(S^\vee \oplus (S^\perp)^\vee)$. We define $\phi\colon L \to S^\vee$ by $\phi(l)(s) = (l,s)$. Then $\ker(\phi) = S^\perp$. Since $L/(S \oplus S^\perp) \cong \phi(L)/S$ we obtain

$$|L/(S \oplus S^\perp)| = |\phi(L)/S| = |\det S|/[S^\vee : \phi(L)],$$

as $|\det S| = [S^\vee : S]$. From the inclusions above

$$|\det S| \cdot |\det S^\perp| = (|\det L|)[\phi(L) : S]^2 = |\det L| \cdot |\det S|^2/[S^\vee : \phi(L)]^2.$$

In the particular case $S = \mathbb{Z}h$ and $L_h = h_L^\perp$ we have $[S^\vee : \phi(L)] = \operatorname{div}(h)$, where

$$\operatorname{div}(h)\mathbb{Z} = (h, L). \tag{7.4}$$

The positive number $\operatorname{div}(h)$ is called the *divisor* of h in L. We have now proved the following lemma.

Lemma 7.5. *Let L be any nondegenerate even integral lattice and let $h \in L$ be a primitive vector with $h^2 = 2d \neq 0$. If L_h is the orthogonal complement of h in L then*

$$|\det L_h| = \frac{|(2d) \cdot \det L|}{\operatorname{div}(h)^2}.$$

We come back to the inclusion (7.3). Following [97] we consider the projections

$$p_S\colon H \to D(S), \qquad p_{S^\perp}\colon H \to D(S^\perp).$$

Using the definitions and the fact that the lattices S and $S^\perp$ are primitive in L one can show (see [97, Prop. 1.5.1]) that these projections are injective and moreover that if $d_S \in p_S(H)$ then there is a unique $d_{S^\perp} \in p_{S^\perp}(H)$ such that $d_S + d_{S^\perp} \in H$. Using these arguments one proves the next lemma (see [50, Lemma 3.2])

Lemma 7.6. *Let S be a primitive sublattice of an even lattice L and denote by $\overline{g}$ the image in $O(D(L))$ of $g \in O(L)$.*

(i) $g \in O(L,S)$ *if and only if* $g(S) = S$, $\overline{g}|_{D(S)} = \mathrm{id}$ *and* $\overline{g}|_{p_{S^\perp}(H)} = \mathrm{id}$.
(ii) $\alpha \in O(S^\perp)$ *can be extended to* $O(L,S)$ *if and only if* $\overline{\alpha}|_{p_{S^\perp}(H)} = \mathrm{id}$.
(iii) *If* $p_{S^\perp}(H) = D(S^\perp)$ *then* $O(L,S)|_{S^\perp} \cong \widetilde{O}(S^\perp)$.

(iv) *Assume that the projection* $O(S^\perp) \to O(D(S^\perp))$ *is surjective. Then*

$$O(L,S)|_{S^\perp}/\widetilde{O}(S^\perp) \cong \{\overline{\gamma} \in O(D(S^\perp)) \mid \overline{\gamma}|_{p_{S^\perp}(H)} = \mathrm{id}\}.$$

Corollary 7.7. *If* $|H| = |\det S^\perp|$ *then* $O(L,S)|_{S^\perp} \cong \widetilde{O}(S^\perp)$.

For example, the condition of Corollary 7.7 is true if L is an even unimodular lattice and S is any primitive sublattice of L (see Example 7.9 below). We recall the following result which we call *Eichler's criterion* (see [30, Section 10], [40, Section 3] and [49]).

Lemma 7.8. *Let* L *be a lattice containing two orthogonal hyperbolic planes. Then the* $\widetilde{SO}(L)$*-orbit of a primitive vector* $l \in L$ *is determined by two invariants: by its length* $l^2 = (l,l)$ *and its image* $l^* + L$*, where* $l^* = l/\operatorname{div}(l)$*, in the discriminant group* $D(L)$.

We note that l^* is a primitive element of the dual lattice $L^\vee$. Therefore $\operatorname{div}(l)$ is a divisor of the exponent of the discriminant group $D(L)$. In particular, $\operatorname{div}(l)$ divides $\det(L)$. Lemma 7.8 can be used to classify all possible vectors of fixed length in different lattices.

Example 7.9. *(The* K3 *lattice.)* $L(K3) = 3U \oplus 2E_8(-1)$ is the even unimodular lattice of signature (3,19): its discriminant group is trivial and all the primitive vectors $h_{2d} \in L_{K3}$ of length 2d form a single $SO(L_{K3})$-orbit. Therefore we can take h_{2d} in the first hyperbolic plane U, so

$$(h_{2d})^\perp_{L_{K3}} \cong L_{2d} = 2U \oplus 2E_8(-1) \oplus \langle -2d \rangle. \tag{7.10}$$

Then according to Corollary 7.7

$$O^+(L_{K3}, h_{2d}) \cong \widetilde{O}^+(h_{2d}^\perp) = \widetilde{O}^+(2U \oplus 2E_8(-1) \oplus \langle -2d \rangle). \tag{7.11}$$

The case of K3 surfaces is an exception, since the K3 lattice is unimodular.

Example 7.12. $K3^{[2]}$*-lattice, split and non-split polarisations.* We consider the Beauville lattice of a deformation $K3^{[n]}$ manifold, which is isomorphic to $L_{K3,2n-2}$ by Proposition 3.2(i). In general there will be several, but finitely many, orbits of primitive polarisation vectors h_{2d}. Let $h_{2d} \in L_{K3,2}$ be a primitive vector of length $2d > 0$. Then $\operatorname{div}(h_{2d})$ divides $|\det L_{K3,2}| = 2$.

All vectors with $\operatorname{div}(h_{2d}) = 1$ constitute a single $\widetilde{SO}^+(L_{K3,2})$-orbit, by Lemma 7.8. Therefore, as in Example 7.9, we obtain

$$(h_{2d})^\perp_{L_{K3,2}} \cong L_{2,2d} = 2U \oplus 2E_8(-1) \oplus \langle -2 \rangle \oplus \langle -2d \rangle. \tag{7.13}$$

We call a polarisation determined by a primitive vector h_{2d} with $\operatorname{div}(h_{2d}) = 1$ a *split polarisation*.

If $h_{2d} \in L_{K3,2}$ and $\operatorname{div}(h_{2d}) = 2$ then we can write h_{2d} as $h_{2d} = 2v + cl_2$, where $v \in 3U \oplus 2E_8(-1)$ and l_2 is a generator of the orthogonal component $\langle -2 \rangle$ in $L_{K3,2}$. The coefficient c is odd because h_{2d} is primitive. Note that $2d = h_{2d}^2 = 4(v,v) - 2c^2$, so $d \equiv -1 \mod 4$. According to Eichler's criterion the $\widetilde{SO}(L_{K3,2})$-orbit of h_{2d} is uniquely determined by the class $h_d^* \equiv l_2/2 \mod L_{K3,2}$. Therefore as in the case of $\operatorname{div}(h_{2d}) = 1$ all vectors with $\operatorname{div}(h_{2d}) = 2$ form only one orbit. We can take a representative in the form $h_{2d} = 2\nu + cl_2 \in U \oplus \langle -2 \rangle$. The orthogonal complement of h_{2d} in $U \oplus \langle -2 \rangle$ can be found by direct calculation. This is an even rank 2 lattice $Q(d)$ of determinant d. It follows that if $d \equiv -1 \mod 4$ then there is only one orbit of vectors h_{2d} with $\operatorname{div}(h_{2d}) = 2$, and the orthogonal complement of h_{2d} is uniquely determined:

$$(7.14) \qquad (h_{2d})^{\perp}_{L_{K3,2}} \cong L_{Q(d)} = 2U \oplus 2E_8(-1) \oplus \begin{pmatrix} -2 & 1 \\ 1 & -\frac{d+1}{2} \end{pmatrix}.$$

We call a polarisation of this kind a *non-split polarisation*. We note that $|\det L_{Q(d)}| = d$ and the discriminant group of $L_{Q(d)}$ is cyclic (see [50, Remark 3.15]).

Remark 7.15. Taking the orthogonal complement of the (-2)-vector in $Q(d)$ we have a split sublattice of index 2

$$\langle -2 \rangle \oplus \langle -2d \rangle < Q(d).$$

Therefore $L_{2,2d} < L_{Q(d)}$ is also a sublattice of index 2 and according to Lemma 7.1

$$(7.16) \qquad \widetilde{O}^+(L_{2,2d}) < \widetilde{O}^+(L_{Q(d)})$$

is a subgroup of finite index. This implies that the modular variety $\mathcal{F}_L(\widetilde{O}^+(L_{2,2d}))$ is a finite covering of $\mathcal{F}_L(\widetilde{O}^+(L_{Q(d)}))$ where $d \equiv -1 \mod 4$. One can calculate this index using the explicit formula for the Hirzebruch–Mumford volume (see [47]).

We gave a classification of all possible type of polarisations for the symplectic varieties of $K3^{[n]}$ type in [50, Proposition 3.6]. Results about the polarisation types of 10-dimensional O'Grady varieties can be found in [51, Theorem 3.1].

The branch divisor (Equation (6.12)) of the modular varieties is defined by $\pm$reflections in the modular groups. Below we give a description of the branch divisors in the cases of the polarised K3 surfaces and polarised holomorphic symplectic varieties of type $K3^{[2]}$.

Let L be a nondegenerate integral lattice and $r \in L$ be a primitive vector. If the reflection is integral, i.e. $\sigma_r \in O(L)$ (see Equations (6.11)) and (7.4)), then

$$(7.17) \qquad \operatorname{div}(r) \mid r^2 \mid 2\operatorname{div}(r).$$

The following general result was proved in [46].

Proposition 7.18. *(i) Let L be a nondegenerate even integral lattice. Let $r \in L$ be primitive. Then $\sigma_r \in \widetilde{O}(L)$ if and only if $r^2 = \pm 2$.*

If $-\sigma_r \in \widetilde{O}(L)$, i.e. $\sigma_r|_{A_L} = -\mathrm{id}$, then we also have

(ii) $r^2 = \pm 2a$ *and* $\operatorname{div}(r) = a \equiv 1 \mod 2$, *or* $r^2 = \pm a$ *and* $\operatorname{div}(r) = a$ *or* $a/2$; *and*

(iii) $A_L \cong (\mathbb{Z}/2\mathbb{Z})^m \times (\mathbb{Z}/a\mathbb{Z})$, *for some* $m \geqslant 0$.

If (iii) holds then

(iv) *If* $r^2 = \pm a$ *and either* $\operatorname{div}(r) = a$ *or* $\operatorname{div}(r) = a/2 \equiv 1 \bmod 2$, *then* $-\sigma_r \in \widetilde{O}(L)$;

(v) *If* $r^2 = \pm 2a$ *and* $\operatorname{div}(r) = a \equiv 1 \bmod 2$, *then* $-\sigma_r \in \widetilde{O}(L)$.

With polarised K3 surfaces in mind, we consider in more detail the lattice L_{2d} (see Equation (7.10)). In this case the ramification divisor has three irreducible components.

Corollary 7.19. *Let* σ_r *be a reflection in* $O^+(L_{2d})$ *defined by a primitive vector* $r \in L_{2d}$. *The reflection* σ_r *induces* $\pm\operatorname{id}$ *on the discriminant group* $L_{2d}^\vee/L_{2d}$ *if and only if* $r^2 = -2$ *or* $r^2 = -2d$ *and* $\operatorname{div}(r) = d$ *or* $2d$. *If* $r^2 = -2$ *then*

$$r^\perp_{L_{2d}} \cong 2U \oplus E_8(-1) \oplus E_7(-1) \oplus \langle -2d \rangle.$$

If $\operatorname{div}(r) = d$ *then either*

$$r^\perp_{L_{2d}} \cong U \oplus 2E_8(-1) \oplus \langle 2 \rangle \oplus \langle -2 \rangle$$

or

$$r^\perp_{L_{2d}} \cong U \oplus 2E_8(-1) \oplus U(2).$$

See the proof in [46, Corollary 3.4 and Proposition 3.6]. Geometrically the three cases in the last proposition correspond to the Néron-Severi group being (generically) U, $U(2)$ or $\langle 2 \rangle \oplus \langle -2 \rangle$ respectively. The K3 surfaces (without polarisation) themselves are, respectively, a double cover of the Hirzebruch surface F_4, a double cover of a quadric, and the desingularisation of a double cover of $\mathbb{P}^2$ branched along a nodal sextic.

We note that in the case of polarised deformation $K3^{[2]}$ manifolds the branch divisor has one main (i.e. $\operatorname{div}(r) = 1$) component, with $r^2 = -2$, and 6 (respectively, 1) additional components ($\operatorname{div}(r) > 1$) for split (respectively, non-split) type (see the proof of Proposition 8.22 below).

8. The quasi pull-back of modular forms

The main aim of this section is to show how we can construct cusp forms of small weight, for example on the moduli spaces of polarised K3 surfaces. We use the method of quasi pull-back of the Borcherds form Φ_{12} which was proposed in [11, pp. 200-201]. This method was successfully applied to the theory of moduli spaces in [14], [76], [46], [50] and [51]. In this section we review this method and prove a new result (Theorem 8.18 below) showing that non-trivial quasi pull-backs are cusp forms.

8.1. Quasi pull-back

First we give a general property of rational quadratic divisors. Let M be a lattice of signature $(2, m)$ and L be a primitive nondegenerate sublattice of signature

$(2, n)$ where $n < m$. Then $L_M^\perp$ is negative definite and we have the usual inclusions $L \oplus L_M^\perp < M < M^\vee < L^\vee \oplus (L^\perp)^\vee$. For $v \in M$ we write

$$v = \alpha + \beta, \qquad \alpha = \mathrm{pr}_{L^\vee}(v) \in L^\vee,\ \beta \in (L^\perp)^\vee. \tag{8.1}$$

Lemma 8.2. *Let* L *and* M *be as above. Then for any* $v \in M$ *with* $v^2 < 0$ *we have*

$$\mathcal{D}_L \cap \mathcal{D}_v(M) = \begin{cases} \mathcal{D}_\alpha(L), & \text{if } \alpha^2 < 0, \\ \emptyset, & \text{if } \alpha^2 \geqslant 0,\ \alpha \neq 0, \\ \mathcal{D}_L, & \text{if } \alpha = 0,\ \text{i.e. } v \in L^\perp. \end{cases}$$

Proof. We have $\mathcal{D}_L \subset \mathcal{D}_M$ because L is a sublattice of M. For any $Z = X + iY \in \mathcal{D}_L^\bullet$ with $X, Y \in L \otimes \mathbb{R}$ we have $(X, Y) = 0$ and $(X, X) = (Y, Y) > 0$. Therefore the quadratic space $\langle X, Y\rangle_{\mathbb{R}}$ is of signature (2,0). Note that $(Z, v) = 0$ is equivalent to $(Z, \alpha) = 0$. The signature of L is equal to $(2, n)$. Analysing the signature of $\langle X, Y\rangle_{\mathbb{R}} \oplus \langle \alpha\rangle_{\mathbb{R}}$ we get that $\mathcal{D}_L \cap \mathcal{D}_v(M)$ is non-empty if and only if $\alpha = \mathrm{pr}_{L^\vee}(v)$ belongs to the negative definite quadratic space $\langle X, Y\rangle_{M\otimes\mathbb{R}}^\perp$. This proves the first two cases of the lemma.

The finite group $H = M/(L \oplus L^\perp)$ is a subgroup of the orthogonal sum of the discriminant groups $D(L) \oplus D(L^\perp)$ where $D(L) = L^\vee/L$. The decomposition (8.1) defines a projection $\mathrm{pr}\colon H \to D(L) \oplus D(L^\perp)$. For a primitive sublattice L the class $(\alpha + L) \in D(L)$ is uniquely determined by the class $\beta + L^\perp$ in $D(L^\perp)$ (see [97, Proposition 1.5.1]). Therefore $\alpha \in L$ if and only if $\beta \in L^\perp$. In particular v is orthogonal to L if and only if $\alpha = 0$. This proves the last assertion of the lemma. □

In the next theorem we explain the main idea of the method of quasi pull-back applied to the strongly reflective modular form Φ_{12} (see [11, pp. 200–201] and [14]).

Theorem 8.3. *Let* $L \hookrightarrow II_{2,26}$ *be a primitive nondegenerate sublattice of signature* $(2, n)$, $n \geqslant 3$, *and let* $\mathcal{D}_L \hookrightarrow \mathcal{D}_{II_{2,26}}$ *be the corresponding embedding of the homogeneous domains. The set of* (-2)*-roots*

$$R_{-2}(L^\perp) = \{r \in II_{2,26} \mid r^2 = -2,\ (r, L) = 0\}$$

in the orthogonal complement is finite. We put $N(L^\perp) = \#R_{-2}(L^\perp)/2$. *Then the function*

$$\Phi|_L = \left.\frac{\Phi_{12}(Z)}{\prod_{r\in R_{-2}(L^\perp)/\pm 1}(Z, r)}\right|_{\mathcal{D}_L} \in M_{12+N(L^\perp)}(\widetilde{O}(L), \det), \tag{8.4}$$

where in the product over r *we fix a finite system of representatives in* $R_{-2}(L^\perp)/\pm 1$. *The modular form* $\Phi|_L$ *vanishes only on rational quadratic divisors of type* $\mathcal{D}_v(L)$ *where* $v \in L^\vee$ *is the orthogonal projection of a* (-2)*-root* $r \in II_{2,26}$ *on* $L^\vee$.

We say that the modular form $\Phi|_L$ is a *quasi pull-back* of Φ_{12} if the set of roots $R_{-2}(L^\perp)$ is non-empty.

Proof. We introduce coordinates $Z = (Z_1, Z_2) \in \mathcal{D}_{II_{2,26}}$ related to the embedding $L \hookrightarrow II_{2,26}$ and the splitting (8.1), namely $Z_1 \in L \otimes \mathbb{C}$ and $Z_2 \in (L \otimes \mathbb{C})^{\perp}$. We have $\widetilde{O}^+(L) < O^+(II_{2,26})$ (see Lemma 7.1) and we denote by $\tilde{g} \in O^+(II_{2,26})$ the extension of $g \in \widetilde{O}^+(L)$ by $\tilde{g}|_{L^\perp} = \mathrm{id}$.

If the root system $R_{-2}(L^\perp)$ is empty, then $\Phi|_L$ is the usual pull-back of Φ_{12} on $\mathcal{D}_L$. Then $\Phi|_L \not\equiv 0$ and we have

$$\det(g)\Phi_{12}(Z_1, Z_2) = \Phi_{12}(\tilde{g}(Z_1, Z_2)) = \Phi_{12}(g \cdot Z_1, Z_2).$$

Therefore the pull-back of Φ_{12} on $\mathcal{D}_L$ is a modular form of weight 12. We note that even in this simple case one obtains interesting reflective modular forms (see [55, Section 4.2]).

If there are roots in $R_{-2}(L^\perp)$ then the pull-back of Φ_{12} on $\mathcal{D}_L$ vanishes identically, and one has to divide by the equations of the rational quadratic divisors, as in Equation (8.4).

According to Lemma 8.2 the order of zero of Φ_{12} along $\mathcal{D}_L$ is equal to $N(L^\perp) = \#R_{-2}(L^\perp)/2$. Therefore the non-zero function $\Phi|_L$ is holomorphic on $\mathcal{D}_L^\bullet$. Moreover it is homogeneous of degree $12 + N(L^\perp)$. Any $g \in \widetilde{O}^+(L)$ acts trivially on $L^\perp$ and $(Z, r) = (Z_1 + Z_2, r) = (Z_2, r)$. Therefore

$$\left.\frac{\Phi_{12}(gZ_1, Z_2)}{\prod_r (Z_2, r)}\right|_{\mathcal{D}_L} = \left.\frac{\Phi_{12}(\tilde{g} \cdot (Z_1, Z_2))}{\prod_r (\tilde{g} \cdot (Z_1, Z_2), r)}\right|_{\mathcal{D}_L} = \det(g) \left.\frac{\Phi_{12}(Z)}{\prod_r (Z, r)}\right|_{\mathcal{D}_L}.$$

It follows that $\Phi|_L$ is modular with respect to $\widetilde{O}^+(L)$ with character det. We finish the proof using Koecher's principle.

The zeros of $\Phi|_L$ can be determined using Lemma 8.2 and the fact that Φ_{12} vanishes along $\mathcal{D}_r(II_{2,26})$ with $r^2 = -2$. □

Remark 8.5. Theorem 8.3 is still true for $n \leqslant 2$. We can show this by computing the Fourier expansions of the quasi pull-back. Moreover we prove in Theorem 8.18 that the quasi pull-back is always a cusp form.

Remark 8.6. The modular group of $\Phi|_L$ might be larger than $\widetilde{O}^+(L)$ (see, for example, [51, Lemma 4.4]).

Remark 8.7. For the applications to the theory of moduli spaces we use the quasi pull-back of Φ_{12}. It is easy to prove an analogue of Theorem 8.3 for an arbitrary modular form whose divisor consists only of rational quadratic divisors (in the style of Theorem 8.18 below).

In [46] we showed that some quasi pull-backs for lattices related to the moduli spaces of polarised K3 surfaces are cusp forms. In this paper we prove a new, more general result: that the quasi pull-back construction always gives a cusp form. The main idea of the proof is to consider the quasi pull-back as a differential operator (see [46, Section 6]).

8.2. Tube domain realisation

We define a Fourier expansion of a modular form F at a 0-dimensional cusp. The Fourier expansion depends on the choice of affine coordinates at a cusp. We consider the general case, following the approach used in [12, Theorem 5.2 and page 542]: for more details see [55, Section 2.3] and [33].

A 0-dimensional cusp of $\mathcal{D}_L$ is defined by an isotropic sublattice of rank 1 or, equivalently, by a primitive isotropic vector $c \in L$ (up to sign: c and $-c$ define the same cusp). The choice of c identifies $\mathcal{D}_L$ with an affine quadric:

$$\mathcal{D}_{L,c} = \{Z \in \mathcal{D}_L^\bullet \mid (Z, c) = 1\} \cong \mathcal{D}_L.$$

The lattice

$$L_c := c^\perp/c = c_L^\perp/\mathbb{Z}c \tag{8.8}$$

is an integral lattice of signature $(1, n-1)$. We fix an element $b \in L^\vee$ such that $(c, b) = 1$. A choice of b gives a realisation of the hyperbolic lattice L_c as a sublattice in L

$$L_c \cong L_{c,b} = L \cap c^\perp \cap b^\perp. \tag{8.9}$$

We have

$$L \otimes \mathbb{Q} = L_{c,b} \otimes \mathbb{Q} \oplus (\mathbb{Q}b + \mathbb{Q}c).$$

Using the hyperbolic lattice $L_c \otimes \mathbb{R}$ we define a positive cone

$$C(L_c) = \{x \in L_c \otimes \mathbb{R} \mid (x, x) > 0\}.$$

We may choose $C^+(L_c)$, one of the two connected components of $C(L_c)$ so that corresponding tube domain, which is the complexification of $C^+(L_c)$

$$\mathcal{H}_c = L_c \otimes \mathbb{R} + i\, C^+(L_c)$$

has an isomorphism $\mathcal{H}_c \to \mathcal{D}_{L,c} \cong \mathcal{D}_L$ by

$$z \mapsto [z] = z \oplus \left(b - \frac{(z, z) + (b, b)}{2} c\right) \quad (z \in \mathcal{H}_c,\ [z] \in \mathcal{D}_{L,c}). \tag{8.10}$$

Using the coordinate $z \in \mathcal{H}_c$ defined by the choice of c and b we can identify an arbitrary modular form F of weight k with a modular form $F_{c,b}$ (or simply F_c) on the tube domain $\mathcal{H}_c$:

$$F(\lambda[z]) = \lambda^{-k} F_{c,b}(z).$$

8.3. Fourier expansion at 0-dimensional cusps

In order to define the Fourier expansion at the cusp c we consider an unipotent subgroup of the stabiliser $\widetilde{O}^+(L)$, the subgroup of the Eichler transvections. For any $a \in c_L^\perp$ the map

$$t'(c, a)\colon v \longmapsto v - (a, v)c \qquad (v \in c_L^\perp)$$

belongs to the orthogonal group $O(c_L^\perp)$. It has the unique orthogonal extension on L which is given by the map

$$t(c,a)\colon v \longmapsto v-(a,v)c+(c,v)a-\frac{1}{2}(a,a)(c,v)c. \tag{8.11}$$

This element is called an *Eichler transvection*: see [30, Section 3] and [49]. We note that $t(c,a) \in \widetilde{SO}^+(L)$ for any $a \in c_L^\perp$, that $t(c,a)(c) = c$, and that for $a,\ a' \in c_L^\perp$

$$t(c,a)t(c,a') = t(c,a+a') \quad \text{and} \quad t(c,a)^{-1} = t(c,-a).$$

We can identify the lattice $L_{c,b}$ with the corresponding group of transvections $E_c(L) = \langle t(c,a) \mid a \in L_c\rangle$. The group $E_c(L)$ is the unipotent radical of the parabolic subgroup associated to c. A direct calculation shows that $t(c,a)$ acts as linear translation in the affine coordinates (8.10):

$$t(c,a)([z]) = [z+a].$$

Let $F \in M_k(\widetilde{SO}^+(L))$. Then $F_c(z+a) = F_c(z)$ for all $a \in L_{c,b}$ and we obtain the Fourier expansion of F at the cusp c:

$$F_c(z) = \sum_{l \in L_{c,b}^\vee} f(l)\exp(2\pi i\,(l,z)). \tag{8.12}$$

The function $F_c(z)$ is holomorphic at the cusp c if the Fourier coefficient $f(l)$ can be different from 0 only if its index $l \in L_{c,b}^\vee$ belongs to the closure of the positive cone $C^+(L_c)$. Another formulation of this fact is $(l,Y) > 0$ for any Y in the positive cone $C^+(L_c)$.

Remark 8.13. In general the Fourier expansion (8.12) depends on the choice of b. For another b' the Fourier coefficients will be different by a factor $\exp(2\pi i(l,b-b'))$ which is a root of unity (see details in [55, §2.3]). In particular the Fourier coefficient f_0 (the value of F at the cusp c) is well-defined.

Remark 8.14. The stabiliser of the cusp c in $\widetilde{O}^+(L)$ is isomorphic to the semi-direct product of $\widetilde{O}^+(L_c)$ and $E_c(L)$.

Remark 8.15. Let $\operatorname{div}_L(c) = N$. Then $\det L = N^2 \det L_c$. If $\operatorname{div}_L(c) = 1$ then c can be completed to a hyperbolic plane in L and $L = U \oplus L_c$. This cusp is called the *simplest cusp*. If $\operatorname{div}(c) > 1$ then $c/\operatorname{div}(c) + L$ is an isotropic element of the discriminant group $D(L)$ and $|D(L)| = |\det L|$ is divisible by $\operatorname{div}(c)^2$. If $D(L)$ contains no non-trivial isotropic elements (in this case the lattice is maximal) then all 0-dimensional cusps are equivalent to the simplest cusp.

Remark 8.16. If L contains two hyperbolic planes then one can use the Eichler criterion (see Lemma 7.8) in order to classify the 0-dimensional cusps with respect to the action of $\widetilde{SO}^+(L)$. The orbit of c, in this case, is uniquely determined by the isotropic class $c/\operatorname{div}(c) + L$ in the discriminant group $D(L)$.

If $D(L)$ does not contain isotropic elements (in particular if $\det(L)$ is square free) then the modular variety $\widetilde{O}^+(L)\setminus\mathcal{D}_L$ has only one 0-dimensional cusp.

Remark 8.17. If $[O(L):\Gamma]<\infty$ and $F\in M_k(\Gamma,\chi)$ then we can define the Fourier expansion using a sublattice $L_c(\Gamma)$ of finite index in L_c corresponding to the group $E_c(L)\cap\ker\chi$.

8.4. Properties of quasi pull-back

We may consider the quasi pull-back of other modular forms, not only Φ_{12}. If L is of signature $(2,n)$ and $r\in L$ is primitive with $(r,r)<0$ then $\mathcal{D}_r(L)=\mathcal{D}_{tr}(L)$ for any $t\in\mathbb{Q}^\times$, and we write $r^\perp=r^\perp_L$ if there is no ambiguity.

Theorem 8.18. *Let L be an integral lattice of signature $(2,n)$. Suppose that the modular form $F\in M_k(\widetilde{SO}^+(L))$ vanishes with order $m>0$ on the rational quadratic divisor $\mathcal{D}_r(L)$ where r is a primitive vector in L with $(r,r)<0$. We define the quasi pull-back of F on the domain $\mathcal{D}_{r^\perp}$ of complex dimension $n-1$ by*

$$F|_{r^\perp}=\left.\frac{F(Z)}{(Z,r)^m}\right|_{\mathcal{D}_{r^\perp}}. \tag{8.19}$$

Then the quasi pull-back is a cusp form of weight $k+m$

$$F|_{r^\perp}\in S_{k+m}(\widetilde{SO}^+(r^\perp)).$$

Proof. There are two parts to the assertion: that $F|_{r^\perp}$ is a modular form, and that it vanishes at every cusp.

For the first part, we have the inclusions

$$r^\perp\oplus\mathbb{Z}r\subset L\subset L^\vee\subset(r^\perp)^\vee\oplus\mathbb{Z}\frac{r}{(r,r)},$$

and $r^\perp$ has signature $(2,n-1)$. We consider the two embeddings $\mathcal{D}_{r^\perp}\hookrightarrow\mathcal{D}_L$ and $\widetilde{SO}^+(r^\perp)\hookrightarrow\widetilde{SO}^+(L)$. We note that any element $g\in\widetilde{SO}^+(r^\perp)$ extends to $\tilde{g}\in\widetilde{SO}^+(L)$ by acting trivially on r (see Lemma 7.1). Therefore any $\tilde{g}$ preserves $\mathcal{D}_{r^\perp}$ and $\tilde{g}\cdot r=r$. The function $F|_{r^\perp}$ is a holomorphic function on $\mathcal{D}_{r^\perp}$ and it is homogeneous of degree $k+m$. For any $g\in\widetilde{SO}^+(r^\perp)$ we have

$$\left.\frac{F(\tilde{g}Z)}{(\tilde{g}Z,r)^m}\right|_{\mathcal{D}_{r^\perp}}=\left.\frac{F(Z)}{(Z,r)^m}\right|_{\mathcal{D}_{r^\perp}}.$$

Therefore the quasi pull-back $F|_{r^\perp}$ is $\widetilde{SO}^+(r^\perp)$-invariant. If instead F has character χ then $F_{r^\perp}$ transforms according to the induced character $\chi|_{\widetilde{SO}^+(r^\perp)}$. If $n>3$ then using Koecher's principle we conclude that

$$F|_{r^\perp}\in M_{k+m}(\widetilde{SO}(r^\perp)).$$

If $n\leqslant 3$ we cannot apply Koecher's principle and instead we must use the Fourier expansion to check that the quasi pull-back is a modular form as well as to show the vanishing at the cusps.

We calculate the Fourier expansion of the quasi pull-back at an arbitrary 0-dimensional cusp of $r^\perp$. Let $c \in r^\perp$ be a primitive isotropic vector in $r^\perp$ (if there are any: if not, there is nothing more to prove). We fix a vector $b \in (r^\perp)^\vee$. Then we can define the two homogeneous domains $\mathcal{H}_c(r^\perp)$ and $\mathcal{H}_c(L)$ because the vector c defines also a 0-dimensional cusp of L. We write $Z \in \mathcal{H}_c(L)$ in the form $Z = Z_1 + zr$ where

$$Z_1 \in r^\perp \otimes \mathbb{C},\ z = x + iy \in \mathbb{C},\ \ (\operatorname{Im} Z_1, \operatorname{Im} Z_1) + (r,r)y^2 > 0.$$

If $[Z]$ is the image of Z in $\mathcal{D}_c(L)$ (see (8.10)) then $([Z], r) = (r,r)z$. Therefore the equation of the divisor $\mathcal{D}_r(L)$ in the affine coordinates $Z = Z_1 \oplus zr \in \mathcal{H}_c$ is $z = 0$. The quasi pull-back $F|_{r^\perp}$ is equal, up to a constant, to the first non-zero coefficient in the Taylor expansion of $F_{c,b}(Z)$ in z.

Consider the Fourier expansion

$$F_{c,b}(Z) = \sum_{l \in L^\vee_{c,b}} f(l) \exp\left(2\pi i\,(l, Z)\right).$$

The modular form $F_{c,b}(Z)$ is holomorphic at the boundary. Therefore l belongs to the closure of the dual cone, in particular, $(l,l) \geqslant 0$. The vectors c and b are orthogonal to r and the lattice $(r^\perp)_{c,b} \oplus \mathbb{Z}r$ is a sublattice of finite index in the lattice of translations $L_{c,b}$. For any $Z_1 \in \mathcal{H}_c(r^\perp) \subset r^\perp \otimes \mathbb{C}$ we consider $z = x + iy$ such that $Z = Z_1 \oplus zr \in \mathcal{H}_c(L)$. (If y is small enough then $(\operatorname{Im} Z_1, \operatorname{Im} Z_1) > -(r,r)y^2 > 0$.) Therefore we can rewrite the Fourier expansion using this parametrisation

$$(8.20)\quad F_{c,b}(Z_1 + zr) = \sum_{l_1 \in (r^\perp)^\vee_{c,b},\ l_2 \in \mathbb{Z}r/(r,r)} f(l_1 + l_2) \exp\left(2\pi i\,(l_1, Z_1) + z(l_2, r)\right)$$

where $(l,l) = (l_1 + l_2, l_1 + l_2) = (l_1, l_1) + (l_2, l_2) \geqslant 0$. The Fourier coefficients of $(F|_{r^\perp})_{c,b}$ are proportional to the Fourier coefficients of the m-th Taylor coefficient

$$\left.\frac{\partial^m F_{c,b}(Z_1 + zr)}{(\partial z)^m}\right|_{z=0}.$$

The derivatives of the terms in the Fourier expansion (8.20) vanish if $l_2 = 0$. If $(l_2, l_2) < 0$ then $(l_1, l_1) > 0$, so nonzero Fourier coefficients occur only when the index l_1 has positive square. We have proved this, which is much stronger than just the vanishing of the zeroth coefficient, for an arbitrary 0-dimensional cusp of $r^\perp$, so we have shown that $F|_{r^\perp}$ is holomorphic at the boundary even for $n = 1$ or $n = 2$. Moreover the value of $F|_{r^\perp}$ at an arbitrary 0-dimensional cusp is zero.

The Baily-Borel compactification of $\widetilde{SO}^+(r^\perp) \setminus \mathcal{D}_{r^\perp}$ contains only boundary components of dimension 0 and 1 (the latter only if $r^\perp$ contains a totally isotropic sublattice of rank two). The Fourier expansion at a 1-dimensional cusp E is called Fourier–Jacobi expansion (see [4], [107], [40]). The value of a modular form G on the boundary component E is given by the Siegel operator $\Phi_E(G)$ (see [5]). This is the zeroth coefficient of the Fourier–Jacobi expansion which is a modular form with respect to a subgroup of $SL(2, \mathbb{Z})$.

The boundary of a 1-dimensional cusp E is a union of some 0-dimensional cusps. We consider the Fourier expansion of $\Phi_E(F|_{r^\perp})$ at a 0-dimensional cusp c associated to E as a part of the Fourier expansion of $(F|_{r^\perp})_{c,h}$ (see [74, Section 5.2] for the Fourier expansion of Fourier–Jacobi coefficients of modular forms). The indices of the Fourier coefficients of $\Phi_E(F|_{r^\perp})$ are of hyperbolic norm 0. Therefore $\Phi_E(F|_{r^\perp}) \equiv 0$ because, as shown above, all such coefficients in $(F|_{r^\perp})_{c,h}$ are equal to zero. Therefore the quasi pull-back $F|_{r^\perp}$ is a cusp form. □

Using Theorem 8.18 we prove that the quasi pull-back defined in Theorem 8.3 is a cusp form.

Corollary 8.21. *Let* $L \hookrightarrow II_{2,26}$ *be a nondegenerate sublattice of signature* $(2,n)$, $n \geqslant 1$. *We assume that the set* $R_{-2}(L^\perp)$ *of* (-2)*-roots in* $L^\perp$ *is non-empty. Then the quasi pull-back* $\Phi|_L \in S_{12+N(L^\perp)}(\widetilde{O}(L), \det)$ *of the Borcherds form* Φ_{12} *is a cusp form.*

Proof. To prove the corollary we divide the procedure of the quasi pull-back of Theorem 8.3 into finitely many steps.

First, we take a root $r_1 \in R_{-2}(L^\perp) \neq \emptyset$ and define $M_1 = (r_1)^\perp_{II_{2,26}}$. The form Φ_{12} has a zero of order 1 on $\mathcal{D}_{r_1}(II_{2,26})$. According to Theorem 8.18 we have

$$\Phi|_{M_1} = \left.\frac{\Phi_{12}(Z)}{(r_1, Z)}\right|_{M_1} \in S_{13}(\widetilde{O}^+(M_1), \det).$$

We note that the cusp form $\Phi|_{M_1}$ might have divisors different from the (-2)-divisors of Φ_{12}. If $s \in R_{-2}(L^\perp)$ such that $(r_1, s) \neq 0$ then $s_1 = \mathrm{pr}_{M_1^\vee}(s)$ has negative norm $-2 < (s_1, s_1) < 0$ and $\Phi|_{M_1}$ vanishes along $\mathcal{D}_{s_1}(M_1)$ according to Lemma 8.2. Therefore the divisors of $\Phi|_{M_1}$ are rational quadratic divisors defined by some vectors in $v \in M_1^\vee$. But $\mathcal{D}_v(M_1) = \mathcal{D}_{tv}(M_1)$ and we can fix t in order to have a primitive vector $tv \in M_1$.

Second, we consider the lattice $L^\perp \cap M_1$. Suppose first that there is no vector $v \in L^\perp \cap M_1$ such that $\Phi|_{M_1}$ vanishes on $\mathcal{D}_v(M_1)$. In this case $\Phi|_L$ is equal to the pull-back of $\Phi|_{M_1}$ on $\mathcal{D}_L$. The pull-back of a cusp form is a cusp form and the proof is finished.

Otherwise, if $r_2 \in L^\perp \cap M_1$ is a vector such that $\Phi|_{M_1}$ vanishes on $\mathcal{D}_{r_2}(M_1)$, then we define $M_2 = (r_2)^\perp_{M_1} = \langle r_1, r_2\rangle^\perp_{II_{2,26}}$. As in Theorem 8.18 we have

$$\Phi|_{M_2} = \left.\frac{\Phi|_{M_1}}{(r_2, Z)^m}\right|_{M_2} \in S_{13+m}(\widetilde{O}^+(M_2), \det)$$

where $m \geqslant 1$ is the degree of zero of $\Phi|_{M_1}$ on $\mathcal{D}_{r_2}(M_1)$.

The function $\Phi|_{M_2}$ is a modular form vanishing along some rational quadratic divisors. We can repeat the procedure described above for M_2. After a finite number of steps we get the cusp form $\Phi|_L$. □

Proposition 8.22. *Let* L *be one of the lattices* L_{2d}, $L_{2,2d}$ *or* $L_{Q(d)}$ *defined in* (7.10)–(7.14). *We assume that there exists an embedding* $L \hookrightarrow II_{2,26}$ *such that the weight of the* *quasi pull-back $\Phi|_L$ is smaller than the dimension of $\mathcal{D}_L$. Then $\Phi|_L$ vanishes along the ramification divisor of the modular projection*

$$\pi\colon \mathcal{D}_L \to \widetilde{O}^+(L) \setminus \mathcal{D}_L.$$

Proof. We give here a proof which works for all cases including the moduli spaces of polarised holomorphic symplectic O'Grady varieties (see [51, Corollary 4.6]).

The components of the branch divisor of π are $\mathcal{D}_r(L)$ where $r \in L$ and σ_r or $-\sigma_r$ is in $\widetilde{O}^+(L)$ (see [46, Corollary 2.13] and Equation (6.12) above). If $\sigma_r \in \widetilde{O}^+(L)$, then $\Phi|_L$ vanishes along $\mathcal{D}_r(L)$ because $\Phi|_L$ is modular with character det. We have to prove that $\Phi|_L$ vanishes also on $\mathcal{D}_r(L)$ with $-\sigma_r \in \widetilde{O}^+(L)$. To prove this we use the transitivity of the quasi pull-back construction. Let $r \in L \hookrightarrow II_{2,26}$ and $L_r = r_L^\perp$. Then

$$(\Phi|_L)|_{L_r} = \Phi|_{L_r}.$$

We have to consider three cases.

1) Let $r \in L = L_{2d}$. Then $\operatorname{rank} L_r = 18$, and, according to Corollary 7.19, $\det|(L_r)^\perp_{II_{2,26}}|$ is either 1 or 4. In [23, Table I] one can find all classes of the indecomposable lattices of small rank and determinant. Analysing that table we find three classes of lattices of rank 8 of determinant 1, 2 or 4:

$$E_8\ \ (|R(E_8)| = 240),\qquad E_7 \oplus A_1\ \ (|R(E_7 \oplus A_1)| = 128),\qquad D_8\ \ (|R(D_8)| = 112).$$

Therefore Φ_{12} has a zero of order at least 56 along $\mathcal{D}_r(L)$. The modular form $\Phi|_L$ is of weight $k < 19$. Therefore $R(L^\perp_{II_{2,26}})$ has at most 12 roots. Therefore $\Phi|_L$ vanishes on $\mathcal{D}_r(L_{2d})$ with order at least 50.

2) Let $r \in L = L_{2,2d}$. Then $\operatorname{rank} L_r = 19$. We described reflective vectors r with $-\sigma_r \in \widetilde{O}^+(L)$ in Proposition 7.18. There are three possible cases:

$$r^2 = 2d,\ \operatorname{div}(r) = d \text{ or } 2d \quad \text{and} \quad r^2 = d = \operatorname{div}(r)\ (d \text{ is odd}).$$

According to Lemma 7.5, $\det L_r = 2, 4$ or 8. Therefore the rank 7 lattice $(L_r)^\perp_{II_{2,26}}$ has the same determinant. According to [23, Table I] there are six possible classes of such lattices:

$$E_7,\ D_7,\ D_6 \oplus A_1,\ A_7,\ [D_6 \oplus \langle 8\rangle]_2,\ [E_6 \oplus \langle 24\rangle]_3,$$

where $[M]_n$ denotes an overlattice of order n of M. Any of these lattices contains at least 60 roots ($|R(D_6)| = 60$). The modular form $\Phi|_L$ is of weight $k < 20$. Therefore $R(L^\perp_{II_{2,26}})$ has at most 14 roots and $\Phi|_L$ vanishes on $\mathcal{D}_r(L_{2,2d})$ with order at least 23.

3) Let $r \in L = L_{Q(d)}$. In this case $d \equiv 3 \mod 4$, and the discriminant group $D(L_{Q(d)})$ is cyclic of order d. Using Proposition 7.18 and Lemma 7.5 we obtain that only one class for $(L_r)^\perp_{II_{2,26}}$ is possible. This is the lattice E_7. We finish the proof as above. □

9. Arithmetic of root systems

In this section we finish the proof of Theorem 6.1. To prove it we use the low weight cusp form trick (see the discussion in the proof of Theorem 6.13) by using the quasi pull-back of the Borcherds modular form Φ_{12} to construct cusp forms of small weight with large divisor.

According to Theorem 8.3, Theorem 8.18 and Proposition 8.22 the main point for us is the following. We want to know for which $2d > 0$ there exists a vector

$$l \in E_8,\ l^2 = 2d,\ l \text{ is orthogonal to at least 2 and at most 12 roots.} \tag{9.1}$$

To solve this problem we use the combinatorial geometry of the root system E_8 together with the theory of quadratic forms. First we give some properties of the lattice E_8 and we show how one can construct the first polarisations of general type in Theorem 6.1. After that we outline the answer to the question in (9.1).

9.1. Vectors in E_8 and E_7

By definition, the lattice D_8 is an even sublattice of the Euclidean lattice $\mathbb{Z}^8$

$$D_8 = \{l = (x_1, \dots, x_8) \in \mathbb{Z}^8 \,|\, x_1 + \dots + x_8 \in 2\mathbb{Z}\}.$$

The determinant of D_8 is equal to 4. We denote by $e_1, \dots, e_8$ the Euclidean basis of $\mathbb{Z}^8$ ($(e_i, e_j) = \delta_{ij}$). The lattice E_8 is the double extension of D_8:

$$E_8 = D_8 \cup \left(\frac{e_1 + \dots + e_8}{2} + D_8\right).$$

We consider the Coxeter basis of simple roots in E_8 (see [17])

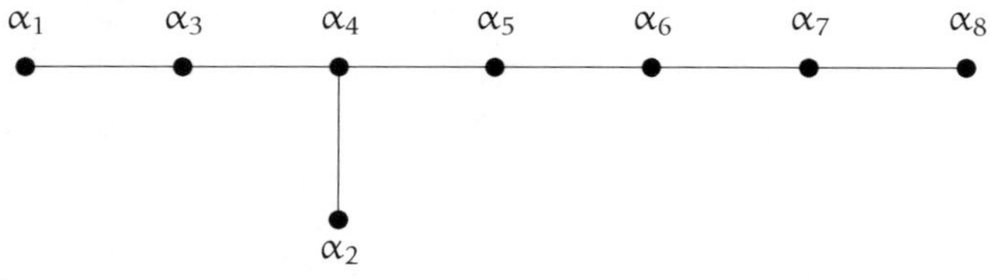

where

$$\alpha_1 = \frac{1}{2}(e_1 + e_8) - \frac{1}{2}(e_2 + e_3 + e_4 + e_5 + e_6 + e_7),$$
$$\alpha_2 = e_1 + e_2, \quad \alpha_k = e_{k-1} - e_{k-2} \ (3 \leqslant k \leqslant 8)$$

and $E_8 = \langle \alpha_1, \dots \alpha_8 \rangle_{\mathbb{Z}}$. The lattice E_8 contains 240 roots. We recall that any root is a sum of simple roots with integral coefficients of the same sign. The fundamental weights ω_j of E_8 form the dual basis, so $(\alpha_i, \omega_j) = \delta_{ij}$. The formulae for the weights are given in [17, Tabl. VII]. We shall use the Cartan matrix of the dual basis

$$(9.2) \qquad ((\omega_i, \omega_j)) = \begin{pmatrix} 4 & 5 & 7 & 10 & 8 & 6 & 4 & 2 \\ 5 & 8 & 10 & 15 & 12 & 9 & 6 & 3 \\ 7 & 10 & 14 & 20 & 16 & 12 & 8 & 4 \\ 10 & 15 & 20 & 30 & 24 & 18 & 12 & 6 \\ 8 & 12 & 16 & 24 & 20 & 15 & 10 & 5 \\ 6 & 9 & 12 & 18 & 15 & 12 & 8 & 4 \\ 4 & 6 & 8 & 12 & 10 & 8 & 6 & 3 \\ 2 & 3 & 4 & 6 & 5 & 4 & 3 & 2 \end{pmatrix}.$$

Let us assume that $l_{2d} \in E_8$ and $(l_{2d})^{\perp}_{E_8}$ contains exactly 12 roots. We consider two cases

$$R((l_{2d})^{\perp}_{E_8}) = A_2 \oplus 3A_1 \quad \text{or} \quad R((l_{2d})^{\perp}_{E_8}) = A_2 \oplus A_2.$$

There are four possible choices of the subsystem $A_2 \oplus 3A_1$ inside the Dynkin diagram of E_8. There are four choices for a copy of A_2:

$$A_2 = \langle \alpha_1, \alpha_3 \rangle, \ \langle \alpha_2, \alpha_4 \rangle, \ \langle \alpha_5, \alpha_6 \rangle, \ \langle \alpha_7, \alpha_8 \rangle.$$

Then the three pairwise orthogonal copies of A_1 are defined automatically. For example, if $A_2^{(1)} = \langle \alpha_5, \alpha_6 \rangle$ then $3A_1^{(1)} = \langle \alpha_2 \rangle \oplus \langle \alpha_3 \rangle \oplus \langle \alpha_8 \rangle$. The sum $A_2^{(1)} \oplus 3A_1^{(1)}$ is the root system of the orthogonal complement of the vector $l_{5,6} = \omega_1 + \omega_4 + \omega_7 \in E_8$. In fact, if $r = \sum_{i=1}^{8} x_i \alpha_i$ is a positive root ($x_i \geqslant 0$) then $(r, l_{5,6}) = x_1 + x_4 + x_7 = 0$. Therefore $x_1 = x_4 = x_7 = 0$ and r belongs to $A_2^{(1)} \oplus 3A_1^{(1)}$ (see the Dynkin diagram of E_8 above). Using the matrix (9.2) we find $l_{5,6}^2 = 46$. Doing similar calculations with

$$l_{1,3} = \omega_4 + \omega_6 + \omega_8, \quad l_{2,4} = \omega_3 + \omega_5 + \omega_7, \quad l_{7,8} = \omega_1 + \omega_4 + \omega_6$$

we obtain polarisations for $d = 50$, 54 and 57

$$l_{1,3}^2 = 2 \cdot 50, \qquad l_{2,4}^2 = 2 \cdot 54, \qquad l_{7,8}^2 = 2 \cdot 57.$$

To get a good vector for $d = 52$ we consider the lattice $M = A_2 \oplus A_2 = \langle \alpha_3, \alpha_4 \rangle \oplus \langle \alpha_6, \alpha_7 \rangle$ in E_8. Then M is the root system of the orthogonal complement of $l_{104} = \omega_1 + \omega_2 + \omega_5 + \omega_8$ and $l_{104}^2 = 2 \cdot 52$.

In this way we construct the first five polarisations of general type from the list of Theorem 6.1. These elementary arguments, similar to the arguments of Kondo in [76], do not give the whole list but only a few early cases. A sufficient condition for existence of vectors satisfying (9.1) is given in the theorem below.

Theorem 9.3. *A vector l satisfying* (9.1) *does exist if the inequality*

$$(9.4) \qquad 4N_{E_7}(2d) > 28N_{E_6}(2d) + 63N_{D_6}(2d)$$

is valid, where $N_L(2d)$ denotes the number of representations of $2d$ by the lattice L.

Proof. We use bouquets of copies of A_2 in $E_8 \setminus E_7$. The root system $R(E_8)$ is a disjoint union of 126 roots of E_7 and the bouquet of 28 copies of A_2 centred in A_1. This fact explains the coefficients 28 and 63 in the right hand side of (9.4). See [46, Section 7] for more details. □

In [46] we found explicit formulae for the numbers of representations and we proved that

$$N_{E_7}(2m) > \frac{24\pi^3}{5\zeta(3)}, \qquad 21m^2 > N_{D_6}(2d), \qquad 103.69m^2 > N_{E_6}(2d).$$

These inequalities give a finite set of d for which (9.4) is not valid. Analysing the corresponding theta series we found the set of all such d, containing 131 numbers. The five tables in [46, Section 7] and the argument above for $d = 52$ give the list of polarisations of general type of Theorem 6.1.

The geometric genus of $\mathcal{F}_{2d}$ is positive if there exists a cusp form of canonical weight 19. For each d not in the general type list but satisfying $d \geqslant 40$ and $d \neq 41$, 44, 45 or 47, there exists a vector $h_{2d} \in E_8$ of length 2d orthogonal exactly to 14 roots. The corresponding quasi pull-back is a cusp forms of canonical weight. For $d = 42$, 43, 51, 53 and 55 such cusp forms were constructed by Kondo in [76]. For other d see [46].

A similar arithmetic method applied to E_7 (see [50, Section 4]) gives the proof of Theorem 6.2. There is a significant technical difficulty in the case of E_7. The proof involves estimating the number of ways of representing certain integers by various root lattices of odd rank. In [50, Section 5]) we gave a new, clear, explicit version of Siegel's formula for this number in the odd rank case. It may be expressed either in terms of Zagier L-functions or in terms of the H. Cohen numbers. For example we obtained a new, very short formula for the number of representations by 5 squares (see [50, Section 4] for the details).

9.2. Binary quadratic forms in E_8

We saw in Example 4.5 that the results of Debarre and Voisin [25] imply that the moduli space $\mathcal{M}_{2\cdot 11}^{[2],\text{non-split}}$ of Beauville degree $d = 11$ is unirational. We note that $\mathcal{M}_{2\cdot 11}^{[2],\text{split}}$, which is a finite covering of $\mathcal{M}_{2\cdot 11}^{[2],\text{non-split}}$ by (7.16), has non-negative Kodaira dimension. Theorem 6.2 shows that there can be at most 11 exceptional split polarisations of non general type. Proposition 9.5 hints that one can expect a theorem for the non-split case, which we hope to prove in the future, similar to Theorem 6.2.

We recall that the Beauville degree $d \equiv -1 \mod 4$ if the polarisation h_{2d} is of non-split type. For small d we can calculate the class of the orthogonal complement of $Q(d)$. According to [23, Table 0] the rank 6 lattice $Q(d)^{\perp}_{E_8}$ of determinant d contains at least 24 roots for $d = 3$, 7, 11 and 15. One can continue this analysis but we propose below a simple algorithm which gives us the first "good" embeddings of $Q(d)$ in E_8.

Proposition 9.5. *The moduli spaces* $\mathcal{M}_{2\cdot 39}^{[2],non\text{-}split}$ *and* $\mathcal{M}_{2\cdot 47}^{[2],non\text{-}split}$ *are of general type.*

Proof. Let $Q(d)(-1) = \begin{pmatrix} 2 & -1 \\ -1 & 2c \end{pmatrix}$, where $c = (d+1)/4$, be the binary quadratic form associated to a non-split polarisation of degree $2d$ (see (7.14)). To make the notation simpler we denote this binary form by $Q(d)$. We have to embed $Q(d)$ in E_8 so as to satisfy

$$2 \leqslant |R(Q(d)^{\perp}_{E_8})| \leqslant 14.$$

We describe below a method (we call it $\frac{1}{2}(++)$-algorithm) which gives many such embeddings. We take the following realisation of the binary quadratic form of determinant d in E_8:

$$Q(d) = \langle e_2 - e_1, v_{2c} \rangle = \langle e_2 - e_1, \frac{1}{2}(e_1 - e_2 + x_3 e_3 + \cdots + x_8 e_8) \rangle,$$

where $x_i \geqslant 0$ and $2c = (2 + x_3^2 + \cdots + x_8^2)/4$. The second generator v_{2c} of the binary quadratic form is a half-integral element of E_8. According to the definition of E_8 given above

$$v_{2c} + \frac{1}{2}(e_1 + \cdots + e_8) = e_1 + \frac{x_3+1}{2} e_3 + \cdots + \frac{x_8+1}{2} e_8 \in D_8.$$

Therefore $x_3 \equiv \cdots \equiv x_8 \mod 2$ and $x_3 + \cdots + x_8 \equiv 0 \mod 4$. Now we can find all roots orthogonal to $Q(d)$.

First, we consider the integral roots of E_8. The roots $\pm(e_1 + e_2)$ are orthogonal to $Q(d)$ therefore $R(Q(d)^{\perp}_{E_8})$ is not empty. Then the integral roots $\pm(e_i \pm e_j)$ $(i, j > 2)$ are orthogonal to $Q(d)$ if and only if $x_i = \mp x_j$.

Second, a half-integral root orthogonal to $Q(d)$ has the form

$$\pm\frac{1}{2}(e_1 + e_2 + \sum_{i=3}^{8} (-1)^{\varepsilon_i} e_i)$$

where the number of $-$ signs in the sum is even. Therefore v_{2c} is orthogonal to a half-integral root if and only if we can divide the vector of coefficients $(x_3, \ldots, x_8)$ in two parts containing even number of terms and with the same sum. If this is possible there are two pairs of roots orthogonal to v_{2c}. For example, if there is a root of type $\pm(++;++++--)$ then we also have $\pm(++;----++)$. The rules described above give the number (it is always positive) of roots orthogonal to $Q(d)$.

To finish the proof of Proposition 9.5 we consider two vectors

$$v_{20} = \frac{1}{2}(1,1;5,5,3,3,3,1) \quad \text{and} \quad v_{24} = \frac{1}{2}(1,1;7,5,3,3,1,1).$$

The quadratic forms $Q(39)$ and $Q(47)$ are orthogonal to exactly 14 roots in E_8. As above, using Theorem 8.3, Theorem 8.18 and Proposition 8.22 we see that the quasi pull-back of Φ_{12} on the modular variety defined by the lattice $L_{Q(d)}$ is a cusp form of weight 19 vanishing on the ramification divisor. We finish the proof using Theorem 6.13. □

References

[1] A. Apostolov. *Moduli spaces of polarized irreducible symplectic manifolds are not necessarily connected.* arXiv:1109.0175, 17 pp. ← 480

[2] M. Artin. Supersingular K3 surfaces. *Ann. Scient. Éc. Norm. Sup. (4)*, **7**, 543–568, 1974. ← 462

[3] A. Ash, D. Mumford, M. Rapoport, and Y. Tai. *Smooth compactification of locally symmetric varieties.* Second edition: Cambridge University Press, Cambridge, 2010. ← 487, 488, 489, 494, 498

[4] W. L. Baily. Fourier-Jacobi series, in: Algebraic Groups and Discontinuous Subgroups, 296–300. *Proc. Sympos. Pure Math.*, **9**, Amer. Math. Soc., Providence, 1966. ← 496, 511

[5] W. L. Baily Jr. and A. Borel. Compactification of arithmetic quotients of bounded symmetric domains. *Ann. of Math. (2)*, **84**, 442–528, 1966. ← 469, 483, 485, 486, 511

[6] W. Barth, K. Hulek, C. Peters, and A. Van de Ven. Compact complex surfaces. Second Edition. *Ergebnisse der Mathematik und ihrer Grenzgebiete (3)*, **4**, Springer, Berlin, 2004. ← 461, 463, 464, 465, 466, 468

[7] J. Bauermann. Lokale Tensorfelder auf arithmetischen Quotienten symmetrischer beschränkter Gebiete. *J. Reine Angew. Math.*, **428**, 111–138, 1992. ← 496

[8] A. Beauville. Variétés Kähleriennes dont la première classe de Chern est nulle. *J. Differential Geom.*, **18**, 755–782, 1983. ← 462, 473, 474, 475

[9] F. Bogomolov. The decomposition of Kähler manifolds with a trivial canonical class. *Mat. Sb. (N.S.)*, **93(135)**, 573–575, 630, 1974. ← 462, 473, 475

[10] F. Bogomolov. Hamiltonian Kählerian manifolds. *Dokl. Akad. Nauk SSSR*, **243**, 1101–1104, 1978. ← 473

[11] R. E. Borcherds. Automorphic forms on $O_{s+2,2}(\mathbb{R})$ and infinite products. *Invent. Math.*, **120**, 161–213, 1995. ← 495, 499, 500, 505, 506

[12] R. E. Borcherds. The moduli space of Enriques surfaces and the fake monster Lie superalgebra. *Topology*, **35**, 699–710, 1996. ← 508

[13] R. E. Borcherds. Automorphic forms with singularities on Grassmannians. *Invent. Math.*, **132**, 491–562, 1998. ← 497

[14] R. E. Borcherds, L. Katzarkov, T. Pantev, and N. I. Shepherd-Barron. Families of K3 surfaces. *J. Algebraic Geom.*, **7**, 183–193, 1998. ← 505, 506

[15] A. Borel. Some metric properties of arithmetic quotients of symmetric spaces and an extension theorem. *J. Differential Geom.*, **6**, 543–560, 1972. ← 472

[16] A. Borel and L. Ji. Compactifications of symmetric and locally symmetric spaces. *Mathematics: Theory & Applications*, Birkhäuser, Boston, 2006. ← 484, 485, 486

[17] N. Bourbaki. *Groupes et Algèbres de Lie, Chapitres 4, 5 et 6*. Hermann, Paris, 1968. ← 514, 515

[18] R. Brussee. The canonical class and the C^∞ properties of Kähler surfaces. *New York J. Math.*, **2**, 103–146, 1996. ← 461

[19] N. Buchdahl. On compact Kähler surfaces. *Ann. Inst. Fourier (Grenoble)*, **49**, 287–302, 1999. ← 461

[20] D. Burns and M. Rapoport. On the Torelli problem for Kählerian K3 surfaces. *Ann. Sc. ENS.*, **8**, 235–274, 1975. ← 461, 467

[21] H. Cartan. Quotient d'un espace analytique par un groupe d'automorphismes, in: Algebraic geometry and topology, a symposium in honor of S. Lefschetz, 90–102. Princeton University Press, Princeton, 1957. ← 490

[22] C. Chevalley. Invariants of finite groups generated by reflections. *Amer. J. Math.*, **77**, 778–782, 1955. ← 483

[23] J. H. Conway and N. J. A. Sloane. Low-dimensional lattices I: quadratic forms of small determinant. *Proc. Roy. Soc. London Ser. A*, **418**(1854), 17–41, 1988. ← 513, 516

[24] O. Debarre. Un contre-exemple au théorème de Torelli pour les variétés symplectiques irréductibles. *C. R. Acad. Sci. Paris Sér. I Math.*, **299**, 681–684, 1984. ← 462, 475

[25] O. Debarre and C. Voisin. Hyper-Kähler fourfolds and Grassmann geometry. *J. Reine Angew. Math.*, **649**, 63–87, 2010. ← 482, 516

[26] I. Dolgachev. Mirror symmetry for lattice polarized K3 surfaces. *J. Math. Sci.*, **81**, 2599–2630, 1996. ← 471

[27] I. Dolgachev, B. van Geemen, and S. Kondo. A complex ball uniformization of the moduli space of cubic surfaces via periods of K3 surfaces. *J. Reine Angew. Math.*, **588**, 99–148, 2005. ← 481

[28] I. Dolgachev and S. Kondo. Moduli of K3 surfaces and complex ball quotients, in: Arithmetic and geometry around hypergeometric functions, 43–100. *Progr. Math.*, **260**, Birkhäuser, Basel, 2007. ← 481

[29] M. Dürr. *Seiberg-Witten theory and the C^∞-classification of complex surfaces*. Ph.D. thesis, Zürich, 2002. ← 461

[30] M. Eichler. Quadratische Formen und orthogonale Gruppen. *Grundlehren der mathematischen Wissenschaften*, **63**, Springer, Berlin–New York, 1974. ← 503, 509

[31] G. Faltings and C.-L. Chai. Degeneration of abelian varieties.*Ergebnisse der Mathematik und ihrer Grenzgebiete (3)*, **22**. Springer, Berlin, 1990. ← 490, 494

[32] E. Freitag. Siegelsche Modulfunktionen. *Grundlehren der mathematischen Wissenschaften*, **254**. Springer, Berlin–Göttingen–Heidelberg, 1983. ← 496, 498

[33] E. Freitag. Some modular forms related to cubic surfaces. *Kyungpook Math. J.*, **43**, 433–462, 2003. ← 508

[34] R. Friedman. A new proof of the global Torelli theorem for K3 surfaces. *Ann. of Math. (2)*, **120**, 237–269, 1984. ← 461, 482

[35] R. Friedman and J. W. Morgan. Algebraic surfaces and Seiberg-Witten invariants. *J. Algebraic Geom.*, **6**, 445–479, 1997. ← 461

[36] A. Fujiki. On the de Rham cohomology group of a compact Kähler symplectic manifold, in: Algebraic geometry (Sendai 1985), 105–165. *Adv. Stud. Pure Math.*, **10**, North-Holland, Amsterdam, 1987. ← 474

[37] A. Garbagnati and A. Sarti. Elliptic fibrations and symplectic automorphisms on K3 surfaces. *Comm. Algebra*, **37**, 3601–3631, 2009. ← 481

[38] V. Gritsenko. Jacobi functions of n-variables. *Zap. Nauk. Sem. LOMI*, **168**, 32–45, 1988. English translation in *J. Soviet Math.*, **53**, 243–252, 1991. ← 496

[39] V. Gritsenko. Irrationality of the moduli spaces of polarized Abelian surfaces. *Internat. Math. Res. Notices*, **6**, 235–243, 1994. ← 496

[40] V. Gritsenko. Modular forms and moduli spaces of Abelian and K3 surfaces. *Algebra i Analiz*, **6**, 65–102, 1994; English translation in: *St. Petersburg Math. J.*, **6**, 1179–1208, 1995. ← 495, 496, 498, 503, 511

[41] V. Gritsenko. Reflective modular forms in algebraic geometry, 28 pp. `arXiv:1005.3753` ← 500, 501

[42] V. Gritsenko and K. Hulek. Appendix to the paper "Irrationality of the moduli spaces of polarized abelian surfaces". Abelian varieties, Egloffstein 1993, 83–84. Walter de Gruyter, Berlin, 1995. ← 498

[43] V. Gritsenko and K. Hulek. Minimal Siegel modular threefolds. *Math. Proc. Cambridge Philos. Soc.*, **123**, 461–485, 1998. ← 476, 496

[44] V. Gritsenko and K. Hulek. Commutator coverings of Siegel threefolds. *Duke Math. J.*, **94**, 509–542, 1998. ← 496

[45] V. Gritsenko and K. Hulek. Uniruledness of orthogonal modular varieties. 14 pp., `arXiv:1202.3361`. ← 501

[46] V. Gritsenko, K. Hulek, and G. K. Sankaran. The Kodaira dimension of the moduli of K3 surfaces. *Invent. Math.*, **169**, 519–567, 2007. ← 461, 490, 493, 494, 495, 497, 504, 505, 507, 513, 516

[47] V. Gritsenko, K. Hulek, and G. K. Sankaran. The Hirzebruch-Mumford volume for the orthogonal group and applications. *Documenta Math.* , **12**, 215–241, 2007. ← 504

[48] V. Gritsenko, K. Hulek, and G. K. Sankaran. Hirzebruch-Mumford proportionality and locally symmetric varieties of orthogonal type. *Documenta Math.*, **13**, 1–19, 2008. ← 470, 499

[49] V. Gritsenko, K. Hulek, and G. K. Sankaran. Abelianisation of orthogonal groups and the fundamental group of modular varieties. *J. of Algebra*, **322**, 463–478, 2009. ← 469, 496, 500, 503, 509

[50] V. Gritsenko, K. Hulek, and G. K. Sankaran. Moduli spaces of irreducible symplectic manifolds. *Compositio Math.*, **146**, 404–434, 2010. ← 462, 472, 477, 478, 479, 480, 495, 502, 504, 505, 516

[51] V. Gritsenko, K. Hulek, and G. K. Sankaran. Moduli spaces of polarised symplectic O'Grady varieties and Borcherds products. *J. Differential Geom.*, **88**, 61–85, 2011. ← 462, 478, 495, 504, 505, 507, 513

[52] V. Gritsenko and V. Nikulin. K3 surfaces, Lorentzian Kac–Moody algebras and mirror symmetry. *Math. Res. Lett.*, **3**, 211–229, 1996. ← 471

[53] V. Gritsenko and V. Nikulin. Siegel automorphic form correction of some Lorentzian Kac–Moody Lie algebras. *Amer. J. Math.*, **119**, 181–224, 1997. ← 496

[54] V. Gritsenko and V. Nikulin. Automorphic forms and Lorentzian Kac-Moody algebras. I. *International J. Math.*, **9**(2), 153–200, 1998. ← 499

[55] V. Gritsenko and V. Nikulin. Automorphic forms and Lorentzian Kac-Moody algebras. II. *International J. Math.*, **9** (2), 201–275, 1998. ← 496, 497, 507, 508, 509

[56] V. Gritsenko and V. Nikulin. The arithmetic mirror symmetry and Calabi–Yau manifolds. *Comm. Math. Phys.*, **200**, 1–11, 2000. ← 471

[57] V. Gritsenko and G. K. Sankaran. Moduli of abelian surfaces with a $(1, p^2)$ polarisation. *Izv. Ross. Akad. Nauk Ser. Mat.*, **60**, 19–26, 1996 ; reprinted in: *Izv. Math.*, **60**, 893–900, 1996. ← 498

[58] B. Hassett and Y. Tschinkel. Moving and ample cones of holomorphic symplectic fourfolds. *Geometric and Functional Analysis*, **19**, 1065–1080, 2009. ← 478

[59] B. Hassett and Y. Tschinkel. Flops on holomorphic symplectic fourfolds and determinantal cubic hypersurfaces. *J. Inst. Math. Jussieu*, **9**, 125–153, 2010. ← 478

[60] S. Helgason. Differential geometry, Lie groups, and symmetric spaces. *Graduate Studies in Mathematics*, **34**, Amer. Math. Soc., Providence, 2001. ← 485

[61] E. Horikawa. Surjectivity of the period map of K3 surfaces of degree 2. *Math. Ann.*, **228**, 113–146, 1977. ← 461

[62] E. Horikawa. On the periods of Enriques surfaces I. *Proc. Japan Acad.*, **53**, 124–127, 1977. ← 481

[63] D. Huybrechts. Compact hyper-Kähler manifolds: basic results. *Invent. Math.*, **135**, 63–113, 1999. ← 462, 475

[64] D. Huybrechts. *A global Torelli theorem for hyperkähler manifolds [after Verbitsky]*. Séminaire BOURBAKI 63ème année, n 1040, 2010–2011. ← 462, 476

[65] A. Iliev and K. Ranestad. K3 surfaces of genus 8 and varieties of sums of powers of cubic fourfolds. *Trans. Amer. Math. Soc.*, **353**, 1455–1468, 2001. ← 482

[66] A. Iliev and K. Ranestad. Addendum to K3 surfaces of genus 8 and varieties of sums of powers of cubic fourfolds. *C. R. Acad. Bulgare Sci.*, **60**, 1265–1270, 2007. ← 482

[67] D. Kaledin, M. Lehn, and C. Sorger. Singular symplectic moduli spaces. *Invent. Math.*, **164**, 591–614, 2006. ← 473

[68] Y. Kawamata. Unobstructed deformations. A remark on a paper of Z. Ran. *J. Algebraic Geom.*, **1**, 183–190, 1992. ← 474

[69] D. Kazhdan. On the connection of the dual space of a group with the structure of its closed subgroups. *Funkcional. Anal. i Pril.*, **1**, 71–74, 1967. ← 496

[70] M. Kneser. Erzeugung ganzzahliger orthogonaler Gruppen durch Spiegelungen. *Math. Ann.*, **255**, 453–462, 1981. ← 480

[71] K. Kodaira. On the structure of compact complex analytic surfaces. I. *Amer. J. Math.*, **86**, 751–789, 1964. ← 461

[72] K. Kodaira and D. C. Spencer. On deformations of complex analytic structures. III. Stability theorems for complex structures. *Ann. of Math. (2)* , **71**, 43–76, 1960. ← 467

[73] J. Kollár and T. Matsusaka. Riemann-Roch type inequalities. *Amer. J. Math.*, **105**, 229–252, 1983. ← 478

[74] S. Kondo. On the Kodaira dimension of the moduli space of K3 surfaces. *Compositio Math.*, **89**, 251–299, 1993. ← 461, 494, 512

[75] S. Kondo. The rationality of the moduli space of Enriques surfaces. *Compositio Math.*, **91**, 159–173, 1994. ← 481

[76] S. Kondo. On the Kodaira dimension of the moduli space of K3 surfaces. II. *Compositio Math.*, **116**, 111–117, 1999. ← 505, 515, 516

[77] V. Kulikov. Degenerations of K3 surfaces and Enriques surfaces. *Izv. Akad. Nauk SSSR Ser. Mat.*, **41**, 1008–1042, 1199, 1977. ← 461

[78] A. Lamari. Courants kählériens et surfaces compactes. *Ann. Inst. Fourier (Grenoble)*, **49**, 263–285, 1999. ← 461

[79] D. Lieberman and D. Mumford. Matsusaka's big theorem, in: Algebraic Geometry, Arcata 1974, 513–530. *Proc. Sympos. Pure Math.*, **29**, Amer. Math. Soc., Providence, 1975. ← 472

[80] E. Looijenga. A Torelli theorem for Kähler-Einstein K3 surfaces. *Lecture Notes in Math.*, **894**, 107–112. Springer, Berlin–New York, 1981. ← 461

[81] E. Looijenga and C. Peters. Torelli theorems for Kähler K3 surfaces. *Compositio Math.*, **42**, 145–186, 1981. ← 461

[82] S. Ma. The unirationality of the moduli spaces of 2-elementary K3 surfaces. With an appendix by K. Yoshikawa. *Proc. London Math. Soc.*, doi: 10.1112/plms/pds008, 2012. ← 481

[83] E. Markman. On the monodromy of moduli spaces of sheaves on K3 surfaces II , 76 pp. math.AG/0305043 ← 476, 479

[84] E. Markman. On the monodromy of moduli spaces of sheaves on K3 surfaces. *J. Algebraic Geom.*, **17**, 29–99, 2008. ← 476

[85] E. Markman. Integral constraints on the monodromy group of the hyperkahler resolution of a symmetric product of a K3 surface. *Internat. J. Math.*, **21**, 169–223, 2010. ← 476

[86] E. Markman. A survey of Torelli and monodromy results for holomorphic-symplectic varieties, in: Complex and Differential Geometry, Hannover 2009, 257–322. *Springer Proceedings in Mathematics*, **8**, Springer, Berlin, 2011. ← 462, 476, 477, 478, 479

[87] T. Matsusaka. Polarized varieties with a given Hilbert polynomial. *Amer. J. Math.*, **94**, 1027–1077, 1972. ← 472, 478

[88] D. Morrison. Some remarks on the moduli of K3 surfaces, in: Classification of algebraic and analytic manifolds, 303–332. *Progr. Math.*, **39**, Birkhäuser, Boston, 1983. ← 470

[89] S. Mukai. Curves, K3 surfaces and Fano 3-folds of genus $\leqslant 10$, in: *Algebraic geometry and commutative algebra*, Vol. **I**, 357–377. Kinokuniya, Tokyo, 1988. ← 461, 481

[90] S. Mukai. Polarized K3 surfaces of genus 18 and 20, in: Complex projective geometry, Trieste 1989/Bergen 1989, 264–276. *London Math. Soc. Lecture Note Ser.*, **179**, Cambridge Univ. Press, Cambridge, 1992. ← 461, 481

[91] S. Mukai. Curves and K3 surfaces of genus eleven, in: Moduli of vector bundles, Sanda 1994/Kyoto 1994, 189–197. *Lecture Notes in Pure and Appl. Math.*, **179**, Dekker, New York, 1996. ← 461, 481

[92] S. Mukai. Polarized K3 surfaces of genus thirteen, in: Moduli spaces and arithmetic geometry, Kyoto 2004, 315–326, *Adv. Stud. Pure Math.*, **45**, Math. Soc. Japan, Tokyo, 2006. ← 461, 481

[93] S. Mukai. Polarized K3 surfaces of genus sixteen. Oberwolfach Report 02/2010. DOI: 10.4171/OWR/2010/02 ← 461, 481

[94] D. Mumford. Hirzebruch's proportionality theorem in the noncompact case. *Invent. Math.*, **42**, 239–272, 1977. ← 499

[95] Y. Namikawa. Counter-example to global Torelli problem for irreducible symplectic manifolds. *Math. Ann.*, **324**, 841–845, 2002. ← 462, 476

[96] V. V. Nikulin. Finite automorphism groups of Kähler K3 surfaces. *Trudy Moskov. Mat. Obshch.*, **38**, 75–137, 1979; English translation in *Trans. Mosc. Math. Soc.*, **2**, 71-135, 1980. ← 471

[97] V. V. Nikulin. Integral symmetric bilinear forms and some of their applications. *Izv. Akad. Nauk SSSR Ser. Mat.*, **43**, 111–177, 1979; English translation in *Math. USSR, Izvestiia*, **14**, 103–167, 1980. ← 469, 502, 506

[98] N. O. Nygaard. The Torelli theorem for ordinary K3 surfaces over finite fields, in: Arithmetic and geometry, Volume I. *Progr. Math.*, **36**, 267–276, Birkhäuser, Boston 1983. ← 462

[99] K. O'Grady. Desingularized moduli spaces of sheaves on a K3. *J. Reine Angew. Math.*, **512**, 49–117, 1999. ← 462, 473, 495

[100] K. O'Grady. A new six-dimensional irreducible symplectic variety. *J. Algebraic Geom.*, **12**, 435–505, 2003. ← 462, 473

[101] K. O'Grady. Irreducible symplectic 4-folds and Eisenbud-Popescu-Walter sextics. *Duke Math. J.*, **134**, 99–137, 2006. ← 482

[102] A. Ogus. A crystalline Torelli theorem for supersingular K3 surfaces, in: Arithmetic and geometry, Volume II, 361–394. *Progr. Math.*, **36**, Birkhäuser, Boston 1983. ← 462

[103] C. Okonek and A. Van de Ven. Stable bundles, instantons and C^∞-structures on algebraic surfaces, in: Several complex variables VI, 197–249. *Encyclopaedia Math. Sci.*, **69**, Springer, Berlin, 1990. ← 461

[104] M. Olsson. Semistable degenerations and period spaces for polarized K3 surfaces. *Duke Math. J.*, **125**, 121–203, 2004. ← 482

[105] U. Persson and H. Pinkham. Degeneration of surfaces with trivial canonical bundle. *Ann. of Math. (2)*, **113**, 45–66, 1981. ← 461

[106] A. Peterson and G. K. Sankaran. On some lattice computations related to moduli problems. With an appendix by V. Gritsenko. *Rend. Semin. Mat. Univ. Politec. Torino*, **68**, 289–304, 2010. ← 462

[107] I. Piatetskii-Shapiro. *Automorphic functions and the geometry of classical domains.* Gordon and Breach, New York, 1969. ← 496, 511

[108] I. Piatetskii-Shapiro and I. Shafarevich. A Torelli theorem for algebraic surfaces of type K3. *Izv. Akad. Nauk SSSR, Ser. Mat.*, **35**, 530–572, 1971; English translation in: *Math. USSR, Izv.*, **5** (1971), 547–588, 1972. ← 461

[109] Z. Ran. Deformations of manifolds with torsion or negative canonical bundle. *J. Algebraic Geom.*, **1**, , 279–291, 1992. ← 474

[110] A. Rapagnetta. Topological invariants of O'Grady's six dimensional irreducible symplectic variety. *Math. Z.*, **256**, 1–34, 2007. ← 474

[111] A. Rapagnetta. On the Beauville form of the known irreducible symplectic varieties. *Math. Ann.*, **340**, 77–95, 2008. ← 474

[112] M. Reid. Young person's guide to canonical singularities, in: Algebraic geometry, Bowdoin 1985, 345–414. *Proc. Sympos. Pure Math.*, **46**, Part 1, Amer. Math. Soc., Providence, 1987. ← 490

[113] B. Saint-Donat. Projective models of K3 surfaces. *Amer. J. Math.*, **96**,602–639, 1974. ← 468, 472

[114] I. Satake. Algebraic structures of symmetric domains. *Kano Memorial Lectures*, **4**, Iwanami Shoten, Tokyo; Princeton University Press, Princeton, 1980. ← 469, 486

[115] F. Scattone. On the compactification of moduli spaces of algebraic K3 surfaces. *Mem. Amer. Math. Soc.*, **70**(374), 1987. ← 482, 494

[116] M. Schütt. Arithmetic of K3 surfaces. *Jahresber. Deutsch. Math.-Verein.*, **111**, 23–41, 2009. ← 462

[117] J. Shah. Surjectivity of the period map in the case of quartic surfaces and sextic double planes. *Bull. Amer. Math. Soc.*, **82**, 716–718, 1976. ← 461

[118] J. Shah. A complete moduli space for K3 surfaces of degree 2. *Ann. of Math. (2)*, **112**, 485–510, 1980. ← 461, 482

[119] J. Shah. Degenerations of K3 surfaces of degree 4. *Trans. Amer. Math. Soc.*, **263**, 271–308, 1981. ← 461, 482

[120] N. I. Shepherd-Barron. The rationality of some moduli spaces of plane curves. *Compositio Math.*, **67**, 51–88, 1988. ← 481

[121] N. I. Shepherd-Barron. Apolarity and its applications. *Invent. Math.*, **97**, 433–444, 1989. ← 481

[122] Y. T. Siu. Every K3 surface is Kähler. *Invent. Math.*, **73**, 139–150, 1983. ← 461, 463

[123] H. Sterk. Lattices and K3 surfaces of degree 6. *Linear Algebra Appl.*, **226/228**, 297–309, 1995. ← 482

[124] Y. Tai. On the Kodaira dimension of the moduli space of abelian varieties. *Invent. Math.*, **68**, 425–439, 1982. ← 491

[125] G. Tian. Smoothness of the universal deformation space of compact Calabi-Yau manifolds and its Petersson-Weil metric, in: Mathematical aspects of string theory, 629–646. *Advanced Series in Mathematical Physics*, **1**, World Scientific, Singapore, 1987. ← 474

[126] A. N. Todorov. The Weil-Petersson geometry of the moduli space of SU($n \geqslant 3$) (Calabi-Yau) manifolds I. *Comm. Math. Phys.*, **126**, 325–346, 1989. ← 461, 468, 474

[127] M. Verbitsky. A global Torelli theorem for hyperkahler manifolds , 46 pp. arXiv:0908.4121v7 ← 462, 476, 477, 478

[128] E. Viehweg. Quasi-projective moduli for polarized manifolds. *Ergebnisse der Mathematik und ihrer Grenzgebiete (3)*, **30**, Springer, Berlin, 1995. ← 471

[129] C. Voisin. Théorème de Torelli pour les cubiques de $\mathbb{P}^5$. *Invent. Math.*, **86**, 577–601, 1986. ← 462, 482

[130] C. Voisin. *Géométrie des espaces de modules de courbes et de surfaces K3 [d'après Gritsenko-Hulek-Sankaran, Farkas-Popa, Mukai, Verra. . .]*. Séminaire BOURBAKI 59ème année, 2006–2007, n 981. ← 481

[131] A. Weil. *Oeuvres scientifiques – Collected papers*, Volumes I, II, III, Springer, New York–Heidelberg, 1979. ← 460

[132] M. Zowislok. On moduli spaces of sheaves on K3 or abelian surfaces, 24 pp. arXiv:1104.5629, ← 473

Université Lille 1, Laboratoire Paul Painlevé, F-59655 Villeneuve d'Ascq, Cedex, France
E-mail address: `Valery.Gritsenko@agat.univ-lille1.fr`

Institut für Algebraische Geometrie, Leibniz Universität Hannover, D-30060 Hannover, Germany
E-mail address: `hulek@math.uni-hannover.de`

Department of Mathematical Sciences, University of Bath, Bath BA2 7AY, England
E-mail address: `G.K.Sankaran@bath.ac.uk`

Normal functions and the geometry of moduli spaces of curves

Richard Hain

Contents

1. Introduction

A normal function on a complex manifold T is a special kind of holomorphic section of a bundle $\mathcal{J}(\mathbb{V}) \to T$ of compact complex tori constructed from a weight -1 variation of Hodge structure $\mathbb{V}$ over T. Normal functions in their modern formulation arose in the work [12] of Griffiths as a tool for understanding algebraic cycles in a complex projective manifold.

In this paper we give two examples to illustrate how normal functions might be a useful, if unconventional, tool for understanding the geometry of moduli spaces of curves. The first is to give a partial answer to a question of Eliashberg, which arose in symplectic field theory [10]. Namely, we compute the class in rational cohomology of the pullback of the 0-section of the universal jacobian $\mathcal{J}^c_{g,n} \to \mathcal{M}^c_{g,n}$ over the

Supported in part by grants DMS-0103667 and DMS-1005675 from the National Science Foundation.

moduli space of n-pointed, stable projective curves of compact type of genus g along the section defined by

$$F_{\mathbf{d}} : [C; x_1, \ldots, x_n] \mapsto \sum_{j=1}^{n} d_j [x_j] \in \operatorname{Jac} C,$$

where $\mathbf{d} = (d_1, \ldots, d_n) \in \mathbb{Z}^n$ satisfies $\sum_{j=1}^{n} d_j = 0$. This section is a normal function.

The second application is to give an alternative and complementary approach to the slope inequalities of the type discovered by Moriwaki [32, 33] such as his result that the divisor class

$$M := (8g+4)\lambda_1 - g\delta_0 - 4 \sum_{h=1}^{\lfloor g/2 \rfloor} h(g-h)\delta_h$$

on $\overline{\mathcal{M}}_g$ has non-negative degree on all complete curves in $\overline{\mathcal{M}}_g$ that do not lie in the boundary divisor $\Delta := \overline{\mathcal{M}}_g - \mathcal{M}_g$.[1] (Here the δ_h $(0 \leqslant h \leqslant g/2)$ denote the classes of the components of Δ.) This alternative approach leads to actual and conjectural strengthening of his inequalities: we show that the Moriwaki divisor M has non-negative degree on all complete curves in $\mathcal{M}_g^c$ and conjecture that M has non-negative degree on all complete curves in $\overline{\mathcal{M}}_g$ that do not lie in the boundary divisor $\Delta_0 := \overline{\mathcal{M}}_g - \mathcal{M}_g^c$.

Each normal function section ν of $J(\mathbb{V}) \to T$ determines a class $c(\nu)$ in $H^1(T, \mathbb{V})$. When $H^0(T, \mathbb{V}_{\mathbb{Q}})$ vanishes, the normal function ν is determined, mod torsion, by its characteristic class $c(\nu)$. In such cases, there is a rigid relationship between normal functions and cohomology. When $T = \mathcal{M}_{g,n}^c$, $g \geqslant 3$, and $\mathbb{V}$ is a variation of Hodge structure corresponding to a non-trivial rational representation of Sp_g that does not contain the trivial representation, the group of normal function sections of $J(\mathbb{V})$ is finitely generated and is known modulo torsion, [15, §8]. This result is recalled in Appendix A. When $\mathbb{V}$ corresponds to the fundamental representation of Sp_g, $J(\mathbb{V})$ is the universal jacobian $\mathcal{J}_{g,n}^c$ over $\mathcal{M}_{g,n}^c$ and the class of the pullback $F_{\mathbf{d}}^* \eta_g$ of the zero section of $\mathcal{J}_{g,n}^c$ can be expressed in terms of the classes of certain basic normal functions defined on $\mathcal{M}_{g,n}^c$.

All variations of Hodge structure $\mathbb{V}$ of geometric origin have a *polarization*; that is, an invariant inner product $S : \mathbb{V}^{\otimes 2} \to \mathbb{Q}$ that satisfies the Riemann-Hodge bilinear relations on each fiber. When V has odd weight, the polarization is skew symmetric. Each invariant, skew-symmetric inner product $S : \mathbb{V}^{\otimes 2} \to \mathbb{Q}$ gives rise to a class $S \circ c(\nu)^2$ in $H^2(T, \mathbb{Q})$. It is the image of $c(\nu)^{\otimes 2}$ under the composition of the cup product with the map induced by S:

$$H^1(T, \mathbb{V})^{\otimes 2} \xrightarrow{\smile} H^2(T, \mathbb{V}^{\otimes 2}) \xrightarrow{S_*} H^2(T, \mathbb{Q}).$$

The class $S \circ c(\nu)^2$ has a natural de Rham representative which is a non-negative (1,1)-form when S is a polarization. Moriwaki's inequality for complete curves in

[1] A weaker version of this inequality had been proved previously by Cornalba and Harris in [7].

$\mathcal{M}_g^c$ is an immediate consequence of this semi-positivity and the fact that the class of the Moriwaki divisor equals the square $S \circ c(\nu)^2$ of the class of the most fundamental normal function over $\mathcal{M}_g$ — viz., the normal function associated to the cycle $C - C^-$ in $\operatorname{Jac} C$ that was first studied by Ceresa [6].

When $\mathbb{V}$ is a weight -1 polarized variation of Hodge structure over T, there is a naturally metrized line bundle $\mathcal{B}$ over $J(\mathbb{V})$, which is called the *biextension line bundle*. The curvature of its pullback along a normal function $\nu : T \to J(\mathbb{V})$ is the natural de Rham representative of $S \circ c(\nu)^2$. This line bundle extends naturally to any compactification $\overline{T}$ of T, even if $\mathbb{V}$ (and hence ν) does not extend to $\overline{T}$. This extension is characterized by the property that the metric extends across the codimension 1 strata of $\overline{T} - T$. Surprisingly, this extension is *not* natural under pullback to a smooth variety as the metric on the extended line bundle may be singular on strata of codimension $\geqslant 2$ of $\overline{T} - T$. This curious phenomenon is called *height jumping*. Height jumping and its relevance to refined slope inequalities is discussed in Section 14.

The classes $c(\nu)$ form part of a larger structure. When $T = \mathcal{M}_{g,n}$ they should be regarded as twisted tautological cohomology classes. To explain this, we need to introduce a certain graded commutative algebra associated to $\mathcal{M}_{g,n}$. Denote the coordinate ring of the symplectic group Sp_g by $\mathcal{O}(\mathrm{Sp}_g)$. Left and right multiplication induce commuting left and right actions of the mapping class group $\pi_1(\mathcal{M}_{g,n}, *)$ on it via the standard representation $\pi_1(\mathcal{M}_{g,n}, x_o) \to \mathrm{Sp}_g(\mathbb{Q})$. Using the right action, one obtains a local system $\mathbb{O}$ of $\mathbb{Q}$-algebras over $\mathcal{M}_{g,n}$. Since $\mathbb{O}$ is a local system of commutative $\mathbb{Q}$-algebras, its cohomology

$$A^\bullet_{g,n} := H^\bullet(\mathcal{M}_{g,n}, \mathbb{O})$$

is a graded commutative $\mathbb{Q}$-algebra. The left Sp_g-action on $\mathbb{O}$ gives it the structure of a graded commutative algebra in the category of Sp_g-modules.

The algebraic analogue of the Peter-Weyl Theorem implies that there is an $\mathrm{Sp}_g \times \mathrm{Sp}_g$-equivariant isomorphism

$$\mathcal{O}(\mathrm{Sp}_g) \cong \bigoplus_\lambda \operatorname{End}(V_\lambda)^* \cong \bigoplus_\lambda V_\lambda \boxtimes V_\lambda^*,$$

where $\{V_\lambda\}$ is a set of representatives of the isomorphism classes of irreducible Sp_g-modules. There is thus an isomorphism

$$A^\bullet_{g,n} \cong \bigoplus_\lambda H^\bullet(\mathcal{M}_{g,n}, \mathbb{V}_\lambda) \otimes V_\lambda^*$$

where $\mathbb{V}_\lambda$ denotes the local system over $\mathcal{M}_{g,n}$ that corresponds to V_λ.

The classes $c(\nu)$ are more fundamental than their squares $S \circ c(\nu)^2$, which are known to be tautological classes. For this reason, we define the tautological subalgebra $T^\bullet_{g,n}$ of $A^\bullet_{g,n}$ as the graded subalgebra generated by the classes $c(\nu) \otimes V_\lambda^*$ of the normal function sections ν of the $J(\mathbb{V}_\lambda)$.[2] It is finitely generated. The classification of normal functions [15] over $\mathcal{M}_{g,n}$, and the work of Kawazumi and Morita [25] imply that the ring of Sp_g-invariants $(T^\bullet_{g,n})^{\mathrm{Sp}_g}$ is Faber's tautological ring $R_{g,n}$ (in cohomology) [11] of $\mathcal{M}_{g,n}$. The computations of Morita [31], Kawazumi and Morita [25], and those of this paper, may be regarded as computations in $T^\bullet_{g,n}$. For this reason, we propose that the ring $T^\bullet_{g,n}$ is more fundamental than its Sp_g-invariant part $R_{g,n}$. It would be interesting to define and study a Chow analogue of $T^\bullet_{g,n}$. The significance of this algebra and its relation to normal functions is discussed in Appendix B.

Advice to the reader: Although normal functions have long been a part of algebraic geometry (examples were first considered by Poincaré), they are not currently part of the standard repertoire of modern algebraic geometry. Their modern definition (Definition 5.2), in terms of extensions of variations of Hodge structure, requires an understanding of variations of (mixed) Hodge structure. However, if the reader is prepared to believe that the local systems associated to locally topologically trivial families of algebraic varieties are motivic, and so are variations of mixed Hodge structure, then the definition should be natural.

We assume the reader is familiar with the basic definitions and constructions of Hodge theory. In particular, the reader should know the definition of Hodge structures, mixed Hodge structures, and variations of Hodge structure. The book [35] by Peters and Steenbrink is a good source of basic material on these topics. The paper [18] contains a brief exposition of Schmid's work [40] on the asymptotic properties of variations of Hodge structure that emphasizes the case of degenerations of curves. Finally, the recent survey of normal functions [26] by Kerr and Pearlstein should be a useful supplement, although its emphasis is quite different from that of this article.

Acknowledgments: I am grateful to Yasha Eliashberg for posing his question to me in 2001 and to Gavril Farkas for his interest, which resulted in this paper seeing the light of day. Theorem 11.1 was proved during visits to the University of Sydney and the Université de Nice during the author's sabbatical in 2002–03. Many thanks to both institutions for their support, and to my respective hosts, Gus Lehrer and Arnaud Beauville, for their hospitality. I am especially grateful to Samuel Grushevsky, Robin de Jong and Dmitry Zharkov for their interest in this work and for their constructive comments and corrections. Sam and Dmitry pointed out an error in the coefficient of δ_h^P in Theorem 11.1; Sam and Robin isolated the error, which was a missing term in the formula Theorem 10.4.

I would also like to thank Gregory Pearlstein for his numerous constructive comments on the manuscript, and also Renzo Cavalieri for communicating his

[2]Each $\mathbb{V}_\lambda$ is the local system that underlies a polarized variation of Hodge structure over $\mathcal{M}_{g,n}$. It is unique up to Tate twist. The only $\mathbb{V}_\lambda$ that admit non-torsion normal functions have weight -1.

related results [5] with Steffen Marcus. The section on the genus 1 case of Eliashberg's problem was written as a result of correspondence with him. Finally, I would like to thank the referee for helpful comments.

2. Notation and conventions

All varieties (and stacks) will be defined over the complex numbers. Denote the moduli space of stable n-pointed curves of genus g by $\overline{\mathcal{M}}_{g,n}$. This is defined when $2g-2+n>0$ and will be viewed as a stack or as a complex analytic orbifold. As such, it is smooth. Denote the Zariski open subset corresponding to the set of n-pointed smooth curves by $\mathcal{M}_{g,n}$ and by $\mathcal{M}_{g,n}^c$ the Zariski open subset consisting of n-pointed curves of compact type. Note that $\mathcal{M}_{g,n}^c = \overline{\mathcal{M}}_{g,n} - \Delta_0$, where Δ_0 denotes the boundary divisor of $\overline{\mathcal{M}}_{g,n}$ whose generic point is an irreducible, geometrically connected curve with one node.

The moduli stack of principally polarized abelian varieties of dimension g will be denoted by $\mathcal{A}_g$. The universal curve of compact type will be denoted by $\mathcal{C}_g^c \to \mathcal{M}_g^c$ and its restriction to $\mathcal{M}_g$ by $\mathcal{C}_g$. All of these moduli spaces are globally the quotient of a smooth variety by a finite group, [29, 1].

Vector bundles, variations of (mixed) Hodge structure, etc on a stack T that is the quotient of a smooth variety S by a finite group G, are G-invariant bundles, variations of (mixed) Hodge structure, etc, over S. Since all stacks that occur in this paper are of this form, we will not distinguish between stacks and varieties, as working on one of these stacks is working equivariantly on a finite cover that is a variety.

The category of $\mathbb{Z}$-mixed Hodge structures by MHS. For $d \in \mathbb{Z}$, the Hodge structure of type $(-d,-d)$ whose underlying lattice is isomorphic to $\mathbb{Z}$ will be denoted by $\mathbb{Z}(d)$. We shall denote by ΓV the set $\mathrm{Hom}_{\mathsf{MHS}}(\mathbb{Z}(0), V)$ of Hodge classes of type (0,0) of the mixed Hodge structure V . The category of admissible variations of mixed Hodge structure over a smooth variety X will be denoted by $\mathsf{MHS}(X)$.

All cohomology groups will be with $\mathbb{Q}$ coefficients unless otherwise stated. Similarly, the Chow group of codimension d cycles on a stack X, *tensored with* $\mathbb{Q}$, will be denoted by $CH^d(X)$.

3. Eliashberg's problem

To motivate the discussion of normal functions and related topics in subsequent sections, we begin with a brief discussion of the universal jacobian and Eliashberg's problem. Some readers may prefer to begin with Sections 4 and 5. Recall that a stable curve C of genus g is of *compact type* if its dual graph is a tree. This condition is equivalent to the condition that its jacobian $\mathrm{Jac}\, C := \mathrm{Pic}^0 C$ be an abelian variety.

3.1. The universal jacobian

We begin with a review of the transcendental construction of the jacobian of the universal curve over $\mathcal{M}_g$. This is a special case of Griffiths' construction of families of intermediate jacobians and normal functions, which are reviewed in Section 5.

First recall the transcendental construction of the jacobian of a smooth projective curve C, which we recast in the language of Hodge structures. It will be generalized in Section 5 where Griffiths intermediate jacobians are introduced. The Hodge Theorem implies that

$$H^1(C,\mathbb{C}) \cong H^{1,0}(C) \oplus H^{0,1}(C) \tag{3.1}$$

where $H^{1,0}(C)$ denotes the space $H^0(C,\Omega^1_C)$ of holomorphic 1-forms on C and $H^{0,1}(C)$ its complex conjugate, the space of anti-holomorphic 1-forms. The first integral cohomology group $H^1(C,\mathbb{Z})$ endowed with the decomposition (3.1) is the prototypical Hodge structure of weight 1. Its dual

$$H_1(C,\mathbb{C}) = H^{-1,0}(C) \oplus H^{0,-1}(C)$$

is a Hodge structure of weight -1, where $H^{-p,-q}(C)$ is defined to be the dual of $H^{p,q}(C)$. The Hodge filtration

$$H_1(C,\mathbb{C}) = F^{-1}H_1(C) \supset F^0H_1(C) \supset F^1H_1(C) = 0.$$

of $H_1(C)$ is defined by

$$F^pH_1(C) = \bigoplus_{\substack{s\geqslant p\\ s+t=-1}} H^{s,t}(C).$$

The projection onto $H^{1,0}(C)$ induces an isomorphism

$$H_1(C,\mathbb{C})/F^0 \cong H^{-1,0}(C) \cong H^0(C,\Omega^1_C)^* := \mathrm{Hom}_{\mathbb{C}}(H^0(C,\Omega^1_C),\mathbb{C}).$$

The composite

$$H_1(C,\mathbb{Z}) \hookrightarrow H_1(C,\mathbb{C}) \to H_1(C,\mathbb{C})/F^0 \cong H^0(C,\Omega^1_C)^*$$

is the map that takes the homology class of the 1-cycle γ to the functional

$$\int_\gamma := \left\{\omega \mapsto \int_\gamma \omega\right\} \in H^0(C,\Omega^1_C)^*.$$

It is injective and its image is a lattice. The jacobian of C is the quotient

$$\operatorname{Jac} C := H^0(C,\Omega^1_C)^*/H_1(C,\mathbb{Z}) \cong H_1(C,\mathbb{C})/(H_1(C,\mathbb{Z}) + F^0H_1(C)).$$

Every divisor D of degree 0 on C can be written as the boundary $D = \partial\gamma$ of a real 1-chain γ. The Abel-Jacobi mapping

$$\{\text{divisors of degree 0 on C}\}/\text{rational equivalence} \to \operatorname{Jac} C$$

is defined by taking the divisor class of the boundary of the 1-chain γ to the functional $\int_\gamma$. Abel's Theorem implies that it is a group isomorphism. This construction works equally well when C is a curve of compact type. In this case, the Hodge structure on

$H_1(C)$ is the direct sum of the Hodge structures of its irreducible components. This construction can also be carried out for families of complete curves, where each fiber is either smooth or of compact type. Below we carry this out for the universal curve of compact type. The family of Hodge structures associated to such a family is an example of a *variation of Hodge structure*.

To the universal curve $\pi : \mathcal{C}_g^c \to \mathcal{M}_g^c$ of compact type we associate the variation of Hodge structure

$$\mathbb{H} := R^1\pi_*\mathbb{Z}$$

and the corresponding holomorphic vector bundle $\mathcal{H} := \mathbb{H} \otimes_{\mathbb{Z}} \mathcal{O}_{\mathcal{M}_g^c}$. The fiber of $\mathcal{H}$ over the moduli point of C is $H^1(C, \mathbb{C})$. This bundle has a flat holomorphic connection ∇. The local monodromy transformations about the divisor $\Delta_0 := \overline{\mathcal{M}}_{g,n} - \mathcal{M}_{g,n}^c$ are given by the Picard-Lefschetz formula and are therefore unipotent. Consequently, $\mathcal{H}$ has a canonical extension (in the sense of Deligne [8]) to a vector bundle $\overline{\mathcal{H}}$ over $\overline{\mathcal{M}}_g$.[3] It is characterized by the property that the connection is regular and that its residue at each smooth point of Δ_0 is nilpotent. Since the monodromy of $\mathbb{H}$ about Δ_0 is non-trivial, the local system $\mathbb{H}$ does not extend across Δ_0. Consequently, $\mathcal{M}_{g,n}^c$ is the maximal Zariski open subset of $\overline{\mathcal{M}}_{g,n}$ to which $\mathbb{H}$ extends.

The *Hodge bundle* $\mathcal{F} := \mathcal{F}^1\mathcal{H}$ is the sub-bundle of $\mathcal{H}$ whose fiber over the moduli point of C is $H^{1,0}(C)$. It is holomorphic and extends, by a result of Deligne [8], to a holomorphic sub-bundle of $\overline{\mathcal{F}}^1$ of $\overline{\mathcal{H}}$.[4] (See [18, §4] for an exposition.) There is a natural projection $\check{\mathcal{F}} \to \mathcal{J}_g$ from the dual of the Hodge bundle to the universal jacobian, which is a covering map on each fiber. The kernel of the projection

$$\check{\mathcal{F}}_C \to \mathcal{J}_{g,C}$$

at the moduli point of the stable curve C is the image of $H_1(C - C^{\text{sing}}, \mathbb{Z})$ under the integration map $H_1(C - C^{\text{sing}}, \mathbb{Z}) \to \check{\mathcal{F}}_C$. The restriction of $\mathcal{J}_g$ to $\overline{\mathcal{M}}_g - \Delta_0^{\text{sing}}$ is a Hausdorff complex analytic orbifold, a fact which follows, for example, from [44, Prop. 2.9]. It is the analytic orbifold associated to the restriction of the universal Pic^0 stack over $\overline{\mathcal{M}}_g$ to $\overline{\mathcal{M}}_g - \Delta_0^{\text{sing}}$, which is constructed in [3].

Observe that the fiber of $\mathcal{J}_g \to \overline{\mathcal{M}}_g$ over the moduli point of a stable curve C is an abelian variety if and only if C is of compact type. From the construction, it is clear that the normal bundle of the zero section of $\mathcal{J}_g$ is the dual $\check{\mathcal{F}}$ of the Hodge bundle.

---

[3] See [18] for a concise exposition.

[4] The fiber $\mathcal{F}_C$ of the Hodge bundle over the stable curve C can be described as follows. Denote the normalization of C by $\nu : \tilde{C} \to C$. Let $D \subset \tilde{C}$ be the inverse image of the double points of C. Then $\mathcal{H}_C$ is the subset of $H^0(\tilde{C}, \Omega^1_{\tilde{C}}(D))$ consisting of those w such that $\text{Res}_P\, w + \text{Res}_Q\, w = 0$ whenever $P \neq Q$ and $\nu(P) = \nu(Q)$. It is naturally isomorphic to $F^1H^1(C_{\vec{v}})$ where $H^1(C_{\vec{v}})$ denotes the limit MHS on the first order smoothing $C_{\vec{v}}$ of C associated to a tangent vector $\vec{v}$ of $\overline{\mathcal{M}}_g$ at $[C]$ which is not tangent to the boundary divisor.

3.2. Eliashberg's question

Suppose that $2g - 2 + n > 0$. Given an integer vector $\mathbf{d} = (d_1, \ldots, d_n)$ with $\sum_j d_j = 0$, we have the *rational* section F_d

$$\begin{array}{ccc} & & \mathcal{J}_g \\ & \overset{F_d}{\nearrow} & \downarrow \\ \overline{\mathcal{M}}_{g,n} & \longrightarrow & \overline{\mathcal{M}}_g \end{array}$$

of the universal jacobian defined by

$$F_d : [C; x_1, \ldots, x_d] \mapsto \left[\sum_{j=1}^{n} d_j x_j\right] \in \operatorname{Jac} C$$

when C is smooth. It is holomorphic over $\overline{\mathcal{M}}_{g,n} - \Delta_0^{\mathrm{sing}}$, the complement of the singular locus of Δ_0.

Denote the class of the zero section of $\mathcal{J}_g$ in $H^{2g}(\mathcal{J}_g, \mathbb{Q})$ by η_g.

Problem 3.2 (Eliashberg). Compute the class in $H^{2g}(\overline{\mathcal{M}}_{g,n} - \Delta_0^{\mathrm{sing}})$ of the pullback $F_d^*\eta_g$ of the zero section of $\mathcal{J}_g$.

Denote the jth Chern class of the Hodge bundle by λ_j. When all d_j are zero, the section F_d is defined on all of $\overline{\mathcal{M}}_{g,n}$. Since the normal bundle of the zero section is $\check{\mathcal{F}}$, the dual of the Hodge bundle, we have:

Proposition 3.3. *If* $\mathbf{d} = 0$, *then* $F_d^*\eta_g = (-1)^g\lambda_g \in H^{2g}(\overline{\mathcal{M}}_{g,n}, \mathbb{Q})$.

Remark 3.4. This result also holds in the Chow ring.

4. Families of compact tori

The restriction of the universal jacobian to $\mathcal{M}_{g,n}^c$ is a family of compact tori. This section is a discussion of some general properties of families of compact tori.

Definition 4.1. A family of compact (real) r-dimensional tori is a smooth fiber bundle $f : T \to B$, each of whose fibers is a compact, connected abelian Lie group. This bundle is locally (but typically not globally) trivial as a bundle of Lie groups.

We shall assume throughout that B is connected. The identity section will be denoted by $s : B \to T$. Denote the fiber of T over $b \in B$ by T_b.

For a coefficient ring R, denote the local system over B whose fiber over b is $H_1(T_b, R)$ by $\mathbb{H}_R$. The following assertion is easily proved.

Proposition 4.2. *If* $f : T \to B$ *is a family of compact tori, then there is a natural bijection*

$$T \to \mathbb{H}_{\mathbb{R}}/\mathbb{H}_{\mathbb{Z}}$$

which commutes with the projections to B *and is a group homomorphism on each fiber.* □

The flat structure on $\mathbb{H}_\mathbb{R}$ descends to a flat structure on $T = \mathbb{H}_\mathbb{R}/\mathbb{H}_\mathbb{Z}$.

Corollary 4.3. *Every bundle of compact tori has a natural flat structure in which the torsion multi-sections are leaves. Equivalently, a bundle of compact real tori has a natural trivialization over each contractible subset of* B *in which the torsion sections are constant.* □

4.1. The class of a section

Each section s of a family $T \to B$ of compact tori determines a class $c(s) \in H^1(B, \mathbb{H}_\mathbb{Z})$. We review three standard constructions of this class.

The first is sheaf theoretic. Denote the sheaf of C^∞ real-valued functions on B by $\mathcal{E}_B$. The flat vector bundle associated to $\mathbb{H}_\mathbb{R}$ has sheaf of sections $\mathcal{H} := \mathbb{H}_\mathbb{R} \otimes_\mathbb{R} \mathcal{E}_B$. Denote the sheaf of smooth sections of $T \to B$ by $\mathcal{T}$. Then one has a short exact sequence

$$0 \to \mathbb{H}_\mathbb{Z} \to \mathcal{H} \to \mathcal{T} \to 0$$

of sheaves. Taking cohomology yields the exact sequence

$$0 \to H^0(B, \mathbb{H}_\mathbb{Z}) \to H^0(B, \mathcal{H}) \to H^0(B, \mathcal{T}) \xrightarrow{c} H^1(B, \mathbb{H}_\mathbb{Z}) \to 0.$$

The connecting homomorphism is well defined up to a sign. With the appropriate choice it takes a section s to its characteristic class. Note that the vanishing of $c(s)$ implies that s is homotopic to the zero section. Thus $H^1(B, \mathbb{H}_\mathbb{Z})$ can be identified with the group of homotopy classes of smooth sections of $T \to B$.

The second description is obtained by regarding $H^1(B, \mathbb{H}_\mathbb{Z})$ as congruence classes of extensions

$$0 \to \mathbb{H}_\mathbb{Z} \to \mathbb{E} \to \mathbb{Z}_B \to 0$$

of local systems over B. Given a section s of $T \to B$, we can construct such an extension $\mathbb{E}$ as the local system whose fiber over $b \in B$ is $H_1(\widehat{T}_b, \mathbb{Z})$, where

$$\widehat{T}_b := T_b \cup_h [0,1]$$

where $h(0) = 0$ and $h(1) = s(b)$. There is a short exact sequence

$$0 \to H_1(T_b) \to H_1(\widehat{T}_b) \to \mathbb{Z} \to 0$$

in which $H_1([0,1], \{0,1\})$ is identified with $\mathbb{Z}$ by taking the generator to be the class of a path from 1 to 0.

When $s(b) \neq 0$, $H_1(\widehat{T}_b) \cong H_1(T_b, \{0, s(b)\}) \cong H^{r-1}(T_b - \{0, s(b)\})$. The first description of c is determined only up to a sign. This description fixes the sign.

The third description uses de Rham cohomology. Each $b \in B$ has an open neighbourhood U where s lifts to a section $\tilde{s} : U \to \mathcal{H}$ of the flat vector bundle $\mathcal{H}$. When U is connected, such a lift is unique up to translation by a local section of $\mathbb{H}_\mathbb{Z}$. The 1-form $d\tilde{s}$ on U with values in $\mathbb{H}_\mathbb{R}$ is therefore independent of the choice of the lift $\tilde{s}$. The de Rham representative of $c(s)$ is the class that is locally represented by $d\tilde{s}$.

4.1.1. Equivalence of these constructions Here is a quick sketch of the equivalence of these definitions. Choose an open covering $\mathcal{U} = \{U_\alpha\}$ of T such that $U_{\alpha_0} \cap \cdots \cap U_{\alpha_q}$ is contractible for all multi-indices $(\alpha_0, \dots, \alpha_q)$. Such an open covering can be constructed by taking each U_α to be a geodesically convex ball with respect to some riemannian metric on T. The complex $C^\bullet(\mathcal{U}, \mathcal{F})$ of Cech cochains with coefficients in $\mathcal{F}$ computes $H^\bullet(T, \mathcal{F})$ when $\mathcal{F}$ is $\mathbb{H}_\mathbb{Z}$, $\mathbb{H}_\mathbb{R}$, $\mathcal{H}$ and $\mathcal{T}$. Suppose that s is a smooth section of $T \to B$. Its restriction to U_α can be lifted to a section s_α of $\mathcal{H}$. The difference $c_{\alpha\beta} := s_\beta - s_\alpha$ is a section of $\mathbb{H}_\mathbb{Z}$ over $U_\alpha \cap U_\beta$. The class of $c(s)$ is represented by the cocycle $(c_{\alpha\beta}) \in C^1(\mathcal{U}, \mathbb{H}_\mathbb{Z})$.

The class of an extension

$$0 \to \mathbb{H}_\mathbb{Z} \to \mathbb{E} \to \mathbb{Z}_T \to 0.$$

is computed by choosing sections e_α of $\mathbb{E} \to \mathbb{Z}_T$ over each U_α. The class of the extension is represented by the cocycle $(e_\beta - e_\alpha)_{\alpha\beta} \in C^1(\mathcal{U}, \mathbb{H}_\mathbb{Z})$. In the case above, one can take the local section e_α to be the class in $H_1(T_b, \{0, s(b)\})$ of the image of the path in the universal covering $\mathbb{H}_{\mathbb{R},b}$ of T_b that goes from 0 to $s_\alpha(b)$. Then $e_\beta - e_\alpha = c_{\alpha\beta}$, as required.

Denote the de Rham sheaf of smooth $\mathbb{R}$-valued forms on T by $\mathcal{E}^\bullet_T$. Standard arguments imply that the inclusions

$$C^\bullet(\mathcal{U}, \mathbb{H}_\mathbb{R}) \hookrightarrow C^\bullet(\mathcal{U}, \mathcal{E}^\bullet_T \otimes \mathbb{H}_\mathbb{R}) \hookleftarrow \mathcal{E}^\bullet_T \otimes \mathbb{H}_\mathbb{R}$$

induce isomorphisms on homology. The standard zig-zag argument implies that the class of the cocycle $(c_{\alpha\beta})$ is represented by the element of

$$E^\bullet(T, \mathbb{H}_\mathbb{R}) := H^0(T, \mathcal{E}^\bullet_T \otimes \mathbb{H}_\mathbb{R})$$

whose restriction to U_α is ds_α.

4.2. Invariant cohomology classes

The flat structure of a family of compact tori $T \to B$ can be used to construct natural de Rham representatives of cohomology classes on T. We will say that a differential form w on a manifold M with a foliation $\mathcal{L}$ is *parallel with respect to* $\mathcal{L}$ if the Lie derivative of w with respect to each vector field tangent to $\mathcal{L}$ vanishes. A family of tori is foliated as it is a flat family of tori.

The following lemma is proved in [23, Lemma 5.1].

Lemma 4.4. *If* $f : T \to B$ *is a family of compact tori, there is a natural mapping*

$$\sigma : H^0(B, R^k f_* \mathbb{R}) \to H^k(T, \mathbb{R})$$

whose composition with the projection

$$H^k(T, \mathbb{R}) \to H^0(B, R^k f_* \mathbb{R})$$

is the identity. Moreover, for each $u \in H^0(B, R^k f_* \mathbb{R})$, *the extended class* $\sigma(u)$ *has a natural differential form representative* w_u *whose restriction to each fiber is a closed, translation-invariant differential form, and which is parallel with respect to the flat structure. This class has the property that its restriction to every leaf (such as the zero section and every torsion multi-section) is zero.* □

4.3. The Poincaré dual of the zero section

Let r be the real dimension of the fiber of $f : T \to B$. If B and T are oriented and B is connected, then

$$H^0(B, R^r f_* \mathbb{R}) \cong \mathbb{R}.$$

Let u be the element of this group whose value on one (and hence all) fibers is 1. Denote the class $\sigma(u) \in H^r(T, \mathbb{R})$ by ψ.

Proposition 4.5. *If the base* B *is a compact manifold (possibly with boundary), then the Poincaré dual of the zero section is* ψ.

Proof. Set $d = \dim_{\mathbb{R}} B$. For $e \in \mathbb{Z}$ define $[e] : T \to T$ to be the map whose restriction to each fiber is multiplication by e. Since $[e]$ induces multiplication by e^k on $R^k f_* \mathbb{Q}$, it follows that the Leray spectral sequence degenerates at E_2. It also follows that the eigenvalues of the induced mapping $[e]^*$ on $H^k(T)$ lie in $\{1, e, \dots, e^k\}$. Since none of these eigenvalues is zero when $e \neq 0$, $[e]^*$ is invertible on rational cohomology when $e \neq 0$. Similar assertions hold for the homology spectral sequence

$$H_s(B, \partial B; \mathbb{H}_t) \implies H_{s+t}(T, \partial T),$$

where $\mathbb{H}_t$ denotes the local system whose fiber over $b \in B$ is $H_t(T_b)$. Since

$$H_{r+d}(T, \partial T) = H_d(B, \partial B; \mathbb{H}_r)$$

it follows that $[e]_*[T] = e^r[T]$, where $[T]$ denotes the fundamental class of $(T, \partial T)$.

Now assume that $e > 1$. Then the collapsing of the Leray spectral sequence implies that the dimension of the e^r-eigenspace of $H^r(T)$ is one. Since the form w_ψ that naturally represents ψ has the property that

$$[e]^* w_\psi = e^r w_\psi,$$

and since ψ is non-trivial (as it has non trivial integral over a fiber), it follows that ψ spans this eigenspace.

Note that the class $[Z] \in H_d(T, \partial T)$ of the zero-section is an eigenvector of $[e]_*$ with eigenvalue 1. Denote the Poincaré dual of the zero section by η_Z. It is characterized by the property that

$$[T] \cap \eta_Z = [Z] \in H_d(T, \partial T),$$

where $\cap$ denotes the cap product [41, p. 254]

$$\cap : H_{d+r}(T, \partial T) \otimes H^r(T) \to H_d(T, \partial T).$$

Since $e_*[Z] = [Z]$, standard properties of the cap product [41, Assertion 16, p. 254], we have

$$[e]_*([T] \cap [e]^*\eta_Z) = ([e]_*[T]) \cap \eta_Z = e^r[T] \cap \eta_Z = [e]_*([T] \cap e^r\eta_Z).$$

Since $[e]_*$ and capping with $[T]$ are both isomorphisms, it follows that $[e]^*\eta_Z = e^r\eta_Z$. Since η_Z and ψ both lie in the e^r-eigenspace and agree on each fiber, they are equal. □

Since the normal bundle of the zero section Z is the flat bundle $\mathbb{H}_{\mathbb{R}}$, the Euler class of the normal bundle of Z vanishes in rational cohomology.

Corollary 4.6. *The restriction of the Poincaré dual of the zero section Z of a family of compact tori to Z vanishes in rational cohomology.* □

Combined with Proposition 3.3, this implies the well-known property of the top Chern class of the Hodge bundle.

Corollary 4.7. *The restriction of λ_g to $\mathcal{M}_g^c$ vanishes in rational cohomology.*

4.4. The class $S \circ c(s)^2$

Suppose that $f : T \to B$ is a flat family of tori and that $\mathbb{H}$ is the corresponding local system such that $T = \mathbb{H}_{\mathbb{R}}/\mathbb{H}_{\mathbb{Z}}$. A flat, skew-symmetric inner product $S : \mathbb{H}_{\mathbb{R}} \otimes \mathbb{H}_{\mathbb{R}} \to \mathbb{R}$ gives an element of $H^0(T, R^2f_*\mathbb{R}_T)$. Lemma 4.4 implies that S determines a closed 2-form ϕ_S on T. It is characterized by the properties:

(i) its restriction to the fiber T_b of T is the translation invariant 2-form on T_b that corresponds to S,
(ii) it is parallel with respect to the flat structure on T,
(iii) its restriction to the zero-section of T is zero.

The following result is easily proved using the de Rham description of $c(s)$.

Proposition 4.8. *If s is a holomorphic section of $T \to B$, then the form $s^*\phi_S$ represents the cohomology class $S \circ c(s)^2 \in H^2(T, \mathbb{R})$.* □

5. Normal functions

Normal functions are our primary tool. They are holomorphic sections of families of intermediate jacobians that satisfy certain infinitesimal and asymptotic conditions. In this section, we recall Griffiths construction of intermediate jacobians and of the normal function associated to a family of homologically trivial algebraic cycles in a family of smooth projective varieties.

5.1. Intermediate jacobians

Suppose that Y is a compact Kähler manifold and that Z is an algebraic d-cycle in Y where $0 \leqslant d < \dim Y$. One has the exact sequence

$$0 \to H_{2d+1}(Y) \to H_{2d+1}(Y, Z) \to H_{2d}(|Z|) \to H_{2d}(Y) \to \cdots$$

of integral homology groups associated to the pair $(Y, |Z|)$, where $|Z|$ denotes the support of Z. It is an exact sequence of mixed Hodge structure. The class of the cycle Z defines a morphism of mixed Hodge structures

$$c_Z : \mathbb{Z}(d) \to H_{2d}(|Z|),$$

where $\mathbb{Z}(d)$ denotes the Hodge structure of type $(-d, -d)$ whose underlying lattice is isomorphic to $\mathbb{Z}$. If Z is null homologous, we can pull back the above sequence along c_Z to obtain an extension

$$0 \to H_{2d+1}(Y) \to E_Z \to \mathbb{Z}(d) \to 0$$

in MHS, the category of mixed Hodge structures. Tensoring with $\mathbb{Z}(-d)$ gives an extension

$$0 \to H_{2d+1}(Y, \mathbb{Z}(-d)) \to E_Z(-d) \to \mathbb{Z}(0) \to 0$$

and thus a class e_Z in

$$\mathrm{Ext}^1_{\mathrm{MHS}}(\mathbb{Z}(0), H_{2d+1}(Y, \mathbb{Z}(-d))).$$

Note that, since $H_{2d+1}(Y)$ has weight $-(2d+1)$, $H_{2d+1}(Y, \mathbb{Z}(-d))$ has weight -1.

Suppose that V is a Hodge structure of weight -1 whose underlying lattice $V_{\mathbb{Z}}$ is torsion free. The associated jacobian

$$J(V) := V_{\mathbb{C}}/(V_{\mathbb{Z}} + F^0 V_{\mathbb{C}})$$

is a compact complex torus. In general, $J(V)$ is not an abelian variety. When V is the weight -1 Hodge structure $H_{2d+1}(Y, \mathbb{Z}(-d))$ (mod its torsion), $J(V)$ is the dth Griffiths intermediate jacobian of Y.

There is a natural isomorphism (see [4] or [35, §3.5], for example)

$$\mathrm{Ext}^1_{\mathrm{MHS}}(\mathbb{Z}(0), V) \cong J(V).$$

The class e_Z of a homologically trivial d-cycle Z in Y can thus be viewed as a class

$$e_Z \in J(H_{2d+1}(Y, \mathbb{Z}(d)))$$

in the dth Griffiths intermediate jacobian. This class can be described explicitly by Griffiths' generalization [12] of the Abel-Jacobi construction , which we now recall. First observe that the standard pairing between $H_{2d+1}(Y)$ and $H^{2d+1}(Y)$ induces an isomorphism

$$H_{2d+1}(Y, \mathbb{Z}(-d))/F^0 \cong \mathrm{Hom}_{\mathbb{C}}(F^{d+1}H^{2d+1}(Y), \mathbb{C}).$$

The natural mapping $H_{2d+1}(Y, \mathbb{Z}(-d)) \to H_{2d+1}(Y, \mathbb{Z}(-d))/F^0$ corresponds to the integration mapping

$$H_{2d+1}(Y, \mathbb{Z}) \to \mathrm{Hom}_{\mathbb{C}}(F^{d+1}H^{2d+1}(Y), \mathbb{C})$$

that takes the homology class z to $\xi \mapsto \int_z \xi$. A homologically trivial d-cycle Z in Y can be written as the boundary $\partial\Gamma$ of a (topological) $(2d+1)$-chain Γ. Note that Γ is well defined up to the addition of an integral $(2d+1)$-cycle. Classical Hodge theory implies that each element u of $F^{d+1}H^{2d+1}(Y)$ can be represented by a closed C^∞ form ξ_u in the $(d+1)$st level of the Hodge filtration on the de Rham complex of Y and that any two such forms differ by the exterior derivative of a form in the same level F^{d+1} of the de Rham complex. The point e_Z is represented by the element

$$\int_\Gamma : u \mapsto \int_\Gamma \xi_u.$$

of

$$\mathrm{Hom}_{\mathbb{C}}(F^{d+1}H^{2d+1}(Y), \mathbb{C})/H_{2d+1}(Y, \mathbb{Z}) \cong J(H_{2d+1}(Y, \mathbb{Z}(-d))).$$

Stokes' Theorem implies that the image of this functional in the intermediate jacobian depends only on Z and not on the choice of Γ or ξ_u.

This construction generalizes the classical construction for 0-cycles on curves that was sketched in Section 3. More generally, it generalizes the classical construction for 0-cycles, where $J(H_1(Y)) \cong \mathrm{Alb}\, Y$, and for divisors, where $J(H_{2d-1}(Y)) \cong \mathrm{Pic}^0 Y$ and $d = \dim Y$.

5.2. Normal functions

Suppose that $\overline{X}$ is a complex projective manifold and that $X = \overline{X} - D$ where D is a normal crossings divisor in $\overline{X}$. Suppose that $\mathbb{V}$ is a variation of Hodge structure over X of weight -1. Denote by $J(\mathbb{V}) \to X$ the corresponding bundle of intermediate jacobians; the fiber over $x \in X$ is $J(V_x)$, where V_x is the fiber of $\mathbb{V}$ over x. It is a family of compact tori.

The discussion of Section 4.1 implies that a holomorphic section $\nu : X \to J(\mathbb{V})$ determines a cohomology class $c(\nu) \in H^1(X, \mathbb{V})$ and a local system $\mathbb{E} \to X$ which is an extension

$$0 \to \mathbb{V} \to \mathbb{E} \to \mathbb{Z}_X \to 0.$$

The point $\nu(x) \in J(V_x) \cong \mathrm{Ext}^1_{\mathsf{MHS}}(\mathbb{Z}(0), V_x)$ determines a mixed Hodge structure on the fiber E_x of $\mathbb{E}$ over $x \in X$ so that

$$0 \to V_x \to E_x \to \mathbb{Z}(0) \to 0$$

is an extension in MHS. That ν is holomorphic implies that this family of MHSs varies holomorphically with $x \in X$.

Example 5.1. Families of homologically trivial algebraic cycles give rise to such extensions. Suppose that $\overline{Y}$ is a complex projective manifold and that $f : \overline{Y} \to \overline{X}$ is a morphism whose restriction to X is a family $Y \to X$ of projective manifolds. Suppose that that Z is an algebraic d-cycle in Y such that the restriction Z_x of Z to the fiber over each $x \in X$ is a homologically trivial d-cycle in Y_x. Applying the construction of the previous section fiber-by-fiber, one obtains an extension $\mathbb{E}_Z$ of $\mathbb{Z}_X(0)$ by the variation of Hodge structure $\mathbb{V}$ of weight -1 whose fiber over $x \in X$ is $V_x = H_{2d+1}(Y_x, \mathbb{Z}(-d))$. The family $\{E_x\}_{x\in X}$ of extensions of MHS corresponds to a holomorphic section ν of the bundle $J(\mathbb{V}) \to X$ of intermediate jacobians.

The section ν is not an arbitrary holomorphic section. It satisfies the *Griffiths infinitesimal period relation*[5] at each $x \in X$ and also satisfies strong (and technical) conditions as Y_x degenerates. A succinct way to state these conditions is to say that the corresponding family $\mathbb{E}$ of MHS is an *admissible variation of MHS* in the sense Steenbrink-Zucker [42] and Kashiwara [24]. (The definition and an exposition of admissible variations of MHS can be found in [35, §14.4.1].)

All "naturally defined local systems" over a smooth variety X that arise from families of varieties over X, such as the one constructed in Example 5.1, are admissible variations of MHS.[6] The admissibility conditions axiomatize the infinitesimal and asymptotic properties satisfied by such geometrically defined local systems.

Definition 5.2 (Hain [15, §6], Saito [39]). A section ν of a family $J(\mathbb{V}) \to X$ of intermediate jacobians is a *normal function* when the corresponding family of MHS $\mathbb{E}$ is an admissible variation of MHS.

The preceding discussion implies that the section ν of $J(\mathbb{V}) \to X$ associated to a family of null homologous cycles is a normal function. Several concrete examples of normal functions over moduli spaces of curves will be given in Section 5.5. A detailed discussion of normal functions can be found in [26, §2.11], where they are called *admissible* normal functions.

Denote the category of admissible variations of mixed Hodge structure over a smooth variety X by MHS(X). It is abelian. The definition of normal functions implies that normal function sections of $J(\mathbb{V}) \to X$ correspond to elements of $\operatorname{Ext}^1_{\mathsf{MHS}(X)}(\mathbb{Z}_X(0), \mathbb{V})$. In the appendix to [19] it is proved that one has an exact sequence

$$(5.3) \qquad 0 \to \operatorname{Ext}^1_{\mathsf{MHS}}(\mathbb{Z}(0), H^0(X, \mathbb{V}_\mathbb{Z})) \xrightarrow{j} \operatorname{Ext}^1_{\mathsf{MHS}(X)}(\mathbb{Z}_X(0), \mathbb{V}_\mathbb{Z}) \xrightarrow{\delta} H^1(X, \mathbb{V}_\mathbb{Z})$$

where δ takes a normal function ν to its class $c(\nu)$. An immediate consequence is the following rigidity property of normal functions.

Proposition 5.4. *If* $H^0(X, \mathbb{V}_\mathbb{Q}) = 0$, *then the group* $\operatorname{Ext}^1_{\mathsf{MHS}(X)}(\mathbb{Z}(0)_X, \mathbb{V}_\mathbb{Z})$ *is finitely generated and each normal function section* ν *of* $J(\mathbb{V})$ *is determined, up to a torsion section, by its class* $c(\nu) \in H^1(X, \mathbb{V})$.

5.3. Extending normal functions

The following result guarantees that the normal functions that we will define over $\mathcal{M}_{g,n}$ extend to $\mathcal{M}^c_{g,n}$. A proof can be found, for example, in [15, Thm. 7.1].

Proposition 5.5. *Suppose that* $\mathbb{V}$ *is a variation of Hodge structure of weight* -1 *over a smooth variety* X. *If* ν *is a normal function section of* $J(\mathbb{V})$ *that is defined on the complement*

[5]This states that if $\tilde{\nu}$ is a local holomorphic lift of a normal function ν to a section of $\mathbb{E} \otimes \mathcal{O}_X$, then its derivative $\nabla\tilde{\nu}$ is a section of $\mathcal{F}^{-1}(\mathbb{E} \otimes \mathcal{O}_X) \otimes \Omega^1_X$.

[6]These results are due to Steenbrink-Zucker [42], Navarro et al [13] and Saito [38]. Precise statements can be found in [35, Thm. 14.51].

$X - Y$ *of a closed subvariety* Y *of* X, *where* $Y \neq X$, *then* ν *extends across* Y *to a normal function section of* $J(\mathbb{V})$ *over* X.

Note that, in this result, the variation $\mathbb{V}$ is defined over X, not just over $X - Y$. The proposition asserts that normal functions defined generically on X extend to X. It does not assert that if $\mathbb{V}$ is defined only over $X - Y$, then a normal function section of $J(\mathbb{V})$ over $X - Y$ will extend to X. Before discussing this problem, one first has to construct an extension of $J(\mathbb{V})$ to X. The existence of such extensions is discussed in [44] and [26], for example.

5.4. Some variations of Hodge structure

Denote the moduli stack of principally polarized abelian varieties of dimension g by $\mathcal{A}_g$. Here we introduce three variations of Hodge structure over $\mathcal{A}_g$ whose pullbacks to $\mathcal{M}_{g,n}^c$ have a special geometric significance. Throughout we suppose that $g \geqslant 2$.

Denote the universal abelian variety over $\mathcal{A}_g$ by $f : \mathcal{X} \to \mathcal{A}_g$. The local system

$$\mathbb{H} := R^1 f_* \mathbb{Z}_{\mathcal{X}}(1)$$

is a variation of Hodge structure over $\mathcal{A}_g$ of weight -1. Its pullback to $\mathcal{M}_{g,n}^c$ along the period mapping $\mathcal{M}_{g,n}^c \to \mathcal{A}_g$ will also be denoted by $\mathbb{H}$. The corresponding family of intermediate jacobians $J(\mathbb{H})$ over $\mathcal{A}_g$ is naturally isomorphic to $\mathcal{X}$, the universal principally polarized abelian variety.

The construction of the universal jacobian $\mathcal{J}_g$ in Section 3 implies that its restriction to $\mathcal{M}_g^c$ is a bundle of intermediate jacobians.

Proposition 5.6. *If* $g \geqslant 2$, *then* $J(\mathbb{H}) \to \mathcal{A}_g$ *is naturally isomorphic to the universal abelian variety and its pullback to* $\mathcal{M}_g^c$ *is isomorphic to the universal jacobian* $\mathcal{J}_g^c$.

The canonical polarization defines a morphism $S_H : \mathbb{H} \otimes \mathbb{H} \to \mathbb{Z}(1)$ into the constant variation of Hodge structure $\mathbb{Z}(1)$. It can be regarded as a section of $(\Lambda^2 \mathbb{H})(-1)$, which we shall also denote by S_H.

Denote by $\mathbb{L}$ the variation of Hodge structure $(\Lambda^3 \mathbb{H})(-1)$. It has weight -1. Wedging with S_H defines an inclusion

$$\mathbb{H} \hookrightarrow \mathbb{L}, \qquad x \mapsto x \wedge S_H$$

of variations of Hodge structure. Set $\mathbb{V} = \mathbb{L}/\mathbb{H}$. Note that when $g = 2$, $\mathbb{L} \cong \mathbb{H}$ and $\mathbb{V} = 0$. Denote the fibers of $\mathbb{H}$, $\mathbb{L}$ and $\mathbb{V}$ over the moduli point of $\operatorname{Jac} C$ by H_C, L_C and V_C, respectively.

Remark 5.7. The variation $\mathbb{L}$ over $\mathcal{M}_g^c$ is isomorphic to the variation $R^3\pi_*\mathbb{Z}(2)$ where $\pi : \mathcal{J}_g^c \to \mathcal{M}_g$ denotes the projection. Its fiber over $[C]$ is $H^3(\operatorname{Jac} C, \mathbb{Z}(2))$. The twist $\mathbb{Z}(2)$ lowers the weight from 3 to -1. Its quotient $\mathbb{V}$ is an integral form of the $\mathbb{Q}$-variation of Hodge structure whose fiber over $[C]$ is the primitive part of $H^3(\operatorname{Jac} C, \mathbb{Q}(2))$.

5.5. Normal functions over $\mathcal{M}_g^c$ and $\mathcal{M}_{g,n}^c$

Suppose that $g \geqslant 1$. An irreducible representation V of Sp_g determines a variation of Hodge structure $\mathbb{V}$ over $\mathcal{M}_{g,n}$ that is unique up to Tate twist.[7] Since every such variation of Hodge structure $\mathbb{V}$ extends to $\mathcal{M}_{g,n}^c$, Proposition 5.5 implies that every normal function section of $J(\mathbb{V})$ defined over $\mathcal{M}_{g,n}$ extends to a normal function on $\mathcal{M}_{g,n}^c$.

By Proposition 5.6, the bundle of intermediate jacobians that corresponds to the fundamental representation of Sp_g is the pullback of the universal jacobian $J(\mathbb{H}) = \mathcal{J}_g$ to $\mathcal{M}_{g,n}^c$. Its group of sections $\mathrm{Ext}^1_{\mathsf{MHS}(\mathcal{M}_{g,1})}(\mathbb{Z}(0), \mathbb{H})$ is free of rank 1. Its generator $\mathcal{K}$ is the normal function that takes the moduli point $[C, x]$ of the pointed curve (C, x) to

$$\mathcal{K}([C, x]) := (2g-2)[x] - K_C \in \mathrm{Jac}(C)$$

where K_C denotes the canonical divisor class of C.

When $n \geqslant 1$, we can pull back $\mathcal{K}$ along the jth projection $\mathcal{M}_{g,n} \to \mathcal{M}_{g,1}$ to obtain the normal function

$$\mathcal{K}_j : \mathcal{M}_{g,n} \to J(\mathbb{H}) \qquad j = 1, \dots, n.$$

Explicitly

$$\mathcal{K}_j([C, x_1, \dots, x_n]) = (2g-2)[x_j] - K_C \in \mathrm{Jac}(C).$$

When $n \geqslant 2$ we also have the normal functions

$$\mathcal{D}_{j,k} : \mathcal{M}_{g,n} \to J(\mathbb{H}) \qquad 1 \leqslant j < k \leqslant n$$

defined by

$$\mathcal{D}_{j,k}([C, x_1, \dots, x_n]) = [x_j] - [x_k] \in \mathrm{Jac}(C).$$

Proposition 5.8 (cf. [15, Thm. 12.3]). *If $g \geqslant 3$ and $n \geqslant 2$, then the group of normal function sections (indeed, all sections) of $J(\mathbb{H}) \to \mathcal{M}_{g,n}$ is torsion free and is generated by $\mathcal{K}_1, \dots, \mathcal{K}_n$ and the $\mathcal{D}_{j,k}$ where $1 \leqslant j < k \leqslant n$.*

The most interesting normal function over $\mathcal{M}_{g,n}$ is constructed from the Ceresa cycle in the universal jacobian. Suppose that C is a smooth projective curve of genus $g \geqslant 3$ and that $x \in C$. Then one has the imbedding

$$\mu_x : C \to \mathrm{Jac}\, C$$

that takes y to $[y] - [x]$. Its image is an algebraic 1-cycle in $\mathrm{Jac}\, C$ that we denote by C_x. Let i be the involution $u \mapsto -u$ of $\mathrm{Jac}\, C$. Set $C_x^- = i_* C_x$. Since $i^* : H^1(\mathrm{Jac}\, C) \to H^1(\mathrm{Jac}\, C)$ is multiplication by -1, it follows that $i^* : H^k(\mathrm{Jac}\, C) \to H^k(\mathrm{Jac}\, C)$ is multiplication by $(-1)^k$. This implies that the 1-cycle $C_x - C_x^-$, called the *Ceresa cycle*, is homologically trivial. It therefore determines a class

$$\nu_x(C) \in J(H_3(\mathrm{Jac}\, C, \mathbb{Z}(-1)))$$

[7] This is very well-known. A proof can be found, for example, in [15, §9].

and a normal function:

$$J(\mathbb{L}) \xrightarrow{} \mathcal{M}_{g,1} \quad (\text{section } \tilde{\nu})$$

whose value at $[C, x]$ is $\nu_x(C)$. The inclusion $\mathbb{H} \hookrightarrow \mathbb{L}$ induces an inclusion

$$j : \mathcal{J}_{g,1} = J(\mathbb{H}) \hookrightarrow J(\mathbb{L}).$$

It is proved in [36] that if $x, y \in C$, then

$$\tilde{\nu}_x(C) - \tilde{\nu}_y(C) = 2j([x] - [y]) \in J(L_C)$$

so that the image $\nu(C)$ of $\nu_x(C)$ in $J(V_C)$ does not depend on the choice of $x \in C$. This implies that $\tilde{\nu}$ is pulled back from a normal function

$$J(\mathbb{V}) \xrightarrow{} \mathcal{M}_g. \quad (\text{section } \nu)$$

We will abuse notation and also denote its pullback to $\mathcal{M}_{g,n}$ by ν.

Proposition 5.9 ([15, Thm. 8.3]). *If $g \geqslant 3$ and $n \geqslant 0$, then the group of normal function sections of $J(\mathbb{V}) \to \mathcal{M}_{g,n}$ is freely generated by ν.*

Proposition 5.5 implies that the normal functions $\nu, \mathcal{K}_j, \delta_{j,k}$ extend canonically to normal functions over $\mathcal{M}^c_{g,1}$.

6. Biextension line bundles

This section is a brief review of facts about biextension line bundles from [14], [28], and [23]. It is needed to prove that the square of the class of a normal function extends naturally to a class on $\overline{\mathcal{M}}_{g,n}$ even though the normal function itself does not extend.

Suppose that $\mathbb{U}$ is a variation of Hodge structure over an algebraic manifold X of weight -1 endowed with a flat inner product S that satisfies the condition

$$S \in \mathrm{Hom}_{\mathsf{MHS}}(U_x^{\otimes 2}, \mathbb{Z}(1)) \text{ for all } x \in X.$$

Equivalently,

$$S(U_x^{p,q}, \overline{U_x^{r,s}}) = 0 \text{ unless } p = r \text{ and } q = s.$$

Set $\check{\mathbb{U}} := \mathrm{Hom}_{\mathbb{Z}}(\mathbb{U}, \mathbb{Z}_X(1))$. This also a variation of Hodge structure of weight -1. There is a natural isomorphism

$$\mathrm{Ext}^1_{\mathsf{MHS}(X)}(\mathbb{Z}_X(0), \check{\mathbb{U}}) \cong \mathrm{Ext}^1_{\mathsf{MHS}(X)}(\mathbb{U}, \mathbb{Z}_X(1)).$$

The *biextension line bundle* $\mathcal{B}$ is a line bundle over $J(\mathbb{U}) \times_X J(\check{\mathbb{U}})$. Denote the associated $\mathbb{G}_m$-bundle by $\mathcal{B}^*$. The fiber of the projection

$$\mathcal{B}^* \to J(\mathbb{U}) \times_X J(\check{\mathbb{U}}) \to X \tag{6.1}$$

over $x \in X$ is the set of all mixed Hodge structures whose weight graded quotients are $\mathbb{Z}(0)$, V_x and $\mathbb{Z}(1)$ via a fixed isomorphism. These are called *biextensions*. The projection (6.1) takes the biextension B to the pair of extensions $B/\mathbb{Z}(1)$ and $W_{-1}B$. A detailed exposition of the construction of $\mathcal{B}$ is given in [23, §7].

A (Hodge) biextension is a section β of (6.1) that corresponds to an admissible variation of MHS $\mathbb{B}$ over X with weight graded quotients $\mathbb{Z}_X(0)$, $\mathbb{U}$ and $\mathbb{Z}_X(1)$. Its fiber over $x \in X$ is the biextension $\beta(x)$. The composite of a biextension β with the projection $\mathcal{B}^* \to J(\mathbb{U}) \times_X J(\check{\mathbb{U}})$ is a pair of normal functions that determines the extension $\mathbb{B}/\mathbb{Z}(1)$ of $\mathbb{Z}_X(0)$ by $\mathbb{U}$ and the extension $W_{-1}\mathbb{B}$ of $\mathbb{U}$ by $\mathbb{Z}_X(1)$. The biextension line bundle has a canonical metric $|\ |_{\mathcal{B}}$. A biextension β thus determines the real-valued function $\log|\beta|_{\mathcal{B}} : X \to \mathbb{R}$.

The pairing S also defines a morphism $\mathbb{U} \to \check{\mathbb{U}}$ of variations of Hodge structure over X, and therefore a map $i_S : J(\mathbb{U}) \to J(\check{\mathbb{U}})$. Pulling back the line bundle $\mathcal{B}$ along the map

$$(\mathrm{id}, i_S) : \mathcal{J}(\mathbb{U}) \to J(\mathbb{U}) \times_X J(\check{\mathbb{U}})$$

we obtain a metrized line bundle $\widehat{\mathcal{B}} \to J(\mathbb{U})$. By [23, Prop. 7.3], the curvature of $\widehat{\mathcal{B}}$ is the translation-invariant, parallel 2-form $2\omega_S$ on $J(\mathbb{U})$ that corresponds to the bilinear form 2S. Points of the associated $\mathbb{C}^*$-bundle $\widehat{\mathcal{B}}^*$ correspond to "symmetric biextensions".

Denote by $\phi_S \in H^2(J(\mathbb{U}))$ the class of ω_S. Since $2\omega_S$ represents $c_1(\widehat{\mathcal{B}})$, the class $2\phi_S$ is integral.

Suppose now that $X = \overline{X} - Y$, where $\overline{X}$ is smooth and Y is a subvariety. Each normal function section ν of $J(\mathbb{U})$ thus determines a metrized holomorphic line bundle $\nu^*\widehat{\mathcal{B}}$ over X. One can ask whether it extends as a metrized line bundle to $\overline{X}$. Lear's thesis [28] implies that a power of it extends to a *continuously* metrized holomorphic line bundle over $X - Y^{\mathrm{sing}}$.

Theorem 6.2 (Lear [28]). *If* $\dim X = 1$ *and* ν *is a normal function section of* $J(\mathbb{U}) \to X$, *then there exists an integer* $N \geqslant 1$ *such that the metrized holomorphic line bundle* $\nu^*\widehat{\mathcal{B}}^{\otimes N}$ *over* X *extends to a holomorphic line bundle over* $\overline{X}$ *with a continuous metric. Moreover, if* β *is a biextension section defined over* X *that projects to* ν, *and if* $\mathbb{D}$ *is a disk in* $\overline{X}$ *with holomorphic coordinate* t *that is centered at a point of* $\overline{X}-X$, *then there is a rational number* p/q, *which depends only on the monodromy about the origin of* $\mathbb{D}$ *of the VMHS over* $\mathbb{D}^*$ *corresponding to* β, *such that*

$$\Big|\log|\beta(t)|_{\mathcal{B}} - \frac{p}{q}\log|t|\Big| \tag{6.3}$$

is bounded in a neighbourhood of $t = 0$.

Note that the continuity of the metric ensures that the extension is uniquely determined. Since $\overline{X}$ is smooth, every line bundle over $\overline{X} - Y^{\mathrm{sing}}$ extends uniquely to a line bundle over $\overline{X}$. Lear's Theorem thus implies the following result in the case $\dim X \geqslant 1$.

Corollary 6.4. *If* ν *is a normal function section of* $J(\mathbb{U}) \to X$*, then there exists an integer* $N \geqslant 1$ *such that the metrized holomorphic line bundle* $\nu^*\widehat{\mathcal{B}}^{\otimes N}$ *over* X *extends to a holomorphic line bundle* $\overline{\mathcal{B}}_{N,\nu}$ *over* $\overline{X}$ *whose metric extends continuously over* $\overline{X} - Y^{\text{sing}}$.

This result implies that the class $\nu^*\phi_S \in H^2(X)$ of $\nu^*\omega_S$ has a natural extension to a class in $H^2(\overline{X})$; namely $c_1(\overline{\mathcal{B}}_{N,\nu})/2N$.

Corollary 6.5. *If* ν *is a normal function section of* $J(\mathbb{U}) \to X$*, then the class* $\nu^*\phi_S$ *has a natural extension to a class* $\widehat{\nu^*\phi_S} \in H^2(\overline{X})$.

Lear's Theorem implies that the multiplicity of each boundary divisor in $\widehat{\nu^*\phi_S}$ is determined by the asymptotics (6.3) of the restriction of the biextension to a disk transverse to the boundary divisor.

The previous result suggests that $\nu^*\omega_S$, regarded as a current on $\overline{X}$, is a natural representative of $\widehat{\nu^*\phi_S}$.

Conjecture 6.6. If X is a curve, then the 2-form $\nu^*\omega_S$ is integrable on X and

$$\int_X \nu^*\omega_S = \frac{1}{2N}\int_{\overline{X}} c_1(\overline{\mathcal{B}}_{N,\nu}).$$

It is known that, in general, the metric does not extend continuously over Y^{sing} due to the phenomenon of "height jumping" which we shall discuss in Section 14 and which has been explained by Brosnan and Pearlstein in [2].

7. Polarizations

Polarizations play an important and subtle (if sometimes neglected) role in Hodge theory due to their positivity properties.

7.1. Polarizations

A *polarization* on a Hodge structure H of weight k is a $(-1)^k$ symmetric bilinear form S on $H_{\mathbb{Q}}$ satisfying the Riemann-Hodge bilinear relations:

(i) $S(H^{p,q}, \overline{H^{r,s}}) = 0$ unless $p = r$ and $q = s$;

(ii) $i^{p-q}S(v, \overline{v}) > 0$ when $v \in H^{p,q}$ and $v \neq 0$.

A bilinear form S is a *weak polarization* on H if it satisfies the first condition and the weaker version $i^{p-q}S(v,\overline{v}) \geqslant 0$ for all $v \in H^{p,q}$ of the second.

Suppose that Y is a smooth projective variety of dimension n. Denote the hyperplane class by w. For $k \leqslant n$, define a bilinear form S on $H^k(Y)$ by

$$S(u, v) = \int_Y u \wedge v \wedge w^{n-k}. \tag{7.1}$$

This is a non-degenerate, $(-1)^k$ symmetric bilinear form. However, it is not a polarization in general. The Riemann-Hodge bilinear relations imply that the restriction of $(-1)^{k(k-1)/2}S$ to $PH^k(Y)$, the primitive part of $H^k(Y)$, is a polarization. These provide the principal examples of polarized Hodge structures.

A variation of Hodge structure $\mathbb{V}$ over a base X is *polarized* by S if S is a flat bilinear form on the variation which restricts to a polarization on each fiber.

7.2. Some polarized variations of Hodge structure over $\mathcal{A}_g$

The variations $\mathbb{H}$, $\mathbb{L}$ and $\mathbb{V}$ defined in Section 5.4 have natural polarizations. The Riemann bilinear relations imply that the variation $\mathbb{H}$ over $\mathcal{A}_g$ is polarized by the inner product S_H introduced in Section 5.4. The corresponding polarization is easily described on the pullback of $\mathbb{H}$ to $\mathcal{M}_g$. In this case, the fiber H_C of $\mathbb{H}$ over the moduli point $[C]$ of a smooth projective curve C is $H^1(C, \mathbb{Z}(1))$. Under this isomorphism, S_H corresponds to the inner product

$$S(u, v) = \int_C u \wedge v.$$

on $H^1(C)$.

The intersection form S_H extends to the skew symmetric bilinear form

$$S_L : \mathbb{L} \otimes \mathbb{L} \to \mathbb{Z}(1)$$

defined by

$$S_L(x_1 \wedge x_2 \wedge x_3, y_1 \wedge y_2 \wedge y_3) = \det(S_H(x_i, y_j)).$$

Note that this is *not* dual to the inner product S on $H^3(\operatorname{Jac} C)(1)$ defined in equation (7.1) above as is easily seen by a direct computation.

Denote the fiber $\Lambda^3 H_C$ of $\mathbb{L}$ over $[C]$ by L_C and the fiber of $\mathbb{V}$ over $[C]$ by V_C. Define $c : L_C \to H_C$ by

$$c(x \wedge y \wedge z) = S_H(y, z)x + S_H(z, x)y + S_H(x, y)z. \tag{7.2}$$

Regard S_H as an element of $\Lambda^2 H_L$. The projection $p : L_C \to V_C$ has a canonical Sp_g-invariant splitting j. It is defined by

$$j(p(x \wedge y \wedge z)) = x \wedge y \wedge z - S_H \wedge c(x \wedge y \wedge z)/(g-1).$$

A skew symmetric bilinear form on $\mathbb{V}$ can be defined by

$$S_V(u, v) = (g-1)S_L(j(u), j(v)).$$

This form is integral and primitive.

Proposition 7.3. *The variations $(\mathbb{H}, S_H)$, $(\mathbb{L}, S_L)$ and $(\mathbb{V}, S_V)$ are polarized variations of Hodge structure over $\mathcal{A}_g$, as are their pullbacks to $\mathcal{M}_{g,n}^c$.*

Proof. We have already seen that $(\mathbb{H}, S_H)$ is a polarized variation of Hodge structure. For the rest, it suffices to show that $\Lambda^3 H_1(C)$ is polarized by S_L. To do this, choose a basis $u_1, \dots, u_g$ of $H^{-1,0}$ that is orthonormal under the positive definite hermitian inner product

$$(u, v) = i^{-1} S_H(u, \overline{v}).$$

Then, for example,

$$i^{-3-0} S_L(u_1 \wedge u_2 \wedge u_3, \overline{u}_1 \wedge \overline{u}_2, \overline{u}_3) = \det(i^{-1} S_H(u_j, \overline{u}_k)) = 1 > 0.$$

and

$i^{-2-(-1)}S_L(u_1 \wedge u_2 \wedge \overline{u}_3, \overline{u}_1 \wedge \overline{u}_2, u_3) = -i^{-1}i^3 \det(i^{-1}S_H(u_j, \overline{u}_k)) = -i^2 = 1 > 0.$

The remaining computations follow by taking complex conjugates. □

The contraction (7.2) induces a projection $c : \mathbb{L} \to \mathbb{H}$ of variations of Hodge structure. The canonical quotient mapping $p : \mathbb{L} \to \mathbb{V}$ is also a morphism of variation of Hodge structure. The polarizations S_H of $\mathbb{H}$ and S_V of $\mathbb{V}$ can be pulled back along these projections to obtain the invariant inner product $c^*S_H + p^*S_V$ on $\mathbb{L}$. For later use, we record the following fact:

Lemma 7.4 ([22, Prop. 18]). *If* $g \geqslant 2$, *then* $c^*S_H + p^*S_V = (g-1)S_L$. □

8. Cohomology classes

By Lemma 4.4 each invariant inner product on a variation of MHS $\mathbb{U}$ over a smooth variety T gives rise to a parallel, translation invariant 2-form ω on the associated bundle of intermediate jacobians $J(\mathbb{U})$ and its cohomology class $\phi \in H^2(J(\mathbb{U}))$.

The polarizations S_H, S_L and S_V of the variations of Hodge structure $\mathbb{H}$, $\mathbb{L}$ and $\mathbb{V}$ over $\mathcal{A}_g$ therefore give rise to cohomology classes

$$\phi_H \in H^2(J(\mathbb{H})), \quad \phi_L \in H^2(J(\mathbb{L})) \text{ and } \phi_V \in H^2(J(\mathbb{V})).$$

Denote their parallel, canonical translation invariant representatives by ω_H, ω_L and ω_V.

Recall that $J(\mathbb{H}) = \mathcal{J}_g^c$, the universal jacobian over $\mathcal{M}_g^c$ and that $\eta_g \in H^{2g}(\mathcal{J}_g^c)$ denotes the Poincaré dual of the 0-section Z_g of $\mathcal{J}_g^c$. A standard and elementary computation shows that if C is a smooth projective curve of genus g, then

$$\int_{\mathrm{Jac}\, C} \omega_H^g = g!$$

Combining this with Proposition 4.5 yields the following result, which can also be deduced from [43, Cor. 2.2].

Proposition 8.1. *The class* η_g *of the zero section of* $\mathcal{J}_g^c$ *in* $H^{2g}(\mathcal{J}_g^c)$ *is* $\phi_H^g/g!$ □

Denote the restriction of $\mathcal{J}_g$ to $\overline{\mathcal{M}}_g - \Delta_0^{\mathrm{sing}}$ by $\mathcal{J}'_g$. Zucker's Theorem [44] implies that every normal function section μ of $\mathcal{J}_g$ defined over $\mathcal{M}_{g,n}^c$ extends to a section (also denoted μ) of $\mathcal{J}'_g$ defined over $\overline{\mathcal{M}}_{g,n} - \Delta_0^{\mathrm{sing}}$.

Proposition 8.2. *The class* $\phi_H \in H^2(\mathcal{J}_g^c)$ *extends naturally to a class* $\widehat{\phi}_H \in H^2(\mathcal{J}'_g)$. *It is characterized by the property*

$$[e]^*\widehat{\phi}_H = e^2\widehat{\phi}_H \textit{ for all } e \in \mathbb{Z}$$

and has the property that $\widehat{\mu^*\phi_H} = \mu^*\widehat{\phi}_H$ *for all normal function sections* μ *of* $\mathcal{J}_g^c$ *defined over* $\mathcal{M}_{g,n}^c$.

Sketch of Proof. The first rational cohomology of the smooth finite orbi-covering $\overline{\mathcal{M}}_{g-1,2}$ of Δ_0 vanishes. The Gysin sequence thus gives an exact sequence

$$\mathbb{Q}\delta_0 \to H^2(\mathcal{J}'_g) \to H^2(\mathcal{J}^c_g) \to 0$$

of rational cohomology. For each integer $e > 1$, the endomorphism $[e]$ of $\mathcal{J}'_g$ induces an action on this sequence. It acts trivially on the left-hand term. Since $[e]^*\phi_H = e^2\phi_H$ in $H^2(\mathcal{J}^c_g)$, it follows that ϕ_H has a unique lift $\widehat{\phi}_H$ to $H^2(\mathcal{J}'_g)$ with the property that $[e]^*\widehat{\phi}_H = e^2\widehat{\phi}_H$. The restriction of $\widehat{\phi}_H$ to the zero section $\overline{\mathcal{M}}_g - \Delta_0^{\text{sing}}$ of $\mathcal{J}'_g$ vanishes as $[e]$ preserves the zero section and acts trivially on $H^2(\overline{\mathcal{M}}_g - \Delta_0^{\text{sing}})$, and since $[e]^*\widehat{\phi}_H = e^2\widehat{\phi}_H$.

To complete the proof, we need several facts about biextension line bundles. Suppose that $f : Y \to X$ is a morphism of smooth varieties and that $\mathbb{U}$ is a VHS over X polarized by S. Then the constructions of [14] imply that one has a commutative diagram

$$\begin{array}{ccc}
\widehat{\mathcal{B}}_Y & \xrightarrow{f_{\mathcal{B}}} & \widehat{\mathcal{B}}_X \\
\downarrow & & \downarrow \\
J(f^*\mathbb{U}) & \xrightarrow{J(f)} & J(\mathbb{U}) \\
\downarrow & & \downarrow \\
Y & \xrightarrow{f} & X
\end{array}$$

of biextension line bundles, where $f_{\mathcal{B}}$ is a morphism of metrized line bundles. This implies that $J(f)^*\phi_S = \phi_{f^*S}$. The next fact, which follows from Lear's Theorem, is that if X' and Y' are smooth varieties in which X and Y are Zariski dense, and where $X' - X$ is a smooth divisor in X', and if δ is a normal function section of $J(\mathbb{U}) \to X$, then

$$(8.3) \qquad \widehat{(f^*\delta)^*\phi_{f^*S}} = f^*(\widehat{\delta^*\phi_S}) \in H^2(Y').$$

We now apply this with $X = \mathcal{J}^c_g$, $X' = \mathcal{J}'_g$, $Y = \mathcal{M}^c_{g,n}$ and $Y' = \overline{\mathcal{M}}_{g,n} - \Delta_0^{\text{sing}}$. The variation $\mathbb{U}$ is the standard variation $\mathbb{H}$, so that $\mathcal{J}(\mathbb{U}) = \mathcal{J}^c_g \times_{\mathcal{M}^c_g} \mathcal{J}^c_g$. Important here is the fact that $\mathcal{J}^c_g$ is an algebraic variety. The normal function δ will be the diagonal section of $\mathcal{J}^c_g \times_{\mathcal{M}^c_g} \mathcal{J}^c_g \to \mathcal{J}^c_g$. Finally, $f : Y \to X$ will be a normal function $\mu : \mathcal{M}^c_{g,n} \to \mathcal{J}_g$, which is a morphism as $\mathcal{J}^c_g \to \mathcal{M}^c_g$ is a family of (semi)-abelian varieties:

$$\begin{array}{ccc}
\mathcal{J}'_{g,n} & \xrightarrow{(\Delta,\mu)} & \mathcal{J}'_g \times_{\overline{\mathcal{M}}_g} \mathcal{J}'_g \\
\mu \uparrow\downarrow & & \pi_1 \downarrow\uparrow \Delta \\
\overline{\mathcal{M}}_{g,n} - \Delta_0^{\text{sing}} & \xrightarrow[\mu]{} & \mathcal{J}'_g
\end{array}$$

The class $\phi_S \in H^2(\mathcal{J}'_g \times_{\overline{\mathcal{M}}_g} \mathcal{J}'_g)$ is $\pi_2^*\phi_H$, where π_1 and π_2 are the two projections $\mathcal{J}'_g \times_{\overline{\mathcal{M}}_g} \mathcal{J}'_g \to \mathcal{J}'_g$. It is now a tautology that $\widehat{\Delta^*\phi_S} = \widehat{\phi}_H \in H^2(\mathcal{J}'_g)$. This and the naturality statement (8.3) now imply that $\widehat{\mu^*\phi_H} = \mu^*(\widehat{\Delta^*\phi_S}) = \mu^*\widehat{\phi}_H$. □

It is important to note that Proposition 8 1 does not hold over $\overline{\mathcal{M}}_g$. This is because the restriction of $\widehat{\phi}_H$ to the zero section vanishes, whereas the conormal bundle of the zero section is the Hodge bundle, whose top Chern class is non-trivial.

We shall also need the invariant inner product Δ on $\mathbb{H} \oplus \mathbb{H}$ that is defined by

$$S_\Delta\big((u_1, v_1), (u_2, v_2)\big) = S_H(u_1, v_2) - S_H(u_2, v_1)$$

Even though it preserves the Hodge filtration, it is not a weak polarization as can be seen by restricting it to the diagonal of $H \oplus H$ (where it is positive) and anti-diagonal (where it is negative). Denote the associated cohomology class in $H^2(\mathcal{J}(\mathbb{H} \oplus \mathbb{H}))$ by ϕ_Δ. The class ϕ_Δ extends naturally to a class $\widehat{\phi}_\Delta \in H^2(\mathcal{J}'_g \times_{\overline{\mathcal{M}}_g} \mathcal{J}'_g)$. The proof is similar to that of Proposition 8.2 and is left to the reader.

8.1. The classes $\mathcal{K}^*\phi_H$, $\nu^*\phi_V$, $\tilde{\nu}^*\phi_L$ and $(\mathcal{K} \times \mathcal{K})^*\phi_\Delta$

Pulling back the classes ϕ_H, ϕ_V, ϕ_L and ϕ_Δ along the normal functions $\mathcal{K}$, ν, $\tilde{\nu}$ and the normal function section $\mathcal{K} \times \mathcal{K}$ of

$$J(\mathbb{H} \oplus \mathbb{H}) = J(\mathbb{H}) \times_{\mathcal{M}_{g,2}} J(\mathbb{H}) \to \mathcal{M}_{g,2}$$

defined by

$$\mathcal{K} \times \mathcal{K} : [C; x_1, x_2] \mapsto (\mathcal{K}(x_1), \mathcal{K}(x_2)) \in \mathrm{Jac}(C) \times \mathrm{Jac}(C).$$

we obtain rational cohomology classes $\mathcal{K}^*\phi_H \in H^2(\mathcal{M}^c_{g,1})$, $\nu^*\phi_V \in H^2(\mathcal{M}^c_g)$, $\tilde{\nu}^*\phi_L \in H^2(\mathcal{M}^c_{g,1})$, and $(\mathcal{K} \times \mathcal{K})^*\phi_\Delta \in H^2(\mathcal{M}^c_{g,2})$. Lear's Theorem implies (via Cor. 6.5) that they extend naturally to classes $\widehat{\nu^*\phi_V} \in H^2(\overline{\mathcal{M}}_g)$, $\widehat{\tilde{\nu}^*\phi_L} \in H^2(\overline{\mathcal{M}}_{g,1})$, and $\widehat{(\mathcal{K} \times \mathcal{K})^*}\phi_\Delta \in H^2(\overline{\mathcal{M}}_{g,1})$.

Proposition 8.4. *If* $g \geqslant 2$, *then* $(g-1)\widehat{\tilde{\nu}^*\phi_L} = \widehat{\nu^*\phi_V} + \widehat{\mathcal{K}^*\phi_H} \in H^2(\overline{\mathcal{M}}_{g,1})$.

Proof. This follows immediately from Lemma 7.4 and the fact [36] that $c \circ \tilde{\nu} = \mathcal{K}$ and $p \circ \tilde{\nu} = \nu$. □

Our next task is to compute each of these classes First we need to fix notation for the natural classes in $H^2(\overline{\mathcal{M}}_{g,n})$.

9. Divisor classes

Denote the set $\{x_1, x_2, \ldots, x_n\}$ of marked points by I. Denote the relative dualizing sheaf of the universal curve $\pi : \mathcal{C} \to \overline{\mathcal{M}}_{g,n}$ over $\overline{\mathcal{M}}_{g,n}$ by w. Its pushforward $\pi_*\omega$ is locally free of rank g. Recall that λ_1 denotes the first Chern class of the Hodge bundle π_*w. The classes ψ_j, $x_j \in I$ are defined by

$$\psi_j := x_j^* c_1(w).$$

When $n = 1$, we will denote ψ_1 by ψ. Note that this definition is different from the standard definition. With this definition the ψ classes are natural with respect to the forgetful maps $\overline{\mathcal{M}}_{g,n+1} \to \overline{\mathcal{M}}_{g,n}$.

Each component of the boundary divisor of $\mathcal{M}_{g,n}$ in $\overline{\mathcal{M}}_{g,n}$ has as its generic point a stable n-pointed curve of genus g with exactly one node. These components are:

- Δ_0: the generic point is an irreducible, geometrically connected n-pointed curve with one node;
- Δ_0^P, where P is a subset of I with $|P| \geqslant 2$: the generic point is a reducible curve with two geometrically connected components joined at a single node, one of which has genus 0 — the points in P lie on the genus 0 component minus its node, the remaining points $P^c := I - P$ lie on the other (genus g) component minus the node;
- Δ_h^P, where $0 < h < g$ and $P \subseteq I$ (possibly empty): the generic point is a reducible curve with exactly one node and two geometrically connected irreducible components, one of genus h, the other of genus $g-h$; the points in P lie on the genus h component minus the node, and the rest P^c lie on the other component minus the node. Note that $\Delta_{g-h}^{P^c} = \Delta_h^P$.

Denote the classes of the divisors

$$\Delta_0,\ \Delta_0^P\ (P \subseteq I,\ |P| \geqslant 2),\ \Delta_h^P\ (0 < h < g,\ P \subseteq I)$$

by

$$\delta_0,\ \delta_0^P\ (P \subseteq I,\ |P| \geqslant 2),\ \delta_h^P\ (0 < h < g,\ P \subseteq I),$$

respectively. It is well-known that the classes

$$\lambda_1,\ \psi_j\ (x_j \in I),\ \delta_0,\ \delta_0^P\ (P \subseteq I,\ |P| \geqslant 2),\ \delta_h^P\ (0 < h \leqslant g/2,\ P \subseteq I)$$

comprise a basis of $H^2(\overline{\mathcal{M}}_{g,n})$.

One also has the Miller-Morita-Mumford classes

$$\kappa_j := \pi_* c_1(w)^{j+1} \in H^{2j}(\overline{\mathcal{M}}_g)$$

which are defined for $j \geqslant 1$. It follows from Grothendieck-Riemann-Roch (cf. [34]) that $\kappa_1 = 12\lambda_1 - \delta$, where

$$\delta = \sum_{j=0}^{\lfloor g/2 \rfloor} \delta_j \in H^2(\overline{\mathcal{M}}_g).$$

Define $\kappa_j, \delta \in H^\bullet(\overline{\mathcal{M}}_{g,n})$ to be the pullbacks of $\kappa_j, \delta \in H^\bullet(\overline{\mathcal{M}}_g)$ under the natural morphisms $\overline{\mathcal{M}}_{g,n} \to \overline{\mathcal{M}}_g$. We thus have the alternate basis

$$\kappa_1,\ \psi_j\ (x_j \in I),\ \delta_0,\ \delta_0^P\ (P \subseteq I,\ |P| \geqslant 2),\ \delta_h^P\ (1 \leqslant h \leqslant g/2,\ P \subseteq I)$$

of $H^2(\overline{\mathcal{M}}_{g,n})$.

Remark 9.1. All of these divisor classes can be regarded as classes in $H^2(\overline{\mathcal{M}}_{g,n})$ or in $CH^1(\overline{\mathcal{M}}_{g,n})$. Note that these two groups are isomorphic, so that any relation between divisor classes that holds in cohomology also holds in the Chow ring.

10. Formulas for $\nu^*\phi_V$ and $\mathcal{K}^*\phi_H$

The computation of $F_d^*\eta_g$ will be reduced to the computation of the pullbacks of the classes ϕ_H and ϕ_Δ along the normal functions $\mathcal{K}_j$ and $\delta_{i,j}$. In this section we compute these basic classes. The formulas reflect the structure of Torelli groups.

The Moriwaki divisor is the class $\widehat{\nu^*\phi_V}$:

Theorem 10.1 ([23, Thm. 1.3]). *If* $g \geqslant 2$, *then*

$$2\widehat{\nu^*\phi_V} = (8g+4)\lambda_1 - g\delta_0 - 4\sum_{h=1}^{[g/2]} h(g-h)\delta_h \in H^2(\overline{\mathcal{M}}_g, \mathbb{Z}).$$

The case $g \geqslant 3$ is [23, Thm. 1.3]. When $g = 2$, $\nu = 0$ and the result follows from Mumford's computation [34] that $10\lambda_1 = \delta_0 + 2\delta_1$. Suitably interpreted, it holds in genus 1 as $12\lambda_1 = \delta_0$ in $H^2(\overline{\mathcal{M}}_{1,1}, \mathbb{Z})$.

Proposition 8.2 implies that $\widehat{\mathcal{K}^*\phi_H} = \mathcal{K}^*\widehat{\phi}_H$ and $\widehat{(\mathcal{K}\times\mathcal{K})^*\phi_\Delta} = (\mathcal{K}\times\mathcal{K})^*\widehat{\phi}_\Delta$.

Theorem 10.2. *If* $g \geqslant 2$, *then*

$$2\mathcal{K}^*\widehat{\phi}_H = 4g(g-1)\psi - \kappa_1 - \sum_{h=1}^{g-1}(2h-1)^2\,\delta_{g-h}^{\{x\}} \in H^2(\overline{\mathcal{M}}_{g,1}, \mathbb{Z}).$$

This result holds trivially in genus 1 as $\mathcal{K} \equiv 0$ and because $\kappa_1 = 12\lambda_1 - \delta_0 = 0$.

Sketch of Proof. Since $\kappa_1 = 12\lambda_1 - \delta$,

$$\kappa_1 + \sum_{h=1}^{g-1}(2h-1)^2\delta_{g-h}^{\{x\}} = 12\lambda_1 - \delta_0 + \sum_{h=1}^{g-1} 4h(h-1)\delta_{g-h}^{\{x\}}.$$

So it suffices to prove that

$$2\mathcal{K}^*\widehat{\phi}_H = 4g(g-1)\psi - 12\lambda_1 + \delta_0 - \sum_{h=1}^{g-1} 4h(h-1)\delta_{g-h}^{\{x\}}.$$

This formula, modulo boundary terms, was proved by Morita [30, (1.7)]. Another proof is given in [22, Thm. 1].

The coefficient of $\delta_h^{\{x\}}$ is computed using the method of [23, §11]. The Torelli group T_g is replaced by the Torelli group $T_{g,1}$ associated to a 1-pointed surface. Instead of taking $V = \Lambda^3 H/(\theta \wedge H)$, we take it to be H. The quadratic form q (which is denoted S_V in this paper) is replaced by c^*S_H, where $c : \Lambda^3 H \to H$ is the contraction (7.2). We sketch the monodromy computation using the notation of [23, §11].

The coefficient of $\delta_h^{\{x\}}$ is $-\hat{\tau}(\sigma_h)$, where σ_h is a Dehn twist about a separating simple closed curve that divides a pointed, genus g reference surface into a surface of genus h (that does not contain the point) and a surface of genus $g-h$, and where $\hat{\tau}$ is a representation of $T_{g,1}$ into the Heisenberg group associated to (H, S_H).

There is a symplectic basis $a_1, b_1, \ldots, a_g, b_g$ of $H_{\mathbb{Z}}$ such that $a_1, b_1, \ldots, a_h, b_h$ is a basis of H', the first homology of the genus h subsurface. Set $\omega' = a_1 \wedge b_1 + \cdots + a_h \wedge b_h$, the symplectic form of H'. If $u \in H$, then $c(u \wedge \omega') = (h-1)u$. Thus

$$\hat{\tau}_h(\sigma_h) = \frac{8}{2h-2} \sum_{j=1}^{h} S_H(c(a_j \wedge \omega'), c(b_j \wedge \omega'))$$

$$= \frac{8(h-1)^2}{2h-2} \sum_{j=1}^{h} S_H(a_j, b_j) = 4h(h-1).$$

It remains to compute the coefficient of δ_0. The most direct way to compute it is by restricting to a curve in the hyperelliptic locus. First note that if C is a hyperelliptic curve and $x \in C$ is a Weierstrass point, then $\mathcal{K}(C, x) = 0$. Call such a pair (C, x) a *hyperelliptic pointed curve.* Suppose that T is a smooth, complete curve and that $f : T \to \overline{\mathcal{M}}_{g,1}$ is a morphism where $f(t)$ is the moduli point of an irreducible hyperelliptic pointed curve for each $t \in T$.[8] The normal function $f^*\mathcal{K}$ vanishes identically on T, which implies that $f^*\mathcal{K}^*\widehat{\mathcal{B}}$ is trivial as a metrized line bundle over $T - f^{-1}\Delta_0$. Its extension as a metrized line bundle to T is therefore trivial. This implies the vanishing of

$$f^*\mathcal{K}^*\widehat{\phi}_H \in H^2(T).$$

On the other hand, standard techniques can be used to show that

$$f^*(8\lambda_1 + 4g\psi - \delta_0) = 0.$$

The Cornalba-Harris relation [7, Prop. 4.7] implies that

$$f^*\big((8g+4)\lambda_1 - g\delta_0\big) = 0 \in \operatorname{Pic} T.$$

It follows that

$$f^*(4g(g-1)\psi - 12\lambda_1 + \delta_0) = (g-1)f^*(8\lambda_1 + 4g\psi - \delta_0) - f^*\big((8g+4)\lambda_1 - g\delta_0\big) = 0.$$

These two facts together imply that the coefficient of δ_0 in $\mathcal{K}^*\widehat{\phi}_H$ is 1. □

Since $\kappa_1 = 12\lambda_1 - \delta$, Theorems 10.1 and 10.2 and Proposition 8.4 imply the following result when $g \geqslant 2$. The case $g = 1$ follows from the fact that $\psi = \lambda_1$ and the well-known relation $\delta_0 = 12\lambda_1$ in $\operatorname{Pic} \overline{\mathcal{M}}_{1,1}$.

An immediate consequence of Lemma 7.4 and the two previous results is the following formula for $\widehat{\tilde{\nu}^*\phi_L}$.

[8] For example, we can take $T = \mathbb{P}^1$ and f the morphism associated to the family

$$v^2 = (u-t)u \prod_{j=1}^{2g} (u - a_j),$$

where $t \in \mathbb{C}$ and the a_j are distinct non-zero complex numbers. A section of Weierstrass points is given by $x = (0,0)$.

Corollary 10.3. *For all* $g \geqslant 1$,

$$2\widehat{\tilde{\nu}^*\phi_L} = 8\lambda_1 + 4g\psi - \delta_0 - 4\sum_{h=1}^{g-1} h\delta_{g-h}^{\{x\}} \in H^2(\overline{\mathcal{M}}_{g,1}, \mathbb{Z}).$$

The next result is needed in the solution of Eliashberg's problem.

Theorem 10.4. *If* $g \geqslant 2$, *then in* $H^2(\overline{\mathcal{M}}_{g,2}, \mathbb{Z})$ *we have*

$$(\mathcal{K} \times \mathcal{K})^*\widehat{\phi}_\Delta = (2g-2)(\psi_1 + \psi_2) - \kappa_1 - (2g-2)^2\delta_0^{\{x_1,x_2\}}$$
$$- \sum_{h=1}^{g-1} (2h-1)^2\,\delta_{g-h}^{\{x_1,x_2\}} + (2h-1)(2(g-h)-1)\left(\delta_h^{\{x_1\}} + \delta_{g-h}^{\{x_2\}}\right)/2.$$

Note that in this and subsequent formulas, we will often sum from $h = 1$ to $h = g-1$ and over all subsets P of I. Because of this, some terms will appear twice as $\delta_h^P = \delta_{g-h}^{P^c}$. We do this to emphasize the symmetry of the formulas and to facilitate later computations.

Proof. Modulo the coefficients of the $\delta_h^{\{x_1\}}$, this formula can be computed

(i) by restricting $(\mathcal{K} \times \mathcal{K})^*\phi_\Delta$ to any fiber $C \times C$ and
(ii) from Theorem 10.2 by restricting to the diagonal $\overline{\mathcal{M}}_{g,1} \to \overline{\mathcal{M}}_{g,2}$, noting that the restriction of $\widehat{\phi}_\Delta$ to the diagonal $\mathcal{J}_g$ of $\mathcal{J}_g \times_{\mathcal{M}_{g,2}^c} \mathcal{J}_g$ is $2\widehat{\phi}_H$.

These computations are straightforward, once one notes that the divisor $\Delta_0^{\{x_1,x_2\}}$ is the "diagonal" $\mathcal{M}_{g,1}^c \to \mathcal{M}_{g,2}^c$ in $\mathcal{M}_{g,2}^c$ and that the Chern class of its normal bundle is ψ. If we restrict to a single curve C, then in $H^2(C \times C)$

$$\begin{aligned}(\mathcal{K} \times \mathcal{K})^*\widehat{\phi}_\Delta &= (2g-2)^2 \sum_{j=1}^{g} \left(a_j^{(1)} \wedge b_j^{(2)} - b_j^{(1)} \wedge a_j^{(2)}\right)\\ &= (2g-2)^2\left([\text{point}]^{(1)} + [\text{point}]^{(2)} - [\text{diagonal}]\right)\\ &= (2g-2)(\psi_1 + \psi_2) - (2g-2)^2\delta_0^{\{x_1,x_2\}}.\end{aligned}$$

Here $a_1, \ldots, b_g$ is a symplectic basis of $H_1(C)$ and, for $x \in H^1(C)$, $x^{(k)}$ denotes the pullback of x under the kth projection $p_k : C^2 \to C$.

It remains to compute the coefficient of $\delta_h^{x_1}$ when $0 < h < g$. This we do using a test curve suggested by Sam Grushevsky. Since $\widehat{\phi}_\Delta$ is invariant when the two factors of $\mathcal{J}_g^c \times_{\mathcal{M}_{g,2}^c} \mathcal{J}_g^c$ are swapped, it follows that the formula for $(\mathcal{K} \times \mathcal{K})^*\widehat{\phi}_\Delta$. is symmetric in x_1 and x_2. Since $\delta_h^{\{x_1\}} = \delta_{g-h}^{\{x_2\}}$, the formula is also invariant when h is replaced by $g-h$. We therefore conclude that

$$(\mathcal{K} \times \mathcal{K})^*\widehat{\phi}_\Delta = (2g-2)(\psi_1 + \psi_2) - \kappa_1 - (2g-2)^2\delta_0^{\{x_1,x_2\}} - \sum_{h=1}^{g-1} (2h-1)^2\,\delta_{g-h}^{\{x_1,x_2\}} + \sum_{h=1}^{g-1} c_h\delta_h^{\{x_1\}} \tag{10.5}$$

where $c_h = c_{g-h}$.

Suppose that $0 < h < g$. Fix pointed smooth projective curves (C', P') and (C'', P'') with $g(C') = h$ and $g(C'') = g - h$. Let C be the nodal genus g curve with three components C', C'' and $\mathbb{P}^1$, where C' is attached to $\mathbb{P}^1$ by identifying $P' \in C'$ with $0 \in \mathbb{P}^1$ and $P'' \in C''$ with $\infty \in \mathbb{P}^1$. For $t \in \mathbb{P}^1 - \{0, 1, \infty\}$ let C_t be the stable 2-pointed curve $(C; 1, t)$. The closure T of the curve

$$\mathbb{P}^1 - \{0, 1, \infty\} \to \mathcal{M}^c_{g,2}, \quad t \mapsto [C_t]$$

is a copy of $\overline{\mathcal{M}}_{0,4} \cong \mathbb{P}^1$ imbedded in $\mathcal{M}^c_{g,2}$.

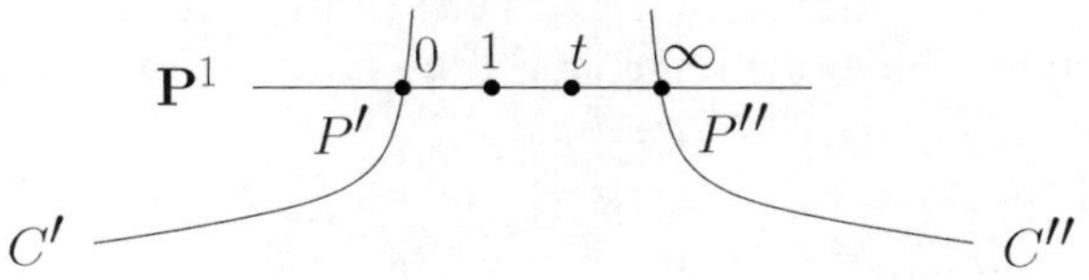

Figure 1. The 2-pointed curve C_t

The restriction of $\mathcal{K} \times \mathcal{K}$ to T takes the constant value

$$((2h - 2)P' - K_{C'}, (2(g - h) - 2)P'' - K_{C''}) \in \operatorname{Jac} C' \times \operatorname{Jac} C'' \cong \operatorname{Jac} C,$$

which implies that $(\mathcal{K} \times \mathcal{K})^* \phi_\Delta = 0$. The coefficient c_h is computed by evaluating the right hand side (RHS) of (10.5) on T.

The curve T is contained in $\Delta_h^{\{x_1,x_2\}} \cap \Delta_{g-h}^{\{x_1,x_2\}}$ and intersects the three boundary divisors $\Delta_0^{\{x_1,x_2\}}$, $\Delta_h^{\{x_1\}}$ and $\Delta_h^{\{x_2\}}$ transversely in three distinct points. These are the three boundary points of $\overline{\mathcal{M}}_{0,4} \cong T$. It does not intersect any other boundary divisors. Consequently,

$$\int_T \delta_0^{\{x_1,x_2\}} = \int_T \delta_h^{\{x_1\}} = \int_T \delta_h^{\{x_2\}} = 1.$$

The projection formula can be used to evaluate the other terms of the RHS of (10.5) on T. Let

$$q : \overline{\mathcal{M}}_{g,2} \to \overline{\mathcal{M}}_g, \quad p_j : \overline{\mathcal{M}}_{g,2} \to \overline{\mathcal{M}}_{g,1} \quad j = 1,2$$

denote the natural projections, where $p_j([C; x_1, x_2]) = [C, x_j]$. Note that q and the p_j collapse T to a point. Since $\psi_j = p_j^* \psi$ and $\kappa_1 = q^* \kappa_1$, the projection formula implies that

$$\int_T \kappa_1 = \int_T q^* \kappa_1 = \int_{q_* T} \kappa_1 = 0 \text{ and } \int_T \psi_j = \int_T p_j^* \psi = \int_{p_j * T} \psi = 0.$$

Since $p_1^* \delta_h^{\{x\}} = \delta_h^{\{x_1,x_2\}} + \delta_h^{\{x_1\}}$, the projection formula implies that

$$\int_T \delta_h^{\{x_1,x_2\}} = -\int_T \delta_h^{\{x_1\}} = -1.$$

Similarly, $\int_T \delta_{g-h}^{\{x_1,x_2\}} = -1$.

Evaluating the expression (10.5) on T we obtain

$$0 = 0 + 0 - (2g-2)^2 + (2h-1)^2 + \big(2(g-h)-1\big)^2 + c_h + c_{g-h}.$$

Since $c_h = c_{g-h}$, this implies that $c_h = (2h-1)\big(2(g-h)-1\big)$. □

11. Solution of Eliashberg's Problem over $\mathcal{M}^c_{g,n}$ when $g > 1$

In this section, we solve Eliashberg's problem over $\mathcal{M}^c_{g,n}$ when $g > 2$. A complete solution in the genus 1 case is given in the following section. The solution in genus > 1 is a direct consequence of Proposition 8.1 and Theorem 11.5 below. Related work on Eliashberg's problem has been obtained independently by Cavalieri and Marcus [5] via Gromov-Witten theory.

Fix an integral vector $\mathbf{d} = (d_1, \ldots, d_n)$ with $\sum_j d_j = 0$. As in the introduction, we have the section

$$F_{\mathbf{d}} : \mathcal{M}^c_{g,n} \to \mathcal{J}_g$$

of the universal jacobian defined by

$$F_{\mathbf{d}} : (x_1, \ldots, x_d) \mapsto \left[\sum_{j=1}^{n} d_j x_j\right] \in \operatorname{Jac} C.$$

For each subset P of $\{x_1, \ldots, x_n\}$, set $d_P = \sum_{j \in P} d_j$. Since $d_P + d_{P^c} = 0$, $d_P^2 \delta_h^P = d_{P^c}^2 \delta_{g-h}^{P^c}$.

Theorem 11.1. *If $g \geqslant 2$, then in $H^{2g}(\mathcal{M}^c_{g,n})$ we have*

$$F_{\mathbf{d}}^* \eta_g = \frac{1}{g!}\left(\sum_{j=1}^{n} d_j^2 \psi_j/2 - \sum_{P \subseteq I} \sum_{\{x_j, x_k\} \subseteq P} d_j d_k \, \delta_0^P - \frac{1}{4} \sum_{P \subseteq I} \sum_{h=1}^{g-1} d_P^2 \, \delta_h^P\right)^g.$$

Recall that our definition of the ψ_j differs from the commonly used one. Here $\psi_j := x_j^* c_1(\omega)$, where ω is the relative dualizing sheaf of the universal curve. Note, too, that since $\delta_h^P = \delta_{g-h}^{P^c}$ and since we are summing from $h = 1$ to $h = g-1$ in this and other results in this section, some boundary divisors occur twice in this expression.

I do not know if this formula holds in the Chow ring. Although this formula makes sense in $H^{2g}(\overline{\mathcal{M}}_{g,n} - \Delta_0^{\text{sing}})$, it does not hold there. For example, when $\mathbf{d} = 0$ the left hand side is the non-trivial class $(-1)^g \lambda_g$, whereas the right-hand side vanishes. A result of Ekedahl and van der Geer [9] implies that, in $CH^g(\overline{\mathcal{M}}_g - \Delta_0^{\text{sing}})$, λ_g is $(-1)^g \zeta(1-2g)$ times a natural class, where $\zeta(s)$ denotes the Riemann zeta function. This suggests that the class

$$F_{\mathbf{d}}^* \eta_g - \frac{1}{g!}(F_{\mathbf{d}}^* \widehat{\phi}_H)^g \in CH^g(\overline{\mathcal{M}}_{g,n} - \Delta_0^{\text{sing}})$$

should be interesting. In particular, does Proposition 8.1 hold in $CH^g(\mathcal{J}_g^c)$? This makes sense, as $\phi_H = \theta - \lambda_1/2$ in $\operatorname{Pic} \mathcal{J}_g^c$. (Cf. the proof of Theorem 11.6 below.)

11.1. The approach and reduction

Denote the pullback of $\mathcal{J}_g$ to $\overline{\mathcal{M}}_{g,n}$ by $\mathcal{J}_{g,n}$ and its restriction to $\mathcal{M}^c_{g,n}$ by $\mathcal{J}^c_{g,n}$. We will consider a more general situation. Namely, we'll assume that $\mathbf{d} \in \mathbb{Z}^n$ and $m \in \mathbb{Z}$ satisfy

$$\sum_{j=1}^{n} d_j = (2g-2)m.$$

Then one has the section

$$F_{\mathbf{d}} : [C; x_1, \dots, x_n] \mapsto \sum_{j=1}^{n} d_j x_j - mK_C \in \operatorname{Jac} C$$

of $\mathcal{J}^c_{g,n}$ over $\mathcal{M}^c_{g,n}$. Our goal is to compute $F^*_{\mathbf{d}}\phi_H$.

Set

$$\mathcal{J}^m_{g,n} := \underbrace{\mathcal{J}_{g,n} \times_{\overline{\mathcal{M}}_{g,n}} \times \mathcal{J}_{g,n} \times_{\overline{\mathcal{M}}_{g,n}} \cdots \times_{\overline{\mathcal{M}}_{g,n}} \mathcal{J}_{g,n}}_{m}.$$

Denote its restriction to $\mathcal{M}^c_{g,n}$ by $\mathcal{J}^{c,m}_{g,n}$. Let

$$\mathcal{K}_n : \overline{\mathcal{M}}_{g,n} - \Delta_0^{\text{sing}} \to \mathcal{J}^n_{g,n}$$

be the nth power of $\mathcal{K}$ — that is, the section of $\mathcal{J}^{c,n}_{g,n} \to \mathcal{M}^c_{g,n}$ defined by

$$\mathcal{K}_n : (C; x_1, \dots, x_n) \mapsto \big(\mathcal{K}(x_1), \mathcal{K}(x_2), \dots, \mathcal{K}(x_n)\big) \in \big(\operatorname{Jac} C\big)^n.$$

Note that $\mathcal{K}_2 = \mathcal{K} \times \mathcal{K}$.

Proposition 11.2. *Define* $\mathbf{d} : \mathcal{J}^n_{g,n} \to \mathcal{J}^n_{g,n}$ *by*

$$\mathbf{d} : (u_1, \dots, u_n) \mapsto (d_1 u_1, \dots, d_n u_n)$$

and $\operatorname{tr}_n : \mathcal{J}^n_{g,n} \to \mathcal{J}_{g,n}$ *by*

$$\operatorname{tr}_n : (u_1, \dots, u_n) \mapsto u_1 + \cdots + u_n.$$

Then the mapping

$$\overline{\mathcal{M}}_{g,n} - \Delta_0^{\text{sing}} \xrightarrow{\mathcal{K}_n} \mathcal{J}^n_{g,n} \xrightarrow{\mathbf{d}} \mathcal{J}^n_{g,n} \xrightarrow{\operatorname{tr}_n} \mathcal{J}_{g,n}$$

equals $(2g-2)F_{\mathbf{d}}$. □

This formula allows the reduction of the computation of $F^*_{\mathbf{d}}\phi_H$ to more basic computations. Denote the pullback of ϕ_H under the jth projection

$$p_j : \mathcal{J}^{c,n}_{g,n} \to \mathcal{J}^c_{g,n}$$

by $\phi_{j,j}$. For $j \neq k$, denote the pullback of ϕ_Δ under the (j,k)th projection

$$p_{j,k} : \mathcal{J}^{c,n}_{g,n} \to \mathcal{J}^{c,2}_{g,n} \quad (u_1, \dots, u_n) \mapsto (u_j, u_k)$$

by $\phi_{j,k}$.

Lemma 11.3. *With notation as above,*

$$\mathbf{d}^*\phi_{j,k} = d_j d_k\, \phi_{j,k} \text{ and } \operatorname{tr}_n^* \phi_H = \sum_{j\leqslant k} \phi_{j,k}.$$

Proof. Since all of the classes $\phi_{j,k}$ are represented by parallel, translation invariant forms, to prove the result, it suffices to prove the result in the cohomology of the jacobian $J := \operatorname{Jac} C$ of a single smooth projective curve C.

Set $J = \operatorname{Jac} C$. Note that the ring homomorphism

$$H^\bullet(J) \to H^\bullet(J)^{\otimes n}$$

induced by the addition map $J^n \to J$ is, in degree 1, given by

$$x \mapsto \sum_{\substack{a+b=n-1\\ a,b\geqslant 0}} 1^{\otimes a} \otimes x \otimes 1^{\otimes b}.$$

The formula for tr_n^* follows using the fact that this is a ring homomorphism.

The formula for $\mathbf{d}^*$ follows as the map $[e] : J \to J$ is multiplication by e on $H^1(J)$, and therefore multiplication by e^k on $H^k(J)$. □

Recall that $\phi_H \in H^2(\mathcal{J}_g^c)$ and $\phi_\Delta \in H^2(\mathcal{J}_g^c \times_{\mathcal{M}_g^c} \mathcal{J}_g)$ extend naturally to classes

$$\widehat{\phi}_H \in H^2(\mathcal{J}_g') \text{ and } \widehat{\phi}_\Delta \in H^2(\mathcal{J}_g' \times_{\overline{\mathcal{M}}_g} \mathcal{J}_g'),$$

where $\mathcal{J}_g'$ denotes the universal jacobian over $\overline{\mathcal{M}}_g - \Delta_0^{\text{sing}}$. Using the scaling by the action of $(e_1, \ldots, e_n) \in \mathbb{Z}^n$ on the cohomology of the nth power of $\mathcal{J}_g' \to \overline{\mathcal{M}}_g - \Delta_0^{\text{sing}}$ one can show that Lemma 11.3 holds when ϕ_H and ϕ_Δ are replaced by the extended classes $\widehat{\phi}_H$ and $\widehat{\phi}_\Delta$. We therefore have:

Corollary 11.4. *If $g \geqslant 2$, then*

$$(2g-2)^2\, F_{\mathbf{d}}^* \widehat{\phi}_H = \sum_{j=1}^{n} d_j^2\, \pi_j^* \mathcal{K}^* \widehat{\phi}_H + \sum_{1\leqslant j<k\leqslant n} d_j d_k\, \pi_{j,k}^* (\mathcal{K} \times \mathcal{K})^* \widehat{\phi}_\Delta \in H^2(\mathcal{M}_{g,n}^c),$$

where $\pi_j : \overline{\mathcal{M}}_{g,n} \to \overline{\mathcal{M}}_{g,1}$ and $\pi_{j,k} : \overline{\mathcal{M}}_{g,n} \to \overline{\mathcal{M}}_{g,2}$ denote the natural projections.

The following result is the main computation of this section.

Theorem 11.5. *If $g \geqslant 2$ and $\mathbf{d} \in \mathbb{Z}^n$ and $m \in \mathbb{Z}$ satisfy $\sum d_j = (2g-2)m$, then in $H^2(\overline{\mathcal{M}}_{g,n})$ we have*

$$\begin{aligned} F_{\mathbf{d}}^* \widehat{\phi}_H = &- m^2\kappa_1/2 + \sum_{j=1}^{n} (d_j m + d_j^2/2)\, \psi_j - \sum_{P\subseteq I} \sum_{\{x_j,x_k\}\subseteq P} d_j d_k\, \delta_0^P \\ &- \frac{1}{4} \sum_{P\subseteq I} \sum_{h=1}^{g-1} \left(d_P - (2h-1)m\right)^2 \delta_h^P. \end{aligned}$$

Recall that $\delta_h^P = \delta_{g-h}^{P^c}$. Note that, in this expression, the coefficients of δ_h^P and $\delta_{g-h}^{P^c}$ are equal as $d_P + d_{P^c} = (2g-2)m$. Note, too, that one recovers Theorem 10.2 when $n = m = 1$ and $d_1 = 2g-2$.

Theorem 11.1 follows directly from Proposition 8.1 and the case $m = 0$. The corresponding genus 1 statement is proved in the following section. The proof below fails in genus 1 as we cannot divide by $2g-2$.

Proof. This follows from Theorems 10.2 and 10.4 and Corollary 11.4. First observe that the coefficient of κ_1 in $(2g-2)^2 F_d^* \widehat{\phi}_H$ is

$$-\sum_j d_j^2/2 - \sum_{j<k} d_j d_k = -(d_1 + \cdots + d_n)^2/2 = -(2g-2)^2 m^2/2.$$

Next, the coefficient of ψ_j in $(2g-2)^2 F_d^* \widehat{\phi}_H$ is

$$\begin{aligned}(2g-2)\sum_{k\neq j} d_j d_k + \frac{1}{2} 4g(g-1)d_j^2 &= (2g-2)\big[(d_1 + \cdots + d_n - d_j)d_j + g\, d_j^2\big] \\ &= (2g-2)^2 d_j m + (2g-2)(g-1)d_j^2 \\ &= (2g-2)^2 (d_j m + d_j^2/2).\end{aligned}$$

Since

$$\pi_{j,k}^* \delta_0^{\{x_1,x_2\}} = \sum_{\{x_j,x_k\}\subseteq P} \delta_0^P, \quad j \neq k$$

the coefficient of δ_0^P in the expression for $(2g-2)^2 F_d^* \widehat{\phi}_H$ is

$$-(2g-2)^2 \sum_{\{x_j,x_k\}\subseteq P} d_j d_k\, \delta_0^P.$$

When $1 \leqslant h \leqslant g-1$ and $j \neq k$

$$\pi_j^* \delta_h^{\{x\}} = \sum_{x_j\in P} \delta_h^P, \quad \pi_{j,k}^* \delta_h^{\{x_1,x_2\}} = \sum_{\{x_j,x_k\}\subseteq P} \delta_h^P, \text{ and } \pi_{j,k}^* \delta_h^{x_1} = \sum_{\substack{x_j\in P\\ x_k\in P^c}} \delta_h^P.$$

Then, computing formally, we see that the coefficient of δ_h^P in $(2g-2)^2 F_d^* \widehat{\phi}_H$ is the coefficient of δ_h^P in

$$\begin{aligned}&-(2(g-h)-1)^2 \sum_j d_j^2\, \pi_j^*\, \delta_h^{\{x\}}/2 - (2(g-h)-1)^2 \sum_{j<k} d_j d_k\, \pi_{j,k}^*\, \delta_h^{\{x_1,x_2\}} \\ &\qquad + (2h-1)(2(g-h)-1) \sum_{j<k} d_j d_k \pi_{j,k}^* (\delta_h^{x_1} + \delta_{g-h}^{x_2})/2,\end{aligned}$$

which is

$$\begin{aligned}&-\frac{(2(g-h)-1)^2}{2}\Bigg(\sum_{x_j\in P} d_j^2 + \sum_{\substack{x_j,x_k\in P\\ x_j\neq x_k}} d_j d_k\Bigg) + \frac{(2h-1)(2(g-h)-1)}{2} \sum_{\substack{x_j\in P\\ x_k\in P^c}} d_j d_k \\ &= -\frac{1}{2}\Big((2(g-h)-1)^2 d_P^2 - (2(g-h)-1)(2h-1) d_P d_{P^c}\Big).\end{aligned}$$

Since $\delta_h^P = \delta_{g-h}^{P^c}$, the coefficient of δ_h^P can be chosen (and will be chosen) to be the average of the formally computed coefficients of δ_h^P and $\delta_{g-h}^{P^c}$, which is

$$\begin{aligned}
&-\frac{1}{4}\Big(\big(2(g-h)-1\big)^2 d_P^2 - \big(2(g-h)-1\big)(2h-1)d_P d_{P^c} + (2h-1)^2 d_{P^c}^2\Big)\\
&= -\frac{1}{4}\Big(\big(2(g-h)-1\big)d_P - (2h-1)d_{P^c}\Big)^2\\
&= -\frac{1}{4}\Big(\big(2(g-h)-1\big)d_P + (2h-1)\big(d_P-(2g-2)m\big)\Big)^2\\
&= -\frac{(2g-2)^2}{4}\big(d_P-(2h-1)m\big)^2
\end{aligned}$$

as $d_P + d_{P^c} = (2g-2)m$. □

11.2. Variants

Theorem 11.5 can be adapted to establish more general results. Suppose that r divides $2g-2$ and suppose that $\mathbf{d} = (d_1, \ldots, d_n) \in \mathbb{Z}^n$ and $m \in \mathbb{Z}$ satisfy

$$\sum_{j=1}^{n} d_j = m(2g-2)/r.$$

Let $f : X \to \overline{\mathcal{M}}_{g,n} - \Delta_0^{\mathrm{sing}}$ be a morphism over which there is a globally defined rth root α of the relative dualizing sheaf. Then one has the section $E_{\mathbf{d}}$ of $\mathcal{J}_{g,n}^c$ over X defined by

$$E_{\mathbf{d}} : x \mapsto -m\alpha + \sum_{j=1}^{n} d_j x_j \in \operatorname{Jac} C.$$

where $f(x) = [C; x_1, \ldots, x_n]$. As above, $d_P = \sum_{x_j \in P} d_j$.

Theorem 11.6. *The class* $E_{\mathbf{d}}^* \widehat{\phi}_H \in H^2(X)$ *is the pullback along* f *of the class*

$$\begin{aligned}
-(m/r)^2\kappa_1/2 + \sum_{j=1}^{n}\big(d_j(m/r) + d_j^2/2\big)\psi_j - \sum_{P\subseteq I}\ \sum_{\{x_j,x_k\}\subseteq P} d_j d_k\, \delta_0^P&\\
-\frac{1}{4}\sum_{P\subseteq I}\sum_{h=1}^{g-1}\big(d_P-(2h-1)(m/r)\big)^2\,\delta_h^P \in H^2(\overline{\mathcal{M}}_{g,n}).&
\end{aligned}$$

Proof. Since the diagram

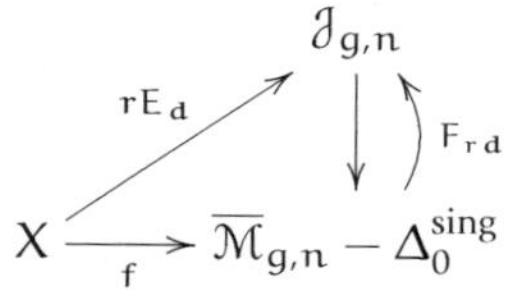

commutes (i.e., $F_{r\mathbf{d}} \circ f = rE_{\mathbf{d}}$), and since the extended class $\widehat{\phi}_H$ satisfies $[e]^*\widehat{\phi}_H = e^2\widehat{\phi}_H$, we have

$$E_{\mathbf{d}}^*\widehat{\phi}_H = r^{-2}E_{r\mathbf{d}}\widehat{\phi}_H = r^{-2}f^*F_{r\mathbf{d}}\widehat{\phi}_H.$$

The result now follows from Theorem 11.5. □

This result can be used to give a partial solution to a problem posed to me by Joe Harris. Suppose that $\mathbf{d} = (d_1, \dots, d_n) \in \mathbb{Z}^n$ satisfies $\sum_{j=1}^n d_j = g - 1$. Then one has the section

$$G_d : [C; x_1, \dots, x_n] \mapsto \sum_j d_j x_j \in \mathrm{Pic}^{g-1} C$$

of the relative Picard bundle

$$\mathcal{P}_{g,n} := \mathrm{Pic}^{g-1}_{\mathcal{C}_{g,n}/\mathcal{M}_{g,n}}$$

over $\mathcal{M}_{g,n}$. This contains the divisor W of effective divisor classes of degree $g-1$. The pullback $G_d^* W$ is the divisor in $\mathcal{M}_{g,n}$ consisting of those $[C; x_1, \dots, x_n]$ where $h^0(C, \sum d_j x_j) > 0$. Denote its closure in $\overline{\mathcal{M}}_{g,n}$ by W_d. Harris' problem is to compute the class of W_d in terms of standard classes. This is a subtle problem as $\mathcal{P}_{g,n}$ is not separated over $\mathcal{M}_{g,n}^c$.

Although we cannot solve this problem, we can solve the following closely related problem. Let $X \to \overline{\mathcal{M}}_{g,n} - \Delta_0^{\mathrm{sing}}$ be any dominant morphism on which there is a globally defined theta characteristic α. Denote the inverse image of $\mathcal{M}_{g,n}^c$ in X by X^c and the inverse image in X of $\mathcal{M}_{g,n}$ by X^o. Denote the universal jacobian over X by $\mathcal{J}_X$ and its restrictions to X^c and X^o by $\mathcal{J}_X^c$ and $\mathcal{J}_X^o$, respectively. Denote the pullback of $\mathcal{P}_{g,n}$ to X^o by $\mathcal{P}_X$. Then α defines an isomorphism of $\mathcal{P}_X$ with $\mathcal{J}_X^o$. Under this isomorphism G_d corresponds to the section

$$F_d : [C; x_1, \dots, x_n] \mapsto -\alpha + \sum_{j=1}^n d_j x_j \in \mathrm{Jac}\, C$$

and the pullback of the divisor W to $\mathcal{P}_X$ corresponds to the divisor Θ_α which is defined locally by the theta function $\vartheta_\alpha(z, \Omega)$ that corresponds to α.

The section F_d extends to a section of $\mathcal{J}_X$ over X and the divisor Θ_α extends to $\mathcal{J}_X$. One therefore has the class $F_d^* \Theta_\alpha$ in $H^2(X)$. Its restriction to $H^2(X^o)$ is the pullback of the class of W_d.

Theorem 11.7. *The class of* $F_d^* \Theta_\alpha$ *in* $H^2(X)$ *is*

$$\delta_0/8 - \lambda_1 + \sum_{j=1}^n (d_j + d_j^2)\psi_j/2 - \sum_{P \subseteq I} \sum_{\{x_j, x_k\} \subseteq P} d_j d_k \delta_0^P$$
$$- \frac{1}{4} \sum_{P \subseteq I} \sum_{h=1}^{g-1} \left(d_P^2 - (2h-1)d_P + h^2 - h\right) \delta_h^P.$$

The class $-\delta_0/8 + F_d^* \Theta_\alpha$ is the pullback of an integral class from $\overline{\mathcal{M}}_{g,n}$ as $\delta_h^P = \delta_{g-h}^{P^c}$ and because each of $d_j + d_j^2$, $d_P^2 - (2h-1)d_P$ and $h^2 - h$ is even.

Proof. The first step is to show that the class θ_α of Θ_α satisfies

$$\theta_\alpha = \hat{\phi}_H + \lambda_1/2 \in H^2(\mathcal{J}_g'), \tag{11.8}$$

where $\mathcal{J}'_g$ denotes the universal jacobian over $\overline{\mathcal{M}}_g - \Delta_0^{\text{sing}}$. Granted this, the result follows from Theorem 11.6 as

$$F_d^*(\theta_\alpha) = F_d^*(\widehat{\phi}_H + \lambda_1/2) = \lambda_1/2 + F_d^*(\widehat{\phi}_H)$$

and

$$\kappa_1 = 12\lambda_1 - \delta = 12\lambda_1 - \delta_o - \frac{1}{2}\sum_{j=1}^{h}\sum_{P\subseteq I}\delta_h^P.$$

To prove (11.8), first note that the relation $\theta_\alpha = \phi_H + \lambda_1/2$ holds in $H^2(\mathcal{J}_g^c)$. This is because the restrictions of θ_α and ϕ_H to each fiber of $\mathcal{J}_g^c \to \mathcal{M}_g^c$ are equal and because the restriction of ϕ_H to the zero section vanishes, while the restriction of θ_α to the zero section has class $\lambda_1/2$ as the corresponding theta null $\vartheta_\alpha(0, \Omega)$ is a modular form of weight $1/2$ for some finite index subgroup of $Sp_g(\mathbb{Z})$. Since $\widehat{\phi}_H$, θ_α and λ_1 are all classes of line bundles, and since $\mathcal{J}_g - \mathcal{J}_g^c$ is the restriction of $\mathcal{J}'_g$ to Δ_0, it follows that

$$\theta_\alpha = \widehat{\phi}_H + \lambda_1/2 + c\delta_0 \in H^2(\mathcal{J}'_g)$$

Restricting both sides to the zero section implies that

$$\lambda_1/2 = 0 + \lambda_1/2 + c\delta_0 \in H^2(\overline{\mathcal{M}}_g - \Delta_0^{\text{sing}}) \cong H^2(\overline{\mathcal{M}}_g),$$

which implies that $c = 0$, as required. □

12. Eliashberg's Problem in Genus 1

The solution of Eliashberg's problem over $\mathcal{M}_{g,n}^c$ in genus $\geqslant 2$ given in the previous section fails when $g = 1$ as we cannot divide by $2g - 2$. However, a variant of the methods of the previous section gives a complete solution in genus 1.

When $g = 1$, the class of $F_d^*\eta_1$ naturally lives in $H^2(\overline{\mathcal{M}}_{1,n+1})$ as the locus of indeterminacy Δ_0^{sing} of F_d has codimension $\geqslant 2$, whereas $F_d\eta_1$ is the class of a divisor, and thus extends uniquely from $\overline{\mathcal{M}}_{1,n+1} - \Delta_0^{\text{sing}}$ to $\overline{\mathcal{M}}_{1,n+1}$.

Theorem 12.1. *If* $\mathbf{d} = (d_0, \dots, d_n) \in \mathbb{Z}^{n+1}$ *satisfies* $\sum_j d_j = 0$, *then*

$$F_d^*\eta_1 = (-1 + (d_0^2 + \cdots + d_n^2)/2)\lambda_1 - \sum_{P\subseteq I}\sum_{\{x_j,x_k\}\subseteq P} d_j d_k\, \delta_0^P \in H^2(\overline{\mathcal{M}}_{1,n+1}).$$

The restriction of this class to $\mathcal{M}_{1,n+1}$ has been computed independently by Cavalieri and Marcus [5] using different methods.

Denote the universal elliptic curve $\mathcal{J}_1 \to \overline{\mathcal{M}}_{1,1}$ by $\mathcal{E}$. Note that $\mathcal{E} = \overline{\mathcal{M}}_{1,2} - \Delta_0^{\text{sing}}$. Its restriction to $\mathcal{M}_{1,1}$ is the universal elliptic curve $\mathcal{E}^c$ of "compact type". Since $\mathcal{E}$ and $\overline{\mathcal{M}}_{1,2}$ differ in codimension 2, their second cohomology and Picard groups are isomorphic:

$$H^2(\overline{\mathcal{M}}_{1,2}) \cong (\operatorname{Pic}\overline{\mathcal{M}}_{1,2}) \otimes \mathbb{Q} \cong (\operatorname{Pic}\mathcal{E}) \otimes \mathbb{Q} \cong H^2(\mathcal{E}).$$

These groups are 2-dimensional with basis $\delta := \delta_0^{\{x_1,x_2\}}$, the class of the zero section D of $\mathcal{E} \to \overline{\mathcal{M}}_{1,1}$, and δ_0, the class of the fiber over the cusp of $\overline{\mathcal{M}}_{1,1}$. The class λ_1 of the Hodge bundle is $\delta_0/12$.

We will deduce Theorem 12.1 from the corresponding result for the nth power

$$\mathcal{E}^n = \underbrace{\mathcal{E} \times_{\overline{\mathcal{M}}_{1,1}} \cdots \times_{\overline{\mathcal{M}}_{1,1}} \mathcal{E}}_{n}.$$

of the universal elliptic curve over $\overline{\mathcal{M}}_{1,1}$. A point in $\mathcal{E}^n$ corresponds to the isomorphism class of an n-pointed elliptic curve $(E; x_0, x_1, \dots, x_n)$ where $x_0 = 0$, the identity.[9] Let $p : \mathcal{E}^n \to \overline{\mathcal{M}}_{1,1}$ be the canonical projection. For each $\mathbf{d} = (d_0, \dots, d_n)$ as above, there is a section of the pullback $p^*\mathcal{E} \to \mathcal{E}^n$ of the universal elliptic curve defined by

$$F_{\mathbf{d}}(E; x_0, \dots, x_n) = \sum_{j=0}^{n} d_j x_j = \sum_{j=1}^{n} d_j x_j \in E.$$

For $0 \leqslant j < k \leqslant n$ let $D_{j,k}$ be the divisor in $\mathcal{E}^n$ where $x_j = x_k$. Denote its class in $H^2(\mathcal{E}^n)$ by $\delta_{j,k}$. Let Δ_0 be the inverse image of the moduli point of the nodal cubic under the projection $\mathcal{E}^n \to \overline{\mathcal{M}}_{1,1}$. Then

$$H^2(\mathcal{E}^n) \cong (\operatorname{Pic} \mathcal{E}^n) \otimes \mathbb{Q} = \mathbb{Q}\delta_0 \oplus \bigoplus_{0 \leqslant j < k \leqslant n} \mathbb{Q}\delta_{j,k}.$$

The pullback of $\delta_{j,k}$ under the morphism $\pi : \overline{\mathcal{M}}_{1,n+1} - \Delta_0^{\text{sing}} \to \mathcal{E}^n$ is

$$\pi^* \delta_{jk} = \sum_{\{x_j, x_k\} \subseteq P} \delta_0^P \in \operatorname{Pic} \overline{\mathcal{M}}_{1,n+1}.$$

Since $d_0 + \cdots + d_n = 0$, $d_0^2 + \cdots + d_n^2 = -2 \sum_{0 \leqslant j < k \leqslant n} d_j d_k$. Since $F_{\mathbf{d}}^* \lambda_1 = \lambda_1$, to prove Theorem 12.1 it suffices to prove that

$$F_{\mathbf{d}}^*(\delta + \lambda_1) = -\sum_{0 \leqslant j < k \leqslant n} d_j d_k (\delta_{j,k} + \lambda_1) \in H^2(\mathcal{E}^n). \tag{12.2}$$

The key step in the proof of this statement is to show that $\delta + \lambda_1$ is a parallel, translation invariant class. This statement is made precise in the following lemma. For a positive integer e, let $[e] : \mathcal{E} \to \mathcal{E}$ be multiplication by e.

Lemma 12.3. *The class $\phi_H \in H^2(\mathcal{E}^c)$ extends uniquely to a class $\widehat{\phi}_H$ in $H^2(\mathcal{E})$ that vanishes on the zero-section D. It is given by*

$$\widehat{\phi}_H = \delta + \lambda_1 \in H^2(\mathcal{E}).$$

and is characterized by the two properties

$$\int_{\Delta_0} \widehat{\phi}_H = 1 \text{ and } [e]^* \widehat{\phi}_H = e^2 \widehat{\phi}_H \text{ for all integers } e > 1.$$

[9] Note that an n-pointed elliptic curves is an $(n+1)$-pointed genus 1 curve.

Proof. The exact sequence

$$0 \to \mathbb{Q}\delta_0 \to H^2(\mathcal{E}) \to H^2(\mathcal{E}^c) \to 0.$$

is invariant under $[e]^*$, which acts trivially on the kernel and by multiplication by e^2 on the quotient. Since ϕ_H spans the right-hand group, it follows that it has a unique lift $\widehat{\phi}_H$ to $H^2(\mathcal{E})$ with the property that $[e]^*\widehat{\phi}_H = e^2\widehat{\phi}_H$.

Since $[e]_*[D] = [D]$, we have

$$(e^2-1)\int_D \widehat{\phi}_H = \int_D [e]^*\widehat{\phi}_H - \int_D \widehat{\phi}_H = \int_{([e]_*-1)D} \widehat{\phi}_H = 0.$$

Since $e > 1$, this implies the vanishing $\int_D \widehat{\phi}_H = 0$ of $\widehat{\phi}_H$ on D. Since Δ_0 is the class of the fiber of $\mathcal{E} \to \overline{\mathcal{M}}_{1,1}$, $\int_{\Delta_0} \widehat{\phi}_H = 1$. Since δ and δ_0 span $H^2(\mathcal{E})$ and since its intersection pairing is non-singular, these two properties characterize $\widehat{\phi}_H$.

To prove that $\widehat{\phi}_H = \delta + \lambda_1$, it suffices to show that $(\delta+\lambda_1)\cdot\delta = 0$ and that $(\delta+\lambda_1)\cdot\delta_0 = 1$. Since the Chern class of the normal bundle of D in $\mathcal{E}$ is $-\lambda_1$, we have

$$\int_D \delta = D^2 = -\int_D \lambda_1,$$

which implies that $(\delta+\lambda_1)\cdot\delta = 0$. Since δ_0 is the class of a fiber of $\mathcal{E} \to \overline{\mathcal{M}}_{1,1}$, $\delta_0\cdot\lambda_1 = 0$, so that $\delta_0\cdot(\delta+\lambda_1) = \delta_0\cdot\delta = 1$. □

Each unordered pair $\{j,k\}$ of integers in $[0,n]$ determines a parallel class $\widehat{\phi}_{j,k} \in H^2(\mathcal{E}^n)$ as follows. Let $p_{j,k} : \mathcal{E}^n \to \mathcal{E}$ be the projection that takes $[E : x_0, \dots, x_n]$ to $[E; x_j, x_k] = [E; 0, x_k - x_j]$. Observe that $D_{j,k} = p_{j,k}^* D$. Set $\widehat{\phi}_{j,k} = p_{j,k}^*\widehat{\phi}_H$. Lemma 12.3 implies that $\widehat{\phi}_{j,k} = \delta_{j,k} + \lambda_1$. Combining this with (12.2), we deduce that, to prove Theorem 12.1, it suffices to show that

$$F_{\mathbf{d}}^*\widehat{\phi}_H = -\sum_{0\leqslant j<k\leqslant n} d_j d_k \widehat{\phi}_{j,k}. \tag{12.4}$$

Lemma 12.5. *The class $\phi_\Delta \in H^2(\mathcal{E}^c \times_{\mathcal{M}_{1,1}} \mathcal{E}^c)$ extends to a class $\widehat{\phi}_\Delta$ in $H^2(\mathcal{E}^2)$ with the property that $[e]^*\widehat{\phi}_\Delta = e^2\widehat{\phi}_\Delta$. Specifically,*

$$\widehat{\phi}_\Delta = p_{0,1}^*\widehat{\phi}_H + p_{0,2}^*\widehat{\phi}_H - p_{1,2}^*\widehat{\phi}_H \in H^2(\mathcal{E}^2). \tag{12.6}$$

Proof. The class given by (12.6) is clearly an eigenvector of $[e]$ with eigenvalue e^2. To prove the result, it suffices to prove that its restriction to $(\mathcal{E}^c)^2 := \mathcal{E}^c \times_{\mathcal{M}_{1,1}} \mathcal{E}^c$ is ϕ_Δ. Since the restriction to $(\mathcal{E}^c)^2$ of all classes in the formula are represented by parallel, translation invariant forms, it suffices to check the formula for a single smooth elliptic curve E. Let a, b be a symplectic basis of $H^1(E)$. Identify $H^\bullet(E^2)$ with $H^\bullet(E)\otimes H^\bullet(E)$. Then a routine computation shows that the restriction of $p_{1,1}^*\phi_H$ to E^2 is

$$(a\wedge b)\otimes 1 + 1\otimes(a\wedge b) - (a\otimes b - b\otimes a) = (p_{0,1}^*\phi_H + p_{0,2}^*\phi_H - \phi_\Delta)|_{E^2},$$

as required. □

For each $0 \leqslant j < k \leqslant n$ define $\hat{\phi}_\Delta^{j,k} = p_{j,k}^* \hat{\phi}_\Delta$.

Corollary 12.7. *If* $0 < j < k \leqslant n$, *then* $\hat{\phi}_\Delta^{j,k} = \hat{\phi}_{0,j} + \hat{\phi}_{0,k} - \hat{\phi}_{j,k}$.

To prove (12.4), factor $F_d : \mathcal{E}^n \to \mathcal{E}$ as follows:

$$\mathcal{E}^n \xrightarrow{(d_1,\ldots,d_n)} \mathcal{E}^n \xrightarrow{\mathrm{tr}_n} \mathcal{E}, \qquad F_d = \mathrm{tr}_n \circ (d_1,\ldots,d_n)$$

where $(d_1, \ldots, d_n) : [E; 0, x_1, \ldots, x_n] \mapsto [E : 0, d_1 x_1, \ldots, d_n x_n]$. Next, note that the formula

$$\mathrm{tr}_n^* \hat{\phi}_H = \sum_{j=1}^n \hat{\phi}_{0,j} + \sum_{1 \leqslant j < k \leqslant n} \hat{\phi}_\Delta^{j,k}$$

holds in $H^2(\mathcal{E}^n)$ as it holds when restricted to $(\mathcal{E}^c)^n$, because both sides are eigenvectors of $[e]$ with eigenvalue e^2, and because both sides vanish on the divisor $x_1 + \cdots + x_n = 0$ by an argument similar to the one used in the proof of Lemma 12.3.

Adapting the arguments of the previous section to this case, we see that

$$\begin{aligned}
F_d^* \hat{\phi}_H &= \sum_{j=1}^n d_j^2 \hat{\phi}_{0,j} + \sum_{1 \leqslant j < k \leqslant n} d_j d_k \hat{\phi}_\Delta^{j,k} \\
&= \sum_{j=1}^n d_j^2 \hat{\phi}_{0,j} + \sum_{1 \leqslant j < k \leqslant n} d_j d_k (\hat{\phi}_{0,j} + \hat{\phi}_{0,k} - \hat{\phi}_{j,k}) \\
&= \sum_{j=1}^n (d_j^2 + (d_1 + \cdots + d_n - d_j) d_j) \delta_{0,j} - \sum_{1 \leqslant j < k \leqslant n} d_j d_k \hat{\phi}_{j,k} \\
&= \sum_{j=1}^n -d_0 d_j \delta_{0,j} - \sum_{1 \leqslant j < k \leqslant n} d_j d_k \hat{\phi}_{j,k} \\
&= - \sum_{0 \leqslant j < k \leqslant n} d_j d_k \hat{\phi}_{j,k}.
\end{aligned}$$

13. Normal functions and positivity

Suppose that $\mathbb{V}$ is a polarized variation of Hodge structure over X of weight -1 endowed with a weak polarization S.

Theorem 13.1. *If* $\nu : X \to J(\mathbb{V})$ *is a normal function, then* $\nu^* \omega_S$ *is a non-negative* $(1,1)$*-form on* X.

Proof. As previously remarked, $J(\mathbb{V})$ is isomorphic, as a bundle of tori, to $\mathbb{V}_\mathbb{R}/\mathbb{V}_\mathbb{Z}$. The normal function ν thus corresponds to a section $s : X \to \mathbb{V}_\mathbb{R}/\mathbb{V}_\mathbb{Z}$. Locally this lifts to a section (also denoted by s) of $\mathbb{V}_\mathbb{R}$. We can view this as as section of

$$\mathcal{V} := \mathbb{V} \otimes \mathcal{O}_X$$

This is a flat holomorphic vector bundle. Denote its Hodge filtration by

$$\mathcal{V} \supseteq \cdots \supseteq \mathcal{F}^p \supseteq \mathcal{F}^{p+1} \supseteq \cdots$$

Each $\mathcal{F}^p$ is a holomorphic sub-bundle of $\mathcal{V}$. Since $\mathbb{V}$ has weight -1, it splits as the sum

$$\mathcal{V} = \mathcal{F}^0 \oplus \overline{\mathcal{F}}^0,$$

where $\overline{\mathcal{F}}^p$ denotes the complex conjugate of $\mathcal{F}^p$ in $\mathcal{V}$. Note that $\mathcal{F}^0$ is a holomorphic sub-bundle, while $\overline{\mathcal{F}}^0$ is not, in general, holomorphic. Decompose s as

$$s = p + n$$

where p is a smooth section of $\mathcal{F}^0$ and n is a smooth section of $\overline{\mathcal{F}}^0$. Since s is real, p and n are complex conjugates of each other.

We can compute the differentials of s, n and p with respect to the flat structure on $\mathbb{V}_{\mathbb{R}}$. Since ω_S is parallel, we have

$$\nu^*\omega_S = s^*\omega_S = S(ds, ds) = S(\partial s + \overline{\partial} s, \partial s + \overline{\partial} s) = 2S(\partial s, \overline{\partial} s).$$

This is clearly of type (1,1).

Next we prove that $\nu^*\omega_S$ is non-negative. Since a 2-form is positive if and only if it is positive on every holomorphic arc in X, we may assume that X is the unit disk. Let t be a holomorphic coordinate in X.

The Griffiths infinitesimal period relation for normal functions implies that $\partial f/\partial t \in \mathcal{F}^{-1}$ for any smooth local lift $f : X \to \mathcal{V}$ of the normal function. Here, and in what follows, the partial derivatives are taken with respect to the natural flat connection on $\mathcal{V}$. Since n and $p + n$ are both smooth local lifts of ν,

$$\text{(13.2)} \qquad \frac{\partial p}{\partial t} \in \mathcal{F}^{-1} \text{ and } \frac{\partial n}{\partial t} \in \mathcal{F}^{-1}.$$

Since $\mathcal{F}^0$ is a holomorphic sub-bundle of $\mathcal{V}$, $\partial p/\partial \overline{t} \in \mathcal{F}^0$. Since n is the conjugate of p, we have $\partial n/\partial t \in \overline{\mathcal{F}}^0$. Combining this with Griffiths infinitesimal period relation (13.2), we conclude that $\partial n/\partial t \in \mathcal{F}^{-1} \cap \overline{\mathcal{F}}^0$ and, by taking complex conjugates, that $\partial p/\partial \overline{t} \in \mathcal{F}^0 \cap \overline{\mathcal{F}}^{-1}$.

Next compute the pullback of ω_S: as above,

$$\begin{aligned} \nu^*\omega_S &= 2S(\partial s, \overline{\partial} s) \\ &= 2S(\partial n + \partial p, \overline{\partial} n + \overline{\partial} p) \\ &= 2S\left(\frac{\partial n}{\partial t} + \frac{\partial p}{\partial t}, \frac{\partial n}{\partial \overline{t}} + \frac{\partial p}{\partial \overline{t}}\right) dt \wedge d\overline{t} \\ &= 2S(\nu(t), \overline{\nu(t)}) dt \wedge d\overline{t} \end{aligned}$$

where $\nu(t) := \partial n/\partial t + \partial p/\partial t$. Set $t = x + iy$. Since $\nu(t) \in H_t^{-1,0}$ for all t, it follows that $\nu^*\omega_S = 2i^{-1-0}S(\nu, \overline{\nu})dx \wedge dy$, which is non-negative. □

Corollary 13.3. *Suppose that $\mathbb{V}$ is a variation of Hodge structure of weight -1 over a smooth complex algebraic variety X. If S is a weak polarization of $\mathbb{V}$ and ν is a normal function section of $J(\mathbb{V}) \to X$, then for all complete curves T in X*

$$\int_T \nu^* \omega_S \geqslant 0$$

with equality if and only if the infinitesimal invariant of the normal function vanishes on T.

Suppose that $\overline{X}$ is a smooth completion of X. If Conjecture 6.6 holds, then we can conclude that the natural extension of $\nu^*\phi_S$ to a class in $H^2(\overline{X})$ has non-negative degree on all complete curves T in $\overline{X}$ that do not lie in $\overline{X} - X$.

14. Slope inequalities

As an immediate consequence of Corollary 13.3 with the computations in Section 10, we obtain the following versions Moriwaki's inequalities [32, 33]. The second assertion below was obtained independently by the author in the late 1990s (cf. [33, p. 195]).

Theorem 14.1. *If $g \geqslant 2$, then*

(i) *the divisor*

$$(8g+4)\lambda_1 - 4\sum_{h=1}^{[g/2]} h(g-h)\delta_h$$

has non-negative degree on each complete curve in $\mathcal{M}_g^c$,

(ii) *the divisors*

$$4g(g-1)\psi - \kappa_1 - \sum_{h=1}^{g-1}(2h-1)^2\delta_{g-h}^{\{x\}}$$

$$8\lambda_1 + 4g\psi - \delta_0 - \sum_{h=1}^{g-1} h\delta_{g-h}^{\{x\}} \in H^2(\mathcal{M}_{g,1}^c)$$

have non-negative degree on each complete curve in $\mathcal{M}_{g,1}^c$.

The semi-positivity of the 2-forms representing these classes implies that their powers are also semi-positive.

Corollary 14.2. *If $g \geqslant 2$ and $k \geqslant 1$, then*

(i) *the cohomology class*

$$\Big((8g+4)\lambda_1 - 4\sum_{h=1}^{[g/2]} h(g-h)\delta_h\Big)^k \in H^{2k}(\mathcal{M}_g^c)$$

has non-negative degree on each complete k-dimensional subvariety of $\mathcal{M}_g^c$.

(ii) *the cohomology classes*

$$\Big(4g(g-1)\psi - \kappa_1 - \sum_{h=1}^{g-1}(2h-1)^2\delta_h^{\{x\}}\Big)^k$$

and

$$\Big(8\lambda_1 + 4g\psi - \delta_0 - \sum_{h=1}^{g-1} h\delta_h^{\{x\}}\Big)^k$$

in $H^{2k}(\mathcal{M}_{g,1}^c)$ *have non-negative degree on each complete* k*-dimensional subvariety of* $\mathcal{M}_{g,1}^c$.

This result also follows from Kleiman's criterion [27, Thm. 1], as János Kollár pointed out to me.

The statements in Theorem 14.1 are weaker than Moriwaki's in the sense that they apply only to complete curves in $\mathcal{M}_{g,n}^c$ where $n = 0,1$, but stronger than Moriwaki's as his versions apply only to complete curves in $\overline{\mathcal{M}}_{g,n}$ that do not lie in the boundary divisor Δ. These two versions suggest the following stronger version of Moriwaki's inequalities, which would follow from Conjecture 6.6 if it were true.

Conjecture 14.3. For all $g \geqslant 2$:

(i) the divisor

$$M := (8g+4)\lambda_1 - 4\sum_{h=1}^{[g/2]} h(g-h)\delta_h$$

has non-negative degree on each complete curve in $\overline{\mathcal{M}}_g$ that does not lie in Δ_0;

(ii) the divisors

$$W_H := 4g(g-1)\psi - \kappa_1 - \sum_{h=1}^{g-1}(2h-1)^2\delta_{g-h}^{\{x\}}$$

$$W_L := 8\lambda_1 + 4g\psi - \delta_0 - \sum_{h=1}^{g-1} h\delta_{g-h}^{\{x\}} \in H^2(\mathcal{M}_{g,1}^c)$$

have non-negative degree on each complete curve in $\overline{\mathcal{M}}_{g,1}$ that does not lie in Δ_0;

(iii) and for all $\mathbf{d} = (d_1, \dots, d_n) \in \mathbb{Z}^d$ with $\sum d_j = m$, the class

$$-m^2\kappa_1/2 + \sum_{j=1}^{n}(d_j m + d_j^2/2)\,\psi_j - \sum_{P\subseteq I}\sum_{\{x_j,x_k\}\subseteq P} d_j d_k\,\delta_0^P$$

$$-\frac{1}{4}\sum_{P\subseteq I}\sum_{h=1}^{g-1}\big(d_P - (2h-1)m\big)^2\,\delta_h^P$$

has non-negative degree on all complete curves in $\overline{\mathcal{M}}_{g,n}$ that do not lie in Δ_0.

Other slope inequalities of this type may be proved more directly, rather than deducing them from the fundamental slope inequalities above. For example, since F_d is a normal function, the class $F_d^*\phi_H$ is represented by a non-negative (1,1)-form on $\mathcal{M}_{g,n}^c$. Theorem 11.5 thus implies the following positivity statement:

Proposition 14.4. *If* $g \geqslant 2$ *and* $(d_1, \dots, d_n) \in \mathbb{Z}^n$ *satisfies* $\sum_{j=1}^n d_j = (2g-2)m$, *then for each* $k \geqslant 1$,

$$\Big(-m^2\kappa_1/2 + \sum_{j=1}^{n} (d_j m + d_j^2/2)\,\psi_j - \sum_{P \subseteq I} \sum_{\{x_j, x_k\} \subseteq P} d_j d_k\, \delta_0^P \\ - \frac{1}{4} \sum_{P \subseteq I} \sum_{h=1}^{g-1} \big(d_P - (2h-1)m\big)^2 \delta_h^P \Big)^k$$

has non-negative degree on all complete k*-dimensional subvarieties of* $\mathcal{M}_{g,n}^c$.

When $k = g$ and $m = g$, this implies that the pullback $F_d^*\eta_g$ of the zero section has non-negative degree on all complete, g-dimensional subvarieties of $\mathcal{M}_{g,n}^c$.

14.1. The jumping divisor

To better understand the behaviour of the degree of the Moriwaki divisors on curves in $\overline{\mathcal{M}}_{g,n}$ that pass through Δ_0^{sing} but are not contained in Δ_0, we need to consider the phenomenon of "height jumping" and the associated notion of a *jumping divisor.*

Suppose that $X = \overline{X} - D$ where $\overline{X}$ is a smooth projective variety and D is a normal crossings divisor in $\overline{X}$. Suppose that $\mathbb{U}$ is a weight -1 variation of Hodge structure over X that is polarized by S. Suppose that ν is a normal function section of $J(\mathbb{U})$. Then one has the symmetric biextension line bundle $\nu^*\widehat{\mathcal{B}}$ over X. Lear's Theorem implies that a positive power $\widehat{\mathcal{B}}_{N,\nu}$ of $\nu^*\widehat{\mathcal{B}}$ extends naturally to a line bundle over $\overline{X}$. It is characterized by the property that the metric on $\nu^*\widehat{\mathcal{B}}$ extends to a continuous metric on this line bundle over $\overline{X} - D^{\mathrm{sing}}$. For clarity of exposition, we suppose that the power is 1, so that $\nu^*\widehat{\mathcal{B}}$ itself extends.[10] Denote the extended line bundle $\widehat{\mathcal{B}}_{1,\nu}$ by $\mathcal{B}_{\overline{X}}$.

Now suppose that $f : T \to \overline{X}$ is a morphism from a smooth projective curve to $\overline{X}$ whose image is not contained in D. Set $T' = T - f^{-1}(D)$. Applying Lear's Theorem to the normal function section $f^*\nu$ of $J(f^*\mathbb{U}) \to T'$, we obtain the Lear extension $\mathcal{B}_T$ of $(\nu \circ f)^*\widehat{\mathcal{B}}$ to T. If the image of T avoids D^{sing}, then $f^*\mathcal{B}_{\overline{X}} \cong \mathcal{B}_T$. This is because $\mathcal{B}_{\overline{X}}$ is metrized over $\overline{X} - D^{\mathrm{sing}}$, so that its pullback to T is the unique metrized extension to T of $(\nu \circ f)^*\widehat{\mathcal{B}}$. In general, there is a 0-cycle J on T, supported on $T - T'$, such that

$$f^*\mathcal{B}_{\overline{X}} \cong \mathcal{B}_T(J).$$

[10] Otherwise, replace it by the power that does extend. In all known examples the power is 1.

If $f(T) \cap D^{sing}$ is empty, then $J = 0$. We will call J the *jumping divisor of* ν *on* T. It is not always trivial, as we shall explain below. The jumping divisor encodes "height jumping".

14.2. Height jumping

Set $d = \dim X$. Assume that $\mathbb{D}^d$ is a polydisk in $\overline{X}$ with coordinates $(t_1, \ldots, t_d)$, and that its intersection with D is the divisor $t_1 \ldots t_m = 0$. Assume that the monodromy of $\mathbb{U}$ about the branch $t_j = 0$ of D is unipotent for each $j \in \{1, \ldots, m\}$.[11] Suppose that

$$\beta \in H^0\big((\mathbb{D}^*)^m \times \mathbb{D}^{d-m}, \nu^*\widehat{\mathcal{B}}\big)$$

is a biextension section of $\nu^*\widehat{\mathcal{B}}$ defined over the punctured polydisk. The associated *height function* $(\mathbb{D}^*)^m \times \mathbb{D}^{d-m} \to \mathbb{R}^+$ is the function

$$(t_1, \ldots, t_d) \mapsto \log|\beta(t_1, \ldots, t_d)|.$$

Suppose that $P \in T$ and that $f(P)$ is the origin of $\mathbb{D}^d$. Suppose that $\mathbb{D}$ is a disk in T with coordinate t with $t(P) = 0$. The restriction of f to $\mathbb{D}$ is a holomorphic arc $f : \mathbb{D} \to \mathbb{D}^d$. Set

$$r_j := \operatorname{ord}_{t=0} f^* t_j, \quad j = 1, \ldots, m.$$

There is a rational number $q(r_1, \ldots, r_m)$, which depends only on the exponents r_j, such that

$$\log|\beta(f(t))|_{\mathcal{B}} \sim q(r_1, \ldots, r_m) \log|t|.$$

One might expect that $q(r_1, \ldots, r_m)$ is linear. Surprisingly, this is not the case.

To better understand this, write

$$q(r_1, \ldots, r_m) = q_0(r_1, \ldots, r_m) + j(r_1, \ldots, r_m)$$

where q_0 is the linear function

$$q_0(r_1, \ldots, r_m) = \sum_{j=1}^{m} r_j q(\mathbf{e}_j)$$

and $\mathbf{e}_1, \ldots, \mathbf{e}_m$ is the standard basis of $\mathbb{Z}^m$. We shall call j the *jump function* of β at P. When j vanishes, the height behaves as expected. Surprisingly, the height can jump. I first observed this when trying to understand Moriwaki's inequality, as explained in the following example. Although I was aware of height jumping through this example, I had no explanation for it. Recently Brosnan and Pearlstein [2] have given a complete explanation of this phenomenon.

For the curve $f : T \to \overline{X}$ and $P \in T$, define $j_P = j(r_1, \ldots, r_m)$. The jumping function determines the jumping divisor.

---

[11] This condition is satisfied by the variations $\mathbb{H}$, $\mathbb{L}$ and $\mathbb{V}$ over $\mathcal{M}_{g,n}$.

Proposition 14.5. *The jumping divisor* J *of* ν *on* T *associated to* $f : T \to \overline{X}$ *is the* 0*-cycle*

$$J = \sum_{P \in T} j_P P$$

on T.

This is proved, using techniques similar to those described in [23, §8], by considering the asymptotics as one approaches P of the length

$$|s|_{\overline{X}}/|s|_T$$

of a section s of $f^*\mathcal{B}_{\overline{X}} \otimes \mathcal{B}_T^{-1}$ that trivializes it over T'. Here $|\ |_{\overline{X}}$ denote the natural metric of $(\nu \circ f)^*\widehat{\mathcal{B}}$ over T' and $|\ |_T$ denotes the natural metric on $\mathcal{B}_T$.

14.3. An example of height jumping

In this example, $g \geqslant 3$, $X = \mathcal{M}_g^c$, $\mathbb{U}$ is the variation $\mathbb{V}$ and ν is the normal function associated to the Ceresa cycle. Denote the Lear extension of $\nu^*\widehat{\mathcal{B}}$ to $\overline{\mathcal{M}}_g$ by $\mathcal{B}_{\overline{\mathcal{M}}}$. The main result of [23] implies that $c_1(\mathcal{B}_{\overline{\mathcal{M}}})$ is the Moriwaki divisor

$$M := (8g+4)\lambda_1 - g\delta_0 - \sum_{h=1}^{\lfloor g/2 \rfloor} 4h(g-h)\delta_h$$

Let $\overline{\mathcal{H}}_g := \{[C] \in \overline{\mathcal{M}}_g : C \text{ is hyperelliptic}\}$ be the locus of hyperelliptic curves in $\overline{\mathcal{M}}_g$. Set $\mathcal{H}_g^c = \overline{\mathcal{H}}_g \cap \mathcal{M}_g^c$. The normal function ν vanishes identically on $\mathcal{H}_g^c$. This implies that the line bundle $\nu^*\widehat{\mathcal{B}}$ over $\mathcal{H}_g^c$ is trivial as metrized holomorphic line bundle. Consequently, its Lear extension to $\overline{\mathcal{H}}_g$, which we denote by $\mathcal{B}_{\overline{\mathcal{H}}}$, is a trivial as a metrized holomorphic line bundle over $\overline{\mathcal{H}}_g$.

Recall that the boundary $\overline{\mathcal{H}}_g - \mathcal{H}_g$ of $\overline{\mathcal{H}}_g$ is a union of divisors

$$\Delta_h, \quad 1 \leqslant h \leqslant g/2 \text{ and } \Xi_h, \quad 0 \leqslant h \leqslant (g-1)/2,$$

where Δ_h is the restriction of the boundary divisor Δ_h of $\overline{\mathcal{M}}_g$ to $\overline{\mathcal{H}}_g$; where Ξ_0 is the divisor whose generic point is an irreducible, geometrically connected hyperelliptic curve with one node; and where Ξ_h, $(h \neq 0)$ is the locus whose generic point is a hyperelliptic curve with two nodes that are conjugate under the hyperelliptic involution and whose normalization is the union of two smooth, geometrically connected hyperelliptic curves, one of genus h and the other of genus $g-h-1$. Denote the class of Ξ_h in $H^2(\overline{\mathcal{H}}_g)$ by ξ_h.

Since $\mathcal{B}_{\overline{X}}$ is metrized over $\overline{\mathcal{M}}_g - \Delta_0^{\text{sing}}$, the restrictions of $\mathcal{B}_{\overline{\mathcal{M}}}$ and $\mathcal{B}_{\overline{\mathcal{H}}}$ to

$$\overline{\mathcal{H}}_g - \bigcup_{h>0} \Xi_h$$

are isomorphic as metrized line bundles. This implies that

$$j^*\mathcal{B}_{\overline{\mathcal{M}}} \otimes \mathcal{B}_{\overline{\mathcal{H}}}^{-1} \cong \mathcal{O}(J)$$

where j denotes the inclusion $\overline{\mathcal{H}}_g \hookrightarrow \overline{\mathcal{M}}_g$ and where J is a linear combination of the Ξ_h, $h > 0$. Note that, since $\mathcal{B}_{\overline{\mathcal{H}}}$ is trivial,

$$\mathcal{O}(J) \cong j^*\mathcal{B}_{\overline{\mathcal{M}}} \cong \mathcal{O}(M)|_{\overline{\mathcal{H}}_g} \in \operatorname{Pic}\overline{\mathcal{H}}_g.$$

The restriction of the Moriwaki divisor to $\overline{\mathcal{H}}_g$ is easily seen to be

$$j^*M = (8g+4)\lambda_1 - g\xi_0 - \sum_{h=1}^{\lfloor (g-1)/2 \rfloor} 2g\xi_h - 4\sum_{h=1}^{\lfloor g/2 \rfloor} h(g-h)\delta_h.$$

On the other hand, Cornalba and Harris [7] have shown that

$$(8g+4)\lambda_1 - g\xi_0 - \sum_{h=1}^{\lfloor (g-1)/2 \rfloor} 2(h+1)(g-h)\xi_h - 4\sum_{h=1}^{\lfloor g/2 \rfloor} h(g-h)\delta_h = 0$$

in $\operatorname{Pic}\overline{\mathcal{H}}_g$. Together these imply that

$$J = \sum_{h=1}^{\lfloor (g-1)/2 \rfloor} 2h(g-h-1)\Xi_h.$$

It is now easy to construct an example of height jumping. Suppose that $f : T \to \overline{\mathcal{M}}_g$ is a curve whose image lies in the hyperelliptic locus and is not contained in Δ_0. If $h > 0$ and if the image of f intersects Ξ_h transversely at a smooth point $f(P)$, then the computations above imply that

$$j_P = 2(h+1)(g-h) - 2g = 2h(g-h-1) > 0.$$

This implies that

$$\deg_T M = \deg_T J > 0.$$

Moriwaki's inequality and positivity in Hodge theory suggest that the jumping divisor associated to any curve in $\overline{\mathcal{M}}_g$ should be effective. Denote by J_T the jumping divisor associated to a morphism $f : T \to \overline{\mathcal{M}}_g$ whose image is not contained in Δ_0.

Conjecture 14.6 (weak form). For all projective curves $f : T \to \overline{\mathcal{M}}_g$ whose image is not contained in Δ_0, the jumping divisor J_T is effective.

This and Conjecture 6.6, if true, imply a stronger version of Moriwaki's inequalities as, for example,

$$\deg_T M = \deg_T \mathcal{B}_T + \deg_T J_T \geqslant \deg_T J_T \geqslant 0.$$

This would imply that the degree of Moriwaki's divisor on most curves not contained in Δ_0 that pass through Δ_0^{sing} would be strictly positive as $\deg_T M$ would be bounded below by the degree of its jumping divisor.

Similarly, one can conjecture that for all projective curves $f : T \to \overline{\mathcal{M}}_{g,1}$ whose image is not contained in Δ_0, the jumping divisor associated to the biextension line bundle associated to the normal function section $\mathcal{K}$ of $J(\mathbb{H})$ is always effective.

In general, one might hope that in the situation described in Paragraph 14.1, the jumping divisor J associated to a curve $f : T \to \overline{X}$ whose image is not contained in D, is effective.

A. Normal functions over $\mathcal{M}_{g,n}$

For completeness, we state the classification (mod torsion) of normal functions over $\mathcal{M}_{g,n}$ associated to variations of Hodge structure whose monodromy representation factors through a rational representation of Sp_g. It follows quite directly from results proved in [15, §8].[12]

The isomorphism classes of irreducible rational representations of the $\mathbb{Q}$-group Sp_g are indexed by partitions λ

$$n = \lambda_1 + \cdots + \lambda_h, \quad \lambda_1 \geqslant \lambda_2 \geqslant \cdots \geqslant \lambda_h, \quad h \leqslant g$$

of integers n into at most g parts. Denote the local system over $\mathcal{A}_g$ that corresponds to the partition λ by $\mathbb{V}_\lambda$. It underlies a $\mathbb{Q}$-variation of Hodge structure of weight $-|\lambda|$, where

$$|\lambda| := \lambda_1 + \cdots + \lambda_h.$$

These can be pulled back to variations of Hodge structure over $\mathcal{M}_{g,n}$ along the period map. Note that $\mathbb{H} = \mathbb{V}_{[1]}$ and that $\mathbb{V} = \mathbb{V}_{[1^3]}(-1)$.

Recall that ΓA denotes the set of rational $(0,0)$ classes of a $\mathbb{Q}$-Hodge structure A. If A is polarized, then $A = \Gamma A \oplus (\Gamma A)^\perp$ in the category of $\mathbb{Q}$-Hodge structures.

Theorem A.1. *Suppose that* $2g-2+n > 0$. *If* $\mathbb{U}$ *is a polarized variation of Hodge structure over* $\mathcal{M}_{g,n}$ *whose monodromy representation factors through a rational representation of* Sp_g, *then there is an isomorphism of variations of* $\mathbb{Q}$*-Hodge structure*

$$\mathbb{U} \cong \bigoplus_\lambda A_\lambda \otimes_\mathbb{Q} \mathbb{V}_\lambda,$$

where A_λ *is the Hodge structure* $H^0(\mathcal{M}_{g,n}, \mathrm{Hom}(\mathbb{V}_\lambda, \mathbb{U}))$. *Moreover, if* A *is a polarized Hodge structure, then*

(i) $\mathrm{Ext}^1_{\mathsf{MHS}(\mathcal{M}_{g,n})}(\mathbb{Q}(0), A \otimes \mathbb{V}_\lambda)$ *vanishes unless* $\lambda = [1]$ *or* $\lambda = [1^3]$,
(ii) $\mathrm{Ext}^1_{\mathsf{MHS}(\mathcal{M}_{g,n})}(\mathbb{Q}(0), A \otimes \mathbb{H})$ *vanishes when* $\Gamma A = 0$.
(iii) $\mathrm{Ext}^1_{\mathsf{MHS}(\mathcal{M}_{g,n})}(\mathbb{Q}(0), \mathbb{H})$ *has basis* $\mathcal{K}_1, \dots, \mathcal{K}_n$,
(iv) $\mathrm{Ext}^1_{\mathsf{MHS}(\mathcal{M}_{g,n})}(\mathbb{Q}(0), A \otimes \mathbb{V})$ *vanishes when* $\Gamma A = 0$,
(v) $\mathrm{Ext}^1_{\mathsf{MHS}(\mathcal{M}_{g,n})}(\mathbb{Q}(0), \mathbb{V})$ *is one-dimensional, spanned by* ν.

Crudely stated, this result says that all normal functions over $\mathcal{M}_{g,n}$ associated to variations of Hodge structure that are representations of Sp_g can be expressed, modulo torsion, as rational linear combinations of the basic normal functions ν and $\mathcal{K}_1, \dots, \mathcal{K}_n$. For example, $(2g-2)\delta_{j,k} = \mathcal{K}_j - \mathcal{K}_k$.

---

[12] Note that there are two typos on page 121. Line 4 should begin $\dim \Gamma H^1(\Gamma^n_{g,r}, V(\lambda))$, and there is a 2 missing from the right-hand side of line -7.

B. The Big Picture

The philosophy behind this work is that a significant amount of the geometry of $\overline{\mathcal{M}}_{g,n}$ is encoded in the category $\mathsf{MHS}_{g,n}$ of those admissible variations of $\mathbb{Z}$-MHS over $\overline{\mathcal{M}}_{g,n}$ whose weight graded quotients are subquotients of Tate twists $\mathbb{H}^{\otimes m}(d)$ of tensor powers of the fundamental local system

$$\mathbb{H} := R^1\pi_*\mathbb{Z}(1)$$

associated to the universal curve $\pi : \mathcal{C} \to \mathcal{M}_{g,n}$. The normal functions discussed in this paper are objects of $\mathsf{MHS}_{g,n}$.

The pure objects of $\mathsf{MHS}_{g,n}$ are the variations of Hodge structure that correspond to the irreducible representations of the group GSp_g of symplectic similitude.[13] These are all pulled back from variations of Hodge structure over $\mathcal{A}_g$ along the period mapping $\mathcal{M}_{g,n} \to \mathcal{A}_g$. Theorem A.1 implies that the only non-trivial extensions between these (mod torsion) are of the form

$$\text{(B.1)} \qquad 0 \to \mathbb{V}_b \to \mathbb{E} \to \mathbb{V}_a \to 0$$

where the weights w_a and w_b of the pure variations $\mathbb{V}_a$ and $\mathbb{V}_b$ satisfy $w_a = 1 + w_a$. This is because the extension (B.1) is determined by the extension

$$0 \to \mathrm{Hom}_{\mathbb{Z}}(\mathbb{V}_a, \mathbb{V}_b) \to \mathbb{E}' \to \mathbb{Z}_{\mathcal{M}_{g,n}}(0) \to 0$$

obtained by applying $\mathrm{Hom}_{\mathbb{Z}}(\mathbb{V}_a, \ )$ to (B.1) and then pulling back along the identity $\mathbb{Z}(0) \to \mathrm{Hom}_{\mathbb{Z}}(\mathbb{V}_a, \mathbb{V}_a)$. That is, the 1-extensions in $\mathsf{MHS}_{g,n}$ correspond to normal functions (mod torsion) over $\mathcal{M}_{g,n}$.

The question arises as to how one can understand $\mathsf{MHS}_{g,n}$. A first approximation is to understand the monodromy representations of objects of $\mathsf{MHS}_{g,n}$. The (orbifold) fundamental group of $\mathcal{M}_{g,n}$ is the mapping class group:

$$\pi_1(\mathcal{M}_{g,n}, [C; P]) \cong \Gamma_{C,P} := \pi_0 \,\mathrm{Diff}^+(C, P),$$

where $P = \{x_1, \dots, x_n\}$ and $\pi_0 \,\mathrm{Diff}^+(C, P)$ denotes the group of connected components of the group of orientation preserving diffeomorphisms of C that fix P pointwise.[14] The action of $\Gamma_{C,P}$ on $H_1(C, \mathbb{Z})$ induces a homomorphism

$$\rho : \Gamma_{C,P} \to \mathrm{Aut}\big(H_1(C, \mathbb{Z}), \text{intersection form}\big) =: \mathrm{Sp}(H_1(C, \mathbb{Z})) \cong \mathrm{Sp}_g(\mathbb{Z})$$

which is well known to be surjective. Its kernel is the *Torelli group*, $T_{C,P}$.

Because each object $\mathbb{V}$ of $\mathsf{MHS}_{g,n}$ is filtered by its weight filtration

$$\cdots \subseteq W_{j-1}\mathbb{V} \subseteq W_j\mathbb{V} \subseteq W_{j+1}\mathbb{V} \subseteq \cdots$$

where each $\mathrm{Gr}^W_j \mathbb{V} := W_j\mathbb{V}/W_{j-1}\mathbb{V}$ is a pure variation of Hodge structure, the Zariski closure (over $\mathbb{Q}$) of its monodromy representation

$$\rho_V : \Gamma_{C,P} \to \mathrm{Aut}(V_{C,P})$$

[13] This is well known. An explanation can be found in [19, §8.1].

[14] This group depends only on g and n and is often denoted by $\Gamma_{g,n}$.

is an extension

$$1 \to U_V \to G_V \to \mathrm{Sp}(H_1(C)) \to 1$$

of algebraic $\mathbb{Q}$-groups where U_V is unipotent. It is thus natural to consider all Zariski dense representations $\Gamma_{C,P} \to G(\mathbb{Q})$ where G is a $\mathbb{Q}$ algebraic group that is an extension of $\mathrm{Sp}(H_1(C))$ by a unipotent group and where $\Gamma_{C,P} \to G \to \mathrm{Sp}(H_1(C))$ is the standard representation ρ. These form an inverse system. Their inverse limit is known as the *completion of* $\Gamma_{C,P}$ *relative to* ρ; it is studied in [16].

The completion of $\Gamma_{C,P}$ relative to ρ is an extension

$$1 \to \mathcal{U}_{C,P} \to \mathcal{G}_{C,P} \to \mathrm{Sp}(H_1(C)) \to 1$$

of affine $\mathbb{Q}$-groups, where $\mathcal{U}_{C,P}$ is prounipotent. There is a canonical homomorphism $\Gamma_{C,P} \to \mathcal{G}_{C,P}(\mathbb{Q})$. Denote the corresponding sequence of Lie algebras by

$$0 \to \mathfrak{u}_{C,P} \to \mathfrak{g}_{C,P} \to \mathfrak{sp}(H_1(C)) \to 0. \tag{B.2}$$

It is proved in [17] and [16] that for each choice of a base point $[C, P]$ of $\mathcal{M}_{g,n}$, the sequence (B.2) is a short exact sequence of Lie algebras in MHS.

Define a *Hodge representation* of $\Gamma_{C,P}$ to be a MHS V and a pair of representations $\phi_{\mathbb{Z}} : \Gamma_{C,P} \to \mathrm{Aut}_{\mathbb{Z}} V$ and $\phi : \mathcal{G}_{C,P} \to \mathrm{Aut}\, V$ such that the diagram

$$\begin{array}{ccc} \Gamma_{C,P} & \xrightarrow{\phi_{\mathbb{Z}}} & \mathrm{Aut}_{\mathbb{Z}} V \\ \downarrow & & \downarrow \\ \mathcal{G}_{C,P}(\mathbb{Q}) & \xrightarrow{\phi} & \mathrm{Aut}_{\mathbb{Q}} V \end{array}$$

commutes and the induced homomorphism $d\phi : \mathfrak{g}_{C,P} \to \mathrm{End}\, V$ is a morphism of MHS. The main result of [21] implies that $\mathsf{MHS}_{g,n}$ is equivalent to the category of Hodge representations of $\mathfrak{g}_{C,P}$, a statement informally conjectured by Deligne. Because of this, extensions in $\mathsf{MHS}_{g,n}$ are determined up to isogeny by Lie algebra cohomology via the following isomorphisms (cf. [20, Cor. 3.7]):

$$\mathrm{Ext}^{\bullet}_{\mathsf{MHS}_{g,n}}(\mathbb{Q}(0), \mathbb{V}) \cong H^{\bullet}(\mathcal{G}_{C,P}, V_{C,P}) \cong H^{\bullet}(\mathfrak{u}_{C,P}, V_{C,P})^{\mathfrak{sp}(H_1(C))}.$$

Although Theorem A.1 was originally proved by a direct argument, it is most natural to regard it as a consequence of this general result and known facts about the cohomology of mapping class and Torelli groups.

The relevance of the algebra $A^{\bullet}_{g,n}$ defined in the introduction is that there is a natural Sp_g-equivariant algebra homomorphism

$$H^{\bullet}(\mathfrak{u}_{C,P}) \to A^{\bullet}_{g,n}$$

which is an isomorphism in degrees 0 and 1 and injective in degree 2. The construction is described in [17] in a more general context. The algebra $T^{\bullet}_{g,n}$ is simply the subalgebra of $A^{\bullet}_{g,n}$ generated by the image of $H^1(\mathfrak{u})$ — that is, by normal functions.

The coefficients of the boundary component δ^P_h, $(h > 0)$ in the formulas in Section 10 are determined by the image in the second weight graded quotient of

$u_{C,P}$ of the Dehn twist corresponding to a loop about Δ_h^P. The coefficients of δ_0^P can be determined similarly, although one has to work with the appropriate relative weight filtration.

One remaining question is whether more information can be obtained by considering other natural categories of MHS over $\mathcal{M}_{g,n}$ (possibly with a level structure), such as the category of variations of MHS obtained from the Prym construction. At present not enough is know about the topology of the Prym construction to understand this problem, although recent progress has been made by Putman [37].

References

[1] M. Boggi and M. Pikaart. Galois covers of moduli of curves. *Compositio Math.*, **120**, 171–191, 2000. ← 531

[2] P. Brosnan and G. Pearlstein. Jumps in the archimedean height. manuscript, March, 2006. ← 546, 570

[3] L. Caporaso. Neron models and compactified Picard schemes over the moduli stack of stable curves. *Amer. J. Math.*, **130**, 1–47, 2008. ← 533

[4] J. Carlson. Extensions of mixed Hodge structures, Journées de Géometrie Algébrique d'Angers, Juillet 1979, Sijthoff & Noordhoff, Alphen aan den Rijn, 1980, 107–127. ← 539

[5] R. Cavalieri, S. Marcus, and J. Wise. Polynomial families of tautological classes on $\mathcal{M}_{g,n}^{rt}$, *J. Pure Appl. Algebra*, **216**, 950–981, (2012). ← 531, 556, 562

[6] G. Ceresa. C is not algebraically equivalent to C^- in its Jacobian. *Ann. of Math.*, **117**, 285–291, 1983. ← 529

[7] M. Cornalba and J. Harris. Divisor classes associated to families of stable varieties, with applications to the moduli space of curves. *Ann. Sci. École Norm. Sup.*, **21**, 455–475, 1988. ← 528, 553, 572

[8] P. Deligne. Équations différentielles à points singuliers réguliers. *Lecture Notes in Mathematics*, **163**, Springer-Verlag, 1970. ← 533

[9] T. Ekedahl and G. van der Geer. Cycles representing the top Chern class of the Hodge bundle on the moduli space of abelian varieties. *Duke Math. J.*, **129**, 187–199, 2005. ← 556

[10] Y. Eliashberg, A. Giventhal, and H. Hofer. Introduction to Symplectic Field Theory, GAFA 2000 (Tel Aviv, 1999), *Geom. Funct. Anal. 2000*, Special Volume, Part II, 560–673. ← 527

[11] C. Faber. A conjectural description of the tautological ring of the moduli space of curves, in: Moduli of curves and abelian varieties, 109–129, *Aspects Math.*, **E33**, Vieweg, Braunschweig, 1999. ← 530

[12] P. Griffiths. Periods of integrals on algebraic manifolds, II. Local study of the period mapping. *Amer. J. Math.*, **90**, 805–865, 1968. ← 527, 539

[13] F. Guillén, V. Navarro Aznar, P. Pascual Gainza, and F. Puerta. Hyperrésolutions cubiques et descente cohomologique. Papers from the Seminar on Hodge-Deligne Theory held in Barcelona, 1982. *Lecture Notes in Mathematics*, **1335**, Springer-Verlag, New York, 1988. ← 541

[14] R. Hain. Biextensions and heights associated to curves of odd genus. *Duke Math. J.*, **61**, 859–898, 1990. ← 544, 549

[15] R. Hain. Torelli groups and Geometry of Moduli Spaces of Curves, in:Current Topics in Complex Algebraic Geometry (C. H. Clemens and J. Kollar, eds.), 97–143. *MSRI publications*, no.**28**, Cambridge University Press, Cambridge, 1995. ← 528, 530, 541, 543, 544, 573

[16] R. Hain. Infinitesimal presentations of Torelli groups. J. *Amer. Math. Soc.*, **10**, 597–651, 1997. ← 575

[17] R. Hain. Hodge-de Rham theory of relative Malcev completion. *Ann. Sci. École Norm. Sup.*, **31**, 47–92, 1998. ← 575

[18] R. Hain. Periods of limit mixed Hodge structures. *Current developments in mathematics, 2002*, Int. Press, Somerville, MA, 113–133, 2003, ← 530, 533

[19] R. Hain. Rational Points of Universal Curves. *J. Amer. Math. Soc.*, **24**, 709-769, 2011. ← 541, 574

[20] R. Hain. Remarks on non-abelian cohomology of proalgebraic groups. *J. Algebraic Geom.*, to appear. ← 575

[21] R. Hain, M. Matsumoto, G. Pearlstein, and T. Terasoma. Tannakian fundamental groups of categories of variations of mixed Hodge structure, in preparation. ← 575

[22] R. Hain and D. Reed. Geometric proofs of some results of Morita. *J. Algebraic Geom.*,**10**, 199–217, 2001. ← 548, 552

[23] R. Hain and D. Reed. On the Arakelov Geometry of Moduli Spaces of Curves. *J. Differential Geom.*, **67**, 195-228, 2004. ← 536, 544, 545, 552, 571

[24] M. Kashiwara. A study of variation of mixed Hodge structure. *Publ. Res. Inst. Math. Sci.*, **22**(5), 991–1024, 1986. ← 541

[25] N. Kawazumi and S. Morita. The primary approximation to the cohomology of the moduli space of curves and cocycles for the stable characteristic classes. *Math. Res. Lett.*, **3**, 629–641, 1996. ← 530

[26] M. Kerr and G. Pearlstein. An exponential history of functions with logarithmic growth, in: Topology of Stratified Spaces, 281–374 *Math. Sci. Res. Inst. Publ.*, **58**, Cambridge Univ. Press, New York, 2011. ← 530, 541, 542

[27] S. Kleiman. Toward a numerical theory of ampleness. *Ann. of Math.*, **84**,293–344, 1966. ← 568

[28] D. Lear. Extensions of Normal Functions and Asymptotics of the Height Pairing, Ph.D. Thesis, University of Washington, 1990. ← 544, 545

[29] E. Looijenga. Smooth Deligne-Mumford compactifications by means of Prym level structures. *J. Algebraic Geom.*, **3**, 283–293, 1994. ← 531

[30] S. Morita. Families of jacobian manifolds and characteristic classes of flat bundles, II. *Math. Proc. Camb. Phil. Soc.*, **105**, 79–101, 1989. ← 552

[31] S. Morita. Generators for the tautological algebra of the moduli space of curves. *Topology*, **42**, 787–819, 2003. ← 530

[32] A. Moriwaki. Relative Bogomolov's inequality and the cone of positive divisors on the moduli spaces of stable curves. *J. Amer. Math. Soc.*, **11**, 569–600, 1998. ← 528, 567

[33] A. Moriwaki. Nef divisors in codimension one on the moduli space of stable curves. *Compositio Math.*, **132**, 191–228, 2002. ← 528, 567

[34] D. Mumford. Towards and enumerative geometry of the moduli space of curves, in: Arithmetic and Geometry, Artin and Tate editors, 271–328, Birkhauser, 1983. ← 551, 552

[35] C. Peters and J. Steenbrink. Mixed Hodge Structures. *Ergebnisse der Mathematik und ihrer Grenzgebiete*, **52**, Springer-Verlag, 2008. ← 530, 539, 541

[36] M. Pulte. The fundamental group of a Riemann surface: mixed Hodge structures and algebraic cycles. *Duke Math. J.*, **57**, 721–760, 1988. ← 544, 550

[37] A. Putman. Abelian covers of surfaces and the homology of the level L mapping class group. *J. Topol. Anal.*, to appear. arXiv:0907.1718 ← 576

[38] M. Saito. Mixed Hodge modules. *Publ. Res. Inst. Math. Sci.*, **26**, 221–333, 1990. ← 541

[39] M. Saito. Admissible normal functions. *J. Algebraic Geom.*, **5**, 235–276, 1996. ← 541

[40] W. Schmid. Variation of Hodge structure: the singularities of the period mapping. *Invent. Math.*, **22**, 211–319, 1973. ← 530

[41] E. Spanier. *Algebraic Topology*, McGraw-Hill, 1966. ← 537, 538

[42] J. Steenbrink and S. Zucker. Variation of mixed Hodge structure, I. *Invent. Math.*, **80**, 489–542, 1985. ← 541

[43] C. Voisin. Chow rings and decomposition theorems for families of K3 surfaces and Calabi-Yau hypersurfaces. *Geom. Topol.*, **16**, 433-473, (2012). ← 548

[44] S. Zucker. Generalized intermediate Jacobians and the theorem on normal functions. *Invent. Math.*, **33**, 185–222, 1976. ← 533, 542, 548

Department of Mathematics, Duke University, Durham, NC 27708-0320
E-mail address: hain@math.duke.edu

Colophon

The *Handbook of Moduli* was designed by Ian Morrison in LaTeX using a variant of the standard style files of the Higher Education Press. The text of the *Handbook* is set in ITC Giovanni and the mathematics in AMS Euler.

Giovanni was designed by Robert Slimbach in 1989 for ITC and was one of the early faces that earned him the Prix Charles Peignot—the Fields Medal of type design—awarded "to a designer under the age of 35 who has made an outstanding contribution to type design." Giovanni combines the basic proportions of traditional old-style designs with the more even color and higher x-height of modern digital fonts, to produce an inconspicuous but legible typeface.

Euler was designed in 1981 by Hermann Zapf, a major figure in twentieth century type design and a pioneer in digital typography, working in close cooperation with Donald Knuth, as an upright, cursive symbol font that would give the effect of mathematics handwritten on a blackboard. In 2008, Zapf reshaped many of the glyphs, with the assistance of Hans Hagen, Taco Hoekwater, and Volker R.W. Schaa, in order to harmonize the designs and bring them into line with contemporary standards of digital typography.

Made in United States
Orlando, FL
04 November 2026

90201248R00328